METHODS IN PHARMACOLOGY AND TOXICOLOGY

Series Editor
Y. James Kang
University of Louisville
School of Medicine
Prospect, Kentucky, USA

For further volumes:
http://www.springer.com/series/7653

Optimization in Drug Discovery

In Vitro Methods

Second Edition

Edited by

Gary W. Caldwell and Zhengyin Yan

CREATe Analytical Sciences, Janssen Research & Development, LLC, Spring House, PA, USA

Humana Press

Editors
Gary W. Caldwell
CREATe Analytical Sciences
Janssen Research & Development, LLC
Spring House, PA, USA

Zhengyin Yan
CREATe Analytical Sciences
Janssen Research & Development, LLC
Spring House, PA, USA

ISSN 1557-2153 ISSN 1940-6053 (electronic)
ISBN 978-1-62703-741-9 ISBN 978-1-62703-742-6 (eBook)
DOI 10.1007/978-1-62703-742-6
Springer New York Heidelberg Dordrecht London

Library of Congress Control Number: 2013953234

© Springer Science+Business Media New York 2014
This work is subject to copyright. All rights are reserved by the Publisher, whether the whole or part of the material is concerned, specifically the rights of translation, reprinting, reuse of illustrations, recitation, broadcasting, reproduction on microfilms or in any other physical way, and transmission or information storage and retrieval, electronic adaptation, computer software, or by similar or dissimilar methodology now known or hereafter developed. Exempted from this legal reservation are brief excerpts in connection with reviews or scholarly analysis or material supplied specifically for the purpose of being entered and executed on a computer system, for exclusive use by the purchaser of the work. Duplication of this publication or parts thereof is permitted only under the provisions of the Copyright Law of the Publisher's location, in its current version, and permission for use must always be obtained from Springer. Permissions for use may be obtained through RightsLink at the Copyright Clearance Center. Violations are liable to prosecution under the respective Copyright Law.
The use of general descriptive names, registered names, trademarks, service marks, etc. in this publication does not imply, even in the absence of a specific statement, that such names are exempt from the relevant protective laws and regulations and therefore free for general use.
While the advice and information in this book are believed to be true and accurate at the date of publication, neither the authors nor the editors nor the publisher can accept any legal responsibility for any errors or omissions that may be made. The publisher makes no warranty, express or implied, with respect to the material contained herein.

Printed on acid-free paper

Humana Press is a brand of Springer
Springer is part of Springer Science+Business Media (www.springer.com)

Preface

The discovery and commercialization of new drugs for humans is extremely complex. Typically, pharmaceutical companies have approached this pharmacology challenge by dividing the problem into three stages. These stages in a pharmaceutical drug development process are the discovery of drug candidates interacting at a particular therapeutic target using whole-animal models, the development of these drug candidates into new chemical entities (NCEs) using human subjects, and the commercialization of NCEs into medicines. The process requires an enormous financial investment since a decade or longer is typically required to transform a drug candidate into a medicine. In addition and probably most important is that the process requires a large interdisciplinary team of scientists and support staff working closely together with a focused management team to be successful.

Whole-animal *in vivo* pharmacology models, which are required by drug regulatory agencies, are the gold standard for biopharmaceutic, pharmacokinetic, toxicokinetic, and pharmacodynamic late stage drug candidate predictions; however, for many biological pathways and mechanisms, they do not provide a good extrapolation to humans. To address this issue, pharmaceutical scientists have used an *in vitro* and *in situ* surrogate assay reductionisms approach to understand biopharmaceutic, pharmacokinetic, toxicokinetic, and pharmacodynamic properties and thus to select drug candidates that have a high probability of becoming an NCE and eventually a medicine (Fig. 1). These surrogate assays provide more representative methods to rule out adverse effects early in the screening process for new drug candidates and to provide a knowledge platform for the correlation of whole-animal *in vivo* pharmacology results to humans. In this strategy, surrogate assays have been developed to understand the biopharmaceutics of drug candidates including the solid-state characteristics of the drug in physiological fluids; that is, dissolution rates, dissociation constants, ionization potential, lipophilic partition coefficients, hydrophobicity, stability, solubility, formulation, and permeability. The pharmacokinetics and toxicokinetics of drug candidates have been addressed by examining individual physiological processes such as absorption (i.e., passive, active, efflux transport of drugs), distribution (i.e., tissue, protein, and cell drug binding), metabolism (i.e., cytochrome P450 (CYP) and UDP-glucuronosyltransferases (UGT)), and drug excretion mechanisms (i.e., metabolism, renal and bile). The overall pharmacodynamic predictions of the drug candidates have been rationalized by receptor binding and functional assays and safety assessment assays including CYP inhibition and induction, drug–drug interactions via assessment of reactive metabolites, hERG (the human *Ether-à-go-go*-Related Gene), DNA damage, genotoxicity, and mutagenicity assays.

Thus, based on this reductionism approach, the pharmacology of a drug can be understood and examined by studying its parts, such as biopharmaceutics, pharmacokinetics, toxicokinetics, and pharmacodynamics. As previously mentioned, each sub-part contained in Fig. 1 can be further subdivided into *in vitro* and *in situ* surrogate assays. Due to the

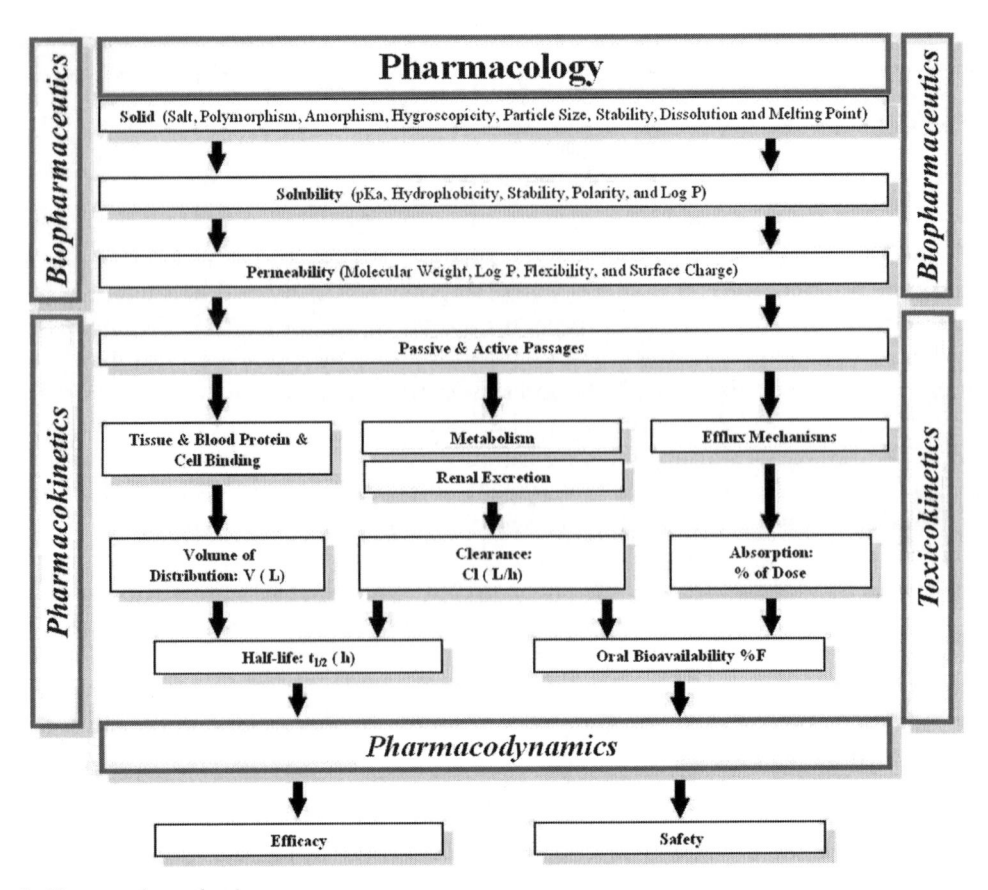

Fig. 1 Pharmacology of a drug

large number of drug candidates that need to be tested, drug discovery and development groups in the pharmaceutical industry have adopted an assay tiered approach toward selecting potential new drug candidates with superior drug properties from large compound collections; that is, funneling thousands of compounds through a series of high-throughput capacity assays to lower capacity assays, which reveal more and more detailed information on a particular sub-part of the reductionism scheme. Using this approach, "drug-like" characteristics in addition to efficacy properties and good safety profiles are achieved. With this process in mind, the book, *Optimization in Drug Discovery: In Vitro Methods*, first published in 2004, presented a compilation of detailed experimental protocols necessary for setting up a variety of *in vitro* and *in situ* assays important in the selection of drug candidates based on the reductionism scheme outlined in Fig. 1. Each chapter contained Introductions, Materials, Methods, Notes, and References sections providing scientists with important background information on the assay, a list of all the equipment and reagents necessary to carry out the assay, a step-by-step protocol, information on dealing with common and unexpected experimental problems in the assay and finally, a listing of important supplementary readings.

We now have compiled a second edition following the same format as the first edition, which contains updated variations on previously reported assays and many new protocols. A total of 34 chapters have been contributed by experts covering a wide spectrum of subjects including formulation, plasma binding, absorption and permeability, cytochrome P450 (CYP) and UDP-glucuronosyltransferases (UGT) metabolism, CYP inhibition and induction, drug transporters, drug–drug interactions via assessment of reactive metabolites, genotoxicity, and chemical and photo-mutagenicity assays.

Since biopharmaceutic, pharmacokinetic, toxicokinetic, and pharmacodynamic are all interrelated, it has been long recognized that a series of *in vitro* and *in situ* assays are required to understand how to develop "drug-like" characteristics in new drug candidates. For example, biopharmaceutic parameters influence the transfer of a drug across cell membranes, and thus affect absorption and distribution of the drug, which in turn affects pharmacokinetic properties, which in turn affects pharmacodynamic properties and so on. *Chapters 1–3* of the new second edition provide experimental methods for preparing an optimal drug formulation, measuring protein binding and red blood cell binding. When combined with measuring pK_a, solubility, lipophilicity, and plasma protein binding techniques from Chaps. 1, 8, and 9 in the first edition, the most fundamental physicochemical properties of a drug candidate can be determined.

Having a good absorption profile for new drug candidates is another important requirement for a drug to be effective. Drug absorption is primarily governed by solubility properties of the solid neat drug, permeability, and influx and efflux transport mechanisms. *Chapters 4* and *5* are included in the second edition to address different issues on this aspect using a 5-day cultured Caco-2 cell model and an *in situ* single pass perfused rat intestinal model. Combining these assays with absorption models described in Chaps. 2–5 in the first edition, the most commonly used assays to investigate drug absorption mechanisms are available to research scientists.

Optimal metabolic stability of new drug candidates is one property that is necessary for a drug to have a long systemic half-life in the body and thus, lasting pharmacological effects on the action site. There are many different *in vitro* metabolic stability assays that can be used to understand the metabolism fate of new drug candidates, identify potent metabolites with better "drug-like" properties, and for using metabolic stability information to guide new synthesis and generate more stable drug candidates. In *Chaps. 6* and *7* in the second edition, the assessment of CYPs and UGTs metabolism is determined from incubations with either hepatocytes or microsomes. *Chapters 8* and *9* in the second edition outline assays to determine the CYP and UGT phenotyping. When these assays are combined with metabolic stability assays from Chaps. 10–12 in the first edition, an arsenal of assays are available with clear advantages and objectives to address most metabolic stability questions.

Drug–drug interactions (DDIs) are defined when one drug alters the pharmacokinetics or pharmacodynamics of another drug. Since biopharmaceutic, pharmacokinetic, toxicokinetic, and pharmacodynamic are all interrelated, in many cases, one drug generally alters the metabolism or transport of a second drug. Most DDIs involve alternations in the metabolic pathways within the CYP system. There are two mechanisms involve for the CYPs; that is, through the process of CYP induction which increases drug clearance that causes a decline or loss of therapeutic efficacy or when one drug inhibits metabolism of another drug. Drug-CYP induction is typically caused by activation of gene transcription via ligand-activated specific receptor which eventually leads to an increase in CYP enzyme expression. The three most commonly involved nuclear hormone receptors are (1) PXR, which up-regulates expression of the CYP3A, CYP2B, and CYP2C, (2) CAR, which also results in enhanced expression of the CYP3A, CYP2B, and CYP2C, and (3) AhR, which results in enhanced expression of the CYP1A and 1B enzymes. Nuclear hormone receptor activation assays using stable cell lines are described in *Chaps. 10–13* in the second edition. CYP induction can be evaluated using human hepatocytes as described in *Chap. 14* in the second edition or Chap. 13 in the first edition. Each system has its own advantages and limitations, and the decision to use a particular approach depends upon the goal of the drug evaluation. Drug-inhibition of CYPs is typically caused by reversible and irreversible inhibition mechanisms. In Chaps. 14 and 15 in the first edition, high-throughput screening assays for 13 individual CYPs by using fluorescent substrates and cDNA-expressed enzymes, and 6 individual CYPs using specific substrate probes and human liver microsomes were described to measure reversible inhibition mechanisms, respectively. In the second edition, several assays are described to measure irreversible inhibition mechanisms. *Chapters 15* and *16* in the second edition outline assays to measure irreversible inhibition (i.e., time-dependent inhibition) using plated and suspended human hepatocytes while *Chaps. 17–19* use human liver microsomes combined with novel detection methods. A systemic approach is given in Chap. 16 in the first edition to identify mechanism-based CYP inhibitors. Thus, a complete set of assays are available to address many DDI questions that occur due to alterations in the metabolic CYP system pathways.

The ATP binding cassette (ABC) superfamily and the solute carrier (SLC) family of transporters play a major role in influencing the pharmacokinetics and toxicokinetics of drugs since they are responsible for the efflux of a plethora of therapeutic drugs, peptides, and endogenous compounds across biological membranes. The ABC subfamily contains nine transporters which have different intracellular localizations, substrate specificities, and structures. In *Chaps. 20–22* in the second edition, *in vitro* methods for discovering substrates and inhibitors for the P-glycoprotein (P-gp/ABCB1), the breast cancer resistance protein (BCRP), and the multidrug resistance-associated protein 2 (MRP2; ABCC2) are discussed in detail, respectively. The organic anion transporting polypeptides (OATPs) are members of the SLC family of transporters. In *Chaps. 23* and *24* in the second edition, *in vitro* assays for discovering substrates and inhibitors of OATP1B1 and OATP1B3, which are predominantly expressed at the sinusoidal membrane of hepatocytes, and OAT1 (SLC22A6), OAT3 (SLC22A8), and OCT2 (SLC22A2), which are primarily expressed in the proximal tubule epithelial cells of the kidney, are discussed in detail, respectively. When these assays are combined with transporter assays from Chaps. 6 and 7 in the first edition, an arsenal of assays are available to understand the major role that transporters play in influencing the drug's pharmacodynamics.

In *Chaps. 25–27* in the second edition, a variety of assays are presented dealing with establishing good *in vitro* LC/MS/MS assays, LC/MS/MS methods for the identification of metabolites in biological fluids, and the detection of endogenous and xenobiotic compounds from biological fluids using LC/MS/MS and dried blood spot techniques, respectively.

The most important clearance pathways for most drugs in humans involve drugs being metabolized by CYP enzymes to more polar compounds that are eventually eliminated in urine. However, CYP enzyme-mediated metabolism can also lead to drug bioactivation resulting in the formation of reactive metabolites that can potentially induce idiosyncratic toxicity by covalently binding to endogenous proteins and nucleic acids before being eliminated from the body. Because reactive metabolites are not stable, direct detection and characterization of them is not technically feasible. Therefore, many assay strategies have been developed to study the bioactivation liability of drug candidates by using trapping reagents that result in the formation of stable adducts which are subsequently characterized by tandem mass spectrometry. In *Chaps. 28–30* in the second edition, a variety of assays are presented dealing with drug bioactivation including the utilization of trapping reagents that results in the formation of stable adducts, quantitative methods for detecting reactive metabolites using radioactive and non-radioactive reagents, and screening assays for determining the reactivity of acyl glucuronides. When these assays are combined with reactive metabolites assays from Chaps. 24 and 25 in the first edition, a set of assays are available with clear advantages and objectives to help medicinal chemists to optimize lead compounds at an early stage of drug discovery.

The failure of NCEs in both clinical development and aftermarket launch for toxicity reasons is still a major concern for pharmaceutical companies. Therefore, toxicity assays that can provide information at an early stage of drug discovery are of major concerns for medicinal chemists to optimize lead compounds. Interaction of drugs with DNAs potentially results in DNA damage or covalent modifications which may lead to genotoxicity. In *Chaps. 31* and *32* in the second edition, a method for detecting DNA damage at the level of individual eukaryotes induced by drugs using the traditional *in vitro* Comet assay (neutral and alkaline) is presented and a system based in eukaryotic yeast cells that utilize an endogenous DNA damage-responsive gene promoter and a reporter gene fusion to assess the ability of the drugs to damage DNA is presented, respectively. Here the authors provide examples of these assays with detailed procedures used in their laboratory for the analysis and interpretation of assay data. Combining these new versions with DNA damage assays from Chaps. 17–20 in the first edition, a set of assays are available to medicinal chemists to provide a path forward in early stage drug discovery lead optimization programs. Although the Ames test has long been used to detect mutagens and possible carcinogens, an improved version of the assay is given in *Chap. 33* in the second edition as compared to the version in Chap. 21 in the first edition. Also, a new version of the mouse lymphoma assay (MLA) is outlined in *Chap. 34* in the second edition as compared to the version in Chap. 22 in the first edition. In Chap. 23 in the first edition, a high through *in vitro* hERG channel assay was presented.

Finally, we want to express our tremendous gratitude to all the authors that contributed chapters to this book. Without their time and energy, the second edition of Optimization in Drug Discovery: In Vitro Methods would not have been possible.

Spring House, PA, USA   *Gary W. Caldwell*
Spring House, PA, USA   *Zhengyin Yan*

Contents

Preface..		*v*
Contributors..		*xv*
1	Small Molecule Formulation Screening Strategies in Drug Discovery........ *Gary W. Caldwell, Becki Hasting, John A. Masucci, and Zhengyin Yan*	1
2	Assessment of Drug Plasma Protein Binding in Drug Discovery............ *Dennis Kalamaridis and Nayan Patel*	21
3	Drug Partition in Red Blood Cells..................................... *Dennis Kalamaridis and Karen DiLoreto*	39
4	Permeability Assessment Using 5-day Cultured Caco-2 Cell Monolayers..... *Gary W. Caldwell, Chrissa Ferguson, Robyn Buerger, Lovonia Kulp, and Zhengyin Yan*	49
5	In Situ Single Pass Perfused Rat Intestinal Model .... *Maria Markowska and L. Mark Kao*	77
6	Metabolic Stability Assessed by Liver Microsomes and Hepatocytes......... *Kevin J. Coe and Tatiana Koudriakova*	87
7	Metabolic Assessment in Alamethicin-Activated Liver Microsomes: Co-activating CYPs and UGTs .... *Gary W. Caldwell and Zhengyin Yan*	101
8	Phenotyping UDP-Glucuronosyltransferases (UGTs) Involved in Human Drug Metabolism: An Update .... *Michael H. Court*	117
9	In Vitro CYP/FMO Reaction Phenotyping............................ *Carlo Sensenhauser*	137
10	Human Pregnane X Receptor (hPXR) Activation Assay in Stable Cell Lines.. *Judy L. Raucy*	171
11	Characterization of Constitutive Androstane Receptor (CAR) Activation..... *Caitlin Lynch, Haishan Li, and Hongbing Wang*	195
12	DNA Binding (Gel Retardation Assay) Analysis for Identification of Aryl Hydrocarbon (Ah) Receptor Agonists and Antagonists.............. *Anatoly A. Soshilov and Michael S. Denison*	207
13	Cell-Based Assays for Identification of Aryl Hydrocarbon Receptor (AhR) Activators... *Guochun He, Jing Zhao, Jennifer C. Brennan, Alessandra A. Affatato, Bin Zhao, Robert H. Rice, and Michael S. Denison*	221

14	In Vitro CYP Induction Using Human Hepatocytes .... *Monica Singer, Carlo Sensenhauser, and Shannon Dallas*	237
15	Assessment of CYP3A4 Time-Dependent Inhibition in Plated and Suspended Human Hepatocytes.................... *J. George Zhang and David M. Stresser*	255
16	Evaluation of Time-Dependent CYP3A4 Inhibition Using Human Hepatocytes............................... *Yuan Chen and Adrian J. Fretland*	269
17	Rapidly Distinguishing Reversible and Time-Dependent CYP450 Inhibition Using Human Liver Microsomes, Co-incubation, and Continuous Fluorometric Kinetic Analyses .... *Gary W. Caldwell and Zhengyin Yan*	281
18	Identification of Time-Dependent CYP Inhibitors Using Human Liver Microsomes (HLM) .... *Kevin J. Coe, Judith Skaptason, and Tatiana Koudriakova*	305
19	CYP Time-Dependent Inhibition (TDI) Using an IC_{50} Shift Assay with Stable Isotopic Labeled Substrate Probes to Facilitate Liquid Chromatography/Mass Spectrometry Analyses .... *Gary W. Caldwell and Zhengyin Yan*	315
20	Screening for P-Glycoprotein (Pgp) Substrates and Inhibitors .... *Qing Wang and Tina M. Sauerwald*	337
21	In Vitro Characterization of Intestinal Transporter, Breast Cancer Resistance Protein (BCRP)....................... *Chris Bode and Li-Bin Li*	353
22	In Vitro Characterization of Intestinal and Hepatic Transporters: MRP2 ..... *Ravindra Varma Alluri, Peter Ward, Jeevan R. Kunta,* *Brian C. Ferslew, Dhiren R. Thakker, and Shannon Dallas*	369
23	In Vitro Characterization of Hepatic Transporters OATP1B1 and OATP1B3.. *Blair Miezeiewski and Allison McLaughlin*	405
24	In Vitro Characterization of Renal Transporters OAT1, OAT3, and OCT2 . . . *Ying Wang and Nicole Behler*	417
25	General Guidelines for Setting Up an In Vitro LC/MS/MS Assay .... *John A. Masucci and Gary W. Caldwell*	431
26	Metabolite Identification in Drug Discovery........................ *Wing W. Lam, Jie Chen, Rongfang Fran Xu, Jose Silva,* *and Heng-Keang Lim*	445
27	Drug, Lipid, and Acylcarnitine Profiling Using Dried Blood Spot (DBS) Technology in Drug Discovery...................... *Wensheng Lang, Jenson Qi, and Gary W. Caldwell*	461
28	In Vitro Trapping and Screening of Reactive Metabolites Using Liquid Chromatography-Mass Spectrometry................... *Zhengyin Yan and Gary W. Caldwell*	477

29	Quantitative Assessment of Reactive Metabolites............................	489
	Jie Chen, Rongfang Fran Xu, Wing W. Lam, Jose Silva, and Heng-Keang Lim	
30	In Vitro Assessment of the Reactivity of Acyl Glucuronides................	505
	Rongfang Fran Xu, Wing W. Lam, Jie Chen, Michael McMillian, Jose Silva, and Heng-Keang Lim	
31	In Vitro Comet Assay for Testing Genotoxicity of Chemicals..............	517
	Haixia Lin, Nan Mei, and Mugimane G. Manjanatha	
32	Assessing DNA Damage Using a Reporter Gene System..................	537
	Michael Biss and Wei Xiao	
33	Improved AMES Test for Genotoxicity Assessment of Drugs: Preincubation Assay Using a Low Concentration of Dimethyl Sulfoxide......	545
	Atsushi Hakura	
34	Methods for Using the Mouse Lymphoma Assay to Screen for Chemical Mutagenicity and Photo-Mutagenicity.............	561
	Nan Mei, Xiaoqing Guo, and Martha M. Moore	
Index...		593

Contributors

ALESSANDRA A. AFFATATO • *Division of Oncology, Clinical Research Office, Infermi Hospital, Rimini, Italy*
RAVINDRA VARMA ALLURI • *Eshelman School of Pharmacy, University of North Carolina, Chapel Hill, NC, USA*
NICOLE BEHLER • *Absorption Systems, Exton, PA, USA*
MICHAEL BISS • *Department of Microbiology and Immunology, University of Saskatchewan, Saskatoon, SK, Canada*
CHRIS BODE • *Absorption Systems, Exton, PA, USA*
JENNIFER C. BRENNAN • *Department of Environmental Toxicology, University of California, Davis, CA, USA*
ROBYN BUERGER • *Absorption Systems, Exton, PA, USA*
GARY W. CALDWELL • *CREATe Analytical Sciences, Janssen Research & Development, LLC, Spring House, PA, USA*
JIE CHEN • *Drug Safety Sciences, Drug Metabolism and Pharmacokinetics, Janssen Research & Development, LLC, Spring House, PA, USA*
YUAN CHEN • *Early Development, Drug Metabolism and Pharmacokinetics, Genentech Research, South San Francisco, CA, USA*
KEVIN J. COE • *Discovery Sciences, Drug Metabolism and Pharmacokinetics, Janssen Research & Development, LLC, La Jolla, CA, USA*
MICHAEL H. COURT • *Department of Veterinary Clinical Sciences, College of Veterinary Medicine, Washington State University, Pullman, WA, USA*
SHANNON DALLAS • *Drug Safety Sciences, Drug Metabolism and Pharmacokinetics, Janssen Research & Development, LLC, Spring House, PA, USA*
MICHAEL S. DENISON • *Department of Environmental Toxicology, University of California, Davis, CA, USA*
KAREN DILORETO • *Discovery Sciences, Drug Metabolism and Pharmacokinetics, Janssen Research & Development, LLC, Spring House, PA, USA*
CHRISSA FERGUSON • *Absorption Systems, Exton, PA, USA*
BRIAN C. FERSLEW • *Eshelman School of Pharmacy, University of North Carolina, Chapel Hill, NC, USA*
ADRIAN J. FRETLAND • *Roche Pharma Research and Early Development, Non-Clinical Safety, Roche-Nutley, Nutley, NJ, USA*
XIAOQING GUO • *Division of Genetic and Molecular Toxicology, National Center for Toxicological Research (NCTR), Food and Drug Administration (FDA), Jefferson, AR, USA*
ATSUSHI HAKURA • *Global Drug Safety, Drug Safety Japan, Eisai Co., Ltd, Ibaraki, Japan*
BECKI HASTING • *CREATe Analytical Sciences, Janssen Research & Development, LLC, Spring House, PA, USA*

GUOCHUN HE • *Department of Environmental Toxicology, University of California, Davis, CA, USA*

DENNIS KALAMARIDIS • *Drug Safety Sciences, Drug Metabolism and Pharmacokinetics, Janssen Research & Development, LLC, Spring House, PA, USA*

L. MARK KAO • *Drug Safety Sciences, Drug Metabolism and Pharmacokinetics, Janssen Research & Development, LLC, Raritan, NJ, USA*

TATIANA KOUDRIAKOVA • *Discovery Sciences, Drug Metabolism and Pharmacokinetics, Janssen Research & Development, LLC, La Jolla, CA, USA*

LOVONIA KULP • *Absorption Systems, Exton, PA, USA*

JEEVAN R. KUNTA • *Discovery Sciences, Drug Metabolism and Pharmacokinetics, Janssen Research & Development, LLC, Spring House, PA, USA*

WING W. LAM • *Drug Safety Sciences, Drug Metabolism and Pharmacokinetics, Janssen Research & Development, LLC, Spring House, PA, USA*

WENSHENG LANG • *CREATe Analytical Sciences, Janssen Research & Development, LLC, Spring House, PA, USA*

HAISHAN LI • *Research Center for Import–Export Chemicals Safety, Chinese Academy of Inspection and Quarantine, Beijing, China*

LI-BIN LI • *Absorption Systems, Exton, PA, USA*

HENG-KEANG LIM • *Drug Safety Sciences, Drug Metabolism and Pharmacokinetics, Janssen Research & Development, LLC, Spring House, PA, USA*

HAIXIA LIN • *Division of Genetic and Molecular Toxicology, National Center for Toxicological Research (NCTR), Food and Drug Administration (FDA), Jefferson, AR, USA*

CAITLIN LYNCH • *Department of Pharmaceutical Sciences, School of Pharmacy, University of Maryland, Baltimore, MD, USA*

MUGIMANE G. MANJANATHA • *Division of Genetic and Molecular, Toxicology National Center for Toxicological Research (NCTR), Food and Drug Administration (FDA), Jefferson, AR, USA*

MARIA MARKOWSKA • *Drug Safety Sciences, Drug Metabolism and Pharmacokinetics, Janssen Research & Development, LLC, Spring House, PA, USA*

JOHN A. MASUCCI • *CREATe Analytical Sciences, Janssen Research & Development, LLC, Spring House, PA, USA*

ALLISON MCLAUGHLIN • *Absorption Systems, Exton, PA, USA*

MICHAEL MCMILLIAN • *Drug Safety Sciences, Drug Metabolism and Pharmacokinetics, Janssen Research & Development, LLC, Spring House, PA, USA*

NAN MEI • *Division of Genetic and Molecular Toxicology, National Center for Toxicological Research (NCTR), Food and Drug Administration (FDA), Jefferson, AR, USA*

BLAIR MIEZEIEWSKI • *Absorption Systems, Exton, PA, USA*

MARTHA M. MOORE • *Division of Genetic and Molecular Toxicology, National Center for Toxicological Research (NCTR), Food and Drug Administration (FDA), Jefferson, AR, USA*

NAYAN PATEL • *Drug Safety Sciences, Bioanalytical, Janssen Research & Development, LLC, Spring House, PA, USA*

JENSON QI • *Cardiovascular Metabolism, Janssen Research & Development, LLC, Spring House, PA, USA*

JUDY L. RAUCY • *Puracyp, Inc., Carlsbad, CA, USA*

ROBERT H. RICE • *Department of Environmental Toxicology, University of California, Davis, CA, USA*

TINA M. SAUERWALD • *Absorption Systems, Exton, PA, USA*
CARLO SENSENHAUSER • *Drug Safety Sciences, Drug Metabolism and Pharmacokinetics, Janssen Research & Development, LLC, Spring House, PA, USA*
JOSE SILVA • *Drug Safety Sciences, Drug Metabolism and Pharmacokinetics, Janssen Research & Development, LLC, Spring House, PA, USA*
MONICA SINGER • *Drug Safety Sciences, Investigative Toxicology, Janssen Pharmaceutical Research & Development, Spring House, PA, USA*
JUDITH SKAPTASON • *Discovery Sciences, Drug Metabolism and Pharmacokinetics, Janssen Research & Development, LLC, La Jolla, CA, USA*
ANATOLY A. SOSHILOV • *Department of Environmental Toxicology, University of California, Davis, CA, USA*
DAVID M. STRESSER • *Gentest Contract Research Services, Corning Life Sciences, Becton Dickinson, Woburn, MA, USA*
DHIREN R. THAKKER • *Eshelman School of Pharmacy, University of North Carolina, Chapel Hill, NC, USA*
HONGBING WANG • *Department of Pharmaceutical Sciences, School of Pharmacy, University of Maryland, Baltimore, MD, USA*
QING WANG • *Absorption Systems, Exton, PA, USA*
YING WANG • *Absorption Systems, Exton, PA, USA*
PETER WARD • *Drug Safety Sciences, Drug Metabolism and Pharmacokinetics, Janssen Research & Development, LLC, La Jolla, CA, USA*
WEI XIAO • *Department of Microbiology and Immunology, University of Saskatchewan, Saskatoon, SK, Canada; College of Life Sciences, Capital Normal University, Beijing, China*
RONGFANG FRAN XU • *Drug Safety Sciences, Drug Metabolism and Pharmacokinetics, Janssen Research & Development, LLC, Spring House, PA, USA*
ZHENGYIN YAN • *CREATe Analytical Sciences, Janssen Research & Development, LLC, Spring House, PA, USA*
J. GEORGE ZHANG • *Gentest Contract Research Services, Corning Life Sciences, Becton Dickinson, Woburn, MA, USA*
JING ZHAO • *Department of Environmental Toxicology, University of California, Davis, CA, USA*
BIN ZHAO • *State Key Laboratory of Environmental Chemistry and Ecotoxicology, Research Center for Eco-Environmental Sciences, Chinese Academy of Sciences, Beijing, China*

Chapter 1

Small Molecule Formulation Screening Strategies in Drug Discovery

Gary W. Caldwell, Becki Hasting, John A. Masucci, and Zhengyin Yan

Abstract

The correct formulation of new drug candidate compounds in drug discovery is mandatory since the majority of go/no-go decisions to advance candidates are based on *in vitro* ADME, receptor binding, *in vivo* pharmacokinetic and efficacy screens. For this reason, having a rapid formulation screen would be a valuable tool for chemists and biologists working in drug discovery. This chapter will describe a rapid solubilization screen that consumes minimal amounts of compound using an HPLC detection method to measure the solubility of drug candidates in various formulations. Using the pKa and Log P of the compound, formulation selection for drug candidates are based on a decision tree approach to guide the user in the selection of appropriate formulations.

Key words Drug discovery formulation, Solubilization techniques, Buffers, Cosolvents, Surfactants Complexants, Lipids

1 Introduction

The main goal of drug discovery research is to select drug candidates that are worthy of becoming preclinical candidates [1]. These preclinical candidates receive more extensive and time-consuming development in the hope of entering clinical testing. From clinical testing, medicines emerge which are commercialized. This pharmaceutical drug discovery/development process requires an enormous financial investment since a decade or longer is required to take a drug candidate to commercialization [2]. In addition, it requires a large interdisciplinary team of scientists and support staff working seamlessly together with a focused management team.

There is a high attrition rate of drug candidates in preclinical and clinical development due primarily to insufficient efficacy, safety issues, and/or economic reasons. Efficacy and safety deficiencies can be related in part to poor oral absorption, distribution, metabolism and excretion (ADME) properties, pharmacokinetics (PK), toxicokinetics (TK), and formulation issues [3].

Biopharmaceutics

Solid (Salt type, Polymorphism, Amorphism, Hygroscopicity, Melting Point, Particle Size, Surface Area, and Stability)
Solubility (Dissolution Rates, Ksp, pKa, Log P/Log D, Hydrophobicity, and Stability)
Permeability (Molecular Weight, Log P, polar surface area, and Flexibility)

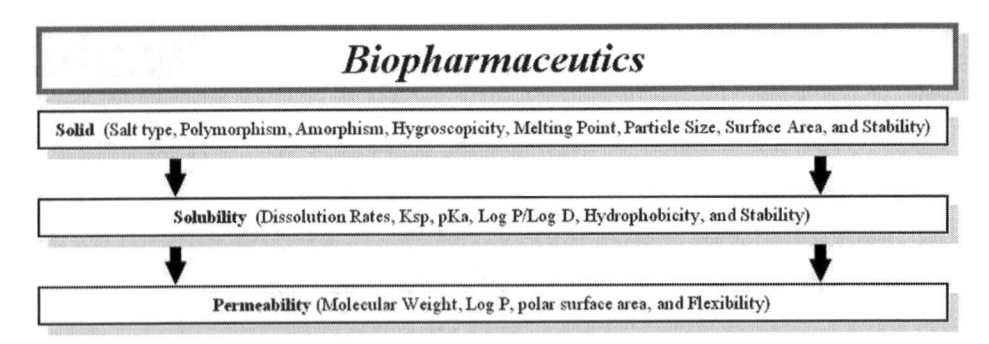

Fig. 1 Factors involved in biopharmaceutics

Therefore, most pharmaceutical companies today use panels of well-characterized ADME/PK, toxicity and formulation screens in parallel with *in vivo* efficacy and safety assays to identify drug candidates that have the potential of becoming preclinical & clinical candidates [4, 5]. In this chapter, we will describe a biopharmaceutics strategy to understand formulation issues and use this information to create formulation screens that can be used at the early stages of drug discovery research to de-risk drug candidates before entering preclinical development. The selections of appropriate formulations are based on a decision tree approach that is used by many pharmaceutical companies.

There are many excellent books and research papers that cover the theoretical [6–11] and practical [12–17] aspects of biopharmaceutics including drug formulation in the pharmaceutical industry or related industries. From a drug discovery biopharmaceutics point of view, physicochemical parameters of drug candidates including understanding the solid-state characteristics of the drug, solubility and permeability need to be evaluated with the highest degree of accuracy in the shortest amount of time using the least amount of drug compound (Fig. 1). Since the majority of drug candidates in drug discovery are solids at room temperatures, solid-state characteristics typically involve the investigation of salt type, polymorphism/amorphism tendencies, melting point, hygroscopicity, particle size distribution, specific surface area and stress stability. The characteristics of a drug candidate in solution involve dissolution rates, dissociation constants (Ksp), ionization potential (pKa), lipophilic partition coefficients (Log P) as a function of pH (Log D), hydrophobicity, and stability in solution. Molecular factors important to the permeability of a drug candidate include its molecular weight (MW), lipophilicity/hydrophobicity tendency, polar surface area (PSA) and the number of rotatable bonds (flexibility).

The relationship between solubility and permeability has been described by the Biopharmaceutical Classification System (BCS)

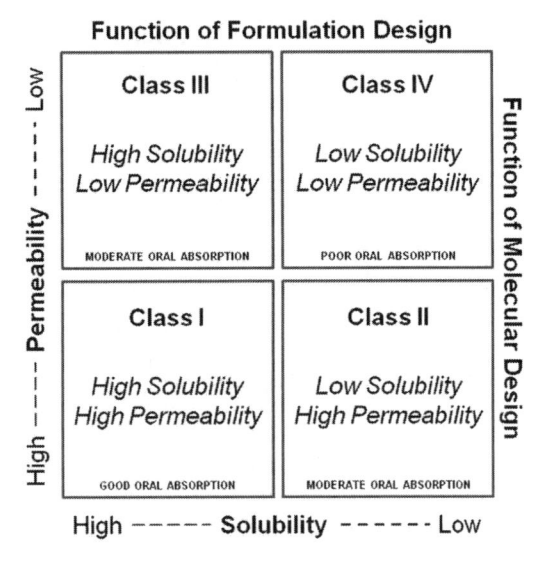

Fig. 2 Factors involved in oral absorption

classification (Fig. 2) [18, 19]. While the BCS was originally proposed to provide a scientific basis for an FDA biowaiver for conducting human bioavailability and bioequivalence studies, it is also a useful system to understand the formulation and molecular optimization needs of drug candidates [20, 21]. That is, oral absorption for small molecule drug candidates (i.e., compounds with MW <1,000 Da) is a dynamic process that involves the transfer of drug molecules from the stomach to the gastrointestinal lumen followed by transfer of the drug molecule across the apical intestinal epithelium membrane followed by diffusion through the cytoplasm and finally exiting through the basolateral membrane into the portal blood system (i.e., transcellular passage). The transcellular drug flux across the intestinal membrane is a product of the drug concentration (i.e., solubility) in the luminal fluid and the rate that the drug travels from the apical side of the epithelium cell to the basolateral side (i.e., permeability). Thus, a qualitative and quantitative understanding of solubility and permeability is essential in drug discovery to de-risk drug candidates against having poor oral absorption characteristics. Consider Fig. 2, a drug is classified as Class I if it has high solubility and high permeability; that is, good oral absorption characteristics. Hydrophobic drugs are typically classified as Class IV since they have low solubility and low permeability or, in other words, poor oral absorption characteristics. Class II and Class III represents drugs that have moderate oral absorption characteristics due to either having high or low solubility or permeability. Low and high cutoff values for solubility and permeability are typically based on animal or human physiological parameters.

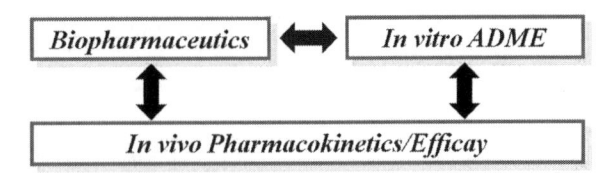

Fig. 3 Integrated "drug developability" assessment of drug candidates

By combining the biopharmaceutical profile as outlined in Figs. 1 and 2 with the results of early *in vitro* absorption, distribution, metabolism and excretion (eADME) assays [3, 4] leads to an integrated "drug developability" assessment of drug candidates for *in vivo* pharmacokinetic and efficacy studies (Fig. 3). Rapid and simple methods for classifying drug candidates according to Fig. 2 have been developed. However, a universally agreed upon choice of assays for solubility and permeability that provide an adequate biopharmaceutics assessment of drug candidates at the drug discovery stage has not been established. The primary reason for this situation is that drug candidate compounds, which are synthesized in small batches in drug discovery medicinal chemistry groups, are often only available in limited amounts (i.e., 5–20 mg) with varying degrees of purity (i.e., 70–95 %) from batch to batch. In addition, these candidate compounds may be available only as dimethyl sulfoxide (DMSO) stock solutions (i.e., 5–20 mM), which limit the range of assay options. As with all assays or combination of assays, the availability and purity of the drug candidate compound dictates the choice of assay used and the reliability of the data.

There have been more or less two approaches to biopharmaceutical profiling in a drug discovery environment: (a) using high-throughput screens (HTS) where drug candidate compounds are in DMSO stock solutions with varying degrees of purity and accuracy in stock concentrations; (b) using lower throughput screens where drug candidate compounds are initially solids with a higher degree of purity. Some pharmaceutical companies use one approach or the other while others use both approaches in a tiered strategy; that is, thousands of compounds are funneled through a series of high-throughput capacity solubility and permeability assays to lower capacity assays, which reveal more and more detailed information. A typical HTS approach to solubility might be the use of a kinetic or semi-equilibrium solubility assay [10, 22]. Here drug candidate compounds typically start as an inaccurate stock concentration DMSO solution that is both added directly to a buffer [23] or the DMSO is first evaporated and then buffer is added [24]. If the solubility measurement is taken in a short amount of time (i.e., 1 h), it is referred to as a kinetic measurement [25]; if the solution is allowed to equilibrate for a few hours to a day, the method is referred to as a semi-equilibrium measurement [24]. The buffer used in these HTS approaches is typically a phosphate buffer at pH 6.5 or 7.4 [26]. The parallel artificial membrane assay (PAMPA) has been used as a HTS approach to permeability [27, 28]. In this type of assay, a lipophilic microfilter is impregnated with 10 % wt/vol egg lecithin dissolved in n-dodecane to create a filter-immobilized artificial membrane. The filter-immobilized artificial membrane is used in a chamber apparatus to separate an aqueous buffer solution containing the drug candidate compound at a known concentration from an aqueous buffer solution containing no compound. The kinetics of transport by diffusion across the artificial membrane is measured and a permeability parameter is calculated. While these HTS approaches to solubility and permeability provide approximate values, they are not accurate enough for conclusive biopharmaceutical profiling of drug candidates and many pharmaceutical companies have selected to stop using them. Balbach and Korn [29] have designed a series of lower throughput assays whereby the solid-state, and solubility characteristics of drug candidates can be evaluated in about 4-weeks using no more than 100 mg of highly purified drug compound. In their approach, the following solid-state characteristics of drug candidates are measured: the dissolution rates, dissociation constants (Ksp), ionization potential (pKa), lipophilic partition coefficients (Log P) as a function of pH (Log D), hydrophobicity, particle size distribution, polymorphism tendency, and stress stability at solid-state. The solubility and stability in pH 1.2–8.0 in fed-state simulating intestinal fluid (FeSSIF) and fasted-state simulating intestinal fluid (FaSSIF) are measured for each drug candidate. The human colorectal carcinoma intestinal cell line (Caco-2) is a cell culture model that is used to measure the permeability of drug candidates [30, 31]. Caco-2 cells spontaneously differentiate on microporous filter membranes into polarized monolayers with tight cellular junctions. The Caco-2 membrane is used in a chamber apparatus to separate an aqueous pH 6.4 buffer solution containing the drug candidate compound at a known concentration from an aqueous pH 7.4 buffer solution initially containing no compound. Drug molecule diffusion across the Caco-2 membrane, from the apical side to the basolateral side, with the permeability parameter being calculated based on the amount drug molecule reaching the basolateral side [32]. The Caco-2 cell model is designed to emulate transcellular drug flux across the intestinal membrane. The more prudent approach for an ideal biopharmaceutical profile of drug candidates would be some combination of moderate throughput solubility and permeability screens using the shortest amount of time (1–2 weeks), and the least amount of drug compounds (5–10 mg) with a high degree of purity (90–95 %).

Once the classification of a drug candidate has been determined, a formulation strategy for *in vitro* and *in vivo* assays can be established (Figs. 2 and 3). For example, in some cases, the solubility of a drug can be improved, from low to high or from low to moderate, by optimizing its formulation. Thus, formulation strategies can be used to re-classify drugs; that is, change a Class II drug to a Class I drug. Formulation strategies for Class III or Class IV drugs may not improve their oral absorption characteristics since they are still compromised by their low membrane permeability. The best strategies for Class III and Class IV drugs are to make molecular design changes to improve their permeability based on physicochemical parameters outlined in Fig. 1. It should be understood that creating sufficient lipophilicity for membrane permeability and receptor binding, while polar enough for aqueous solubility, is not a trivial medicinal chemistry problem. Therefore, the correct choice of formulation of drug candidates in drug discovery is mandatory since many go/no-go decisions for the advancement of candidates are based on *in vitro* ADME and *in vivo* pharmacokinetic screens.

There are more or less three formulation approaches that are either used individually or in combination with each other to enhance the solubility of poorly soluble drug candidate compounds (i.e., <10 μg/mL) in a drug discovery environment [33–35]: (a) in some cases, the drug formulation strategy is oriented toward the creation of suspensions (i.e., supersaturating solutions) using polymers including methyl cellulose (MC), hydroxylethyl cellulose (HEC) or hydroxypropyl cellulose (HPC) [36]; (b) solubilization techniques using enhancers to aqueous media such as, buffers, cosolvents, surfactants (micellar system), and complexants are used to increase solubility [11]; (c) lipid-based formulations including lipid solutions, lipid emulsions, lipid dispersions, self-emulsifying drug delivery systems (SEDDS) and self-microemulsifying drug delivery systems (SMEDDS) have been investigated [37, 38]. In addition, rapid dissolving solid state formulations using drug particle engineering to enhance drug is also applied. These formulations include solid dispersions, nanoparticles and co-ground mixtures [39].

It is imperative that formulations need to be *in vitro* and *in vivo* biocompatible and stable. That is, the primary purpose of enhancing the solubility of poorly soluble drug candidates is to acquire sufficient *in vitro* and *in vivo* exposure without interfering with the experimental interpretation of the data. For example, using suspension formulations can lead to miss interpretation of the experimental results since the dissolution rate of the solid is typically not measured. In addition, physical stability of the drug candidate compound is a major issue of suspension formulations and short-term storage. Extreme pH conditions and cosolvents can have biocompatibility issues due to tissue irritation and drug precipitation in the lumen of the gastrointestinal tract. While cyclodextrins have an acceptable safety profile, there are still concerns of nephrotoxicity [40] especially at high acute concentration or in chronic studies. Some surfactants have systemic toxicity including histamine release and adverse cardiovascular effects and are poorly tolerated in chronic studies [34]. In general, nephrotoxicity is a concern for lipid-based formulations. In addition, formulations should not mask the pharmacological effect being studied, such as, avoiding ethanol formulations when investigating CNS behavioral effects or dextrose-based formulation in diabetic animal models. While many formulations based on cosolvents, surfactants, and complexants will be acceptable for animal studies, they may not be acceptable in human studies.

We will focus on formulation approaches using buffers, cosolvents, surfactants (micellar system), and complexants that are either used individually or in combination with each other. Various solubilization techniques have been developed to alter the solubility and dissolution rates of small molecules in aqueous media [11–16]. These techniques range from simple methods such as the addition of 0.9 % sodium chloride (NaCl) or 5 % dextrose (D5W) to water for intravenous (i.v.) dosing to more complex strategies based on the addition of enhancers to aqueous media such as, buffers, cosolvents, surfactants (micellar system), complexants and lipids. A short list of common solubilizing enhancers is shown in Table 1 [33] where the recommended percent of the enhancer is given along with the route of administration of the dose. These solubilizing enhancers are either used individually or in combination with other enhancers or other solid-state particle size reduction techniques. For example, control of the pH of the aqueous media using buffers and particle size reduction using a mortar and pestle are the most common methods of enhancing the solubility of drug candidates that are weak acids or bases that ionize at physiological pH 2–9. Individual or combination formulations can be created using, cosolvents, surfactants, and complexants in combination with pH adjustments for weak electrolytes drug candidates along with solid-state nanoparticle size reduction methods [39].

Based upon these solubilization techniques (i.e., buffers, cosolvents, surfactants, and complexants) several formulation decision tree strategies have been reported in the literature. For example, Lee and coworkers [41] have applied these solubilization techniques to i.v. formulations for 317 drug candidates and were able to formulate over 80 % of the compounds. Gopinathan and coworkers [42] have applied these solubilization techniques to oral formulations for 26 drug candidates. Using 54 formulation conditions, all drug candidates could be formulated. The formulation conditions can be based on a decision tree approach using the pKa and Log P of the compound to guide the user in the selection of appropriate formulations (Fig. 4). Following the decision tree in Fig. 4, if the drug candidate molecule has acceptable aqueous solubility, then an aqueous formulation is selected. If the aqueous solubility is unacceptable and the drug candidate molecule has ionizable functional groups, a buffer formulation is attempted; if acceptable, a pH-based formulation is selected. If changing the pH is unacceptable and the Log P <3, then cosolvents are tried to improve the solubility. If this

Table 1
Buffers, cosolvents, surfactants, complexants and lipids for altering solubility [33]

Buffers[a] (pH range) (route)	Cosolvents[b] (% range)[f] (route)	Surfactants[c] (% range)[f] (route)	Complexants[d] (% range)[f] (route)	Lipids[e] (% range)[f] (route)
Acetic acid (pH 4–6) (oral, i.v.)	DMA (10–30 %) (i.v.)	Cremophor EL (5–10 %) (oral, i.v.)	HPβCD (20–40 %) (oral, i.v.)	Soybean oil (50–100 %) (oral)
Na citric (pH 2.5–7) (oral, i.v.)	Ethanol (5–10 %) (oral, i.v.)	Gelucire 44/14 (20–50 %) (oral)	SBEβCD (Captisol) (20–40 %) (oral, i.v.)	Labrafil 1944CS (30–60 %) (oral, i.v.)
Lactic acid (pH 3–4.5) (oral, i.v.)	NMP (5–20 %) (oral, i.v.)	Lecithin (20–50 %) (oral, i.v.)		Capmul MCM (30–60 %) (oral, i.v.)
Maleic acid (pH 2–3) (oral, i.v.)	PG (20–60 %) (oral, i.v.)	Pluronic F68 (20–50 %) (oral, i.v.)		
Na bicarbonate (pH 4–9) (oral, i.v.)	PEG-400 (20–100 %) (oral, i.v.)	Solutol HS (20–50 %) (oral, i.v.)		
Na phosphate (pH 6–8) (oral, i.v.)	Transcutol (5–30 %) (oral)	Na cholate (10–20 %) (oral, i.v.)		
Tris (pH 8–9) (oral, i.v.)		Tween 80 (5–10 %) (oral, i.v.)		

[a]Adjustment of pH can also be accomplished by using hydrochloride acid or sodium hydroxide. Tris (tris (hydroxymethyl) aminomethane)
[b]DMSO (dimethyl sulfoxide), DMA (N,N-dimethylacetamide), NMP (N-methylpyrrolidon) PG (Propylene Glycol), PEG (polyethylene glycol), Transcutol (diethylene glycol monoethyl ether)
[c]Cremophor El (polyoxyl-35 castor oil), Gelucire 44/14 (lauroyl macrogol-32 glycerides), Lecithin (phosphatidylcholine), Pluronic F68 (81 % PEG and 19 % poly-PG), Solutol HS (macrogol-15-hydroxystearate), Na Cholate (sodium cholate), Tween 80 (polyoxyethylene-sorbitanmonooleate 80)
[d]HPβCD (hydroxyl-β-cyclodextrin), SBEβCD (sulfobutylether-β-cyclodextrin)
[e]Labrafil 1944CS (polyoxyethyllated oleic glycerides), Capmul MCM (medium chain mono- and diglycerides)
[f]% range for rat or mouse for a 10 mL/kg dose

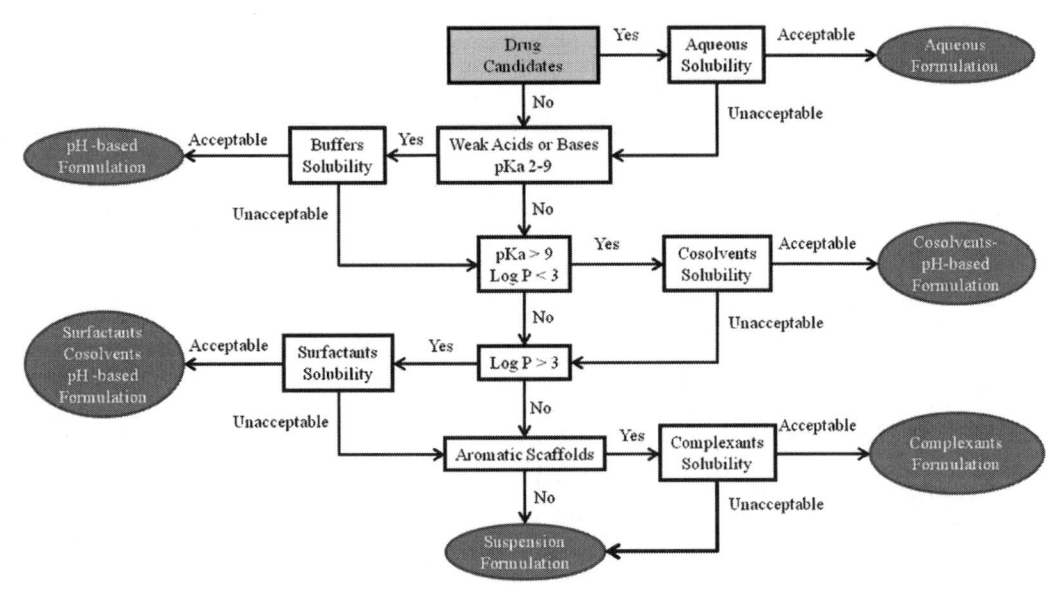

Fig. 4 Formulation decision tree

fails and the Log P >3, then surfactants can be selected. If the drug candidate molecule has aromatic scaffolds, then complexants such as cyclodextrins can be selected. If all soluble formulation conditions fail, a suspension-based formulation can be used.

Since there is only a limited amount of compound synthesized in drug discovery (i.e., typically 5–20 mg), it is not possible to try all formulation combinations listed in Table 1 using a formulation decision tree. However, there are many suggestions in the literature on individual or combinations of buffers, cosolvents, surfactants, and complexants that appear to work for many different types of chemical scaffolds [33–35, 41, 42]. For example, following the decision tree, one would select:

- pH 4 (Citrate) or 8 (Tris) for ionizable compounds
- 5 % NMP (pH 4 or 8)
- 30 % PEG-400
- 30 % PG
- 40 % PEG-400/10 % EtOH
- 10 % DMA/10 % EtOH/20 % PEG-400
- 10 % DMA/10 % EtOH/20 % PG
- 10 % DMA/20 % PG/40 % PEG-400
- 10 % Cremophor EL
- 10 % Solutol (pH 4 or 8)
- 10 % Cremophor EL/20 % PEG-400

- 10 % Cremophor EL/10 % EtOH
- 20 % HPβCD (pH 4 or 8)
- 20 % Captisol (pH 4 or 8)
- and so on

Thus, if the drug candidate molecule has an ionizable group, then using a citrate or Tris buffer would be selected. If the buffers failed to produce an acceptable solubility (i.e., 1 mg/mL) and the drug candidate molecule had a Log P <3, then a cosolvent such as PEG or PG would be selected. If these cosolvents failed to give an acceptable solubility, combinations of cosolvents would be selected. In the above list, there are four suggested cosolvent combinations. If the drug candidate molecule has a Log P >3, then a surfactant such as Cremophor EL or Solutol or combinations with cosolvents PEG-400 and ethanol would be selected. If the drug candidate molecule has aromatic scaffolds, then complexants such as HPβCD or Captisol at pH 4 and 8 would be selected. While combinations of cyclodextrins and cosolvents are acceptable, combinations between cyclodextrins and surfactants should be avoided primarily due to potential nephrotoxicity [40].

A general rule of thumb is that typical drug discovery PK studies in small animal models require a solubility range of 1–2 mg/mL for doses ranging from 1 to 10 mg/kg. For toxicokinetics studies, a solubility range of 25–50 mg/mL may be required to support future studies.

In this chapter, we have provided a step-by-step protocol of a 96-well plate liquid chromatography-based assay used for rapid identification of formulation conditions for poorly soluble drug candidates using buffers, cosolvents, surfactants and complexants [43]. Briefly, a pre-weighed drug candidate compound is dissolved in an organic solvent, aliquoted into a 96-well plate and dried to obtain accurate sub-milligram quantity of the compound in each well. A panel of buffers, cosolvents, surfactants, and complexants is selected for solubility determination in these different formulations. The formulations are shaken for 24 h and analyzed by an HPLC/UV method after precipitated compound is removed by centrifugation. This type of preformulation data is useful in minimizing the challenges in late stage drug development efforts where stable biopharmaceutically suitable drug dosage forms are required.

2 Materials

2.1 Equipment

Specific equipment used in this assay is described; however any model of comparable capability can be easily substituted.

1. Hot plate (VWR, Bridgeport, NJ)
2. Vacuum oven (VWR, Bridgeport, NJ)

3. Centrifuge (VWR, Bridgeport, NJ)
4. pH meter (VWR, Bridgeport, NJ)
5. Multi-channel (8) automatic pipettor (50–1,200 µL) (VWR, Bridgeport, NJ)
6. 96-well plate cover plate lids (VWR, Bridgeport, NJ)
7. Sonicator (Branson 2510, Thomas Scientific, Swedesboro, NJ)
8. 96 well plates (VWR, Bridgeport, NJ; Cat# 46600-670)
9. Hoefer Red Rotor Shaker (Harvard Bioscience, Inc. Holliston, Massachusetts)
10. HPLC (Agilent 1100) (Agilent Technologies, Inc., Santa Clara, CA, USA)

2.2 Reagents

1. Solutol (Sigma-Aldrich, St. Louis, MO; Cat# 42966).
2. 100 mM Citrate buffer, pH 4.0 (VWR, Bridgeport, NJ; Cat# 100219-854)—dilute this stock solution to 30 mM buffer.
3. 1 M Tris buffer pH- 8.0 (Mediatech, Inc. A Corning Subsidiary, Manassas, VA; Cellgro Cat# 46031CM)—dilute this stock solution to 30 mM buffer.
4. Hydroxy propyl beta cyclodextrine (HPβCD) (Sigma-Aldrich, St. Louis, MO; Cat# H107).
5. Captisol (SBEβCD) (CyDex, Inc. La Jolla, CA; Cat# CY03A).
6. HPLC grade acetonitrile (EMD Chemicals, Darmstadt, Germany; Cat# EM-AX0145).
7. Trifluoroacetic acid (TFA) (EMD Chemicals, Darmstadt, Germany; Cat# EM-TX1276-6).
8. Test Compounds: Betaestradiol, Verapamil, Propranolol, Ketoconazole, Diethylstilbestrol, 4,5 Diphenylimidazole, Itraconazole, Griseofulvin, and Testosterone (Sigma-Aldrich, St. Louis, MO) (*see* **Note 1**).

2.3 Formulation Solubility Procedure

Prepare the following six formulations to be tested using combinations of buffers, cosolvents, surfactants, and complexants (Table 1) and Fig. 4 (*see* **Note 2**).

1. **10 % Solutol (pH 4):** Melt the solutol with a hot plate in a beaker and mix 10 mL of solutol with 90 mL of 30 mM citrate buffer at pH 4. Sonicate final volume.
2. **10 % Solutol (pH 8)** Melt the solutol with a hot plate in a beaker and mix 10 mL of solutol with 90 mL of 30 mM Tris buffer at pH 8. Sonicate final volume.
3. **20 % HPβCD (pH 4)** Weigh 20 g of HPβCD and dissolve it in 100 mL of 30 mM sodium citrate buffer. Sonicate final volume.

4. **20 % HPβCD (pH 8)** Weigh 20 g of HPβCD and dissolve it in 100 mL of 30 mM Tris buffer. Sonicate final volume.

5. **20 % Captisol (pH 4)** Weigh 20 g of Captisol and dissolve it in 100 mL of 30 mM sodium citrate buffer. Sonicate final volume.

6. **20 % Captisol**—Weigh 20 g of Captisol and dissolve it in distilled water (pH 7) to 100 mL final volume and sonicate.

2.4 Test Compound Solubility Procedure

Prepare 3 mg/mL solutions of the test and control compounds.

1. Weigh out 5 mg of each compound
2. Dissolve compounds in 1.67 mL of acetonitrile or methanol (*see* **Note 3**)

3 Methods

3.1 Assay Procedure

1. Place 100 μL of control in all wells in column 1 except G1 of a 96 well plate (Fig. 5) (*see* **Note 4**).
2. Place 100 μL of test compound(s) in the same manner down a column, leaving G empty.

	1	2	3	4	5	6	7	8	9	10	11	12
A 10% Solutol pH 4.0	Control	Compound 1	Compound 2	Compound 3	Compound 4	Compound 5	Compound 6	Compound 7	Compound 8	Compound 9	Compound 10	Compound 11
B 10% Solutol pH 8.0	Control	Compound 1	Compound 2	Compound 3	Compound 4	Compound 5	Compound 6	Compound 7	Compound 8	Compound 9	Compound 10	Compound 11
C 20%HPβCD pH 4.0	Control	Compound 1	Compound 2	Compound 3	Compound 4	Compound 5	Compound 6	Compound 7	Compound 8	Compound 9	Compound 10	Compound 11
D 20 %HPβCD pH 8.0	Control	Compound 1	Compound 2	Compound 3	Compound 4	Compound 5	Compound 6	Compound 7	Compound 8	Compound 9	Compound 10	Compound 11
E 20 % Captisol pH 4.0	Control	Compound 1	Compound 2	Compound 3	Compound 4	Compound 5	Compound 6	Compound 7	Compound 8	Compound 9	Compound 10	Compound 11
F 20 % Captisol	Control	Compound 1	Compound 2	Compound 3	Compound 4	Compound 5	Compound 6	Compound 7	Compound 8	Compound 9	Compound 10	Compound 11
G												
H Organic Solvent	Control	Compound 1	Compound 2	Compound 3	Compound 4	Compound 5	Compound 6	Compound 7	Compound 8	Compound 9	Compound 10	Compound 11

Fig. 5 Plate layout

Formulation Screening in Drug Discovery 13

3. Evaporate the organic via a low flow nitrogen gas (30 min at 25 °C) (*see* **Note 5**).
4. Place the six formulations in the water bath for at least 20 min before adding them to the plate. This can be done while evaporating off organic solvents.
5. Place 100 μL of each formulation in rows A-F (1 into A, 2 into B etc.). At this point there is 3 mg/mL in each well.
6. No formulations are placed into row H at this time.
7. Cover with plate lid and put on rotor shaker for 24 h.
8. After 24 h, spin in centrifuge for 10 min at 3,000 rpm.
9. Prepare the standards at 1.5 mg/mL by adding 200 μL of acetonitrile or methanol to row H and mix well.
10. Transfer 80 μL of the test solutions to a new 96-well plate (ROWS A-F), and 160 μL of the 1.5 mg/mL standards (ROW H).
11. Cover with a plate lid, analyze the samples by HPLC with UV/VIS detection (*see* **Notes 6** and **7**).

3.2 HPLC Method

3.2.1 HPLC Conditions

1. Software: Agilent OpenLAB CDS CS Workstation C.01.04 (ENG).
2. Instrument: Agilent 1100 pump.
3. Auto sampler: Agilent 1100 autosampler.
4. Detector: Agilent Diode array detector (DAD).
5. Column: Inertsil ODS-3, particle size 3 μ 2.1 × 50 mm, P/N5020-04412.
6. Detector Setting: 220–320 nm depending on the compound.
7. Injection volume: 1 μL.
8. Flow Rate: 1.0 mL/min.
9. Column Temperature: 60 °C.
10. Mobile Phase: A. Water with 0.1 % TFA; B. Acetonitrile with 0.1 % TFA.

3.2.2 HPLC Gradient

The HPLC gradient was set up as follows:

Time (min)	%B
0	5
0.5	5
3.5	95
4.5	95
5.0	5
6.5	5

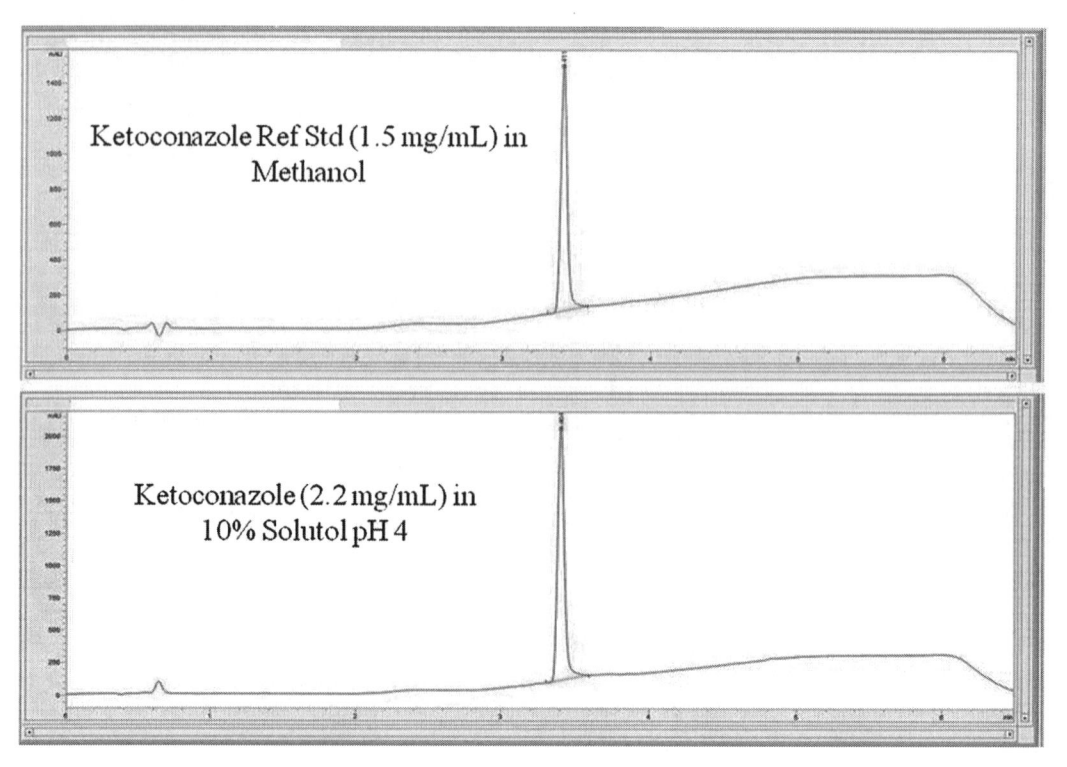

Fig. 6 HPLC traces of ketoconazole

3.2.3 Analysis Time and Injection Sequence

The analysis time was 6.5 min per sample. The entire experiment requires about 10 h with a 96-well plate containing 11 test compounds and one positive control and six formulation conditions. The 1.5 mg/mL standard samples are injected three times followed by the six formulations (*see* **Note 8**).

3.2.4 Data Processing

To calculate the concentration of the test compounds (Fig. 6), the average peak area of the three standard injections is calculated. This single point calibration value is used to determine the concentration of the test compound using the following formula:

$$Conc_{Test} = \left(\frac{Conc_{std}}{Area_{std}} \right) Area_{Test} \qquad (1)$$

3.3 Marker Compounds

The concentration (mg/mL) of nine test compounds was measured in six different formulations on five consecutive days (Table 2). Note that the upper limit of solubility in this assay was 3 mg/mL. As shown in Table 2, the upper limit on solubility was not obtained for any of the formulations tested. The %CV was typically 15 % or less for all compounds except for Itraconazole which was about 20 %. The experiment was repeated if there was more than 30 % average deviation between the replicates.

Table 2
Solubility of test compounds in various formulations

	10 % Solutol pH 4 (mg/mL) (%CV)	10 % Solutol pH 8 (mg/mL) (%CV)	20 % HPβCD pH 4 (mg/mL) (%CV)	20 % HPβCD pH 8 (mg/mL) (%CV)	20 % Captisol (mg/mL) (%CV)	20 % Captisol pH 4 (mg/mL) (%CV)
Betaestradiol	ND	ND	2.5 (14 %)	2.4 (12 %)	2.4 (12 %)	2.4 (13 %)
Verapamil	2.5 (5 %)	2.6 (7 %)	2.6 (6 %)	2.6 (6 %)	2.6 (7 %)	2.5 (6 %)
Propranolol	2.0 (4 %)	2.0 (5 %)	2.0 (2 %)	2.0 (3 %)	2.0 (3 %)	2.1 (4 %)
Ketoconazole	2.2 (10 %)	2.1 (7 %)	2.3 (11 %)	2.3 (11 %)	2.3 (10 %)	2.3 (9 %)
Diethylstilbestrol	1.5 (16 %)	1.9 (11 %)	2.4 (13 %)	2.4 (9 %)	2.3 (13 %)	2.3 (9 %)
4,5 Diphenylimidazole	1.4 (10 %)	0.7 (10 %)	1.5 (11 %)	0.3 (12 %)	0.8 (9 %)	0.2 (9 %)
Itraconazole	ND	ND	1.7 (20 %)	1.7 (20 %)	0.8 (23 %)	1.3 (20 %)
Griseofulvin	0.2 (8 %)	0.3 (13 %)	ND	ND	ND	ND
Testosterone	0.7 (10 %)	0.8 (9 %)	0.2 (8 %)	0.2 (9 %)	0.2 (9 %)	0.2 (9 %)

3.4 Validation Studies

Since all the test compounds were available commercially in high purity, we wanted to see how real drug candidate compounds synthesized in small batches would behave in the assay. We were particularly interested in seeing if the buffering capacity of the citrate (pH 4) (Table 3) and Tris (pH 8) (Table 4) would be maintained. As shown in Tables 3 and 4, on average the pH of the formulations were maintained using compounds with unknown salt content, hydration, and trifluoroacetic acid contamination. However, there were a few exceptions particularly Compound 9 in Table 3 and Compound 2 in Table 4 (*see* **Note 9**). The reason for this pH change is not known.

4 Conclusion

It is important to study the formulation behavior upon oral and i.v. administration. Upon oral administration, aqueous dilution of the formulation occurs in the gastrointestinal lumen and upon i.v. bolus injection, dilution of the formulation occurs in the systemic system. Therefore, a major limitation of using cosolvents is potential compound precipitation in the lumen once the formulation is dosed. Surfactant formulations are the least likely to precipitate upon dilution. There are a number of *in vitro* methods available to assess compound precipitation in formulations [33–35]. In addition, with formulations that are stored longer than 24 h before use, stability studies should also be performed to ensure the quality of the formulations.

Table 3
pH values for real drug discovery compounds using a citrate buffer

	30 % PEG-400 pH 4 (pH)	30 % PEG-400 5 % NMP pH 4 (pH)	30 % PG pH 4 (pH)	30 % PG 5 % NMP pH 4 (pH)	20 % HPβCD pH 4 (pH)	20 % HPβCD 5 % NMP pH 4 (pH)
Compound 1	4.3	4.5	4.0	4.1	4.5	4.2
Compound 2	3.5	3.5	3.3	3.4	3.1	3.2
Compound 3	4.2	4.8	3.9	4.1	3.3	3.5
Compound 4	4.2	4.1	3.4	3.4	3.2	3.2
Compound 5	5.2	5.0	4.5	4.5	4.6	4.5
Compound 6	4.5	4.6	3.8	3.9	3.8	3.7
Compound 7	3.7	3.5	3.3	3.4	3.0	3.2
Compound 8	4.2	4.6	4.0	4.7	3.9	3.7
Compound 9	6.5	6.6	7.9	7.8	7.5	7.6
Compound 10	4.3	4.0	4.4	4.5	4.3	4.0
Compound 11	4.2	4.4	4.1	4.2	4.1	3.8

It is possible to test different salt forms of free bases in our assay by using *in situ* salt formation techniques [44]. Briefly, depending on the functional group present in the drug candidate molecule, an acid is selected (e.g., HCl or Tartaric acid) that is 2 pH units lower than the pKa of the free base. The acid is added in excess to a solution containing a known amount of the free base and the solvent is removed after 24 h. Different formulations can be added to this new salt form and the solubility measured using the procedure outlined above. A key requirement to ensure the successful generation of appropriate salts is the need for the free base to be highly pure. Compounds that are generated in drug discovery by medicinal chemistry groups are sometimes contaminated with trifluoroacetic acid (TFA), since TFA is a common acidic modifier in preparative reversed-phase HPLC methods [45].

5 Notes

1. Testosterone is a schedule III controlled substance in the United States. A controlled substance license, from the Drug Enforcement Agency, is required to purchase with this material.
2. Any combination in Table 1 can be used in this formulation screen. Here we are using six formulations to illustrate the assay.

Table 4
pH values for real drug discovery compounds using a Tris buffer

	30 % PEG-400 pH 8 (pH)	30 % PEG-400 5 % NMP pH 8 (pH)	30 % PG pH 8 (pH)	30 % PG 5 % NMP pH 8 (pH)	20 % HPβCD pH 8 (pH)	20 % HPβCD 5 % NMP pH 8 (pH)
Compound 1	7.3	7.6	8.1	8.2	7.8	7.8
Compound 2	5.4	5.3	7.5	7.7	7.1	7.2
Compound 3	7.4	7.0	8.8	8.8	8.0	8.4
Compound 4	6.4	6.9	7.9	7.9	7.7	7.3
Compound 5	7.3	7.9	8.3	8.5	8.1	8.0
Compound 6	7.0	7.3	7.9	8.3	8.5	8.8
Compound 7	7.2	8.0	7.2	7.8	7.6	7.7
Compound 8	7.6	7.5	7.7	7.8	7.5	7.5
Compound 9	8.3	8.5	8.5	8.9	8.5	8.8
Compound 10	7.2	7.3	7.9	8.0	7.3	7.4
Compound 11	7.5	7.7	7.9	8.1	7.8	7.9

3. This step sets the upper limit for solubility testing. In this case, it is 3 mg/mL; however, this can be increased or decreased depending on the situation. For example, to increase the upper limits for solubility testing to 6 mg/mL decrease the volume of the organic solvent to 835 µL. A smaller amount of compound can be utilized as long as it can be accurately weighed out. We have used 2.5 mg in 835 µL (i.e., 3 mg/mL); however, it is time consuming to accurately weigh out these amounts. Finally, make sure the compounds tested go into solution in the organic solvent prior to beginning the experiment, since the quantitative results depend upon this step.

4. We are only using six formulations to demonstrate the method. However, a maximum of seven formulations can be used per 96-well plate with a maximum of 11 test compounds. One positive control (i.e., ketoconazole) is used in each experiment.

5. When evaporating the compounds make sure the nitrogen flow isn't up too high, or that the plate isn't too close to the nitrogen supply. This can cause splashing resulting in non-accurate amounts of compound in the wells. Once the organic solvent has been evaporated each well will contain 300 µg of compounds. This solvent casting method will produce accurate amounts of NEAT compounds in the wells as long as the starting stock solution concentrations are correct.

6. Watch out for evaporation of standards while running the HPLC. Standard samples are primarily organic solvent and will evaporate quicker than the aqueous samples being tested.

7. Tips that can help with evaporation of standards:

    (a) Inject the standards in triplicate before each unknown, pay attention to the area of the peaks of each standard from the first sample through the last. The area should remain consistent. If it deviates dramatically evaporation could be an issue.

    (b) Using a larger volume, helps to minimize evaporation.

    (c) The LC run time can be shorter as long as the chromatography is still accurate.

8. Make sure the integration parameters are optimized. The ChemStation software has difficulty in integrating up a "slope". Try to narrow the specific time, and set a lower limit for an acceptable area.

9. If the buffering capacity is compromised, this may lead to a suboptimal formulation. The pH of the wells can be measured if necessary.

Acknowledgements

The views expressed here are solely those of the author and do not reflect the opinions of Janssen Research & Development, LLC.

We thank Dr. Andrew Mahan and Malini Dasgupta for expert advice regarding HPLC method development.

References

1. Caldwell GW, Yan Z, Masucci JA, Hageman W, Leo G, Ritchie DM (2003) Applied pharmacokinetics in drug development: an overview of drug discovery. Pharm Dev Regul 1(2):117–132
2. Kaitin KI, DiMasi JA (2011) Pharmaceutical innovation in the 21st century: new drug approvals in the first decade, 2000–2009. Clin Pharmacol Ther 89(2):183–188
3. Caldwell GW (2000) Compound optimization in early- and late-phase drug discovery: acceptable pharmacokinetics properties utilizing combined physicochemical, in vitro and in vivo screens. Curr Opin Drug Discov Devel 3:30–41
4. Caldwell GW, Ritchie DM, Masucci JA, Hageman W, Yan Z (2001) The new pre-preclinical paradigm: compound optimization in early and late phase drug discovery. Curr Top Med Chem 1(5):353–366
5. Kerns EH, Di L (2002) Multivariate pharmaceutical profiling for drug discovery. Curr Top Med Chem 2(1):87–98
6. Hsieh C-M, Wang S, Lin S-T, Sandler SI (2011) A predictive model for the solubility and octanol-water partition coefficient of pharmaceuticals. J Chem Eng Data 56(4):936–945
7. Jouyban A (2008) Review of the cosolvency models for predicting solubility of drugs in water-cosolvent mixtures. J Pharm Pharm Sci 11(1):32–58
8. Jouyban-Gharamaleki A (1998) The modified Wilson model and predicting drug solubility in water-cosolvent mixtures. Chem Pharm Bull 46(6):1058–1061
9. Adjei A, Newburger J, Martin A (1980) Extended Hildebrand approach: solubility of caffeine in dioxane-water mixtures. J Pharm Sci 69(6):659–661

10. Avdeef A (2001) Physicochemical profiling (solubility, permeability and charge state). Curr Top Med Chem 1(4):277–351
11. Yalkowsky SH (1999) Solubility and solubilization in aqueous media. Oxford University Press, New York, NY, p 464
12. Trivedi JS (2008) Solubilization using cosolvent approach. In: Liu R (ed) Water-insoluble drug formulation, 2nd edn. CRC Press, Boca Raton, FL, pp 161–194
13. Steele G (2009) Preformulation as an aid to product design in early drug development. In: Gibson M (ed) Pharmaceutical preformulation and formulation, 2nd edn. Informa Healthcare, New York, NY, pp 188–246
14. Guo J, Elzinga, Paul A, Hageman MJ, Herron JN (2008) Rapid throughput solubility screening method for BCS class II drugs in animal GI fluids and simulated human GI fluids using a 96-well format. J Pharm Sci 97(4):1427–1442
15. Gibson M (2009) Pharmaceutical preformulation and formulation: a practical guide from candidate drug selection to commercial dosage form, 2nd edn, Drugs and the pharmaceutical sciences. Informa Healthcare, New York, NY, p 541
16. Tong W-Q (2007) Practical aspects of solubility determination in pharmaceutical preformulation, vol 6. Springer, New York, NY, pp 137–149
17. Heinrich Stahl P, Wermuth CG (2008) Handbook of pharmaceutical salts: properties, selection, and use. Hevetica Chimica Acta, Zurich, p 374
18. Amidon GL, Lennernas H, Shah VP, Crison JR (1995) A theoretical basis for a biopharmaceutic drug classification: the correlation of in vitro drug product dissolution and in vivo bioavailability. Pharm Res 12(3):413–420
19. Chen M-L, Amidon GL, Benet LZ, Lennernas H, Yu LX (2011) The BCS, BDDCS, and regulatory guidances. Pharm Res 28(7):1774–1778
20. Wu C-Y, Benet LZ (2005) Predicting drug disposition via application of BCS: transport/absorption/elimination interplay and development of a biopharmaceutics drug disposition classification system. Pharm Res 22(1):11–23
21. Butler JM, Dressman JB (2010) The developability classification system: application of biopharmaceutics concepts to formulation development. J Pharm Sci 99(12):4940–4954
22. Di L, Kerns EH (2007) Solubility issues in early discovery and HTS. In: Augustijns P, Brewster M (eds) Solvent systems and their selection in pharmaceutics and biopharmaceutics. Springer, New York, NY, pp 111–136
23. Bevan CD, Lloyd RS (2000) A high-throughput screening method for the determination of aqueous drug solubility using laser nephelometry in microtiter plates. Anal Chem 72:1781–1787
24. Alelyunas YW, Liu R, Pelosi-Kilby L, Shen C (2009) Application of dried-DMSO rapid throughput 24-h equilibrium solubility in advancing discovery candidates. Eur J Pharm Sci 37:172–182
25. Fligge TA, Schuler A (2006) Integration of a rapid automated solubility classification into early validation of hits obtained by high throughput screening. J Pharm Biomed Anal 42:449–454
26. Di L, Fish PV, Mano T (2012) Bridging solubility between drug discovery and development. Drug Discov Today 17(9/10):486–495
27. Kansy M, Senner F, Gubernator K (1998) Physicochemical high throughput screening: parallel artificial membrane permeation assay in the description of passive absorption processes. J Med Chem 41(7):1007–1010
28. Ruell JA, Avdeef A (2004) Absorption screening using the PAMPA approach. In: Yan Z, Caldwell GW (eds) Optimization in drug discovery. Humana Press, Totowa, NJ, pp 37–64
29. Stefan B, Christian K (2004) Pharmaceutical evaluation of early development candidates "the 100 mg-approach". Int J Pharm 275(1–2):1–12
30. Hidalgo IJ (2001) Assessing the absorption of new pharmaceuticals. Curr Top Med Chem 1(5):385–401
31. Hu M, Ling J, Lin H, Chen J (2004) Use of Caco-2 cell monolayers to study drug absorption and metabolism. In: Yan Z, Caldwell GW (eds) Optimization in drug discovery. Humana Press, Totowa, NJ, pp 37–64
32. Caldwell GW, Easlick SM, Gunnet J, Masucci JA, Demarest K (1998) In vitro permeability of eight β-blockers through Caco-2 monolayers utilizing liquid chromatography/electrospray ionization mass spectrometry. J Mass Spectrom 33(7):607–614
33. Ping L, Luwei Z (2007) Developing early formulations: practice and perspective. Int J Pharm 341(1–2):1–19
34. Pouton CW (2006) Formulation of poorly water-soluble drugs for oral administration: physicochemical and physiological issues and the lipid formulation classification system. Eur J Pharm Sci 29(3–4):278–287
35. Williams HD, Trevaskis NL, Charman SA, Shanker RM, Charman WN, Pouton CW, Porter CJH (2013) Strategies to address low drug solubility in discovery and development. Pharmacol Rev 65(1):315–499

36. Stillhart C, Cavegn M, Kuentz M (2013) Study of drug supersaturation for rational early formulation screening of surfactant/co-solvent drug delivery systems. J Pharm Pharmacol 65(2):181–192
37. Sakai K, Obata K, Yoshikawa M, Takano R, Shibata M, Maeda H, Mizutani A, Terada K (2012) High drug loading self-microemulsifying/micelle formulation: design by high-throughput formulation screening system and in vivo evaluation. Drug Dev Ind Pharm 38(10):1254–1261
38. Chen X-Q, Gudmundsson OS, Hageman MJ (2012) Application of lipid-based formulations in drug discovery. J Med Chem 55(18):7945–7956
39. Merisko-Liversidge E, Liversidge GG (2011) Nanosizing for oral and parenteral drug delivery: a perspective on formulating poorly-water soluble compounds using wet media milling technology. Adv Drug Deliv Rev 63(6):427–440
40. Irie T, Uekama K (1997) Pharmaceutical applications of cyclodextrins. III. Toxicological issues and safety evaluation. J Pharm Sci 86(2):147–162
41. Lee Y-C, Zocharski PD, Samas B (2003) An intravenous formulation decision tree for discovery compound formulation development. Int J Pharm 253(1–2):111–119
42. Gopinathan S, Nouraldeen A, Wilson AGE (2010) Development and application of a high-throughput formulation screening strategy for oral administration in drug discovery. Future Med Chem 2(9):1391–1398
43. Shanbhag A, Rabel S, Nauka E, Casadevall G, Shivanand P, Eichenbaum G, Mansky P (2008) Method for screening of solid dispersion formulations of low-solubility compounds-miniaturization and automation of solvent casting and dissolution testing. Int J Pharm 351(1–2):209–218
44. Tong W-Q, Whitesell G (1998) In situ screening—a useful technique for discovery support and preformulation studies. Pharm Dev Technol 3(2):215–223
45. Reilly J, Wright P, Larrow J, Mann T, Twomey J, Grondine M, Dodd S, Capacci-Daniel C, Bilotta J, Shah L et al (2012) Counterion quantification for discovery chemistry and preformulation salt screening. J Liq Chrom Relat Tech 35(8):1001–1010

Chapter 2

Assessment of Drug Plasma Protein Binding in Drug Discovery

Dennis Kalamaridis and Nayan Patel

Abstract

Determination of plasma protein binding is important in drug discovery and development for developing PK-PD relationships, and projecting clinical doses because the free drug concentration at the site of action is responsible for the pharmacological activity. This chapter will describe various methods commonly used for the *in vitro* determination of plasma protein binding including classical equilibrium dialysis (CED), rapid equilibrium dialysis (RED), and ultrafiltration Nonspecific binding and mass balance methods used to recover or prevent the loss of compound due to binding to the wall and/or membrane of the apparatus will be briefly discussed as well. Finally, the techniques will be discussed in relation to the type of study required and the method of analysis of both the bound (plasma) and free-fraction (buffer) components and the analytical procedures needed for the quantification of concentrations in these matrices. Techniques using both radiolabeled and cold compounds will be presented as both are usually available at the early or late discovery phases and employ different methods of analysis.

Key words Plasma protein binding, Equilibrium dialysis, RED device, Ultrafiltration, Non-specific binding

1 Introduction

In vitro plasma protein binding experiments that determine the fraction of protein-bound drug are frequently used in drug discovery to guide structure design in order to prioritize compounds for *in vivo* studies. Binding of small molecule drugs to plasma proteins has a major clinical significance *in vivo* for both the pharmacokinetics and pharmacodynamics of a drug. If a drug exhibits low binding to serum proteins, its tissue distribution may be limited. This relationship is especially relevant to poorly hydrosoluble ligands, where binding to serum proteins enhances their solubilization and subsequent tissue distribution. However, when a drug binds too tightly to plasma proteins, the level of free drug may never reach the therapeutic concentration. Therefore, determination of the drug binding properties early during the development process will allow identification of desirable drug candidates.

Fraction unbound (fu) is a critical parameter in estimating therapeutic index, estimating drug-drug interaction potential, developing pharmacokinetics–pharmacodynamics (PK-PD) relationships, and projecting clinical doses because the free drug concentration at the site of action is responsible for the pharmacological activity based on the free drug hypothesis [1].

The degree of protein binding of a therapeutic agent should be considered when interpreting the PK profile of a compound. High plasma protein binding will lead to more drug being present in the central blood compartment and therefore a lower volume of distribution. This relationship might not always hold true if the drug also shows a high degree of binding to tissue proteins. Low protein binding means more drug is free to partition into tissues and will therefore result in a high volume of distribution. Since only free drug is available to exert a pharmacological effect, but in most PK studies it is the total plasma concentration which is measured, the knowledge of the unbound fraction in blood is essential to the correlation of the observed total plasma concentrations with activity [2].

High protein binding of a compound can have important clinical effects with regard to drug toxicity. A small change in the extent of binding of a very highly protein bound drug will equate to a very large change in drug concentration in plasma. These alterations in degree of drug binding can result from drug-drug interactions (usually with concomitantly administered drugs) or disease states such as hepatic or renal disease [2].

The most common proteins in plasma and serum which bind drugs are albumin and α_1-acid glycoprotein (AAG). Albumin is the most significant contributor to the binding of therapeutic drugs but AAG (high binding capacity for basic/lipophilic drugs) and lipoproteins also are contributors to this observed binding. Since differential binding to these plasma proteins is sometimes important when evaluating potential drug-drug interactions during the course of certain disease states, methods used for these analyses can be included in your assay plans. More recent innovations in the accurate determination of plasma protein binding requiring long incubation times, specifically the use of a static 5 % CO_2 environment as a means of controlling pH throughout the entire incubation period has proved to be a significant optimization tool as a means of preventing the over-estimation of binding [3].

Currently, there are three widely used and accepted methods for the plasma protein binding evaluation of drugs in Discovery and Early Development (Table 1). Equilibrium dialysis methods (Classical Equilibrium Dialysis; CED and Rapid Equilibrium Dialysis; RED) and ultrafiltration methods (Centrifree® Ultrafiltration devices; EMD Millipore Corp., Billerica, MA.) will be presented along with the analytical methods used to support these assays. Methods using radiolabeled (as ^3H or ^{14}C-labeled) and non-radiolabeled, cold, drugs will be discussed in the context of assay design and sample analysis techniques.

Table 1
Summary of equipment and reagents used in the different analysis procedures for plasma protein binding

Method	Apparatus/reagent description	Manufacturer	Part number	Special considerations
CED—classical equilibrium dialysis	Spectrum/Dianorm rotating dialysis units (with Teflon cells and holders)	Spectrum/Dianorm; Spectrum Laboratories, Inc. (Rancho Dominguez, CA)		
	SpectaPor 2 dialysis membranes discs	Spectrum Laboratories, Inc. (Rancho Dominguez, CA)	132480 (50/pk)	MWCO = 10–14 kDa; cellulose
RED—rapid equilibrium dialysis	RED device reusable base plate	Thermo Fisher Scientific Inc. (Rockford, IL)	89811	48 well base plate capable of holding 48, 2-sided membrane inserts
	RED device inserts, 8 K MWCO	Thermo Fisher Scientific Inc. (Rockford, IL)	89809 (50/pk)	MWCO = 8 kDa; cellulose
Ultrafiltration—Centrifree®	Centrifree® ultrafiltration devices	Merck-Millipore Corp. (Billerica, MA)	4104 (50/ok)	
	Plasma (from all species)	Bioreclamation, Inc. (Hicksville, NY)	Various	Plasma received frozen and freshly harvested

2 Materials

Specific devices and equipment used in this assay are described in Tables 1 and 2.

3 Methods

3.1 Equilibrium Dialysis: CED Methods Using Spectrum\Dianorm Equilibrium Dialyzing Apparatus and Spectra/Por® Dialysis Membranes

Equilibrium dialysis (ED) is the preferred method to determine the extent of plasma protein binding or free-fraction analysis because it is less susceptible to potential experimental artifacts such as non-specific binding [4]. There are two widely used methods of ED which have different methods of analysis and throughput. The classical equilibrium dialysis (CED) methods of analysis use sealed paired Teflon cells (1 mL volume per side) with buffer and plasma separated by a regenerated cellulose membrane (Spectra/Por®;

Table 2
Instrumentation and chemistries used in the bioanalytical analysis of plasma protein binding study samples using equilibrium dialysis techniques

Instrument/device	Manufacturer/vendor	Model
Mass spectrometer		
API 5000 LC/MS/MS system	PE Sciex	API 5000
Shimadzu SIL-HTC autosampler	Shimadzu	SIL-HTC
Shimadzu LC-20AD prominence pump "A"	Shimadzu	LC-20AD
Shimadzu LC-20AD prominence pump "B"	Shimadzu	LC-20AD
Shimadzu LC-20AD prominence pump "C"	Shimadzu	LC-20AD
Shimadzu DGU-20A$_3$ prominence degasser	Shimadzu	DGU-20A$_3$
Divert valve	Valco Instruments Co. Inc.	Two position actuator control model
Sample preparation		
96-well 2 mL square top, tapered v-bottom	Analytical Sales & Services, Inc.	59623-23
Advantage 1 mL protein precipitation crash plate	Analytical Sales & Services, Inc.	60513
Phenomenex®, Strata impact protein precipitation 2 mL	Phenomenex	CE0-7565
Square well filter plate		
Item	Descriptor	
Analytical column	Zorbax SB-C8, 4.6×75 mm, 3.5 µm (Agilent Technologies)	
Guard column	Zorbax SB-C8, 4.6×12.5 mm, 5 µm (Agilent Technologies)	
Mobile phase "A"	5 mM ammonium acetate in water	
Mobile phase "B"	Acetonitrile	
Mobile phase "C"	50/50 HPLC grade acetonitrile/millipore water (v/v)	
Needle solvent	50/50 HPLC grade acetonitrile/millipore water (v/v) with 0.2 % TFA	

Spectrum Laboratories Inc., Rancho Dominguez, CA.) containing a molecular weight cut-off of 12–14 kDa. These sealed cells are loaded into metal flanges and rotated in motorized dialyzers (Spectrum/Dianorm apparatus; Spectrum Laboratories Inc., Rancho Dominguez, CA) either in a water bath or temperature-controlled incubator at 37 °C for a predetermined equilibration period. The advantages of this technique are the observed low incidence of non-specific binding due to the inert nature of the surrounding Teflon cell containing your analyte of interest and the relative ease of use [5]. A disadvantage of this method is the relatively large volumes of plasma needed (0.7–0.9 mL) to fill the plasma side of the 1 mL cell. Another disadvantage includes potential volume shifts as a result of the long incubation times (4–12 h) and differences in osmotic pressure between the plasma and buffer sides. These osmotic pressure differences can be controlled through the use of a high ionic strength buffer such as Sorensen's Buffer (125 mM sodium/potassium phosphate, pH 7.4) or standard phosphate buffered saline (PBS) solutions adjusted to 290–300 mOsm/L (using an osmometer) in order to be isotonic with the dialyzing plasma compartment. Other disadvantages of the CED technique involve the time consuming setup procedure as well as the limited number of sample replicates that can be run in each dialyzing unit (20 cells) leading itself to low throughput analysis which might not be acceptable for most Discovery and Early Development assay schemes.

The procedure below represents a typical assay design which investigates plasma protein binding using CED techniques in multiple species using freshly harvested plasma collected in tubes containing an anticoagulant such as sodium heparin or potassium EDTA (K_2EDTA). Concentrations chosen for these assays are usually set at 1 or 10 μM when the compound is in early Discovery screening but when the compound transitions into Early Development and early PK studies are performed in multiple species, the concentrations chosen for these assays are usually around the observed C_{max}. Many investigators prefer to run three concentrations (i.e., 100, 500, and 1,000 ng/mL) for each species in order to show concentration-dependent changes in plasma protein binding *in vitro*. The CED method is widely used where radioactive compounds are readily available, as in Early Development. The volume of plasma and buffer used for these assays (0.7–0.9 mL) is very convenient for radioactive assays where the free fraction is very low (<0.5 %) and the large volume recovered and counted by liquid scintillation counting from the buffer side should yield a radioactivity concentration (as dpm/mL) which is many fold higher than background. When non-radiolabeled material is available, these large volumes recovered are also good for analytical techniques, as will be described later, which require liquid-liquid or solid phase extraction procedures needed for sample preparation for LC-MS/MS analysis.

3.1.1 Equilibrium Dialysis: CED Methods

1. Blood samples are collected from randomly chosen male and female rats, dogs, mice, monkeys, rabbits, and from healthy drug-free male and female human volunteers into tubes containing a suitable anticoagulant.

2. Plasma is harvested by centrifugation, pooled by species, and spiked with an aliquot of the test solution (for radiolabeled studies this is usually a mixture of hot and cold material) to a stock plasma concentrations predetermined by the assay design. These plasma stock solutions are further diluted with blank plasma to yield the final concentrations needed for the assay.

3. Equilibrium dialysis experiments are performed with a Specta/Por® (or Dianorm) Equilibrium Dialyzer using 1 mL/side Teflon® dialysis cells containing SpectraPor® 4 dialysis membranes (Molecular Weight Cutoff of 12–14 kDa). Dialysis membranes are prepared by soaking in distilled water for at least 15 min, followed by 30 min of soaking in buffer then mounted between the Teflon® half-cells, and each cell is loaded into the driving flange of the Spectra/Por (or Dianorm) apparatus.

4. Five replicate aliquots (0.7–0.9 mL) of each plasma pool, spiked and diluted to the final assay concentrations are loaded into one half-cell and an equal volume of buffer will be loaded into the opposing half-cell. All cells are stoppered and the flanges loaded in the dialysis unit.

5. The dialysis unit is either immersed in a 37 °C water bath or placed in a temperature controlled incubator and the cells rotated at 20 rpm for a pre-determined equilibrium period.

6. At the end of the equilibration period, each half-cell containing either plasma or buffer is emptied into 2 mL polypropylene vials for aliquoting or stored frozen for analysis.

When using radiolabeled materials (^{3}H or ^{14}C-labeled) the method of analysis of both plasma and buffer samples recovered from each half-cell will be liquid scintillation spectrometry. Usually 50–100 µL of each matrix is added to scintillation vials containing cocktail (Ultima Gold®, Perkin Elmer, Shelton, CT) and each sample collected is counted in duplicate in a liquid scintillation analyzer (Packard TriCarb 3100TR, Perkin Elmer, Shelton, CT). As mentioned earlier, the large volume of buffer collected using this method is advantageous for the determination of fraction unbound where the free fraction is very low (i.e., <0.5 %) as larger volumes can be counted in order to obtain sufficient counts for the calculation of f_u.

```
┌─────────────────────────────────────────────────────────────────────┐
│       Prepare Spectrum/Dianorm cells and apparatus for sample assay │
└─────────────────────────────────────────────────────────────────────┘
                                  ↓
┌─────────────────────────────────────────────────────────────────────┐
│   Spike analyte into plasma from each species to appropriate final  │
│                       concentrations                                │
│                   (using hot or cold stocks)                        │
└─────────────────────────────────────────────────────────────────────┘
                                  ↓
┌─────────────────────────────────────────────────────────────────────┐
│  Pipet 700-900 µL of plasma (spiked) and buffer (blank) into        │
│            opposite sides of the Teflon cell housings               │
└─────────────────────────────────────────────────────────────────────┘
                                  ↓
┌─────────────────────────────────────────────────────────────────────┐
│  Load the dialyzing units into the apparatus and spin at 20 rpm at  │
│           37°C for a predetermined equilibration time               │
└─────────────────────────────────────────────────────────────────────┘
                                  ↓
┌─────────────────────────────────────────────────────────────────────┐
│ Following the incubation time, empty the contents of each cell side │
│         (plasma or buffer) into separate tubes for analysis         │
└─────────────────────────────────────────────────────────────────────┘
```

Sample Analysis Phase:

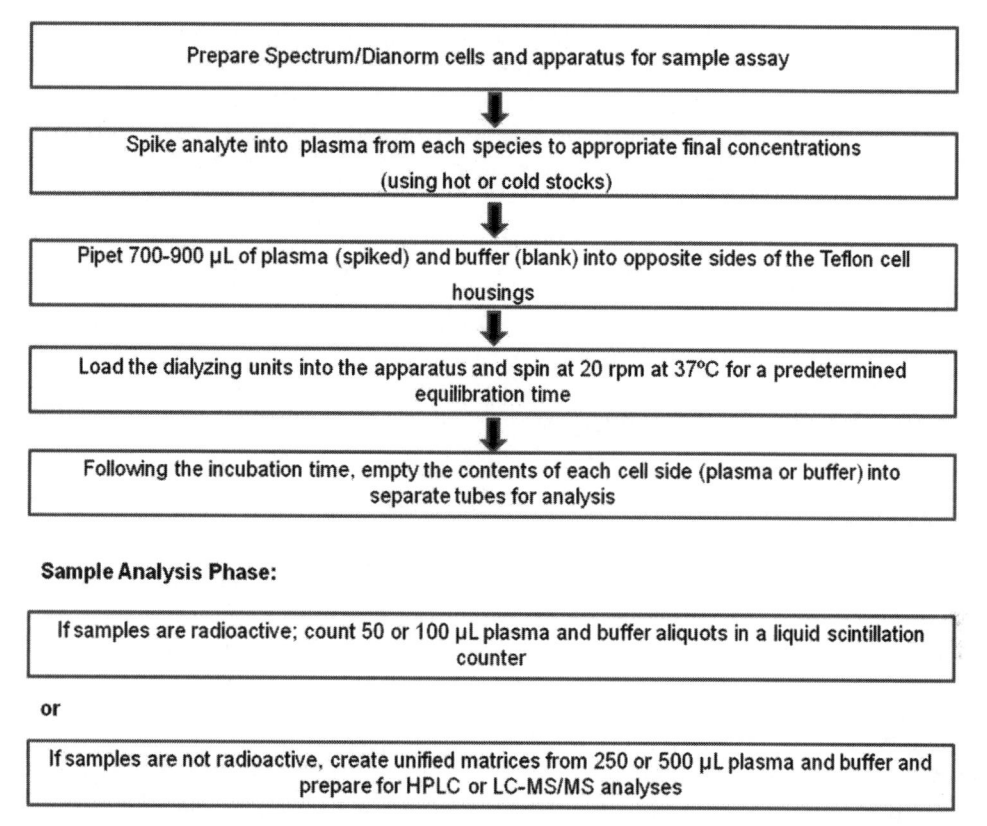

```
┌─────────────────────────────────────────────────────────────────────┐
│  If samples are radioactive; count 50 or 100 µL plasma and buffer   │
│            aliquots in a liquid scintillation counter               │
└─────────────────────────────────────────────────────────────────────┘
```

or

```
┌─────────────────────────────────────────────────────────────────────┐
│  If samples are not radioactive, create unified matrices from 250   │
│   or 500 µL plasma and buffer and prepare for HPLC or LC-MS/MS      │
│                            analyses                                 │
└─────────────────────────────────────────────────────────────────────┘
```

Fig. 1 Flow chart for plasma protein binding assay initiation, execution, and sample analysis using Classical Equilibrium Dialysis (CED) techniques

When non-radiolabeled material is used as is the case in early to late Discovery stages of development, analytical methods usually include LC-MS/MS analysis proceeded by spiking and sample preparation using a suitable internal standard. The buffer (free fraction) and plasma (bound fraction) matrices are collected from each side of the cells and each matrix is aliquoted (usually as 250 or 500 µL volumes) and combined with the same volume of the corresponding opposite matrix (i.e., 250 µL of free fraction buffer is combined with 250 µL of blank plasma, and vice versa) to create unified matrices for each sample in preparation for the subsequent sample preparation for bioanalysis. For the analysis of these unified matrices from the assay, standards and QC's will be prepared in the same matrix format (as 50:50, plasma:buffer) using human plasma as the contributing and validated plasma matrix as this will prevent the need for preparing, running, and validating large numbers of QC's and standard curves for assays which involve plasma protein binding determinations in multiple species. The techniques used for the Bioanalytical analysis of these samples from the above-mentioned CED procedure will be outlined in a separate section (*see* Fig. 1).

3.1.2 Calculations of Plasma Protein Binding and Free Fraction

The data will be analyzed to determine the fraction of drug bound to plasma proteins and the corresponding free fraction (as f_u). The following equations will be used to determine the extent of plasma protein binding of total radioactive equivalents when labeled material is used:

$$Percent\ free\ drug = 100 \times \frac{(dpm\ in\ dialysate\ buffer\ aliquot)}{dpm\ in\ dialysate\ plasma\ aliquot} \quad (1)$$

$$Percent\ bound\ drug = 100 - \%\ free\ drug \quad (2)$$

For non-radiolabeled material where parent drug is assayed for in the unified matrices created, the following equation will be used for the calculations of % bound and % free:

$$Percent\ free\ drug = 100 \times \frac{(conc.in\ dialysate\ buffer\ aliquot\ used\ to\ create\ the\ unified\ matrix)}{(conc.in\ plasma\ sample\ side\ aliquot\ used\ to\ create\ the\ unified\ matrix)} \quad (3)$$

$$Percent\ bound\ drug = 100 - \%\ free\ drug \quad (4)$$

3.2 Equilibrium Dialysis: Rapid Equilibrium Dialysis (RED) Methods Using RED Device Baseplate (48-Well) and Inserts

A rapid equilibrium dialysis (RED) device has recently become commercially available (Pierce Biotechnology, Thermo Fisher Scientific, Waltham, MA) offering the potential for reduced preparation and equilibration times. Each RED device Teflon base plate holds up to 48 RED device inserts and has a standard 96-well plate footprint with 9×9 mm well spacing to provide compatibility with automated liquid handling systems. Each insert contains a buffer and a plasma compartment separated by a semipermeable membrane with a molecular weight cutoff of approximately 8 kDa. The RED device insert is filled with buffer and plasma and the unique design of this side-by-side chamber system has a vertical cylinder of dialysis membrane with a high membrane surface area-to-volume ratio [6]. Some advantages of this method is that it offers shorter incubation times (usually 2–4 h), low non-specific binding as well as ease of use. Some disadvantages include low plasma sample volumes, usually from 100 to 500 μL (Table 3) which might present a problem with radiolabeled drugs where sensitivity of the free fraction analysis might be compromised for highly protein bound drugs where the free fraction is <0.2 %. As will be described later, this is the method of choice when requiring a higher throughput for sample analysis as needed for a Discovery setting where multiple compounds are being screened in plasma from many species. The advantage of this system from an Early Development perspective is that multiple concentrations in plasma can be assayed for each species

Table 3
Summary of suggested volume pairings for both plasma and buffer compartments for the RED device inserts

Plasma sample chamber (µL)	Buffer chamber (µL)
100	300
200	350
300	500
400	600
500	750

for a given new chemical entity (NCE) and concentration-dependent changes in plasma protein binding can be investigated. All techniques discussed here have a wide range of potential applications as well as experimental short-comings and liabilities so the investigator will need to choose each one based on the circumstances at hand [4].

The procedure below represents a typical assay design which investigates plasma protein binding using the Rapid Equilibrium Dialysis (RED device) method in multiple species using freshly harvested plasma collected in tubes containing an anticoagulant such as sodium heparin or potassium EDTA (K_2EDTA). As mentioned in the other method, concentrations chosen for these assays are generally set at 1 or 10 µM when the compound is in early Discovery screening but when the compound transitions into Early Development and early PK studies are performed in multiple species, the concentrations chosen for these assays are usually around the observed C_{max}.

The ability to assay low volumes of plasma (as low as 100 µL) makes this method particularly amenable to reduced blood volumes collected from some species where harvested plasma volumes might be very limited and pooling is not an option. The working paired plasma and buffer sample volumes suggested below by the manufacturer shows the flexibility of this method.

This method is widely used where radioactive compounds are readily available as in late Discovery phases or Early Development. As with that observed for the CED method, the volume of buffer used for these assays (0.3–0.75 mL) is very convenient for radioactive assays where if the free fraction is very low (<0.5 %) the large volume recovered and counted by liquid scintillation counting from the buffer side should yield a radioactivity concentration (as dpm/mL) which is many fold higher than background thus being above the limit of detection. When non-radiolabeled material is available, these large volumes recovered are also favorable for Bioanalytical techniques, as will be described later, which require liquid-liquid or solid phase extraction procedures for sample preparation for LC-MS/MS analysis.

3.2.1 Equilibrium Dialysis- RED Procedure

1. Rinse the Teflon® base plate in 20 % ethanol for 10 min followed by rinsing twice in ultrapure water. Dialysis membrane inserts will be rinsed three times in distilled water for 10 min. The pre-soaked inserts will be blotted to remove excess water and added to the pre-soaked base plate. Each dialysis insert contains two chambers. One chamber is red and will contain 100–500 µL plasma solution while the second chamber is white and will contain 300–750 µL buffer (see paired volumes from Table 2 above).

2. Prepare the buffer solution. Dialysis buffer solution is normal PBS (phosphate buffered saline) containing 100 mM sodium phosphate and 150 mM sodium chloride (pH 7.4). This buffer is commercially available as powder packets from Thermo Scientific (part no. 28372) and should be prepared fresh prior to each assay.

3. Plasma sample working solutions from each species being investigated, will be prepared by adding aliquots (30 µL) of the test compound stock solutions to blank plasma (3 mL) to achieve the anticipated nominal final plasma concentrations for the assay. When radioactivity is used as tracer, 10 µL of the radiolabeled stock solution will be added to each species working solution stocks. The contribution of mass from the radiolabeled material to the final concentration, if not negligible, will need to be calculated and appropriate adjustments to the mass of the cold material might be necessary.

4. When radioactivity is available, linearity with time in order to determine the equilibration period for the length of time needed for the definitive study can be conducted in human plasma spiked to the nominal assay concentrations. Sets of four replicates per concentration will be assayed and each replicate set will correspond to incubation sampling times of 1, 2, 3, 4, or 5 h. Equilibrium will be determined as the time where the % bound does not change between two successive timepoints following the sampling (50 µL) and counting of both plasma and buffer samples from all sets.

5. The possibility of non-specific binding will be monitored by adding buffer, spiked with the test compound concentrations to both sides of the RED device insert and incubated and analyzed in a separate assay prior to running the definitive study.

6. Four replicates at each concentration in the plasma solutions will be added to the plasma (red) chamber of the insert while buffer will be added to the buffer (white) chamber of the RED device according to the worklog. After sealing the RED device with a clear film, dialysis will be completed at 37 °C with shaking at 100 rpm for the pre-determined equilibration time.

7. In order to verify assay integrity, aliquots of cold and/or radiolabeled acetaminophen, warfarin, and lidocaine can be prepared in human plasma at concentrations of 1 μg/mL and run as positive controls.

8. At the end of the assay incubation period 50–100 μL will be removed from the sample and buffer compartments of each insert and analyzed using either liquid scintillation counting (for all radiolabeled materials) or LC-MS/MS analysis, if non-radiolabeled, following the creation of the unified matrices (mentioned below) and the appropriate sample work-up.

When using radiolabeled materials (^{3}H or ^{14}C-labeled) the method of analysis of both plasma and buffer samples recovered from each half-cell will be liquid scintillation spectrometry. Usually 50–100 μL of each matrix is added to scintillation vials containing cocktail (Ultima Gold®, Perkin Elmer, Shelton, CT) and each sample collected is counted in duplicate in a liquid scintillation analyzer (Packard TriCarb 3100TR, Perkin Elmer, Shelton, CT). As mentioned earlier, the large volume of buffer collected using this method is advantageous for the determination of fraction unbound where the free fraction is very low (i.e., <0.5 %) as larger volumes can be counted in order to obtain sufficient counts for the calculation of f_u.

When non-radiolabeled material is used as is the case in early to late Discovery stages of development, analytical methods usually include LC-MS/MS analysis proceeded by spiking and sample preparation using a suitable internal standard. The buffer (free fraction) and plasma (bound fraction) matrices are collected from each side of the cells and each matrix is aliquoted (usually as 250 or 500 μL volumes) and combined with the same volume of the corresponding opposite matrix (i.e., 250 μL of free fraction buffer is combined with 250 μL of blank plasma, and vice versa) to create unified matrices for each sample in preparation for the subsequent sample preparation for bioanalysis. For the analysis of these unified matrices from the assay, standards and QC's will be prepared in the same matrix format (as 50:50, plasma:buffer) using human plasma

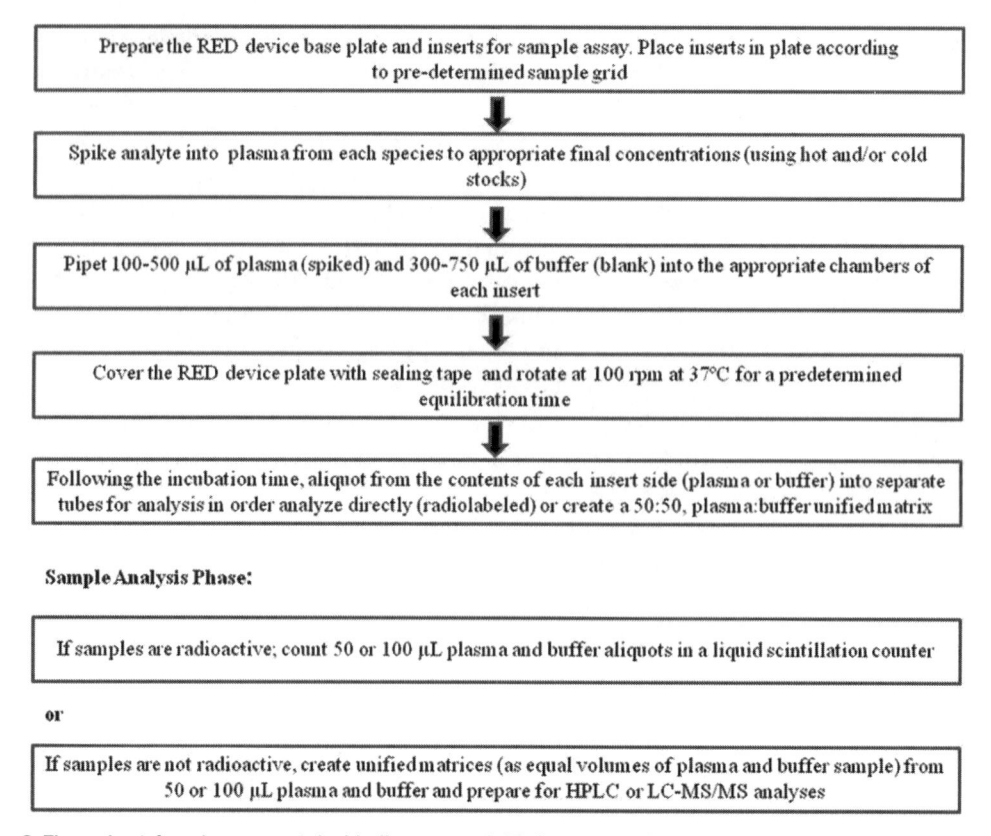

Fig. 2 Flow chart for plasma protein binding assay initiation, execution, and sample analysis using Rapid Equilibrium Dialysis (RED device) techniques

as the contributing and validated plasma matrix as this will prevent the need for preparing, running, and validating large numbers of QC's and standard curves for assays which involve plasma protein binding determinations in multiple species. The techniques used for the Bioanalytical analysis of these samples from the above-mentioned RED procedure will be outlined in a separate section (*see* Fig. 2).

3.2.2 Calculations of Plasma Protein Binding and Free Fraction from RED Device Samples

The data will be analyzed to determine the fraction of drug bound to plasma proteins and the corresponding free fraction (as f_u). Equations 1 and 2 (Sect. 3.1.1) will be used to determine the extent of plasma protein binding of total radioactive equivalents when labeled material is used. Equations 3 and 4 (Sect. 3.1.1) will be used for non-radiolabeled material where parent drug is assayed for in the unified matrices.

3.3 Ultrafiltration Methods of Plasma Protein Binding- Centrifree® Ultrafiltration Device

Much attention has been paid to the equilibrium dialysis methods of plasma protein binding from a Discovery setting with automated, and medium to high-throughput analysis. Centrifree® devices rapidly and efficiently separate free from protein-bound analytes in small volumes (0.5–1.0 mL) of serum, and plasma using a method called ultrafiltration. Accurate partitioning occurs in minutes without dilution, change in physiologic pH, ion composition, or unbound analyte concentration. The Centrifree® ultrafiltration device provides maximum efficiency for multiple sample processing. The procedure for these analyses usually involves addition of the aliquot of plasma to the device and subsequent centrifugation at $1,000-2,000 \times g$ in a fixed angle rotor for approximately 10–20 min. Each unit consists of a membrane and O-ring permanently sealed between the sample reservoir and support base. A removable filtrate cup is attached to the base. The volume of filtrate (as plasma water) usually ranges from 160 to 170 µL for a volume of 1 mL of human serum centrifuged at $1,000 \times g$ for 10 min. This method is well suited for quick analysis where stability of the analyte might be an issue. While ultrafiltration has the main advantage of requiring a shorter assay time (usually 10–20 min depending on the volume of plasma or serum used), the main drawback to the accuracy of this technique is non-specific binding (NSB) of test compounds to the filtration apparatus (usually polycarbonate plastic in nature) [5]. Due to the nature of the material used in the design of these devices there is the possibility of high non-specific binding with some ligands.

3.3.1 Use of the Centrifree® Ultrafiltration Devices

1. Before using a Centrifree® device, remove the red reservoir cap. Hold reservoir angled at approximately 45° with pipette tip touching reservoir wall. Add sample solution smoothly in one even flow to avoid air locking.

2. Cap sample reservoir, then place device in a fixed-angle centrifuge rotor with 17×100 mm adapters. Counterbalance centrifuge with a similar device.

3. Equilibrate device and sample to temperature required by your application.

4. Spin device at $1,000-2,000 \times g$ for required time to obtain desired filtrate volume (10–20 min).

5. Carefully remove device from centrifuge rotor and disconnect filtrate cup containing the ultrafiltrate. Cover filtrate cup with supplied cap until ultrafiltrate can be analyzed.

This technique is well suited for the analysis of radiolabeled materials where filtrate sample volumes collected (150–170 µL from a 1 mL plasma volume) for free fraction analysis should be adequate for the analysis of analytes which have a very low free fraction concentration.

3.3.2 Data Analysis from Ultrafiltration Technique Using Radiolabeled Material

Equations 1 and 2 (Sect. 3.1.1) will be used to determine the extent of plasma protein binding of total radioactive equivalents when labeled material is used (ultrafiltrate was used as free fraction volume). Note that stock plasma aliquot is from the original plasma stock solution not the remaining plasma in the device after centrifugation.

3.4 Practical Aspects for Using LC-MS/MS Analysis in Equilibrium Dialysis Protein Binding Studies

As mentioned earlier, for protein binding studies, experimental plasma samples from multiple species are usually tested with known concentrations of the analyte. Both free and bound forms of the analyte can exist in plasma. The unbound fraction, also known as the "free fraction", can be isolated by equilibrium dialysis techniques as described earlier. Depending upon the extent of protein binding and the concentrations planned for the study, a broad concentration range of calibration standards may be needed for measuring both unbound and bound analytes. Therefore, to avoid analytical pitfalls, such as non-linear dynamic range and carryover, it can be practical to use two concentration ranges, lower and higher. For highly protein bound analytes, the unbound fraction can be quite low in concentration, and the bound fraction can remain very high.

Before conducting an experiment, it is important to test carryover, matrix effect, recovery, and stability. Non-specific binding of the apparatus should also be investigated prior to collecting study samples. Immediately after sample collection, the plasma fraction is typically mixed with an equal volume of buffer prepared at physiological pH, and likewise, the buffer samples are mixed with an equal volume of plasma from the respective species. In this way, a unified matrix is established for both the plasma and buffer fractions. Therefore, samples from either the plasma or the buffer fractions are expected to experience identical matrix effects for a given species. Calibration standards and quality control samples (QCs) can also be prepared by fortifying control unified matrix. Assuming plasma from all experimental species experiences similar matrix effects, calibration curves can be prepared from one species to analyze samples from multiple species. With careful advance planning, the plasma and buffer fractions can be analyzed in a single analytical run using a calibration curve that covers the range of concentration of those fractions.

In general, a generic LC-MS/MS method can be used with a common internal standard and a protein precipitation extraction procedure (Table 3). This provides fast turnaround, efficient set-up, and cost-effective bioanalytical support (Figs. 3, 4 and 5). Weighted linear regression is typically used to generate the calibration curve.

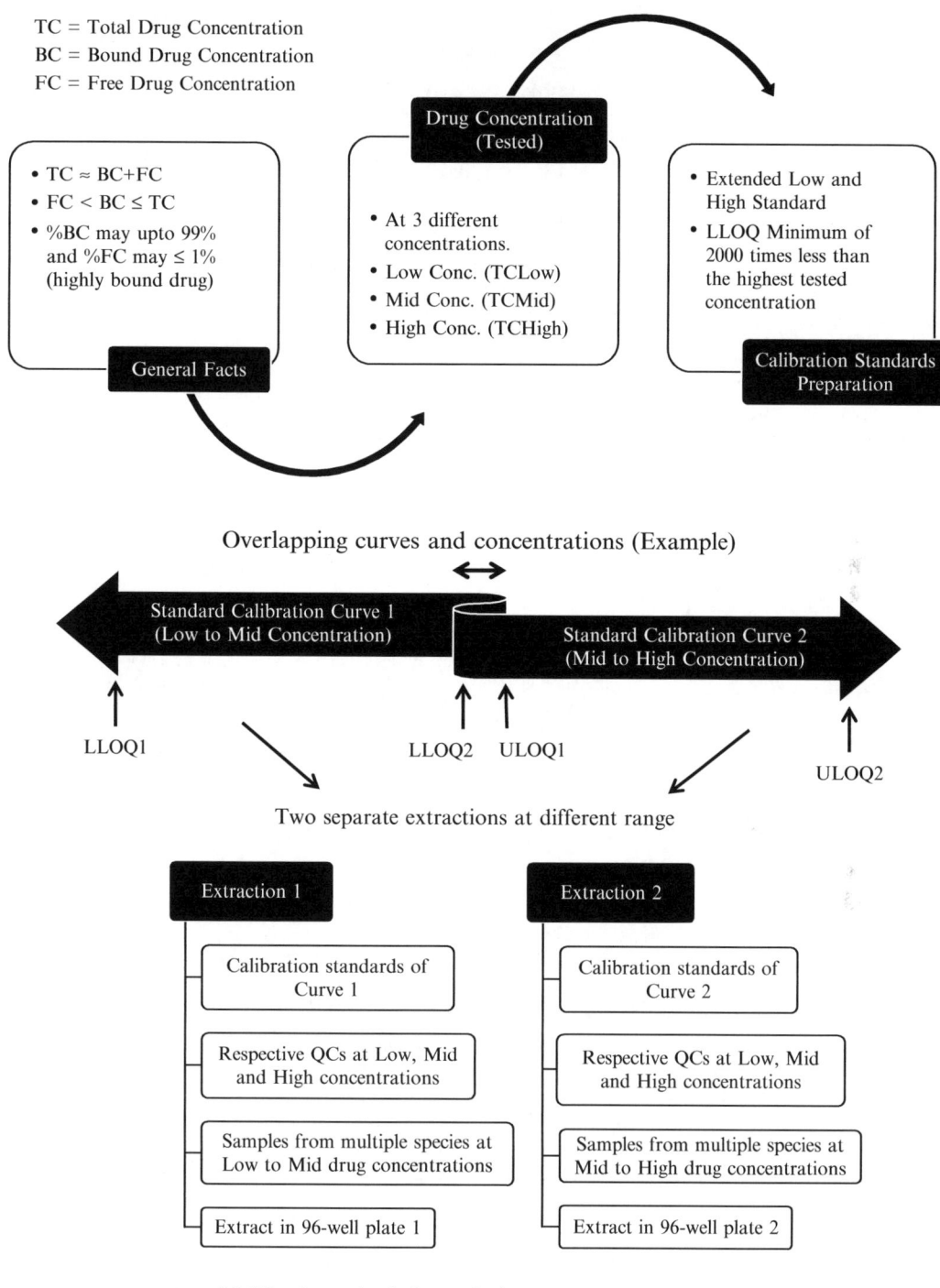

Plasma Protein Binding   35

TC = Total Drug Concentration
BC = Bound Drug Concentration
FC = Free Drug Concentration

General Facts
- TC ≈ BC+FC
- FC < BC ≤ TC
- %BC may upto 99% and %FC may ≤ 1% (highly bound drug)

Drug Concentration (Tested)
- At 3 different concentrations.
- Low Conc. (TCLow)
- Mid Conc. (TCMid)
- High Conc. (TCHigh)

Calibration Standards Preparation
- Extended Low and High Standard
- LLOQ Minimum of 2000 times less than the highest tested concentration

Overlapping curves and concentrations (Example)

Standard Calibration Curve 1 (Low to Mid Concentration) ← → Standard Calibration Curve 2 (Mid to High Concentration)

LLOQ1   LLOQ2 ULOQ1   ULOQ2

Two separate extractions at different range

Extraction 1
- Calibration standards of Curve 1
- Respective QCs at Low, Mid and High concentrations
- Samples from multiple species at Low to Mid drug concentrations
- Extract in 96-well plate 1

Extraction 2
- Calibration standards of Curve 2
- Respective QCs at Low, Mid and High concentrations
- Samples from multiple species at Mid to High drug concentrations
- Extract in 96-well plate 2

- LLOQ = Lower level of quantitation.
- ULOQ = Upper level of quantitation
- LLOQ1, ULOQ2 may vary during method development due to extent of plasma protein binding

Fig. 3 Standard, QC preparation and sample analysis

	1	2	3	4	5	6	7	8	9	10	11	12
A	STD1	STD2	STD3	STD4	STD5	STD6	STD7	STD8	Blank			
B	Low QC	Mid QC	High QC	Blank								
C	Rat Sample 001	Rat Sample 002	Rat Sample 003	Rat Sample 004	Rat Sample 005	Rat Sample 006						
D	Rabbit Sample 007	Rabbit Sample 008	Rabbit Sample 009	Rabbit Sample 010	Rabbit Sample 011	Rabbit Sample 012						
E	Dog Sample 013	Dog Sample 014	Dog Sample 015	Dog Sample 016	Dog Sample 017	Dog Sample 018						
F	Human Sample 019	Human Sample 020	Human Sample 021	Human Sample 022	Human Sample 023	Human Sample 024						
G	Mouse Sample 025	Mouse Sample 026	Mouse Sample 027	Mouse Sample 028	Mouse Sample 029	Mouse Sample 030						
H	Low QC	Mid QC	High QC	Blank								

Fig. 4 Example of typical 96 well plate format for sample analysis

Pipette Experimental Samples

Pipette Stds/QCs prepared in plasma

Buffer fraction (Free Fraction)

Plasma fraction (Bound Fraction)

Pipette equal volume of buffer

Pipette equal volume of plasma

Pipette equal volume of buffer

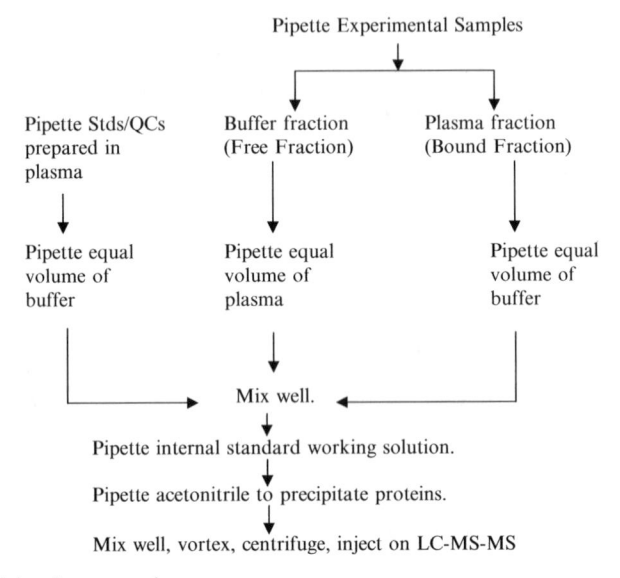

Mix well.

Pipette internal standard working solution.

Pipette acetonitrile to precipitate proteins.

Mix well, vortex, centrifuge, inject on LC-MS-MS

Fig. 5 Extraction procedure

4 Notes

1. Although methods were presented for both equilibrium dialysis methods of plasma protein binding, the sample handling (as unified matrix) are the same for both methods using bioanalytical LC-MS/MS analysis.

2. The use of radiometric analysis of plasma and buffer samples represents total radioactivity concentrations and is not specific for the parent compound. Radioprofiling techniques might be

used in the case where degradants might have been produced as a result of the incubation conditions or inherent instability of the molecule.

3. For all ED techniques, non-specific binding will need to be assessed in order to ensure high (>90 %) recovery in the free fraction buffer component. Even if recovery is high for the buffer component collected, a small amount of albumin (HSA) can be added to each collection vessel to ensure that the solute remains in solution throughout the post-incubation, pre-analysis aliquoting process.

4. The equilibration times for the ED techniques need to be established prior to running the definitive assays. As mentioned earlier, CED equilibration times will probably be longer than those observed for the RED device assays. This is due to the membrane surface area per unit volume contained within the RED device sample insert. For some compounds, the incubation times might be critical due to inherent compound stability issues at 37 °C.

5. The relevance of experimental conditions such as pH control through increased buffer strength or the use of a CO_2 incubator has been shown [3] to improve the reproducibility of fraction unbound data by controlling and optimizing conditions during the ED procedures.

Acknowledgements

The views expressed here are solely those of the author and do not reflect the opinions of Janssen Research & Development, LLC.

References

1. Bowers W, Fulton S, Thompson J (1984) Ultrafiltration vs equilibrium dialysis for determination of free fraction. Clin Pharmacokinet 9(suppl 1):49–60
2. Smith DA, Kerns EH (2010) The effect of plasma protein binding on in vivo efficacy: misconceptions in drug discovery. Nat Rev Drug Discov 9:929–939
3. Curan R, Claxton C, Hutchinson L, Harridine P, Martin I, Littlewood P (2011) Control and measurement of plasma pH in equilibrium dialysis: influence on drug plasma protein binding. Drug Metab Dispos 39(3):551–557
4. Kariv I, Cao H, Oldenburg K (2000) Development of a high throughput equilibrium dialysis method. J Pharm Sci 90(5):580–587
5. Di L, Umland J, Trapa P, Maurer T (2012) Impact of recovery on fraction unbound using equilibrium dialysis. J Pharm Sci 101(3):1327–1335
6. Waters N, Jones R, Williams G, Sohal B (2008) Validation of a rapid equilibrium dialysis approach for the measurement of plasma protein binding. J Pharm Sci 97(10):4586–4595

Chapter 3

Drug Partition in Red Blood Cells

Dennis Kalamaridis and Karen DiLoreto

Abstract

This chapter will focus on the techniques used in the evaluation of red blood cell partitioning/binding (RCB) of drugs in discovery and development. Certain therapeutic compounds have a high degree of affinity for the red blood cell fraction of whole blood and have large RBC-to-plasma concentration ratios. Knowledge of *in vitro* RBC partitioning of compounds is important to the interpretation and understanding of the compounds pharmacokinetic profile and distribution *in vivo*. The goals of this chapter will be to examine some of the current methods used in the evaluation of red cell partitioning and the determination of blood-to-plasma ratios at the discovery level of development where species comparisons can be an important determinant in the design and interpretation of some observed pharmacokinetic study results. Methods employing the use of both radiolabeled and cold materials will also be discussed using a variety of analytical tools.

Key words Red cell partitioning, Blood to plasma ratio, Hematocrit

1 Introduction

The determination of the extent of distribution or partitioning of drugs in RBCs from different species in Discovery phases is sometimes not fully studied and might lead to missed opportunities in predicting the kinetics of the compound at these early stages of development. Knowledge of RBC partitioning of compounds enables: (a) physiologically meaningful referencing of pharmacokinetic parameters of drugs to concentrations in either whole blood, plasma, or serum; (b) *in vitro* prediction of drug distribution *in vivo*; (c) the calculation of blood to plasma ratio for the determination of the relevance of plasma clearance in conjunction with the plasma protein binding of drugs; and (d) a rational choice of appropriate biological fluid, either whole blood, plasma, or serum, for assay, especially in the cases of high partitioning in red cells. Among the cellular constituents of blood, the RBCs represent, by far, the largest population both in number and cell size. Since RBCs make up more than 99 % of the total cellular space of blood in humans [1], knowing the extent and nature of this partitioning can also be used to predict or understand potential haematoxicity issues observed in preclinical species early on in drug development which might assist in the drug candidate selection process.

At the Discovery phases of new chemical entity (NCE) selection, rate and extent of RBC partitioning of drugs is determined using both *in vitro* and *ex vivo* methodologies [2]. This chapter will focus mainly on these *in vitro* procedures using both radiolabeled (if available) and cold, unlabeled materials. In the *ex vivo* procedure, drug is administered (usually intravenously) to the preclinical species, a series of blood samples are taken at timed intervals post-dose, and, following centrifugal separation, the drug concentrations are measured in the RBCs and plasma. In these *in vitro* procedures, drug is added to whole blood, and after mixing, the drug's concentration is measured in RBCs and plasma following centrifugal separation [3]. Determinations using radiolabeled materials are quick and easy requiring the quantitation of total radioactivity in aliquots of whole blood and prepared plasma using liquid scintillation spectroscopic analysis. For determining concentrations in RBCs and plasma using unlabeled or cold materials, differential centrifugation of whole blood and the subsequent analysis of the plasma and red cell volumes are afforded using lysis and protein precipitation of aliquots from the individual fractions and LC-MS/MS analysis of the extracts using a suitable internal standard. The reasons which prefer the *in vitro* procedure are such that since the rate of partitioning is fast, and distribution equilibrium is reached within a few seconds to minutes, and only the *in vitro* procedure enables determination of the rate of partitioning to obtain meaningful results. The extent of drug partitioning into RBCs should be determined under steady-state equilibrium conditions. Because the *in vitro* method uses a closed system, these steady state equilibrium conditions can be easily established through timed incubation and better control over the acquisition of a desirable equilibrium time-course. However, with more slowly equilibrating drugs, the erythrocyte uptake process is often more difficult to separate from the other kinetic events, such as tissue distribution and elimination from the body, which are often occurring simultaneously using the *ex vivo* approach. This is not an issue with the *in vitro* procedure [4] as presented in this chapter where multiple timepoints (usually 0, 15, 30, and 60 min) are investigated.

To summarize, the significance of the blood to plasma ratio:

- Parameters usually determined using only plasma data may be misleading if concentrations of drug differ between plasma and red blood cells as a consequence of differential binding to a specific component in the blood.

- Pharmacokinetic parameters are normally determined by analysis of drug concentrations in plasma rather than whole blood. The blood to plasma ratio determines the concentration of the drug in whole blood compared to plasma and provides a comparative indication of drug binding to erythrocytes.
- When the blood to plasma ratio is >1or 2 (usually as a consequence of the drug distributing into the erythrocyte), the plasma clearance significantly overestimates blood clearance and could exceed hepatic blood flow.

Blood to plasma ratio is an important parameter, in conjunction with other ADME and physicochemical properties, for predicting whole body pharmacokinetics which is important in comparative NCE selection and subsequent dose selection for the preclinical species when evaluating comparative exposure [5]. The following sections will discuss methods used for the evaluation of red cell partitioning which is usually performed using non-radiolabeled cold material and blood to plasma ratios (as Kb/p) where radiolabeled material is used in the determination of the observed partitioning of total radioactive analytes in both the whole blood and plasma matrices and serves as an estimation of total plasma clearance. At blood to plasma ratios of greater than 1 (usually as a consequence of the drug distributing into the erythrocyte), the plasma clearance significantly overestimates blood clearance and could exceed hepatic blood flow.

2 Red Blood Cell Partitioning Protocol Using Unlabeled (Cold) Material

2.1 Materials (See Table 1)

1. Fresh whole blood obtained from human and preclinical species (Human, NHP, Dog, Rat, Mouse, or Guinea Pig)
2. Test compound (as neat powder)
3. DMSO (Sigma-Aldrich, St. Louis, MO)
4. Acetonitrile (EMD Millipore Corp. Omnisolv® High Purity Solvents)

2.2 Method

2.2.1 Study Design

Three milliliter of whole blood is pre-incubated to 37 °C for approximately 30 min prior to study. Three microliter of a 10 mM DMSO stock solution containing the test article is added to obtain a final concentration of 10 µM (1 and 10 µM can be assayed). Whole blood is vortexed gently for initial mixing and 500 µL aliquot is taken (this will be 0 timepoint). Aliquots of blood are taken at each of the following time points after the initial 0, 15, 30, and 60 min. Blood is returned to the incubator between time points.

Aliquots are centrifuged for about 3 min at 7,000 rpm to separate plasma and red blood cells. 100 µL of plasma layer is removed from the red blood cells and placed in a separate vial; remaining plasma layer is carefully removed and discarded or kept in a separate

Table 1
Instrumentation and chemistries used in the bioanalytical analysis of red blood cell partitioning using unlabeled (cold) method

Instrument/device	Manufacturer/vendor	Model
Mass spectrometer		
API 4000 QTRAP LC/MS/MS system	PE Sciex	API 4000 QTRAP
Shimadzu LC-20AD prominence pump "A"	Shimadzu	LC-20AD
Shimadzu LC-20AD prominence pump "B"	Shimadzu	LC-20AD
Leap HTS PAL autosampler	Leap Technologies	HTS PAL
Item	*Descriptor*	
Analytical column	C18, 100A, 2×50 mm, 5 U (Princeton Chromatography)	
Mobile phase "A"	0.1 % formic acid in HPLC grade water	
Mobile phase "B"	Acetonitrile	

vial as a backup sample (*see* **Note 1**). 400 µL of acetonitrile spiked with an internal standard is added for protein precipitation and samples are vortexed and placed in the centrifuge for 3 min at 7,000 rpm as generated.

From the initial vial, 100 µL of red blood cells are transferred to a new vial, placed on dry ice and stored in a −80 °C freezer overnight (lysis step). Any remaining red blood cells are also discarded or kept as a backup sample. On the next day, red blood cell samples are thawed and 400 µL of acetonitrile spiked with an internal standard is added for protein precipitation. Samples are vortexed and placed in the centrifuge for 3 min at 7,000 rpm.

2.2.2 Standard Curve Preparation

Additional blank whole blood for each species is centrifuged for 3 min at 7,000 rpm to separate the plasma and red blood cells. The plasma layer is carefully transferred to a separate vial leaving only blank red blood cells. Remaining blank red blood cells are stored in −80 °C freezer overnight with red blood cell samples generated during the assay. A standard curve in plasma is prepared at the following concentrations: 25, 10, 5, 2.5, 1, and 0.5 µM. 400 µL of acetonitrile spiked with an internal standard is added for protein precipitation. Standard curves for the red blood cells are prepared after thawing under the same conditions as stated for plasma above.

2.2.3 Sample Analysis

Samples were analyzed by LC-MS/MS (API4000 QTRAP). Data analysis and sample quantitation was done with vendor software package (Analyst 1.4.2, AB SCIEX, Framingham, MA).

Distribution between the RBC and plasma phases was calculated based on an estimated hematocrit of 44 % for human, 42 % for dog, 41 % for monkey, 45 % for mouse and 46 % for rat [6].

3 Red Blood Cell Partitioning (as Blood to Plasma Ratio) Protocol Using Radiolabeled Material (^{14}C or ^{3}H-Labeled)

3.1 Materials

1. Fresh whole blood obtained from human and preclinical species (Bioreclamation Inc., Hicksville, NY; Human, NHP, Dog, Rat, Mouse, or Guinea Pig)
2. Radiolabeled test compound (^{14}C-labeled or ^{3}H-labeled material; usually dissolved in ethanol)
3. Cold test compound (as neat powder)
4. Acetonitrile, methanol, or ethanol for spiking solution (EMD Millipore Corp. Omnisolv® High Purity Solvents)

3.2 Methods

3.2.1 Study Design

1. Blood samples are collected from randomly chosen male and female rats, dogs, mice, monkeys, and from healthy drug-free male and female human volunteers into tubes containing a suitable anticoagulant (BD Biosciences, San Jose, CA; the most commonly used anticoagulants are sodium heparin and potassium EDTA; K$_2$EDTA).
2. Blood from each species are then pooled and divided into two 10 mL (one for each concentration usually 1 and 10 µM as a mixture of hot and cold material) aliquots, and placed into large scintillation vials. These vials are placed in a chamber (Barnstead/Labline MaxQ 4000 shaking incubator, Melrose, IL) at 37 °C and pre-incubated for a period of 15 min prior to spiking.
3. Following this pre-incubation period, the vials are still maintained at 37 °C with gentle shaking at 20 rpm, and 100 µL is then removed from each of the whole blood vials (10 mL) for each species and replaced with 100 µL of the radiolabeled stock solutions (prepared above) to achieve the corresponding final concentrations with the [^{14}C or ^{3}H]-labeled radiolabeled stock solutions.
4. Immediately upon spiking, duplicate sets of two aliquots of 0.9 mL of whole blood from each species is then removed and placed in tubes on ice (these served as the '0' time sample) and one of these sets are then spun in a centrifuge at 14,000 rpm for 6 min to prepare plasma. The resulting plasma fractions are then aliquoted and placed into two separate tubes on ice. Following the '0' time sampling for each species, duplicate sets of two separate aliquots are removed at 30 min and 1 h from each of the stock whole blood vials and placed in separate tubes

for processing. The processing procedures for these samples at 30 min and 1 h are identical as that performed for the '0' time samples above. Following each collection interval, all whole blood and plasma samples are stored at 5 °C pending analysis.

3.2.2 Sample Analysis

Whole blood and plasma samples are analyzed for [^{14}C or ^{3}H]-labeled dpm values using liquid scintillation counting. Whole blood samples are aliquoted as 50–100 μL aliquots and combusted in a Packard System 307 sample oxidizer (Perkin Elmer, Shelton, CT) and counted on a Packard 3100TR Liquid Scintillation Analyzer (Perkin Elmer, Shelton, CT). Plasma samples are aliquoted as 50–100 μL aliquots and counted directly on a Packard 3100TR Liquid Scintillation Analyzer (Perkin Elmer, Shelton, CT). Alternative methods for directly counting aliquots of whole blood can be performed according to solubilization procedures. The successful preparation of blood samples for LSC can often be technically difficult, and successful digestion can be largely dependent on the practical experience of the researcher. The source of blood and the correct choice of solubilizer also influence the results of digestion. In general, most of the sample preparation problems occur with blood samples from smaller animals such as rats and mice. In this case, it may be necessary to consider smaller sample volumes, and even then, the end result is usually color quench problems. Suggested sample preparation methods include solubilization and sample combustion. Direct sample addition is not recommended due to color quench and sample/cocktail incompatibility [7]. Also, even with the correction for color quench which is performed quite nicely with the newer software corrections and photomultiplier tube sensitivities present within some of the more recent liquid scintillation analyzers, sensitivity issues might exist which might confound the accurate interpretation of the data.

A few manufacturers have developed quite impressive solubilization methods for counting biological samples such as whole blood and are paired up with a few cocktails which have shown to be very compatible with these samples and solubilization procedures. In this chapter discussion of the methods using Perkin Elmer products will be presented (*see* **Note 2**).

3.2.3 Solubilization Techniques

Solubilization methods [8] are given below for both Soluene®-350 and SOLVABLE™ (PerkinElmer, Shelton, CT).

Soluene®-350 Method

1. Add a maximum of 0.4 mL of blood to a glass scintillation vial.
2. Add, while swirling gently, 1.0 mL of a mixture of Soluene-350 and isopropyl alcohol (IPA; Sigma-Aldrich, St Louis, MO) at 1:1 or 1:2 ratio. Ethanol (Sigma-Aldrich, St Louis, MO) may be substituted for the IPA if desired.

3. Incubate at 60 °C for 2 h. The sample at this stage will be reddish-brown.

4. Cool to room temperature.

5. Add 0.2–0.5 mL of 30 % hydrogen peroxide (Sigma-Aldrich, St Louis, MO) dropwise or in small aliquots. Foaming will occur after each addition; therefore, gentle agitation is necessary. Keep swirling the mixture until all foaming subsides and then continue swirling until all of the hydrogen peroxide has been added. Hydrogen peroxide treatment helps reduce the amount of color present and thus reduces color quench in the final mixture.

6. Allow to stand for 15–30 min at room temperature to complete the reaction.

7. Cap the vial tightly and place in an oven or water bath at 60 °C for 30 min. The samples at this stage should now have changed to pale yellow.

8. Cool to room temperature and add 10–15 mL of either Hionic-Fluor™, Pico-Fluor™ 40 or Ultima Gold™ (Perkin Elmer, Shelton, CT). If color is present use 15 mL cocktail, as this reduces color quench by diluting the color.

9. Temperature and light adapt for 1 h before counting.

SOLVABLE™ Method

1. Add a maximum of 0.5 mL blood to a glass scintillation vial.

2. Add 1.0 mL SOLVABLE™.

3. Incubate the sample at 55–60 °C for 1 h. Sample at this stage will be brown/green in appearance.

4. Add 0.1 mL of 0.1 M EDTA-di-sodium salt solution which helps reduce foaming when the subsequent hydrogen peroxide (Sigma-Aldrich, St Louis, MO) is added.

5. Add 0.3–0.5 mL of 30 % hydrogen peroxide in 0.1 mL aliquots. Gently agitate between additions to allow reaction foaming to subside. Hydrogen peroxide treatment helps reduce the amount of color present, and thus reduces color quench in the final mixture.

6. Allow to stand for 15–30 min at room temperature to complete the reaction.

7. Cap the vial tightly and place in an oven or water bath at 55–60 °C for 1 h. The color will change from brown/green to pale yellow.

8. Cool to room temperature and add 10–15 mL of either Ultima Gold, Opti-Fluor™, Hionic-Fluor or Pico-Fluor 40 (Perkin Elmer, Shelton, CT). If color is present, use 15 mL cocktail, as this reduces color quench by diluting the color.

9. Temperature and light adapt for 1 h before counting.

4 Data Analysis for Both Labeled and Unlabeled Methods

4.1 Radiolabeled

The blood to plasma ratio (often referred to as Kb/p) is the ratio of the concentration of drug in whole blood (i.e. contains both red blood cells and plasma) to the concentration of drug in plasma, namely C_B/C_P. For radiolabeled compounds the blood to plasma ratio is calculated as the ratio of the observed radioactivity detected (as dpm values following Liquid Scintillation Analysis of equivalent volumes of whole blood and plasma; observed dpm in whole blood/observed dpm in plasma) and represents total radioactivity concentrations. This method also does not directly represent the concentration of parent drug or other metabolites within the RBCs as the result of partitioning or binding to erythrocyte membranes or carbonic anhydrase. These determinations would need to be performed using other analytical techniques as described above for cold non-radioactive material where direct harvesting, lysis and subsequent analysis of the packed red cell volume can yield more concise information on direct partitioning and/or binding to erythrocytes.

When looking for comparisons across species for red cell binding or partitioning, sometimes adjustments for hematocrit may be used in order to normalize the data in order to see if there is a definite species difference in the time course and extent of red cell partitioning. The following equation is used for these comparisons when correcting for species hematocrit (H):

$$Blood\ to\ plasma\ ratio = (Kb/p \times H) + (1-H) \quad (1)$$

Using radioactivity and measuring total radioactivity in both whole blood and plasma corrected for hematocrit:

$$C_B/C_P = (dpm\ whole\ blood/dpm\ plasma \times H) + (1-H) \quad (2)$$

4.2 Non-radiolabeled Data Analysis

Both fractions of the whole blood (plasma and red blood cells) are analyzed by LC-MS/MS alongside the reference samples.

The red blood cell partition coefficient (often referred to as Ke/p) is the ratio of the concentration of drug in the red blood cells (i.e. not including plasma) to concentration of drug in plasma i.e. C_{RBC}/C_P [5].

$$Ke/p = \frac{I_{RBC}}{I_{RBC}^{REF}} \Big/ \frac{I_{PL}}{I_{PL}^{REF}} \quad (3)$$

where

Ke/p is the red blood cell to plasma partition coefficient

I_{PL} is the LC-MS/MS response (peak area ratio to an internal standard) for the plasma fraction

I_{PL}^{REF} is the LC-MS/MS response (peak area ratio to an internal standard) for the reference plasma

I_{RBC} is the LC-MS/MS response (peak area ratio to an internal standard) for the red blood cell fraction

I_{RBC}^{REF} is the LC-MS/MS response (peak area ratio to an internal standard) for the reference red blood cells

5 Notes

1. When analyzing red blood cells, the additional issue of plasma trapped between the cells needs to be considered. To minimize this issue, increasing centrifugation speeds might be attempted

2. Solubilization methods for handling the detection of radioactivity in whole blood samples are involved and quite time consuming. Sample combustion using an oxidizer is usually more reliable but not always available.

Acknowledgements

The views expressed here are solely those of the author and do not reflect the opinions of Janssen Research & Development, LLC.

References

1. Diem K, Lentner C (1975) Documenta Geigy scientific tables, 7th edn. Geigy Pharmaceuticals (Ciba-Geigy Ltd.), Basel, pp 617–618
2. Hinderling PH (1997) Red blood cells: a neglected compartment in pharmacokinetics and pharmacodynamics. Pharmacol Rev 49(3):279–295
3. Hinderling PH (1984) Kinetics of partitioning and binding of digoxin and its analogues in the subcompartments of blood. J Pharm Sci 73:1042–1053
4. Wallace SM, Riegelman S (1977) Uptake of acetazolamide by human erythrocytes. J Pharm Sci 66:729–731
5. Yu S et al (2005) A novel liquid chromatography/tandem mass spectrometry based depletion method for measuring red blood cell partitioning of pharmaceutical compounds in drug discovery. Rapid Commun Mass Spectrom 19(2):250–254
6. Davies B, Morris T (1993) Physiological parameters in laboratory animals and humans. Pharm Res 10(7):1093–1095
7. Moore PA (1981) Preparation of whole blood for liquid scintillation counting. Clin Chem 27(4):609–611
8. "LSC Sample Preparation by Solubilization" LSC Application Note (1996) Perkin Elmer

Chapter 4

Permeability Assessment Using 5-day Cultured Caco-2 Cell Monolayers

Gary W. Caldwell, Chrissa Ferguson, Robyn Buerger, Lovonia Kulp, and Zhengyin Yan

Abstract

In this chapter, we have provided a step-by-step protocol of an accelerated differentiation 5 day culturing Caco-2 (HTB-37) model in a 24-well plate format. The 5-day accelerated Caco-2 assay is based on previous literature protocols where inserts are coated with collagen protein and the culturing cell media contains sodium butyrate to modify differentiation and growth properties of the cell line. Protocol conditions for the accelerated 5 day Caco-2 and conventional 21 day Caco-2 assays are outlined in the chapter. Permeability values and inhibitory efflux ratios for drug candidate compounds are measured using both Caco-2 models using a liquid chromatography-mass spectrometer as the analytical detection method. A comparison between the 5 and 21 day Caco-2 models revealed that a reasonable correlation was demonstrated between the models for ranking permeability values of drug candidate compounds. Thus, permeability studies for large number of drug candidate compounds using the 5 day accelerated Caco-2 model is more convenient and productive than the 21 day model. Using the 5 day model for assessing whether a drug is a substrate of P-gp, revealed efflux ratios lower than those using a 21 day model. The data indicated that the 5 day monolayer expressed P-gp transporters and these transporters were functioning in the proper orientation with presumably a lower expression. While the 5 day Caco-2 model can be used as a high throughput screen for permeability and P-gp efflux measurements, it is recommended to use the 5 and 21 day Caco-2 assays in a tiered approach; that is, large number of compounds in a rapid manner can be funneled through the 5 day assay for both permeability and P-gp data, however, compounds that are positive or borderline substrates of P-gp should be retested in the 21 day assay for a more definitive answer.

Key words Drug discovery, Caco-2, 5-day cultured, Cell monolayers

1 Introduction

The oral absorption of small drug molecules (i.e., drug candidate compounds with molecular weights <1,000 Da) is a dynamic process that involves the transfer of drug molecules from the stomach to the gastrointestinal lumen followed by transfer of the drug molecule across the apical intestinal epithelium membrane followed by diffusion through the cytoplasm and finally the drug molecules exiting through the basolateral membrane into the portal blood system. This transport process across an epithelium membrane is referred to as a passive transcellular passage. There are other mechanisms for the transport of drug molecules across epithelium membranes including paracellular, carrier-mediated and transcytotic pathways. The passive transcellular drug flux across the intestinal membrane is a product of the drug concentration (i.e., solubility) in the luminal fluid and the rate that the drug travels from the apical side of the epithelium cell to the basolateral side (i.e., permeability). Therefore, *in vitro* solubility and permeability assays have been established in drug discovery Drug Metabolism/Pharmacokinetic (DMPK) groups as surrogate markers for small drug molecule oral absorption; thus, de-risking drug candidates as they move from drug discovery into pre-clinical and clinical testing [1–5].

The human intestinal colon carcinoma (Caco-2) cell line has been extensively utilized and studied over the last 2 decades as an *in vitro* human intestinal permeability model [6–10]. In fact, it is the most commonly used assay in a drug discovery research setting for screening drug candidate compounds for human intestinal permeability. The Caco-2 cell line is an excellent *in vitro* model for human intestinal permeability studies since it shares many characteristics with the human small intestinal epithelium including morphology, polarity, and enterocytic differentiation. For example, when Caco-2 cells are cultured on semi-permeable membranes they undergo spontaneous enterocytes differentiation in cell culture that leads to the formation of a monolayer of cells that are cylindrically polarized with tight cell junctions between adjacent cells, microvilli on the apical side of the membrane, biotransformation activities, expression of several carrier-mediated nutrients, and efflux transporters [11]. Thus, it is clear that the Caco-2 model provides a platform for permeability screening, transport mechanism studies and biopharmaceutical assessment.

The parental ATCC Caco-2 cell line (HTB-37) is composed of a heterogeneous population of clones. Some of these subpopulation clones have been isolated and characterized over the years including Caco-2/TC7, Caco-2/BBE 1&2, Caco-2/15 and Caco-2/AQ [11]. These clones have shown to have different morphologies than the parental cell line with Caco-2/TC7 being the most widely studied and utilized for small drug molecules to measure permeability values and to investigate efflux transporters [12]. The heterogeneity of the parental ATCC Caco-2 cell line (HTB-37) makes it susceptible to variations in cultural protocols such as the number of passages, and cell culture conditions (i.e., time of culture, cell feeding media and frequency, cell density, type of semi-permeable membranes, transport media buffers, and so on). It is typically assumed that variations in different subpopulation clones and culturing protocols are the main reasons for laboratory to laboratory variation in Caco-2 experimental permeability values

[9]. Even with variations in cell replication, senescence and differentiation, the Caco-2 permeability cell model is considered to be a reliable *in vitro* method for the estimation of human *in vivo* permeability [13]. In fact, Caco-2 permeability measurements are accepted by the Food and Drug Administration (FDA) [14, 15] as a surrogate for human intestinal permeability measurements as part of the Biopharmaceutical Classification System biowaiver guidance [16].

There are several potential issues that hamper the interpretation of Caco-2 data [17]. While the major epithelium membrane efflux transporters including the permeability glycoprotein (P-gp), the breast cancer resistant protein (BCRP) and the multidrug resistance (MDR) are expressed in Caco-2 cells and the small human intestine, many other efflux transporters are either under expressed or absent in Caco-2 cell monolayers [6–11]. This difference in transporter levels can lead to false negative results; that is, *in vitro* Caco-2 data indicating a drug candidate compound is not permeable, however, the same compound being well absorbed in humans. Other issues around this point concern compounds that diffuse primarily via a paracellular route. It is speculated that the Caco-2 cell junctions are smaller and thus, less permeable than those in the small human intestine. This difference in tight cell junctions can lead to false negative results. Hydrophobic drug candidate compounds can have significant non-specific binding to plastic surfaces and the cell membrane. This problem can lead to false negative results since the permeability of the compound is underestimated by the Caco-2 devise itself and therefore, are not adequately evaluated. Since many of the drug candidate compounds synthesized in drug discovery groups have hydrophobic characteristics, it is highly recommended that a solubility test, using transport media and buffering conditions of the Caco-2 assay, be used before performing the Caco-2 assay to eliminate compounds with poor solubility (i.e., <10 µg/mL) and help with the interpretation of low permeability compounds. To summarize, drug candidate compounds with high permeability values (i.e., >100 nm/s) in the Caco-2 model are typically well absorbed *in vivo* while low permeability results (i.e., <10 nm/s) may not indicate that a drug candidate compound will be poorly absorbed *in vivo*.

The preparation of fully differentiated confluent Caco-2 monolayers grown on semi-permeable membranes typically requires a 21–27 day culture period with 7–20 days of continuous cell feeding [8, 9]. Due to this long culturing period, laborious cell feeding schedules, and the need to measure the Caco-2 permeability values for a larger number of compounds in a short period of time (i.e., 5–10 days), accelerated differentiation 3–7 day culturing Caco-2 models have been developed [18–26]. The conventional 21 day Caco-2 assay in a 24-well device typically starts with seeded cells with:

- Passage numbers between 20 and 60
- Polycarbonate (PB) or polyethyleneterephthalate (PET) or other semi-permeable membranes
- Semi-permeable membrane inserts with a surface size of 0.3 cm^2
- Seeded cell grown in DMEM media supplemented with 10 % fetal bovine serum (FBS) plus other growth factors, antibiotics and metabolites
- Cell seeding densities in the range of 10^5 cells/cm^2
- Media changed every 48 h for 21 days under 37 °C, 90 % humidity and 5 % CO^2 culturing conditions
- Transport medium of HBSS/HEPES at 37 °C and pH of 6.8 or 7.4 for permeability and efflux studies

The primary differences between the conventional 21 day Caco-2 assay and the accelerated 3–7 day assays are the semi-permeable membranes utilized, cell seeding density, extracellular protein coating on the membrane inserts to facilitate cell adhesion, and replacement of 10 % fetal bovine serum (FBS) with butyric acid. A summary of accelerated differentiation 3–7 day culturing Caco-2 protocols are shown in Table 1. Many of the accelerated 3–7 day Caco-2 assays use sodium butyrate to modify differentiation and growth properties of the cell line [27]. It is assumed that butyrate reduces c-*myc* mRNA levels and thus, induces differentiation of intestinal epithelial cells *in vitro* via down regulation of c-*myc* expression.

In this chapter, we have provided a step-by-step protocol of an accelerated differentiation 5 day culturing Caco-2 model in a 24-well plate format using liquid chromatography-mass spectrometry for rapid assessment of permeability and efflux values for drug candidate compounds. The 5-day accelerated Caco-2 assay is based on previous literature protocols [18–26]. Briefly, the Caco-2 cells (HTB-37) are cultivated in medium supplemented with salts, non-essential amino acids, serum, growth factors, antibiotics and metabolites. These cells are split several times by trypsinization creating passages typically numbered between 20 and 50. For permeability and efflux transport studies, cells are seeded on specially designed donor/receiver devices that contain a semi-permeable membrane insert between two chambers. The insert is coated with collagen protein to help cell adhesion and the culturing media contains sodium butyrate to modify differentiation and growth properties of the cell line. Fully differentiated confluent Caco-2 monolayers form in 5 days on semi-permeable membrane inserts with approximately 2 days of cell feeding. In drug permeability studies, drugs are added to apical (mucosal) side and the time resolved drug appearance on the basolateral (serosal) side is measured. From this data, the rate (i.e., velocity or permeability) of

Table 1
Accelerated Caco-2 assay protocols

Parameters	Ref. [18]	Ref. [19]	Ref. [20]	Ref. [21]	Ref. [22]	Ref. [22]	Ref. [23]
Cell source (kit)	ATCC (HTB-37) BioCoat™	ATCC (HTB-37)	ATCC (HTB-37)	Private (TC7) BioCoat™	ATCC (HTB-37) BioCoat™	ATCC (HTB-37)	ATCC (HTB-37)
Cells seeded (passage number)	35–37	30–50	20–50	NA	NA	NA	50–72
Cell seeding density (cells/insert area cm^2)	2.0×10^6	2.0×10^5	2.0×10^6	2.0×10^6	2.0×10^5	7.2×10^4	9.3×10^4
Cell seeding media	DMEM MITO™	DMEM	DMEM	DMEM MITO™	MEM 10 % FBS	MEM 10% FBS	DMEM 10 % FBS heat inactivated
Insert wells (type)	6-wells (PET)	6-wells (PC)	6-wells (PC)	6-wells (PET)	24-wells (PC)	96-wells (PC)	96-wells (PC)
Insert area (pore size)	4.19 cm^2 (1 μm)	4.71 cm^2 (3 μm)	4.71 cm^2 (3 μm)	4.19 cm^2 (1 μm)	0.31 cm^2 (NA)	0.11 cm^2 (NA)	0.11 cm^2 (0.4 μm)
Extra-cellular protein coating on insert	Fibrillar collagen	None	None	Fibrillar collagen	Fibrillar collagen	Fibrillar collagen	None
Cell differentiation inducing media (culturing time)	Entero-STIM™ (3-days)	DMEM/F12 2 % sCS (5 days)	DMEM/F12 2 % sCS (4-days)	Entero-STIM™ (3-days)	Entero-STIM™ MITO™ (3-days)	Entero-STIM™ MITO™ (4-days)	None
Transport medium (temp) (pH)	MHBS HEPES (37 °C) (7.4)	HBSS NA (37°)	HBSS HEPES (37 °C) (6.8)	HBSS HEPES (37 °C) (7.4)	HBSS HEPES (37 °C) (7.4)	HBSS HEPES (37 °C) (7.4)	HBSS HEPES (37 °C) (7.4)

NA not available, *ATCC* American type culture collection, *BioCoat*™ intestinal epithelium differentiation environment—fibrillar collagen cell culture inserts, *DMEM* serum free Dulbecco's modification of eagle's medium, *MITO*™ serum extender is a concentrated, fully defined formulation of hormones, growth factors (EGF and FGF), and other metabolites (insulin and steroid hormones), *Entero-STIM*™ serum free DMEM containing butyric acid, *MHBS* modified hanks balance salt solution, *HEPES* 4-(2-hydroxyethyl)-1-piperazineethanesulfonic acid, *DMEM/F12 sCS* a 1:1 mixture of Dulbecco's modified eagle's medium (DME) and Ham's F-12 nutrient mixture containing 2 % (v/v) iron-supplemented calf serum, *FBS* fetal bovine serum, *PC* polycarbonate, *PET* polyethyleneterephthalate

Table 2
Accelerated and conventional Caco-2 assay protocols used in this chapter

Parameters	Accelerated Caco-2 assay	Conventional Caco-2 assay
Cell source	ATCC (HTB-37)	ATCC (HTB-37)
Cells seeded (passage number)	20–50	20–50
Cell seeding density (cells/insert area cm^2)	4.6×10^5	6.0×10^4
Cell seeding media	DMEM complete: For one 500 mL bottle of DMEM (without sodium pyruvate, with glucose, and phenol red) add 5.6 mL of Pen strep, 5.6 mL of MEM, 5.6 mL of sodium pyruvate, and 56 mL of FBS	MEM 10 % FBS
Insert wells (type)	24-wells (PC)	24-wells (PC)
Insert area (pore size)	0.33 cm^2 (0.4 µm)	0.33 cm^2 (0.4 µm)
Extra-cellular protein coating on insert	Rat tail collagen type I	Rat tail collagen type I
Cell differentiation inducing media (culturing time)	STIM complete: For one 250 mL bottle of Entero-STIM™ media add 12.5 mL FBS, and 2.5 mL Pen strep (5 days)	MEM 10 % FBS (21-days)
Transport medium	HBSSg: HBSS/Glucose/HEPES	HBSSg: HBSS/Glucose/HEPES
Permeability	HBSSg buffer with 10 µM CSA	HBSSg buffer with 10 µM CSA
P-gP (temp) (pH)	(37 °C) (7.4)	(37 °C) (7.4)

See Table 1 for definition of abbreviations
MEM minimum essential medium, *HBSS* hanks balance salt solution, *Pen-Strep* penicillin streptomycin, *CSA* Cyclosporin A

passage through the monolayer is measured. The transport process across the Caco-2 monolayer can be passive transcellular, passive paracellular, carrier-mediated and/or transcytotic. To investigate transport mechanisms, carrier-mediated inhibitors are added to the basolateral side and compared to data where no inhibitor is used. Transport experiments are carried out at known temperatures and pH values under sink conditions in order to avoid back diffusion through the cell monolayer. The protocol conditions are summarized in Table 2 for the accelerated and conventional Caco-2 assays used in this chapter. Data for both Caco-2 assays are collected and compared with a discussion of advantages and disadvantages of the accelerated model.

2 Materials

2.1 Equipment

Specific equipment used in this assay is described; however, any model of comparable capability can be easily substituted.

1. Microscope (Olympus, Center Valley, PA; Model 215078)
2. Epithelial Voltohmmeter for Trans Epithelial Electric Resistance (TEER) Measurements (World Precision Instruments, Sarasota, FL; Model # EVOM2 with Model # STX-100 Electrode)
3. Thermo Forma Steri-cult CO_2 incubator (Thermo Scientific, Asheville, NC; Model# 3307)
4. Centrifuge (Beckman Coulter, Inc, Brea, CA; GPR centrifuge)
5. −80 °C freezer (Baxter Scientific Products, Deerfield, IL; Cryo-Fridge Model # U2186 ABA)
6. Liquid Nitrogen Cell Storage (Thermo Scientific, Asheville, NC; Thermolyne Locator JR. Cryo Biological storage system Model# 5810)
7. Envision plate reader: (Perkin Elmer, Milltown, NJ; Envision 2102 multilabel reader)
8. Cryogenic Freezing vials (blue lid) (Thermo Scientific, Asheville, NC; Nalgene Cryo 1 ° C Freezing container Cat # 5100-0001)
9. Corning 24-well receiver plates (Corning Cat # 3526, VWR, Bridgeport, NJ; Cat# 29443-952)
10. Corning 24 well HTS plates with cell culture inserts (Absorption Systems, Exton, PA; VWR, Bridgeport, NJ; Cat # 29442-106) (*see* **Note 1**)
11. BD Falcon 96-well black cytofluor plates(BD Biosciences, Franklin Lakes, NJ; Cat # 353241)
12. 24-well 10 mL deep well plate (VWR, Bridgeport, NJ; Cat # 89080-532)
13. Transfer Pipets (VWR, Bridgeport, NJ; Cat # 14670-200)
14. Water Bath 37 °C (Fisher Scientific, Bridgewater, NJ; ISO TEMP 220)
15. Thermo shaker (Thermo Scientific, West Palm Beach, FL; Model Max Q 4450)
16. Tissue culture T-75, T-150 and T-225 flasks (Sigma-Aldrich, St. Louis, MO)
17. Plate Loc Velocity 11 plate sealer (Hamilton Robotics, Inc. Reno, NV)
18. Sciex API 4000 Triple Quad LC/MS/MS (AB SCIEX Ontario, Canada)

19. SPHER-100 LC C18 column 100 Å, 5 μ, 50×2.1 mm (Princeton Chromatography Inc, Cranbury Township, NJ; Cat # 050021-03501)
20. Stop watch
21. pH meter
22. Multi-channel (8) automatic and pipettor (50–1,200 μL)
23. Plastic waste dishpan
24. Paper towels

2.2 Reagents

1. Caco-2 wild type cells (HTB-37) (American Type Culture Collection (ATCC) Manassas, VA, USA) (*see* **Note 2**)
2. Hanks Balance Salt Solution (HBSS) 500 mL bottle (Cellgro, Manassas, VA; Cat #: 21-023CV with Ca^{+2} and Mg^{+2})
3. Dulbecco's modification of Eagle's medium (DMEM) 1× with 4.5 g/L glucose, phenol red and L-glutamine without sodium pyruvate. (Cellgro, Manassas, VA; Cat #10-017-CV) 500 mL bottle
4. Dulbecco's Phosphate Buffered Saline with Ca^{+2} and Mg^{+2} (DPBS) (Cellgro, Manassas, VA; Cat # 21-030-CV)
5. Trypsin with EDTA 1× (Cellgro, Manassas, VA; Cat # 25-053-Cl)
6. 4-(2-hydroxyethyl)-1-piperazineethanesulfonic acid (HEPES) Buffer 1 M solution (Cellgro, Manassas, VA; Cat # 25-060-Cl)
7. Sodium Hydroxide (NaOH) (Sigma-Aldrich, St. Louis, MO)
8. D-(+)- Glucose (Sigma-Aldrich, St. Louis, MO; Cat # G-7520)
9. Cyclosporin A (CSA; 1202.61 g/mol) (Sigma-Aldrich, St. Louis, MO; Cat # C3662)
10. Digoxin (780.94 g/mol) (Sigma-Aldrich, St. Louis, MO; Cat # D-6003)
11. (±) Atenolol (266.34 g/mol) (Sigma-Aldrich, St. Louis, MO; Cat # A-7655)
12. Chloramphenicol (323.13 g/mol) (Sigma-Aldrich, St. Louis, MO; Cat # C-0378)
13. (±) Propranolol (295.80 g/mol) (Sigma-Aldrich, St. Louis, MO; Cat # P0884)
14. Dimethyl Sulfoxide (DMSO) (Sigma-Aldrich, St. Louis, MO; Cat # 472301-100 mL)
15. Lucifer Yellow CH dipotassium salt (457.25 g/mol) (Sigma-Aldrich, St. Louis, MO; Cat # L0144-100 mg)
16. Enterocyte differentiation medium 250 mL bottle (Entero-STIM™) (Becton Dickinson Labware, Franklin Lakes, NJ; BD Biosciences-Biocoat Cat # 05496)

17. Fetal Bovine Serum (FBS) (Cellgro, Manassas, VA; Cat # 35-101-CV)

18. Penicillin Streptomycin (Pen-Strep) solution 100×; 10,000 I.U./mL Penicillin, 10,000 μg/mL Streptomycin (Cellgro, Manassas, VA; Cat # 30-002-cl)

19. Minimum Essential Medium (MEM) Nonessential Amino Acids 100× solution (Cellgro, Manassas, VA; Cat # 25-025-cl)

20. Sodium Pyruvate 100 mM solution (Cellgro, Manassas, VA; Cat # 25-000-cl)

2.3 Media Solution Procedures

All media should be stored in the dark to avoid possible deterioration by light and should not be frozen. In addition, media supplements should be added under aseptic conditions without filtering.

1. Hanks Balance Salt Solution with glucose (HBSSg): To prepare this solution add 2.25 g of D-(+)-Glucose and 5 mL of 1 M HEPES buffer to a 500 mL bottle of HBSS. Adjust the pH to 7.4 (±0.2) with NaOH.

2. Prepare a 6.2 mg/mL (11.9 mM) working stock solution of Lucifer Yellow (521.57 g/mol) by dissolving 100 mg of Lucifer yellow in 4 mL of DMSO and 12.129 mL of water (33 % DMSO).

3. Preparation of STIM complete: For one 250 mL bottle of Entero-STIM™ media add 12.5 mL FBS, and 2.5 mL Pen strep. One 250 mL bottle of STIM media is enough for six plates (*see* **Note 3**).

4. Preparation of DMEM complete: For one 500 mL bottle of DMEM (without sodium pyruvate, with glucose, and phenol red) add 5.6 mL of Pen strep, 5.6 mL of MEM, 5.6 mL of sodium pyruvate, and 56 mL of FBS.

5. Prepare 10 mM DMSO stock solution of Digoxin and Atenolol. Add 5 mg of Digoxin to 640 μL of DMSO to prepare a 10 mM stock solution. Add 5 mg of Atenolol to 1.88 mL of DMSO to prepare a 10 mM stock solution.

6. Prepare HBSSg buffer containing 10 μM of CSA. Add 5 mg of CSA to 416 μL of DMSO to prepare a 10 mM stock solution. Take 100 mL of HBSSg buffer and add 100 μL of the CSA 10 mM DMSO stock solution. This solution contains 0.1 % DMSO. The CSA compound is a known inhibitor of the P-gp transporter.

7. Preparing test compound solutions (10 μM) and control solutions (10 μM): Add 5 μL of 10 mM DMSO stock solution of test compounds to 4.995 mL of HBSSg (or HBSSg containing 10 μM of CSA) in a 24-well 10 mL deep well plate. The percent DMSO is 0.1 % in this solution (*see* **Note 4**).

2.4 Test Compound Solubility Procedure

All test compounds are prepared from 10 mM DMSO stock solutions. The stock solutions are diluted into transport buffer media to a final concentration of 10 µM (*see* **Note 5**).

2.5 Thawing Cells

The following procedure is used to remove cells from liquid nitrogen storage, thaw cells and to prepare them for plating onto semiporous membrane inserts. All cell culture procedures need to be performed under aseptic conditions with incubations being done at 37 °C, 5 % CO_2 and 90 % humidity.

1. Remove two vials of cells from the liquid nitrogen storage tank and gently shake while submerging the vials in a 37 °C water bath.
2. Once the cells are thawed, immediately remove cells from vials and combine them in a T-75 Flask with 18 mL of pre warmed DMEM complete media (*see* **Notes 6** and **7**).
3. Label the Caco-2 cells as HTB-37, the date thawed and one passage up from the passage they were frozen at ATCC.
4. Change the media the day after thawing the cells and remove any damaged cells (*see* **Note 8**).
5. Check the cells under the microscope daily, once they reach 70 % confluence, split the cells (*see* **Note 9**).
6. Wash the cells 2× with DPBS, add 1.5 mL Trypsin and incubate the cells for 5 min and check every few minutes until cells have detached.
7. Once the cells have detached, add 10.5 mL DMEM complete to the cells to neutralize the trypsin.
8. Place 18 mL DMEM complete into a new T-75 flask, add 3–5 mL of the cell suspension to this flask, up the passage number by 1 and label appropriately.
9. Repeat process when cells are again 70 % confluent, do not let the cells grow for more than 7 days, split them even if they are less than 70 % confluent.
10. Continue this process until the cells have been passed ten times from their initial passage number (*see* **Note 10**).
11. After ten passages, trypsinize cells for about 5 min and then add 10.5 mL DMEM complete to the cells to neutralize the trypsin. Remove 1 mL of cell suspension and seed into a T-225 flask containing 30 mL of DMEM complete. Incubate the cells until 70 % confluent changing the media every 2–3 days. These cells will be used in Sect. 3.1 *Cell Splitting*.

2.6 Cryopreservation of Cell Culture

The following procedure is used to cryopreserve cells for future studies if necessary.

1. Check the status of the cell monolayer flask (s) that will be cryopreserved. They should be about 80 % confluent.

2. Pre-warm DMEM complete, DPBS and trypsin reagents.
3. Depending on the number of flasks, estimate the number of cryogenic vials needed. If freezing at three million cells per vial about six vials per T-150 is a rough estimate.
4. Label the cryogenic storage vials with the cell line, passage number, date, and preparer's initials.
5. Dissociate the cells using 3 mL trypsin for a T-150 and neutralizing with 12 mL DMEM complete for a total of 15 mL suspension.
6. Remove an aliquot and count the cells. Remove a volume of cell suspension to achieve 1–3 million cells per vial. Vials can have up to 2 mL in them (*see* **Note 11**).
7. Collect the appropriate amount of cell suspension in 15 or 50 mL tubes. Gently centrifuge at 1,500 rpm for 5 min to pellet cells.
8. While centrifuge is running prepare freezing medium: 5 % DMSO in DMEM complete or media appropriate for cell culture. Volume of freezing solution depends on volume of cells spun down.
9. Re-suspend the cells in freezing medium.
10. Aliquot 1–2 mL into each vial. Place vials in cell freezing container.
11. Cells are frozen slowly (1 °C per minute) place freezing container in −80 °C freezer.
12. Transfer the vials from the freezer after 48–72 h to the liquid nitrogen storage container.

3 Methods

The following procedure is used to create a cell monolayer on collagen coated 24-well semi-permeable polycarbonate inserts. Caco-2 cells spontaneously undergo enterocytic differentiation in culture. All cell culture procedures are completed under aseptic conditions and incubations are done at 37 °C, 5 % CO_2 and 90 % humidity.

3.1 Cell Splitting and Growth Procedure

1. Take a flask of cells prepared in Sect. 2.5.11 *Thawing Cells*.
2. Aspirate media from the T-225 flasks. Wash cells once with approximately 20 mL DPBS and aspirate DPBS off. Repeat the washing step one more time.
3. Add 3 mL of trypsin with EDTA per flask. Swirl to make sure cells are covered. Incubate for no more than 5 min. Continue to check if cells are detached throughout the 5 min and remove flasks from the incubator as soon as they detach.

4. Add 12 mL of DMEM complete media per flask to neutralize the trypsin. Combine the cell suspension into a new T-225 flask.

5. Count the cells and make a suspension at 3.0×10^5 cells/mL (*see* **Note 11**).

6. Seed 3–5 T-225 flasks with approximately 2.0×10^6 cells per flask.

7. Check the flasks seeded every day for 6 days to ensure that cells are <70 % confluent.

8. On the seventh day, splits the cells (i.e. trypsinize the cells for about 5 min and then add 10.5 mL DMEM complete to the cells to neutralize the trypsin) and count the cells (*see* **Note 11**).

3.2 Cells Seeding Procedure

The following procedure is used for a 5-day culturing of Caco-2 cell onto 24 well plates with collagen coated semi-permeable polycarbonate inserts. Use cells prepared in Sect. 3.1.8 *Cell Splitting* for this 5 day culture. All incubations are done at 37 °C, 5 % CO_2 and 90 % humidity. The experiment is designed that each 24-well assay plate can measure one positive control compound and three test compounds for transport from the apical to the basolateral side (A-B) and basolateral to apical side (B-A) in triplet.

1. The seeding density required for each insert is 460,000 cells/cm^2. Each insert has an area of 0.33 cm^2; therefore, 151,800 cells are required. Since each plate has 24-wells with a volume of 0.5 mL per well, approximately 3×10^5 cells/mL is required. The total volume of cell suspension needed for the 24-well plate is 12 mL.

2. To dilute the cell suspension solution in Sect. 3.1.8 *Cell Splitting* to 460,000 cells/cm^2 use the following equations to calculated dilution factor:

$$Dilution\ volume(mL) = \frac{(12\ mL)(3.0 \times 10^5\ cells\ /\ mL)}{Cells\ Counted\ in\ step\ 5} \quad (1)$$

$$Volume\ of\ DMEM\ complete = 12mL - Dilution\ volume(mL) \quad (2)$$

3. Add the volume of DMEM complete calculated in equation (2) to the cell suspension solution in Sect. 3.1.8 *Cell Splitting*.

4. Add 0.5 mL of the solution generated in step 3 to each well in the top insert plate (see Fig. 1).

5. Add 25 mL DMEM complete media to the bottom reservoir plate (*see* Fig. 1). Submerge the top insert plate into the bottom reservoir place and place a cover on the top plate. Incubate these plates for 72 h.

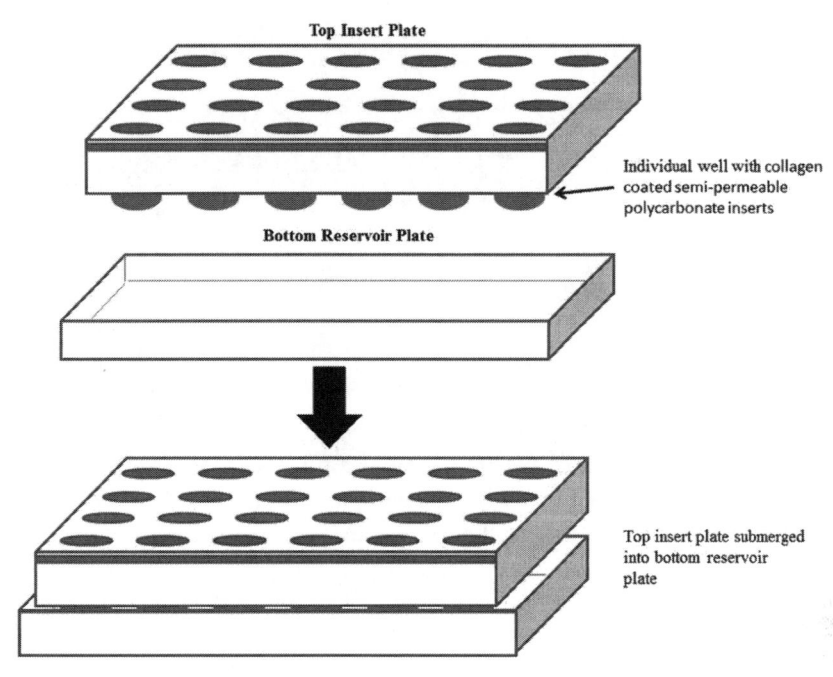

Fig. 1 Diagram of 24-well seeding plate

6. After the 72 h incubation, aspirate DMEM complete media from the top insert plate; add 0.5 mL STIM complete media to top insert plate, 25 mL into the bottom reservoir plate, submerge top plate into bottom plate and incubate for 48 h (*see* **Note 12**).

3.3 Cell Culturing and Dosing Procedures

1. Gather all needed test compounds.
2. Prepare HBSSg complete solution (*see* Sect. 2.3.1).
3. Label all plates.

3.3.1 On the Day Before the Caco-2 Assay is to be Performed

4. Make sure internal standards are prepared/prepare a 5 µM Propranolol and 10 µM Chloramphenicol solution in water. This is the LC/MS/MS internal standards.

3.3.2 Prepare Top Insert Plate on the Day the Assay is to be Performed

1. Place HBSSg into the 37 °C water bath and allow warming to 37 °C (*see* **Note 7**).
2. Make sure the thermo shaker is on and set temperature to 37 °C.
3. Randomly check some monolayers in the insert wells for TEER values using the probe and EVOM. An acceptable reading is approximately >300 $\Omega \cdot cm^2$ (*see* **Note 13**).
4. Place the cell plates in the thermo shaker until ready to use (*see* **Note 14**).

Add 600 µL Buffer (Basolateral Solution)   Add 600 µL Compound (Basolateral Solution)

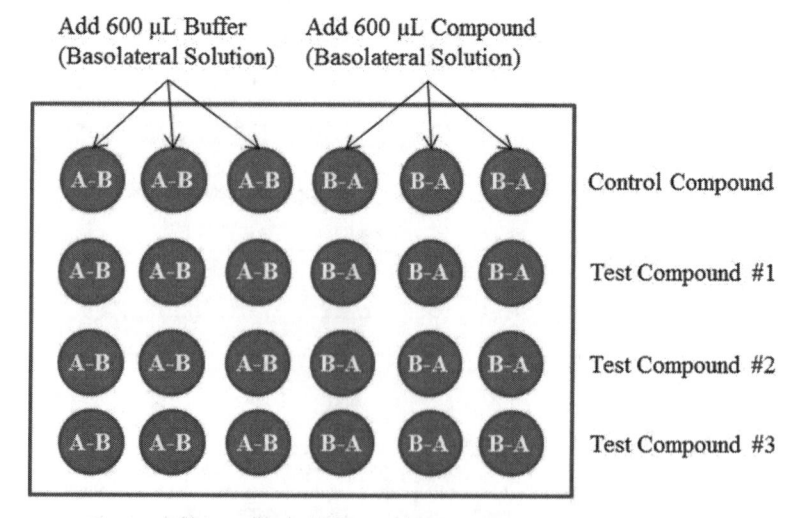

Bottom Plate with Individual Reservoirs

Fig. 2 Caco-2 basolateral plate showing layout of test and control compounds

3.3.3 Preparation of Dosing Solutions for Permeability Studies

1. ***Permeability Measurement:*** Prepare samples according to Sect. 2.3.7.
2. ***Monolayer Integrity Assessment Markers:*** Add 5 µL of 10 mM DMSO stock solution of Digoxin (transcellular marker) and Atenolol (paracellular marker) to a glass bottle and add 4.990 mL of HBSSg. The percent DMSO is 0.1 % in this solution.

3.3.4 Prepare Bottom Plate (Basolateral Side) with Individual Reservoirs for Assay (see Fig. 2)

1. Obtain a bottom plate with individual reservoirs.
2. Add 600 µL of HBSSg buffer to the wells of the A-B receivers.
3. Add 600 µL of test compounds or control to the wells of the B-A donors.
4. Place plates in the 37 °C thermo shaker.

3.3.5 Wash Cells in Top Insert Plate (Apical Side) (see Fig. 1)

1. Take plate from Sect. 3.3.2.4 out of the thermo shaker
2. Invert the plate and dump off solution (STIM complete) into a paper towel, and allow the inverted plate to rest on a paper towel to drain.
3. Dump the bottom reservoir plate into a flask and discard plate
4. Use a disposable pipette and rinse cells in top insert plate with HBSS buffer and also rinse back of the plate to remove STIM complete
5. Invert again and rinse again with HBSS buffer
6. Place the top insert plate into a holder.

Fig. 3 Caco-2 apical plate showing layout of test and control compounds

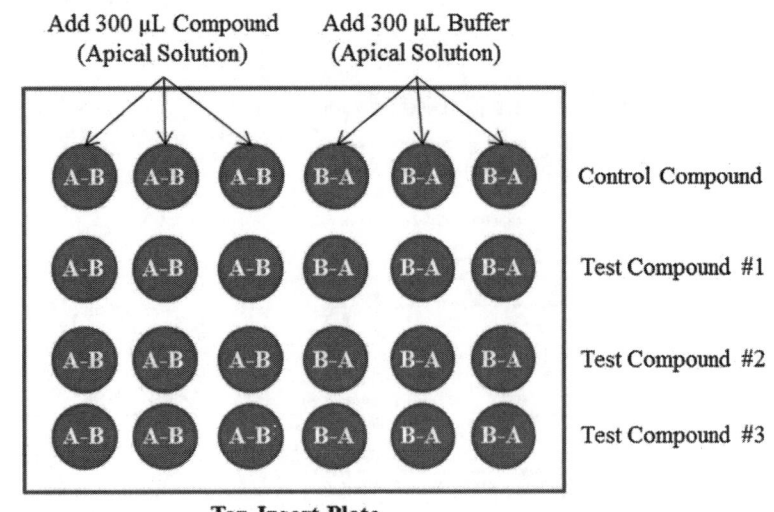

3.3.6 Prepare Top Insert Plate (Apical Side) for Assay (see Fig. 3)

1. Add 300 μL of test compounds or control to the inserts of the A-B (*see* Sect. 3.3.3.1).
2. Add 300 μL of HBSSg buffer to the insert of the B-A receivers.
3. Add 5 μL of a Lucifer Yellow stock solution (6.2 mg/mL; 13.6 mM) to all inserts. The concentration of Lucifer Yellow in each insert (apical side) is 195 μM at 0.5 % DMSO.
4. Start experiment: Submerge the top plate (Fig. 3) into the bottom plate (Fig. 2) and start the timer (Fig. 4).
5. Place the plate in the thermo shaker at 37 °C and shake at speed 32 rpm.
6. Wait 90 min and remove plates.

3.3.7 Preparation of Dosing Solutions for P-gp Studies

1. ***P-gp Measurement***: Prepare all testing compounds (10 μM) and controls (10 μM) using HBSSg buffer with 10 μM CSA. The CSA compound is a known inhibitor of the P-gp transporter. Add 5 μL of 10 mM DMSO stock solution of test compounds to 4.995 mL of HBSSg buffer with 10 μM CSA in a 24-well 10 mL deep well plate.
2. ***Monolayer Integrity Assessment Markers***: Add 5 μL of 10 mM DMSO stock solution of Digoxin and Atenolol to a glass bottle and add 4.990 mL of HBSSg buffer with 10 μM CSA. The percent DMSO in these solutions is 0.1 % (*see* **Note 15**).
3. ***Prepare plate for P-gp Studies***: Follow steps in Sects. 3.3.4, 3.3.5, and 3.3.6 replacing HBSSg with HBSSg containing 10 μM of CSA (*see* **Note 15**).

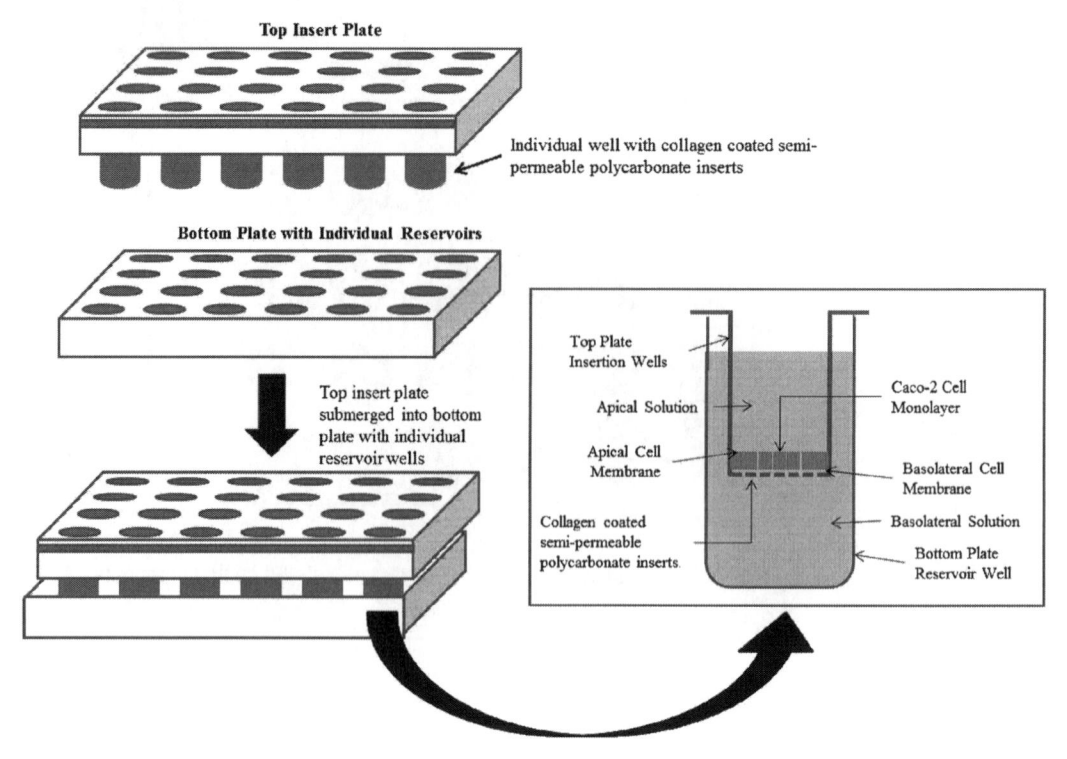

Fig. 4 Caco-2 dosing plate

3.4 Preparing Plates for Lucifer Yellow and LC/MS/MS Analyses

At 90 min remove the plates from the thermo shaker and prepare a 96-well plate for Lucifer Yellow analyses and two 96-well plates for LC/MS/MS analyses.

1. Remove insert plate and place in empty reservoir.
2. Remove 100 µL from each well of the bottom assay plate (i.e., A-B receiver and B-A donor; basolateral solution, *see* Fig. 2) to a black cytofluor 96-well *Lucifer Yellow Sample plate* to assess the integrity of the monolayers (*see* **Note 16**).
3. Remove 125 µL from the A-B receiver (i.e., basolateral solution; *see* Fig. 2) and add to the corresponding wells of the 96-well *MS-Receiver plate*.
4. Remove 125 µL from the B-A receiver (i.e., apical solution; *see* Fig. 3) and add to the corresponding wells of the 96-well *MS-Receiver plate*.
5. Remove 25 µL of the A-B donor (i.e., top plate; *see* Fig. 3) and the B-A donor (i.e., bottom plate; *see* Fig. 2). Add to 100 µL of HBSSg buffer in corresponding wells of another 96-well *MS-Donor plate*. (*see* **Note 15**)
6. Add 20 µL of 5 µM Propranolol and of 10 µM Chloramphenicol solution diluted 1:4 in water to each well on the MS plates as internal standards (*see* **Note 17**).

7. Add 125 μL ACN to all wells on the MS plates
8. Mix all 96-well MS plates well.
9. Seal all plates on the Plate Loc Velocity 11 plate sealer for future analyses

4 Analytical Methods

4.1 Lucifer Yellow Analyses

4.1.1 Standard Plate

1. Add 100 μL of HBSSg medium to each well of the first three rows in the black 96-well cytofluor plate label *Standard Lucifer Yellow plate* (*see* **Note 16**) (*see* Fig. 5).
2. Add 200 μL HBSSg to the first column of the *Standard Lucifer Yellow plate*.
3. Add 5 μL of the Lucifer Yellow working solution (6.2 mg/mL; Sect. 2.3.2) to the well in the first column to make the Lucifer Yellow concentration 100 μg/mL.
4. Transfer 100 μL from the first column to the second column and mix and second column to the third column and mix and so on. This process will create a serial dilution plate as shown in Fig. 5.
5. Leave last column blank. This well will contain only HBSSg medium.

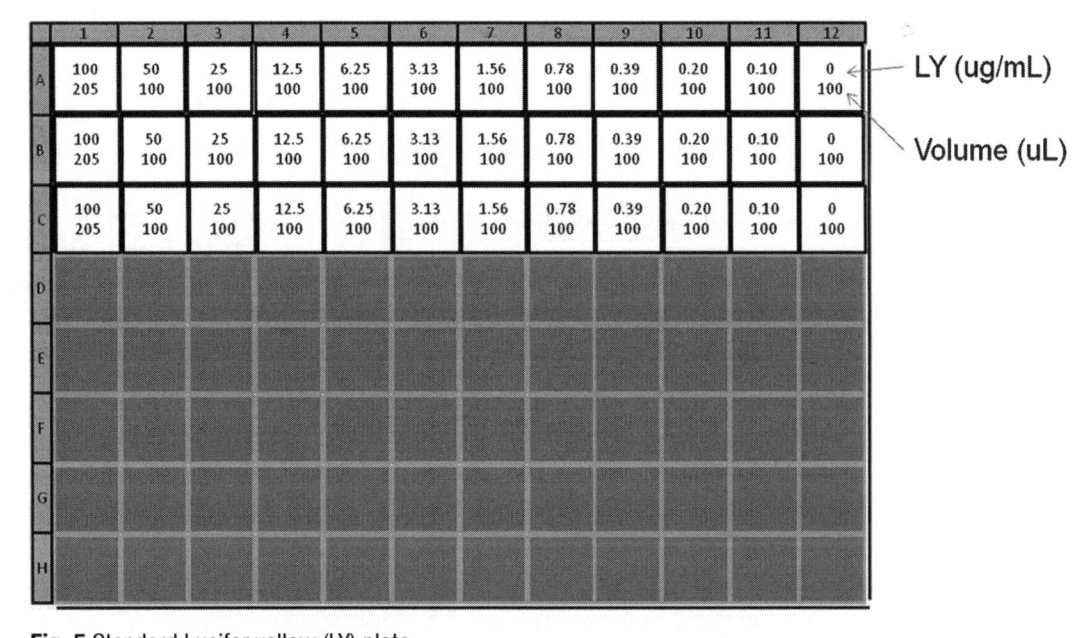

	1	2	3	4	5	6	7	8	9	10	11	12
A	100 / 205	50 / 100	25 / 100	12.5 / 100	6.25 / 100	3.13 / 100	1.56 / 100	0.78 / 100	0.39 / 100	0.20 / 100	0.10 / 100	0 / 100
B	100 / 205	50 / 100	25 / 100	12.5 / 100	6.25 / 100	3.13 / 100	1.56 / 100	0.78 / 100	0.39 / 100	0.20 / 100	0.10 / 100	0 / 100
C	100 / 205	50 / 100	25 / 100	12.5 / 100	6.25 / 100	3.13 / 100	1.56 / 100	0.78 / 100	0.39 / 100	0.20 / 100	0.10 / 100	0 / 100

Top values: LY (ug/mL); Bottom values: Volume (uL)

Fig. 5 Standard Lucifer yellow (LY) plate

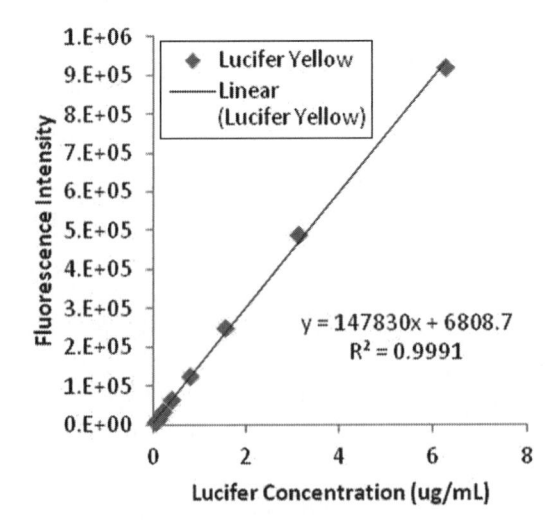

Fig. 6 Standard Lucifer yellow (LY) calibration curve

4.1.2 Sample Plate

1. Analyze both the 96 well **Standard Lucifer Yellow plate** and the **Lucifer Yellow Sample plate** on the Envision 2021 multi-label reader (*see* **Note 17**).

2. The raw readings, from the Envision 2021 multilabel reader, are then pasted into a spreadsheet software package where known concentrations of Lucifer Yellow (see Fig. 5) are plotted against raw fluorescence intensity values that have been corrected for background fluorescence. Typically, the fluorescence intensity data for the 100, 50, and 12.5 µg/mL standards are excluded since at these concentrations the fluorescence signal is saturated. The equation for a straight line is calculated from the remaining data and used to calculate the unknown concentrations in the **Lucifer Yellow Sample plate** (*see* Fig. 6).

3. For the **Lucifer Yellow Sample plate**, first subtracts out the blank florescence intensity values, and then calculate the concentration (µg/mL) of the basolateral solutions using the straight line equation generated in step 7. The amount of Lucifer Yellow transported (µg) through the cell monolayer is calculated by multiplying the calculated concentrations (µg/mL) of the basolateral solutions by the volume in the well (600 µL). The % Lucifer Yellow passage and P_{app} (cm/s) are calculated using (3) and (4), respectively:

$$\% \text{ Lucifer Yellow passage} = \frac{(F_{test} - F_{Blank})}{(F_0 - F_{Blank})} \quad (3)$$

where the fluorescence intensity of Lucifer Yellow in the basolateral solution is F_{test}, the fluorescence intensity of Lucifer

Yellow in the apical solution is F_0, and the fluorescence intensity of the blank HBSS solution is F_{Blank}.

$$P_{app} = \frac{\left(\frac{dQ}{dt}\right)}{(A)(C_D)} \quad (4)$$

In (4), dQ/dt is the rate of appearance of Lucifer Yellow on the basolateral side of the monolayer (μg/s), A is the area of the semi-permeable membrane insert (0.33 cm^2), and C_D is the initial Lucifer Yellow concentration on the apical side of the monolayer (100 μg/mL). This equation assumes sink conditions.

4. Typical % Lucifer Yellow passage values are < 0.1 % which corresponds to P_{app} values of < 10^{-7} cm/s. These types of values indicate that the integrity of the cell monolayer is acceptable; that is, it has tight cell junctions between adjacent cells. If the P_{app} reading of Lucifer Yellow is greater than 1.0 × 10^{-6} cm/s the monolayer fails. The integrity of the monolayer is considered questionable and any data associated with it should be eliminated (*see* **Note 18**).

4.2 LC/MS/MS Analyses

4.2.1 Standard Plate

1. Add 25 μL dosing solution in triplicates to a 96 well **Standard-MS plate** (first three rows of the plate) (i.e., 10 μM test compounds in HBSSg buffer; see 3.3.3.1 Preparation of dosing solutions)
2. Add 100 μL HBSSg buffer to all wells (i.e., standards at 0.1 μM).
3. Add 125 μL HBSSg buffer to the fourth column. This will serve as a blank.
4. Add 125 μL ACN to all wells (i.e., standards at 0.05 μM).
5. Add 20 μL internal standards (1:4 diluted) to the first three columns. Do not add internal standards to the fourth column (blank 50:50 HBSSg and ACN) (*see* **Note 19**)
6. Mix well.
7. Label and seal the plate on the Plate Loc Velocity 11 plate sealer.

4.2.2 Sample Plate

All samples are analyzed with a Sciex API 4000 Triple Quad (LC/MS/MS). The HPLC method may need to change for compounds with extended retention time and the electrospray ionization mode may need to change for compounds that are acids (negative) and bases (positive). Quantitative mass spectrometry is performed using the multiple-reaction-monitoring (MRM) technique (*see* **Note 20**).

1. Analyze the 96 well **Standard-MS plate, the MS-Receiver plate** and the **MS-Donor plate** on the mass spectrometer.
2. LC/MS/MS **Standard-MS plate** Analysis: Automaton software from AB Sciex is used for tuning and determining MRM mass transitions for each unknown compound (Table 3).

Table 3
Auto-tune method and MRM transition method creation (for conditions, see foot notes[a-e])

Time	Module	Events	Parameter
0.10	Pumps	Pump B concentration	A = 5 %/B = 95 %
1.10	System controller	Stop	

[a]Flow injection
[b]Mobile phase: A = 0.1 % formic acid plus 1 g ammonium acetate in a 4 L bottle of water
[c]Mobile phase: B = 0.1 % formic acid plus 1 g ammonium acetate in a 4 L bottle of 70 % MeOH/30 % ACN
[d]Flow rate = 0.2 mL/min
[e]Mass spec: scan type—Q1 MS, start mass 100 amu, stop mass 900 amu, time 0.5 s)

Table 4
Sciex LC method (for conditions, see foot notes[a-e])

Time	Module	Events	Parameter
0.25	Pumps	Pump B concentration	A = 5 %/B = 95 %
0.45	Pumps	Pump B concentration	A = 10 %/B = 90 %
1.25	Pumps	Pump B concentration	A = 10 %/B = 90 %
1.30	Pumps	Pump B concentration	A = 5 %/B = 95 %
1.60	System controller	Stop	

[a]Column: SPHER-100:C18, 100 Å, 5 μ, 50 × 2.1 mm
[b]Column oven temperature: 40 °C
[c]Mobile phase: A = 0.1 % formic acid plus 1 g ammonium acetate in a 4 L bottle of water
[d]Mobile phase: B = 0.1 % formic acid plus 1 g ammonium acetate in a 4 L bottle of 70 % MeOH/30 % ACN
[e]Flow rate = 0.2 mL/min

3. LC/MS/MS *Sample-MS plates* Analysis: Using automation software from AB Sciex and the tune and MRM conditions generated in step 2, analyze the 96 well *Standard-MS plate*, *the MS-Receiver plate* and the *MS-Donor plate*. The mobile phase conditions are shown in Table 4.
4. The AB Sciex standard quantitative software package will quantitate all samples from the 96 well plates.
5. Using the quantitative information, the A-B permeability and the B-permeability for each compound is calculated using (4). The A-B and B-A % mass balances are calculated using (5).

$$\% \, Mass \, Balance = \frac{C_{Basolateral} + C_{Apical}}{C_0} \times 100 \qquad (5)$$

Table 5
Permeability and P-gp efflux data for the 5 day Caco-2 model

Compound	A-B P$_{app}$ (10^{-6} cm/s)	B-A P$_{app}$ (10^{-6} cm/s)	A-B Mass balance (%)	B-A Mass balance (%)	B-A/A-B Efflux ratio
Atenolol	1.4 ± 1.1 %CV = 78 n = 191	2.1 ± 1.9 %CV = 91 n = 191	97 ± 12 %CV = 12 n = 191	100 ± 14 %CV = 14 n = 191	1.7 ± 1.1 %CV = 65 n = 191
Atenolol P-gp inhibitor CSA	1.8 ± 2.2 %CV = 121 n = 191	2.4 ± 2.9 %CV = 125 n = 191	97 ± 12 %CV = 12 n = 191	100 ± 14 %CV = 14 n = 191	1.9 ± 1.2 %CV = 64 n = 191
Digoxin	1.4 ± 0.8 %CV = 53 n = 191	16.4 ± 4.2 %CV = 26 n = 191	92 ± 12 %CV = 12 n = 191	99 ± 13 %CV = 13 n = 191	13.0 ± 4.7 %CV = 36 n = 191
Digoxin P-gp inhibitor CSA	3.6 ± 2.3 %CV = 65 n = 191	5.7 ± 2.7 %CV = 48 n = 191	92 ± 12 %CV = 12 n = 191	99 ± 14 %CV = 14 n = 191	2.0 ± 1.0 %CV = 50 n = 191

where $C_{Basolateral}$ is the concentration of the drug candidate compound on the basolateral side of the monolayer, C_{Apical} is the concentration of the drug candidate compound on the apical side of the monolayer, and C_0 is the initial concentration of the drug candidate compound.

4.2.3 Examples of Permeability and P-gp Efflux Data for Controls

Atenolol is used as a passive paracellular marker for examining the 5 day Caco-2 monolayer while digoxin is used as a passive transcellular marker for examining the P-gp transporter. As shown in Table 5, atenolol is a non P-gp substrate and the efflux ratios are the same in the absence of a P-gp substrate and in the presence of a known substrate of the P-gp transporter (CSA). Digoxin is a P-gp substrate and the efflux ratios are different in the absence of a P-gp substrate and in the presence of a known substrate of the P-gp transporter (CSA). The data indicates that the 5 day Caco-2 model expresses P-gp transporters and these transporters were functioning in the proper orientation.

4.2.4 Examples of Permeability and Efflux Data for Drug Candidate Compounds Using the 21 and 5 Day Models

As shown Fig. 7, the apparent permeability for 12 drug candidate compounds using the 21 and 5 day Caco-2 models have been measured in the apical to basolateral (A-B) direction and the basolateral to apical (B-A) direction. These data have been correlated to each other. It is clear that the correlation between the 21 and 5 day models is significant with a R^2 value of 0.9226 for the A-B comparison.

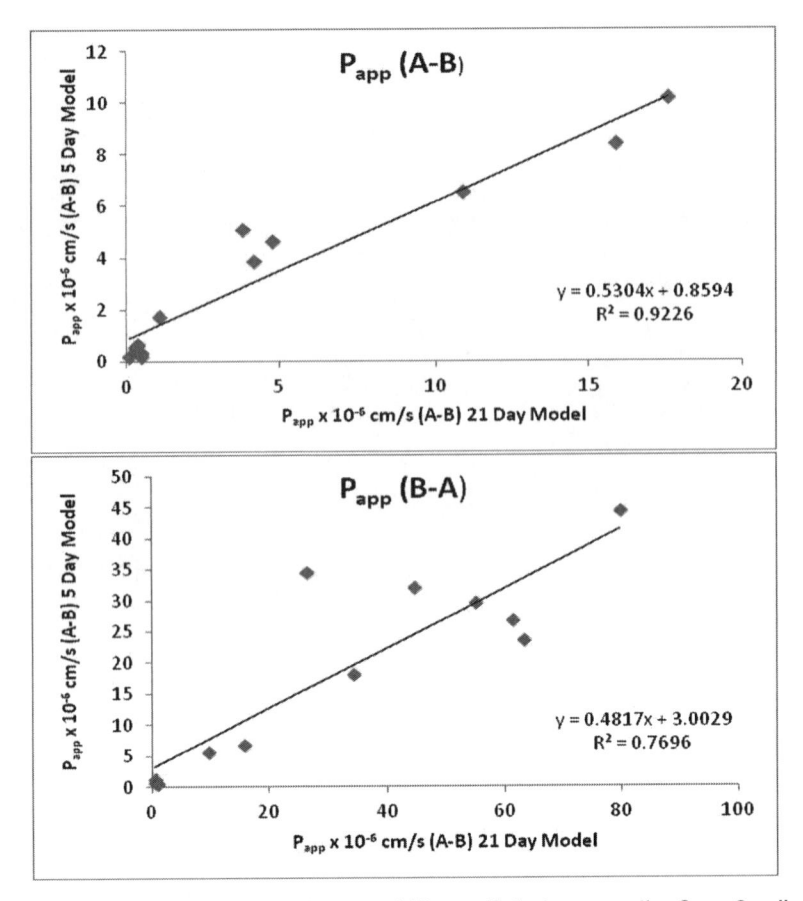

Fig. 7 Correlation of apparent permeability coefficients across the Caco-2 cell monolayer between the conventional 21 day model and the accelerated 5 day model in a 24 well format (n = 3)

The B-A correlation was somewhat less but still significant. The same absorption potential for the 12 compounds for both the 21 day and the 5 day Caco-2 models can be classified as either Low or High using the following criteria for the apical-to-basolateral (A to B) direction:

- Low: Papp < 1.0×10^{-6} cm/s
- High: Papp ≥ 1.0×10^{-6} cm/s

It should be noted that the % mass balance values for these studies ranged from 90 to 120 % for all measurements in Fig. 7.

For passive transcellular and paracellular compounds, the bi-directional P_{app} should be similar in both directions and an efflux ratio, defined as P_{app} (B-to-A)/P_{app} (A-to-B), near unity. However, substrates of efflux transporters expressed on the apical surface of the cells are transported more rapidly in the B-to-A direction and their efflux ratio will be greater than unity. The efflux ratios for the two models are shown in Fig. 8 for 12 drug candidate compounds

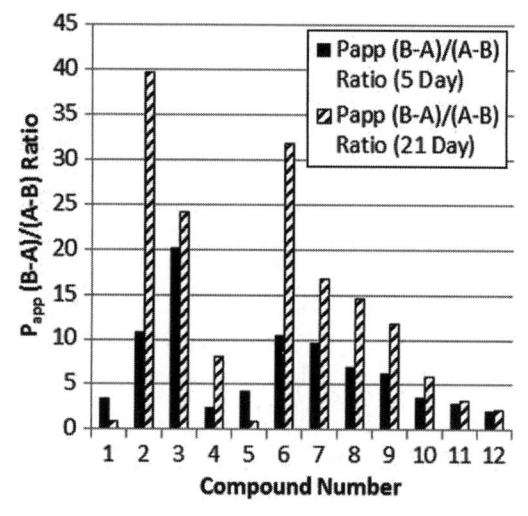

Fig. 8 Efflux ratio (B-A)/(A-B) across the Caco-2 cell monolayer for the conventional 21 day model and the accelerated 5 day model in a 24 well format (n = 3)

utilizing data from Fig. 7. If the two models were identical the 21 and 5 day bars would be of equal height. It is clear that the 5 day accelerated Caco-2 model expressed transporters that are functioning in the proper orientation, but are 40–70 % lower than the conventional 21 day model. However, if we use the criteria that the efflux is classified as significant when only when:

- Efflux ≥ 3.0 and P_{app} (B to A) $\geq 1.0 \times 10^{-6}$ cm/s

We see that compounds 2, 3 and 6–11 are classified as having a significant efflux by both the 5 and 21 day models. In addition, we see that compounds 1, 4 and 12 are classified as not having a significant efflux by both the 5 and 21 day models. For compound 5, the 5 day model classified it as having a significant efflux while the 21 day model classified it as not having a significant efflux.

5 Notes

1. The 24-wells Transswell® Permeable plate have semi-permeable polycarbonate inserts (6.5 mm) with a pore size of 0.4 μm and a well area of 0.33 cm². The inserts have been treated with rat tail collagen, type I, under conditions that allow *in situ* formation of large collagen fibrils. Briefly, rat tail collagen type I solution (4.7 g/L) was diluted with a 0.02 M acetic acid solution to a final concentration of 0.7 g/L. An aliquot of 225 μL containing 1.65 μg of protein was used to cover the insert and allowed to stand for 1 h at 25 °C. The collagen solution was aspirated from the insert and rinsed three times with a PBS

(pH 7.4) buffer. The procedure yielded a uniform thin layer of fibrillar collagen on the inserts.

2. A license from ATCC is required to order cells. When you receive the cells from ATCC they will arrive in dry ice, immediately transfer them to the liquid nitrogen container (−80 °C). They will arrive with a product information sheet; this sheet will contain the passage of the frozen cells.

3. Enterocyte differentiation medium (Entero-STIM™ media) is used in the assay to promote the rapid differentiation of Caco-2 cells. This media contains butyric acid which induces differentiation of intestinal epithelial cells *in vitro* via down regulation of *c-myc* expression [27].

4. Many of the drug candidate compounds tested in a Caco-2 assay are prepared as 10 mM DMSO stock solutions. The DMSO is added at this step to keep the DMSO content consistent in the experiment. If neat test compounds are used without using DMSO to solubilizing the sample, then the DMSO can be eliminated from this step.

5. It should be noted that dissolved test compounds from library collections have inaccurate concentrations. From our own experience, when library 10 mM DMSO stock solutions were measured they were discovered to be closer to 6–7 mM.

6. Ensure that the cells are the same passage and lot number before combining them into one T75 flask. The passage number refers to the number of times the cell line has been grown to confluency (i.e., re-plated with cells adhering to the flask and each other as they grow.) and trypsinized. Thus, add 1 to the passage number each time the cell line is trypsinized and re-plated. It should be noted that the number of passages can influence different functions and activities of the cell line such as active transporters and CYP3A4 metabolism.

7. Avoid repeatedly warming and cooling cycles of the medium since this process will affect the medium bioactivity.

8. After the trypsinization process, cells will be in suspension and appear rounded. The live cells will begin to attach to the flask and each other while the dead or damaged cells will not attach. When the media is changed the dead cells can be aspirated off leaving only attached healthy cells. Note that excessive aspiration may remove live cells.

9. The percent confluence can be subjectively estimated using a microscope by comparing the amount of space covered by cells with the unoccupied spaces.

10. Usually these cells can be used up to 60–70 passages from their original ATCC passage number (i.e., 6–7). The best way to determine if a passage is acceptable is to run positive controls

(i.e., digoxin and atenolol) and reject passages that continuously fail. Once the positive control fails, remove more cells from liquid nitrogen storage and begin the process again.

11. To determine the cell dilution, you must count the cells. To count the cells, load a hemocytometer with 15uL cell solution, which is obtained from the flask of cells that was previously trypsinized and neutralized. Count the cells in the four quadrants of the 4×4 square of the hemocytometer. Take the average of the four quadrants. Correct for the hemocytometer (10,000) by moving the decimal place one to the left. Calculate how much cell suspension you need to add to the DMEM complete to achieve 460,000 cells per cm^2.

12. Caution at this step is advised since damage to the fibrillar collagen coating and/or the Caco-2 cell monolayer can be damaged by excessive aspiration.

13. Transepithelial electrical resistance (TEER) measurements are used to assess the integrity of monolayers (i.e., tight paracelluar cell junctions). TEER is determined by ion flux through paracelluar space. A sharp increase in TEER value indicates the confluence of the cellular monolayer while a drop the TEER values indicate that the monolayer is leaking. It should be noted that TEER measurements are somewhat variable and caution is advised in their interpretation for integrity of the monolayers. Typically, TEER values in the range of 300–500 Ω cm^2 indicate tight paracelluar cell junctions. In addition, changes in temperature are known to have strong effects on TEER measurements; therefore, for comparison purposes, TEER should be measured at the same temperature. The flux of an impermeable marker such as Lucifer yellow is used as an indicator of the size tight junctions in addition to TEER values.

14. The Caco-2 monolayer is agitated using a shaker to reduce the effects of the aqueous boundary layer adjacent to the cell membrane. Without agitation during the experiment, permeability values for rapidly transported test compounds will be significantly under estimated.

15. One set of compounds are for permeability studies and other set is for P-gp study, where 10 µM of CSA is added to all compounds. Make sure to use 10 µM CSA in HBSSg on basolateral and apical sides.

16. Lucifer Yellow is a fluorescent dye used to test cell monolayer integrity; in other words, it answers the question are the cell junctions between adjacent cells tight. Due to the polar characteristics of Lucifer Yellow, it is considered to be a zero permeability marker. The permeability of Lucifer Yellow can be significantly altered by the presence of organic solubilizing solvents, such as, polyethylenglycol, propyleneglycol, dimethyl

sulfoxide, acetonitrile and methanol, since these solvents disrupt cell junctions between adjacent cells.

17. The Envision 2021 multilabel reader should be set such that a photometric 450 excitation filter and a FITC 535 emission filter are used.

18. The % passage of Lucifer Yellow was in the range of 0.1 % or less for many of the monolayers formed using the 5 day model accelerated differentiation model. The rejection rate of monolayers based on % passage of Lucifer Yellow data for the 5 day accelerated differentiation model was no greater than for the conventional 21 day Caco-2 model.

19. Propranolol (5 μM) and Chloramphenicol (10 μM) solutions are used as internal standards for the LC/MS/MS analyses (i.e., injection to injection variability). Propranolol is used in the positive ionization mode and Chloramphenicol is used in the negative ionization mode. Data are normalized to these values.

20. The MRM mode consists of two stages of mass filtering and one stage of collisional fragmentation. Briefly, for a LC/MS/MS triple quadrupole (Q1Q2Q3) mass spectrometer, the m/z cation or anion of interest is generated in the ion source (i.e., electrospray) and preselected in the first stage of mass filtering Q1 and transferred to Q2 where it is induced to fragment by collisional excitation with a neutral gas in a pressurized collision cell. In the second stage of mass filtering Q3, only the fragment ions (transition ions in Q2) are mass analyzed. The MRM mode of detection is a very sensitive technique being able to detect ions in the pictogram range and has approximately a dynamic range of five orders of magnitude [28, 29].

Acknowledgements

The views expressed here are solely those of the author and do not reflect the opinions of Janssen Research & Development, LLC.

We thank Malini Dasgupta for expert advice regarding assay method development.

References

1. Caldwell GW (2000) Compound optimization in early- and late-phase drug discovery: acceptable pharmacokinetics properties utilizing combined physicochemical, *in vitro* and *in vivo* screens. Curr Opin Drug Discov Devel 3:30–41
2. Caldwell GW, Ritchie DM, Masucci JA, Hageman W, Yan Z (2001) The new preclinical paradigm: compound optimization in early and late phase drug discovery. Curr Top Med Chem 1(5):353–366
3. Caldwell GW, Yan Z, Masucci JA, Hageman W, Leo G, Ritchie DM (2003) Applied pharmacokinetics in drug development: an overview of drug discovery. Pharm Dev Regul 1(2):117–132

4. Caldwell GW, Ritchie DM (2004) Is early absorption screening useful in the selection of drug candidates? Preclinical 2(6):399–401
5. Caldwell GW, Yan Z (2006) Screening for reactive intermediates and toxicity assessment in drug discovery. Curr Opin Drug Discov Devel 9(1):47–60
6. Hidalgo IJ, Raub TJ, Borchardt RT (1989) Characterization of the human colon carcinoma cell line (Caco-2) as a model system for intestinal epithelial permeability. Gastroenterology 96(3):736–749
7. Borchardt RT, Hidalgo IJ, Hillgren KM, Hu M (1991) Pharmaceutical applications of cell culture: an overview. In: Wilson G, Davis SS, Illum L (eds) Pharmaceutical applications of cell and tissue culture to drug transport, vol 218, NATO ASI series, series A: life sciences. Plenum, New York, pp 1–14
8. Hidalgo IJ (1996) Cultured intestinal epithelial cell models. Pharm Biotechnol 8:35–50, Models for Assessing Drug Absorption and Metabolism
9. Hidalgo IJ (2001) Assessing the absorption of new pharmaceuticals. Curr Top Med Chem 1(5):385–401
10. Weinstein K, Kardos P, Strab R, Hidalgo IJ (2004) Cultured epithelial cell assays used to estimate intestinal absorption potential. In: Borchardt RT, Kerns EH, Lipinski CA, Thakker DR, Wang B (eds) Biotechnology: pharmaceutical aspects, vol 1, Pharmaceutical profiling in drug discovery for lead selection. AAPS Press, Arlington, pp 217–234
11. Sambuy Y, De Angelis I, Ranaldi G, Scarino ML, Stammati A, Zucco F (2005) The Caco-2 cell line as a model of the intestinal barrier: influence of cell and culture-related factors on Caco-2 cell functional characteristics. Cell Biol Toxicol 21(1):1–26
12. Turco L, Catone T, Caloni F, Di Consiglio E, Testai E, Stammati A (2011) Caco-2/TC7 cell line characterization for intestinal absorption: how reliable is this in vitro model for the prediction of the oral dose fraction absorbed in human? Toxicol In Vitro 25(1):13–20
13. Avdeef A, Tam KY (2010) How well can the Caco-2/Madin-Darby canine kidney models predict effective human jejunal permeability? J Med Chem 53(9):3566–3584
14. US Department of Health and Human Services; Guidance for Industry (2000) Waiver of in vivo bioavailability and bioequivalence studies for immediate-release solid oral dosage forms based on a biopharmaceutics classification system. http://www.fda.gov/downloads/Drugs/GuidanceComplianceRegulatoryInformation/Guidances/UCM070246.pdf
15. Larregieu CA, Benet LZ (2013) Drug discovery and regulatory considerations for improving in silico and in vitro predictions that use Caco-2 as a surrogate for human intestinal permeability measurements. AAPS J 15(2):483–497
16. Amidon GL, Lennernaes H, Shah VP, Crison JR (1995) A theoretical basis for a biopharmaceutic drug classification: the correlation of in vitro drug product dissolution and in vivo bioavailability. Pharm Res 12(3):413–420
17. Cheng K-C, Li C, Uss AS (2008) Prediction of oral drug absorption in humans—from cultured cell lines and experimental animals. Expert Opin Drug Metab Toxicol 4(5):581–590
18. Chong S, Dando SA, Morrison RA (1997) Evaluation of Biocoat intestinal epithelium differentiation environment (3-day cultured Caco-2 cells) as an absorption screening model with improved productivity. Pharm Res 14(12):1835–1837
19. Liang E, Chessic K, Yazdanian M (2000) Evaluation of an accelerated Caco-2 cell permeability model. J Pharm Sci 89(3):336–345
20. Lentz KA, Hayashi J, Lucisano LJ, Polli JE (2000) Development of a more rapid, reduced serum culture system for Caco-2 monolayers and application to the biopharmaceutics classification system. Int J Pharm 200(1):41–51
21. Da Violante G, Zerrouk N, Richard I, Frendo JL, Zhiri A, Li-Khuan R, Tricottet V, Provot G, Chaumeil JC, Arnaud P (2004) Short term Caco-2/TC7 cell culture: comparison between conventional 21-d and a commercially available 3-d system. Biol Pharm Bull 27(12):1986–1992
22. Uchida M, Fukazawa T, Yamazaki Y, Hashimoto H, Miyamoto Y (2009) A modified fast (4 day) 96-well plate Caco-2 permeability assay. J Pharmacol Toxicol Methods 59(1):39–43
23. Galkin A, Pakkanen J, Vuorela P (2008) Development of an automated 7-day 96-well Caco-2 cell culture model. Pharmazie 63(6):464–469
24. Yamashita S, Konishi K, Yamazaki Y, Taki Y, Sakane T, Sezaki H, Furuyama Y (2002) New and better protocols for a short-term Caco-2 cell culture system. J Pharm Sci 91(3):669–679
25. Alsenz J, Haenel E (2003) Development of a 7-day, 96-well Caco-2 permeability assay with high-throughput direct UV compound analysis. Pharm Res 20(12):1961–1969
26. Gupta V, Doshi N, Mitragotri S (2013) Permeation of insulin, calcitonin and exenatide across Caco-2 monolayers: measurement using a rapid, 3-day system. PLoS One 8(2):e57136
27. Souleimani A, Asselin C (1993) Regulation of c-myc expression by sodium butyrate in the colon carcinoma cell line Caco-2. FEBS Lett 326(1–3):45–50

28. Caldwell GW, Easlick SM, Gunnet J, Demarest K, Masucci JA (1998) In-vitro permeability of eight β-blockers through a Caco-2 monolayer utilizing electrospray ionization mass spectrometry. J Mass Spectrom 33:607–614

29. Wang Z, Hop CECA, Leung KH, Pang J (2000) Determination of in vitro permeability of drug candidates through a Caco-2 cell monolayer by liquid chromatography/tandem mass spectrometry. J Mass Spectrom 35(1):71–76

Chapter 5

In Situ Single Pass Perfused Rat Intestinal Model

Maria Markowska and L. Mark Kao

Abstract

Compound solubility and permeability play an important role in oral absorption of drug candidates and ultimately their Biopharmaceutics Classification System (BCS) classification. BCS classification of drug candidates can influence the drug development path, specifically around formulation development, and the potential route used for drug administration. The *in situ* single pass perfused rat intestinal model has been shown to be highly predictive of human absorption of marketed drugs and thus, it is a useful tool for assessing the intestinal permeability of drug candidates during early stages of drug development. With the perfused rat intestinal model, the permeability of drug candidates is calculated using a macroscopic mass balance approach and a complete radial mixing model. This model is an effective approach used for selecting drug candidates with a desired BCS classification, which ultimately could improve the success rate in the selection of new chemical entities for eventual clinical use.

Key words In situ single rat perfusion, Solubility, Stability, Permeability, Jejunum, BCS-biopharmaceutical classification system

1 Introduction

Intestinal permeability of test compounds can be determined by multiple methods including *in vitro*, *ex vivo* and *in situ* models [1]. *In situ* intestinal models have significant advantages over some *in vivo* and *in vitro* models. That is, using the *in situ* model approach site-specific absorption and metabolism studies can be integrated whereby physiological and physicochemical factors influencing absorption are studied at specific segments of the intestine [2]. In addition, bypassing the stomach ensures that acidic compounds are not likely to precipitate, and dissolution rates do not affect drug solubility in the intestine and ultimately plasma exposures [3]. Since mesenteric blood flow is intact this provides for a dynamic distribution of absorbed drug which is similar to that observed under *in vivo* conditions. However, caution must be taken with the choice of anesthetic used in this procedure as it has been demonstrated that anesthesia can have significant effects on intestinal drug absorption [2].

Gary W. Caldwell and Zhengyin Yan (eds.), *Optimization in Drug Discovery: In Vitro Methods*, Methods in Pharmacology and Toxicology, DOI 10.1007/978-1-62703-742-6_5, © Springer Science+Business Media New York 2014

The *in situ* single-pass perfused rat intestinal model has been shown to be highly predictive of human absorption of marketed drugs and thus is a useful tool for investigating the intestinal permeability of drug candidates during early stages of drug discovery and development [1, 4]. Importantly, it is one method recommended by the FDA to be used for BCS classification of drug candidates, and for obtaining a waiver for conducting human bioavailability and bioequivalence studies [5]. The model was first developed by Amidon at the University of Michigan using isolated rat intestine [6, 7]. Subsequent optimization of this method has been described [8]. Variations of the methodology have also been reported, including site-specific studies on different segments of the intestine. The *in situ* rat jejunum model provides excellent correlation with human absorption potential [9] and is also useful in evaluating potential drug-drug interactions, and in the facilitation of pro-drug development [4, 10]. In conjunction with the use of knock-out mice models, it may also be a useful tool for understanding transporters involved in the absorption of test compounds at the level of the intestine [11]. The model herein describes intestinal perfusion studies for one region of the GI-tract of the rat, i.e., the jejunum.

2 Materials

Reagents and equipment for the study can be found in Tables 1 and 2 (**Note 1**).

Table 1
List of reagents needed for *in situ* perfusion studies (Note 2)

Reagents	Vendor	Catalog number
NaCl	Sigma-Aldrich (St. Louis, MO)	S5886
KCl	Sigma-Aldrich (St. Louis, MO)	P5405
2-(N-morpholino) ethanesulfonic acid (Mes)	Sigma-Aldrich (St. Louis, MO)	M3058
NaOH	Sigma-Aldrich (St. Louis, MO)	S5881
PEG-4000	Sigma-Aldrich (St. Louis, MO)	81240
Saline	Abbott (Abbott Park, IL)	009861
Ketamine	Sigma-Aldrich (St. Louis, MO)	K2753
Acepromazine maleate	Sigma-Aldrich (St. Louis, MO)	A7111
Xylazine	Sigma-Aldrich (St. Louis, MO)	X1251
DMSO	Sigma-Aldrich (St. Louis, MO)	D1251
^{14}C-PEG-4000 (Note 2)	Perkin Elmer (Shelton, CT)	NEC8260
Ultima Gold	Perkin Elmer (Shelton, CT)	6013329

Table 2
List of equipment needed for *in situ* perfusion studies

Equipment	Vendor	Catalog number
Harvard infusion pump	Harvard Apparatus (Holliston, Mass.)	704501
Tri-Carb Liquid Scintillation Counter	Perkin Elmer (Shelton, CT)	3110TR B291000
Tygon Laboratory tubing	VWR (Radnor, PA)	63010
Water bath	VWR (Radnor, PA)	89132-162
Heating pad pillow	Sears (Chicago, IL)	SPM7551663610
Surgical instruments	Roboz Surgical Instruments (Gaithersburg, MD)	
16–100 mm glass tubes	VWR (Radnor, PA)	47729-576
Eppendorf tubes 1.5 mL	VWR (Radnor, PA)	53511-997
Scintillation vials 7 mL	VWR (Radnor, PA)	66009-254
Teflon tubing	VWR (Radnor, PA)	63015-336
60 cm^3 syringes	BD Gentest (San Jose, CA)	309654
Parafilm	VWR (Radnor, PA)	21020-238
Q-tips	VWR (Radnor, PA)	15141-356
Suture 20 vicro silk	Roboz Surgical Instruments (Gaithersburg, MD)	
Electric razor-cordless	Wahl (Sterling, IL)	
Scissors, micro-scissor, forceps, 1 in. square parafilm	VWR (Radnor, PA)	

3 Methods

3.1 Animals

Male or female Sprague Dawley rats, 250–390 g, can be purchased from Charles River Laboratories (Wilmington, MA), or another appropriate vendor. They should be housed and handled in a facility which is in compliance with the Association for the Assessment and Accreditation of Laboratory Animal Care (AAALAC) and require an acclimatization period of 1 week prior to use.

3.2 Anesthesia and Buffers Preparation

Approved protocol by Institutional Animal Care and Use Committee (IACUC) is mandatory before any study is initiated.

3.2.1 Anesthesia

Anesthetic doses contain ketamine HCl (100 mg/mL), acepromazine maleate (10 mg/mL), and xylazine (20 mg/mL), in combination. A mixture of all three anesthetic agents should be prepared based on rat weight. The final dose of 75 mg/kg ketamine HCl, 2.5 mg/kg of acepromazine maleate and 5 mg/kg xylazine should be prepared fresh on the day the perfusion study is conducted.

3.2.2 Perfusion Buffer Preparation

The constituents of the perfusion buffer are: Mes buffer, pH 6.5 (Note 3) with NaCl (110–145 mM) range, KCl (5 mM), Mes (10 mM), NaOH (5 mM) and PEG-4000 (0.01 %).

1. Weight out each chemical required (final concentrations as above), and dissolve in a 950 mL beaker using deionized water.
2. Adjust the pH to 6.5 with 2 N NaOH and then add H_2O to achieve a final volume of 1,000 mL using a volumetric flask.
3. Calibrate the Micro-Osmometer with Reference Standard Solution (290 ± 2 mOsm) and measure the osmotic concentration of the buffer.
4. The osmotic concentration of the final perfusion buffer should 300 ± 10 mOsm/L; adjust to final osmotic concentration with NaCl, sterilize via filtration using a 0.45 μm (Millipore) filter.
5. Add the test article to its final anticipated concentration into the perfusion buffer (Note 4) and measure the osmotic concentration, re-adjusting if necessary to 300 mOsm/L.
6. Final adjustments (osmotic concentration and pH) of the test article solution should be performed on the day of the study, to avoid compound decomposition.

3.3 Drug Solubility Assessment

1. Obtain physiochemical properties such as molecular weight, batch molecular weight, and other pertinent information including log P, permeability in Caco-2 cells and solubility data, if available.
2. Prepare drug stock solution (100 mM) in perfusion buffer, DMSO or other another suitable solvent (**Notes 4** and **5**).
3. Run a preliminary solubility test from 10 μM to 100 mM on the test article at the desired final concentration in perfusion buffer prior to the initiation of the study (**Note 6**).
4. The final test article concentration in the perfusion studies should be 10–100 μM (**Note 7**). Make sure that the compound is stable in the perfusion buffer (for 24 h) to ensure no precipitation during the course of the study. The final organic content of the compound in perfusion buffer can range from 0.01 to 0.1 %.
5. The solubility of the test article will also need to be determined in simulated intestinal fluid (SIF, pH 7.5) (**Note 8**). 50–200 mg of compound is added to 2–5 mL of SIF (pH 7.5). The samples are incubated at room temperature for 24 h in order to attain saturation equilibrium. After incubation, a 1 mL sample is removed, filtered (0.45 μm), and analyzed for drug concentration by LC-MS or another suitable analytical procedure.

3.4 Drug Stability

1. Determine test article stability (10–100 μm) in rat intestinal perfusate. Extra perfusate can be collected from satellite animals and frozen until needed (**Note 9**). Test article should be incubated at 37 °C in a shaking water bath.

2. During the incubation, 100 μL aliquots are withdrawn at 0, 30, 60 and 120 min after the addition of the test article.

3. Samples are quenched with 100 μL acetonitrile solution maintained at 4 °C, vortexed and centrifuged at 1,000 g for 10 min and the resulting supernatants collected and analyzed immediately by LC-MS or another suitable analytical procedure (**Note 10**). If drug stability in the perfusate is acceptable, proceed with perfusion study.

3.5 In Situ Single Rat Perfusion

1. A day before the study, rats are fasted overnight with free access to water. Label glass tubes, microtube and scintillation vials. Prepare stock solutions of the internal permeability markers propranolol and terbutaline at 100 mM in DMSO.

2. On the day of the perfusion, prepare initial dosing solution in perfusion buffer. This will contain the internal permeability markers at final concentrations of 100 μM, and the test article at 10–100 μM. Adjust final osmotic concentration to 300 ± 10 mOsm/L, if required. Aliquot 1 mL of initial drug solution (without ^{14}C-PEG 4000 addition, next step) to prepare standard curve for subsequent bio-analytical measurements of the test compound and controls.

3. Add ^{14}C-PEG 4000 (used as a non-absorbable marker for measuring water flux) to the dosing solution to achieve a final concentration of 10 nM (*see* **Note 11**) in 100 mL of drug solution. Remember to aliquot 1 mL of the dosing solution at the beginning and at the end of the perfusion study for subsequent analysis.

4. Warm saline and perfusion buffer (200 mL) to 37°.

5. Weigh rats, record the weights and inject the sedatives intramuscularly based on the weight. Have 10 cm suture, scissors, micro-scissors, forceps, 1 in. square parafilm, mesh, gauze and saline (37 °C) ready.

6. Shave the abdominal cavity of each rat with an electric razor. Working on each individual rat, make a midline abdominal incision of 3–4 cm. Handling the GI tract delicately, locate the jejunum and isolate by ligation two suitable jejunal segments using sutures of approximately 10–12 cm each. Cannulate at both ends of each segment and keep the jejunum moist with gauze saturated with saline at 37 °C.

7. Perfuse the selected jejunum segment with perfusion buffer at 37 °C at a flow rate of 0.5 mL/min to wash out the residual mucin and bile contents.

8. During the wash period, warm the drug solutions containing the test article, internal markers propranolol and terbutaline and ^{14}C-PEG-4000 to 37 °C, and fill 4×60 cm^3 syringes with 45 mL of drug solution.

9. Perfuse the drug solution at 0.5 mL/min for 10 min, at 0.25 mL/min for 30 min. After 30 min, start to collect the perfusate in 10 min intervals for 90 min, keeping the collecting tubes on wet ice. At the end of the perfusion, euthanize the rat.

10. Immediately after euthanizing the rat, remove the jejunum and measure length immediately.

11. Pipette 1 mL of perfusion sample into 7 mL scintillation vial (pre-filled with 5 mL of scintillation cocktail), and count dpm values using a liquid scintillation counter.

12. Pipette another 1 mL of perfusion sample into a microcentrifuge tube (1.5 mL), and freeze for subsequent analysis (**Note 12**).

3.6 Sample Analysis

1. Samples from step 12 above, are diluted tenfold with perfusion buffer to bring the radioactivity to a background level (20–25 dpm) and drug concentration is analyzed by LC/MS-MS or another appropriate analytical procedure.

2. All standards are assayed in duplicate. Analyte concentrations are calculated from standard curves of the peaks versus standard concentration.

3.7 Data Analysis

Using a macroscopic mass balance approach and a complete radial mixing model, the effective permeability (P_{eff}) is determined from equation (1),

$$Peff = \frac{-Q \cdot \ln\left(\frac{Cout'}{Cin'}\right)}{2\pi RL} \quad (1)$$

where Q is the perfusion buffer flow rate (mL/min), C_{out}' is the outlet concentration (μg/mL) that has been adjusted for water transport equation (2) after passing through the intestinal segment, C_{in}' is the inlet or starting concentration (μg/mL), R is the radius (cm) of the intestinal segment (set to 0.2 cm) and L is the length of the intestinal segment (cm).

The ratio of concentrations (C_{out}/C_{in}) is adjusted to account for water transport that may have occurred during the perfusion. To correct for this, a non-absorbed radioactive tracer, ^{14}C-PEG 4000 is included in the perfusion buffer. The C_{out}/C_{in} ratio is then corrected for water transport according to equation (2):

$$\frac{Cout'}{Cin'} = \frac{Cout}{Cin} \times \frac{Ain - A_0}{Aout - A_0} \quad (2)$$

where A_{out}-A_0 is equal to the radioactivity counts (dpm) in the outlet sample minus background (dpm), and A_{in}-A_0 is equal to the radioactivity counts (dpm) in the inlet sample minus background (dpm). The C_{out}'/C_{in}' is used to determine the permeability according to equation (1) [6, 12, 13].

% Water Transport along the intestine is also calculated for each perfusion sample according to equation (3).

$$W.T. = \frac{((Ain - A_0) - (Aout - A_0))}{Aout - A_0} \times 100 \qquad (3)$$

This water transport value is divided by the length of the perfused segment (cm) to determine W.T./cm. The value must be within ±0.5 %/cm W.T. for the acceptance of the data for permeability assessment.

All experiments should be undertaken with a minimum number of four intestinal segments.

3.8 Sample Permeability and Solubility Determination

Sample replicate permeability data is given in Table 3, with final perfusion data calculated and presented in Table 4 and graphically in Fig. 1.

Upon determination of the P_{eff} from the *in situ* rat perfusion, the suitability of the permeability model should be established by comparing the P_{eff} values of the internal markers propranolol and terbutaline, with reference values of 5.87×10^{-5} and 0.41×10^{-5} cm/s, respectively [12, 13].

Table 3
Intestinal permeability ($\times 10^{-5}$ cm/s) of compounds determined by *in situ* rat perfusion

Compound	Segment 1	Segment 2	Segment 3	Segment 4	Mean	SD
JNJ-1	6.40	5.50	5.0	5.80	5.68	0.59
Propranolol	4.56	4.32	4.38	4.15	4.35	0.17
JNJ-2	2.51	3.15	3.5	2.85	3.0	0.42
Terbutaline	0.38	0.32	0.45	0.40	0.39	0.05
JNJ-3	0.12	0.22	0.23	0.19	0.19	0.05

Table 4
Intestinal permeability ($\times 10^{-5}$ cm/s) and solubility of test compounds

Compound	JNJ-1	Propranolol	JNJ-2	Terbutaline	JNJ-3
SIF solubility (mg/mL)	0.85	NA	1.70	NA	4.98
Permeability ($\times 10^{-5}$ cm/s)	5.68	4.35	3.05	0.39	0.19

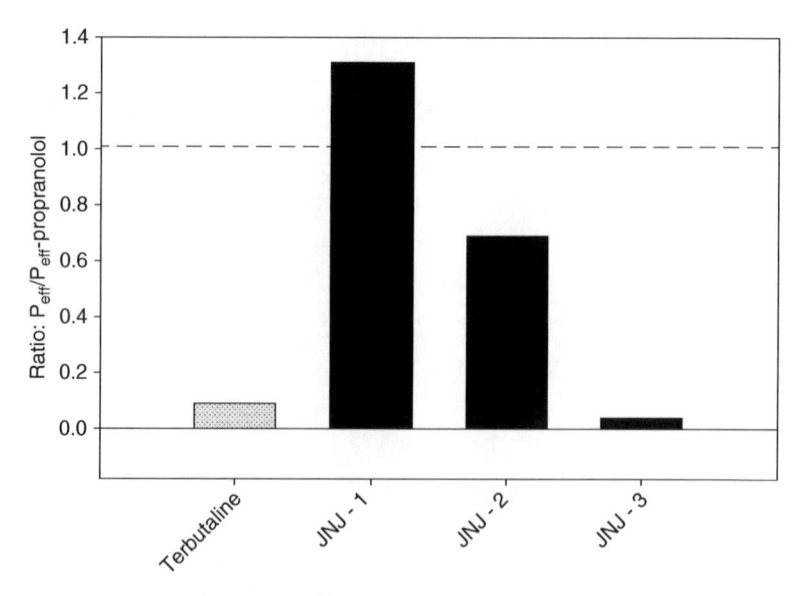

Fig. 1 Summary of the permeability ratio of compounds to the permeability of co-perfused internal standard propranolol. The *horizontal line* at a ratio of approximately 1 indicates the limit for compounds with high permeability

According to the FDA guideline [5] classification, the data obtained classifies JNJ-1 into a BCS II, based on assessment of high permeability and low solubility (Table 4), JNJ-2 falls into a BCS I, based on assessment of high permeability and high solubility, and JNJ-3, falls into a BCS III, based on assessment of low permeability and high solubility (Table 4).

The permeability ratio of compounds to the internal permeability standard propranolol is shown in Fig. 1, indicating JNJ-1 with high permeability and JNJ-3 with very low permeability. This approach of comparing permeability ratios provides an effective means in screening and ranking of compound absorption [14].

4 Notes

1. Chemicals and equipment indicated in Tables 1 and 2, respectively, are an immediate reference. Equivalent chemicals and equipment can be obtained from other commercial sources.
2. In order to conduct *in situ* rat perfusion, the laboratory has to be certified to use radioactive material, in accordance with appropriate state and federal laws.
3. The perfusion buffer (pH 6.5) can be prepared prior to the study and stored refrigerated.
4. If compound cannot be dissolved in perfusion buffer or DMSO (see Drug Solubility Assessment, Sect. 2.4), use other solvents

such as ethanol, acetonitrile, water, or perfusion buffer with solubility enhancers such as bovine serum albumin (BSA).

5. Conduct the initial compound stock solution solubility test first in perfusion buffer, pH 6.5, at 100 μM. If the compound does not dissolve perform sequential dilutions with perfusion buffer to attain the highest concentration possible.

6. If the compound cannot be dissolved to the final concentration desired in perfusion buffer, try altering the stock solution organic solvent content (as indicated in Note 4).

7. Typically the final drug concentration is tested at 10–100 μM due to limit of quantitation from a bio-analytical perspective.

8. Solubility of the test article is also performed under physiologically relevant conditions, such as in SIF, as detailed in the FDA guidance. These values will be used during interpretation of the study as detailed (Table 4).

9. Test article stability is conducted in the perfusate buffer to rule out any degradation in the jejunum segment of the gastrointestinal tract (GIT).

10. If samples are not being analyzed immediately, freeze at −20 °C until assay availability. Test article freeze-thaw stability should be checked prior.

11. Final concentration of ^{14}C-PEG-4000 solution (specific activity 45 mCi/mmol, concentration 0.05 mCi/mL) should result (200–250 dpm).

12. Samples are quantitated by an LC-MS/MS procedure using a Sciex API-5000 (Applied Biosystems, Foster City, CA), or similar system.

Acknowledgements

The views expressed here are solely those of the author and do not reflect the opinions of Janssen Research & Development, LLC.

The authors would like to thank Shannon Dallas and Dennis Kalamaridis for reviewing the chapter.

References

1. Volpe DA (2010) Application of method suitability for drug permeability classification. AAPS J 12(4):670–678
2. Ferrec EL, Chesne C, Artusson P, Brayden D, Fabre G, Gires P, Guillou F, Rousset M, Rubas W, Scarino M-L (2001) In vitro models of the intestinal barrier: *the report and recommendations of ECVAM Workshop 46*. Altern Lab Anim 29:649–688
3. Holst B, Williamson G (2004) Methods to study bioavailability of phytochemicals. In: Bao Y, Fenwick R (eds) Phytochemicals in health and disease. Marcel Dekker, New York, p 40
4. Chaturvedi PR, Decker CJ, Odinecs A (2001) Prediction of pharmacokinetic properties using experimental approaches during early drug discovery. Curr Opin Chem Biol 5(4):452–463

5. US-FDA (2000) Guidance for industry: waiver of in vivo bioavailability and bioequivalence studies for immediate-release solid oral dosage forms based on a biopharmaceutics classification system
6. Amidon GL et al (1980) Analysis of models for determining intestinal wall permeabilities. J Pharm Sci 69:1369–1373
7. Sinko PJ, Amidon GL (1988) Characterization of the oral absorption of beta-lactam antibiotics. I. Cephalosporins: determination of intrinsic membrane absorption parameters in the rat intestine in situ. Pharm Res 5(10): 645–650
8. Jeong EJ, Liu Y, Lin H, Hu M (2004) In situ single-pass perfused rat intestinal model for absorption and metabolism. In: Yan Z, Caldwell W (eds) Optimization in drug discovery: in vitro methods. Humana Press Inc., Spring House, pp 65–76
9. Peternel L, Kristan K, Petruševska M, Rižner TL, Legen I (2012) Suitability of isolated rat jejunum model for demonstration of complete absorption in humans for BCS-based biowaiver request. J Pharm Sci 101:1436–1449
10. Nozawa T, Imai T (2011) Prediction of human intestinal absorption of the prodrug temocapril by in situ single-pass perfusion using rat intestine with modified hydrolase activity. Drug Metab Dispos 39(7):1263–1269
11. Shugarts S, Benet LZ (2009) The role of transporters in the pharmacokinetics of orally administered drugs. Pharm Res 26(9): 2039–2054
12. Amidon GL, Sinko PJ, Fleisher D (1988) Estimating human oral fraction dose absorbed: a correlation using rat intestinal membrane permeability for passive and carrier-mediated compounds. Pharm Res 5(10):651–654
13. Fagerholm U, Johansson M, Lennernas H (1996) Comparison between permeability coefficients in rat and human jejunum. Pharm Res 13(9):1336–1342
14. Kim J-S, Mitchell S, Kijjek P, Tsume Y, Hilfinger J, Amidon GL (2006) The suitability of an in situ perfusion model for permeability determinations: utility for BCS class I biowaiver requests. Mol Pharm 3(6):686–694

Chapter 6

Metabolic Stability Assessed by Liver Microsomes and Hepatocytes

Kevin J. Coe and Tatiana Koudriakova

Abstract

Hepatic metabolism is often a major contributor to drug clearance from the body, highlighting the utility of *in vitro* liver systems to address chemotypes that undergo extensive metabolism. Drug metabolism can be assessed in a variety of *in vitro* test systems, including microsomes, cytosol or S9 fractions, hepatocytes (suspension or plated), and isolated liver slices. In drug discovery, metabolic stability is typically determined by measuring the depletion of test compound over time in a relevant *in vitro* system. Most commonly used *in vitro* systems are liver microsomes and suspension hepatocytes, which are amenable to automated, high-throughput assay formats. Test compound is quantified in samples using LC/MS/MS, and its metabolic half-life in the *in vitro* test system is derived from percent remaining vs. time data. Half-life data can be used to rank order compounds according to stability, derive SAR to improve metabolic stability for chemotypes suffering from high turnover, and predict *in vivo* clearance. The latter is of particular importance to understand *in vitro-in vivo* correlations and select compounds with appropriate pharmacokinetics in humans. This chapter provides details on the methods used to investigate metabolic stability of test compounds in liver microsomes (for both cytochrome P450s and UDP-glucuronosyltransferases) and in suspension hepatocytes.

Key words Cytochrome P450, Hepatocytes, Microsomes, Metabolism, Metabolic stability, UDP-glucuronosyltransferases

1 Introduction

Hepatic metabolism is a primary elimination pathway for a majority of drugs [1]. Hepatic clearance can impact the oral bioavailability through first pass metabolism and the observed half-life of a drug thereby influencing its dose and frequency of administration. Thus, hepatic *in vitro* test systems to determine the metabolic stability of discovery compounds are routinely employed during lead optimization oftentimes in automated, high throughput assays. Liver microsomes and suspension hepatocytes serve as the most common *in vitro* metabolic stability systems. Although single-time point incubations, broadly utilized in the past, are amenable to rank-ordering compounds based on stability (e.g. high, moderate, and low turnover), time-course incubations offer the advantage of determining a compound's rate of metabolism (or metabolic half-life) necessary to calculate the intrinsic and predicted hepatic clearance of a test compound. The most conventional means of determining the metabolic half-life is through measuring the percent compound remaining over time using LC/MS/MS detection. Using this approach, compounds can be selected based on metabolic stability for pre-clinical pharmacokinetic studies and the utility of the *in vitro* test system can be evaluated by comparing the predicted to observed clearance values.

1.1 Selection of In Vitro Test Systems of Metabolic Stability Assay

Microsomes are typically the test system of choice due to their relatively inexpensive cost, robustness and ease of use for high throughput applications, low lot-to-lot variability due to pooling multiple donors, and wide availability from commercially reputable vendors. Microsomes are prepared by homogenizing fresh or frozen livers with subsequent isolation of the endoplasmic reticulum subcellular fraction via centrifugation. A number of drug metabolizing enzymes are located in the endoplasmic reticulum membrane (*see* **Note 1**), most notably cytochrome P450 (CYP), a superfamily of enzymes that is responsible for the metabolism for a majority of marketed drugs [2], and UDP-glucuronosyltransferases (UGT), an enzyme family that conjugates the sugar molecule UDPGA to alcohols, acids, and basic amines [3]. Liver microsomes require the addition of either reduced nicotinamide adenine dinucleotide phosphate (NADPH) or uridine-diphosphate-glucuronic acid (UDPGA) to support the catalytic cycle of CYP [4] and UGT enzymes, respectively.

In certain instances, hepatocytes may serve as a more appropriate test system particularly if multiple enzyme systems are involved in metabolism. Hepatocytes contain the full complement of Phase I and II drug metabolizing enzymes, including the cytosolic enzymes such as aldehyde oxidase, xanthine oxidase, sulfotransferases, methyltransferases, *N*-acetyl-transferases, and glutathione transferases. In addition, hepatocytes do not require the addition of co-factors to initiate enzymatic reactions. Cryopreserved hepatocytes are widely available from commercial vendors to minimize the inter-experimental variability often observed from freshly isolated hepatocytes. Compound stability studies can be conducted in plated hepatocytes, which can be cultured for 1 week and are of particular use for low turnover compounds. Metabolic stability studies are more commonly conducted in suspension hepatocytes due to their improved throughput and capability for automation as well as lower potential for non-specific binding of test compound to plastic material relative to plated hepatocytes.

The decision to employ microsomes or hepatocytes for stability studies is often dependent on the major route of metabolism for a particular chemotype. Thus, early metabolite identification studies in both microsomes and hepatocytes may guide selection of the appropriate test system. Typically, liver microsomes represent the pragmatic system of choice given the prevalence of CYPs and UGTs in drug metabolism. However, given the recent appreciation for non-microsomal enzyme systems, such as aldehyde oxidase [5] as the major route of metabolism for certain chemotypes, metabolic stability studies may warrant use of hepatocytes.

1.2 Overview of Metabolic Stability Assay

Prior to initiating the reaction, liver microsomes or hepatocytes need to be properly thawed and diluted with the incubation buffer to concentrations chosen for conducting stability studies.

Metabolic stability reactions are typically started by either adding the appropriate co-factor (NADPH or UDPGA) to microsomes containing test compound or by spiking test compound into hepatocytes. The reactions samples are briefly mixed and incubated at 37 °C for the duration of the reaction. At pre-defined times, the reaction is terminated by the addition of organic solvent containing an internal standard to precipitate proteins. After centrifugation, the supernatant can be further diluted in water, and the test compound is semi-quantified by LC/MS/MS. The percent parent remaining is calculated relative to 0 min. The half-life is derived from the initial linear portion of the slope of the natural log (ln) of the percent compound remaining over time. As illustrated in Fig. 1, verapamil, cerivastatin, and warfarin are clearly demarcated as high, moderate, and low turn-over compounds in rat liver microsomes.

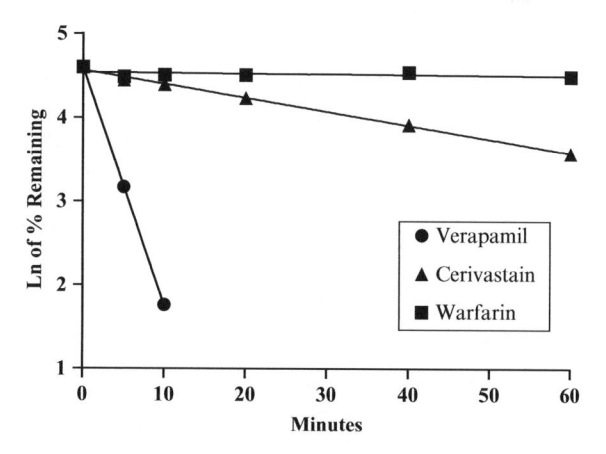

Fig. 1 The metabolic stability of three drugs (verapamil, cerivastatin, and warfarin) were studied in rat liver microsomes using 0.5 mg/mL microsomal protein fortified with NADPH in order to illustrate the utility of the assay to classify high, moderate, and low turn-over compound profiles and to derive the metabolic half-life from the slope (k)

Such information is invaluable to medicinal chemists to help derive SAR to address chemotypes suffering from metabolic instability. In addition, the half-life can be used to calculate a compound's intrinsic and predicted hepatic clearance. Thus, the assay has predictive power to help select compounds for *in vivo* studies and allows for the building of *in vitro–in vivo* correlations for a particular chemotype.

1.3 Defining Critical Experimental Variables

Defining conditions for key experimental variables is essential to optimize the assay for front-end stability testing. Such variables include the *in vitro* system composition, time-point selection, test compound concentration, organic solvent content, and assay automation.

1. *In Vitro System Composition:* Microsomal incubations are traditionally carried out in potassium phosphate buffer and Tris–HCl for CYPs and UGTs, respectively (*see* **Note 2**) and magnesium chloride; however, the microsomal concentration can range from 0.1 to 1 mg/mL. A starting concentration of 0.5 mg/mL is suggested for CYP stability studies during lead optimization phase to flag high turnover compounds likely unsuitable for PK studies, derive SAR around metabolic instability, and identify moderate to low turn-over compounds. Incubations using greater microsomal protein content, such as 1 mg/mL, are often used to enhance the turnover of advanced stage compounds, which typically are more stable, in order to derive a reliable hepatic clearance value from *in vitro* data. UGT activity is oftentimes less robust than CYPs in microsomes due to compound restriction from the UGT active site, which faces the luminal side of microsomal membranes (unlike CYPs). As such, greater microsomal concentrations of 1 mg/mL and the addition of detergents such as alamethicin are recommended for UGT stability assays to improve test compound access to the UGT active site. Microsomal concentrations >2 mg/mL are not recommended due to a potential negative impact of increased non-specific binding on the rate of metabolism.

   Hepatocyte incubations are typically conducted in Krebs-Henseleit Buffer (KHB) or Williams E Media. Similar to microsomes, a range of concentrations of hepatocytes can be employed from 0.2 to 1.5×10^6 cells/mL. We recommend a cell density of 0.5×10^6 cells/mL as a starting point (*see* **Note 3**) for similar reasons above.

2. *Time-Point Selection:* At least four time-points (in addition to the 0 min time-point) provide sufficient data to characterize the rate of reaction. Since the assay in principal is designed to address compounds that demonstrate high turnover to guide medicinal chemists during lead optimization, a bias towards earlier time-points is recommended, such as 5, 10 and 20 min for microsomal incubations. Later time-points, such as 40 and 60 min, aide in improving the microsomal half-life determination for moderate—low turnover compounds (*see* **Note 4**).

For hepatocyte suspensions, where cells remain viable up to ~5–6 h after thawing, time-points of 0.5, 1, 2, 3, and 4 h are recommended.

3. *Test Compound Concentration*: In order for the *in vitro* reaction to obey apparent first-order kinetics based on the Michaelis-Menten model, the test compound concentration must be beneath its K_M. For the majority of compounds, a test concentration of 1 µM is assumed to be lower than its K_M value. However, "non-linear" kinetics may occur if a test compound has a K_M value equal to or greater than 1 µM (*see* **Note 5**). In this instance, re-testing metabolic stability at lower compound concentration, such as 0.1 µM, is recommended (*see* **Note 6**).

4. *Organic Solvent Composition*: DMSO is most often used as the organic solvent of choice to dissolve discovery compounds, where stock concentrations of 10 or 20 mM DMSO stocks are often possible. However, high concentrations of DMSO in the test reaction can inhibit CYP activity [6]. DMSO at concentrations as low as 0.2 % (v/v) can attenuate the activity of CYP2C19, 2E1, and 3A4/5, while acetonitrile and methanol demonstrate inhibition of CYPs at concentrations >0.5 % (v/v). Thus, it is advisable to reduce the organic solvent composition in the test reaction to $\leq$0.1 % DMSO and $\leq$0.5 % methanol or acetonitrile in order to minimize inhibition of CYP activity.

5. *Automation*: Although the methodology herein describes benchtop handling, assay automation on robotic plate-decks, such as the Beckman Coulter Biomek FX™ or Tecan Freedom EVO™, is recommended if greater compound throughput from the bench-top assay is desired.

1.4 Overview of LC/MS/MS Analyses of Samples from Metabolic Stability Incubations

Before compounds are analyzed by LC/MS/MS, a MS tune method is first determined for each test compound. Tuning includes establishing a select ion monitoring (SIM) transition particular for each test compound as well as optimizing the collision energy, declustering potential, and collision exit potential to maximize test compound quantification. Although this information can be manually derived, most LC/MS software permits facile selection of tune methods for a large set of test compounds. This information is then used to create the MS methodology for test compound detection.

The LC run-times are often short, usually 3–4 min in duration, sufficient to separate the test compound from the sample matrix and amenable to multiple injections to generate the kinetic profile for a battery of compounds.

In its simplest form, the LC/MS/MS method can be set up to detect the test compound and the assay internal standard only; however, compound pooling can help maximize time and improve throughput. Since LC/MS instruments have scan times amenable to detecting 2–10 analytes simultaneously, while still retaining sensitivity to detect low analyte levels, compounds can be pooled post-reaction from similar reaction time-points and analyzed together (*see* **Note 7**).

2 Materials

2.1 Metabolic Stability Using Microsomal Incubations

2.1.1 Preparation of Test Compound

1. Test compound of known molecular weight
2. DMSO, anhydrous (ARCO)
3. HPLC grade acetonitrile (JT Baker, Phillipsburg, NH)
4. HPLC grade water (JT Baker, Phillipsburg, NH)

2.1.2 Reagents for Microsomal Incubations

1. Potassium phosphate buffer, 0.5 M, pH 7.4 (BD Gentest, San Jose, CA)
2. Magnesium chloride (Sigma, St. Louis, MO)
3. β-Nicotinamide adenine dinucleotide phosphate, reduced form (NADPH) (Sigma, St. Louis, MO)
4. Tris–HCl buffer, 1 M, pH 7.7 at 37 °C (Sigma, Saint Louis, MO)
5. Ethylenediaminetetraacetic acid (EDTA) (Sigma, St. Louis, MO)
6. Uridine-diphosphate-glucuronic acid (UDPGA) (Sigma, St. Louis, MO)
7. Alamethicin (Sigma, St. Louis, MO)
8. Rat or human liver microsomes, 20 mg/mL (BD Gentest, San Jose, CA)
9. HPLC grade methanol (JT Baker, Phillipsburg, NH)

2.1.3 Materials for Microsomal Incubation

1. 1.2 and 2 mL—96-well polypropylene reaction plates and seals
2. Plate mixing apparatus
3. Water bath or incubator set to 37 °C
4. Ice tray and crushed ice

2.1.4 Preparation of Samples for LC/MS/MS Analyses

1. Selected internal standard (e.g. phenytoin)
2. Refrigerated centrifuge capable of $10,000 \times g$
3. 1.2 mL—96-well polypropylene blocks and seals

2.2 Metabolic Stability Using Suspension Hepatocyte Incubations

2.2.1 Thawing of Cryopreserved Hepatocytes

1. Cryopreserved rat and human hepatocytes, 5 million cells per mL (Celsis-IVT, Baltimore, MD)
2. Thawing hepatocyte media (Celsis-IVT, Baltimore, MD)
3. Krebs-Henseleit Buffer (KHB) (Celsis-IVT, Baltimore, MD)
4. Centrifuge capable of 40–60 g
5. Rocking tray

6. 37 °C culture incubator at relative humidity (95 %) and CO_2 (5 %)
7. Trypan Blue (Sigma, St. Louis, MO)
8. Hemacytometer
9. 1.5 mL racked polypropylene trays

2.3 Materials for Data Analyses from Samples Generated from Incubations

2.3.1 LC/MS/MS Materials

1. API 4000 LC/MS/MS system (Applied Biosystems, Concord, Ontario, Canada)
2. Agilent 1100 Series HPLC (Santa Clara, CA)
3. Autosampler (LEAP PAL, Carrboro, NC)
4. Zorbax™ SB-Phenyl column, 2 × 50 mm, 5 µm (Agilent, Santa Clara, CA)

3 Methods

3.1 Metabolic Stability Using Liver Microsomes

3.1.1 CYP Metabolic Stability Reagent Preparation

1. Potassium phosphate, 0.5 M, pH 7.4, stock buffer is diluted to 0.1 M in HPLC-grade water.
2. Magnesium chloride is prepared as a 30 mM stock solution in 0.1 M potassium phosphate buffer.
3. NADPH is prepared as a 10 mM stock concentration in 0.1 M potassium phosphate buffer and pre-incubated in a 37 °C water bath.
4. The organic solvent mixture is composed of a 3:1 mixture of acetonitrile:methanol containing an appropriate internal standard (e.g. 0.05 µg/mL phenytoin). The solution is kept cold to facilitate reaction termination and protein precipitation.

3.1.2 UGT Metabolic Stability Reagent Preparation

1. Alamethicin is dissolved in HPLC-grade methanol as a 5 mg/mL stock solution.
2. Tris–HCl, 1 M pH 7.7 at 37 °C, stock buffer is diluted to 0.1 M in HPLC-grade water.
3. EDTA is prepared as a 1 M stock solution in 0.1 M Tris–HCl buffer.
4. UDPGA is prepared as an 80 mM stock concentration in 0.1 M Tris–HCl buffer and pre-incubated in a 37 °C water bath.
5. The organic solvent mixture is identical to that employed above for the CYP metabolic stability reactions.

3.1.3 CYP Microsomal Reaction Mix Preparation

1. Rat or human liver microsomes, 20 mg/mL, are thawed within 30 min from the reaction initiation and kept on ice.
2. The Reaction Mix is prepared by dispensing 0.45 mL of 20 mg/mL liver microsomes into a 50 mL tube followed by the addition of 13.95 mL 0.1 M potassium phosphate buffer,

and 1.8 mL 30 mM magnesium chloride. Upon the addition of NADPH (occurring just prior to reaction), this yields a final reaction condition of 0.5 mg/mL liver microsomes, 1 mM NADPH, 100 mM potassium phosphate, and 3 mM magnesium chloride. This is sufficient to maximize the use of one vial of liver microsomes, typically provided as 0.5 mL aliquots by commercial vendors, as freeze-thaw cycles of microsomes may reduce enzymatic activity.

3.1.4 UGT Microsomal Reaction Mix Preparation

1. Alamethicin is pipetted onto the bottom of a 20 mL glass scintillation vial, 20 mL, in order to yield a final concentration of 25 µg of alamethicin/mg of microsomal protein (e.g. 45 µL alamethicin solution to 450 µL microsomal protein).

2. Alamethicin solution is dried over a gentle stream of nitrogen to evaporate off methanol.

3. Rat or human liver microsomes, 20 mg/mL, are thawed and 0.45 mL is pipetted to the bottom of the alamethicin-coated scintillation vial.

4. Microsomes are gently mixed and pre-incubated on ice for 15 min prior to preparation of the Reaction Mix.

5. The Reaction Mix is prepared by adding 6.75 mL of 0.1 M Tris–HCl buffer, 0.9 mL of 30 mM magnesium chloride, and 0.018 mL of 0.5 M EDTA. Upon the addition of UDPGA (occurring just prior to reaction), this yields a final reaction condition of 1 mg/mL liver microsomes, 100 mM Tris–HCl phosphate, 8 mM UDPGA, 3 mM magnesium chloride, and 1 mM EDTA.

3.1.5 Preparation of Test Compound

1. Dissolve test compound in DMSO at a stock concentration of 10 mM.

2. Transfer 5 µL of 10 mM DMSO stock solution to 495 µL of 1:1 acetonitrile:water to make a working concentration of 100 µM test compound in a 2 mL deep-well polypropylene 96-well plate.

3. Seal plate and mix thoroughly.

3.1.6 Microsomal Incubation Procedure

The following procedure describes the bench-top microsomal stability assay using a final reaction volume of 500 µL, where samples are excised for each time-point from a lone reaction plate (*see* **Note 8**). It is recommended that in addition to test compounds, a high, moderate, and low turnover control compound should be included in the incubation to help calibrate and validate the assay run.

1. Dispense 450 µL of Reaction Mix into a 96-well, 1.2 mL-deep polypropylene plate, which serves as the reaction plate.

2. Spike the reaction plate with 5 µL of test compound and mix.

3. Pre-incubate the reaction plate for 5 min in a 37 °C shaking water-bath.

4. Initiate the reaction by the addition of 50 μL of 10 mM NADPH or 80 mM UDPGA to the reaction plate for CYP and UGT metabolic stability studies, respectively.

5. Mix the reaction plate quickly (~5–10 s).

6. Excise 50 μL from the reaction plate and transfer into a new plate containing 200 μL of 3:1 acetonitrile:methanol spiked with an appropriate internal standard (e.g. 0.05 μg/mL phenytoin) to terminate the reaction. This represents the 0 min time-point.

7. Transfer the reaction plate back to a 37 °C shaking water bath or incubator.

8. At each designated time-point (e.g. 5, 10, 20, 40, and 60 min), excise 50 μL from the reaction plate and terminate the reaction similar to the 0 min time-point (*see* **Step 6**).

9. Process the terminated plates by vortexing vigorously for 15 min followed by centrifugation at 14,000 g for 10 min at 4 °C.

10. Remove 100 μL of supernatant from the termination plate and mix with 300 μL water for subsequent LC/MS/MS analyses.

11. If compound pooling is desired, mix the diluted samples from step 10 with compounds from similar time-points.

3.2 Metabolic Stability Using Suspension Hepatocytes

3.2.1 Thawing Cryopreserved Hepatocytes

1. It is recommended to follow the procedure provided by the commercial vendor for thawing cryopreserved hepatocytes as thawing protocols may vary between vendors.

3.2.2 Measuring Cell Density and Viability Using Trypan Blue Exclusion

1. From a homogenous solution of cells, excise 100 μL of solution and mix with 700 μL KHB buffer and 200 μL of Trypan blue solution into a 1.5 mL tube.

2. Invert the tube ten times and after 1 min of incubation, pipet 10 μL to each side of a hemacytometer. Count five squares per each side of the hemacytometer, and total the number of live cells (clear) and dead cells (blue).

3. Determine the cell viability by dividing the total number of dead cells by the total number of cells counted (*see* **Note 9**). A cell viability of >80 % is considered acceptable for metabolic stability experiments.

4. Determine the cell concentration by taking the average of live cells per square (total live cells divided by 10) and multiplying by 10 (the dilution factor) and then by 10,000 (the hemacytometer factor). Add KHB media to the cell suspension in order to obtain the desired cell density for the reaction (e.g. 0.5 million cells/mL).

3.2.3 Hepatocyte Incubation Procedure

The following procedure describes the bench-top hepatocyte stability assay, where the reaction is initiated by first spiking the test compound into 600 μL of hepatocytes followed by aliquoting 100 μL into separate plates for time-point. It is recommended that in addition to test compounds, a Phase I and II metabolic control compound be incubated as well to help calibrate and validate the assay run.

1. From a 10 mM DMSO stock concentration, add 5 μL of the test compound into 995 μL of 1:1 acetonitrile:water to make a 50 μM working stock solution.

2. Dispense 600 μL of hepatocytes into 1 mL 96-well, racked test tubes.

3. Add 12 μL from the working test compound solution to hepatocytes to make a final reaction concentration of 1 μM.

4. Cap cells and gently invert several times before uncapping and dispensing 100 μL of hepatocytes into fresh 1 mL 96-well, racked test tubes (*see* **Note 10**).

5. Incubate hepatocytes for 0, 0.5, 1, 2, 3, and 4 h in a 37 °C tissue culture incubator at relative humidity (95 %) and CO_2 (5 %).

6. At each time point, tubes should be removed from the incubator and the reaction stopped by the addition of 200 μL ice-cold 3:1 acetonitrile:methanol containing internal standard.

7. Process each terminated reaction immediately after reaction by capping the incubation plate and vortexing for 10 min followed by incubating on ice for 15 min.

8. Centrifuge samples at $14,000 \times g$ for 10 min at 4 °C and transfer supernatant to a new plate.

9. Once all reaction plates have been terminated and processed, mix 100 μL supernatant with 100 μL water for LC/MS/MS analyses.

3.3 LC/MS/MS Analyses of Metabolic Stability Samples

Diluted supernatants from either microsomal or hepatocyte stability studies are loaded onto 2×50 mm C8, C18, or SB-Phenyl columns, and test compounds and the internal standard are eluted using a generic reverse-phase HPLC method at 0.8 mL/min flow rate to permit short run-times (often less than 4 min). Acetonitrile spiked with 0.1 % formic acid is typically used as the mobile phase, where separation is often achieved during a 1.5 min gradient from 5 to 90 % organic. The test compounds are quantified on a 4000 triple quadruple MS/MS instrument by single ion monitoring, where mass spectral counts for each test compound are normalized to the response of the internal standard.

3.3.1 Determination of Percent Remaining and Half-Life

From the LC/MS/MS data, the amount of test compound observed at a particular time-point is divided by that observed at 0 min, and this value is converted to a percentage. The *in vitro* metabolic half-life is calculated from the slope (k) of the log-linear regression plot of percent test compound remaining over time (*see* **Note 11**), where:

$$t1/2 = \ln(2)/k$$

Once half-life data is generated, the determination of a test compound's intrinsic clearance and predicted hepatic clearance can be made employing data either from microsomes [7, 8] or hepatocytes [9–11]. Establishing such *in vitro–in vivo* correlations in hepatic clearance can improve confidence in predicting human clearance from in vitro test systems [12–14].

3.4 Conclusions

The prominent role of hepatic metabolism in the clearance of drugs underscores the requirement for throughput and robust metabolic stability assays in drug discovery. Metabolic stability assays conducted in liver microsomes or hepatocytes are invaluable tools for medicinal chemists to derive SAR to reduce metabolic turnover and improve the oral bioavailability and half-life of lead molecules. In addition, the utility of such assays to predict hepatic clearance is instrumental for the DMPK investigator to establish *in vitro–in vivo* correlations and increase confidence in human pharmacokinetic predictions.

4 Notes

1. Microsomes are enriched with a number of other drug metabolizing enzymes in addition to CYPs and UGTs, including flavin-containing mono-oxygenases (FMOs) and esterases. These enzymes may contribute to turnover in NADPH-supplemented microsomes.

2. Optimal CYP activity is achieved in potassium phosphate buffer as certain CYP isoforms, particularly CYP2Cs, are less active in alternative buffers such as Tris–HCl. In contrast, UGT activity is higher in Tris–HCl buffer relative to potassium phosphate buffer [15].

3. Increasing liver microsomal content or hepatocyte cell density can be used as a means to increase the rate of turnover of stable compounds. Although a doubling in the concentration of the *in vitro* system theoretically should double the rate of turnover, this is not always the case in practice as non-specific binding of the test compound to endogenous components present in microsomes or hepatocytes (e.g. lipids, proteins) may slow the rate of metabolism.

4. During the catalytic cycle, CYPs can generate reactive oxygen species such as superoxide anion and hydrogen peroxide. These species can accumulate during the course of the reaction and inactivate CYPs. Therefore, longer incubations in microsomes (>60 min) are often compromised due to less robust CYP activity and offer little additional value to half-life determination.

5. There are other explanations for "non-linear" kinetics such as the formation of metabolites that either inhibit (auto-inhibition) or enhance (auto-activation) the rate of reaction. A close inspection of the half-life curve will aide in the detection of such scenarios.

6. "Non-linear" kinetics results in a slower observed rate of turn-over than compared to apparent first-order kinetics. If the half-life of the test compound at 1 µM is similar to that observed employing lower concentrations, such as 0.1 µM or less, then the reaction likely follows first order kinetics.

7. Compound pooling should be used with some caution, since test compounds with similar mass or with isotopic patterns may confound post-acquisition analyses. It is recommended to pool compounds that are from different chemotypes and at least bear a m/z difference of >4 Da. Furthermore, pooling of too many compounds together may dilute the MS detection of test compounds. For such scenarios, pooling may either not be possible or should be minimized. Pragmatically, the pooling of four compounds together balances the requirement to maximize the LC/MS/MS acquisition time and to provide sufficient MS signal to adequately detect all four compounds.

8. Alternatively, upon the addition of NADPH to initiate the reaction, the mixed reaction can be dispensed at the start of the assay into separate plates, where each plate corresponds to a different time-point. At each designated time-point, the plate is removed from the incubation and quenched in organic solvent containing the internal standard.

9. Cell viability of 80 % or greater should be used for hepatocyte stability studies. Poor cell viability (i.e. an excess of lysed hepatocytes) may compromise the assay by either improving access of test compound to drug metabolizing enzymes (due to the lack of test compound penetration across a cell membrane) or deleteriously impact the viable cells during the time-course through presence of potentially toxic cellular debris.

10. Vigorously mixing hepatocytes may cause cell lyses and is not recommended to prepare a homogenous cell suspension. In addition, hepatocytes readily settle to the bottom of test incubation vials making it challenging to excise a homogenous

cell suspension from a common reaction plate. Instead, it is recommended to first gently mix hepatocytes upon addition of compound followed by dispensing hepatocytes to separate time-point plates (similar to the description in **Note 6**).

11. Half-life determination may not be possible for test compounds that demonstrate either extremely high turnover or little to no turnover. In such instances, assigning a value of <4 or >180 min may be more appropriate. However, if a half-life value is considered important, adjustments in the amount of microsomes or hepatocytes may improve half-life determination.

Acknowledgements

The views expressed here are solely those of the author and do not reflect the opinions of Janssen Research & Development, LLC.

References

1. Wienkers LC, Heath TG (2005) Predicting in vivo drug interactions from in vitro drug discovery data. Nat Rev Drug Discov 4(10):825–833
2. Guengerich FP (2006) Cytochrome P450s and other enzymes in drug metabolism and toxicity. AAPS J 8(1):E101–E111
3. Fisher MB et al (2001) The role of hepatic and extrahepatic UDP-glucuronosyltransferases in human drug metabolism. Drug Metab Rev 33(3–4):273–297
4. Guengerich FP (2001) Common and uncommon cytochrome P450 reactions related to metabolism and chemical toxicity. Chem Res Toxicol 14(6):611–650
5. Hutzler JM et al (2013) Strategies for a comprehensive understanding of metabolism by aldehyde oxidase. Expert Opin Drug Metab Toxicol 9(2):153–168
6. Chauret N, Gauthier A, Nicoll-Griffith DA (1998) Effect of common organic solvents on in vitro cytochrome P450-mediated metabolic activities in human liver microsomes. Drug Metab Dispos 26(1):1–4
7. Obach RS (1999) Prediction of human clearance of twenty-nine drugs from hepatic microsomal intrinsic clearance data: an examination of in vitro half-life approach and nonspecific binding to microsomes. Drug Metab Dispos 27(11):1350–1359
8. Houston JB (1994) Utility of in vitro drug metabolism data in predicting in vivo metabolic clearance. Biochem Pharmacol 47(9):1469–1479
9. Lau YY et al (2002) Development of a novel in vitro model to predict hepatic clearance using fresh, cryopreserved, and sandwich-cultured hepatocytes. Drug Metab Dispos 30(12):1446–1454
10. Naritomi Y et al (2003) Utility of hepatocytes in predicting drug metabolism: comparison of hepatic intrinsic clearance in rats and humans in vivo and in vitro. Drug Metab Dispos 31(5):580–588
11. Brown HS, Griffin M, Houston JB (2007) Evaluation of cryopreserved human hepatocytes as an alternative in vitro system to microsomes for the prediction of metabolic clearance. Drug Metab Dispos 35(2):293–301
12. Chao P, Uss AS, Cheng KC (2010) Use of intrinsic clearance for prediction of human hepatic clearance. Expert Opin Drug Metab Toxicol 6(2):189–198
13. Klopf W, Worboys P (2010) Scaling in vivo pharmacokinetics from in vitro metabolic stability data in drug discovery. Comb Chem High Throughput Screen 13(2):159–169
14. Chiba M, Ishii Y, Sugiyama Y (2009) Prediction of hepatic clearance in human from in vitro data for successful drug development. AAPS J 11(2):262–276
15. Walsky RL et al (2012) Optimized assays for human UDP-glucuronosyltransferase (UGT) activities: altered alamethicin concentration and utility to screen for UGT inhibitors. Drug Metab Dispos 40(5):1051–1065

Chapter 7

Metabolic Assessment in Alamethicin-Activated Liver Microsomes: Co-activating CYPs and UGTs

Gary W. Caldwell and Zhengyin Yan

Abstract

The methods and materials described in this chapter are for a medium throughput screening assay for the study of parallel CYP- and UGT-mediated metabolic pathways in microsomes. Alamethicin, a pore-forming peptide, was used to activate UGTs in human liver microsomes. An alamethicin-microsomal activated system in which both CYPs and UGTs were active can be used for studies of metabolic stability and *in vitro* metabolite profiling. For compounds with minor or no glucuronidation, the metabolic stability remained similar between the co-activating CYPs and UGTs microsomal system and the conventional CYPs microsomal incubation procedure. However, for compounds in which glucuronidation is possible, the microsomal stability of the co-activating CYPs and UGTs microsomal system and the conventional CYPs microsomal incubation procedure are completely different. Literature validation studies addressing if the presence of CYP and UGT in microsomes induce experimental artifacts were summarized and indicated no major issues. Results clearly suggest that the co-activating CYPs and UGTs microsomal system using alamethicin is a valuable *in vitro* model in drug discovery for the study of parallel CYP- and UGT-mediated metabolic pathways.

Key words Metabolism, Co-activating, CYPs, UGTs, Alamethicin, Microsomes

1 Introduction

During the past decade, there has been continued pressure on the pharmaceutical industry to correctly predict the success of drug candidates as early as possible in the drug discovery process. This pressure has led to the development of *in vitro* assays using human/animal tissues, cells, or fluids that aid medicinal chemists and biologists in their go/no-go decision-making process [1–5]. *In vitro* tiered assays have been developed to indicate liabilities in structural scaffolds of drug candidates relating to the absorption, distribution, metabolism and excretion (ADME), drug transporter interactions, drug-drug interactions and toxicity of new drug candidates [6, 7]. Many of these *in vitro* methods have been applied to screening, classification and ranking of potential drug candidates with the hope that these assays will lead to the understanding of human *in vivo* pharmacokinetic and toxicity mechanisms.

In vitro drug metabolism studies have been routinely performed in the early stage of drug discovery and development. Because hepatic metabolism represents the major elimination route for the majority of drugs, these *in vitro* metabolism assays aim at predicting hepatic clearance [8–10]. Hepatic human/animal microsomes are the most common *in vitro* model used for this purpose. Microsomes are derived from liver cells that are largely comprised of the endoplasmic reticulum (ER) that contains the major phase I drug metabolism enzymes including cytochrome P450s (CYPs), and the major phase II enzymes such as UDP glucuronosyltransferases (UGTs). The CYPs enzymes can be classified into families and subfamilies based on the homology of amino acid sequences. Members of the same family exhibit about 40 % amino acid sequence identity, and the members of the same subfamily possess greater than 55 % identity. Three CYP families CYP1, CYP2 and CYP3 account for about 70 % of human hepatic microsomes CYPs with CYP3 accounting for approximately 30 %. These CYPs are the major ones responsible for the metabolism of most marketed drugs. Note also that the four CYP families CYP1, CYP2, CYP3 and CYP4 account for about 99 % of rat hepatic microsomes with CYP2 accounting for approximately 60 %. The CYPs play an important role in drug metabolism since they display very broad substrate specificities [11, 12]. That is, they are capable of converting various xenobiotic and endogenous molecules to more polar metabolites using NADPH as the cofactor. The UGT superfamily of enzymes that catalyze the conjugation of D-glucuronic acid to various endo- and xeno-biotics can be divided into several families and subfamilies. Enzymes in each family are at least 50 % homologous in their cDNA sequences, whereas enzymes in each subfamiliy are more than 60 % homologous. Human UGTs belong to two subfamilies UGT1 and 2 which are predominantly involved in glucuronidation. The subfamily of enzymes UGT1A1, 1A4, 1A9, 2B7, and 2B15 appear to be of greatest significance in phase II elimination. The UGTs catalyzes the glucuronidation of compounds by transferring a glucuronic acid moiety from the cofactor uridine 5′-diphosphoglucuronic acid (UDPGA) to substrates and thus, forming glucuronides which are water-soluble and readily excreted via urine or bile [13, 14]. While compounds containing a glucuronic acid accepting group such as hydroxyl, carboxyl, carbonyl, sulfuryl, amino and imino tend to be a substrate for UGTs, it is more common that glucuronidation occurs after xenobiotics are metabolized by phase I enzymes such as CYPs. With some exceptions, such as morphine and retinoic acids, xenobiotics usually lose their therapeutic potency after glucuronidation and therefore, it is important to understand the metabolism fate of drug candidates [15].

Although both CYPs and UGTs are predominantly located in the ER membrane of liver cells, a major difference is that the active site of CYPs is exposed to cytosol but UGTs largely reside at the luminal face of ER [16, 17]; therefore, in contrast to CYPs, access to UGTs is limited by a membrane barrier for both substrates and the cofactor UDPGA. It is generally believed that, in intact cells, access of substrates and UDPGA to the active site of UGTs is carried out via a transporter-mediated mechanism; however, this transporting mechanism is not active in hepatic microsomes. As a result, UGTs largely remain latent in a conventional microsomal incubation, and only oxidative enzymes such as CYPs are activated by the addition of NADPH. Although UGT activity in the microsomal membrane can be liberated by sonication [18] or membrane-disrupting agents [19], such treatments lead to alteration in CYP activity [20]. Therefore, glucuronidation and CYP-catalyzed metabolism are not simultaneously studied in microsomes. While this situation has some advantages for developing screens to understand CYP metabolism without the interference of UGTs, it is clear that using hepatic human/animal microsomes for *in vitro-in vivo* extrapolation of hepatic clearance will underestimate the metabolism of drug candidates metabolized by phase I and II pathways. Several studies have reported a 10- to 30-fold underprediction of clearance [21].

Studies have shown that alamethicin, a pore forming peptide, activates UGTs in microsomes [22]. It appears that alamethicin-activated microsomes can enhance UGT activity without having any detrimental effect on P450 activity [23]. Thus, alamethicin-activated microsomes can be used for predicting the clearance for compounds with parallel CYP- and UGT-mediated metabolic pathways [14, 23–25]. This model may help to better estimate the metabolism of compounds metabolized by multiple pathways in *in vitro-in vivo* extrapolation studies. In the following chapter, we provide a comprehensive step-by-step instruction for the application of the metabolic assessment of compounds in liver microsomes by co-activating cytochrome P450s and UDP-glycosyltransferases. The materials and methods described below are for medium throughput screening of parallel CYP- and UGT-mediated metabolic pathways. However, both materials and methods can be easily miniaturized for high throughput screening to accommodate a larger number of compounds.

2 Materials

2.1 Equipment

Some specific equipment used in this assay is described; however, any model of comparable capability can be easily substituted.

1. Assay Plates: 96-deep well plates (1.2 mL/well) and 96-shallow well plates with conical bottoms (0.3 mL/well).
2. HPLC vials and microfuge tubes.
3. Multi (8)-channel automatic and pipettor (50–1,200 μL).

4. Variable Temperature Water bath set at 37 °C.
5. Variable Temperature tabletop centrifuge for cell preparation.
6. pH meter.
7. LC/MS/MS: Micromass (Manchester, UK) *Quattro Micro* triple quadrupole mass spectrometer interfaced to a HPLC system.
8. Agilent 1100 HPLC system or a similar instrument with an autosampler interfaced to the electrospray apparatus of a triple quadrupole mass spectrometer.
9. Agilent Zorbax SB C18 column (2.1×100 mm) was used for the chromatographic separations.

2.2 Reagents

2.2.1 Buffer Reagents

1. Monobasic potassium phosphate, KH_2PO_4 (136.1 g/mol) (EM SCIENCE, Gibbstown, NJ)
2. Dibasic potassium phosphate, trihydrate $K_2HPO_4 \cdot 3H_2O$ (228.29 g/mol) (Sigma-Aldrich, St. Louis, MO)

2.2.2 Reagents for the CYPs-NADPH System

1. Dihydronicotinamide adenine dinucleotide phosphate tetrasodium salt, $NADPH\ Na_4$ (833.35 g/mol) (Sigma-Aldrich, St. Louis, MO)
2. Glucose 6-phoshate (Sigma-Aldrich, St. Louis, MO)
3. Glucose-6-phosphate-dehydrogenase (G6PDH) (Sigma-Aldrich, St. Louis, MO)
4. Sodium citrate, tribasic (Sigma-Aldrich, St. Louis, MO)
5. $MgCl_2 \cdot 6H_2O$ (203.31 g/mol) (Sigma-Aldrich, St. Louis, MO)

2.2.3 Reagents for UGT-UDPGA System

1. Alamethicin (1964.31 g/mol) (Sigma-Aldrich, St. Louis, MO, USA).
2. Uridine-diphosphate-glucuronic acid trisodium salt (UDPGA) (646.23 g/mol) (Sigma-Aldrich, St. Louis, MO, USA).

2.2.4 Other Reagents

1. All other reagents (HPLC grade acetonitrile, methanol, acetic acid and DMSO) were purchased from Sigma-Aldrich (St. Louis, MO, USA).

2.2.5 Microsomes, CYPs and UGTs

1. All human (20 mg/mL) and rat (20 mg/mL) microsomes were purchased from BD Gentest Corp (Woburn, MA, USA). Human liver microsomes were prepared from livers with mixed gender pool of 50 donors, and rat liver microsomes were prepared from male Sprauge-Dawley rats. Protein content was determined by the supplier.
2. Supersomes™ containing individual CYPs (1A2, 2C9, 2C19, 2D6 and 3A4) were derived from baculovirus insect cells

transfected with specific CYP cDNAs. Specific CYP contents in derived microsomes were determined by the supplier (BD Gentest Corp.).

3. UGT1A1 and 1A4 were expressed in baculovirus insect cells transfected with corresponding UGT cDNAs and prepared as membrane fractions (Supersomes™) from BD Gentest Corp.

2.2.6 Control Compounds

1. 3-(α-Acetonylbenzyl)-4-hydroxycoumarin, tranylcypromine, imipramine, nicardipine, testosterone, 11β-hyroxyetiocholanolone, clotrimazole, 7-hydroxyl-coumarin and naphthol were obtained from Sigma-Aldrich (St. Louis, MO, USA).

2.3 Solutions

2.3.1 Buffers and Cofactors

1. Buffer A (50 mM potassium phosphate, KH_2PO_4, monobasic): Dissolve 3.4 g of KH_2PO_4 in 450 mL of deionized water and bring the final volume to 500 mL with deionized water. Filter buffer using 0.2 μm Nalgene filter flask unit and store at 4 °C (**Note 1**).

2. Buffer B (50 mM potassium phosphate, K_2HPO_4, dibasic): Dissolve 5.7 g of $K_2HPO_4·3H_2O$ in 450 mL of deionized water and bring the final volume to 500 mL with deionized water. Filter buffer using 0.2 μm Nalgene filter flask unit and store at 4 °C (**Note 1**).

3. Buffer C 50 mM potassium phosphate at pH 7.4: Dissolve 1.52 g of KH_2PO_4, and 8.78 g of $K_2HPO_4·3H_2O$ in 900 mL of deionized water. Use 50 mM K_2HPO_4 if pH below 7.4 or KH_2PO_4 if pH above 7.4. Bring the final volume to 1 L with deionized water. Filter buffer using 0.2 μm Nalgene filter flask unit and store at 4 °C (**Note 1**).

4. CYP Cofactor: Dissolve 333.3 mg of NADPH Na_4 (833.35 g/mol) in 15 mL deionized water. Adjust the final volume to 20 mL to produce a 20 mM solution. Aliquot and store at −20 °C (**Note 2**).

5. Cofactor: Dissolve 427 mg of $MgCl_2·6H_2O$ (203.31 g/mol) in 20 mL deionized water. Adjust the final volume to 30 mL to produce a 70 mM solution. Aliquot and store at −20 °C (**Note 3**).

6. UGT Cofactor: Dissolve 258.5 mg UDPGA (646.23 g/mol) in 15 mL deionized water. Adjust the final volume to 20 mL to produce a 20 mM solution. Aliquot and store at −20 °C.

7. Cofactor: Alamethicin (1964.31 g/mol) 5 mg was dissolved in 40 mL of MeOH/water (50:50) to create a 0.125 mg/mL (64 μM) stock aliquot and store at 4 °C (**Note 4**).

8. Stop solution: acetonitrile containing 1.0 μM propranolol as an internal standard for LC-MS/MS analysis (**Note 5**).

2.3.2 "Standard" CYP Reaction Solution

A 40 mL solution containing 5 mM NADPH and 17.5 mM MgCl$_2$ which provides the necessary cofactors to catalyze a CYP450 enzyme reaction. To prepare this solution:

1. Transfer 10 mL of NADPH (Sect. 2.3.1.4) to a 50 mL tube
2. Transfer 10 mL MgCl$_2$ of (Sect. 2.3.1.5) to the same 50 mL tube
3. Add 20 mL of 50 mM potassium phosphate (Sect. 2.3.1.3) to the tube
4. Solution may be aliquoted and stored at –20 °C

2.3.3 "Standard" UGT Reaction Solution

A 40 mL solution containing 5 mM UDPGA and 17.5 mM MgCl$_2$ which provides the necessary cofactors to catalyze a UGT enzyme reaction. To prepare this solution:

1. Transfer 10 mL of UDPGA (Sect. 2.3.1.6) and 10 mL MgCl$_2$ of (Sect. 2.3.1.5) to a 50 mL tube
2. Add 20 mL of 50 mM potassium phosphate (Sect. 2.3.1.3) to the tube
3. Solution may be aliquoted and stored at –20 °C

2.3.4 "Dual" CYP and UGT Reaction Solution

A 40 mL solution containing 5 mM NADPH, 5 mM UDPGA and 17.5 mM MgCl$_2$ which provides the necessary cofactors to catalyze CYP and UGT enzyme reactions. To prepare this solution:

1. Transfer 10 mL of NADPH (Sect. 2.3.1.4), 10 mL of UDPGA (Sect. 2.3.1.6) and 10 mL MgCl$_2$ of (Sect. 2.3.1.5) to a 50 mL tube
2. Add 10 mL of 50 mM potassium phosphate (Sect. 2.3.1.3) to the tube
3. Solution may be aliquoted and stored at –20 °C

2.3.5 Microsomal Dilutions

Assuming that microsomes are available as 20 mg/mL protein in 0.5 mL, dilute to 2.5 mg/mL protein using phosphate buffer (Sect. 2.3.1.3) (**Note 6**)

1. To make 2.5 mg/mL microsomal solution, add 0.5 mL of microsomes (1 vial @ 0.5 mL) to 3.5 mL of 50 mM potassium phosphate (Sect. 2.3.1.3) to a 15 mL tube.
2. Invert the tube repeatedly to mix all components and keep on ice.

2.3.6 Test Compound Stock Solution Preparation

The solubility and water or organic solvent stability of test compounds are usually not known at the time of the metabolic stability assay and, in many cases, are pre-dissolved in DMSO at known concentrations (**Note 7**). The total organic content of the final reaction solution should be less than 0.1 % since organic solvents such as acetonitrile, methanol, and DMSO are weak to moderate inhibitors of CYPs. Assuming that test compounds are available as 10 mM stocks in DMSO:

1. Prepare 100 µL of a 0.4 mM stock solution at 4 % DMSO by combining 4 µL of the 10 mM DMSO stock solution to 96 µL of acetonitrile (ACN)

2. In a 96 deep well plate prepare 1 mL of a 2.5 µM working solution (0.02 % DMSO and 0.48 % ACN) of the test compound by adding 6.25 µL of the 0.4 mM stock solution in 993.75 µL of 50 mM phosphate buffer (Sect. 2.3.1.3) to corresponding wells using a multi-channel pipette

3 Methods

The CYPs and UGTs "dual-activity" microsomal stability assay is designed to follow the loss of the test compound (i.e., drug candidates) over time under CYP- and UGT-mediated metabolic pathways. For microsomes (0.5 mg/mL), the test compound is added at a low µM concentration to a phosphate buffer system in the presence of alamethicin (50 µg/mg microsomal protein), NADPH (1 mM) and UDPGA (1 mM). The incubation is maintained at 37 °C in a static water bath and the reaction is stopped by the addition of ice-cold acetonitrile methanol. The samples are then centrifuged and the supernant is transferred into a HPLC vial or 96-well plate for LC-MS/MS analysis. Inactivated microsomes or benchmark compounds are used as controls to assist with the interpretation of the results.

The "dual-activity" microsomal stability assay can be conducted manually or robotically, depending upon the throughput requirement. The current method is a manual version, but it can be easily modified to run the assay robotically using a liquid handler. The experiment is conducted twice in triplet (i.e., one set of three controls and one set of three measurements). Thus, 16 test compounds can be measurement per 96- well plate.

3.1 CYP and UGT Activity Assay in Human Microsomes

1. Using a 96-shallow well plate with conical bottoms, dispense 40 µL of the 2 µM test compound (Sect. 2.3.6.2) to each of the wells using a multi-channel pipette (**Note 8**).

2. Dispense 20 µL diluted HLM (2 mg protein/mL) (Sect. 2.3.5) to each well using a multi-channel pipette.

3. Add 100 µL of ice-cold stop solution with internal standard to all control wells to stop reaction (Sect. 2.3.1.8) (**Note 9**).

4. Dispense 20 µL of alamethicin (25 µg/mL) (Sect. 2.3.1.7) to all wells using a multi-channel pipette.

5. Pre-warm plate in 37 °C water bath for 5 min.

6. With a multi-channel pipette, dispense 20 μL of the *"Dual"* **CYP and UGT generating solution** (Sect. 2.3.4) to each well to initiate reaction (**Note 10**).

7. Incubate the plate in 37 °C water bath for 15 min.

8. Add 100 μL of ice-cold stop solution with internal standard to all test compound measurement wells to stop reaction (Sect. 2.3.1.8). Do not add this solution to the control wells.

9. Cover the plate with an adhesive cover.

10. Centrifuge the plate at 4 °C for 30 min at 5000×g to pellet down proteins.

11. Transfer 150 μL of supernants into a 96-well plate or HPLC vials for LC-MS/MS analysis as described in Sect. 3.3.

3.2 LC-MS/MS Analysis

LC-MS/MS analyses were performed on a Micromass (Manchester, UK) *Quattro Micro* triple quadrupole mass spectrometer interfaced to an Agilent 1100 HPLC system. LC-MS/MS analyses were conducted using electrospray ionization using select ion monitoring (SIM). The capillary voltage was 3.1 kV, and the cone voltage was 20 V. The source temperature was set at 120 °C, and the desolvation temperature was 300 °C. The collision gas used was nitrogen. An Agilent Zorbax SB C18 column (2.1×100 mm) was used for the chromatographic separations. The starting mobile phase consisted of 95 % water (0.5 % acetic acid), and analytes were eluted using a linear gradient of 95 % water to 95 % acetonitrile over 15 min at a flow rate of 0.3 mL/min. At 12 min, the column was flushed with 95 % acetonitrile for 3 min before re-equilibration at initial conditions. During the run, the divert valve was activated to divert the HPLC eluant to waste for the first minute of elution, and then switched to the mass spectrometer for analysis. LC-MS/MS analyses were carried out on 10 μL aliquots from incubations. Data were processed using the *Masslynx* v3.5 software from Micromass (Manchester, UK) (**Note 11**).

3.3 Data Analysis

All results are presented as mean±S.D. (standard deviation) of three replicates. Enzyme activity is expressed as percentages relative to the control.

3.4 Control Compounds

A comparison study was performed to investigate the feasibility of this approach using a list of control compounds (Table 1) selected to represent a diversity of structures and metabolism pathways [12, 23, 25, 26]. The experimental procedure outlined above was used for nine control compounds. In the comparison study, metabolic stability was examined by the addition of NADPH to activate only CYPs in human liver microsomes. In contrast, both CYPs and UGTs were activated in the "dual-activity" system. As shown in Table 2, compounds such as 3-(α-acetonylbenzyl)-4-

Table 1
Analyses of control compounds and glucuronides

Control compounds	Ionization mode	Molecular ions detected by SIM Parent	Molecular ions detected by SIM Glucuronide
3-(α-Acetonylbenzyl)-4-Hydroxycoumarin	Positive	308.3	484.3
Tranylcypromine	Positive	391.6	576.6
Imipramine	Positive	281.5	475.5
Nicardipine	Positive	480.6	656.6
Testosterone	Positive	289.4	465.4
11β-Hyroxy etiocholanolone	Positive	307.4	483.2
Clotrimazole	Positive	345.8	521.8
7-Hydroxyl-coumarin	Negative	161.1	337.1
Naphthol	Negative	143.2	319.1

Table 2
Microsomal stability of control compounds in the presence of NADPH and UDPGA

Control compounds	Compound remained (%) after 15 min CYP activation	Compound remained (%) after 15 min CYP-UGT activation	Glucuronidation
3-(α-Acetonylbenzyl)-4-Hydroxycoumarin	103.6 ± 4.6	99.3 ± 3.2	Not detected
Tranylcypromine	93.5 ± 4.1	96.0 ± 6.7	Not detected
Imipramine	98.3 ± 3.9	97.8 ± 2.8	Not detected
Nicardipine	26.8 ± 5.4	26.8 ± 5.9	Not detected
Testosterone	32.9 ± 5.7	36.3 ± 5.1	O-glucuronide, minor
11β-Hyroxy etiocholanolone	83.9 ± 1.7	66.7 ± 6.2	O-glucuronide, significant
Clotrimazole	76.6 ± 4.2	52.1 ± 3.9	O-glucuronide, significant
7-Hydroxyl-coumarin	53.8 ± 7.0	3.5 ± 2.7	N-glucuronide, significant
Naphthol	60.6 ± 6.9	29.5 ± 5.1	O-glucuronide, significant

hydroxycoumarin, tranylcypromine, imipramine, and nicardipine did not form glucuronide conjugates, and therefore showed similar metabolic profiles in both the "dual-activity" system and the conventional incubation. Although a testosterone glucuronide was detected in the "dual-activity" system, the extent of conjugation was low and did not affect the stability profile compared to the conventional incubation. A striking difference in stability was revealed for compounds with significant glucuronidation, which include 11β-hyroxyetiocholanolone, clotrimazole, 7-hydroxyl-coumarin, and naphthol when comparing the "dual activity" system and the conventional incubation in which only CYPs were active. This suggests that it was possible to establish a "dual-activity" microsomal system in which both CYPs and UGTs are activated so that compounds can be evaluated for metabolic stability.

A comparison between compounds remained for CYP activation and CYP-UGT activation are different. Thus, the ranking order for the standard microsomal stability assay and the "dual-activity" microsomal stability assay is different.

3.5 Supplemental Data

Several validation studies have been conducted in the literature to demonstrate if the presences of CYP and UGT in microsomes induce experimental artifacts with the above microsomal stability assay [12, 23, 25].

3.5.1 Effects of Alamacine on UGT Activity

Initial glucuronidation assay was carried out using Supersomes™ containing cDNA expressed UGTs [23]. Preliminary results indicated that trifluoperazine and acetaminophen are good substrates for UGTlAl and 1A4 respectively. It was also found that glucuronide formation was linear with respect to incubation time for 30 min in the presence of 1 mM UDPGA and 3.5 mM $MgCl_2$. Since trifluoperazine and acetaminophen formed N- and O-glucuronides respectively by different UGT isoforms, these two compounds were chosen as marker substrates to investigate the effect of alamethicin on UGT activity in human liver microsomes. Preliminary studies also suggested that a 5 min pre-incubation of alamethicin with human liver microsomes was necessary to obtain the optimal stimulatory effect on UGT activity and improve data reproducibility. In human liver microsomes, alamethicin increased the glucuronidation of both substrates in a dose-dependent fashion. Relative to the control (without alamethicin), an approximate threefold increase in glucuronide formation was observed for both substrates in the presence of 20–50 mg/mL alamethicin. The stimulatory effect of alamethicin was observed in both human and rat liver microsomes for many other compounds. The data presented was consistent with results reported by others [12, 22, 23, 25], suggested that the latency of UGTs can be removed by the addition of alamethicin, and the enhancement of UGT activity is not dependent on a specific substrate or a UGT isoform. In contrast, alamethicin did not have such effects on UGTlAl and UTGlA4 expressed in baculovirus insect cells transfected with specific UGT cDNAs. The difference in the effect of alamethicin on UGT activity is likely due to different localization of UGTs in liver microsomes and supersomes. Unlike that of liver microsomes, the active site of UGTs is fully accessible for UDPGA and substrates in supersomes; thus UGT activity was not enhanced by alamethicin.

3.5.2 Effects of Alamethicin on CYP Activity

It has been shown by many that alamethicin does not have significant effects on the five major CYPs such as CYP1A2, 2C9, 2C19, 2D6 and 3A4 [12, 22, 23, 25]. Results suggested that the pore-forming peptide alamethicin does not alter the interaction of CYPs with substrates and cofactors, as evidenced by the fact that CYP marker activities in microsomes were not affected by treatment with alamethicin.

3.5.3 Effects of CYP Co-factors on UGT Activity

The effect of CYP cofactors on UGT activity has been examined in the literature [23]. To summarize their results, supersomes containing cDNA expressed UGT1A1 and 1A4 respectively were selected using acetaminophen and trifluoperazine as marker compounds. All CYP cofactors (NADPH or an NADPH regenerating system comprising of NADP+, glucose-6-phosphate (G-6-P) and glucose-6-phosphate-dehydrogenase (G-6-P-D)) were tested for their effect on UGT activity. The cofactors did not have significant effects on glucuronide formation of acetaminophen glucuronidation and trifluoperazine glucuronidation.

3.5.4 Effects of UDPGA on CYP Activity

The effect of the cofactor UDPGA on CYP activity has been examined using supersomes containing individual cDNA expressed CYP1A2, 2C9, 2C9, 2D6 and 3A4 [23]. To summarize their results, the marker activity of all five CYPs was not significantly affected when the concentration of UDPGA was below 2 mM. However, UDPGA at 5 mM showed inhibition of CYP2C9 and CYP2C19 to some degree. Non-specific interactions may contribute to inhibition, although the exact mechanism is not clear at present.

3.5.5 Effects on Metabolism

Tramadol has been studied to demonstrate that the co-activating CYPs and UGTs in a microsomal system do not affect metabolite profiling in an unexpected manner [23]. Tramadol can be metabolized by both phase I and phase II enzymes in man including N-dealkylation, O-dealkylation, N-oxidation and hydroxylation. In the standard CYP microsomal assay seven metabolites were identified. The same seven metabolites were detected in the "dual-activity" system in which both CYPs and UGTs were simultaneously activated by NADPH and UDPGA. In addition, tramadol glucuronide was detected in the "dual-activity" system, but was not found in the conventional incubation. The results suggested that activation of UGTs did not have adverse effects on CYP-mediated oxidative metabolism pathways.

4 Conclusion

The use of liver microsomal data to predict *in vivo* metabolic clearance has been demonstrated by several groups [1–11]. However, the success of this approach is limited to those compounds primarily eliminated by oxidative metabolism. For those compounds with significant glucuronidation, results can be misleading when the *in vitro* metabolic data are extrapolated to predict pharmacokinetic parameters. In this chapter, we have shown that an alamethicin-microsomal activated system in which both CYPs and UGTs were active can be used for studies of metabolic stability and *in vitro* metabolite profiling. This "dual-activity" system has advantages for metabolic stability assessment, as indicated by the fact that compounds with little or no glucuronidation showed similar metabolic stability to the conventional CYP microsomal activated system, while compounds that are highly glucuronidated showed significant glucuronidation. It is reasonable to expect that data from the "dual-activity" system is more comprehensive for ranking compounds based on metabolic stability when compared to that from the conventional microsomal incubation.

In conclusion, the "dual activity" assay is a more reliable model than the conventional CYP liver microsome assay; the "dual activity" assay does not require additional effort or resources, since the reaction was carried out in a single incubation under similar experimental conditions; the "dual-activity" assay is a favorable *in vitro* model for early preclinical metabolism studies, and can be applied to evaluate various compounds that primarily undergo hepatic metabolism via either oxidation, glucuronidation or both. Obviously, this "dual-activity" assay is still an over-simplified *in vitro* metabolism model for the prediction of *in vivo* pharmacokinetics parameters. However, recent results have been favorable for the general application of the "dual-activity" assay for *in vitro-in vivo* extrapolation for the assessment of hepatic clearance for compounds [12, 25].

5 Notes

1. It should be remembered that some enzymatic reactions are markedly impaired by even small changes in the pH and/or the ionic strength of the solution. Therefore, in most *in vitro* enzymatic assays, it is necessary to add a buffer to the medium to stabilize the pH and inorganic salts to stabilize the ionic strength of the solution. If these conditions are changed during the experiment, the results may also be altered. The buffer conditions used in our assay do have effects on enzyme activity. It is interesting to note that phosphate is an inhibitor of various

enzymes such as kinases and dehydrogenases and magnesium chloride in HLM can result in increased glucuronidation activity of UGTs [14]. Other factors to keep in mind are that the pH may also vary with temperature and the water in which the buffer substances are dissolved should be of the highest quality.

2. NADPH is an electron donor in CYP-catalyzed oxidation reactions. Reduced NADPH at 1 mM can be added directly to microsomal mixture to initiate the reaction. More commonly, NADPH is generated from NADP+ by glucose-6-phosphate dehydrogenase.

3. Magnesium chloride is extremely hydroscopic. Even new bottles of $MgCl_2 \cdot 6H_2O$ that are not opened may be hydrated more than the label indicates. Because water adds weight, it is impossible to obtain an accurate concentration of magnesium chloride by weighing it out. One trick is to prepare the entire lot of new bottles of $MgCl_2 \cdot 6H_2O$ as a 100× stock solution. Of course, we are assuming that the manufacturer known quantity of the material is correct.

4. Alamethicin (Ac-2-MeAla-L-Pro-2-MeAla-L-Ala-2-MeAla-L-Ala-L-Glu(NH_2)-2-MeAla-L-Val-2-MeAla-Gly-L-Leu-2-MeAla-L-Pro-L-Val-2-MeAla-2-MeAla-L-Glu-L-Glu(NH_2)-phenylalaninol, 1964.31 g/mol) is a monovalent cation ionophore which strongly induces formation of alpha-helical structures in membranes.

5. As an internal standard for LC-MS/MS analysis, the compound must be highly stable under experimental conditions, show good sensitivity in MS analysis and minimal loss in sample preparation process.

6. The thawed microsomes should be vortexed briefly and gently to maintain a homogenous suspension without disrupting the integrity of the membrane.

7. It should be noted that dissolved test compounds from library collects have inaccurate concentrations. From our own experience, library stock solution of 10 mM in DMSO were measured and discovered to be closer to 6–7 mM in many cases.

8. The final concentrations in the well are: test compound (1 μM), NADPH (1 mM), UDPGA (1 mM), 0.5 mg of microsome protein/mL, alamethicin (25 μg/mL) and $MgCl_2$ (3.5 mM). The total organic solvent is 0.002 % DMSO and 0.048 % ACN.

9. In the control experiments, we have added a large amount of organic to de-nature the microsomes before incubation. Since these control samples were handled in the same manner as the fully incubated samples, they can serve as a zero time point for the experiment and the 15 min incubated samples can be normalized against them.

10. By changing the generating solutions different experiments can be preformed. For example, if you want a standard microsomal CYP assay replace the 20 μL of the "Dual" CYP and UGT generating solution (Sect. 2.3.4) with the "Standard" CYP generating solution (Sect. 2.3.2). If you want a standard microsomal UGT assay replace the 20 μL of the "Dual" CYP and UGT generating solution (Sect. 2.3.4) with the "Standard" UGT generating solution (Sect. 2.3.3).

11. The LC run time can be shortened significantly with a newer LC system such as an Agilent 1200.

Acknowledgements

The views expressed here are solely those of the author and do not reflect the opinions of Janssen Research & Development, LLC.

References

1. Tarbit MH, Berman J (1998) High-throughput approaches for evaluating absorption, distribution, metabolism and excretion properties of lead compounds. Curr Opin Chem Biol 2: 411–416
2. Smith DA, Van de Waterbeemd H (1999) Pharmacokinetics and metabolism in early drug discovery. Curr Opin Chem Biol 3:373–378
3. Caldwell GW (2000) Compound optimization in early- and late-phase drug discovery: acceptable pharmacokinetic properties utilizing combined physicochemical, in vitro and in vivo screens. Curr Opin Drug Discov Devel 3: 30–41
4. Caldwell GW, Ritchie DM, Masucci JA, Hageman W, Yan Z (2001) The new pre-preclinical paradigm: compound optimization in early and late phase drug discovery. Curr Top Med Chem 1(5):353–366
5. Caldwell GW, Yan Z, Tang W, Dasgupta M, Hasting B (2009) ADME optimization and toxicity assessment in early- and late-phase drug discovery. Curr Top Med Chem 9(11): 965–980
6. Caldwell GW (ed) (2012) Pharmaceutical R&D: a knowledge-intensive multidisciplinary approach. Curr Top Med Chem 12(11):85
7. Yan Z, Caldwell GW (eds) (2004) Optimization in drug discovery: in vitro methods. Method in pharmacology and toxicology. Humana Press Inc., Totowa, NJ
8. Bertrand M, Jackson P, Walther R (2000) Rapid assessment of drug metabolism in the drug discovery process. Eur J Pharm Sci 11(suppl 2):S61–S72
9. Thompson TN (2001) Optimization of metabolic stability as a goal of modern drug design. Med Res Rev 21(5):412–449
10. Yan Z, Caldwell GW (2001) Metabolism profiling, and cytochrome P450inhibition & induction in drug discovery. Curr Top Med Chem 1(5):403–425
11. Venkatakrishnan K, Von Moltke LL, Greenblatt DJ (2001) Human drug metabolism and the cytochromes P450: application and relevance of in vitro models. J Clin Pharmacol 41(11): 1149–1179
12. Zhou Y, Liu A, Xie H, Cheng GQ, Dai R (2012) Metsbolism of fibrates by cytochrome P450 and UDP glycosyltransferases in rat and human liver microsomes. Chin Sci Bull 57: 1142–1149
13. Lin JH, Wong RK (2002) Complexities of glucuronidation affecting in vitro-in vivo extrapolation. Curr Drug Metab 3(6):623–646
14. Walsky RL, Bauman JN, Bourcier K, Giddens G, Lapham K, Negahban A, Ryder TF, Obach RS, Hyland R, Goosen TC (2012) Optimized assays for human UDP-glucuronosyltransferase (UGT) activities: altered alamethicin concentration and utility to screen for UGT inhibitors. Drug Metab Dispos 40:1051–1065
15. Ethell RT, Beaumont K, Rance DJ, Burchell R (2001) Use of cloned and expressed human UDP-glucuronosyltransferases for the assessment of human drug conjugation and identification

of potential drug interactions. Drug Metab Dispos 29(1):48–53

16. Yokota H, Yuasa A, Sato R (1992) Topological disposition of UDP-glucuronyltransferase in rat liver microsomes. J Biochem 112(2):192–196

17. Meech R, Mackenzie PI (1998) Determinants of UDP glucuronosyltransferase membrane association and residency in the endoplasmic reticulum. Arch Biochem Biophys 356(1): 77–85

18. Ethell RT, Anderson GD, Beaumont K, Rance DJ, Burchell R (1998) A universal radiochemical high-performance liquid chromatographic assay for the determination of UDP-glucuronosyltransferase activity. Anal Biochem 255(1):142–147

19. Fulceri R, Banhegyi G, GamberucciA GR, Mandl J, Benedetti A (1994) Evidence for the intraluminal positioning of p-nitrophenol UDP-glucuronosyltransferase activity in rat liver microsomal vesicles. Arch Biochem Biophys 309(1):43–46

20. Trapnell CR, Klecker RW, Jamis-Dow C, Collins JM (1998) Glucuronidation of 3′-azido-3′-deoxythymidine (zidovudine) by human liver microsomes: relevance to clinical pharmacokinetic interactions with atovaquone, fluconazole, methadone, and valproic acid. Antimicrob Agents Chemother 42(7):1592–1596

21. Miners JO, Knights KM, Houston JB, Mackenzie PI (2006) In vitro-in vivo correlation for drugs and other compounds eliminated by glucuronidation in humans: pitfalls and promises. Biochem Pharmacol 71: 1531–1539

22. Fisher MR, Campanale K, Ackermann RL, Vandenbranden M, Wrighton SA (2000) In vitro glucuronidation using human liver microsomes and the pore-forming peptide alamethicin. Drug Metab Dispos 28(5):560–566

23. Yan Z, Caldwell GW (2003) Metabolic assessment in liver microsomes by co-activating cytochrome P450s and UDP-glycosyltransferases. Eur J Drug Metab Pharmacokinet 28(3): 223–232

24. Soars MG, Ring BJ, Wrighton SA (2003) The effect of incubation conditions on the enzyme kinetics of UDP-glucuronosyltransferases. Drug Metab Dispos 31:762–767

25. Kilford PJ, Stringer R, Sohal B, Houston JB, Galetin A (2009) Prediction of drug clearance by glucuronidation from *in vitro* data: use of combined cytochrome P450 and UDP-glucuronosyltransferase cofactors in alamethicin activated human liver microsomes. Drug Metab Dispos 37:82–89

26. Yuan R, Madani S, Wei XX, Reynolds K, Huang SM (2002) Evaluation of cytochrome P450 probe substrates commonly used by the pharmaceutical industry to study in vitro drug interactions. Drug Metab Dispos 30(12): 1311–1319

Chapter 8

Phenotyping UDP-Glucuronosyltransferases (UGTs) Involved in Human Drug Metabolism: An Update

Michael H. Court

Abstract

Glucuronidation, catalyzed by the UDP-glucuronosyltransferases (UGT), is a major drug clearance mechanism in humans and other mammalian species. UGT reaction phenotyping involves determining which of the 19 known human UGTs are primarily responsible for glucuronidation of a particular drug. This approach is commonly used during the drug development process for drugs that are clearly primarily by glucuronidation, thereby enabling rational predictions of potential drug interactions and pharmacogenomic variation. An integrated approach to phenotyping is described using recombinant expressed UGTs, comparative enzyme kinetic analysis, correlations with UGT selective probe activities, relative activity factor normalization, and chemical inhibition. Updated protocols are provided that overcome several newly discovered model limitations, including endogenous fatty acid inhibition of UGT2B7 and UGT1A9 activities.

Key words Glucuronidation, UDP-glucuronosyltransferases, Potential drug interactions, Phenotyping, Recombinant expressed UGTs, Comparative enzyme kinetic analysis

1 Introduction

1.1 Role of Glucuronidation and the UGTs in Drug Metabolism

Glucuronidation represents one of the major pathways for drug metabolism and clearance in humans and other mammalian species (for recent reviews see [1–3]). This reaction is catalyzed by the UDP-glucuronosyltransferase (UGT) family of enzymes and involves transfer of the sugar group from UDP-glucuronic acid (UDPGA) to a small hydrophobic molecule (aglycone) most commonly containing a hydroxyl, carboxyl, or a nitrogen group (amines and amides). Sulfur and (direct) carbon-linked glucuronidation also occurs, although relatively rarely. Consequently, substrates may include drugs that possess these functional groups or drug metabolites that have had these functional groups generated by other drug metabolizing enzymes, most frequently by cytochrome P450 monooxygenase (CYP). Although, in most instances, glucuronidation results in inactivation of a drug, pharmacological or toxicological activation can occur. Examples include morphine-6-glucuronide, which is a more potent opioid agonist than morphine, and the acyl-glucuronides of various nonsteroidal anti-inflammatory and hypolipidemic drugs, which have the potential for adduct formation. Table 1 compares substrates, possible enzyme inducing agents and tissue distribution for the 18 known human UGT isoforms.

1.2 In Vitro Phenotyping of Drug Metabolizing Enzymes

In vitro reaction phenotyping is now routinely used to identify CYPs responsible for the oxidative metabolism of candidate compounds during the preclinical phase of drug development [5–8]. Such information has proven extremely useful in predicting drug-drug interactions as well as high interindividual variability in drug disposition resulting from genetic polymorphism. Drugs that could be problematic in clinical usage, such as compounds that induce or inhibit CYP3A4, or are metabolized exclusively by the highly polymorphic CYP2D6, can be identified relatively early in the development process. Accumulating evidence indicates that drug-drug interactions and genetic polymorphism may also complicate the clinical utility of drugs that are cleared primarily by glucuronidation [1, 9, 10]. For example, valproate, fluconazole and probenicid have been shown to inhibit the glucuronidation of the antiretroviral drug zidovudine resulting in up to twofold increase in drug area under the curve (AUC) plasma concentrations [10]. On the other hand, the antituberculosis drug rifampicin can reduce zidovudine AUC by about twofold in large part as a consequence of UGT enzyme induction. Severe adverse side effects of irinotecan including neutropenia and diarrhea are more frequently observed in treated colon cancer patients that have a common genetic polymorphism in the gene encoding UGT1A1 [11]. This polymorphism (UGT1A1*28) results in lower glucuronidation and accumulation of SN-38, the active metabolite of irinotecan. Finally, drugs may also interact with endogenous metabolism causing unwanted side effects. Of particular note is the antiretroviral drug atazanavir, which inhibits UGT1A1 mediated glucuronidation of bilirubin resulting in jaundice [12]. This effect is most pronounced in patients with the UGT1A1*28 polymorphism.

1.3 Strategy to Identify UGTs Relevant to In Vitro Glucuronidation of a Drug

The purpose of this chapter is to update the methodology published in the last edition of this text to identify UGT isoforms that are relevant to the metabolism of novel and existing drugs. The primary focus will be identification of the well-characterized hepatic UGTs (UGTs 1A1, 1A4, 1A6, 1A9, 2B7 and 2B15) since the liver is a major site of drug glucuronidation and the research tools are better developed for these isoforms. However, it is clear that the gastrointestinal tract (contributing to first-pass metabolism) and the kidney are also major sites of glucuronidation for many drugs [1, 4, 9]. The strategy used here for UGT phenotyping is based on well established procedures for the CYPs [5, 7, 8] with appropriate

Table 1
Substrates, inducers, and tissue mRNA distribution of the 19 human UGT enzymes

UGT	Some substrates	Possible inducers	Liver	Stomach	Sm. Intestine	Colon	Kidney	Lung	Prostate	Testis	Breast	Ovary	Nasal Epith.	Brain
1A1*	Bilirubin**, ethinylestradiol**, buprenorphine, SN-38, flavopiridol, etoposide	Arylhydrocarbons Omeprazole Antioxidants Phenobarbital Phenytoin Rifampicin Dexamethasone	+	+	+	+	?	?	?	?				
1A3*	Norbuprenorphine, NSAIDs		+	?	+	+	?	?	?	?			+	
1A4*	Tricyclic antidepressants, trifluoperazine**, antipsychotics, antihistamines	Phenobarbital Rifampicin Carbamazepine Phenytoin	+	+	+	+	+	?	?	?	?	?	+	?
1A5	1-Hydroxypyrene		?	+	+	+	+	?	+	+	?			?
1A6*	Acetaminophen, valproic acid, serotonin**	Arylhydrocarbons Antioxidants	+	+	+	+	+	+	?	?			+	+
1A7*	Benzo(α)pyrene		?	?	+	+	+		?				+	
1A8*	Benzo(α)pyrene, mycophenolic acid, raloxifene				+	+	?		?	?	?		+	
1A9*	Propofol**, acetaminophen, salicylic acid, flavopiridol, mycophenolic acid, SN-38	Arylhydrocarbons Antioxidants Fibrates Glitazones	+	?	+	+	+		?	?			+	

(continued)

Table 1 (continued)

UGT	Some substrates	Possible inducers	Liver	Stomach	Sm. Intestine	Colon	Kidney	Lung	Prostate	Testis	Breast	Ovary	Nasal Epith.	Brain
1A10*	Mycophenolic acid, raloxifene, dopamine**		?	+	+	+	+	?	?	?		?	+	
2A1	Multiple, including odorants, menthol, citronellol							?					+	
2A2	Ethinylestradiol				?					?			+	?
2A3	Hyodeoxycholic acid, bile acids	Rifampicin	+	+	+	+	+			+			+	
2B4*	Bile acids, codeine	Bile acids	+		?	?	+	?	?	+	?			
2B7*	Opioids, morphine**, zidovudine**, NSAIDs, epirubicin, catechol estrogens, retinoids, fatty acids, and many other	Phenobarbital	+	+	+	+	+	?	?	?	?	?		+
2B10*	Nicotine, cotinine		+									?		
2B11	Arachidonate metabolites		+	?	+		+			?	+	?	+	?
2B15*	Oxazepam**, lorazepam**, sipoglitazar, androgens	Rifampicin	+	+	+	+			+	+	+	?	+	
2B17*	Vorinostat, exemestane, androgens		+	+	+	+	?		+	?	+	+	+	?
2B28	Estrogens		+								+			

UGTs 1A2, 1A11 and 1A12 are pseudogenes in the human

*Recombinant UGTs available commercially

**Selective substrate for that UGT (only in liver for propofol/UGT1A9)

The following notation is used in Table 1 (Data from [4])

+Detectable mRNA (constitutively)

Blank No detectable mRNA (constitutively)

?Inconsistent findings between studies

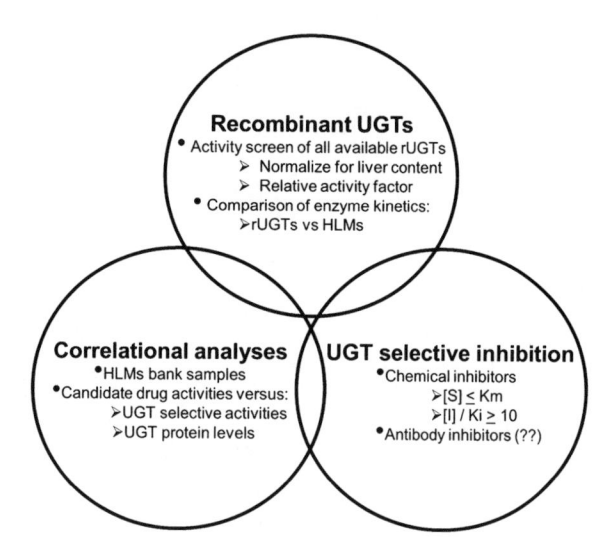

Fig. 1 An integrated approach to identification of UDP-glucuronosyltransferases mediating glucuronidation of a drug in vitro. (??) Indicates that this particular method is not yet practicable because of a lack of appropriate research tools

modifications. Although the available tools for this process are much less well developed, recent work in this and other laboratories have made substantial progress in this regard. There are essentially three components of this strategy, including use of recombinant UGTs (rUGTs), correlation analyses, and isoform selective inhibition, each of which provides complementary and supportive information (see Fig. 1).

2 Material

2.1 In Vitro Glucuronidation Assay

1. Candidate drug and glucuronide.
2. Recombinant expressed UGTs (e.g. BD-Gentest, Woburn, MA).
3. Human liver microsomes—pooled and from individuals (e.g. BD-Gentest, Woburn, MA; Celsis IVT, Baltimore, MD; Xenotech LLC, Lenexa, KS; Gibco, Life Technologies, Grand Island, NY).
4. High performance liquid chromatography (HPLC) system with gradient capability, C18 reverses phase column, and UV absorbance detector (*see* **Note 1**).
5. HPLC mobile phase reagents (see Table 2).
6. Incubation buffer, 50 mM phosphate, pH 7.5 (*see* **Note 2**).
7. UDP-glucuronic acid sodium salt (cat. no.U6751, Sigma-Aldrich, St. Louis, MO).
8. Magnesium chloride solution (50 mM in water).

Table 2
Details of HPLC methods used to assay for glucuronides generated by UGT probe activities

Probe activity	Separation conditions*	UV detection wavelength	Glucuronide R.T.	Substrate R.T.	I.S. R.T.
Estradiol-3-glucuronidation	20–30 % solvent A over 15 min Balance with solvent D	280 nm	9 min-E-3-glu 10 min-E-17-glu	19 min	13 min
Trifluoperazine glucuronidation	10–70 % solvent A over 20 min Balance with solvent C	254 nm	14 min	15 min	7 min
Serotonin glucuronidation	5 % solvent A for 8 min 5–50 % solvent A over 9 min Balance with solvent D	270 nm (225 nm EX/330 nm EM**)	5 min	8 min	14 min
Propofol glucuronidation	20–100 % solvent A over 20 min Balance with solvent D	214 nm	7 min	16 min	14 min
AZT glucuronidation	15 % solvent A for 15 min 15–50 % solvent A over 10 min Balance with solvent B	266 nm	7 min	11 min	8 min
S-oxazepam glucuronidation	25 % solvent A for 15 min 25–60 % solvent A over 10 min Balance with solvent D	214 nm	8 min-R-oxazglu 9 min-S-oxazglu	25 min	17 min

*Flow rate is 1 mL/min
Solvent A: Acetonitrile
Solvent B: 20 mM potassium phosphate buffer in water, pH 2.2
Solvent C: 0.1 % trifluoroacetic acid in water
Solvent D: 20 mM potassium phosphate buffer in water, pH 4.5
**The use of an additional fluorescence detector is optional but provides higher sensitivity and ready identification of serotonin glucuronide. Serially connected UV detector is still needed for quantitation of the internal standard
R.T. Retention time; I.S. Internal standard

9. Alamethicin, 2.5 mg/mL of methanol (cat. no. A4665, Sigma-Aldrich, St. Louis, MO).
10. Saccharolactone; d-saccharic acid 1,4-lactone, 50 mM in water (cat. no. S0375, Sigma-Aldrich, St. Louis, MO).
11. Bovine serum albumin, 20 % in water (cat. no. A7030, Sigma-Aldrich, St. Louis, MO).
12. Escherichia coli β-glucuronidase, 25,000 units in 50 % aqueous glycerol (cat. no.G8162, Sigma-Aldrich, St. Louis, MO).
13. Refrigerated vacuum centrifuge (e.g. SpeedVac concentrator) (*see* **Note 3**).
14. Water bath incubator set at 37 °C (*see* **Note 4**).

2.2 UGT Probe Activities

Substrates

1. Estradiol (cat. no. E1024, Sigma-Aldrich, St. Louis, MO).
2. Trifluoperazine (cat. no. T6062, Sigma-Aldrich, St. Louis, MO).
3. Serotonin; 5-hydroxytryptamine (cat. no. H9523, Sigma-Aldrich, St. Louis, MO).
4. Propofol; 2,6-diisopropylphenol (cat. no. W50, 510-2, Sigma-Aldrich, St. Louis, MO).
5. AZT; 3'-azido-3'-deoxythymidine (cat. no. A2169, Sigma-Aldrich, St. Louis, MO).
6. Oxazepam (cat. no. O5254, Sigma-Aldrich, St. Louis, MO).

Glucuronides

1. Estradiol-3-glucuronide (cat. no. E2127, Sigma-Aldrich, St. Louis, MO)
2. Trifluoperazine N-glucuronide (cat. no. CSTT711, Cachesyn Inc, Mississauga, ON, Canada)
3. Serotonin glucuronide (cat. no. S274990, Toronto Research Chemicals, Toronto, ON, Canada)
4. Propofol glucuronide (cat. no. P829780, Toronto Research Chemicals, Toronto, ON, Canada)
5. AZT-glucuronide (cat. no. A0679, Sigma-Aldrich, St. Louis, MO).
6. Oxazepam glucuronide (cat. no. O845705, Toronto Research Chemicals, Toronto, ON, Canada)

Internal standards

1. Phenacetin (cat. no. A2375, Sigma-Aldrich, St. Louis, MO).
2. Acetaminophen (cat. no. A7085, Sigma-Aldrich, St. Louis, MO).
3. 3-acetamidophenol (cat. no. A4911, Sigma-Aldrich, St. Louis, MO).
4. Thymol (cat. no. T0501, Sigma-Aldrich, St. Louis, MO).

2.3 Chemical Inhibition

1. Hecogenin (cat. no. H2261, Sigma-Aldrich, St. Louis, MO).
2. Diflunisal (cat. no. D3281, Sigma-Aldrich, St. Louis, MO).
3. Niflumic acid (cat. no. N0630, Sigma-Aldrich, St. Louis, MO).

2.4 Data Analyses

1. Graphical computer program capable of nonlinear curve fitting and correlation analyses (e.g. GraphPad Prism 6, GraphPad Software, La Jolla, CA; or Sigmaplot 12, Systat Software, San Jose, CA).

3 Methods

3.1 Development of an In Vitro Glucuronidation Assay for the Candidate Drug

The following is a general approach to developing a HPLC based method to quantify the rate of formation of a glucuronide metabolite using tissue microsomes or recombinant enzyme. A literature search should be conducted prior to starting to determine whether previous assay methods for the substrate and glucuronide have been published and gather any other useful information, such as UV absorbance wavelength maxima (λ max), fluoresce absorbance and emittance wavelength maxima, and/or masses of the parent and possible collision induced fragmentation ions for UV absorbance, fluorescence and mass detectors, respectively. Although not essential, the process is simplified if a small quantity of glucuronide of the candidate drug is available to assist in identification of the appropriate peak on the HPLC chromatogram and enable accurate quantitation. If a glucuronide standard is not available, it can be identified using the methods described below and quantified by reference to a standard curve using the parent compound (data would be expressed as glucuronide equivalents). This approach to quantitation is generally valid for UV absorbance detection methods where extinction coefficients tend to be minimally altered by glucuronide conjugation. However, glucuronidation tends to have a greater effect on fluorescence and mass detection methods. For accurate quantitation and identification, milligram amounts of the glucuronide can be synthesized biologically with this system and purified using methods previously described [13].

3.1.1 Initial Assay Method Development

1. The following assumes a 100 μL incubation volume but can be scaled to other volumes.
2. Dissolve substrate and glucuronide in 50–100 mL methanol and store in a sealed glass container in a −20 °C freezer (*see* **Note 5**).
3. Set up HPLC apparatus and allow equilibration with 1 % solvent A (acetonitrile) and 99 % Solvent B (20 mM potassium phosphate, pH 2.2) at 1 mL/min flow rate (*see* **Note 6**). Set the UV absorbance detector at the λ max for the glucuronide analyte (see **Note 7**).

4. Prepare UDPGA cofactor solution on ice in a microcentrifuge tube. For each 100 µL incubation volume add:
   (a) 0.645 mg UDPGA (5 mM final).
   (b) 10 µL 50 mM magnesium chloride in water (5 mM final).
   (c) 10 µL 50 mM saccharolactone in water (optional).
   (d) 25 µL 100 mM potassium phosphate buffer, pH 7.5.
   (e) Balance to 50 µL with water and vortex.

5. Add 100 µL of substrate dissolved in methanol to empty incubation tubes (0.5 or 1.5 mL polypropylene microcentrifuge) and dry down in the refrigerated vacuum centrifuge (*see* **Note 8**).

6. Place incubation tubes containing dried substrate on ice and add 50 µg of pooled HLM protein, 2.5 µg alamethicin (2.5 µg/µL methanol; 50 µg alamethicin/mg microsomal protein), 1 µL of substrate dissolved in methanol (at 100 times the desired final incubate concentration), and balance to a volume of 50 µL with 50 mM potassium phosphate buffer, pH 7.5 (0.5 mg protein/mL final concentration).

7. Preincubate tubes at 37° for 5 min.

8. Start reaction by adding 50 µL of UDPGA cofactor solution, mix by gently flicking the tube (do not vortex recombinant UGTs), cap tube, and incubate for up to 6 h.

9. To aid in identification of the glucuronide metabolite peak also include three negative controls that: (a) contain no UDPGA; (b) contain no substrate; or (c) are not incubated (i.e. immediately treated with stop solution, vortexed and centrifuged).

10. Stop reactions by adding 100 µL of ice cold acetonitrile, vortex, and centrifuge at 14,000 × g for 10 min (*see* **Note 9**).

11. Transfer 190 µL to glass HPLC vials, dry down in a refrigerated vacuum centrifuge, and reconstitute with 95 µL of water.

12. Analyze 10–50 µL of the incubate by HPLC using a solvent gradient program that increases solvent A from 1 to 50 % over 20 min and then to 90 % solvent A over the next 5 min (balance with solvent B) (*see* **Note 10**).

13. Chromatogram peaks from the incubate are identified by comparison of peak retention times to reference standards (substrate and glucuronide if available), and negative controls. Glucuronide peaks will be absent in all negative controls (*see* **Note 11**).

14. If an authentic glucuronide standard is not available, the identity of the glucuronide peak should be confirmed by showing sensitivity to glucuronidase treatment and/or by mass determination (HPLC-mass spectroscopy) (*see* **Note 12**).

3.1.2 Confirmation of Metabolite Identity by β-Glucuronidase Treatment

1. Generate glucuronide as in previous section but place on ice without adding the stop solution.
2. To 100 μL of incubate add 100 units of Escherichia coli β-glucuronidase solution and mix briefly.
3. Cap tube and incubate at 37 °C for at least 2 h (to overnight).
4. Continue as per step #10 in previous section and analyze for glucuronide content by HPLC.
5. Confirm glucuronide peak identity by comparison with an untreated matched sample (see **Note 13**).

3.1.3 Optimization of the In Vitro Glucuronidation Assay

Once an assay has been developed it will then be necessary to optimize several parameters to ensure maximal sensitivity while maintaining initial rate conditions. The amount of glucuronide formed is determined using a standard curve generated by measuring a series of known amounts of glucuronide (or substrate, if glucuronide is unavailable) dissolved in incubation buffer. Recovery of glucuronides from microsomes is usually 100 %, but can be checked by comparison of standard curves with and without added microsomes. As a general guideline, less than 10 % of the initial mass of substrate should be consumed in any incubation. Metabolite formation should be verified to be linear with respect to incubation time and protein concentration at the lowest substrate concentration to be used (or at least below the enzyme Km). For some slower activities linearity can be observed for up to 6 h incubation. Compared with CYPs, UGTs generally are much more stable under in vitro incubation conditions. Protein concentrations should be minimized (ideally 0.1 mg/mL or less) to minimize effects of nonspecific binding of substrate to microsomes. Alamethicin is a pore-forming antibiotic that activates UGTs by enhancing substrate access to the enzyme active site at the microsomal interior. The amount of alamethicin added to the incubation (usually 20–100 μg/mg of microsomal protein) should also be confirmed to result in maximal activation (usually a 2 to 3-fold increase for HLMs). Activation is not usually observed with rUGTs [14]. Saccharolactone (2–10 mM) may also be required for some activities to inhibit endogenous β-glucuronidase activity. However, inhibition by saccharolactone has also been observed for some activities [15]. Incubate pH can also affect enzymatic activity, however most investigators tend to use a pH within the physiological range (7.0–7.5). Magnesium and UDPGA are essential cofactors that are usually used at saturating concentrations (2–20 mM). An internal standard should also be used to enhance HPLC assay precision and accuracy (see **Note 14**).

3.2 Glucuronidation by Recombinant Expressed UGTs (rUGTs)

Currently out of the 19 known enzymatically functional human UGTs there are 13 rUGTs available through commercial sources including the majority of the UGTs expressed in hepatic tissue (Table 1). Hepatic isoforms that are currently not available commercially include UGT2B11, which is somewhat restricted in substrate specificity to endogenous arachidonic acid metabolites [16], UGT2B28, which may be limited in importance because of aberrant mRNA splicing [17], and UGT2A3, which so far has only been found to efficiently glucuronidate bile acids [18].

3.2.1 Activity Screen with rUGTs

Initially all available rUGTs should be screened for glucuronidation of the candidate drug using the method developed in the previous section. At least two substrate concentrations should be used, one concentration approximating the Km value for HLMs, and one concentration 10 times the Km value. The use of two concentrations will provide preliminary information with regard to the relative affinities of each UGT. Ideally only one UGT is identified that is capable of glucuronidating the candidate drug, with a Km value that is identical for both rUGT and HLM preparations (see next section), thereby simplifying the identification process. We have shown this to be the case for serotonin glucuronidation by HLMs [19]. In most instances multiple UGTs will show activity and it will be necessary to try and identify the major isoform responsible for the activity.

Unfortunately direct comparisons of rUGT activities (normalized to the amount of expressed protein) can be misleading since the abundance of the UGTs in an average human liver differs between isoforms, meaning that the contribution of a lowly expressed UGT would be overestimated, while a highly expressed UGT would be underestimated. A comprehensive study of UGT mRNA expression in 47 human livers by quantitative PCR [4] indicates that the relative content of UGT mRNA is 2B4 > [1A9 = 2B7 = 1A4 = 1A1 = 2B10 = 1A6] > [2B15 = 2B11 = 1A3] > [2A3 = 2B17 = 2B28]. UGTs 1A5, 1A7, A8, 1A10, 2A1, and 2A2 were measured but were either quite low or not detected. A more recent study of UGT protein levels in 17 human livers by quantitative mass spectrometry gave levels of 114, 84, 62, 33, 26, and 17 pmoles/mg microsomal protein for UGTs 1A6, 2B7, 2B15, 1A1, 1A9, 1A3, respectively [20]. Another limitation of extrapolating recombinant enzyme data is that the system most frequently used to produce UGTs commercially (baculovirus-insect cells), while capable of producing large amounts of recombinant protein, much of the protein produced is catalytically inactive.

3.2.2 Relative Activity Factor Normalization of rUGT Activities

One approach that has been used to overcome some of the limitations of extrapolating rUGT activity data to human liver is relative activity factor (RAF) normalization [21–23]. UGT isoform selective probe activities are measured under the same conditions using both the rUGT (that selectively glucuronidates the probe) and pooled HLMs (usually from at least 10 donors). The ratio of rUGT to HLM activities for each isoform selective probe is the RAF value for that rUGT. Glucuronidation activities of the candidate drug for

Table 3
Specificity of six glucuronidation activities evaluated using recombinant UGTs

UGT	Estradiol-3-glucuron-idation	Trifluo perazine glucuron-idation	Serotonin glucuron-idation	Propofol glucuron-idation	AZT glucuron-idation	S-oxazepam glucuron-idation
Vector	0	0	0	0	0	0
1A1*	1,069	0	0	0	0	<1
1A3	210	0	0	0	0	0
1A4*	0	350	0	0	0	0
1A6*	0	0	2,200	0	0	<1
1A7	0	0	0	301	0	0
1A8	306	0	0	61	0	0
1A9*	0	0	0	1,110	0	0
1A10	114	0	0	84	0	0
2B4	–	–	–	–	19	0
2B7*	0	0	0	0	107	0
2B15*	0	0	0	0	0	10
2B17	–	–	–	–	17	0
References	**	[26]	[19]	**	[29]	[30]

*These are considered important hepatic UGT isoforms with regard to drug metabolism
**Unpublished data courtesy of Dr Chris Patten, BD-Gentest, Woburn, MA
"–" Not determined since these isoforms were not available at the time the assays were conducted

each rUGT are then divided by the RAF value for that rUGT, thereby extrapolating the candidate drug rUGT activities to candidate drug activities expected for HLMs. Normalized rUGT activities can then be compared with each other and to HLM activities values for the candidate drug to evaluate the relative contribution of each isoform to total activity. Validated probe activities (see Table 3) for the major drug metabolizing hepatic UGTs include estradiol-3-glucuronidation (UGT1A1), trifluoperazine glucuronidation (UGT1A4), serotonin glucuronidation (UGT1A6), propofol glucuronidation (UGT1A9), AZT glucuronidation (UGT2B7), and S-oxazepam glucuronidation (UGT2B15). One weakness of this approach is that probe activities have not yet been identified for all of the UGTs.

3.2.3 HLM and rUGT Enzyme Kinetic Comparison

Comparison of enzyme kinetic parameters for rUGTs (the most active and those with at least 10 % of the most active) with parameters measured for HLMs under identical experimental conditions also assists in isoform identification. Intrinsic clearance values (Vmax/Km) can be calculated and compared, however the same stipulations with regard to relative isoform abundance differences between recombinant enzymes and liver apply. Direct comparison of Km values will help to exclude low affinity isoforms (Km for rUGT >HLMs) that are unlikely to contribute to HLM activity substantially, and identify high affinity isoforms (Km for rUGT <HLMs) that may contribute significantly at low (clinically relevant) substrate concentrations. The shape of kinetic plots may also assist in identification in that atypical kinetics (such as homotropic activation or substrate inhibition) may be observed for HLMs and also for one of the rUGTs evaluated. High nonspecific binding of substrate to microsomes can be a cause of atypical kinetics and differences in Km values between HLMs and rUGTs [7]. This is more likely to occur with basic and highly lipophilic compounds and at higher protein concentrations (over 0.1 mg/mL) and should be evident as nonlinearity of protein concentration versus glucuronide formation rate during initial assay optimization steps (prior section). If higher protein concentrations cannot be avoided or significant microsomal binding is suspected, equilibrium dialysis or spin column techniques can be used to measure unbound substrate concentrations. Catalytic inhibition by endogenous fatty acids found in human liver microsomes but not in recombinant rUGTs can also cause higher Km values in HLMs versus rUGTs for certain drugs [24]. The effect appears to be greatest for substrates of UGT2B7 [24] and UGT1A9 [25], and can be minimized by addition of 2 % (w/v final concentration) bovine serum albumin to the incubation mix, which binds the inhibitory fatty acids.

UGT Enzyme Kinetic Parameter Determination

1. At least ten different substrate concentrations should be used spanning the Km value determined in preliminary experiments. UDPGA concentration should be saturating (20 mM).

2. Determine glucuronidation activities using pooled HLMs and rUGTs with the optimized assay method developed in the previous section (*see* **Note 15**).

3. Evaluate plots of reaction velocity versus substrate concentration (Michaelis-Menten plot) and of reaction velocity divided by substrate concentration versus reaction velocity (Eadie-Hofstee plot) to determine which kinetic models should be used to fit the data. Typical kinetic models include the Michaelis-Menten (1), Hill (2), uncompetitive substrate inhibition (3) and two-enzyme (4) models.

$$V = V_{max} \times S / (K_m + S) \quad (1)$$

$$V = V_{max} \times S^n / \left(S_{50}^n + S^n\right) \qquad (2)$$

$$V = V_{max} \times S / \left(K_m + S \times (1 + S/K_s)\right) \qquad (3)$$

$$V = V_{max1} \times S / \left(K_{m1} + S\right) + V_{max2} \times S / \left(K_{m2} + S\right) \qquad (4)$$

where V is reaction velocity, S is substrate concentration, V_{max} is maximal reaction velocity, K_m and S_{50} are the substrate concentrations at half maximal velocity, n is an exponent indicative of the degree of curve sigmoidicity, and K_s is an inhibition constant.

4. Fit the kinetic model parameters to the data by nonlinear least squares regression.

5. Evaluate the goodness of fit of the kinetic model to the data by overlaying a curve connecting predicted data points with the observed data points. If a model other than (1) is used, then choice of that model (over 1) needs to be justified by an objective method such as the F test ($P < 0.05$) or the Akaike Information Criterion (AIC) which takes into account model complexity.

3.3 Correlation Analysis

A second approach to UGT phenotyping is to utilize the intrinsic variability in expression of different UGT isoforms in a bank of HLMs. Isoform-selective probe activities for each of the hepatic UGT isoforms (Table 3) are measured using the HLM bank and then correlated to the glucuronidation activities for the candidate drug measured using the same set of HLMs. The highest correlation should be with the probe activity for the relevant UGT isoform. Although correlations may be observed with HLM banks containing as few as ten individuals, larger size HLM banks (>20 individuals) are more useful for this purpose. Additional supportive evidence may also be provided by correlation to immunoquantified UGT protein content determined by Western blotting, although as yet form-specific antibodies are only available commercially for UGT1A1, 1A6 and 2B7. However, as indicated above, protein levels of most of the UGTs expressed in human liver can now be determined using quantitative mass spectrometry [20]. The main limitation with correlation analysis is significant co-regulation of expression of different UGT isoforms. Indeed, a recent study in this laboratory suggests that many of the UGT1A isoforms may be co-regulated [1].

3.3.1 Correlation of Candidate Activities with UGT Probe Substrate Activities

1. Measure glucuronidation activities for the candidate drug using individual HLMs from a HLM bank with the assay method developed in the previous section. The substrate concentration should approximate the Km value of the drug for pooled HLMs.

Table 4
Details of in vitro incubation methods used for UGT probe activities

Probe activity	Substrate concentration	Protein concentration (mg/mL)	Incubation time (min)	Internal standard
Estradiol-3-glucuronidation	100 µM	0.25	30	Phenacetin
Trifluoperazine glucuronidation	200 µM	0.25	30	Acetaminophen
Serotonin glucuronidation	4 mM	0.05	30	Acetaminophen
Propofol glucuronidation	100 µM	0.25	30	Thymol
AZT glucuronidation	500 µM	0.5	120	3-Acetamidophenol
S-Oxazepam glucuronidation	100 µM	0.5	120	Phenacetin

2. Measure UGT probe activities in the HLM bank using the incubation parameters given in Table 4 and the HPLC assay methods outlined in Table 2 (*see* **Note 16**).

3. Correlate the candidate drug activities with data generated for each of the probe activities using an appropriate computer program. Non-parametric Spearman correlation analysis is preferred over the parametric Pearson correlation method since data frequently are not normally distributed. Significant correlations are indicated by Spearman correlation coefficients (r_s) greater than 0.5 and *P* values less than 0.001.

3.4 Isoform Selective Inhibition

The final approach to UGT phenotyping is to use UGT isoform-selective inhibitors of HLM activity. Few of the chemical inhibitors currently used for the UGTs have been rigorously evaluated for isoform selectivity. Inhibitors that have shown good selectivity so far include hecogenin (50 µM) for UGT1A4, and diflunisal (50 µM) or niflumic acid (2.5 µM) for UGT1A9 [14, 26, 27]. Caution should be exercised when evaluating the glucuronidation of low turnover drugs using inhibitors that are also high turnover glucuronidation substrates since accumulation of UDP from rapid glucuronidation of the inhibitor can result in nonspecific end-product inhibition of the candidate drug [28].

Inhibition of glucuronidation of the candidate drug should be performed using both HLMs and rUGTs. Inhibitor effects on selected UGT probe activities in HLMs and rUGTs should also be studied in parallel to verify inhibitor potency and selectivity.

Immuno-inhibition, although theoretically possible, is not feasible at present because of a lack of commercially available inhibitory antibodies.

3.4.1 Chemical Inhibition Study

1. Prepare incubation tubes containing substrate (control activity) and substrate combined with each of the inhibitors that are being tested. The substrate concentration should approximate the Km value of the drug for pooled HLMs. Several inhibitor concentrations should be chosen that span those concentrations known to inhibit the intended UGT by >50 % but do not inhibit other UGTs (<10 % inhibition).
2. Measure activities as in the previous section (*see* **Note 17**).
3. Calculate reaction velocities as a percentage of control (i.e. no inhibitor) activity.

4 Notes

1. These are minimum requirements. Optional enhancements to the basic HPLC system would include a diode array UV detector, which is useful for glucuronide peak identification; a fluorescence detector, which provides superior sensitivity for fluorescent compounds such as serotonin; or an ion trap or triple quadrupole mass detector that enable accurate quantitation and identification of glucuronides by mass transitions (typically a collision induced dissociation neutral loss of 176 amu). Standard (4.6 mm×250 mm 5 μ C18) columns work well for most of the described assays, while shorter narrower columns (2.1 mm×50 mm 5 μ C18) work best for the lower mobile phase flow rates required for mass detectors. Volatile nonmetallic buffers must also be used in mobile phases for mass detectors.
2. Phosphate buffers need to be refrigerated and checked prior to use for cloudiness, indicative of microbial growth. Tris buffer can be substituted for the phosphate incubation buffer. Slightly higher glucuronidation activities have been reported for Tris versus phosphate [14]. High ionic strengths (>50 mM) should be avoided because of significant inhibition.
3. A simple vacuum oven or nitrogen gas drier could also be used for this purpose. Mild heating to ~45 °C will speed solvent evaporation but should be not be used during initial method development until the heat stability of substrate and glucuronide can be verified.
4. Agitation of the incubation tubes is usually unnecessary unless relatively high incubation volumes (>100 μL) or high protein concentrations (>0.5 mg/mL) are used.
5. Appropriate working concentrations are about 10 times the Km value (if known) for the substrate and about the Km value for the glucuronide. Some glucuronides will not dissolve completely in pure methanol and may require addition of up to 10 % water.

6. This is a general HPLC method that we have found useful for initial analysis of glucuronide metabolites. Modifications that may be needed for some analytes include use of a higher pH (4.5 or 7.0), or use of a different buffer (0.1 % trifluoroacetic acid). For mass spectrometry methods, 0.1 % formic acid in water should be used for Solvent B.

7. The λ max for the glucuronide can be determined either: (a) from published values for the glucuronide (or substrate); (b) by running a UV absorbance scan of the glucuronide (or substrate) with a spectrophotometer; (c) by using the peak spectral capability of a diode array UV absorbance detector.

8. An alternative approach is to prepare the substrate in methanol at 100 times the desired final incubation concentration and then add 1 µL of this to the incubation tubes in the next step with microsomes and alamethicin. This approach would avoid potential problems with poor substrate solubility, but is not feasible for investigation of relatively high substrate concentrations. Higher amounts of methanol than this (i.e. 2 % final methanol concentration from the substrate and the alamethicin) should be avoided to minimize potential enzyme inhibitory effects. Other solvents such as DMSO, ethanol, and acetonitrile can also be used if substrate solubility is problematic, although preliminary studies should be conducted to ensure that the solvent used does not inhibit glucuronidation. As much as 10 % DMSO (final concentration) is required to solubilize resveratrol for use in glucuronidation experiments.

9. Initial studies should also evaluate the effect of including acid (such as 5 % glacial acetic acid) in the stop solution since some glucuronides (e.g. acyl-glucuronides) are stabilized by acids, while some (e.g. N-glucuronides) may be acid labile. The internal standard (once identified) can also be included in the stop solution to minimize pipetting steps.

10. Once the analyte peaks are positively identified the HPLC method can be optimized to provide adequate peak separation while minimizing total run times. The stability of analytes should be verified by repeated injection of the same sample over the course of the study.

11. The chromatogram "overlay" capability of modern HPLC systems is particularly useful for this purpose. Comparison of peak spectra is also helpful if a diode array UV detector is available. The success of this approach is highly dependent on consistency of HPLC peak retention times which should be ensured by proper HPLC equipment (especially pump) maintenance.

12. In addition to mass determination, some structural information can be obtained via mass spectrometry collision-induced fragmentation of glucuronides that typically yield a product

ion with a neutral loss to 176 amu. Treatment with acids or alkalis can also assist in identification in that acyl-glucuronides tend to hydrolyze under alkaline conditions, while some N-glucuronides (especially primary amines) tend to hydrolyze when treated with acids. If multiple potential conjugation sites are present on the substrate, determination of the exact site of conjugation will likely require structural elucidation via NMR.

13. Some glucuronides (such as propofol glucuronide) are resistant to β-glucuronidase treatment. In addition, spontaneous isomerization of some acyl-glucuronides yields a compound that is insensitive to enzymatic hydrolysis.

14. Although it is desirable to use an internal standard that is similar structurally to the analyte, this is usually not necessary if the described direct injection HPLC assay method is used (i.e. no liquid-liquid or solid phase extraction steps). The easiest approach to identify an appropriate internal standard is to optimize the HPLC method for the glucuronide and then screen all available compounds for retention times that are similar to but distinct from the analytes. For the assay the internal standard can be dissolved in the acetonitrile stop solution.

15. The effect of adding 2 % bovine serum albumin on HLMs and rUGTs activities should be evaluated in preliminary studies using substrate concentrations less than the Km value. Increased activity with addition of albumin to HLMs would be indicative of endogenous fatty acid inhibition of the responsible UGT (usually UGT2B7 or UGT1A9). In this instance, kinetic curves should be determined with and without added albumin. Decreased activity might indicate that the substrate is binding to the albumin.

16. Incubations with propofol should be performed in glass vials, since this compound is highly lipophilic and tends to adsorb to plastic containers. Oxazepam glucuronidation yields two glucuronide stereoisomers that can be readily separated by HPLC. The S-oxazepam glucuronide is the major metabolite which elutes immediately after the R-oxazepam glucuronide.

17. Adjustment of the HPLC method may be necessary because of interfering peaks from the inhibitor and inhibitor metabolites.

References

1. Court MH (2010) Interindividual variability in hepatic drug glucuronidation: studies into the role of age, sex, enzyme inducers, and genetic polymorphism using the human liver bank as a model system. Drug Metab Rev 42(1):209
2. Meech R, Miners JO, Lewis BC, Mackenzie PI (2012) The glycosidation of xenobiotics and endogenous compounds: versatility and redundancy in the UDP glycosyltransferase superfamily. Pharmacol Ther 134(2):200–218
3. Miners JO, Mackenzie PI, Knights KM (2010) The prediction of drug-glucuronidation parameters in humans: UDP-glucuronosyltransferase enzyme-selective substrate and

inhibitor probes for reaction phenotyping and in vitro-in vivo extrapolation of drug clearance and drug-drug interaction potential. Drug Metab Rev 42(1):196–208
4. Court MH, Zhang X, Ding X, Yee KK, Hesse LM, Finel M (2012) Quantitative distribution of mRNAs encoding the 19 human UDP-glucuronosyltransferase enzymes in 26 adult and 3 fetal tissues. Xenobiotica 42(3):266
5. Harper TW, Brassil PJ (2008) Reaction phenotyping: current industry efforts to identify enzymes responsible for metabolizing drug candidates. AAPS J 10(1):200–207
6. Zhang H, Davis CD, Sinz MW, Rodrigues AD (2007) Cytochrome P450 reaction-phenotyping: an industrial perspective. Expert Opin Drug Metab Toxicol 3(5):667–687
7. Venkatakrishnan K, Von Moltke LL, Greenblatt DJ (2001) Human drug metabolism and the cytochromes P450: application and relevance of in vitro models. J Clin Pharmacol 41(11):1149
8. Rodrigues AD (1999) Integrated cytochrome P450 reaction phenotyping: attempting to bridge the gap between cDNA-expressed cytochromes P450 and native human liver microsomes. Biochem Pharmacol 57(5):465–480
9. Fisher MB, Paine MF, Strelevitz TJ, Wrighton SA (2001) The role of hepatic and extrahepatic UDP-glucuronosyltransferases in human drug metabolism. Drug Metab Rev 33(3–4):273–297
10. Hirata-Koizumi M, Saito M, Miyake S, Hasegawa R (2007) Adverse events caused by drug interactions involving glucuronoconjugates of zidovudine, valproic acid and lamotrigine, and analysis of how such potential events are discussed in package inserts of Japan, UK and USA. J Clin Pharm Ther 32(2):177–185
11. Marsh S, Hoskins JM (2010) Irinotecan pharmacogenomics. Pharmacogenomics 11(7):1003–1010
12. Zhang D, Chando TJ, Everett DW, Patten CJ, Dehal SS, Humphreys WG (2005) In vitro inhibition of UDP glucuronosyltransferases by atazanavir and other HIV protease inhibitors and the relationship of this property to in vivo bilirubin glucuronidation. Drug Metab Dispos 33(11):1729–1739
13. Soars MG, Mattiuz EL, Jackson DA, Kulanthaivel P, Ehlhardt WJ, Wrighton SA (2002) Biosynthesis of drug glucuronides for use as authentic standards. J Pharmacol Toxicol Methods 47(3):161–168
14. Walsky RL, Bauman JN, Bourcier K, Giddens G, Lapham K, Negahban A, Ryder TF, Obach RS, Hyland R, Goosen TC (2012) Optimized assays for human UDP-glucuronosyltransferase (UGT) activities: altered alamethicin concentration and utility to screen for UGT inhibitors. Drug Metab Dispos 40(5):1051–1065
15. Oleson L, Court MH (2008) Effect of the beta-glucuronidase inhibitor saccharolactone on glucuronidation by human tissue microsomes and recombinant UDP-glucuronosyltransferases. J Pharm Pharmacol 60(9):1175
16. Turgeon D, Chouinard S, Belanger P, Picard S, Labbe JF, Borgeat P, Belanger A (2003) Glucuronidation of arachidonic and linoleic acid metabolites by human UDP-glucuronosyltransferases. J Lipid Res 44(6): 1182–1191
17. Levesque E, Turgeon D, Carrier JS, Montminy V, Beaulieu M, Belanger A (2001) Isolation and characterization of the UGT2B28 cDNA encoding a novel human steroid conjugating UDP-glucuronosyltransferase. Biochemistry 40(13):3869–3881
18. Court MH, Hazarika S, Krishnaswamy S, Finel M, Williams JA (2008) Novel polymorphic human UDP-glucuronosyltransferase 2A3: cloning, functional characterization of enzyme variants, comparative tissue expression, and gene induction. Mol Pharmacol 74(3):744
19. Krishnaswamy S, Duan SX, Von Moltke LL, Greenblatt DJ, Court MH (2003) Validation of serotonin (5-hydroxtryptamine) as an in vitro substrate probe for human UDP-glucuronosyltransferase (UGT) 1A6. Drug Metab Dispos 31(1):133
20. Ohtsuki S, Schaefer O, Kawakami H, Inoue T, Liehner S, Saito A, Ishiguro N, Kishimoto W, Ludwig-Schwellinger E, Ebner T, Terasaki T (2012) Simultaneous absolute protein quantification of transporters, cytochromes P450, and UDP-glucuronosyltransferases as a novel approach for the characterization of individual human liver: comparison with mRNA levels and activities. Drug Metab Dispos 40(1):83–92
21. Kato Y, Nakajima M, Oda S, Fukami T, Yokoi T (2012) Human UDP-glucuronosyltransferase isoforms involved in haloperidol glucuronidation and quantitative estimation of their contribution. Drug Metab Dispos 40(2):240–248
22. Zhu L, Ge G, Zhang H, Liu H, He G, Liang S, Zhang Y, Fang Z, Dong P, Finel M, Yang L (2012) Characterization of hepatic and intestinal glucuronidation of magnolol: application of the relative activity factor approach to decipher the contributions of multiple UDP-glucuronosyltransferase isoforms. Drug Metab Dispos 40(3):529–538
23. Venkatakrishnan K, von Moltke LL, Greenblatt DJ (2001) Application of the relative activity factor approach in scaling from heterologously expressed cytochromes p450 to human liver microsomes: studies on amitriptyline as a model substrate. J Pharmacol Exp Ther 297(1): 326–337

24. Rowland A, Gaganis P, Elliot DJ, Mackenzie PI, Knights KM, Miners JO (2007) Binding of inhibitory fatty acids is responsible for the enhancement of UDP-glucuronosyltransferase 2B7 activity by albumin: implications for in vitro-in vivo extrapolation. J Pharmacol Exp Ther 321(1):137–147
25. Rowland A, Knights KM, Mackenzie PI, Miners JO (2008) The "albumin effect" and drug glucuronidation: bovine serum albumin and fatty acid-free human serum albumin enhance the glucuronidation of UDP-glucuronosyltransferase (UGT) 1A9 substrates but not UGT1A1 and UGT1A6 activities. Drug Metab Dispos 36(6):1056–1062
26. Uchaipichat V, Mackenzie PI, Elliot DJ, Miners JO (2006) Selectivity of substrate (trifluoperazine) and inhibitor (amitriptyline, androsterone, canrenoic acid, hecogenin, phenylbutazone, quinidine, quinine, and sulfinpyrazone) "probes" for human udp-glucuronosyltransferases. Drug Metab Dispos 34(3):449–456
27. Miners JO, Bowalgaha K, Elliot DJ, Baranczewski P, Knights KM (2011) Characterization of niflumic acid as a selective inhibitor of human liver microsomal UDP-glucuronosyltransferase 1A9: application to the reaction phenotyping of acetaminophen glucuronidation. Drug Metab Dispos 39(4):644–652
28. Fujiwara R, Nakajima M, Yamanaka H, Katoh M, Yokoi T (2008) Product inhibition of UDP-glucuronosyltransferase (UGT) enzymes by UDP obfuscates the inhibitory effects of UGT substrates. Drug Metab Dispos 36(2):361–367
29. Court MH, Krishnaswamy S, Hao Q, Duan SX, Patten CJ, Von Moltke LL, Greenblatt DJ (2003) Evaluation of 3′-azido-3′-deoxythymidine, morphine, and codeine as probe substrates for UDP-glucuronosyltransferase 2B7 (UGT2B7) in human liver microsomes: specificity and influence of the UGT2B7*2 polymorphism. Drug Metab Dispos 31(9):1125
30. Court MH, Duan SX, Guillemette C, Journault K, Krishnaswamy S, Von Moltke LL, Greenblatt DJ (2002) Stereoselective conjugation of oxazepam by human UDP-glucuronosyltransferases (UGTs): S-oxazepam is glucuronidated by UGT2B15, while R-oxazepam is glucuronidated by UGT2B7 and UGT1A9. Drug Metab Dispos 30(11):1257

Chapter 9

In Vitro CYP/FMO Reaction Phenotyping

Carlo Sensenhauser

Abstract

Reaction phenotyping is the process of identifying *in vitro* the drug-metabolizing enzymes involved in the clearance of a drug in order to predict whether the drug might be susceptible to changes in its exposure or to cause changes in the exposure of concomitantly administered drugs. The semi-quantitative assessment of the relative contributions of these pathways can provide early insight to the potential for drug-drug interactions or to possible affinities to functionally polymorphic enzymes which can cause inter-subject variability in the clinic. This chapter presents an initial strategy that can be implemented at the drug discovery or early development stage of a drug candidate to evaluate the extent of *in vitro* clearance, identify possible metabolic liabilities and predict potential interaction with other relevant drugs.

Key words Drug-metabolizing enzymes, Phenotyping, Drug-drug interactions, Drug clearance, Metabolic liabilities

1 Introduction

Identification of metabolic pathways and the drug-metabolizing enzymes (DME) involved in the metabolic clearance of a drug is an important step in the early development of a new molecular entity (NME) [1]. It is essential in providing an initial assessment of the potential for drug-drug interactions (DDI) resulting from the effects of a therapeutic agent on the exposure and clearance of a concomitantly administered drug [2]. Metabolic clearance involves the biotransformation of drugs which generally introduces a more polar functional group following oxidation, dealkylation, hydrolysis, etc., converting xenobiotics into more readily excreted hydrophilic compounds. Determining whether a drug candidate has multiple routes of elimination can help mitigate the susceptibility to variable exposure levels. Conversely, identifying the drug as a substrate of specific enzymes can provide valuable information in the assessment of the drug as a potential perpetrator (i.e. inhibitor). Recently, more stringent requirements have been proposed by the FDA and European Medicines Agency (EMA), emphasizing the urgency to identify and exhaustively investigate metabolic pathways that account for more than 25 % of the systemic clearance of an investigational drug in order to gain a more adequate understanding of the drug-drug interaction potential [3, 4].

Cytochrome P450 monooxygenases (CYPs) are a superfamily of diverse, membrane-bound, heme-containing enzymes involved in the metabolism and potential bioactivation of a wide range of substrates. The heme is bound non-covalently to the active site of the enzyme and binds molecular oxygen via its iron atom, to initiate the catalytic cycle in which protons and electrons are transferred from a co-factor, NADPH, and closely associated amino acid side chains to typically yield a monooxygenated product and a mole of water. Although several thousand CYP proteins have been identified across all types of living organisms, only a few dozen isoforms are involved in the metabolism of xenobiotics, triggering the possibility of potentially adverse DDIs in humans. Nowadays, the major CYP enzymes involved in drug metabolism are readily available commercially. These are single recombinant enzyme systems heterologously over-expressed in bacterial, insect or mammalian cell lines, which can be used to determine precisely the affinity of the test compound for a specific CYP isoform. They are prepared with higher than natural levels of NADPH-CYP450 oxido-reductase to achieve higher catalytic rates and usually also with cytochrome b5. They are highly reactive unique systems that provide the advantage of facilitating unequivocal assessment of the affinity of a substrate for a particular CYP isoform.

Flavin-containing monooxygenases (FMOs) are a family of five enzymes, FMO1-5, sharing over 50 % amino acid homology, and are located, as the CYPs, in the endoplasmic reticulum region and expressed to varying levels in the liver and other extrahepatic tissue [5]. Similarly to the CYPs, they are involved in the detoxification of nucleophilic xenobiotics, by typically oxygenating heteroatoms, such as nitrogen and sulfur, and require NADPH as the cofactor. While substrate selectivity is variable among the FMOs, FMO3 has been widely investigated in reaction phenotyping due to its high selectivity and several polymorphic variants [6–8]. Differentiating between CYP and FMO contributions is an important step in the phenotyping of a new drug, rendered more practical by the availability of recombinantly expressed systems of the major FMO isoforms.

In addition to CYP or FMO metabolism, Phase I biotransformations can also be the product of other classes of membrane-bound or cytosolic enzymes. Although less characterized and normalized than CYPs and FMO's, aldehyde oxidases, xanthine oxidases and monoamine oxidases can all play a role in the metabolic clearance of the test compound and should be identified as early as possible in the drug development process. In the present chapter, phenotyping of these non-microsomal enzyme families will not be discussed further.

2 General *In Vitro* Reaction Phenotyping Strategy

Reaction phenotyping assays to identify the enzymes involved in Phase I biotransformation pathways are performed routinely and at the various stages of drug development. However, at the discovery and early development stages, the semi-quantitative prediction of the relative contributions of the identified pathways in humans may be limited due to the absence of radiolabeled or bioanalytical reference standards of parent compound and major metabolites. Nonetheless, a general strategy can be implemented and subsequently fine-tuned in a stepwise, integrated process based on the incremental data generated from previous assays, identification of major metabolites and even knowledge of the primary routes of elimination in preclinical species.

In this chapter, the approach presented allows for an initial *in vitro* elucidation of the primary metabolic clearance pathways involved and an early assessment of the potential risks and liabilities that could lead to a high inter-individual variability, such as susceptibility to functionally polymorphic enzymes (such as CYP2C9 and CYP2D6) or DDIs. The assays discussed are either qualitative, to determine which enzymes are involved, or semi-quantitative, to provide an initial estimate of the degree of involvement of a specific enzyme. Qualitatively, recombinant enzymes can be incubated with the test compound and NADPH-dependent substrate depletion can be monitored to assess the extent of specific CYP involvement. Further tests can be performed to evaluate the possibility of non-CYP-mediated Phase I metabolism. The contribution of an enzyme would then be evaluated by the determination of intrinsic clearance, kinetic constants or the effect of CYP-specific chemical inhibitors or monoclonal antibodies on the turnover of the substrate. In order to do this, initial rate conditions would have to be established. Figure 1 shows a flow diagram strategy that can be implemented as a drug candidate progresses through early drug discovery, lead optimization and subsequently to new molecular entity (NME) status, with the objective of understanding the routes of clearance and identifying possible liabilities.

Although not essential, laboratory automation (e.g. robotic liquid handing instruments, such as Beckman Coulter Biomek or Tecan EVO workstations) can facilitate the implementation and execution of these assays, by increasing throughput and reproducibility. Similarly, tandem liquid chromatography/mass spectroscopy (LC/MS) technology is a valuable tool for these early stage screens,

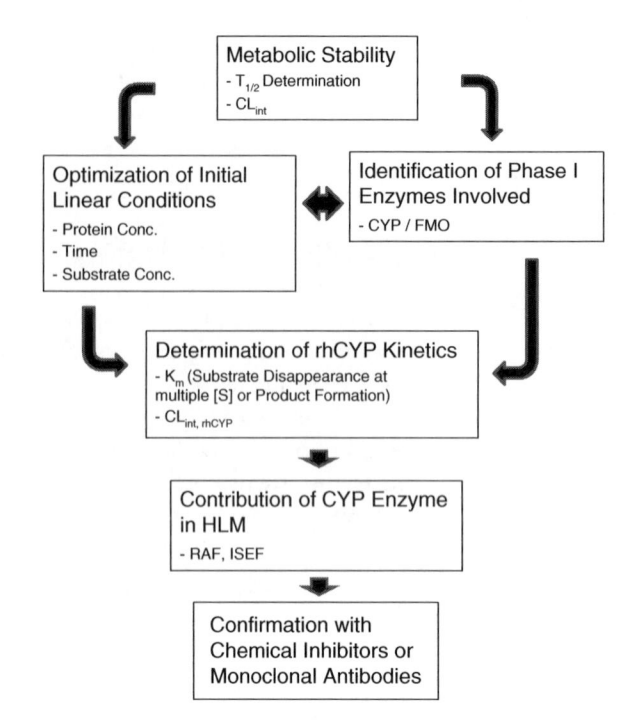

Fig. 1 Reaction phenotyping strategy in drug discovery. Beginning with the assessment of metabolic stability in human liver microsomes, reaction phenotyping of a test compound targets identification of biotransformation pathways, the enzymes involved and the relative contributions to the overall metabolic clearance

when knowledge of the major metabolites formed is still limited and therefore chromatographic separation can be impractical. Monitoring commonly expected biotransformation products (e.g. monooxygenation, dealkylation, carboxylation, etc.) by the appropriate parent mass and possible fragmentation products provides an initial, qualitative indication of potential metabolic pathways. Methodologies in this chapter, have been developed on a Biomek FXp robotic platform, and analyzed on a Sciex API4000 triple quadrupole mass spectrometer, coupled to a Shimadzu Nexera Ultra-high Performance Liquid Chromatography (UPLC) system, but whenever necessary, details and comments on adapting these methods for manual execution or non-LC/MS analysis will be presented in Sect. 5.

3 Phenotyping Assays

3.1 Determination of Microsomal Stability and Intrinsic Clearance (T1/2 Method)

Today, although whole cell systems (e.g. suspension or plated hepatocytes, 3D cultures, etc.) are widely recognized as more predictive and complementary for metabolic profiling of a drug candidate, microsomes are still widely utilized to generate phenotyping, intrinsic clearance and enzyme kinetic data. Their ease of use and full complement of membrane-bound enzymes (CYPs, FMOs, esterases and Phase II conjugation enzymes such as the uridine 5′-diphospho-glucuronosyltransferases, UGTs) make them the system of choice for DDI assessment in drug discovery and early development. Therefore, the first step in reaction phenotyping should be to determine the stability (i.e. the rate of substrate turnover) of a compound in a microsomal incubation. Consideration can be given to optimizing parameters such as substrate concentration and microsomal protein content, but for practical purposes this can be perfected once an initial assessment of the overall microsomal stability is performed. Therefore, by incubating the compound at a concentration that can be reasonably expected to be below the K_m value (e.g. 1 µM) and within a commonly used microsomal protein range (e.g. 0.1–0.4 mg/mL), with and without NADPH and collecting aliquots at various time points, an initial indication of metabolic clearance, linear kinetic range and metabolite formation can be determined in a single assay. Although outside the scope of this chapter, an initial assessment of potential Phase II metabolism can be evaluated by also including incubations with and without UDPGA (in addition to NADPH) in the initial protocol [9] and will only be presented here in Sect. 3.1 (adapted from a Cyprotex Stability assay).

The assay in its simplest form can be set up by incubating the test compound at one concentration (1 µM), with and without co-factors and quenching aliquots of the incubation at various time points.

The incubations can be prepared in a 96-well format as shown in Fig. 2, to test for non-enzymatic degradation, Phase I metabolism, direct and indirect glucuronidation. Aliquots are then transferred to the termination plate containing internal standard in acetonitrile at the selected time points. A separate plate, containing a control compound, such as diclofenac or testosterone can be set up in the same way.

3.1.1 Materials

All reagents, consumables and instrumentation can be obtained from commercial sources. Catalog numbers are given in Tables 1 and 2 (**Note 1**).

3.1.2 Instrumentation

1. Automated liquid handling workstation (e.g. Biomek, Beckman Coulter, Indianapolis, IN or Tecan EVO, Tecan Systems, San Jose, CA)
2. Dubnoff Shaking Water Bath (VWR, Radnor, PA) or Integrated robotic shaking incubator (e.g. Cytomat Automated Incubator, Thermo Scientific, Burlington, ON, Canada)
3. Orion pH Meter with Glass Electrode (Thermo Scientific, Chelmsford, MA)

Incubation Plate

	1	2	3	4	5	6	7	8	9	10	11	12
A	TC - NADPH - UDPGA 1	TC - NADPH - UDPGA 2	TC - NADPH - UDPGA 3	TC + NADPH - UDPGA 1	TC + NADPH - UDPGA 2	TC + NADPH - UDPGA 3	TC - NADPH + UDPGA 1	TC - NADPH + UDPGA 2	TC - NADPH + UDPGA 3	TC + NADPH + UDPGA 1	TC + NADPH + UDPGA 2	TC + NADPH + UDPGA 3

Termination Plate

	1	2	3	4	5	6	7	8	9	10	11	12
A	0 min	0 min	0 min	0 min	0 min	0 min	0 min	0 min	0 min	0 min	0 min	0 min
B	5 min	5 min	5 min	5 min	5 min	5 min	5 min	5 min	5 min	5 min	5 min	5 min
C	10 min	10 min	10 min	10 min	10 min	10 min	10 min	10 min	10 min	10 min	10 min	10 min
D	20 min	20 min	20 min	20 min	20 min	20 min	20 min	20 min	20 min	20 min	20 min	20 min
E	30 min	30 min	30 min	30 min	30 min	30 min	30 min	30 min	30 min	30 min	30 min	30 min
F	40 min	40 min	40 min	40 min	40 min	40 min	40 min	40 min	40 min	40 min	40 min	40 min
G	50 min	50 min	50 min	50 min	50 min	50 min	50 min	50 min	50 min	50 min	50 min	50 min
H	60 min	60 min	60 min	60 min	60 min	60 min	60 min	60 min	60 min	60 min	60 min	60 min

Fig. 2 Plate map for incubation and termination of substrate disappearance assay. Incubations are performed in triplicate, with and without NADPH and with and without UDPGA (optional). At the selected time points, aliquots of the incubated samples are transferred to a termination plate (containing acetonitrile and internal standard). (TC, test compound)

4. Analytical Scale (Mettler Toledo, Columbus, OH)

5. Sciex API 4000™ triple quadrupole mass spectrometer (AB Sciex, Foster City, CA) (**Note 2**)

6. Shimadzu Nexera 30AD UPLC binary pump system (Shimadzu Scientific Instruments, Columbia, MD)

7. Zymark Turbovap Evaporator for 96-well deep well plates (available through Biotage AB, Charlotte, NC)

8. Multi-channel electronic pipettes (e.g. Biohit ProLine, Biohit, Inc., Neptune, NJ)

3.1.3 Assay Procedure

The assay is set up in triplicate as 500 μL incubations at a 0.2 mg/mL microsomal concentration in 100 mM potassium phosphate buffer (KPB), pH 7.4, (**Note 3**) from which 40 μL aliquots will be removed at each time point and quenched in a separate termination plate. A plate map is shown in Fig. 2 to aid the set up of the assay using 96-well plate configuration. Incubations will be carried out with and without cofactors (an NADPH Regenerating System (NRS) (**Note 4**) and, optionally, a UGT Reaction solution (**Note 5**) to screen for Phase II metabolism, see Chap. 8). Samples without cofactor will receive a corresponding volume of blank KPB.

1. Prepare the test compound as a 1 mM stock in a suitable organic solvent (preferably methanol, DMSO, acetonitrile or a combination) (**Note 6**).

2. Prepare the previously selected internal standard stock solution.

Table 1
Common chemicals, reagents and consumables

Chemicals/reagents/consumables	Catalog #	Vendor
Potassium phosphate, Monobasic, 1.0 M	P8709	Sigma (St. Louis, MO)
Potassium phosphate, Dibasic, 1.0 M	D4902	Sigma (St. Louis, MO)
Ethylenediaminetetraacetic acid disodium salt solution, 0.5 M	E7889	Sigma (St. Louis, MO)
Magnesium chloride solution, 1.0 M	M1028	Sigma (St. Louis, MO)
Dimethylsulfoxide	D8779	Sigma (St. Louis, MO)
Glacial acetic acid	A6283	Sigma (St. Louis, MO)
Methanol	MX0488-6	EM Science through VWR (Bridgeport, NJ)
Acetonitrile	AX0142P-1	EM Science through VWR (Bridgeport, NJ)
Deionized water		Millipore Corp. (Billerica, MA)
UltraPool™ HLM Pooled from 150 mixed gender donors	452117	BD Biosciences (San Jose, CA)

NADPH Regenerating System
  Solution A
  Solution B
	
	451220
	451200
		BD Biosciences (San Jose, CA)

NADPH Regenerating System, Sol. B	451200	BD Biosciences (San Jose, CA)
Flexi-tier 96-well Clear Systems (inclusive of base plates and glass inserts—0.5, 1.0 and 1.5 mL volumes)	96FP05-C 96FP10-C 96FP15-C	Analytical Sales and Services (Pompton Plains, NJ)

Biomek FXp Span 8 and 96-channel Tips
  AP96 P250 Tips, Nonsterile
  AP96 P20 Tips, Nonsterile
  Span-8 P1000 Tips, Nonsterile
  Span 8 Universal P50, Nonsterile
  Span 8 Universal P250, Nonsterile
	
	717251
	717254
	987935
	A21578
	379501
		Beckman Coulter (Indianapolis, IN)

Costar 96-square well (2-mL) deep-well Plate	3960	Corning Life Sciences (Tewksbury, MA)

Microcentrifuge tubes
  2 mL
  1.5 mL
	
	89004-296
	89004-288
		VWR (Radnor, PA)

Strata Impact Protein Precipitation Filter Plate, 96-square well,	CE0-7565	Phenomenex Corp. (Torrance, CA)
Kinetex Phenyl-Hexyl 1.7 μm, 2.1 × 150 mm, 100 Å, Analytical Column	00F-4500-AN	Phenomenex Corp. (Torrance, CA)

Table 2
Chemicals, reagents and consumables for T$_{1/2}$ method

Chemicals/reagents/consumables	Catalog #	Vendor
Appropriate analytical internal standard (**Note 2**)		
UGT Reaction Mix (Optional)		BD Biosciences (San Jose, CA)
Solution A	451300	
Solution B	451320	

3. Prepare the termination plate (either Costar 96-deep-well plate or Flexi-Tier 1.0 mL glass inserts) with 300 μL of acetonitrile containing the internal standard. Keep refrigerated until ready to start the assay.

4. Thaw and dilute the human liver microsomes (HLM) with KPB. There are 12 incubated samples, based on the plate map in Fig. 2. Each sample will receive 300 μL of diluted microsomes, spiked with the test compound and 200 μL of buffer and appropriate cofactor solution (*see* step 7). Microsomes should be diluted to yield a final 0.2 mg/mL incubation concentration. In order to have sufficient volume, prepare 4,000 μL of KPB and add 70 μL of HLM.

5. Spike the test compound into the microsomal suspension to yield the 1 μM incubation concentration (i.e. approximately 7 μL of 1 mM stock spiked in 4,000 μL of diluted HLM will result in a 1 μM incubation concentration).

6. Aliquot the spiked microsomes into the incubation vessels (**Note 7**).

7. If Phase II metabolism is included in the assay, add 100 μL of UGT Reaction Mix Solution B, (5× solution of alamethicin, **Note 5**) to the +UDPGA samples. Samples incubated without UDPGA would receive an equivalent volume of blank KPB.

8. Preincubate the samples for 10 min in the shaking incubator or water bath thermostatted at 37 °C.

9. While the samples are pre-incubating, prepare co-factor mixtures to initiate the reactions. The volumes required per sample are shown Table 3. Immediately transfer the time 0 aliquot (40 μL) to the termination plate (**Notes 2** and **8**).

10. Subsequent time points will be quenched in the same manner (**Note 9**).

11. Filter the completed termination plate using the Strata Impact protein precipitation filter plate.

12. Either evaporate the filtrate to dryness under a steady stream of nitrogen gas and then reconstitute the dried samples with mobile phase, or inject directly (after dilution, if necessary) for LC/MS analysis.

Table 3
T$_{1/2}$ method—phase I and phase II cofactor solutions preparation

(μL/sample)	-NADPH -UDPGA	+NADPH -UDPGA	-NADPH +UDPGA	+NADPH +UDPGA
Blank buffer	100	70	60	30
NRS (Soln A + Soln B)		30		30
UGT Reaction Mix, Soln A			40	40

3.1.4 Analytical Method and Data Interpretation

The LC/MS method will be based on the optimized chromatographic and ionization conditions of the parent compound, since it is unlikely that reference standards of the metabolites will be available at the Discovery stage. Nonetheless, formation of the commonly expected biotransformation products can be monitored based on the structure of the test compound (e.g. mass to charge, m/z, differences of +16 for monooxygenations, +30 for carboxylations, +176 for glucuronides, +192 for monooxygenations followed by glucuronidation, dealkylations, etc.).

Metabolic intrinsic clearance (CL$_{int}$) of the test compound can be determined from the slope of the semi-log titration curves of the NADPH-dependent percent loss of substrate as a function of time, compared to the amount at time 0, set to 100 % (Fig. 3). The rate of substrate turnover ($-k$) can be used to derive the half-life (T$_{1/2}$ = ln 2/-k) and the CL$_{int}$ value determined from (1):

$$CL_{int} = \frac{\ln 2}{\ln\ vitro\ T_{1/2}} \times \frac{\text{mL Incubation}}{\text{mg Microsomes}} \quad (1)$$

In cases of rapid turnover, only the initial linear portion of the titration curve should be included in the determination of the slope.

3.2 Linear Conditions Optimization

If results from the previous assay (Sect. 3.1) indicate that the test compound is rapidly metabolized, i.e. more than 50 % of the initial substrate concentration is depleted and that substrate turnover is nonlinear over the time period tested, any further quantitative evaluation (e.g. T$_{1/2}$ or kinetic parameter determination) would require optimization of the initial linear conditions to avoid enzyme saturation. The three parameters that are typically optimized to ensure linear conditions are protein concentration, incubation time and substrate concentration. Although potentially impractical and of lower throughput, this process can be effectively expedited with automation, by setting up an incubation "matrix" in which all parameters can be tested and optimized in one assay incubation.

Linearity is going to be tested at multiple protein and substrate concentrations and over multiple time points. In order to do this, an incubation plate can be set up in which HLM are serially diluted

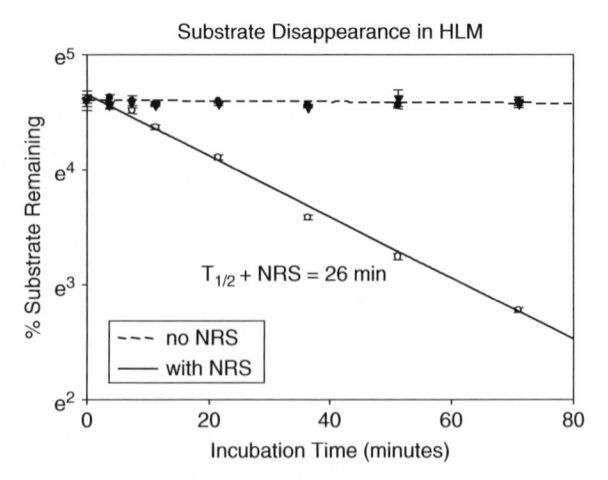

Fig. 3 Determination of intrinsic clearance by substrate disappearance. Test compound is incubated in human liver microsomes, with and without an NADPH Regenerating System. Aliquots are quenched at various time points and substrate depletion is monitored by LC/MS. Turnover is only observed in the presence of NADPH for CYP and FMO mediated reactions. The slope of the titration curve of natural log of percent substrate disappearance as a function of time corresponds to the elimination rate constant from which the half-life ($T_{1/2}$) and the intrinsic clearance ($CL_{int} = \ln 2/T_{1/2}$) can be determined

along the rows of the plate and the substrate is serially diluted along the columns of the plate (see Fig. 4 as a sample layout). In such a way, every microsomal concentration tested is incubated with every substrate concentration, and linearity can be assessed in a simple and straightforward fashion, by taking aliquots at selected time points (e.g. 0, 10, 20, 40 and 60 min) and transferring them to a termination plate, as in the previous assay (Sect. 3.1).

3.2.1 Assay Procedure

The assay is set up by serially diluting the HLM with KPB and the test compound with the same solvent used to prepare the stock solution, i.e. methanol, DMSO, acetonitrile or a combination. Each substrate concentration level is then spiked in each HLM concentration level (**Note 10**). The assay can be scaled up to any particular volume, but similar calculations as in the single substrate concentration assay should be carried out to ensure correct and sufficient volumes.

1. Prepare the test compound (at a 10 mM stock solution concentration), internal standard, termination plates and thaw the HLM as in the previous assay (Sect. 3.1). If automated, each time point will be quenched in individual termination plates.

2. Dilute the HLM to yield a 1.0 mg/mL incubation concentration. Then serially dilute with blank KPB to yield 0.5, 0.2, 0.1 and 0.05 mg/mL concentrations.

CYP/FMO Phenotyping

	1	2	3	4	5	6	7	8	9	10	11	12
Substrate Concentration	Protein Concentration → Increasing											
A	HLM 0.05 mg/mL [S] 0.5 µM	HLM 0.05 mg/mL	HLM 0.1 mg/mL	HLM 0.1 mg/mL	HLM 0.2 mg/mL	HLM 0.2 mg/mL	HLM 0.5 mg/mL	HLM 0.5 mg/mL	HLM 1.0 mg/mL	HLM 1.0 mg/mL		
B	[S] 0.5 µM											
C	[S] 1 µM											
D	[S] 1 µM											
E	[S] 5 µM											
F	[S] 5 µM											
G	[S] 10 µM											
H	[S] 10 µM											

(Substrate Concentration: Increasing ↓)

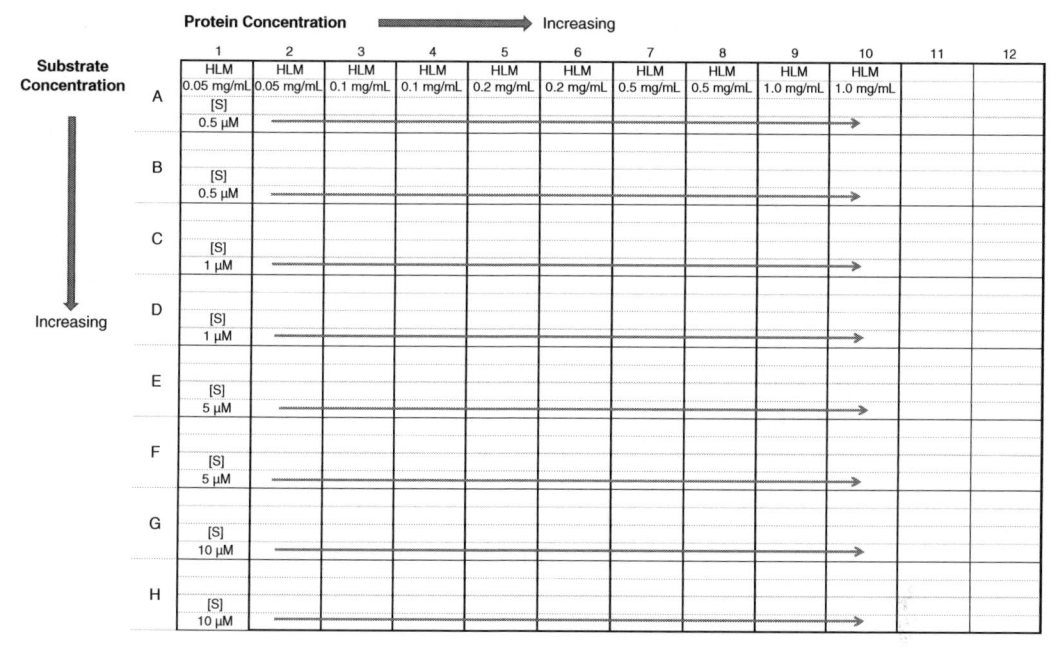

Fig. 4 Plate map for incubation of multiple HLM and substrate concentrations for the optimization of initial linear conditions. Both microsomal protein and substrate concentrations are varied within a 96-well plate and incubated in the presence of NADPH. Aliquots are taken at time points (e.g. 0–60 min) to establish rate of reaction linearity

3. Aliquot each HLM level for each test compound concentration level tested.
4. Dilute the test compound to yield a 10 µM incubation concentration. Then serially dilute with the same organic solvent as the stock solution, to 5, 1 and 0.5 µM concentrations.
5. Aliquot each test compound concentration into each HLM concentration level.
6. Aliquot the spiked HLM into the incubation vessels.
7. Preincubate the samples for 5 min in the shaking incubator or water bath thermostatted at 37 °C.
8. Prepare the NRS mixture to initiate reactions.
9. Initiate the reactions by the addition of the co-factor mixture.
10. Immediately transfer the time 0 aliquot to the termination plate which will contain at least four volumes of acetonitrile spiked with the selected internal standard.
11. Subsequent time points will be quenched in the same manner.
12. Filter and prepare the terminated samples as in the previous assay (Sect. 3.1).

3.2.2 Analytical Method and Data Interpretation

The LC/MS method implemented is the same as in the $T_{1/2}$ method. Linearity optimization will focus on substrate turnover, but some relevant information regarding rates of formation and

Effect of Substrate and Protein Concentration on Initial Linear Conditions

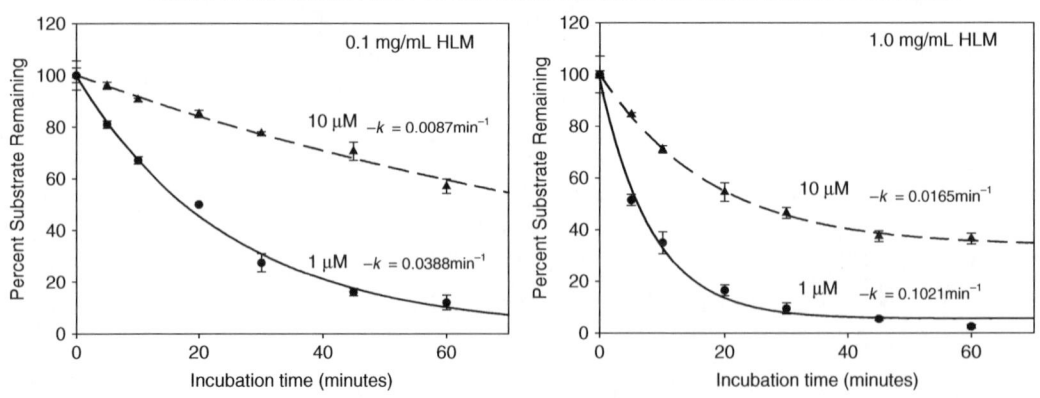

[Left plot: 0.1 mg/mL HLM; 10 µM, $-k = 0.0087\,\text{min}^{-1}$; 1 µM, $-k = 0.0388\,\text{min}^{-1}$]

[Right plot: 1.0 mg/mL HLM; 10 µM, $-k = 0.0165\,\text{min}^{-1}$; 1 µM, $-k = 0.1021\,\text{min}^{-1}$]

Fig. 5 Plots of substrate disappearance as a function of time at different protein and substrate concentrations. Incubations of a test compound in human liver microsomes were monitored for substrate disappearance at time points from 0 to 60 min. Increasing protein concentration will increase initial rates of reaction (i.e. more rapid loss of substrate). The rate constant, k, was determined from the initial linear range of reaction. The difference will be smaller at higher substrate concentrations due to possible enzyme saturation

kinetic profiles of potential biotransformation products can be extracted from the generic method described above.

As before, the rate of substrate disappearance is determined as a percent of the time 0 chromatographic peak area ratio of the substrate to the internal standard. Substrate disappearance as a function of time, for the various protein and substrate concentrations tested (Fig. 5) will provide the range of linear conditions that can be used in future kinetic and semi-quantitative assays.

3.3 Recombinant Human Enzyme Screen

Once the metabolic stability of a test compound has been investigated, CYP and other non-CYP enzymes involved in the metabolism can be readily identified by incubating the commercially available recombinant single human enzyme systems with the compound. Although very little kinetic information is available at this stage, test compounds are incubated at a concentration (typically 1 µM) that allows for accurate detection of depletion and is still reasonably and likely below the K_m value to avoid saturation of enzymatic pathways.

These recombinant systems from cells infected with a virus engineered to express a single enzyme isoform have higher specific activities than human liver microsomes. So while they are extremely useful in qualitative determinations of the isoforms involved in turning over the test compound, it should not be overlooked that they are in effect artificial systems and therefore adequate adjustments (i.e. relative activity factors, RAF or inter-system extrapolation factors, ISEF) need to be made for kinetic parameters derived from these single enzyme systems.

At the drug discovery stage, it is customary to screen the major CYP enzymes (CYPs 1A2, 2B6, 2C8, 2C9, 2C19, 2D6 and 3A4),

CYP/FMO Phenotyping

Plate Map for Recombinant Human Enzyme Screen

	1	2	3	4	5	6	7	8	9	10	11	12
	\multicolumn{6}{c}{INSECT CONTROL}	\multicolumn{6}{c}{CYP2D6}
A	TIME 0	TIME 0	TIME 60 with NADPH	TIME 60 with NADPH	TIME 60 no NADPH	TIME 60 no NADPH	TIME 0	TIME 0	TIME 60 with NADPH	TIME 60 with NADPH	TIME 60 no NADPH	TIME 60 no NADPH
	\multicolumn{6}{c}{CYP1A1}	\multicolumn{6}{c}{CYP2E1}
B	TIME 0	TIME 0	TIME 60 with NADPH	TIME 60 with NADPH	TIME 60 no NADPH	TIME 60 no NADPH	TIME 0	TIME 0	TIME 60 with NADPH	TIME 60 with NADPH	TIME 60 no NADPH	TIME 60 no NADPH
	\multicolumn{6}{c}{CYP1A2}	\multicolumn{6}{c}{CYP3A4}
C	TIME 0	TIME 0	TIME 60 with NADPH	TIME 60 with NADPH	TIME 60 no NADPH	TIME 60 no NADPH	TIME 0	TIME 0	TIME 60 with NADPH	TIME 60 with NADPH	TIME 60 no NADPH	TIME 60 no NADPH
	\multicolumn{6}{c}{CYP2A6}	\multicolumn{6}{c}{CYP3A5}
D	TIME 0	TIME 0	TIME 60 with NADPH	TIME 60 with NADPH	TIME 60 no NADPH	TIME 60 no NADPH	TIME 0	TIME 0	TIME 60 with NADPH	TIME 60 with NADPH	TIME 60 no NADPH	TIME 60 no NADPH
	\multicolumn{6}{c}{CYP2B6}	\multicolumn{6}{c}{FMO1}
E	TIME 0	TIME 0	TIME 60 with NADPH	TIME 60 with NADPH	TIME 60 no NADPH	TIME 60 no NADPH	TIME 0	TIME 0	TIME 60 with NADPH	TIME 60 with NADPH	TIME 60 no NADPH	TIME 60 no NADPH
	\multicolumn{6}{c}{CYP2C8}	\multicolumn{6}{c}{FMO3}
F	TIME 0	TIME 0	TIME 60 with NADPH	TIME 60 with NADPH	TIME 60 no NADPH	TIME 60 no NADPH	TIME 0	TIME 0	TIME 60 with NADPH	TIME 60 with NADPH	TIME 60 no NADPH	TIME 60 no NADPH
	\multicolumn{6}{c}{CYP2C9}	\multicolumn{6}{c}{FMO5}
G	TIME 0	TIME 0	TIME 60 with NADPH	TIME 60 with NADPH	TIME 60 no NADPH	TIME 60 no NADPH	TIME 0	TIME 0	TIME 60 with NADPH	TIME 60 with NADPH	TIME 60 no NADPH	TIME 60 no NADPH
	\multicolumn{6}{c}{CYP2C19}	\multicolumn{6}{c}{HUMAN LIVER MICROSOMES}
H	TIME 0	TIME 0	TIME 60 with NADPH	TIME 60 with NADPH	TIME 60 no NADPH	TIME 60 no NADPH	TIME 0	TIME 0	TIME 60 with NADPH	TIME 60 with NADPH	TIME 60 no NADPH	TIME 60 no NADPH

Fig. 6 Plate map for recombinant human enzyme screen. Test compound is screened against 16 enzyme systems, including human liver microsomes and enzyme insect control. Incubations are carried out with and without NADPH (Regenerating system) and substrate depletion is compared with the Time 0 samples for each enzyme system

since one or a combination of these isoforms is involved in over 75 % of CYP-metabolized marketed drugs [10]. However, in this simple screen some of the other CYPs commonly, albeit to a lesser extent, involved in drug metabolism (CYPs 1A1, 2A6, 2E1 and 3A5) and the three FMO enzymes are included. Samples are incubated in duplicate, for 60 min, with and without NADPH regenerating system, and substrate depletion is compared with the starting amount at time 0. Human liver microsomes for comparison with the previous assays and a negative insect control (no CYP enzyme content), to assess possible non-CYP chemical degradation, are also included in the incubation plate (Fig. 6). A second plate containing a cocktail of well-characterized CYP-specific probe substrates (Table 4) is typically included in the assay. Substrate depletion or CYP-specific metabolite formation can be monitored (see Table 5 for available metabolite reference standards). Automation would significantly facilitate the execution of the assay. LC/MS analysis can monitor substrate depletion and potential product formation, and can provide an indication of which isoforms to investigate in subsequent assays.

Table 4
List of probe substrates (Note 11)

CYP	Substrate	Catalog number	Vendor
1A1	Theophylline[a]	T-1633	Sigma (St. Louis, MO)
1A2	Phenacetin	A-2500	Sigma (St. Louis, MO)
2A6	Coumarin	C-4261	Sigma (St. Louis, MO)
2B6	Bupropion	B-102	Sigma (St. Louis, MO)
2C8	Amodiaquine	A-2799	Sigma (St. Louis, MO)
2C9	Tolbutamide	T-0891	Sigma (St. Louis, MO)
2C19	S-(+)-Mephenytoin	457053	BD Biosciences (San Jose, CA)
2D6	Dextromethorphan	D-9684	Sigma (St. Louis, MO)
2E1	Chlorzoxazone	C-4397	Sigma (St. Louis, MO)
3A4/5	Midazolam Maleate	M-2419	Sigma (St. Louis, MO)
3A4/5	Testosterone	T-1500	Sigma (St. Louis, MO)
FMO	Clozapine	C-6305	Sigma (St. Louis, MO)
FMO	Cimetidine	C-4522	Sigma (St. Louis, MO)

[a]Theophylline is a substrate for both CYP1A1 and CYP1A2. However at higher substrate concentrations (>10 µM), preferential catalysis of the N-3-demethylation reaction pathway by CYP1A1 is observed [11]

Table 5
List of metabolite reference standards

CYP	Substrate	Catalog number	Vendor
1A2	Acetaminophen	A-7085	Sigma (St. Louis, MO)
2A6	7-Hydroxycoumarin	U-7626	Sigma (St. Louis, MO)
2B6	Hydroxybupropion	H-3167	Sigma (St. Louis, MO)
2C8	Desethylamodiaquine	451782	BD Biosciences (San Jose, CA)
2C9	Hydroxytolbutamide	UC160	Sigma (St. Louis, MO)
2C19	4-Hydroxymephenytoin	UC126	Sigma (St. Louis, MO)
2D6	Dextrorphan	UC205	Sigma (St. Louis, MO)
2E1	6-Hydroxychlorzoxazone	UC148	Sigma (St. Louis, MO)
3A4	1'-Hydroxymidazolam	UC430	Sigma (St. Louis, MO)
3A4/5	6ß-Hydroxytestosterone	451012	BD Biosciences (San Jose, CA)
3A5	4-Hydroxymidazolam	UC431	Sigma (St. Louis, MO)
FMO	Clozapine-N-oxide	C-0832	Sigma (St. Louis, MO)

3.3.1 Assay Procedure

1. The assay is set up as 120 µL incubations at concentrations of 40 pmol/mL for the recombinant CYP enzymes, 200 µg/mL for the FMO enzymes (Table 6) and 0.2 mg/mL for the HLM. All enzymes are prepared in 100 mM KPB, pH 7.4.

2. Prepare the test compound as a 1 mM stock in a suitable organic solvent (preferably methanol, DMSO, acetonitrile or a combination)

3. Prepare the assay termination plate containing acetonitrile spiked with internal standard as in Sect. 3.1.

4. Thaw the pre-aliquoted enzymes (**Note 13**) and add of 900 µL of KPB which will yield the proper final incubation concentration.

5. Spike the test compound into the enzyme suspensions to yield the 1 µM incubation concentration (in this step, 1 µL into 900 µL of enzyme from step 3).

6. Aliquot the spiked microsomes into the incubation vessels (100 µL per sample).

7. Preincubate the samples for 5 min in the shaking incubator or water bath thermostatted at 37 °C.

8. Prepare co-factor mixture (NRS) to initiate reactions. Blank KPB will be added to the −NADPH samples

Table 6
List of recombinant CYP and FMO enzymes from BD Biosciences (Note 12)

CYP	Catalog number	With cytochrome b5
1A1	456211	No
1A2	456203	No
2A6	456254	Yes
2B6	456255	Yes
2C8	456252	Yes
2C9*1	456258	Yes
2C19	456259	Yes
2D6*1	456217	No
2E1	456206	Yes
3A4	456202	Yes
3A4/5	456256	Yes
FMO1	456241	No
FMO3	456233	No
FMO5	456245	No
Control	456244	Yes

Clozapine Turnover in rhCYPs and FMOs

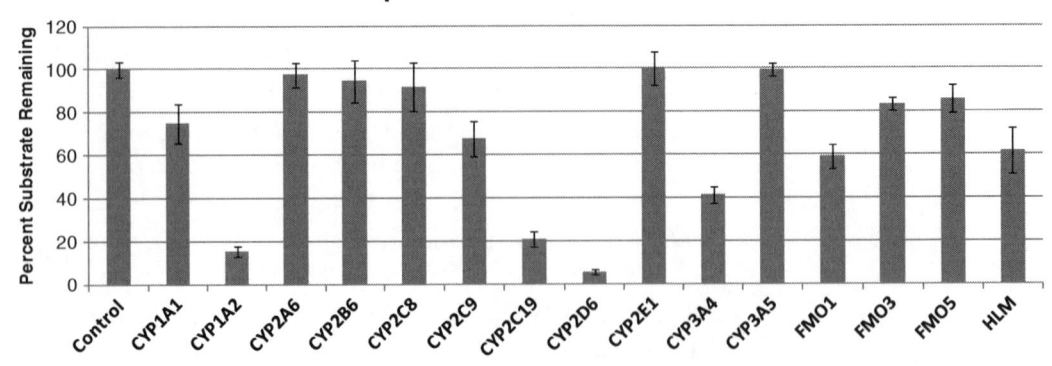

Fig. 7 In vitro reaction phenotyping for clozapine. Clozapine was incubated at 1 μM in recombinant single human enzyme systems (40 pmol/mL CYPs, 200 μg/mL FMOs, 0.2 mg/mL HLM) and substrate depletion was monitored by LC/MS to determine which isoforms are involved in substrate turnover

9. Initiate the reactions by the addition of the NRS mixture or KPB (20 μL).
10. Immediately quench the time zero sample to the termination plate which will contain at least 300 μL acetonitrile spiked with the selected internal standard (**Notes 2** and **8**).
11. After 60 min, quench the remaining samples in the same manner.
12. Filter and prepare the terminated samples as in the previous assay (Sect. 3.1).

3.3.2 Analytical Method and Data Interpretation

Being a qualitative assay, the objective is to determine which enzymes are involved in the metabolism of the test compound. Any NADPH-dependent depletion of substrate can be interpreted as enzymatic turnover and will provide an initial guideline of which enzymes appear to be the major contributors to metabolic clearance and which could be alternative or secondary pathways. Nonetheless, accurate predictions of *in vivo* profiles start by identifying the *in vitro* pathways. At this stage, conclusions as to the contributions of each pathway to the overall systemic clearance of the test compound can be premature. Indeed, in the example in Fig. 7, in vitro reaction phenotyping of clozapine with recombinant enzymes indicates that CYPs 1A1, 1A2, 2C9, 2C19, 2D6, 3A4 and the FMOs are all involved in substrate turnover, with depletion ranging from 15 to 95 % after 60 min. Although CYP2D6 and CYP1A2 appear to be the major enzymes involved using single enzyme systems, the contribution by CYP3A4 was found to be a significant pathway for the formation of both major metabolites in vitro [12] and in vivo [13]. In addition, while FMO1 showed higher turnover among the recombinant FMOs, FMO3, more abundant in native microsomal systems [14] is the primary isoform involved [15].

CYP Versus FMO Metabolism

Nowadays, the availability of recombinant enzyme systems facilitates the task of differentiating between CYP and FMO pathways. However, in case of oxidation of a heteroatom, such as nitrogen of sulfur, some techniques can be implemented to confirm the involvement of FMOs, such as exposure to carbon monoxide, which, unlike CYPs, does not affect the FMO catalytic cycle [16], or incubating at increased pH levels (up to pH 8.8). Loss of FMO-specific activity on the other hand can be accomplished following thermal inactivation at temperatures higher than 45 °C, for 1 min, in the absence of NADPH, [17].

3.4 Recombinant CYP Kinetic Parameters Determination

The isoforms identified in the turnover of the test compound can be evaluated further by establishing the kinetic parameters that provide insight on the potential to saturate or limit the metabolic clearance pathways. These parameters are useful when determining the contribution of the particular enzyme in native systems such as microsomes or hepatocytes. Classic protocols to determine the K_m value involve initial velocity measurements to quantify the amount of product formed as the initial substrate concentration is varied [18]. When the substrate concentration incubated is significantly lower than the K_m ($[S] \ll K_m$) then the Michaelis-Menten equation can be reduced to (2):

$$V = \frac{V_{max}}{K_m} \times [S] \qquad (2)$$

Typically, without an authentic metabolite reference standard, as is usually the case in Drug Discovery, measurements of V_{max}, i.e. the maximum rate of product formation for a particular reaction, are problematic. However, CL_{int} estimates can be obtained from the single substrate concentration method. Since

$$CL_{int} = V_{max} / K_m \qquad (3)$$

once a value for CL_{int} is established from the single concentration substrate depletion method (Sect. 3.1), determination of K_m can then allow for calculation of V_{max}. A variation of the in vitro $T_{1/2}$ method allows for an accurate K_m determination of the overall metabolic clearance of a test compound [19, 20].

3.4.1 Assay Procedure

The assay is set up as the single substrate concentration assay, (Sect. 3.1), but with the variable of incubating multiple, serially-diluted substrate concentrations (either 8 or 12 concentrations, **Note 14**). Incubation volume should be 500 µL at a 0.2 mg/mL microsomal concentration in KPB, from which 40 µL aliquots will be removed at each time point. Reactions will be initiated by the addition of NRS. At least duplicate samples should be incubated.

1. Prepare the test compound as a 10 mM stock in a suitable organic solvent (preferably methanol, DMSO, acetonitrile or a combination) (**Note 15**).
2. Prepare the previously selected internal standard stock solution (**Note 2**). The internal standard should then be spiked in acetonitrile which will be added to the termination plate. The concentration of internal standard should be similar (within twofold) to the expected test compound concentration.
3. Prepare the termination plate (either Costar 96-deep-well plate or Flexi-Tier 1.0 mL glass inserts) with 300 µL of acetonitrile containing the internal standard. Keep refrigerated until ready to start the assay.
4. Thaw and dilute the human liver microsomes (HLM) with KPB. Microsomes should be diluted to yield a final 0.2 mg/mL incubation concentration (proper calculations to ensure the correct final incubation concentration should be carried out, if necessary).
5. Serially dilute the test compound with the same organic solvent used to prepare the stock solution.
6. Spike the test compound concentrations into the microsomal suspension. Mix repeatedly with a pipette to avoid clouding of the microsomal suspension following the spike of the higher test compound concentrations.
7. Aliquot the spiked microsomes into the incubation vessels to yield duplicate samples.
8. Preincubate the samples for 5 min in the shaking incubator or water bath thermostatted at 37 °C.
9. Prepare the NRS mixture to initiate reactions.
10. Initiate the reactions by the addition of the co-factor mixtures.
11. Immediately transfer the time 0 aliquot (40 µL) to the termination plate which will contain at least 300 µL acetonitrile spiked with the selected internal standard (**Notes 2 and 8**).
12. Subsequent time points will be quenched in the same manner.
13. Filter and prepare the terminated samples as in the previous assay (Sect. 3.1).

3.4.2 Analytical Method and Data Interpretation

For each substrate incubation level, percent turnover is plotted on a log-linear scale as a function of time, yielding rates for substrate depletion represented by the negative slope, k_{dep}. The rate values are then plotted as a function of substrate concentration ([S]), yielding a sigmoidal curve as depletion transitions from first order kinetics at low [S] to zero-order kinetics as [S] increases past the K_m value (Fig. 8). The inflection point in the plot corresponds to the K_m value, at which point the rate of depletion is 50 % of the

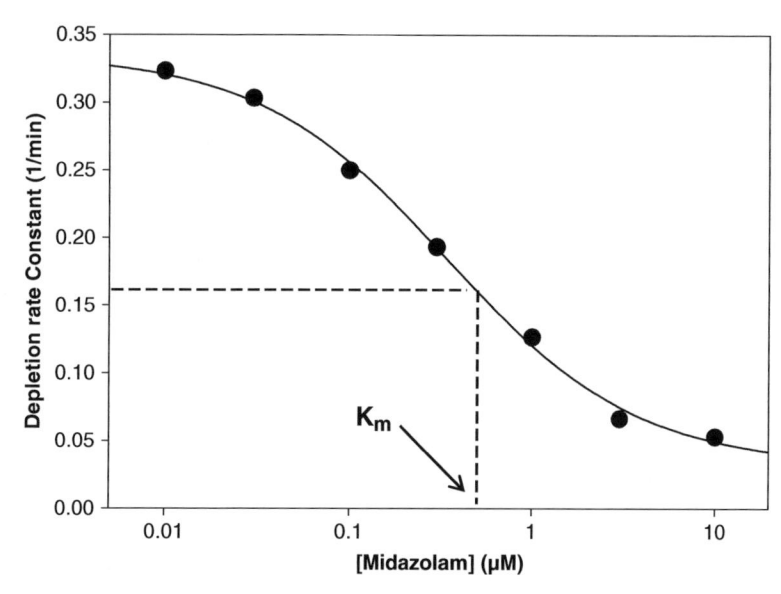

Fig. 8 Determination of K_m of CYP3A4-catalyzed midazolam hydroxylation by substrate depletion. Serially diluted midazolam concentrations were incubated in human liver microsomes. Depletion rate constants, k_{dep}, were determined by the T1/2 method and plotted as a function of substrate concentration to determine the substrate K_m value. Figure design from reference [19]

theoretical maximum rate at an infinitesimally low [S]. Once the K_m is estimated, together with the previously determined CL_{int} value, an estimated V_{max} value can be calculated from (3).

3.5 Relative activity factors (RAF)

Because of the possible differences in the catalytic activity between recombinant enzymes and the corresponding CYP in microsomes, a normalization method can be used to improve quantitative scaling predictions of the relative contributions of specific enzymes in the clearance of a test compound. Mechanistic elucidations of this nature are probably unlikely at the discovery stage of drug development but should be considered as potentially valuable tools in lead optimization when mitigating certain metabolic liabilities may be a critical aspect in advancing a program.

In principle, relative activity factors (RAF) establish a correlation between the activity in recombinant enzymes and the same enzymatic activity observed in a particular pool of microsomes [21]. This correlation is initially measured using a CYP-specific probe substrate and is then extended to the data set acquired for the test compound. RAFs can be determined from the product formed via a specific pathway (e.g. CYP3A4-catalyzed testosterone 6ß-hydroxylation) or from substrate depletion, so special consideration should be given to the selection of the probe substrate to avoid contribution from other enzymes. Activity in either enzyme system is compared at saturating probe substrate concentrations (i.e. V_{max}) and the RAF is derived from the ratio between the two systems as shown in (4), [22].

$$\text{RAF} = \frac{V_{MAX,HLM}}{V_{MAX,rhCYP}} \quad (4)$$

For the test compound, if V_{max} is undetermined, substrate depletion ($T_{1/2}$) CL_{int} values can be substituted [23]. Test compound clearance should be determined in each recombinant single enzyme system previously identified as a metabolic pathway (Sect. 3.3) and in HLM (Sect. 3.1). For the probe substrate, if RAF is determined from V_{max} values, the CL_{int} ratio can be used optionally as a confirmatory tool.

Whether determined from V_{max} values or CL_{int}, values, RAFs will be expressed in units of $pmol_{CYP}$/mg HLM protein as a scaling factor between recombinant enzymes and microsomes. It can then be used to predict the contribution of a particular enzyme to the metabolic clearance of the test compound in microsomes. The RAF is multiplied by the corresponding $CL_{int\,rhCYP}$ value obtained for the test compound. As an example, (5) shows the calculation to determine the contribution by CYP3A4 in the overall microsomal CL_{int} value:

$$CL_{int\,HLM}\,f_{3A4} = RAF_{3A4} \times CL_{int\,rh3A4} \quad (5)$$

where $CL_{int\,HLM}\,f_{3A4}$ represents the contribution by CYP3A4 to the overall intrinsic clearance in HLM.

3.5.1 Assay Procedure

The assay is set up as a two-step determination to include both the test compound (for which typically the CL_{int} value is determined) and the probe substrate (for which either CL_{int} or V_{max} values, or both, can be determined, whether the metabolite reference standard is available).

CL_{int} value determination can be carried out by the substrate depletion method (Sect. 3.1). V_{max} values can be determined by monitoring product formation for a specific metabolic pathway and for which the metabolite reference standard is commercially available for quantification. To determine V_{max}, both incubation time and microsomal protein concentration are kept constant and should be chosen at the lower end of the range established by the determination of initial linear conditions (Sect. 3.2). Under these conditions, substrate concentration is varied and incubated over a range that encompasses the estimated K_m value and can reasonably be expected to achieve saturating kinetics (i.e. velocity or rates of reaction reach a plateau, that is, V_{max}).

Assay Procedure: V_{max} Determination

1. The assay is set up as 120 μL incubations under initial linear conditions (e.g. 0.2 mg/mL HLM or 40 pmol/mL recombinant CYP and 10 min incubations). The substrate will be serially

diluted with organic solvent and then spiked into diluted HLM or recombinant enzyme. The spiked enzymes will then be aliquoted (100 μL) into the incubation vessels. Reactions will be started by the addition of a 20 μL-aliquot of NRS. Samples should be incubated in at least duplicate, preferably triplicate. All enzymes are prepared in 100 mM KPB, pH 7.4.

2. Prepare the test compound as a 10 mM stock in a suitable organic solvent (preferably methanol, DMSO, acetonitrile or a combination).

3. Prepare the assay termination plate containing acetonitrile spiked with internal standard as in Sect. 3.1.

4. Thaw and dilute the HLM and recombinant enzymes with KPB according to the final incubation protein concentration. Aliquot 300 μL (for duplicates) or 400 μL (for triplicates) for each substrate concentration level.

5. Serially dilute the substrate with the same organic solvent with which the stock solution was prepared and spike into the diluted enzymes (no more than 2 μL for each concentration level, in order to keep the organic content below 0.5 %).

6. Aliquot the spiked enzymes into the incubation vessels (100 μL per sample).

7. Preincubate the samples for 5 min in the shaking incubator or water bath thermostatted at 37 °C.

8. Prepare co-factor mixture (NRS) to initiate reactions.

9. Initiate the reactions by the addition of the NRS mixture or KPB (20 μL).

10. Quench the reactions at the appropriate time (e.g. 10 min) by transferring to the termination plate which will contain at least 300 μL acetonitrile spiked with the selected internal standard. (**Note 16**)

11. Filter and prepare the terminated samples as in the previous assay (Sect. 3.1).

3.5.2 Analytical Method and Data Interpretation

Monitoring product formation by LC/MS can be accomplished in a straightforward fashion, especially if metabolite reference standard is available to develop the chromatographic method and optimize ionization conditions. A search of the literature to gather information on recent methodologies may be advisable.

Plotting rates of product formation (in either pmol/min/mg protein for HLM or pmol/min/pmol CYP for the recombinant enzymes) as a function of substrate concentration will yield the classic Michaelis-Menten hyperbolic plot from which V_{max} and K_m can be determined (Fig. 9).

Once the V_{max} values for the probe substrate are determined, the RAF can be calculated and the percent contribution of the

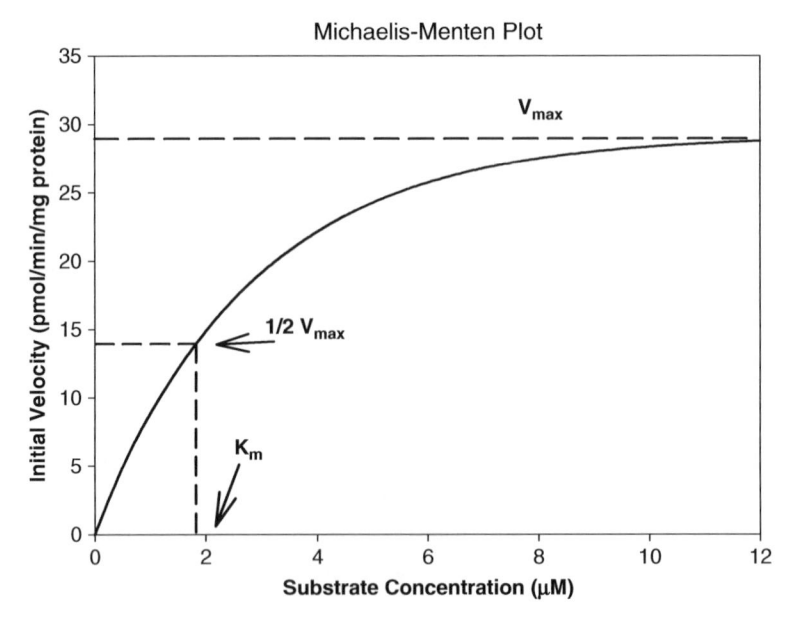

Fig. 9 Classic Michaelis-Menten plot for enzymatic reactions. V_{max} represents the maximum rate of reaction at saturating substrate concentrations. The K_m value (Michaelis constant) represents the substrate concentration yielding half V_{max} and is a measure of the intrinsic affinity of a substrate for an enzyme

same enzyme to the metabolism of the test compound estimated. The following example provides an overview of the calculations involved in assessing the contributions of two pathways to the overall metabolic clearance of a test compound.

Example (**Note 17**) For test compound X, CYP3A4 and CYP2D6 were found to the major CYPs involved and using recombinant enzymes, the following CL_{int} values were obtained:

$$CL_{int} rh3A4 = 50 \ \mu L / min / pmol_{CYP}$$

$$CL_{int} rh2D6 = 40 \ \mu L / min / pmol_{CYP}$$

The probe substrates, testosterone and dextromethorphan were used for CYP3A4 and CYP2D6, respectively. V_{max} values were determined in HLM and the corresponding recombinant enzyme for each substrate.

For CYP3A4 (testosterone 6ß-hydroxylation):

$$V_{max} \text{ in HLM} = 4750 \text{ pmol} / \text{min} / \text{mg protein}$$

$$V_{max} \text{ in rh3A4} = 180 \text{ pmol} / \text{min} / pmol_{CYP}$$

The RAF is calculated from the V_{max} ratio ((4), Sect. 3.5):

$$RAF = 4750 / 180 = 26.4 \text{ pmolCYP3A4} / \text{mg HLM protein}$$

For CYP2D6 (dextromethorphan O-demethylation):

V_{max} in HLM = 350 pmol / min/ mg protein

V_{max} in rh2D6 = 45 pmol / min/ $pmol_{CYP}$

The RAF is calculated from the V_{max} ratio:

RAF = 350 / 45 = 7.8 pmolCYP2D6 / mg HLM protein

These RAF values can then be applied to the CL_{int} values for Compound X, to estimate the CYP3A4 and 2D6 contributions to the overall metabolic clearance in HLM ((5), Sect. 3.5):

For CYP3A4:

26.4 $pmol_{CYP}$ / mg HLM × 50 μL / min/ $pmol_{CYP}$ = 1320 μL / min/ mg HLM

For CYP2D6:

7.8 $pmol_{CYP}$ / mg HLM × 40 μL / min/ $pmol_{CYP}$ = 312 μL / min/ mg HLM

Assuming these two pathways to be the only ones contributing to clearance, it can be estimated that the contribution by CYP3A4 is 1,320/(1,320+312)=81 % of total clearance while the contribution by CYP2D6 is 312/(1,320+312)=19 %.

These estimates can be compared to total clearance measured in HLM and can eventually be confirmed through the use of chemical inhibitors or monoclonal antibodies, as discussed in the next section.

3.5.3 Other Scaling Factors

Other scaling approaches rely on immuno-quantification of each CYP isoforms in a pool of microsomes to determine the relative abundance. The data obtained in recombinant enzymes are then scaled up by multiplying by the relative abundance factor in native microsomes [24, 25]. Table 7 shows mean expression levels and relative abundances of the major human CYPs in both liver and intestine.

More recently, the use of intersystem extrapolation factors (ISEFs) has expanded as population-based pharmacokinetic (PBPK) modeling and simulation programs such as SimCYP and GastroPlus have become widely accepted as sophisticated tools for predicting *in vivo* exposure levels and interactions. ISEFs are dimensionless and include the immuno-quantified abundance of a specific CYP in an HLM sample. Since determination of ISEFs was initially reported from product formation experiments [26] applications in drug discovery were limited. However, a new approach involving CL_{int} determination by substrate depletion [27], provide new possibilities.

3.6 CYP-Specific Chemical Inhibitors and Monoclonal Antibodies

An additional, confirmatory tool for estimating relative contributions of specific metabolic pathways is the use of chemical inhibitors or inhibitory monoclonal antibodies. By selectively inhibiting a pathway, the change in rates of reaction or clearance of a test

Table 7
Range of expression levels and relative abundances of human CYP isoforms in the liver and intestine

CYP	Liver [28] Mean (pmol/mg protein) Min	Max	Mean relative abundance (%) Min	Max	Intestine [29] Range (pmol/mg protein)	Mean RA (%)
CYP1A1					3.6–7.7	7.4
CYP1A2	19	67	7.5	13	BLD	BLD
CYP2A6	14	68	5.5	13	BLD	BLD
CYP2B6	1	45	0.4	8.4	BLD	BLD
CYP2C8	12	64	4.5	12	BLD	BLD
CYP2C9	50	96	20	18	2.9–27	11
CYP2C19	8	20	3.1	3.7	<0.6–3.9	1.3
Total CYP2C	60	64	24	12		
CYP2D6	5	11	2	2.1	<0.2–3.1	0.66
CYP2E1	22	52	8.6	9.8	BLD	BLD
CYP3A4	37	108	15	20	8.8–150	57
CYPA5	1	117	0.4	22	4.9–25	21
Total CYP3A	96	262	38	49		
Total CYP	255	534				

BLD below limit of detection

compound should correspond to the contribution of that particular pathway to the overall clearance. However, for an accurate assessment, particular care has to be given to the selection of inhibitors or antibodies, the concentration levels used and the incubation times (Table 8). Using excessively high concentrations of inhibitor may potentially lead to inhibition of additional or different isoforms, thus affecting the interpretation of the data generated. For example, ketoconazole is a very potent CYP3A4 inhibitor at submicromolar concentrations ($K_i = 20$ nM); at higher concentrations (>5 µM) it can also have a progressively more potent inhibitory effect of several other CYPs, including CYPs 1A2, 2C9 and 2C19 [30]. Identification of a more selective CYP3A4 inhibitor, azamulin, which exhibits less inhibitory cross-reactivity with other isoforms than ketoconazole [31] provides a valid alternative when phenotyping CYP3A4-catalyzed pathways.

Monoclonal antibodies (MAbs), typically ascites based preparations from mice against specific CYP isoforms, offer in some

Table 8
CYP-specific substrates and chemical inhibitors

CYP	Substrate	Inhibitor	Ki (nM)	Inhibitor concentration[a]
1A2	Phenacetin	α-Naphthoflavone	10	500 nM
2A6	Coumarin	Tranylcypromine	300	5 μM
2B6	Bupropion	Ticlopidine[b]	200	1 μM
2C8	Amodiaquine	Montelukast[c]	50	0.1 μM
2C9	Tolbutamide	Sulphaphenazole	300	2 μM
2C19	S-(+)-Mephenytoin	N-3-Benzylnirvanol	250	1 μM
2D6	Dextromethorphan	Quinidine	100	1 μM
2E1	Chlorzoxazone	4-Methylpyrazole	500	10 μM
3A4/5	Testosterone/Midazolam	Ketoconazole	20	500 nM
3A4/5	Testosterone/Midazolam	Azamulin[d]	200	1 μM

All Inhibitors available from Sigma Aldrich (St. Louis, MO)
[a]Concentrations for CYP Reaction Phenotyping to achieve >90 % Inhibition of the particular pathway
[b]Ticlopidine is also a Time-Dependent Inhibitor of CYP2C19
[c]Protein binding is a significant factor with Montelukast. Adjustments should be made depending on the HLM concentration used [33]
[d]Azamulin is a Time-Dependent Inhibitor of CYP3A4. Its inhibitory strength will increase if pre-incubated (10–20 min) with NADPH prior to addition of the CYP3A4 substrate

cases a very potent tool for specific inhibition. Microsomes pre-treated with a specific antibody will lose up to greater than 90 % of the particular CYP activity. A significant advantage of MAbs over chemical inhibitors is the lack of competitive kinetic substrate/inhibitor interactions, which may be susceptible to fine-tuned concentration levels or possible depletion or turnover over time. Treatment with MAbs essentially lowers the quantity of available enzyme, reducing V_{max}, and is generally an irreversible process. Drawbacks of MAbs are the cost, substantially higher than chemical inhibitors, and the fact that in some cases (e.g.CYP3A4) cross-reactivity with similar isoforms (i.e. CYP3A5) will not allow the activity to be quenched beyond a certain limit, which could be less than 80 % inhibition. Without the possibility of driving the equilibrium to increase inhibition, as may be possible with chemical inhibitors, lack of sufficient inhibition of a pathway could lead to uncertainty in the interpretation of phenotyping data.

Whether chemical inhibitors, MAbs or a combination [32] are used in the reaction phenotyping of a test compound, it is advisable to examine the inhibitory efficiency by establishing concentration- and time-dependent titration curves and to test the inhibition of a known, well-characterized marker to gain perspective on the extent of inhibition possible in a particular microsomal pool.

3.6.1 Assay Procedure

Only the procedure for monoclonal antibodies will be presented here. Incubation with chemical inhibitors is either a straightforward substrate depletion assay (Sect. 3.1) or product formation (kinetic) assay (Sect. 3.5.1.1). In either case, the inhibitor can be added at a single concentration or at serially diluted concentrations to obtain an inhibition titration curve. In addition, incubations can be quenched at a single time point, or multiple time points, establishing a time course for either substrate depletion or metabolite formation. The proper negative control (with and without MAb and/or NRS) and probe substrate positive controls should be included in the assay design.

Assay Procedure: Monoclonal Antibodies (Note 18)

1. Incubation volumes will be 500 μL at 0.2 mg/mL HLM and a single substrate concentration. Incubations will be prepared by adding 30 μL of treated HLM (diluted 2-fold after treatment), 440 μL of incubation buffer spiked with the test compound or control substrate and 30 μL of NRS which will be added to start the reactions. Aliquots will be quenched at 0, 10, 20, 30, 45 and 60 min. Incubations will be run in duplicate, with and without MAb and with and without NRS, in 100 mM KPB, pH 7.4, for both the test compound and a probe substrate control (e.g. testosterone). The blank vehicle for the MAb is 25 mM Tris buffer, pH 7.5, which is included in the MAb kit and can also be used to dilute the MAb, if necessary.

2. Microsomes are pre-treated with MAb for 20 min on ice prior to addition of substrate and incubation buffer. The MAb is sold as a 10 mg/mL protein suspension and will be added to HLM (20 mg/mL) at a 2:1 ratio (v/v) to yield a 1:1 weight ratio. In this assay, 100 μL HLM will be added to either 200 μL MAb or 200 μL blank Tris buffer (**Note 19**).

3. Prepare the test compound and control as a 1 mM stocks in suitable organic solvents (preferably methanol, DMSO, acetonitrile or a combination).

4. Prepare the assay termination plate containing acetonitrile spiked with internal standard as in Sect. 3.1.

5. Spike the two substrates in KPB, if possible, to yield 1 μM incubation concentrations, (see Step 2). Prepare sufficient volume to aliquot (440 μL) for each incubated sample. If solubility limitations prevent the compound from dissolving in KPB, add the treated HLM to blank KPB, prior to spiking with the substrate.

6. Aliquot to the incubation vessels (440 μL, without HLM, per sample).

7. Spike the MAb-treated HLM (30 μL) in each incubated sample.

8. Preincubate the samples for 5 min in the shaking incubator or water bath thermostatted at 37 °C.

9. Prepare co-factor mixture (NRS) to initiate reactions.

10. Initiate the reactions by the addition of the NRS mixture or KPB (30 μL).

11. Immediately transfer the time 0 aliquot (40 μL) to the termination plate which will contain at least 300 μL acetonitrile spiked with the selected internal standard (**Notes 2** and **8**).

12. Subsequent time points will be quenched in the same manner.

13. Filter and prepare the terminated samples as in the previous assay (Sect. 3.1).

3.6.2 Data Interpretation

Plotting the substrate depletion or product formation data as a function of incubation time for the samples incubated with and without MAb, the difference in CL_{int}, i.e. the relative contributions of each CYP to the overall clearance of the test compound can be determined. In the case of substrate depletion, the rate constant, k, can be used as a surrogate of CL_{int} in (6) to determine the contribution by CYP3A4 (f_{3A4}) to the overall clearance in HLM:

$$CL_{int\,HLM}\,f_{3A4} = (CL_{int\text{-}MAb3A4} - CL_{int+MAb3A4})\,/\,CL_{int\text{-}MAb3A4} \quad (6)$$

In Fig. 10, for test compound X used in the example in Sect. 3.5.2, (for which the RAF method predicted 81 and 19 % contributions from CYP3A4 and CYP2D6, respectively), incubations with the CYP3A4 MAb reduced CL_{int} by 86 % ((0.076−0.011)/0.076), whereas incubations with the CYP2D6 MAb showed a lesser reduction in CL_{int}, (11 %). Therefore, the predominant contribution by CYP3A4 to the metabolic clearance of compound X, is confirmed, in agreement with the numbers derived from the RAF calculations.

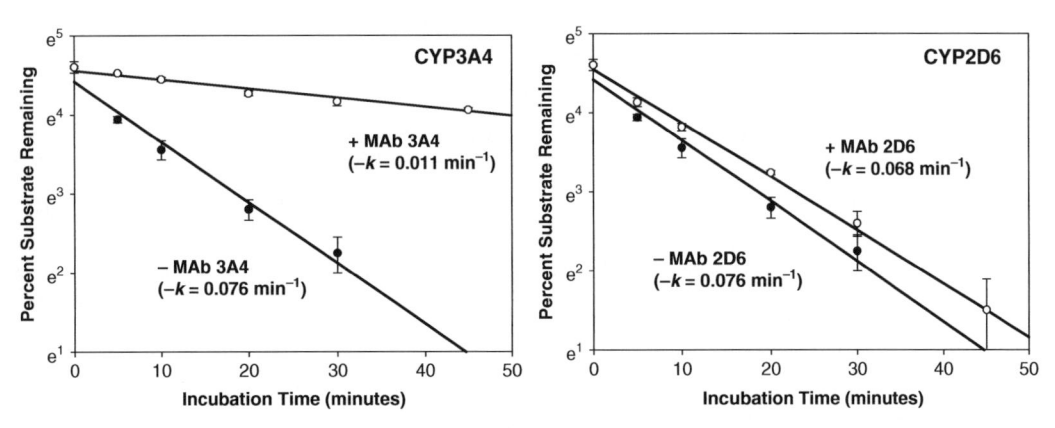

Fig. 10 Incubations of test compound X with human liver microsomes treated with inhibitory monoclonal antibodies against CYP3A4 and CYP2D6. By inhibiting a specific metabolic pathway, the change in rate constant, k (as a surrogate of CL_{int}) represents the estimated contribution of that particular pathway to the overall metabolic clearance of test compound X. Inactivating CYP3A4 leads to a more significant change in k than inactivation of CYP2D6. The −Mab curves are the same for both CYP3A4 and CYP2D6

4 Conclusion

Reaction phenotyping in drug discovery can be limited by the lack of radiolabeled analogs or metabolite reference standards. In addition, information about therapeutic exposure levels, *in vivo* PK profiles and routes of elimination may be partial or not available. Nonetheless, the *in vitro* tools available today allow for a confident screening of drug candidates to identify potential liabilities before progressing into development.

5 Notes

1. The commercial reagents, consumables and equipment listed in this chapter are an immediate example of specific vendor source information that can be used for CYP reaction phenotyping assays.

2. An appropriate analytical internal standard should be identified and used in the LC/MS analysis of the test compound throughout the reaction phenotyping strategy. Ideally, it should be a stable label (i.e. deuterated or 13C) or structural analog of the compound being screened, in order to ensure chromatographic elution and MS ionization consistently similar to the test compound. If an analog is not available, another compound with well-tested solubility and LC/MS properties can be used. Some method development, to evaluate ionization signal strength and variability is recommended. It should be prepared in a suitable organic solvent (methanol, DMSO or acetonitrile) and spiked in acetonitrile which will be added to the assay termination plate, prior to quenching the incubated samples. The final concentration of the internal standard should be similar (within twofold) to the expected test compound concentration in the post-incubated samples.

3. Microsomal or recombinant enzyme incubations are typically carried out at 37 °C in 100 mM phosphate buffer at pH 7.4, supplemented with 5 mM magnesium chloride and 1 mM EDTA. In order to ensure maintaining the proper molarity when adjusting the pH, 100 mM monobasic and dibasic potassium phosphate solutions can be prepared separately and then the monobasic solution can be used to adjust the pH of the starting dibasic solutions (typically around pH 9.2) down to pH 7.4. This can be easily done with a simple pH meter with a standard glass electrode. Once the 100 mM potassium phosphate buffer, pH 7.4, is prepared, it can be transferred to a 1,000 mL volumetric flask containing 5 mL 1.0 M magnesium chloride solution and 1 mL 1.0 M EDTA solution.

This phosphate buffer can be stored at 4 °C for up to 2 months and should be checked at room temperature for cloudiness before use. In addition, it is recommended to sparge buffer with oxygen for several minutes when freshly prepared and before each use.

4. The NADPH Regenerating System referred to here, is purchased from BD Biosciences. It consists of two Solutions: Solution A is a 20× mixture of 26 mM NADP+ and 66 mM Glucose-6-phosphate; Solution B contains 40 U mL of Glucose-6-phosphate dehydrogenase in 5 mM sodium citrate. They are stored at −20 °C or below and prepared in a 5:1 ratio (i.e. 25 μL solution A and 5 μL solution B for a 500 μL incubation volume), yielding incubation concentrations of 1.3 mM NADP+, 3.3 mM Glucose-6-phosphate and 0.4 U.mL of Glucose-6-phosphate dehydrogenase. It can be used to initiate the enzymatic reactions. An alternative to the NRS from BD Biosciences is RapidStart™ from Xenotech, LLC. Compared to adding simple, reduced NADPH to an incubation mix, NADPH regenerating systems provide a more consistent level of NADPH over time, without the risk of depletion, which could ultimately affect the kinetics of a reaction.

5. The UGT Reaction Mix referred to here, is purchased from BD Biosciences. It consists of two solutions; Solution A is a 12.5× concentrate of UDPGA, (for a final 2 mM incubation concentration) in water. UDPGA is the essential cofactor to promote UGT-catalyzed glucuronidation. Solution B is a 5× mixture of 0.125 mg/mL alamethicin, 250 mM Tris-HCl and 40 mM $MgCl_2$. Solution B is typically added 5–10 min prior to addition of UDPGA, to improve rates of glucuronidation via the pore-forming action of alamethicin.

6. The final organic solvent content in a microsomal incubation should not be greater than 0.5 %, (preferably less than 0.3 %). Higher organic content can have an inhibitory effect of the activity of many CYP enzymes [34, 35]. In addition to the absolute level of organic solvent, it is paramount that all samples, especially in kinetic or semi-quantitative assays, have the same organic content, even at varying substrate concentrations.

7. The incubation vessels are Flexi-Tier 96-well inserts from Analytical Sales and Services. They are available in 0.5, 1 and 1.5 mL volumes in glass, silanized glass or polypropylene. The screened compound should be tested for non-specific binding to either glass or plastic prior to the assay.

8. In any time-dependent assay, an accurate measurement of time 0 conditions is essential in order to compare later time points to the initial levels. In CYP phenotyping, quenching the time 0 point immediately after adding the cofactor is preferable to quenching

and then adding the cofactor. Potentially, this could be due to non-specific binding, as the NRS itself contains some protein and adding before denaturing (with the quenching solvent) could affect the rate of binding of the test compound. Quenching the reaction as shortly after adding the cofactor as possible (a few seconds) is therefore preferable and yields more consistent results.

9. While this assay benefits from automation in terms of precision and ease of execution, it can also be carried out manually using a shaker water bath, such as a Dubnoff shaking incubator. The use of multi-channel electronic pipettes (e.g. Biohit) is recommended for transferring aliquots to the termination plate.

10. Automation is paramount for this assay, as a large number of samples are generated by varying three parameters simultaneously. If executed manually, the best approach is to proceed by incubating a single protein concentration with the serially diluted test compound concentrations to establish the linear range of substrate depletion as a function of time. Testing two or three protein concentrations in this fashion should suffice.

    If automated, the most efficient way to set up the assay is to dilute the HLM in the x-direction and the test compound in the y-direction. Aliquots of each HLM concentration are then spiked with the serially diluted substrate. Once started by the addition of NRS, reaction aliquots are transferred to separate termination plates (one per time point). There is no need to include a negative (-NADPH) control since the objective is to establish the linear range. The number of protein and substrate levels screened should be at a minimum 4 in order to assess linearity satisfactorily over at least four time points.

11. Testosterone is a schedule III controlled substance in the United States. A controlled substance license is required to work with this material and must be obtained from the Drug Enforcement Agency.

12. Unless specified, assays presented in this chapter were developed using recombinant CYP Supersomes from BD Biosciences. In addition to BD Biosciences, Cypex Limited (Dundee, Scotland), through Xenotech LLC (Lenexa, KS) in the United States, offers a wide assortment of recombinant human enzymes (Bactosomes).

13. Commercial recombinant enzymes generally are available in 0.5 or 1.0 nmol/0.5 mL aliquots. To avoid repeated freeze/thawing, a vial is pre-aliquoted (45 µL/aliquot) and stored at −80 °C. Addition of 900 µL of KPB yields enough enzyme for the assay

14. In kinetic assays, reaching the highest substrate concentration is essential to attain enzymatic saturation. Prior to the assay, test compound solubility should be tested to establish the

highest potential incubation concentration. A target for the highest concentration would be 100 μM, which would then be serially diluted to approximately 100 nM, a four orders of magnitude range.

15. Stock concentration of the test compound should be higher than that used in the previous assays, since concentration will be varied over a wide enough range to ensure reaching saturation of enzyme.

16. In this particular assay, since all incubations are quenched at one time point, it is optional to add acetonitrile (preferably with the internal standard) directly to the incubation vessel rather than transferring the incubation to the termination plate.

17. Further information on RAF calculations can be found in the BD Biosciences Application Note "Cytochrome P450 Enzyme Mapping in Drug Discovery using BD Supersomes Enzymes" [36].

18. The monoclonal antibody assay is set up using BD Biosciences Inhibitory Monoclonal Antibodies. The vendor protocol calls for treating the HLM with MAbs on ice for 15–20 min. Similar inhibitory MAbs are available from Xenotech, LLC and their protocol calls for treating the HLM at room temperature.

19. Prior to incubating the test compound, a titration curve of varying amounts of MAb (diluted with 25 mM Tris buffer) with a known probe substrate can help determining the maximum observed loss of activity due to inactivation of the isoform.

Acknowledgements

The views expressed here are solely those of the author and do not reflect the opinions of Janssen Research & Development, LLC.

References

1. Williams JA, Hurst SI, Bauman J et al (2003) Reaction phenotyping in drug discovery: moving forward with confidence? Curr Drug Metab 4(6):527–534
2. Emoto C, Murase S, Kl I (2006) Approach to the prediction of the contribution of major cytochrome P450 enzymes to drug metabolism in the early drug-discovery stage. Xenobiotica 36(8):671–683
3. European Medicines Agency (2010) Guideline on the investigation of drug interactions. London. Available at www.ema.europa.eu/docs/en_GB/document_library/Scientific_guideline/2010/05/WC500090112.pdf
4. US Food and Drug Administration, US Department of Health and Human Services, Center for Drug Evaluation and Research, Center for Biologics Evaluation and Research (CBER) (2012) Guidance for industry: drug interaction studies—study design, data analysis, and implications for dosing and labeling. Silver Spring, MD. Available at www.fda.gov/downloads/Drugs/GuidanceCompliance Regulatory Information/Guidances/UCM 292362.pdf
5. Cashman JR, Zhang J (2006) Human flavin-containing monooxygenases. Annu Rev Pharmacol Toxicol 46:65–100

6. Lang DH, Yeung CK, Peter RM et al (1998) Isoform specificity of trimethylamine N-oxygenation by human flavin-containing monooxygenases (FMO) and P450 enzymes: selective catalysis by FMO3. Biochem Pharmacol 56:1005–1012
7. Krueger SK, Williams DE (2005) Mammalian flavin-containing monooxygenases: structure/function, genetic polymorphisms and role in drug metabolism. Pharmacol Ther 106(3):357–387
8. Poetsch M, Caerwinsli M, Wingenfeld L et al (2010) A common FMO3 polymorphism may amplify the effect of nicotine exposure in sudden infant death syndrome (SIDS). Int J Legal Med 124(4):301–306
9. Adapted from Cyprotex in vitro ADME & PK S9 Stability Assay Protocol: http://www.cyprotex.com/admepk/in-vitro-metabolism/s9-stability/
10. Gonzalez FJ, Tukey RH (2005) Drug metabolism. In: Brunton L et al (eds) Goodman and Gilman's the pharmacological basis of therapeutics, 11 edn. The Mc-Graw-Hill Companies, Publisher
11. Fukami T, Nakajima M, Sakai H et al (2007) CYP2A13 metabolizes the substrates of human CYP1A2, phenacetin and theophylline. Drug Metab Dispos 35(3):335–339
12. Eiermann B, Engel G, Johansson I et al (1997) The involvement of CYP1A2 and CYP3A4 in the metabolism of clozapine. Br J Clin Pharmacol 44(5):439–446
13. Zhang WV, D'Esposito F, Edwards RJ et al (2008) Interindividual variation in relative CYP1A2/CYP3A4 phenotype influences susceptibility of clozapine oxidation to cytochrome P450-specific inhibition in human hepatic microsomes. Drug Metab Dispos 36(12):2547–2555
14. Cashman JR (2000) Human flavin-containing monooxygenases substrate specificity and role in drug metabolism. Curr Drug Metab 1(2):181–191
15. Tugnait M, Hawes EM, McKay G et al (1997) N-Oxygenation of clozapine by flavin-containing monoxygenases. Drug Metab Dispos 25(4):524–527
16. Penner N, Woodward C, Prakash C (2012) Drug-metabolizing enzymes and biotransformation reactions. In: Zhang D, Sekhar S (eds) ADME-enabling technologies in drug design and development, 1st edn. John Wiley & Sons, Inc., Publisher
17. Cashman JR (2008) Role of flavin-containing monooxygenases in drug development. Expert Opin Drug Metab Toxicol 4(12):1507–1521
18. Segel IH (1975) Enzyme kinetics: behavior and analysis of rapid equilibrium and steady-state enzyme systems. Wiley, New York
19. Obach RS, Reed-Hagen AE (2002) Measurement of the Michaelis constants for cytochrome P450-mediated biotransformation reactions using a substrate depletion method. Drug Metab Dispos 30(7):831–837
20. Nath A, Atkins W (2006) A theoretical validation of the substrate depletion approach to determining kinetic parameters. Drug Metab Dispos 34(9):1433–1435
21. Venkatakrishnan K, Von Moltke LL, Greenblatt DJ (2001) Application of the relative activity factor approach in scaling from heterologously expressed cytochrome P450 to human liver microsomes: studies on amitriptyline as a model substrate. J Pharmacol Exp Ther 267(1):326–337
22. Crespi CL (1995) Xenobiotic-metabolizing human cells as tools for pharmacological and toxicological research. Adv Drug Res 26:179–235
23. Soars MG, Gelboin HV, Krausz KW et al (2003) A comparison of relative abundance, activity factors and inhibitory monoclonal antibody approaches in the characterization of human CYP enzymology. Br J Clin Pharmacol 55:175–181
24. Rodrigues AD (1999) Integrated cytochrome P450 reaction phenotyping: attempting to bridge the gap between cDNA-expressed cytochrome P450 and native human liver microsomes. Biochem Pharmacol 57(5):465–480
25. Stringer RA, Strain-Damerell C, Nicklin P et al (2009) Evaluation of recombinant cytochrome P450 enzymes as an in vitro system for metabolic clearance predictions. Drug Metab Dispos 37(5):1025–1034
26. Proctor NJ, Tucker GT, Rostami-Hodjegan A (2004) Predicting drug clearance from recombinantly expressed CYPs: intersystem extrapolation factors. Xenobiotica 34:151–178
27. Chen Y, Liu L, Nguyen K et al (2011) Utility of intersystem extrapolation factors in early reaction phenotyping and the quantitative extrapolation of human liver microsomal intrinsic clearance using recombinant cytochromes P450. Drug Metab Dispos 39(3):373–382
28. Zhang H, Davis CD, Sinz MW et al (2007) Cytochrome P450 reaction-phenotyping: an industrial perspective. Expert Opin Drug Metab Toxicol 3(5):667–687
29. Paine MF, Hart HL, Ludington SS et al (2006) The human intestinal cytochrome P450 "pie". Drug Metab Dispos 34(5):880–886
30. Lu C, Miwa GT, Prakash SR et al (2007) A novel model for the prediction of drug-drug interactions in humans based on in vitro cytochrome P450 phenotypic data. Drug Metab Dispos 35(1):79–85

31. Stresser DM, Broudy MI, Ho T et al (2004) Highly selective inhibition of human CYP3A in vitro by azamulin and evidence that inhibition is irreversible. Drug Metab Dispos 32(1): 105–112
32. Rock DA, Foti RS, Pearson JT (2008) The combination of chemical and antibody inhibitors for superior P450 3A inhibition in reaction phenotyping studies. Drug Metab Dispos 36(1):105–112
33. Walsky RL, Obach RS, Gaman EA et al (2005) Selective inhibition of human cytochrome P450 2C8 by montelukast. Drug Metab Dispos 33(3):413–418
34. Chauret N, Gauthier A, Nicoll-Griffith DA (1998) Effect of common organic solvents on in vitro cytochrome P450-mediated metabolic activities in human liver microsomes. Drug Metab Dispos 26(1):1–4
35. Easterbrook J, Lu C, Sakai Y et al (2000) Effects of organic solvents on the activities of cytochrome P450 isoforms, UDP-dependent glucuronyl transferase, and phenol sulfotransferase in human hepatocytes. Drug Metab Dispos 29(2):141–144
36. Stresser DM, Crespi CL, Patten CJ (2009) Cytochrome P450 enzyme mapping in drug discovery using BD supersomes enzymes, application note #467. Available at: http://www.bdbiosciences.com/external_files/dl/doc/tech_bulletin/live/web_enabled/467.pdf

Chapter 10

Human Pregnane X Receptor (hPXR) Activation Assay in Stable Cell Lines

Judy L. Raucy

Abstract

Analysis of pregnane X receptor (PXR, NR1I2) activation to determine induction of drug metabolizing enzymes and transporters and predict drug-drug interactions (DDIs) is a wildly used technique among *in vitro* assays. Direct assessment of PXR activation is a cell-based assay that requires two major components, the PXR and a reporter gene linked to the promoter and enhancer regions of the CYP3A gene. Because of species differences in the ligand binding region of PXR, the receptor from the species of interest should be used when assessing activation. At present, PXR activation determined in stable cell lines can be assessed in medium (96-well) to high throughput (384 to 1,536-well) systems. Assays involving stable cell lines allow for simultaneous detection of PXR activation, CYP3A metabolism and cytotoxicity in a single well of a multi-well plate. In this manner, compounds that are toxic and are both inducers and inhibitors of CYP3A are readily identified. Here, we provide comprehensive step-by-step instructions for the application of screening for human PXR activation using commercially available stable cell lines harboring the PXR and a luciferase reporter gene linked to the promoters of the human CYP3A gene. These instructions provide detailed information on how to thaw, culture, passage and seed the cells in 96 well plates to use for screening of new drug entities to determine their ability to activate PXR. Instructions will also be provided for assessing not only nuclear receptor activation but also cytotoxicity and CYP3A4 metabolism simultaneously in the stable transformants. Finally, methods are provided for interpreting the results generated in the cell lines and a mechanistic model described for predicting clinical drug-drug interactions. The basic protocol provided here for identifying new drugs with the ability to activate human PXR and subsequently cause P450 enzyme induction can be miniaturized for higher throughput and extended to PXR from other species and additional nuclear receptors.

Key words Nuclear receptor, Pregnane X receptor, CYP3A4 induction, DPX2™ cells, PXR, Drug-drug interactions

Abbreviations

AhR	Aryl hydrocarbon receptor
AUC	Area under the concentration curve
CAR	Constitutive androstane receptor
CTF	Cell-Titer Fluor™ cell viability assay
DDIs	Drug-drug interactions

DMSO   Dimethyl sulfoxide
EDTA   Ethylenediaminetetraacetic acid
FBS   Fetal bovine serum
FLU   Fluorescent light units
NME   New molecular entity
NR   Nuclear receptor
PBS   Phosphate buffered saline
PXR   Pregnane X receptor
PXRE   Pregnane x receptor element
RIF   Rifampicin
RIS   Relative induction score
RLU   Relative light units
XREM   Xenobiotic response enhancer module

1 Introduction

Drug-drug interactions remain a leading cause of death and result when one drug alters the pharmacokinetics or pharmacodynamics of another drug. With pharmacokinetic related drug-drug interactions (DDIs) one drug generally alters the metabolism or transport of a second drug. Also at the pharmacokinetic level, dietary supplements, food additives, and even certain foods may alter drug metabolism or transport. Abrupt alterations in metabolism or transport due to drug interactions can change the known safety and efficacy of a drug [1]. With regards to metabolism, there are two mechanisms in which one xenobiotic can alter the metabolic outcome of a second drug. These two mechanisms involve the P450 enzymes and results when one drug inhibits metabolism of another or through the process of enzyme induction resulting from chronic exposure to a drug. The result of P450 enzyme induction is increased clearance that causes a decline or loss of therapeutic efficacy. In some cases, concentrations of the second drug can drop to sub-therapeutic levels. For example, rifampicin can prevent contraception if administered with oral contraceptives [2, 3]. Rifampicin can also precipitate opioid withdrawal symptoms if administered with methadone [4, 5] and cause organ rejection if given with cyclosporine [6, 7]. Other examples of clinically relevant potent enzyme inducers include some anticonvulsants, anti-neoplastics, and anti-retrovirals, drugs that are used in settings where combination therapy is especially common. Alternatively, when a drug is metabolized by a P450 enzyme to a reactive metabolite, induction of that enzyme can lead to increased production of the metabolite and if in a chronic setting, ultimately hepatotoxicity [8]. A well-known example is alcohol-mediated induction of CYP2E1. Increased expression of this enzyme can exacerbate hepatic failure resulting from overdose of acetaminophen [9].

CYP3A is one of the predominant P450 enzymes involved in drug or xenobiotic metabolism and is the primary clearance mechanism for about 50 % of marketed drugs. Furthermore, this enzyme comprises about 30–50 % of the total P450 enzymes in the adult human liver [10, 11]. Of further importance, CYP3A is one of the most highly inducible xenobiotic-metabolizing enzymes. For example, rifampicin can cause between four and tenfold or higher increases in CYP3A4 enzyme activity *in vivo*. The most common mechanism by which drugs can cause P450 enzyme induction is by activation of gene transcription. In general, a NME activates a ligand activated specific receptor, which then causes an increase in the transcription rate of the receptor's target genes. The receptors most commonly involved in regulation of drug metabolizing enzymes are AhR, CAR and PXR. Activation of AhR ultimately results in enhanced expression of the CYP1A and CYP1B P450 enzymes, while activation of PXR and/or CAR results in up-regulation of CYP3A, CYP2B, and CYP2C gene products.

These ligand activated receptors, CAR, PXR, and AhR (commonly designated as "nuclear receptors"), are considered master regulators of drug metabolism. As such, activation of these receptors elicits a plethora of DDIs and adverse drug effects [12, 13]. The capacity to promote these types of effects explains the extensive focus on nuclear receptors in the pharmaceutical/therapeutics arena. These three nuclear receptors bind a wide range of ligands, such as environmental contaminants including polycyclic aromatic hydrocarbons, steroid hormones, bile acids, and small molecules, nearly all of which are lipophilic in nature. Gene expression profiling has demonstrated that PXR and CAR regulate the expression of a multitude of genes including those involved in physiological functions as well as those important in xenobiotic metabolism and transport [14]. PXR alone regulates over 100 genes in rat liver [15] while in humans, there are over 60 genes responsive to PXR [14]. The primary drug-metabolizing enzymes up-regulated by PXR include CYP2B, CYP2C8, CYP3A, and UGT1A1 [14, 16–18]. In addition, the transporter MDR1 that encodes the broad-specificity efflux pump, P-glycoprotein (Pgp), is up-regulated by PXR [14].

Of further importance, X-ray crystallography of the ligand-binding domain of PXR suggests that it is a large, flexible site with the ability to accommodate a variety of structurally un-related ligands [19]. Because of this unique ability, it is difficult to predict the number and nature of compounds that can bind to PXR. It is known that a large number of NMEs can bind and activate the receptor at any given time. For example, chemical library screening suggests that approximately 11 % of compounds exhibit some form of binding to PXR [20, 21]. This promiscuity makes screening for ligands that activate PXR a priority in the drug discovery/development processes to predict clinical DDIs caused by induction of cytochrome P450 enzymes. Because PXR regulates the expression of a majority of the drug metabolizing P450 enzymes, including CYP3A4, CYP3A5, CYP2C8, CYP2C9, CYP2C19 and CYP2B6, it has been estimated that unwanted activation of PXR is responsible for a majority of all DDIs related to enzyme induction [22].

In addition to its role in regulating expression of xenobiotic metabolizing enzymes, PXR sustains endogenous functions. The receptor plays important roles in cholesterol metabolism and lipid homeostasis [23–25]. Many clinical agents that function as PXR ligands are known to increase circulating cholesterol levels. For example, treatment with the PXR ligand, rifampicin, produces hyperlipidemia [26]. Cafestol, a compound found in unfiltered brewed coffee and the most potent cholesterol-elevating agent known in the human diet, is a PXR agonist as well [27]. With regards to lipids, studies in transgenic mice indicate that PXR activation can indeed influence lipid homeostasis. It was previously reported [28, 29] that mice expressing constitutively activated PXR displayed hepatomegaly and marked hepatic steatosis, and that rifampicin treatment of mice possessing a human PXR gene resulted in hepatic triglyceride accumulation [30]. Finally, PXR may be involved, at least in part, in the regulation of gluconeogenesis, as the activation of this receptor suppresses cAMP-dependent induction of glucose 6-phosphatase [28, 31–35].

Due to its important endogenous roles, PXR represents a highly important drug target in terms of potential therapeutic applications [36]. PXR inhibitors and/or antagonists are being sought for various disease states including cancer, and to improve therapeutic efficacy for known activators of this receptor. PXR's regulatory role in bile acid detoxification and transport makes the receptor an attractive target for drug therapy of cholestasis. In fact, rifampicin and phenobarbital have already been employed to reduce serum bile acid concentrations and to treat the pruritus that accompanies cholestasis [37]. The discovery and pharmacological development of new PXR modulators represents an interesting and innovative therapeutic approach to combat various diseases.

Clearly, the role of PXR in DDIs and adverse therapeutic effects makes screening potential drug candidates for their ability to activate and/or inhibit this receptor a priority with most pharmaceutical companies. Several techniques are available for this purpose including ligand binding, transient transfection and mammalian two hybrid assays. However, to illustrate the ease and simplicity of assessing PXR activation to identify potential P450 enzyme and transporter inducers, we focus here on providing comprehensive step-by-step instructions using stable transformants.

The materials and methods described below are for medium throughput screening of PXR activation using 96 well plates. However, both materials and methods can be easily miniaturized for high throughput screening to accommodate 384 or 1,536 well plates.

2 Materials

2.1 Equipment

Specific equipment used in this assay is described; however any model of comparable capability can be easily substituted.

1. Multi-channel automatic and pipettor (50–1,200 µL).
2. Laminar Flow Tissue culture hood
3. Humidified 5 % CO_2/37 °C incubator
4. Variable Temperature Water bath set at 37 °C
5. Luminometer/fluorimeter (BioTek Synergy 2, Winooski, VT)
6. Countess Automated Cell Counter (Life Technologies, Carlsbad CA)
7. Olympus CK40 Inverted microscope
8. Drummond Hood Mate Automatic pipettor

2.2 Reagents and Solutions

1. Sterile PBS
2. Sterile PBS with calcium and magnesium
3. Hyclone fetal bovine serum-characterized
4. 0.25 % trypsin/1 mM EDTA (Life Technologies, Carlsbad CA)
5. 0.4 % Trypan blue
6. DMSO
7. 70 % ethanol or isopropanol
8. Rifampicin (RIF)
9. Cell-Titer Fluor™ cell viability assay (Promega Madison WI, catalogue # G6081)
10. ONE-Glo™ (Promega Madison WI, catalogue # E6110)
11. P450 Glo™ CYP3A4 assay with Luciferin-IPA (Promega Madison WI, catalogue # V9001)
12. Puracyp Culture Medium, stored at 4 °C (Puracyp, Carlsbad CA)
13. Puracyp Dosing Medium, stored at 4 °C (Puracyp, Carlsbad CA)
14. DPX2™ cells, stored in liquid nitrogen (Puracyp, Carlsbad CA)

2.3 Labware/Plasticware

1. Corning 3610 White-clear flat bottom sterile tissue culture treated 96-well plates
2. White-opaque non-sterile non-tissue culture treated 96-well plates
3. Sterile pipettes 2, 5, 10, 25 and 50 mL
4. Sterile pipette tips various sizes ranging from 1 to 1,000 µL
5. Corning 50 mL reagent reservoirs
6. Sterile deep-96 well (2 mL/well) plates (Axygen, Union City CA)

7. 15-mL sterile conical polypropylene centrifuge tubes
8. 2 mL cryogenic vials
9. Tissue culture treated 100×20 mm culture dishes
10. Manual pipettors ranging from 1 to 1,000 µL
11. Countess cell counting chambers (Life Technologies, Carlsbad CA)

3 Methods

3.1 Thawing Cryogenic Vials of Stable Cell Lines to Assess PXR Activation

NOTE: The cell line used here is DPX2™ which have been stably integrated with human PXR and a luciferase (reporter) vector containing human CYP3A4 promoters, XREM and PXRE [38, 39].

Thawing cell vials:

1. Add 10 mL of Puracyp culture medium to a 100 mm culture dish and place in incubator at 37 °C and 5 % CO_2 for 10 min to equilibrate the culture medium.
2. Carefully remove the DPX2™ cells from the liquid nitrogen container.
3. Immediately place the vial in a 37 °C water bath for 5 min to completely thaw the cells.
4. Spray the exterior of the vial with 70 % isopropanol or ethanol to sterilize the surface.
5. Decontaminate the laminar flow hood
6. Wipe any remaining alcohol from the vial and place it in the hood.
7. Remove equilibrated 100 mm culture dish from incubator and place in the hood.
8. Using a 2 mL pipette, gently and aseptically pipette the cells up and down three times to disperse them, and then transfer the cells from the vial to the medium in the dish.
9. Place the dish in the incubator for 24 h at 37 °C and 5 % CO_2. After 24 h, aspirate medium and replace with 10 mL warmed (37 °C) Puracyp culture medium and return to incubator.
10. Twenty-four hours later, check the cells and change medium if any dead (floating) cells are present. If the medium is relatively clear and free of floating cells, there is no need to perform a medium change.

3.2 Culturing DPX2™ Cells

1. Cells are cultured in 100 mm dishes in 10 mL of Puracyp culture medium in a 37 °C and 5 % CO_2 incubator. The incubator should be de-contaminated and **NOT** used for primary cultures.
2. Dishes are checked each day under the microscope to ensure growth and sterility. Medium is aspirated and replaced every 2 days until the cells reach 80–90 % confluency.

3. To split cells, aspirate medium from the culture dish and add 5 mL of sterile PBS. Tilt dish back and forth to ensure complete rinsing of cells.

4. Aspirate PBS and replace with 3 mL Trypsin/EDTA (0.25 % Trypsin, 1 mM EDTA). Tilt dish to ensure all cells are covered with solution. Place dish in incubator at 37 °C and 5 % CO_2 for 3–5 min.

5. Remove dish from incubator and tap gently with hand to detach cells. Add 3 mL of Puracyp culture medium and pipette cells up and down ten times to wash the entire bottom of the dish and detach cells. The medium will also neutralize the trypsin.

6. Once cells are fully dispersed, transfer to 15 mL centrifuge tube and pipette up and down an additional ten times to break-up cell clumps. Pellet cells at 500 rpm for 3 min.

7. Aspirate supernatant and re-suspend cells in 6 mL Puracyp culture medium. Pipette up and down ten times to break up cell clumps.

8. Seed cells into a new 100 mm dish using the split ratio of 1:6; add 1 mL of the 6 mL diluted cells to the dish containing 10 mL of culture medium. Discard used dish.

9. Return dish to incubator at 37 °C and 5 % CO_2. The following day, check cells and change medium. Check dish each day for appearance of floating cells and change medium every other day until monolayer is 70–80 % confluent. If numerous floating cells are evident, replace the medium.

10. Once the cells are 70–80 % confluent in the new dish, use the cells for an assay by seeding in a 96 well plate (Steps under Sect. 3.4), expand the cells into a second 100 mm dish, or freeze an aliquot of cells (Steps under Sect. 3.3).

3.3 Freezing DPX2™ Cells

1. Decontaminate laminar flow hood

2. Label 4 cryogenic vials /100 mm culture dish if all cells are to be frozen, with the catalogue (vial) number, passage number and the date frozen.

3. Once cells are 70–80 % confluent, remove the dish from incubator. Aspirate medium from dish, rinse with 5 mL PBS.

4. Aspirate PBS, and add 3 mL Trypsin/EDTA. Incubate at 37 °C and 5 % CO_2 for 3–5 min.

5. Remove dish and gently tap dish to dislodge cells. Pipette cells up and down to disperse and add 3 mL of room temperature Puracyp culture medium. Transfer to 15 mL centrifuge tube.

6. Pellet cells for 3 min at 500 rpm.

7. Aspirate the supernatant.

8. Re-suspend cell pellet in 3.6 mL of Fetal Bovine Serum. Pipette the mixture up and down ten times to disperse the cell pellet. Check viability and cell number and ensure that viability is greater than 90 % and that cell number is $2-3 \times 10^6$ cells/vial.

9. Add 400 µL neat DMSO (to a final concentration of 10 %) to the 15 mL tube and aliquot 1 mL into each cryogenic vial.

10. Mix each cryogenic vial by inverting 2–3 times.

11. Transfer the vials to a freezing container and store overnight at −80 °C. The following day, transfer cells to liquid nitrogen cryogenic container.

3.4 Assessing Cell Viability and Identifying Compounds That Activate Human PXR

NOTE: Methods provided here are for triplicates of each concentration of test agent and controls. Because of the low variability observed between wells when using DPX2™ cells, duplicates or singlets can be used and the assay revised to accommodate smaller numbers of replicates.

1. Following trypsinization and re-suspension (Steps 4–7 under Sect. 3.2 "Culturing DPX2™ Cells"), remove 10 µL of the cell suspension from the 15 mL centrifuge tube into a microfuge tube containing 10 µL of 0.4 % trypan blue.

2. Mix the 10 µL of the cell suspension and 10 µL of trypan blue by pipetting up and down ten times and remove 10 µL and add to a cell counting chamber.

3. Count cells and determine viability with a Countess Automated Cell Counter. Viability should be 90 % or greater for use in the assay.

4. A 96 well plate is seeded at 2×10^4 cells/well. To achieve this, determine the number of cells/mL and how many mL of cells are needed to seed the 96-well plates. Each 96 well plate requires 10 mL of cell suspension containing two million cells.

5. Dilute the cells in a sterile trough to 2×10^5 cells/mL in the amount of Puracyp culture medium needed for the number of 96 well plates being seeded.

6. Using an automated multichannel pipettor, add 100 µL of the diluted cells to each well of a clear, flat bottom-96 well clear bottom plate. Return plate to 37 °C and 5 % CO_2 incubator.

7. After overnight incubation decontaminate the laminar flow hood.

8. Warm Puracyp dosing medium to room temperature and prepare dosing solutions. For three well replicates, use 1 mL of pre-warmed Puracyp dosing medium for each concentration of test agent made up in neat DMSO and the positive control (RIF). In a 2 mL deep-well plate, aliquot 1 mL of Puracyp dosing medium per concentration of RIF and test articles, and for the solvent (DMSO) control. Here, we use DMSO since

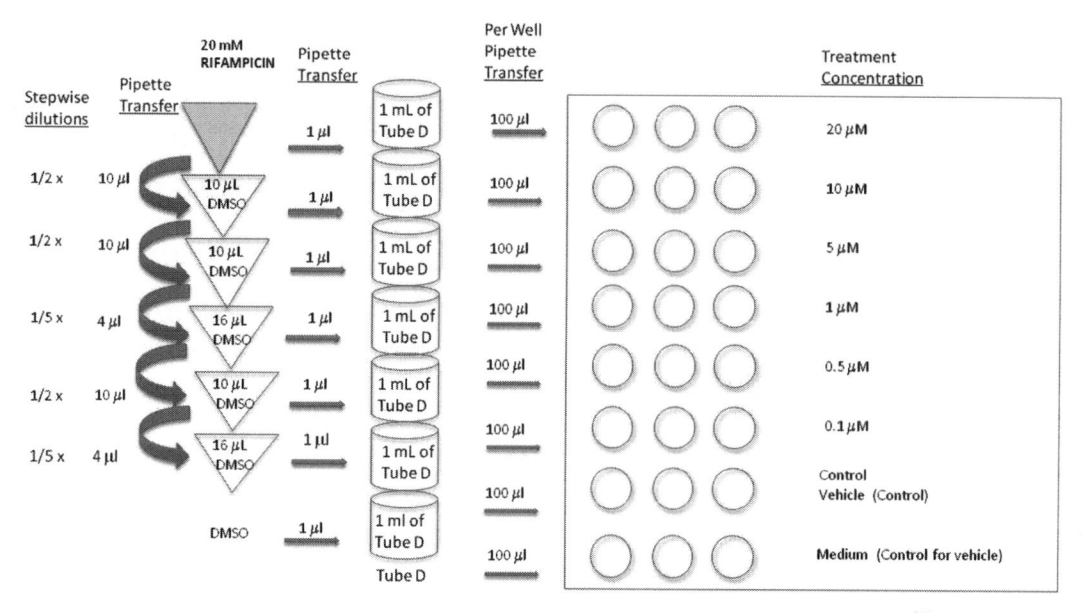

Fig. 1 Dilution scheme for dosing of positive control (Rifampicin): All rifampicin dilutions are made in neat DMSO from a stock concentration of 20 mM, after which the individual solutions are added to 1 mL of Puracyp dosing medium (Tube D) previously aliquoted into 7 wells of a deep 96-well plate (2 mL/well capacity) at a ratio of 1:1,000. Transfer of 100 μL of the dosing solution from the deep well plate into triplicate wells of a 96-well culture plate containing the DPX2™ cells is accomplished with a multichannel automatic pipettor

most compounds are soluble in this agent. However, other solvents can be used.

9. In separate microfuge tubes, prepare the necessary concentrations of Test Articles and RIF in DMSO.

10. The stock concentration, RIF dissolved in 20 mM in DMSO. The stock solution will be diluted 1:2 in DMSO to 10 mM; the 10 mM will then be diluted 1:2 in DMSO to 5 mM; 5 mM will be diluted 1:5 to 1 mM; 1 mM is diluted 1:2 to 0.5 mM; and 0.5 mM diluted 1:5 to 0.1 mM (Fig. 1). If determining EC_{50} values of RIF, use at least six concentrations.

11. An example of a dilution scheme for the test articles is shown in Fig. 2. The test articles (100 mM stock concentration) will be diluted 1:4 in DMSO to 25 mM, 25 mM diluted 1:4 to 6.25 mM; 6.25 mM diluted 1:4 to 1.56 mM; 1.56 mM diluted 1:4 to 0.39 mM; and 0.39 mM diluted 1:4 to 0.098 mM in DMSO (Fig. 2). If determining EC_{50} values of test agent, use at least six concentrations.

12. In the deep well plate containing the dosing medium, transfer 1 μL of the DMSO diluted test articles, RIF, and neat DMSO to separate wells (Figs. 1 and 2). Dilutions should be such that the final DMSO concentration does not exceed 0.1 %. Mix the

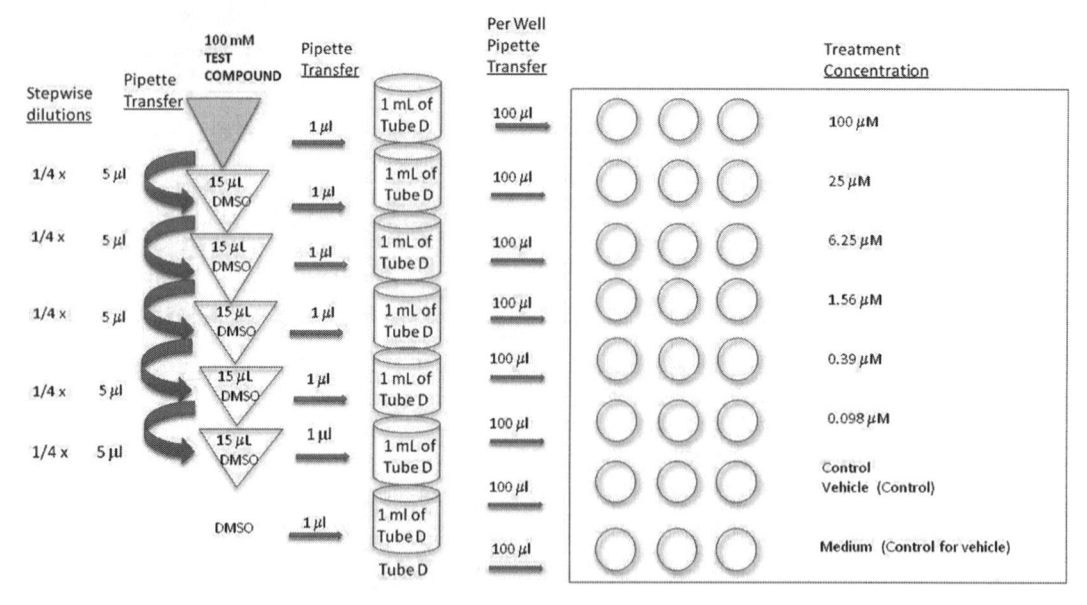

Fig. 2 Dilution scheme for dosing of test articles: All test articles are diluted in neat DMSO from a stock concentration of 100 mM, after which the individual solutions are added to 1 mL of Puracyp dosing medium (Tube D) previously aliquoted into 7 wells of a deep 96-well plate (2 mL/well capacity) at a ratio of 1:1,000 to maximize solubility. Transfer of 100 μL of the dosing solution from the deep well plate into triplicate wells of a 96-well culture plate containing the DPX2™ cells is accomplished with a multichannel automatic pipettor

test articles, DMSO and positive control in the dosing medium by pipetting up and down 2–3 times with an automated pipettor.

13. Following preparation of controls and test articles in dosing medium, remove the 96 well-plate from the incubator.

14. Using a multi-channel pipettor, gently and slowly aspirate culture medium from each well of the 96 well plate and discard. **Be very careful not to dislodge cells when aspirating medium.**

15. Mark the 96 well-plate according to the dosing scheme.

16. Using an automated multi-channel pipettor, transfer 100 μL of Puracyp dosing medium containing the test articles, RIF and DMSO to each designated well of the 96-well plate (Figs. 1 and 2). Be very careful not to dislodge cells when pipetting medium with test articles into the individual wells. Once the entire plate is dosed, return it to the incubator.

17. Incubate the plate overnight (at least 24 h) at 37 °C and 5 % CO_2.

18. After the overnight incubation (24 h or more) decontaminate the laminar flow hood.

19. If performing metabolism and/or inhibition studies re-dose the cells with the test articles by repeating Steps 8–16 after 24 h and return the plate to the incubator for an additional

24 h (48 h total treatment time). If PXR activation only is the final end-point, cells can be treated for 24 or 48 h, depending on the level of activation desired.

20. Following the 24 or 48 h incubation, decontaminate the laminar flow hood.

21. Using a multi-channel pipettor, gently and slowly aspirate the medium containing the test agents from each well of the 96-well plate. **Be very careful not to dislodge cells when aspirating medium.**

22. To normalize the luminescence results based on viable cell number it is recommended that a viability assay be performed in the same well as the induction assay. We use Cell-Titer Fluor™ (CTF) Cytotoxicity Assay Reagent because it measures live cells using fluorescence.

23. For each 96 well plate, add 10 mL room temperature PBS containing magnesium and calcium to a sterile 15 mL conical centrifuge tube. It is important to use the PBS containing these ions because they are necessary for luciferase function.

24. Add 5 μL of CTF and mix the tube by inverting up and down.

25. Add the contents to a medium trough.

26. Using the multi-channel pipettor add 100 μL of the CTF/PBS solution to each well.

27. Return plate to the 37 °C and 5 % CO_2 incubator for 60 min. **The following steps do not require sterile conditions.**

28. Measure resulting fluorescence. Excitation: 380–400 nm; Emission : 505 nm

29. Prepare ONE-Glo™ according to manufacturer's protocol.

30. Using the 12-channel pipettor, add 100 μL of the ONE-Glo™ reagent per well of the 96-well plate and carefully agitate the plate to mix the reagents.

31. Allow the plate to incubate at RT for 5 min.

32. Read luminescence of the individual wells of the 96 well-plate with the luminometer set for a 5 s pre-shake with a 5 s/well read time. A relatively high gain (sensitivity) setting should be used.

33. Collect data in excel spreadsheet.

3.5 Assessing CYP3A4 Metabolism Simultaneously with PXR Activation and Cell Viability

To perform this assay, follow steps 1–19 above. Expose the cells to the dosing media containing the various concentrations of test agents and positive control and DMSO for at least 48 h.

1. Following incubation, de-contaminate the laminar flow hood.
2. Allow CTF, P450 Glo™ CYP3A4 assay components and ONE-Glo™ kit components to reach room temperature.

3. Add 6 μL from P450 Glo™ substrate (Luciferin-IPA) to 6 mL of Puracyp dosing media contained in a sterile 15 mL conical tube. Mix by inverting three times and pour into a sterile medium trough.

4. Remove 96-well plate from the incubator and place into the sterile hood.

5. Carefully remove the liquid from each well using a multichannel pipette and discard.

6. Add 50 μL/well from the medium trough containing Luciferin-IPA reagent to each well.

7. Return plate to incubator for 60 min at 37 °C.

8. During incubation with Luciferin-IPA, pour the contents of P450-Glo™ Buffer from the assay kit into the Luciferin Detection Reagent bottle. Mix by inversion.

9. After the 60 min incubation, remove 96-well plate from the incubator.

10. Carefully transfer 50 μL from each well of the original plate to a corresponding well of a white non-sterile 96-well plate that replicates the format of the original plate. Following transfer of all wells of P450-Glo™ substrate to replicate plate, remove the non-sterile plate from hood.

11. Back in the hood, transfer 10 mL of PBS containing calcium and magnesium to a sterile 15 mL conical tube, and then add 5 μL of CTF. Mix by inversion.

12. Pour into media trough.

13. On the original plate containing the cells, gently add 100 μL of the CTF in PBS reagent into each well using a multi-channel pipettor.

14. Return plate to the incubator for 60 min.

    The following steps do not require sterile conditions.

    NOTE: While the original plate containing the cells and CTF solution are incubating for 60 min, determine CYP3A4 metabolism of Luciferin-IPA.

15. Add 50 μL of P450-Glo™ Buffer/Luciferin Detection Reagent (prepared in step 8 above) to each well of the replicate plate (outside of hood), and incubate at room temperature for 20 min.

16. After the 20 min incubation with Luciferin Detection Reagent, measure the luminescence of the white 96-well plate with the luminometer set for a 1–5 s/well read time. A fairly high gain (increased sensitivity) setting should be used. Collect data in Excel format.

17. Add the contents of ONE-Glo™ Assay Buffer to the ONE-Glo™ Assay Substrate, cap, and mix by inversion.

18. After the 60 min incubation with CTF, remove the original 96-well plate from the incubator, and allow cooling to room temperature.

19. Measure fluorescence of individual wells with a microplate reader in fluorescence mode using an excitation wavelength of 380–400 nm and an emission wavelength of 505 nm. Collect data in Excel format.

20. Pour ONE-Glo™ Assay reagent (prepared in Step #17) into a media trough, and add 100 μl of the reagent into each well of the plate.

21. Carefully agitate the plate to mix the reagents contained in the wells.

22. After 5 min, read the luminescence of the individual wells with the luminometer set for a 5 s pre-shake with a 5 s/well read time. A relatively high gain (sensitivity) setting should be used. Collect data in Excel format.

3.6 Quantitation of PXR Receptor Activation

1. Determine the average Relative Luminescence Units (RLU) of the three replicates for each test compound and RIF at each dosage. Acceptable error limits for the replicates at each concentration and each test article including the positive controls are coefficient of variation (CV) values <20 %. In addition, determine the average Fluorescence Light Units (FLU) of the three replicates for each test compound and RIF at each dosage.

2. Determine the mean RLU and FLU for the 0.1 % DMSO vehicle control replicates. Acceptable error limits for the replicate RLUs of the negative control (DMSO) are CV values < 20 %.

3. Normalize the luciferase activity for cell viability by dividing the mean RLU by the average FLU for each test compound and RIF at each dosage as well as for the vehicle control. Dividing the RLU by the FLU provides a way to normalize PXR activation to the number of viable cells in each well. Moreover, this latter value (FLU) also is an indication of the degree of cytotoxicity produced by each compound at each concentration.

4. PXR receptor activation at the individual test compound doses is calculated by dividing the normalized luciferase activity (RLU/FLU) for the test compound doses by that of the normalized DMSO vehicle control. The final data is expressed as fold activation relative to the vehicle control. The use of 5 or more doses of test compound and positive control allows for the derivation of EC_{50} and E_{MAX} values from nonlinear regression analysis of the log (dose) vs. response (three parameters) curves according to the equation Y=bottom + (top-bottom)/(1+10^(logEC50-X)). Prism V6.0c (Graphpad Software, La Jolla, CA) and SigmaPlot (Systat Software, San Jose, CA) are among the software programs with built-in equations for deriving these PXR activation kinetic parameters.

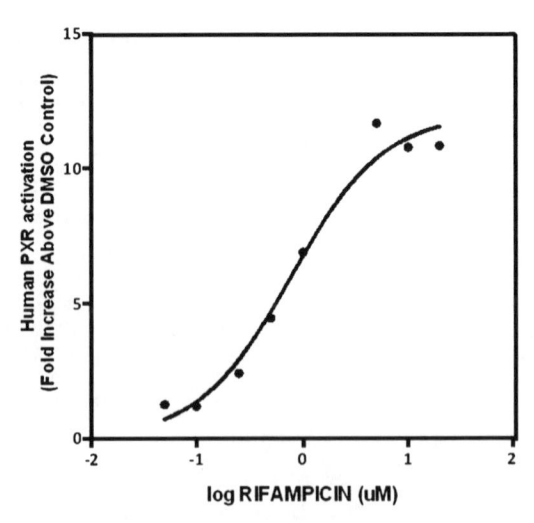

Fig. 3 A typical dose-response curve generated in DPX2™ cells treated with eight concentrations of rifampicin. Cells were seeded in a 96 well culture plate and treated for 24 h with various concentrations of rifampicin ranging from 0.05 to 20 μM. Following treatment, cytotoxicity was determined using Cell-Titer Fluor™ cell viability assay and luciferase activity assessed with ONE-Glo™. Results were the mean of three replicates of luminescence generated in the luciferase assay and normalized to three replicates of fluorescence values produced by assessing cell viability. The capacity to multiplex in this system allows for identification of molecules that not only activate PXR, but also exhibit cellular toxicity. The normalized results, expressed as fold increase above DMSO control, were plotted on a log-concentration vs. response curve and an EC_{50} value of 0.78 μM and an E_{max} of 11.98-fold activation were calculated

5. Test compound values should be compared to those obtained for the positive (rifampicin) control. An example of a typical PXR activation log dose-response curve generated with nonlinear regression analysis described above and using eight concentrations of rifampicin is shown in Fig. 3. Using the built-in equation (shown above in Sect. 3.6, step 4), the calculated EC_{50} value is 0.78 μM and the E_{max} is 11.98 fold above DMSO.

3.7 Quantitation of P450 Substrate Metabolism

1. Receptor activation by test compounds, positive controls, and negative (vehicle) controls are determined as described above.

2. To assess whether the test compounds also increased P450 metabolic activity, determine the average RLU of the three Luciferin-IPA replicates for each test compound and RIF at each dosage. In addition, determine the average RLU of the three replicates for the DMSO (vehicle) controls.

3. Normalize the P450-Glo™ activity for cell viability by dividing the average RLU by the average viability (FLU determined in the CTF assay) for each test compound and RIF at each dosage as well as for the vehicle control.

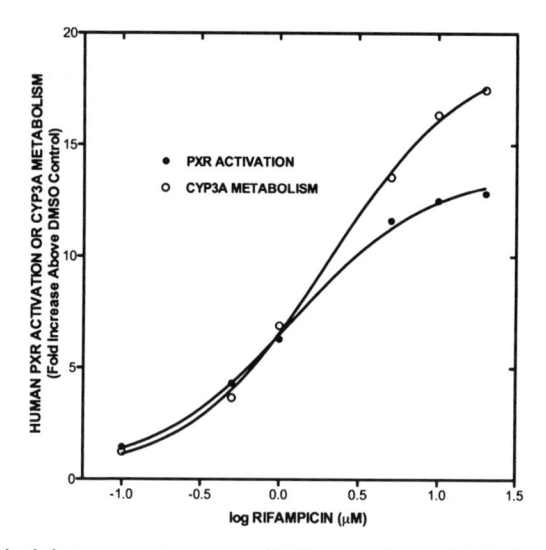

Fig. 4 A typical dose-response curve of PXR activation and CYP3A4 metabolism generated in DPX2™ cells treated with 6 concentrations of Rifampicin: Cells were seeded in a 96 well culture plate and treated for 48 h with various concentrations of rifampicin ranging from 0.1 to 20 μM. Both PXR activation and CYP3A4 metabolism of luciferin-IPA were monitored in a multiplex fashion. Following treatment with rifampicin, cytotoxicity using Cell-Titer Fluor™ reagent, CYP3A4 mediated metabolism of luciferin-IPA (P450-Glo™), and luciferase activity with ONE-Glo™ were assessed. Results were the mean of three replicates of luminescence generated in the luciferase and P450-Glo assays and normalized to three replicates of fluorescence values produced by assessing cell viability. The normalized results for both PXR activation and CYP3A4 metabolism of luciferin-IPA were expressed as fold increase above DMSO control values. As expected, both metabolic activity and PXR activation were enhanced by rifampicin when plotted on a log-concentration vs. response curve

4. The fold induction by the individual doses of test compound and RIF is calculated by dividing the normalized P450-Glo™ luciferase activity (RLU/FLU) for the test compound doses by that of the normalized DMSO vehicle control. The results should be expressed as fold increase (induction) above DMSO-treated cells.

5. If more than five concentrations are used for each compound and RIF, then a log (dose) vs. response curve can be generated from the PXR activation and Luciferin IPA metabolism data. The fold induction produced from the PXR activation and metabolism of Luciferin-IPA for cells exposed to RIF at six concentrations are shown in Fig. 4. Compounds such as RIF that activate the receptor, will exhibit log (dose) vs. response curves with an increase over the concentration range in both the metabolism of Luciferin-IPA and PXR activation (Fig. 4). Nonlinear regression analysis of these typical dose-response curves for PXR activation and induction of CYP3A4 mediated

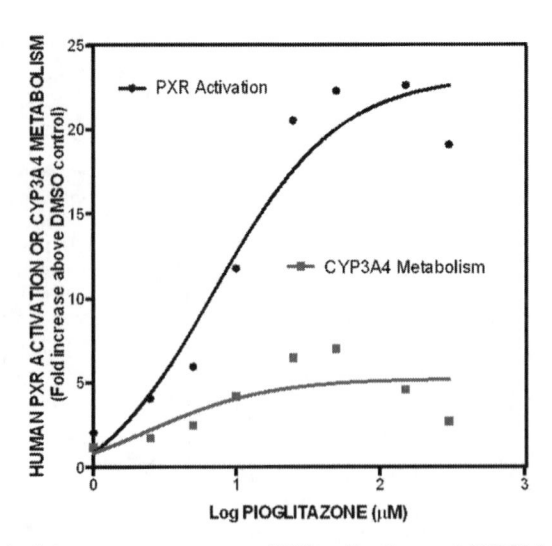

Fig. 5 A typical dose-response curve of PXR activation and CYP3A4 metabolism generated in DPX2™ Cells treated with eight concentrations of Pioglitazone: Cells were seeded in a 96 well culture plate and treated for 48 h with various concentrations of pioglitazone ranging from 1 to 300 µM. Both PXR activation and CYP3A4 metabolism of luciferin-IPA were monitored in a multiplex fashion. Following treatment with pioglitazone, cytotoxicity using Cell-Titer Fluor™ reagent, CYP3A4 mediated metabolism of luciferin-IPA (P450-Glo™), and luciferase activity with ONE-Glo™ were determined. Results represent the mean of three replicates of luminescence generated in the luciferase and P450-Glo assays and normalized to three replicates of fluorescence values produced by assessing cell viability. The normalized results for both PXR activation and CYP3A4 metabolism of luciferin-IPA were expressed as fold increase above DMSO control values and plotted using the log concentration vs. fold increase. The lack of an increase in luciferin-IPA metabolism with a simultaneous increase in PXR activation with increasing doses of pioglitazone suggests that this compound is an inducer and metabolic inhibitor of CYP3A4. It is not uncommon to obtain extensive PXR activation but negligible induction of CYP3A4 substrate metabolism. This allows DPX2™ cells to be used to identify both inducers and inhibitors of CYP3A4 in a single assay

metabolism gave an $EC_{50} = 1.21$ µM and $EC_{50} = 2.0$ µM, respectively. However, compounds that activate the receptor but are metabolic inhibitors of CYP3A4 will exhibit kinetic parameters similar to those shown in Fig. 5. In this example with pioglitazone, compounds that are both inducers and inhibitors of CYP3A4 can be identified simultaneously in this manner. The log (dose) vs. response curve was generated for both PXR activation and luciferin-IPA metabolism. However, since Luciferin-IPA metabolism produced negligible induction and an atypical dose-response curve, an EC_{50} value (7.1 µM) was only calculable for PXR activation by pioglitazone.

4 Interpretation of Results

4.1 Receptor Activation (Induction)

If only 1–4 concentrations are examined then the percent of 10 μM Rif at an equimolar concentration of the test article should be calculated to assess the potential for the test article to cause PXR activation *in vivo* [1, 40–42]. Activation potency is defined as negative, weak, moderate and strong. Negative, weak, moderate and strong activators are those that give <15 %, <40 %, <69 % and >70 %, respectively, of the response produced by 10 μM RIF.

4.2 Receptor Activation Kinetics for Predicting In Vivo Induction of P450 Enzymes

Several algorithms and quantitation approaches have been proposed for studying enzyme induction using *in vitro* data [8, 43–46]. For extrapolating receptor activation results to predictions *in vivo*, we use EC_{50} and E_{MAX} values obtained from PXR activation studies. If five or more concentrations are utilized, these values are calculated for test compounds that exhibit PXR activation using nonlinear regression of typical dose-response curves. A test agent exhibiting either no activation or a non-saturated dose-response curve should not be analyzed in this fashion. To determine these parameters, dose-response data can be fit to a sigmoid 3-parameter function according to the following equation:

$$f = a / \left(1 + \exp\left(-(x - x0) / b\right)\right) \quad (1)$$

where a, b and x0 denote E_{max}, slope, and EC_{50}, respectively.

Once EC_{50} and E_{max} values are obtained from PXR activation studies and the clinical plasma concentrations are known for each NME, the induction score (RIS) can be calculated. RIS is a metric to predict the magnitude of drug interactions from *in vitro* induction data for PXR mediated inducers [47]. This metric is then correlated to clinical effects of known inducers on well-established CYP3A metabolized drugs (midazolam). To calculate RIS the following equation is used:

$$RIS = \frac{E\max * I}{EC_{50} + I} \quad (2)$$

where I denotes the efficacious plasma concentration of an inducer (NME) achieved after a standard therapeutic dose [48]. Values used to estimate the *in vivo* concentration of I are: (a) total systemic steady-state C_{max} concentration or; (b) unbound systemic steady-state C_{max} concentration. RIS can then be used to extrapolate from a curve of induction magnitude versus RIS that is established with known inducers [48]. The magnitude of induction is expressed as the predicted percentage change in midazolam AUC [47]. Because of the low variation in induction response generated with the DPX2™ cells among experiments, one curve could be generated and used for all subsequent studies.

Alternatively, an abbreviated calibration curve (containing two weak and two moderate inducers plus rifampicin) could be generated for each experiment.

The merit in using DPX2™ cells to predict clinical DDIs was established in a recent report [49]. The RIS values for 34 therapeutic agents were determined together with the parameters used to derive the induction scores. The correlation between RIS and clinical DDIs was assessed by plotting RIS values versus the percent decrease in AUC of co-administered midazolam, calculated using published data from clinical DDI trials. Clinical inducers and clinical non-inducers were utilized to construct the RIS correlation plot. A strong correlation ($r^2 = 0.90$) was found between RIS and percent decrease in midazolam AUC, indicating that assessing PXR activation in DPX2™ cells can serve as a predictor of clinical DDIs with reasonably good accuracy. Importantly, no false negatives were observed, and the RIS values obtained with DPX2™ cells did not predict substantial DDIs for several clinical non-inducers [49].

5 Notes

For the most part, the assays described above are easy to perform. However, during the course of the experiments, problems may arise. Here, we describe symptoms or problems, possible causes, and ways to resolve the problems.

5.1 High Well-to-Well Variation

A. *Cause:* The main cause for the variation observed between wells of multi-well plates is that the cells are not evenly seeded among the wells.

B. *Resolution:* Mix cells thoroughly with medium before seeding. To facilitate even plating, ensure the cells have been pipetted up/down at least ten times at each step of the trypsinization process and in the culture medium prior to seeding the cells in the multi-well plate.

5.2 Weak Activation with Potent Positive Control, RIF

A. Causes: There are two potential reasons for weak activation by RIF. One is degraded (oxidized) DMSO or RIF and the other is degraded luciferin reagents.

B. Resolution:

B.1. Ensure that DMSO (or other solvent used to dissolve test compounds) is not degraded. DMSO degradation can be checked by including cells treated with medium alone. Luciferase values obtained with cells treated with medium only should be similar to those obtained with cells treated with 0.1 % DMSO (<10 % difference).

- B.2. RIF and its dilutions should be made fresh for each experiment and the stock powder purchased recently and stored according to the manufacturer's recommendations.
- B.3. Luciferase reagents need to be fresh. The manufacturer does not recommend freeze/thawing of luciferin substrates. Include control wells containing only luciferase assay components (no cells) in each assay, giving the background level produced by these reagents. Wells containing cells that have been treated with DMSO or medium alone should give luciferase values two- to threefold higher than luciferase assay components alone.

5.3 Weak Activation with Test Compounds

A. *Causes:* Two possible reasons exist for weak activation with NMEs. They include poor NME solubility or a decrease in cell viability produced by the NME.

B. *Resolutions:*
- B.1. NMEs dissolved in DMSO and other solvents can precipitate out of solution when added to culture medium. This often occurs at higher concentrations. Weak activation stemming from poor solubility would be indicated by: (a) a peak activation response occurring midway in the dose-response curve, followed by decreasing responses at higher concentrations and; (b) no change in cell viability throughout the entire concentration curve.
- B.2. NME cytotoxicity can result in weak PXR activation. Cell viability should be assessed using CTF, and the fold PXR activation normalized to the viable cell number for a given test compound. A decline in receptor activation and cell viability over the concentration range examined would indicate that the test compound is toxic to the DPX2™ cells. If the viable cell number at the highest concentration declines more than 70 % of the lowest concentration of NME, it is not recommended that the high luciferase values be used in generating a dose response curve.

5.4 False Positives

A. *Cause:* Background Activation by Non-PXR Ligands

B. *Resolution:* The low levels of PXR activation observed with certain compounds may not be indicative of "false positives" if cut-off values are established. Cut-off values, commonly expressed as % of RIF, should be established to differentiate negative, weak, moderate and potent PXR activators. Compounds giving activation <15 % of that observed with 10 µM RIF should be considered negative.

5.5 Potent PXR Activation But Weak Induction of CYP3A Metabolism

A. *Cause:* NME is an Irreversible (Time-Dependent) CYP3A4 Inhibitor

B. *Resolution:* Certain agents elicit potent activation of PXR together with marked induction of CYP3A enzyme levels but fail to give a corresponding increase in CYP3A4-dependent metabolism. This well-documented phenomenon stems from the irreversible, time-dependent inhibition of CYP3A4-mediated metabolism by the inducer.

6 Conclusions

The assays described here represent a rapid, easy, reproducible method to screen drug candidates for the ability to activate PXR and hence, predict drug-drug interactions based on enzyme or transporter induction. The potential for drug candidates to elicit DDIs is attracting more attention during early discovery and development since these interactions can profoundly impact patient safety, concomitant drug administration, dosing schemes, product labeling strategies, and a given drug's overall marketability. Minimizing DDIs through detection of P450 inducers in drug discovery can reduce the costs of development and reduce liabilities associated with DDIs. Furthermore, early detection of PXR activation can provide valuable information regarding structure activity relationships that can direct chemistry towards the synthesis of drugs that lack this ability in the case of DDIs or activate the receptor in the case of producing PXR target drugs. Assays such as those described here can be used in the early stages of drug discovery to minimize development costs.

A clear advantage to stable cell lines over other nuclear receptor assays is that the receptor levels and transporter levels do not deviate from one lot to another, eliminating some of the variability associated with transient cells. Therefore, confidence in prediction is greater for very strong and very weak inducers. In a recent report, Fahmi et al. [49] demonstrated that *in vitro* data may be used to predict the effect of inducer (i.e., perpetrator) drugs on the AUC of object (i.e., victim) drugs using the PXR stable transfectant, DPX2™ cells. The prediction requires three parameters, namely EC_{50}, E_{max} and, the [I] *in vivo* [19]. With this approach a strong correlation ($R^2 = 0.90$) was found between the predicted enzyme induction deduced from receptor activation results and the percent decrease in the victim, midazolam, AUC for 19 clinical inducers and 15 clinical non-inducers [49]. Most importantly, no false negatives were obtained among the 34 therapeutics analyzed in the DPX2™ cell-based transactivation assay, and PXR activation data obtained with the 15 clinical non-inducing agents failed to translate into DDI predictions of significance. Thus, there are many advantages to using PXR activation assays and, in particular, stable cell lines. As with any *in vitro* system, there are caveats in predicting clinical induction including uncertainty in predicting the exact magnitude of drug interactions for more moderate inducers. Although infrequent, other caveats associated with the PXR activation assays are the inability to predict induction of a metabolite when the parent compound does not exhibit this property. Moreover, If the parent compounds and metabolites are inducers and if exposure to these metabolites is significant, the RIS could under predict clinical induction [47].

Considering these caveats, assessing PXR activation remains an important utility for predicting clinical DDIs. An alternative would be to perform all CYP3A4 induction studies *in vivo* and conducting clinical DDI trials for every potential co-administered agent is not practical or cost-effective. Thus, *in vitro* studies, such as those described here, play a significant role in the drug discovery process in several ways. One way is that results from PXR activation studies can provide information regarding the nature and extent of *in vivo* studies that may be required to confirm potential DDIs. Still another way, is that results from these *in vitro* studies serve as a screening mechanism to rule out the need for additional *in vivo* studies [1]. Taken together, applications such as those described here could streamline the drug discovery/development process as it pertains to P450 induction.

References

1. U.S. Department of Health and Human Services, Food and Drug Administration, Center for Drug Evaluation and Research (CDER) (2012) Drug interaction studies-study design, data analysis, implications for dosing and labeling recommendations. Guidance for Industry, pp 15–34
2. Back DJ, Breckenridge AM, Crawford FE, Hall JM, Maciver M, Orme ML, Rowe PH, Smith E, Watts MJ (1980) The effect of rifampicin on the pharmacokinetics of ethynylestradiol in women. Contraception 21:135–143
3. Zhang H, Cui D, Wang B, Han YH, Balimane P, Yang Z, Sinz M, Ridrigues AD (2007) Pharmacokinetic drug interactions involving 17alpha-ethinylestradiol: a new look at an old drug. Clin Pharmacokinet 46:133–157
4. Kreek MJ, Garfield JW, Gutjahr CL, Giusti LM (1976) Rifampicin-induced methadone withdrawal. N Engl J Med 294:1104–1106
5. Raistrick D, Hay A, Wolff K (1996) Methadone maintenance and tuberculosis treatment. BMJ 313:925–926
6. Modry DL, Stinson EB, Oyer PE, Jamieson SW, Baldwin JC, Shumway NE (1985) Acute rejection and massive cyclosporine requirements in heart transplant recipients treated with rifampicin. Transplantation 39:313–314
7. Hebert MF, Roberts JP, Prueksaritanont T, Benet LZ (1992) Bioavailability of cyclosporine with concomitant rifampicin administration is markedly less than predicted by hepatic enzyme induction. Clin Pharmacol Ther 52:453–457
8. Fahmi O, Ripp S (2010) Evaluation of models for predicting drug-drug interactions due to induction. Expert Opin Drug Metab Toxicol 6:1399–1416
9. Black M, Raucy JL (1986) Acetaminophen, alcohol, and cytochrome P450. Ann Intern Med 104:427–429
10. Frye RF (2004) Probing the world of cytochrome P450 enzymes. Mol Interv 4:157–162
11. Shimada T, Yamazaki H, Mimura M, Inui Y, Guengerich FP (1994) Interindividual variations in human liver cytochrome P450 enzymes involved in the oxidation of drugs, carcinogens and toxic chemicals: studies with liver microsomes of 30 Japanese and 30 Caucasians. J Pharmacol Exp Ther 270:414–423

12. Maglich JM, Parks DJ, Moore LB, Collins JL, Goodwin B, Billin AN, Stoltz CA, Kliewer SA, Lambert MH, Willson TM, Moore JT (2003) Identification of a novel human CAR agonist and its use in the identification of CAR target genes. J Biol Chem 278:17277–17283
13. Moore JT, Kliewer SA (2000) Use of the nuclear receptor PXR to predict drug interactions. Toxicology 153:1–10
14. Xie W, Yueh MF, Radominska-Pandya A, Saini SP, Negishi Y, Bottroff BS, Cabrera GY, Tukey RH, Evans RM (2003) Control of steroid, heme, and carcinogen metabolism by nuclear pregnane X receptor and constitutive androstane receptor. Proc Natl Acad Sci USA 100:4150–4155
15. Guzelian J, Barwick JL, Hunter L, Phang TL, Quattrochi LC, Guzelian PS (2006) Identification of genes controlled by the pregnane x receptor by microarray analysis of mRNAs from pregnenolone 16alpha-carbonitrile-treated rats. Toxicol Sci 94:379–387
16. Synold TW, Dussault I, Forman BM (2001) The orphan nuclear receptor SXR coordinately regulates drug metabolism and efflux. Nat Med 7:584–590
17. Sueyoshi T, Kawamoto T, Zelko I, Honkakoski P, Negishi M (1999) The repressed nuclear receptor CAR responds to phenobarbital in activating the human CYP2B6 gene. J Biol Chem 274:6043–6046
18. Raucy JL, Mueller L, Duan K, Allen SW, Strom S, Lasker JM (2002) Expression and induction of CYP2C P450 enzymes in primary cultures of human hepatocytes. J Pharmacol Exp Ther 302:475–482
19. Ekins S, Chang C, Mani S, Krasowski MD, Reschly EJ, Iyer M, Kholodovych V, Ni A, Welsh WJ, Sinz M, Swaan PW, Patel R, Bachmann K (2007) Human pregnane X receptor antagonists and agonists define molecular requirements for different binding sites. Mol Pharmacol 72:592–603
20. Shukla SJ, Nguyen D-T, MacArhur R, Simeonov A, Frazee WJ, Hallis TM, Marks BD, Singh U, Eliason HC, Printen J, Austin CP, Inglese J, Auld DS (2009) Identification of pregnane X receptor ligands using time-resolved fluorescence resonance energy transfer and quantitative high-throughput screening. Assay Drug Dev Technol 7:143–169
21. Shukla SJ, Sakamuru S, Huang R, Moeller TA, Shinn P, Van Leer D, Auld DS, Austin CP, Xia M (2011) Identification of clinically utilized drugs that activate pregnane X receptors. Drug Metab Dispos 39:151–159
22. Kliewer SA, Goodwin B, Willson TM (2002) the nuclear pregnane X receptor: a key regulator of xenobiotic metabolism. Endocr Rev 23:687–702
23. Dai G, He L, Bu P, Wan YJ (2008) Pregnane X receptor is essential for normal progression of liver regeneration. Hepatology 47:1277–1287
24. Roth A, Looser R, Kaufmann M, Blattler SM, Rencurel F, Huang W, Moore DD, Meyer UA (2008) Regulatory cross-talk between drug metabolism and lipid homeostasis: constitutive androstane receptor and pregnane X receptor increase Insig-1 expression. Mol Pharmacol 73:1282–1289
25. Zhou J, Febbraio M, Wada T, Zhai Y, Kuruba R, He J, Lee JH, Khadem S, Ren S, Li S, Silverstein RL, Xie W (2008) Hepatic fatty acid transporter Cd36 is a common target of LXR, PXR and PPARgamma in promoting steatosis. Gastroenterology 134:556–567
26. Bachmann K, Patel H, Batayneh Z, Slama J, White D, Posey J (2004) PXR and the regulation of apoA1 and HDL-cholesterol in rodents. Pharmacol Res 50:237
27. Ricketts ML, Boekschoten MV, Kreeft AJ, Hooiveld GJ, Moen CJ, Muller M, Frants RR, Kasanmoentalib S, Post SM, Princen HMG, Porter JG, Katan MB, Hofker MH, Moore DD (2007) The cholesterol-raising factor from coffee beans, cafestol, as an agonist ligand for the farnesoid and pregnane X receptors. Mol Endocrinol 21:1603–1616
28. Wada T, Gao J, Xie W (2009) PXR and CAR in energy metabolism. Trends Endocrinol Metab 20:273–279
29. Xie W, Radominska-Pandya A, Simon CM, Nelson MC, Ong ES (2001) An essential role for nuclear receptors SXR/PXR in detoxification of cholestatic bile acids. Proc Natl Acad Sci USA 98:3375–3380
30. Guo GL, Lambert G, Negishi M, Ward JM, Brewer HB, Kliewer SA, Gonzalez FJ, Sinal CJ (2003) Complementary roles of farnesoid X receptor, pregnane X receptor and constitutive androstane receptor in protection against bile acid toxicity. J Biol Chem 278:45062–45071
31. Zhou C, Verma S, Blumberg B (2009) The steroid and xenobiotic receptor (SXR), beyond xenobiotic metabolism. Nucl Recept Signal 7:1–21
32. Zhou J, Zhai Y, Mu Y, Gong H, Uppal H, Toma D, Ren S, Evans RM, Xie W (2006) A novel pregnane X receptor-mediated and sterol regulatory element-binding protein-independent lipogenic pathway. J Biol Chem 281:15013–15020
33. Kodama S, Koike C, Negishi M, Yamamoto Y (2004) Nuclear receptors CAR and PXR cross talk with FOXO1 to regulate genes that encode drug-metabolizing and gluconeogenic enzymes. Mol Cell Biol 24:7931–7940
34. Kodama S, Moore R, Yamamoto Y, Negishi M (2007) Human nuclear pregnane x receptor cross-talk with CREB to repress cAMP activation

of the glucose-6-phosphatase gene. Biochem J 407:373–381
35. Yoon JC, Puigserver P, Chen GF, Donovan JL, Wu Z, Rhee J, Adelmant G, Stafford J, Kahn CR, Granner DK, Newgard CB, Spiegelman BM (2001) Control of hepatic gluconeogenesis through the transcriptional coactivator PGC-1. Nature 413:131–138
36. Chen T (2008) Nuclear receptor drug discovery. Curr Opin Chem Biol 12:1–9
37. Zollner G, Trauner M (2009) Nuclear receptors as therapeutic targets in cholestatic liver diseases. Br J Pharmacol 156:7–27
38. Raucy JL, Warfe L, Yueh MF, Allen SW (2002) A cell-based reporter gene assay for determining induction of CYP3A4 in a high volume system. J Pharmacol Exp Ther 303:412–423
39. Raucy JL, Lasker JM (2010) Current in vitro high throughput screening approaches to assess nuclear receptor activation. Curr Drug Metab 11:806–814
40. Bjornsson TD, Callaghan JT, Einolf HJ, Fischer V, Gan L, Grimm S, Kao J, King SP, Miwa G, Ni L, Kumar G, McLeod J, Obach RS, Roberts S, Roe A, Shah A, Snikeris F, Sullivan JT, Tweedie D, Vega JM, Walsh J, Wrighton SA, Pharmaceutical Research and Manufacturers of American (PhRMA) Drug Metabolism/Clinical Pharmacology Technical Working Group; FDA Center for Drug Evaluation and Research (CDER) (2003) The conduct of in vitro and in vivo drug-drug interaction studies: a Pharmaceutical Research and Manufacturers of America (PhRMA) perspective. Drug Metab Dispos 31:815–832
41. Chu V, Einolf HJ, Evers R, Kumar G, Moore D, Ripp SL, Silva J, Sinha V, Sinz M, Skerjanec A (2009) In vitro and in vivo induction of cytochrome P450: a survey of the current practices and recommendations: a pharmaceutical research and manufacturers of America perspective. Drug Metab Dispos 37:1339–1354
42. Sinz M, Kim S, Zhu Z, Chen T, Anthony M, Dickinson K, Rodrigues AD (2006) Evaluation of 170 xenobiotics as transactivators of human pregnane X receptor (hPXR) and correlation to known to CYP3A4 drug interactions. Curr Drug Metab 7:375–388
43. Shou M, Hayashi M, Pan Y, Xu Y, Morrissey K, Xu L, Skiles GL (2008) Modeling, prediction, and in vitro in vivo correlation of CYP3A4 induction. Drug Metab Dispos 36:2355–2370
44. Fahmi OA, Kish M, Boldt S, Obach RS (2010) Cytochrome P450 3A4 mRNA is a more reliable marker than CYP3A4 activity for detecting pregnane X receptor-activated induction of drug-metabolizing enzymes. Drug Metab Dispos 38:1605–1611
45. Almond LM, Yang J, Jamei M, Tucker GT, Rostami-Hodjegan A (2009) Towards a quantitative framework for the prediction of DDIs arising from cytochrome P450 induction. Curr Drug Metab 10:420–432
46. Fahmi OA, Hurst S, Plowchalk D, Cook J, Guo F, Youdim K, Dickins M, Phipps A, Darekar A, Hyland R, Obach RS (2009) Comparison of different algorithms for predicting clinical drug-drug interactions, based on the use of CYP3A4 in vitro data: predictions of compounds as precipitants of interaction. Drug Metab Dispos 37:1658–1666
47. Ripp SL, Mills JB, Fahmi OA, Trevena KA, Liras JL, Maurer TS, deMorais SM (2006) Use of immortalized human hepatocytes to predict the magnitude of clinical drug-drug interactions caused by CYP3A4 induction. Drug Metab Dispos 34:1742–1748
48. Fahmi OA, Boldt S, Kish M, Obach RS, Tremaine LM (2008) Prediction of drug-drug interactions from in vitro induction data. Drug Metab Dispos 36:1971–1974
49. Fahmi O, Raucy J, Ponce E, Hassanali S, Lasker J (2012) The utility of DPX2 cells for predicting CYP3A induction-mediated drug-drug interactions and associated structure-activity relationships. Drug Metab Dispos 40:2204–2211

Chapter 11

Characterization of Constitutive Androstane Receptor (CAR) Activation

Caitlin Lynch, Haishan Li, and Hongbing Wang

Abstract

As the closest relative of the aforementioned PXR, the constitutive androstane/activator receptor (CAR, NR1I3) also governs the transcription of numerous hepatic drug-metabolizing enzymes and transporters in response to various xenobiotic exposures. Unlike most prototypical nuclear receptors, however, CAR can be activated *via* both direct ligand-binding and ligand-independent indirect mechanisms. Moreover, whereas CAR predominantly resides in the cytoplasm of primary hepatocytes in the absence of chemical stimulation, it simultaneously localizes in the nucleus of nearly all immortalized cell lines and is constitutively activated without chemical activation, making in vitro identification of CAR activators extremely challenging. In this chapter, we provide detailed step-by-step instructions for the application of two recently developed in vitro human (h) CAR activation assays: the hCAR1+A-based luciferase assay in HepG2 cells and the adenovirus-based hCAR translocation assay in human primary hepatocytes. In combination, these assays are efficient in the identification of both direct and indirect activators of hCAR in vitro.

Key words CAR, Cytochrome P450, Activation

1 Introduction

The constitutive androstane/active receptor (CAR, NR1I3), originally identified as an orphan nuclear receptor, transcriptionally regulates the expression of a broad spectrum of hepatic genes associated with drug metabolism and transportation, including phase I oxidation enzymes [e.g., cytochrome P450s (CYP)], phase II conjugation enzymes (e.g., UDP-glucuronosyltransferases), as well as phase III efflux transport proteins such as multidrug resistance proteins [1–4]. Along with several other xenobiotic sensors such as PXR and AhR, CAR is predominantly expressed in the liver; the largest metabolic organ in the body. To date, accumulating evidence exhibits that many clinically used drugs that activate CAR can alter the metabolism and clearance of their own or co-administered drugs, leading to pharmacokinetic-related drug-drug interactions [5, 6]. In addition to the well-established role of CAR in drug metabolism/detoxification, recently published results have also implicated that CAR is involved in other hepatic functions, such as energy homeostasis and the development of cancers in rodent animal models [7–13]. As such, efficient identification of CAR activators in vitro is of great interest in both drug development and clinical practice.

Sitting on the same branch of the nuclear receptor superfamily tree, CAR represents the closest relative of PXR. Whereas CAR and PXR share a number of overlapping chemical activators and transcriptional targets, the molecular mechanisms underlying their activation are quite different. Unlike prototypical nuclear receptors such as PXR, CAR exhibits unique subcellular localization and activation patterns between immortalized cell lines and physiologically relevant primary hepatocytes. In an intact liver and primary hepatocytes, CAR localizes in the cytoplasm in the absence of xenobiotic stimulation and translocates into the nucleus after exposure to activators [14, 15]. On the other hand, CAR is simultaneously localized in the nucleus and constitutively activated in immortalized cell lines [15–19]. Furthermore, CAR could be activated *via* either direct ligand binding, such as the selective hCAR agonist 6-(4-chlorophenyl) imidazo[2,1-b][1,3]–thiazole-5-carbaldehyde-O-(3,4-dichlorobenzyl)oxime (CITCO) or ligand-independent (indirect) mechanisms, such as phenobarbital [1, 20]. Together, these distinctive characteristics of CAR significantly limited the application of both the cell-based luciferase assays using wild-type CAR and the typical CAR-ligand binding assays, making in vitro identification of CAR activators an extremely challenging task.

In contrast to the constitutive activation of wild-type CAR, several naturally occurring alternative splicing variants of hCAR, such as hCAR2 (contains a 4-amino acid insertion) and hCAR3 (contains a 5-amino acid insertion), have recently been shown to exhibit low basal but ligand-induced activation in several cell lines [21–23]. Further delineating the functional relevance of these hCAR variants, we identified that the insertion of alanine (A) alone at position 270 is sufficient to switch the constitutively activated wild-type CAR to a xenobiotic-responsive receptor [16]. The generated hCAR1+A construct exhibits robust xenobiotic response over hCAR3 and it maintains the chemical specificities correlated to wild-type hCAR [16, 24]. Thus, cell-based hCAR1+A luciferase assays may represent a sensitive surrogate of hCAR for the identification of hCAR activators in vitro.

Meanwhile, although the underlying molecular mechanism(s) of activator-mediated CAR translocation to the nucleus remains to be elucidated, the unique feature of CAR localization and translocation in primary hepatocytes provides another attractive model for in vitro identification of CAR activators in a potentially high-throughput manner. One of the drawbacks of utilizing primary hepatocyte cultures, however, is the quiescent nature of these cells in vitro, by which the efficiency of chemical-based transfection is extremely low. To circumvent this shortcoming, we have generated a functional adenoviral-enhanced yellow fluorescent protein-tagged hCAR (Ad/EYFP-hCAR) construct that infects human primary hepatocytes with high efficiency and yet maintains hCAR distribution features in a physiologically relevant fashion [15]. Using this experimental system, significant correlations have been observed between the chemical-mediated nuclear accumulation of Ad/EYFP-hCAR in human primary hepatocytes and hCAR activation/target gene induction [15, 24].

These two approaches overcome significant difficulties associated with in vitro CAR activation and represent attractive methodologies for efficient identification of hCAR activators. Nonetheless, several limitations of these assays should also be realized in that the hCAR1+A luciferase assay appears to be more sensitive to direct activators (ligands) over indirect activators [16, 24], while in hepatocyte-based hCAR translocation assays, potent deactivators of hCAR such as clotrimazole and PK11195 also drive Ad-EYFP-hCAR into the nucleus of human primary hepatocytes [15]. Notably, these two methodologies compensate each other and together they offer a valuable avenue for the identification of hCAR activators in vitro. In the following context, we provide comprehensive step-by-step instructions for the application of the hCAR1+A-based luciferase assays in HepG2 cells and the Ad-EYFP-hCAR nuclear translocation assays in human primary hepatocytes.

2 Materials

2.1 Equipment

1. GLOMAX® 20/20 Single-tube Luminometer (Promega, Madison, WI) for dual-luciferase assays.
2. Nikon C1-LU3 confocal laser scanning microscopy (Melville, NY, USA) for imaging.
3. 5430R tabletop centrifuge (Eppendorf AG, Hamburg, Germany) for cell preparation.
4. Milli-Q Synthesis A10 water purification system (Millipore, Billerica, MA) for buffer preparation.
5. 1300 B2 biological safety cabinet (Thermo Fisher Scientific, Asheville, NC) for all cell operation and infection.
6. CO_2 incubator MCO-17AIC (SANYO, Wood Dale, IL) for all cell cultures.

2.2 Reagents and Solutions

1. Seeding medium for HepG2 cells: DMEM (Invitrogen, Carlsbad, CA) 500 mL, add 50 mL fetal bovine serum (FBS) (Sigma, St. Louis, MO), and 5 mL penicillin/streptomycin (100×) (Invitrogen, Carlsbad, CA).

2. Passive lysis buffer (PLB): passive lysis buffer (5×) (Promega, Madison, WI) diluted to 1× with distilled water before use.

3. Opti-MEM® I (1×) reduced-serum medium (Life Technologies, Grand Island, NY).

4. Transfection medium: DMEM 500 mL, add 50 mL fetal bovine serum (FBS) without antibiotics.

5. Plasmids: pGL3-CYP2B6 2.2 kb reporter [25], pCR3-hCAR1+A [16], and pRL-TK (pRenilla Luciferase-Thymadine Kinase) (Promega, Madison, WI).

6. Dual-luciferase® Reporter Assay System (Promega, Madison, WI)

7. Collagen solution: Prepare MCDI (N-Cyclohexyl-N'-(2-morpholinoethyl) carbodiimide metho-p-toluenesulfonate, Sigma, St. Louis, MO) solution (130 µg/mL) in distilled water (**Note 1**). Make a 100 µg/mL solution of collagen (type I from rat tail, BD Biosciences, Bedford, MA) in MCDI solution.

8. 1 M KPO_4: 1 M K_2HPO_4 and 1 M KH_2PO_4 (5.3:1, pH 7.4).

9. Seeding medium for hepatocytes: DMEM (Invitrogen, Carlsbad, CA) 500 mL, add 25 mL fetal bovine serum (FBS) (Sigma, St. Louis, MO), 5 mL penicillin/streptomycin (100×) (Invitrogen, Carlsbad, CA), Insulin 0.5 mL (4 mg/mL), and 50 µL dexamethasone (10 mM) (Sigma, St. Louis, MO).

10. Culture medium for hepatocytes: William's medium 500 mL, add 5 mL ITS (BD Biosciences, Bedford, MA), 5 mL l-glutamine (Invitrogen, Carlsbad, CA), 5 mL penicillin/streptomycin (100×), 5 µL dexamethasone (10 mM).

3 Methods

3.1 hCAR1+A-Based Dual-Luciferase Reporter Assay

1. *Plating HepG2 cells.*

    1.1. The human hepatoma cell line (HepG2) is plated onto a collagen coated plate to produce better monolayer cultures. The plates are coated prior to experiment. Add the appropriate volume of cold collagen solution to each well (Table 1) and swirl to distribute evenly over the well surface (**Note 2**). Incubate plates at 37 °C in a humidified chamber with 5 % CO_2 overnight (**Note 3**). Remove plates and aspirate excess volume. Add PBS to each well. Store all plates at 4 °C until ready to plate cells (**Note 4**).

    1.2. Warm HepG2 seeding medium to 37 °C.

Table 1
Preparation of collagen-coated plates

Number of wells per plate	Collagen solution (mL/well)	1× PBS (mL/well)
6	2	2
12	1	1
24	0.5	0.5
96	0.1	0.1

Table 2
Seeding and harvesting of HepG2 cells

Number of wells per plate	Seeding density (cells/well)	PLB (1×) (µL)
24	1×10^5	100
48	5×10^4	75

1.3. Cells can either be plated in a 24- or 48-well collagen coated plate (Table 2). Cells should be plated using the pre-warmed HepG2 seeding medium (**Note 5**) overnight (in a humidified incubator kept at 37 °C, with 5 % CO_2) or until cells are firmly attached to the bottom of the plate.

2. *Transfecting HepG2 cells.*

   2.1. Make sure cells are about 50–80 % confluent for best transfection results.

   2.2. Combine the appropriate amounts (Table 3) of Opti-MEM® and FuGENE® 6 reagent together and let it sit at room temperature for 5 min (**Note 6**). Then add the reporter, nuclear receptor, and background vectors. After all ingredients are combined, let the mix sit for 20 min at room temperature.

   2.3. Replace seeding medium with 500 and 250 µL of transfection medium for 24- and 48-well plates, respectively.

   2.4. Add specified amounts of the transfection mix into each well. The cells should be transfected overnight before treatment.

Table 3
Transfection mix

	24-well plate (µL)	48-well plate (µL)
Opti-MEM®	7.8	3.9
CYP2B6 (100 ng/µL)	0.6	0.3
hCAR1+A (100 ng/µL)	0.3	0.15
pRL-TK (10 ng/µL)	1	0.5
FuGENE® 6	0.3	0.15
	10 µL/well	5 µL/well

3. *Treatment*

   3.1. Warm HepG2 seeding medium to 37 °C (**Note 7**).

   3.2. Transfected cells will be treated with test compounds, 0.1 % DMSO as negative control, or CITCO (1 µM) as positive control for 24 h. All treatment medium should contain 0.1 % DMSO. Cytotoxicity for new compounds should be tested in preliminary experiments.

4. *Harvesting*

   4.1. Wash the treated cells twice with PBS.

   4.2. After removing PBS, add specified amount of PLB as shown in Table 2 (**Note 8**).

   4.3. Put entire plate into a −20 °C freezer to let the cells go through one freeze cycle before performing the next step.

5. *Dual-luciferase Assay*

   5.1 Make firefly and renilla luciferase assay reagents according to the manufacturers' protocol (**Note 9**).

   5.2 Take out plate from freezer and let it thaw shaking at medium speed.

   5.3 While plate is thawing, get the luminometer ready. Turn on machine and bring up the GLOMAX® SIS program on the accompanying computer.

   5.4 Push the start button on the computer screen and an excel file will be brought up to coincide with the readings on the luminometer.

   5.5 On the luminometer screen, touch "Protocols", "Run Promega Protocol", and finally "DLR-0-INJ". The integration should be set to 10 s. Touch "OK". The luminometer is now ready to read the samples.

   5.6 Once the plate is thawed, carry the rest of the experiment out while keeping the plate on ice.

5.7 Put 25 μL of firefly reagent into a 1.5 mL tube.

5.8 Add 15 μL of cell lysis from one well into the tube and mix. Make sure not to create any bubbles as this may cause an inaccurate reading.

5.9 Put tube into luminometer and touch "Measure Luminescence".

5.10 When the machine is done reading the firefly measurement, it will show a prompt to add the renilla. Add 25 μL of renilla to the same tube and mix up and down, be careful to not create bubbles.

5.11 The machine will then show a firefly, renilla, and ratio reading. The same reading will be on the excel sheet which was opened previously.

5.12 Repeat Steps 5.7–5.11 until all wells are completed.

5.13 Compare all treatments to the negative control by dividing each treatment by the average DMSO ratio value among the three wells. For example:

DMSO #1: 1.039, DMSO #2: 1.056, DMSO #3: 0.998

Average DMSO: $(1.039 + 1.056 + 0.998)/3 = \mathbf{1.031}$

CITCO #1: 13.057, CITCO #2: 13.456, CITCO #3: 12.897

Fold Induction of CITCO:

$[(13.057/1.031) + (13.456/1.031) + (12.897/1.031)]/3 = \mathbf{12.74}$

This means the CITCO activated hCAR 12.74 more times than having no hCAR activation.

5.14 Once all values compared to DMSO are calculated, a comparison of the fold value of CITCO vs. tested compounds should be made to determine whether or not a compound is a possible hCAR activator (**Note 10**).

3.2 Nuclear Translocation of CAR

Ad-EYFP-hCAR infects human primary hepatocytes with high efficiency, and the majority of Ad/EYFP-hCAR (>80 %) is expressed in the cytoplasm of non-induced human primary hepatocytes and is translocated to the nucleus in response to activators and antagonists of human CAR.

1. *Human primary hepatocyte culture*

    1.1. Human primary hepatocytes were prepared by using a modified two-step collagenase digestion as described previously [26] or obtained through commercial sources such as Life Technologies (Durham, NC) or Celsis In Vitro Technologies (Baltimore, MD).

1.2. Cell count and viability assessment:
   1.2.1. Place hepatocyte suspension on ice while the cell count is being carried out.
   1.2.2. Mix 100 µL of cell suspension, 100 µL of Trypan blue and 800 µL of DMEM into a 1.5 mL microcentrifuge tube. Viability of the cells is determined by Trypan blue exclusion.
   1.2.3. Add 10 µL of above mixed hepatocytes to the hemocytometer slide. Count the number of live and dead cells on four separate quadrants. Dead cells are those unable to exclude Trypan blue.
   1.2.4. Example of cell count and calculations:

   Counted 200 live cells and 10 dead cells, total count of 210 cells.

   Viability = 95 % (200/210 = 95 %).

   Total number of viable cells in 40 mL stock = 200×10^6 cells.

   Cell density: $200 \times 10^6 / 40$ mL = 5×10^6/mL.

1.3. Dilute the cell suspension with seeding medium to give the required final cell density as listed in Table 4.
1.4. Ensure that the cell suspension remains homogenous by gently swirling whilst seeding the cells. Seed required volume of cells per collagen-coated dish and place in an incubator.
1.5. Check cell density under microscope, and adjust if required.
1.6. Gently swirl the dishes in a figure eight pattern to ensure formation of an even monolayer.
1.7. Allow hepatocytes to attach for 3–6 h. Then change the seeding medium to hepatocyte culture medium.

Table 4
Seeding density of human primary hepatocytes

Type of dish or well	Seeding density	Volume/dish or well	Total number of viable cell
60 mm dish	1×10^6–1.33×10^6	3 mL	3×10^6–4×10^6
6-well plate	5×10^5–7.5×10^5	2 mL	1×10^6–1.5×10^6
12-well plate	5×10^5–7.5×10^5	1 mL	5×10^5–7.5×10^5
24-well plate	5×10^5–7.5×10^5	0.5 mL	2.5×10^5–3.75×10^5
96-well plate	5×10^5	125 µL	6.25×10^4

2. *Infection and treatment.*
    2.1. Hepatocyte cultures are infected with 2 µL of Ad/EYFP-hCAR for 12 h before treatment with vehicle control (0.1 % DMSO), positive control (phenobarbital 1 mM), or test compounds.
    2.2. After 24 h of treatment, hepatocytes are washed twice with PBS and fixed for 30 min in 4 % buffered paraformaldehyde.
    2.3. The cells were then stained with 4,6-diamidine-2-phenyl-indoledihydrochloride (DAPI) for 30 min.
3. *Imaging and calculation*
    3.1. Laser scanning confocal microscopy is performed on a Nikon C1-LU3 confocal microscope equipped with a Nikon TE2000 inverted microscope. Laser beams with

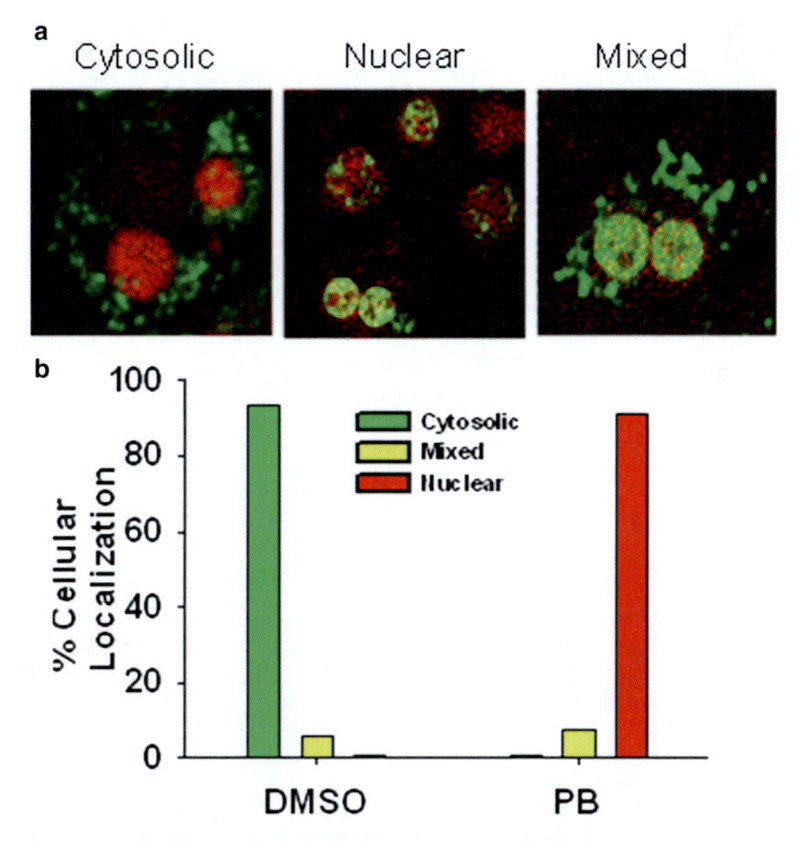

Fig. 1 Localization and translocation of hCAR in human primary hepatocytes. (**a**) Representative images depict EYFP-hCAR expression in cytosolic, nuclear, or mixed localization of human primary hepatocytes. (**b**) Ad/EYFP-hCAR infected human primary hepatocytes were treated with vehicle control (0.1 % DMSO) or phenobarbital (PB, 1 mM) for 24 h. Approximately 120 EYFP-hCAR expressing hepatocytes were classified according to their hCAR localization status. As expected, treatment with PB resulted in significant nuclear accumulation of hCAR

513-, and 340- nm excitation wavelengths were used for EYFP, and DAPI imaging, respectively.

3.2. For each treatment, approximately 100 cells expressing EYFP-hCAR are counted and classified as cytosolic, nuclear, or mixed (cytosolic + nuclear) CAR localization. Figure 1 shows a representative example of EYFP-hCAR localization and phenobarbital induced nuclear accumulation of hCAR in human primary hepatocytes.

4 Notes

1. Recommended expiration date: 6 months at 2–8 °C.
2. Collagen/MCDI solution is freshly prepared before use.
3. If part of the bottom surface of a well dries during storage, the well should not be used.
4. Recommended expiration date of plates: 2 months following coating.
5. PBS should be removed from the well before any medium or cells are added.
6. Once the FuGENE® 6 is added, do not vortex the solution. Gently pipette up and down to mix.
7. HepG2 seeding medium can be used as treatment medium.
8. Make sure to add the PLB evenly to each well. The buffer must touch the entire well so as to get uniform lysis from every cell.
9. All leftover reagents should be kept at −80 °C until the next time the assay is performed.
10. Anything 40 % of the CITCO value is a possible potent inducer of hCAR. If the fold value is between 15 and 40 % of the CITCO value, the compound is a possible mild inducer of hCAR.

References

1. Honkakoski P et al (1998) The nuclear orphan receptor CAR-retinoid X receptor heterodimer activates the phenobarbital-responsive enhancer module of the CYP2B gene. Mol Cell Biol 18(10):5652–5658
2. Moore LB et al (2000) Orphan nuclear receptors constitutive androstane receptor and pregnane X receptor share xenobiotic and steroid ligands. J Biol Chem 275(20):15122–15127
3. Wang HB, LeCluyse EL (2003) Role of orphan nuclear receptors in the regulation of drug-metabolising enzymes. Clin Pharmacokinet 42(15):1331–1357
4. Qatanani M, Moore DD (2005) CAR, the continuously advancing receptor, in drug metabolism and disease. Curr Drug Metab 6(4):329–339
5. Yap KY, Chui WK, Chan A (2008) Drug interactions between chemotherapeutic regimens and antiepileptics. Clin Ther 30(8):1385–1407
6. Ma Q, Lu AY (2008) The challenges of dealing with promiscuous drug-metabolizing enzymes, receptors and transporters. Curr Drug Metab 9(5):374–383
7. Chakraborty S, Kanakasabai S, Bright JJ (2011) Constitutive androstane receptor

agonist CITCO inhibits growth and expansion of brain tumour stem cells. Br J Cancer 104(3):448–459
8. Wada T, Gao J, Xie W (2009) PXR and CAR in energy metabolism. Trends Endocrinol Metab 20(6):273–279
9. Yamamoto Y et al (2004) The orphan nuclear receptor constitutive active/androstane receptor is essential for liver tumor promotion by phenobarbital in mice. Cancer Res 64(20): 7197–7200
10. Huang W et al (2005) Xenobiotic stress induces hepatomegaly and liver tumors via the nuclear receptor constitutive androstane receptor. Mol Endocrinol 19(6):1646–1653
11. Gao J et al (2009) The constitutive androstane receptor is an anti-obesity nuclear receptor that improves insulin sensitivity. J Biol Chem 284(38):25984–25992
12. Dong B et al (2009) Activation of nuclear receptor CAR ameliorates diabetes and fatty liver disease. Proc Natl Acad Sci USA 106(44):18831–18836
13. Maglich JM et al (2004) The nuclear receptor CAR is a regulator of thyroid hormone metabolism during caloric restriction. J Biol Chem 279(19):19832–19838
14. Kawamoto T et al (1999) Phenobarbital-responsive nuclear translocation of the receptor CAR in induction of the CYP2B gene. Mol Cell Biol 19(9):6318–6322
15. Li H et al (2009) Nuclear translocation of adenoviral-enhanced yellow fluorescent protein-tagged-human constitutive androstane receptor (hCAR): a novel tool for screening hCAR activators in human primary hepatocytes. Drug Metab Dispos 37(5):1098–1106
16. Chen T et al (2010) A single amino acid controls the functional switch of human constitutive androstane receptor (CAR) 1 to the xenobiotic-sensitive splicing variant CAR3. J Pharmacol Exp Ther 332(1):106–115
17. Haché RJG et al (1999) Nucleocytoplasmic trafficking of steroid-free glucocorticoid receptor. J Biol Chem 274(3):1432–1439
18. Rowlands JC, Gustafsson J-Å (1997) Aryl hydrocarbon receptor-mediated signal transduction. Crit Rev Toxicol 27(2):109–134
19. Zelko I et al (2001) The peptide near the C terminus regulates receptor CAR nuclear translocation induced by xenochemicals in mouse liver. Mol Cell Biol 21(8):2838–2846
20. Maglich JM et al (2003) Identification of a novel human constitutive androstane receptor (CAR) agonist and its use in the identification of CAR target genes. J Biol Chem 278(19): 17277–17283
21. Auerbach SS et al (2005) Retinoid X receptor-alpha-dependent transactivation by a naturally occurring structural variant of human constitutive androstane receptor (NR1I3). Mol Pharmacol 68(5):1239–1253
22. Auerbach SS et al (2007) CAR2 displays unique ligand binding and RXRalpha heterodimerization characteristics. Drug Metab Dispos 35(3):428–439
23. Arnold KA, Eichelbaum M, Burk O (2004) Alternative splicing affects the function and tissue-specific expression of the human constitutive androstane receptor. Nucl Recept 2(1):1
24. Lynch C et al (2013) Identification of novel activators of constitutive androstane receptor from FDA-approved drugs by integrated computational and biological approaches. Pharm Res 30(2):489–501
25. Wang H et al (2003) A novel distal enhancer module regulated by pregnane X receptor/constitutive androstane receptor is essential for the maximal induction of CYP2B6 gene expression. J Biol Chem 278(16): 14146–14152
26. LeCluyse EL et al (2005) Isolation and culture of primary human hepatocytes. Methods Mol Biol 290:207–229

Chapter 12

DNA Binding (Gel Retardation Assay) Analysis for Identification of Aryl Hydrocarbon (Ah) Receptor Agonists and Antagonists

Anatoly A. Soshilov and Michael S. Denison

Abstract

The gel retardation assay (GRA), also referred to as the electromobility shift assay (EMSA), is commonly used technique to examine DNA binding of transcription factors, including activated nuclear receptors, to their specific DNA recognition sites. GRA of the aryl hydrocarbon receptor (AhR) relies on the *in vitro* ability of the cytosolic AhR protein complex to convert into its high affinity DNA binding form following its interaction with and activation by an AhR agonist and, as such, the GRA can be used for detection of such ligands. In addition, examination of the ability of a chemical to block agonist-dependent DNA binding of the AhR provides an avenue to identify AhR antagonists. Accordingly, this assay allows relatively rapid screening and identification of both AhR agonists and/or antagonists and unlike cell-based or *in vivo* assays, it essentially eliminates the confounding effect of cellular metabolism of the test compounds. The GRA can also be used with nuclear extracts obtained from treated cells to further identify and/or characterize compounds capable of stimulating nuclear translocation and DNA binding of the AhR in intact cells. The methods described here can be applied to cytosolic, nuclear and/or whole cell extracts from various species and tissues.

Key words Ah receptor, DNA binding analysis, Gel retardation assay, GRA, EMSA

1 Introduction

The gel retardation assay (GRA) is a method that allows detection and characterization of protein-DNA interactions. In this assay, the binding of a target protein(s) to a labeled or tagged oligonucleotide that contains the specific DNA recognition sequence of the specific target protein(s) can be detected through separation of protein-bound DNA from protein-free DNA using non-denaturing polyacrylamide gel electrophoresis (PAGE) and visualization of the labeled protein:DNA complexes by autoradiography or imaging methods [1]. For most transcription factors, the specificity of DNA binding of a protein(s) is determined by comparison of protein-DNA complexes formed using cell/tissue extracts containing to those lacking the desired protein. Alternatively, this can be accomplished by a competitive GRA using oligonucleotides that contain the DNA binding sequence and a mutated version of the sequence that eliminates protein binding [1–3]. In contrast, numerous ligand-dependent transcription factors or nuclear receptors (such as those for steroid hormones) only bind to their specific DNA recognition sites following their activation by a ligand or cell signaling event and DNA binding analysis has been used to characterize these receptors [4–6]. While there are technical limitations that prevent the use of GRA analysis for assessment of *in vitro* ligand-dependent DNA binding of steroid hormone receptors (steroid receptors bind to DNA in a ligand-independent manner *in vitro*), one ligand-dependent nuclear receptor that can bind to DNA in a ligand-dependent manner *in vitro* is that of the aryl hydrocarbon receptor (AhR) [7, 8]. The AhR is a ligand-dependent basic-helix-loop-helix-Per-ARNT-Sim (bHLH-PAS)-containing transcription factor that responds to exogenous and endogenous chemicals with the induction/repression of a large battery of genes and production of a diverse spectrum of biological and toxic effects in a wide range of species and tissues [9–14]. The GRA, together with a variety of cell culture-based reporter gene techniques, has been used to detect and characterize numerous agonists and antagonists of the aryl hydrocarbon receptor (AhR) [15–22].

GRA of ligand-dependent AhR DNA binding is typically carried out using cytosolic extracts that contain inactive, ligand-free AhR and these extracts can be easily prepared from a variety of species and tissues although hepatic cytosol is the tissue of choice for GRA given the relatively high AhR concentration and large tissue volume. Following incubation of cytosol with an AhR agonist, the AhR undergoes a process termed transformation, wherein it is converted into its high affinity DNA binding form [13]. In the GRA, the binding of ligand-activated AhR to an oligonucleotide containing its specific DNA binding site, the dioxin-responsive element (DRE), can be resolved by PAGE in native/non-denaturing gel and autoradiography or imaging as a ligand-inducible high molecular weight protein:DNA complex with a slower mobility than that of the protein-free labeled DRE oligonucleotide [2, 3, 7, 8]. Due to the observed change in electrophoretic mobility, this technique is also commonly referred to as the electrophoretic mobility shift assay (EMSA).

The GRA has been used to detect and characterize the AhR agonist or antagonist activity of pure chemicals, mixtures of chemicals and extracts from a wide variety of materials [16, 20–22]. While the prototypical and most potent AhR ligands/agonists are halogenated aromatic hydrocarbons such as 2,3,7,8-tetrachlorodibenzo-p-dioxin (TCDD, dioxin) and related metabolically stable dioxin-like compounds (DLCs), recent studies have shown that the AhR can bind and be activated by a wide variety of structurally diverse compounds including environmental contaminants, pharmaceuticals, natural compounds and chemicals in commercial and consumer products [13, 16, 19, 20]. Persistent AhR activation by metabolically stable ligands (i.e. DLCs) can pose a risk to human health due to AhR-mediated immuno-, dermal and hepatotoxicity, endocrine disruption, carcinogenesis and other effects [9, 10, 12–14]. In contrast, the majority of AhR ligands that have been identified are metabolically labile and do not produce persistent changes in AhR-dependent gene expression or TCDD-like toxic effects. More significantly, the recent identification of a role of the AhR and AhR signaling pathway in endogenous physiological functions, tissue and immune cell development and human disease [9, 10, 23–26] has made the AhR a target for the identification and development of therapeutic agents.

While some compounds reportedly activate the AhR in a ligand-independent manner, the majority of the chemicals that affect the AhR and AhR signaling pathways have been shown to be direct ligands of the AhR agonists and/or antagonists and thus *in vitro* AhR-based assays can be used to identify the majority of AhR ligands. Although AhR ligand-binding assays and structure-activity modeling or docking approaches can be used to assess the ability of a substance to interact with the AhR, these assays provide no insight into the functional activity of the ligands (i.e. whether they are agonists and/or antagonists). Use of the GRA allows direct determination of the ability of a test chemical or extract to exert an agonist/antagonist effect though its ability to stimulate/inhibit AhR transformation and DNA binding. AhR-based GRA bioassays are commonly carried out in combination with cell culture-based AhR-responsive reporter gene assays [15–22] since these latter assays provide information that is complementary to that of the GRA (i.e. reveals information about the metabolic stability of the chemical(s) and their ability to stimulate gene expression in intact cells). Thus, the GRA provides confirmation of the molecular mechanism by which a test chemical can stimulate/inhibit ligand-dependent transformation and DNA binding of the AhR and indirectly assesses its potential to affect AhR-dependent signal transduction.

2 Materials

2.1 Equipment

General purpose equipment can be substituted with comparable models from other companies.

1. Polytron or motorized Teflon-glass homogenizer for tissue homogenization.
2. Sorvall RC-5B Plus centrifuge with Sorvall SS-34 rotor and Beckman L-70 ultracentrifuge with Beckman SW70Ti rotor.

3. Polyacrylamide gel electrophoresis (PAGE) power supply (e.g., EC-105, E-C Apparatus Corporation).
4. Vertical gel running unit (e.g. Vertical Gel Electrophoresis System, Life Technologies), gel plates (for approximately 20×16 cm gels) and spacers (1.5 mm thickness) for PAGE separation.
5. Peristaltic pump for recirculation of gel running buffer.
6. 3 MM blotting paper (Whatman) for gel transfer.
7. Slab gel dryer (e.g., SE1160, Hoeffer Scientific).
8. Fluorescent image analyzer FLA-9000 (Fujifilm) and accompanying cassettes and software to visualize and quantitate dried gels or autoradiographic film developer.
9. Microplate reader or spectrophotometer capable of measuring absorbance at 595 nm for the protein assay.
10. An −80 °C freezer for storage of cytosol samples.
11. Centrifuge with swing-bucket rotor and scintillation counter for preparation of ^{32}P-labeled DRE.
12. Standard benchtop microcentrifuge for quick preparation of nuclear extracts.

2.2 Reagents and Solutions

All buffers should be made using MilliQ-purified water or equivalent. All chemicals should be of Molecular biology grade and are obtained from Sigma (St. Louis, MO) or Fisher Scientific (Pittsburgh, PA) unless indicated otherwise.

1. HEDG buffer: 25 mM Hepes-HCl (pH 7.5), 1 mM EDTA, 1 mM DTT, 10 % (v/v) glycerol; stable at 4 °C for 6 months.
2. HEDGK8 buffer: HEDG, 0.8 M KCl; stable at 4 °C for 6 months.
3. BioRad protein assay kit (BioRad, Hercules, CA).
4. T4 polynucleotide kinase and buffer (New England Biolabs, Ipswich, MA); store at −20 °C.
5. [γ^{32}P]-5′-adenosine triphosphate (6,000 Ci/mmol, 10 mCi/ml; Perkin Elmer, Waltham, Ma) store at −20 °C but use as soon as possible (**Note 1**).
6. DRE oligonucleotides (single stranded and PAGE-purified (**Note 2**)): wild type sense strand (5′-GATCTGGCTCTTCTCACGCAACTCCG-3′) and anti-sense strand (5′-GATCCGGAGTTGCGTGAGAAGAGCCA-3′); mutant sense strand (5′-GATCTGGCTCTTCTCAC**A**CAACTCCG-3′) and anti-sense strand (5′-GATCCGGAGTTG**T**GTGAGAAGAGCCA-3′). The core DRE sequence is underlined and the mutation that eliminates AhR binding to the DRE is indicated in bold type.

7. Oligonucleotide reannealing buffer: 67 mM Tris-HCl (pH 7.5), 1.3 mM $MgCl_2$, 7 mM DTT, 1.3 mM spermidine, 13 mM EDTA; stable at −20 °C for 6 months. Prepare as a 5× solution. Sequenase buffer (from various suppliers) can substitute.
8. TE buffer: 10 mM Tris-HCl (pH 8.0), 1 mM EDTA.
9. Sephadex G-50/TE Quick Spin columns (Roche, Indianapolis, IN).
10. 50× TAE buffer: 2 M Tris-acetate (prepared by adding 242 g Tris base and 57.1 ml glacial acetic acid per 1 l), 0.05 M EDTA.
11. 40 % Acrylamide solution (acrylamide:bis-acrylamide, 37.5:1).
12. 10 % (w/v) Ammonium persulfate (prepared fresh).
13. TEMED (N,N,N′,N′-tetramethylethylenediamine).
14. Poly(dI·dC) (Roche) diluted in HEDG at 10 ng/μl (for cytosolic AhR assays) and 250 ng/μl (for nuclear AhR assays) (**Note 3**).
15. 10× Ficoll GRA gel-loading buffer: 0.25 % (w/v) bromphenol blue, 25 % (w/v) Ficoll 400.
16. Buffer A: 10 mM Hepes-KOH (pH 7.9), 1.5 mM $MgCl_2$, 10 mM KCl, 0.5 mM DTT, 0.5 % (v/v). Protease inhibitor cocktail (Sigma); stable at 4 °C for 6 months.
17. Buffer C: 20 mM Hepes-KOH (pH 7.9), 25 % (v/v) glycerol, 420 mM NaCl, 1.5 mM $MgCl_2$, 0.2 mM EDTA, 0.5 mM DTT, 0.2 mM PMSF, 0.5 % (v/v). Protease inhibitor cocktail (Sigma); stable at 4 °C for 6 months.

3 Methods

Gel Retardation Assay (GRA) of AhR-DNA Binding

The GRA is used to detect the binding of ligand-activated AhR:ARNT complex to its specific DNA binding element, the dioxin responsive element (DRE). It can be performed with hepatic cytosolic samples (**Note 4**) incubated in the presence of AhR agonist ligands (such as TCDD) or solvent control, or alternatively, with nuclear extracts prepared from tissue culture cells incubated with the AhR ligands or solvent control. While sample preparation differs for cytosolic samples and nuclear extracts, the gel retardation protocol is the same for both experimental systems.

3.1. *Preparation of hepatic cytosol (NOTE: perform all steps at 4 °C). This protocol was adapted from [2] and is suitable for use with liver and other tissues from most species.*

3.1.1. Sacrifice the animal using approved procedures. Open the abdominal cavity and cut the vena cava immediately above the liver with scissors. Slowly perfuse the liver with HEDG buffer using a syringe inserted into the hepatic portal vein in order to remove as much blood from the liver as possible (to eliminate blood proteins that can bind AhR ligands).

3.1.2. Place the liver tissue into a beaker and cut it into small pieces with scissors.

3.1.3. Wash the minced tissue with HEDG buffer until the buffer rinse is clear (**Note 5**).

3.1.4. Remove the buffer by pouring the final wash through cheesecloth and weigh the washed tissue in a beaker.

3.1.5. Add 1–1.5 ml HEDG buffer for each gram of tissue.

3.1.6. Homogenize the tissue with a Polytron homogenizer or in a motorized Teflon-glass homogenizer on ice (**Note 6**).

3.1.7. Transfer the homogenate into a 40 ml polypropylene capped centrifuge tube and centrifuge in a Sorvall SS34 rotor for 20 min at $21,000 \times g$, at 4 °C.

3.1.8. Pour the supernatant through a funnel plugged with glass wool.

3.1.9. Transfer the supernatant into a capped 25-ml polycarbonate ultracentrifuge tube and centrifuge in a Beckman SW70Ti rotor for 1 h at $105,000 \times g$, at 4 °C.

3.1.10. Carefully collect the resulting supernatant (hepatic cytosol) from below the surface lipid layer and above the microsomal pellet in the bottom of the tube using Pasteur pipet.

3.1.11. Determine protein concentration of the cytosol using BioRad protein assay and store in small aliquots (200–400 μl) at –80 °C.

3.2. Oligonucleotide annealing and labeling.

3.2.1. Dissolve oligonucleotides in TE (e.g., at 100 μM). Combine 1 μg of each sense and anti-sense oligonucleotide in 20 μl of oligonucleotide annealing buffer. Incubate for 5 min at 90 °C and then cool slowly. The annealed oligonucleotides are dissolved in TE to the desired concentration (e.g., 25 ng/μl) and can be stored at 4 °C for 6 months.

3.2.2. Label 250 ng DRE in a T4 polynucleotide kinase reaction supplemented with 8 μl of [γ^{32}P]ATP (final reaction volume 30 μl).

3.2.3. Purify the resulting radiolabeled DRE using the Sephadex-50 columns (Quick Spin kit).

3.2.4. Count the radioactivity in an aliquot of the radiolabeled DRE in the scintillation counter and dilute the labeled DRE to a final working stock of 25,000 dpm/µl to use as a working stock (**Note 1**).

3.3. *PAGE preparation.*

3.3.1. Pour the native PAGE gels. For each gel to be poured, combine 0.9 ml of 50× TAE, 39.5 ml water and 4.5 ml 40 % acrylamide solution (acrylamide:bis-acrylamide, 37.5:1). Add 0.45 ml 10 % ammonium persulfate and 22.5 µl TEMED. Swirl and pour.

3.3.2. Assemble the PAGE unit, use 1× TAE as running buffer and prerun gels at 130 V for at least 30 min with buffer recycling.

3.4. *Ligand-dependent AhR transformation and DNA binding of hepatic cytosol.*

3.4.1. Dilute cytosol to 8 mg protein/ml in HEDG and incubate in the presence of desired concentrations of AhR ligands or solvent control for 1.5 h at room temperature.

3.4.2. Combine 10 µl of treated cytosol with 4 µl HEDG, 3 µl HEDGK8 and 4 µl of cytosolic poly(dI·dC) (10 ng/µl, **Note 3**). Mix gently and incubate for 10–15 min at room temperature.

3.4.3. Add 4 µl of ^{32}P-labeled oligonucleotide probe (**step 2.4**), mix gently and incubate for 10–15 min at room temperature.

3.4.4. Add 3 µl of 10× Ficoll GRA gel-loading buffer, mix and load 10 µl of the resulting reaction onto the gel.

3.5. *PAGE.*

3.5.1. Turn the power supply and peristaltic pump on.

3.5.2. Run gels at 60 V for 30 min.

3.5.3. Increase voltage to 130 V and run for an additional 1 h to 1 h 30 min. The protein-free [^{32}P]DRE oligonucleotide (25 base pairs) runs slightly faster than the bromophenol blue dye, and thus it is important to not let the dye front get too close to the end of the gel, otherwise the free [^{32}P]DRE oligonucleotide will run off the gel and into the lower chamber of the PAGE unit.

3.5.4. Turn off the power supply and the peristaltic pump and proceed to disassembling the PAGE plates.

3.5.5. Transfer each gel to the 3 MM paper sheet. These gels do not keep shape by themselves. However, when peeling the glass plates apart, the gel usually sticks to one plate. Place the sheet of 3 MM paper over the gel and press slightly. Proceed to drying the gel, layering one extra sheet of 3 MM paper under the first sheet to minimize dryer contamination (**Note 7**).

3.5.6. The running buffer should be checked for ^{32}P contamination and discarded accordingly.

3.5.7. Visualize and analyze dried gels with the Fujifilm FLA-9000 system or similar imaging system or develop the results as an autoradiogram.

3.6. *Quick preparation of nuclear extracts from cells in culture.*

This protocol has been modified from [26, 27] and produces quick and consistent results with cell lines that express high level of AhR, such as the mouse hepatoma (hepa1c1c7) cells.

3.6.1. Treat a confluent 10 cm plate of cells with AhR agonist for 2 h (**Note 8**).

3.6.2. Wash cells with PBS and scrape them into 1.5 ml PBS. Centrifuge in a benchtop microcentrifuge at maximal rpm for 10 s.

3.6.3. Re-suspend the pellet in 400 µl cold Buffer A, incubate on ice for 10 min, then vortex for 10 s. Centrifuge in a benchtop microcentrifuge at maximal rpm for 10 s.

3.6.4. Re-suspend the pellet in 100 µl of cold Buffer C and incubate on ice for 20 min for high salt extraction. Spin down and use the supernatant in the nuclear extract GRA. Generally, this protocol results in 1–1.5 mg/ml protein in the extracts; much lower protein levels may indicate low number of cells or inefficient extraction.

3.7. *DNA binding analysis with nuclear extracts.*

3.7.1. Combine 5 µl nuclear extract (**from step 3.6.4**) with 2.5 µl of poly(dI·dC) solution (250 µg/ml), mix gently and incubate at room temperature for 10 min.

3.7.2. Add 2 µl of working [^{32}P]DRE oligonucleotide stock (**from step 3.2.4**), gently mix and incubate for 10 min.

3.7.3. Load 5–10 µl of sample directly on the gel (without using the dye). Load a control lane with several µl of dye to visually monitor the gel progression.

3.7.4. Run the gel as described in Sect. 3.5.

3.8. Competitive gel retardation analysis of AhR-DNA binding.

This protocol allows confirmation of the sequence specificity of DNA binding. At DNA binding step (3.4.3 or 3.7.2 for cytosolic or nuclear extracts, respectively) add wild type or mutant non-labeled annealed DRE oligonucleotide at a 10–50 M excess relative to [^{32}P]DRE oligonucleotide in 2–4 µl HEDG buffer. Follow by addition of the [^{32}P]DRE

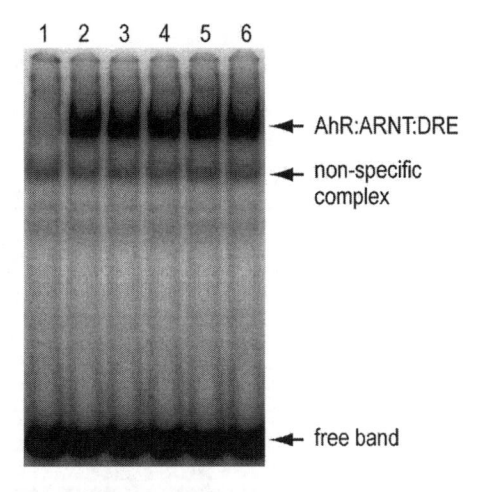

Fig. 1 Transformation/DNA binding analysis with guinea pig hepatic cytosolic extract. Guinea pig cytosolic extract diluted to 8 mg/ml protein in HEDG was incubated in the presence of 1 % v/v DMSO (lane 1), 20 nM TCDD (lane 2), 10 μM YH439 (lane 3), 1 μM β-naphthoflavone (lane 4), 1 μM 3-methylcholanthrene (lane 5) or 1 μM indirubin (lane 6) for 1.5 h at room temperature and DNA binding was analyzed by GRA as described in text. A representative gel is shown. The positions of specific and non-specific complexes, as well as that of the free band, are indicated with *arrows*

oligonucleotide and incubation at room temperature, as described in steps 3.4.3 or 3.7.2, and proceed with the rest of the protocol.

3.9. Interpretation of results.

A typical DNA binding experiment using guinea pig cytosol is presented in Fig. 1. The free band indicates fast-migrating protein-free [^{32}P]DRE oligonucleotide. The slower migrating specific ligand-induced AhR:ARNT:DRE complex is readily apparent as this complex is not present in the same position in the solvent negative control lane (**Note 9**). Non-specific protein-DRE complexes are constitutive (of similar intensity in all lanes) and serve as an important indicator of the overall quality of experiment (i.e. equal loading of protein extract) (**Note 10**).

GRA of nuclear extracts (Fig. 2a) allows detection of ligand-dependent nuclear translocation of the AhR complex in addition to its DNA binding. This assay can be used to confirm AhR-dependent initiation of transcription in cells and can distinguish between agonist and antagonist mode of AhR activation. Thus, in the provided example, co-incubation of cells with TCDD and the AhR antagonists α-naphthoflavone (ANF) or dimethoxynitroflavone (DiMNF), resulted in significantly reduced protein-DNA

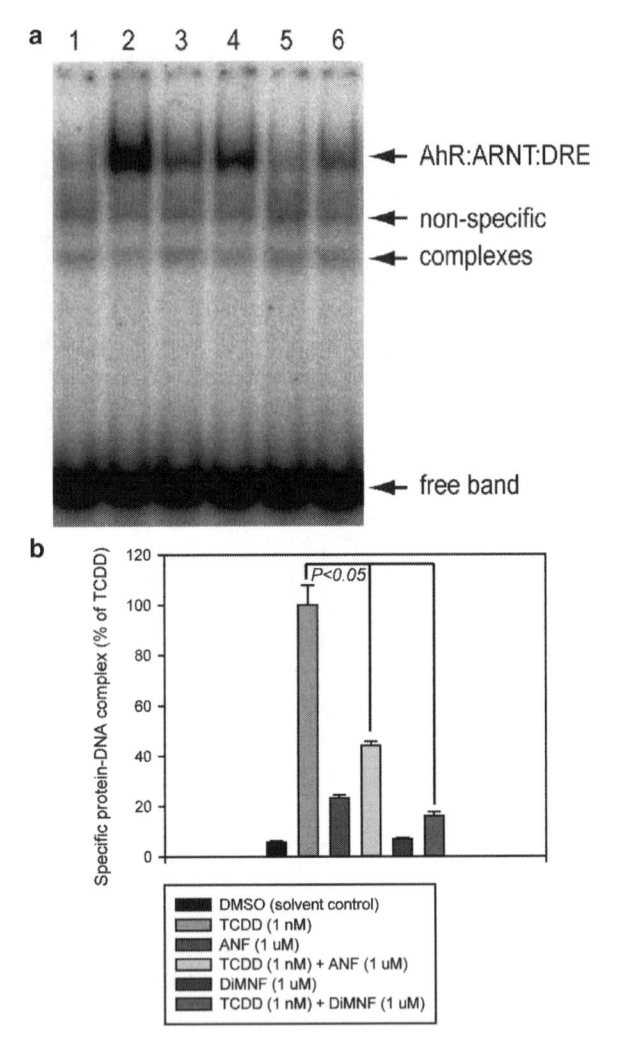

Fig. 2 DNA binding analysis of nuclear extracts. (**a**) Mouse hepatoma (hepa1c1c7) cells were treated with 0.1 % (v/v) DMSO (lane 1), 1 nM TCDD (lanes 2, 4, 6), 1 μM α-naphthoflavone (ANF, lanes 3, 4) or 1 μM dimethoxy-α-naphthoflavone (DiMNF, lanes 5, 6) for 2 h, and the nuclear extracts were prepared and analyzed for DNA binding as described in text. A representative gel is shown. The positions of specific and non-specific complexes, as well as that of the free band, are indicated with *arrows*. (**b**) Quantitation of the experiment shown in part (**a**). Values represent the means ± standard deviations of three independent experiments

complex formation relative to TCDD alone (Fig. 2b). Unlike DiMNF, ANF is a partial agonist of the AhR, demonstrated by its ability to partially activate AhR-dependent DNA binding compared to solvent alone (Fig. 2b).

4 Notes

1. The half-life of ^{32}P is 14 days; therefore, label should be ordered immediately prior to the planned experiments and used promptly. In most cases ^{32}P-labeled oligonucleotides provide good signal for up to 1 month after the reference date of [$\gamma^{32}P$]ATP.

2. PAGE purification is essential for high efficiency of oligonucleotide labeling with T4 polynucleotide kinase.

3. The exact amount of poly(dI·dC) depends on the protein concentration and the specific lot and source of poly(dI·dC) used. With each new lot of poly(dI·dC), the concentration needed for optimal AhR-DRE complex formation and maximal treated/untreated ratio must be established.

4. Although ARNT is a nuclear protein *in vivo*, it is recovered in the cytosolic fraction in the described protocol making formation of the AhR:ARNT complex *in vitro* possible.

5. Tissue should be washed thoroughly to minimize the amount of residual blood as blood proteins have been reported to bind AhR ligands and reduce/interfere with TCDD-dependent AhR:ARNT complex formation [28].

6. It is important to keep samples on ice to prevent tissue from warming up during homogenization.

7. The PAGE gel dryer should be pre-warmed and the gels should be vacuum dried within 45 min to 1 h. Slow or incomplete drying results in the appearance of diffused bands.

8. Optimal treatment time depends on a specific compound. Generally, 2–4 h incubation times works well for most AhR agonists.

9. For accurate quantitative determinations [^{32}P]DRE oligonucleotide should be present in excess (i.e. the intensity of protein-free band should not noticeably change between treated and untreated lanes). If levels of the protein-free [^{32}P]DRE oligonucleotide band changes between treatment lanes, it suggests that each sample lane did not have sufficient [^{32}P]DRE oligonucleotide to produce a maximal amount of protein:DNA binding complex.

10. One of the most common mistakes in reporting gel retardation results is to overlook the inconsistency of non-specific bands between lanes. Dramatic changes in the intensity of non-specific bands between sample lanes may signal serious experimental errors, such as unequal protein or [^{32}P]DRE oligonucleotide concentrations in different sample lanes, and often lead to erroneous interpretations of the results.

References

1. Dey B, Thukral S, Krishnan S et al (2012) DNA-protein interactions: methods for detection and analysis. Mol Cell Biochem 365:279–299
2. Denison MS, Rogers JM, Rushing SR et al (2002) Analysis of the aryl hydrocarbon receptor (AhR) signal transduction pathway. In: Maines M, Costa LG, Reed DJ et al (eds) Current protocols in toxicology. Wiley, New York, pp 4.8.1–4.8.45
3. Yao EF, Denison MS (1992) DNA sequence determinants for binding of transformed Ah receptor to a dioxin responsive element. Biochemistry 31:5060–5067
4. Claessens F, Gewirth DT (2004) DNA recognition by nuclear receptors. Essays Biochem 40:59–72
5. Wan F, Lenardo MJ (2009) Specification of DNA binding activity of NF-kappaB proteins. Cold Spring Harb Perspect Biol 1(4):a000067
6. Zhang DD (2006) Mechanistic studies of the Nrf2-Keap1 signaling pathway. Drug Metab Rev 38:769–789
7. Denison MS, Fisher JM, Whitlock JP Jr (1988) The DNA recognition site for the dioxin-Ah receptor complex. Nucleotide sequence and functional analysis. J Biol Chem 263:17221–17224
8. Denison MS, Yao E (1991) Characterization of the interaction of transformed rat hepatic Ah receptor with a dioxin responsive transcriptional enhancer. Arch Biochem Biophys 284:158–166
9. Hankinson O (1995) The aryl hydrocarbon receptor complex. Annu Rev Pharmacol Toxicol 35:307–340
10. Furness SG, Whelan F (2009) The pleiotropy of dioxin toxicity–xenobiotic misappropriation of the aryl hydrocarbon receptor's alternative physiological roles. Pharmacol Ther 124:336–353
11. Beischlag TV, Morales JL, Hollingshead BD, Perdew GH (2008) The aryl hydrocarbon receptor complex in the control of gene expression. Crit Rev Eukaryot Gene Expr 18:207–250
12. White SS, Birnbaum LS (2009) An overview of the effects of dioxins and dioxin-like compounds on vertebrates, as documented in human and ecological epidemiology. J Environ Sci Health C Environ Carcinog Ecotoxicol Rev 27:197–211
13. Denison MS, Soshilov AA, He G et al (2011) Exactly the same but different: promiscuity and diversity in the molecular mechanisms of action of the aryl hydrocarbon (dioxin) receptor. Toxicol Sci 124:1–22
14. Bock KW (2013) The human Ah receptor: hints from dioxin toxicities to deregulated target genes and physiological functions. Biol Chem 394:729–739
15. Heath-Pagliuso S, Rogers WJ, Tullis K, Seidel SD, Cenijn PH, Brouwer A, Denison MS (1998) Activation of the Ah receptor by tryptophan and tryptophan metabolites. Biochemistry 37:11508–11515
16. Denison MS, Seidel SD, Rogers WJ, Ziccardi M, Winter GM, Heath-Pagliuso S (1999) Natural and synthetic ligands for the Ah receptor. In: Wallace KB, Puga A (eds) Molecular biology approaches to toxicology. Taylor & Francis, Philadelphia, pp 393–410
17. Seidel SD, Li V, Winter GM et al (2000) Ah receptor-based chemical screening bioassays: application and limitations for the detection of Ah receptor agonists. Toxicol Sci 55:107–115
18. Seidel SD, Winters GM, Rogers WJ, Ziccardi MH, Li V, Keser B, Denison MS (2001) Activation of the Ah receptor signaling pathway by prostaglandins. J Biochem Mol Toxicol 15:187–196
19. Nagy SR, Liu G, Lam K, Denison MS (2002) Identification of novel Ah receptor agonists using a high-throughput green fluorescent protein-based recombinant cell bioassay. Biochemistry 41:861–868
20. Denison MS, Nagy SR (2003) Activation of the aryl hydrocarbon receptor by structurally diverse exogenous and endogenous chemicals. Annu Rev Pharmacol Toxicol 43:309–334
21. Zhao B, DeGroot D, Hayashi A, He G, Denison MS (2010) CH223191 is a ligand-selective antagonist of the Ah (dioxin) receptor. Toxicol Sci 117:393–403
22. Zhao B, Bohonowych JES, Timme-Laragy A, Jung D, Affatato AA, Rice RH, Di Giulio RT, Denison MS (2013) Common commercial and consumer products contain activators of the aryl hydrocarbon (dioxin) receptor. PLoS One 8(2):e56860
23. Zhang S, Lei P, Liu X et al (2009) The aryl hydrocarbon receptor as a target for estrogen receptor-negative breast cancer chemotherapy. Endocr Relat Cancer 16:835–844
24. Vondracek J, Umannova L, Machala M (2011) Interactions of the aryl hydrocarbon receptor with inflammatory mediators: beyond CYP1A regulation. Curr Drug Metab 12:89–103
25. Montelone I, MacDonald TT, Pallone F, Montelenoe G (2012) The aryl hydrocarbon receptor in inflammatory bowel disease: linking

the environment to disease pathogenesis. Curr Opin Gastroenterol 28:310–313

26. Anderson G, Beischlag TV, Vinciguerra M, Mazzoccoli G (2013) The circadian clock circuitry and the AHR signaling pathway in physiology and pathology. Biochem Pharmacol 85:1405–1416

27. Andrews NC, Faller DV (1991) A rapid micropreparation technique for extraction of DNA-binding proteins from limiting numbers of mammalian cells. Nucleic Acids Res 19:2499

28. Denison MS, Vella LM, Okey AB (1986) Structure and function of the Ah receptor for 2,3,7,8-tetrachlorodibenzo-p-dioxin. Species difference in molecular properties of the receptors from mouse and rat hepatic cytosols. J Biol Chem 261:3987–3995

Chapter 13

Cell-Based Assays for Identification of Aryl Hydrocarbon Receptor (AhR) Activators

Guochun He, Jing Zhao, Jennifer C. Brennan, Alessandra A. Affatato, Bin Zhao, Robert H. Rice, and Michael S. Denison

Abstract

The Ah receptor (AhR) is a ligand-dependent transcription factor that mediates a wide range of biological and toxicological effects from exposure to structurally diverse synthetic and naturally occurring chemicals. The role of the AhR and its signaling pathway in endogenous physiological functions and its involvement in immune cell development and human diseases has made it a target for development of therapeutic agents. The ability of the AhR to stimulate gene expression in a ligand-specific manner in recombinant mammalian cell lines containing a stably transfected AhR-responsive firefly luciferase or enhanced green fluorescent protein (EGFP) reporter gene permits high throughput chemical screening for AhR activators. The induction of luciferase activity or EGFP fluorescence in these readily available recombinant cell lines occurs in a time-, dose- and AhR-dependent and chemical-specific manner where the magnitude of reporter gene induction is directly proportional to the concentration and potency of the inducing chemical. The AhR agonist activity of positive test chemicals can be confirmed by demonstrating their ability to stimulate expression of CYP1A1, an endogenous AhR-responsive gene, using quantitative real-time PCR. The detailed protocols described here provide step-by-step instructions for detection and characterization of activators of AhR-dependent gene expression that can readily be applied to other appropriate cell lines.

Key words Ah receptor, CALUX, CAFLUX, Luciferase, Green fluorescent protein, Quantitative real time PCR

1 Introduction

The Ah receptor (AhR) is a ligand-dependent transcription factor that not only mediates the induction/repression of expression of a large battery of genes, but it plays a key regulatory role in the production of a broad spectrum of species- and tissue-specific toxic and biological effects of selected AhR ligands [1–6]. Because of their high affinity for the AhR, their metabolic stability and toxicity at extremely low concentrations, halogenated aromatic hydrocarbons such as 2,3,7,8-tetrachlorodibenzo-p-dioxin (TCDD, dioxin)

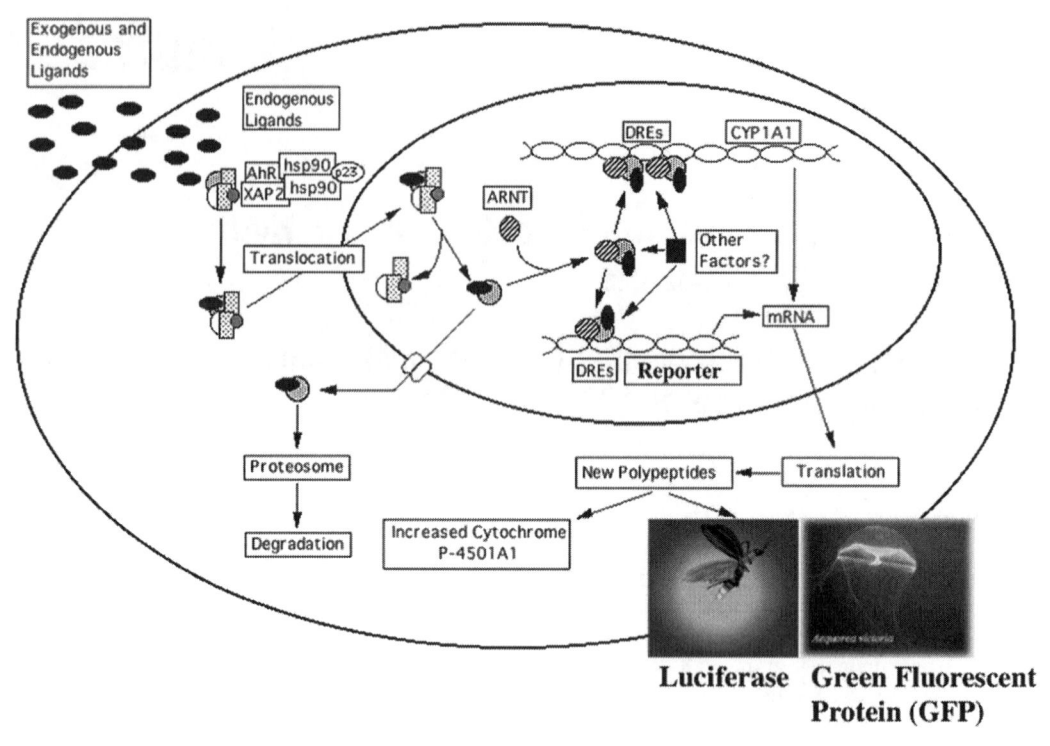

Fig. 1 Mechanisms of Ah receptor (AhR) activation of gene expression and basis for CALUX and CAFLUX cell bioassays. See references [6, 20] for details

and related metabolically stable dioxin-like compounds (DLCs), represent the best studied of all AhR agonists. The toxicity of DLCs appears to be directly related to their metabolic stability and ability to stimulate persistent activation of AhR-dependent gene expression. Consistent with this idea is the recent demonstration that the AhR can bind and be activated by a wide variety of structurally diverse compounds, the majority of which are metabolically labile and do not produce the spectrum of AhR-associated toxic effects [3, 6, 7]. However, all of these compounds stimulate AhR-dependent gene expression by a common mechanistic pathway similar to that of steroid hormones and their receptors [8]. In this mechanism (Fig. 1), the inducing chemical (AhR agonist) diffuses across the plasma membrane of a responsive cell and its specific binding to the cytosolic AhR stimulates nuclear translocation of the AhR protein complex. Once in the nucleus, the AhR is released from its associated proteins following its dimerization with a related nuclear protein called Arnt (AhR nuclear translocator), and formation of the AhR:Arnt heterodimer converts the ligand AhR complex into its high affinity DNA binding form. Binding of the ligand:AhR:Arnt complex to its specific DNA recognition site, the dioxin responsive element (DRE), adjacent to a responsive gene, leads to chromatin disruption and stimulation of transcription of a wide variety of such responsive genes including CYP1A1 and others encoding phase I and phase II drug/xenobiotic metabolizing enzymes [2, 4, 6].

While early studies were focused on the role of the AhR in mediating the biochemical response to xenobiotics (adaptive induction of metabolic enzymes) and the toxic effects of selected AhR ligands, recent studies have identified key endogenous regulatory roles for the AhR in normal human physiology, tissue and immune cell development and disease [2, 9–14]. These and other related studies identify the AhR as a potential target for the identification and development of therapeutic agents for treatment of several maladies including autoimmune disease, various inflammatory conditions and cancer [13, 15–17]. However, while a wide variety of ligands for the AhR have been identified, few chemicals are suitable for development into useful therapeutic agents. Therefore, the identification of activators of the AhR and AhR signaling pathway is critically needed for such development.

The molecular mechanism of the AhR signaling pathway (ligand binding, DNA binding and nuclear translocation) have been used to develop AhR-based bioassays, the most extensively used bioanalytical approach for detection and characterization of AhR ligands [18–22]. In these systems, incubation of cells in culture for varying lengths of time with test chemicals or extracts containing AhR-agonists leads to the induction of AhR-dependent gene expression that occurs in a time-, chemical-, concentration- and AhR-dependent manner. While numerous AhR-based cell bioassays have been described [18–25], the chemically activated luciferase expression (CALUX) and chemically activated fluorescent expression (CAFLUX) bioassays (Fig. 1), which utilize recombinant cell lines that contain stably transfected AhR-responsive firefly luciferase or enhanced green fluorescent protein (EGFP) reporter genes, respectively, have been extensively used [18, 20, 21, 23]. Treatment of these cells with AhR ligands results in induction of reporter gene activity that is directly proportional to the concentration and potency of the inducing chemical (i.e. its AhR agonist activity) [20, 21, 23]. Although these systems use the same AhR-dependent induction, differences in the characteristics of the respective reporters result in bioassay systems with distinct advantages and disadvantages [20, 21, 24]. While the firefly luciferase reporter gene in the CALUX bioassay system is highly sensitive and responsive, primarily due to enzymatic signal amplification, it has limitations with respect to repeated measurement, relatively high cost for reagents and rapidity for high-throughput screening analysis. By contrast, measurement of EGFP reporter gene activity is more rapid, cost effective, amenable to high throughput and repeated analysis of the same cells and the induction response can be measured in "real time" [20, 24].

While the CALUX and CAFLUX bioassays have been used extensively for the relatively rapid screening and assessment of the ability of a chemical(s) to stimulate AhR-dependent gene expression, positive results obtained in these assays are often criticized for not directly demonstrating induction of an endogenous AhR-responsive gene (although we have never identified a CALUX-/CALFUX- positive chemical that does not stimulate expression of an endogenous AhR-responsive gene). Accordingly, to confirm this, additional analysis is often needed to demonstrate that a CALUX- or CAFLUX-positive chemical can direct stimulate CYP1A1 gene expression in cells in culture. This is accomplished by measuring CYP1A1 mRNA levels through quantitative real-type PCR (Q-RT-PCR). Together, the combined results from both CALUX/CAFLUX and Q-RT-PCR assays provide strong evidence for the AhR agonist activity of the positive test chemical. The detailed protocols described here provide step-by-step instructions for rapid screening, detection and characterization of activators of AhR-dependent gene expression using CALUX and CAFLUX cell lines and confirmatory analysis using Q-RT-PCR. Additional analysis of positive test chemicals carried out using in vitro AhR ligand binding [26] or DNA binding (gel retardation) (*see* Chap. 12 and [26]) assays can confirm that the test chemical can directly interact with and activate the AhR.

2 Materials

2.1 Equipment

General purpose equipment can be substituted with comparable models from other companies.

1. Laminar flow hood and CO_2 incubator for cell culture.
2. Inverted microscope to examine cells in tissue culture plates.
3. Centrifuge with swinging-bucket rotor for pelleting cells for plating.
4. Benchtop microcentrifuge for preparation of cell lysates for PCR.
5. Orbital shaker platform (such as a Belly Dancer shaker, Denville Scientific, Denville, NJ) for cell lysis.
6. Microplate Luminometer (Anthos Lucy II, Salzburg, Austria or equivalent) with pumps for automatic addition of luciferin reagent to plate wells.
7. Fluostar Microtiter Plate Fluorometer (Phoenix Research Products, Candler, NC or equivalent) with an excitation and emission wavelengths of 485 and 515 nm, respectively.
8. Applied Biosystems 7500 Fast Sequence Detection System (Life Technologies, Grand Island, NY).

2.2 Reagents and Materials

2.2.1 Luciferase/EGFP Reporter Gene Bioassay

1. Trypsin (1×), tissue culture grade, sterile.
2. Phosphate-buffered saline (PBS; 1×), sterile.
3. Alpha-Minimal Essential Media (MEM; Invitrogen, #12000-063) containing 10 % prescreened (**Note 4.1.1**) fetal bovine serum (FBS; Atlanta Biologicals, #S11150).
4. Tissue culture microplates: Sterile 96-well microplates for cell growth and luciferase analysis (white, clear bottom tissue culture microplate for luciferase (Fisher, #07-200-566)) or EGFP analysis (black, clear bottom tissue culture microplate (Fisher, #07-200-565)).
5. Promega Luciferase Assay Lysis Buffer, 5× stock (Fisher, #PR-E1531).
6. Promega Luciferase Assay System (Fisher, #PR-E1501).
7. Microplate white backing tape (PerkinElmer, #6005199).
8. AhR agonist stock solutions (TCDD in DMSO (**Note 4.1.2**) or other non-toxic AhR agonist in DMSO (**Note 4.1.3**)).

2.2.2 Quantitative Real Time PCR Reagents and Materials

1. Trizol reagent (Invitrogen Life Technologies, Grand Island, NY).
2. High Capacity cDNA Archive Kit (Applied Biosystems, Foster City, CA) or High Capacity cDNA Reverse Transcription Kit.
3. TaqMan Fast Universal PCR Master Mix (no AmpErase UNG) (2×).
4. TaqMan Gene Expression assays (20×).

3 Methods

3.1 Reporter Gene Bioassays

3.1.1 Protocol for Plating Cells into 96-Well Plate

1. General maintenance of cell cultures. Continuous cell lines containing a stably transfected AhR-responsive luciferase reporter gene (mouse H1L6.1c2 (Hepa1c1c7) and human HG2L6.1c3 (HepG2) hepatoma cells [21, 27]) or enhanced green fluorescent protein reporter (EGFP) gene (rat H4G1.1c2 (H4IIe) or mouse H1G1.1c3 (Hepa1c1c7) hepatoma cells [24, 28]) are maintained in alpha-MEM containing 10 % fetal bovine serum. Cells should not exceed 90 % confluence before passaging and cells should be examined prior to trypsinizing and plating.
2. Remove old media from plates by aspiration, rinse cells with PBS, add trypsin solution and incubate for 2–4 min at room temperature or 37 °C (until cells detach from the plates).
3. While the cells are being trypsinized, add a volume of medium/serum to a 50 mL sterile Falcon tube that is at least equal to the volume of trypsin that you will add from the

plates (serum contains trypsin inhibitors). Depending on the number of tissue culture plates, more tubes can be used.

4. Transfer the contents of the culture plate(s) into the Falcon tube(s).

5. Cap and spin cell suspension in a benchtop centrifuge at room temperature for 5 min at 1,100 rpm.

6. In the tissue culture hood, carefully aspirate the media/serum from the centrifuged tubes. Add 10 mL fresh medium/serum to each tube and gently re-suspend the cells (**Note 4.1.4**).

7. An aliquot (10 µL) of re-suspended cells are counted using a hemocytometer or Coulter counter. For the bioassay, the optimal cell density in the 96-well plate format is 750,000 cells/mL, and the counted/re-suspended cells are diluted to this concentration with medium/serum.

8. Add an aliquot of diluted cell suspension (100 µL) to each well of a 96-well plate (clear-bottomed white plate for luciferase analysis or clear-bottomed black plate for EGFP analysis) using a cell trough and multichannel pipette. Cells are typically allowed to attach and grow for 24 h in a tissue culture CO_2 incubator at 37 °C prior to chemical treatment. This typically results in wells containing cells of at least 95 % confluence. Wells should be examined microscopically prior to addition of chemical to ensure that each contains comparable numbers of cells (**Note 4.1.5**).

3.1.2 Protocol for Treating Cells with Test Chemicals

1. In a tissue culture hood, sterilely prepare chemical treatments at the desired concentration at a 1:100 ratio of chemical/sample/control to medium (i.e. 10 µL of test sample is diluted into 990 µL medium) in 7 mL borosilicate glass tubes (autoclaved and baked). Vortex all treatments for several seconds to ensure complete mixing. This volume is sufficient to treat multiple wells, with most analyses carried out in triplicate. In addition to the test samples, each plate contains negative controls (i.e. DMSO and/or solvents used for the test samples), positive controls (i.e. a potent AhR agonist like TCDD (in DMSO) or other AhR agonist) and method blanks (if extracts or mixtures are being examined).

2. Prior to addition of sample treatments, the medium in the 96-well plate(s) can be dumped out with shaking into an appropriate biological waste container containing absorbent material (i.e., bench diaper or paper towels), taking care not to contaminate the cells during this process but to remove as much medium as possible.

3. Carefully fill the appropriate wells of a 96-well microtiter plate with 100 µL of the desired chemical/medium:serum suspension and make sure to note each treatment well.

4. Place the lid(s) back on the treated plate(s) and quickly examine each well microscopically to ensure that cells were not lost during washing and treatment. The plates are placed into a 37 °C incubator for the desired time (4–24 h typically). Alternatively, cells can be incubated with chemicals at 33 °C, which we have found to result in significantly higher luciferase activity and EGFP fluorescence than that obtained at 37 °C ([28]; **Note 4.1.6**).

3.1.3 Measurement of Reporter Gene Activity

Luciferase Activity in Lysed Cells

1. Visually examine each well in the 96-well plate under the microscope for cell toxicity (cells rounding up and detaching from the plate) and for cloudiness or a change in color of the media (i.e., precipitation of the test chemicals or culture components or a change in media pH, respectively). If toxicity is observed, the results should be discarded and test chemicals retested at lower concentrations.

2. Dump medium from 96-well plate(s) into appropriate biological waste container as described in 3.1.2.

3. Carefully rinse the wells twice with 100–150 μL PBS and gently dump the liquid into the waste container.

4. Microscopically examine the cells in each well under the microscope to ensure that they were not lost during PBS washing. Firmly tap the inverted plate onto paper towels to remove any remaining PBS.

5. Add 50 μL of room temperature Promega lysis buffer (1×) to each well. For each 96-well plate, prepare the 1× lysis buffer by mixing 1 mL of 5× lysis buffer with 4 mL of MilliQ water and store in glass.

6. Transfer the plate onto an orbital shaker platform (such as a Belly Dancer table) and shake at a moderate speed for at least 20 min to ensure adequate cell lysis.

7. While the cells are lysing, prepare the luminometer and prime the reagent pumps with luciferase substrate. Mix one bottle of room temperature luciferase buffer with one bottle luciferase substrate (buffer and luciferase substrate (luciferin) are from the Promega Luciferase Assay System) and use it to prime the luminometer pumps. Apply white backing tape to each plate containing lysed cells and insert the plate into the luminometer. The luminescence in each well is measured (integrating luminescence over 10 s after a 10 s delay) following automatic injection of Promega stabilized luciferase reagent.

8. Luciferase activity is typically expressed as a percent of the maximum level of induction in a defined number of cells (**Note 4.1.7**) that is produced by a potent AhR agonist such as TCDD (1 nM for rodent cells and 10 nM for human cells) or other AhR agonist (**Note 4.1.8**).

EGFP Fluorescence in Intact Cells

1. EGFP fluorescence levels are measured in intact cells in each well (without the removal of medium), using a microplate fluorometer with excitation and emission wavelengths of 485 and 515 nm, respectively (**Note 4.1.9**).

2. Following fluorescence measurements, the microplate can be returned to the tissue culture incubator and EGFP levels repeatedly measured in the same cells at later times (**Note 4.1.10**).

3. EGFP activity is typically expressed as a percent of the maximum induction fluorescence produced by a potent AhR agonist such as TCDD (1 nM) or other AhR agonist (**Note 4.1.8**).

3.2 Quantitative Real Time PCR Bioassays

3.2.1 Isolate Total RNA from Cultured Cells (Note 4.2.1)

1. Remove culture medium from plates. Optional: rinse cells with PBS. Tip plates to drain and remove liquid.

2. Add 1 mL Trizol to a 6 cm dish and mix the cells with the Trizol using a scraper (**Note 4.2.2**). If the sample is very viscous and is not sheared well by pipetting up and down through the pipet tip, shear the genomic DNA using a syringe with a 22 gauge needle until the viscosity is reduced (3–4 passes through the needle). Transfer to a microfuge tube and store at −80 °C.

3. Thaw the samples and microfuge them for 10 min at $10,000 \times g$. All centrifugation steps for the rest of protocol are at 4 °C.

4. Transfer the supernatant to a new tube; add 0.2 mL of chloroform and vortex well. Centrifuge for 10 min at $10,000 \times g$.

5. Transfer the top (aqueous) phase to a new tube, taking care to avoid removing any precipitated interphase material. Add 0.5 mL of isopropanol to the aqueous phase, incubate for 10 min at room temperature, and centrifuge for 10 min at $10,000 \times g$.

6. Remove the supernatant with a pipettor or by decanting. Rinse the pellet with 70 % ethanol and re-centrifuging briefly. Remove the supernatant, then re-centrifuge again and remove the remaining liquid with a pipettor. Air-dry the pellet briefly (about 5 min).

7. Dissolve the pellet in 10–50 μL of diethyl pyrocarbonate-treated water. Determine the concentration spectrophotometrically (A^{260}). Store at −80 °C (**Note 4.2.3**).

3.2.2 Prepare the cDNA

1. Mix the following ingredients for one sample: sterile water (4.2 μL), 10× reverse transcriptase buffer (2 μL), 25× dNTP mix (0.8 μL), random primers (2 μL) and reverse transcriptase (1 μL). If you have ten samples, multiply the amounts by 11, enough for each sample plus some extra to cover small pipetting errors.

2. Mix 10 μL aliquots of the above cocktail with 10 μL samples of RNA diluted to 0.2 μg/μL (2 μg total RNA per sample). Incubate for 10 min at 25 °C, then 2 h at 37 °C. Dilute the sample 1:8 with sterile water (140 μL water) and store at −20 °C until needed.

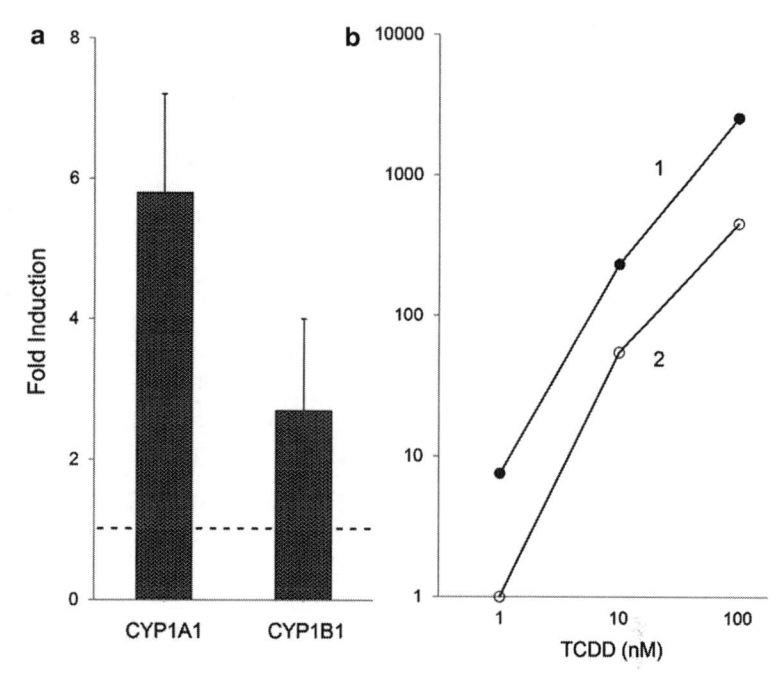

Fig. 2 Illustrated are results for CYP1A1 (#Hs 00153120_m1) and CYP1B1 (#Hs 00164383_m1) using β-actin (#Hs 99999903_m1) as endogenous control. The housekeeping gene GUSB (#Hs 99999908_m1) is also useful in such experiments. (**a**) Human skin organ culture exposed overnight to 10 nM TCDD [29]. Shown are the means ± SDs from three experiments. (**b**) Confluent cultured normal human epidermal cells assayed after overnight TCDD treatment showing a difference in sensitivity in two keratinocyte strains (1 and 2) measured in parallel

3.2.3 Conduct Real Time PCR

1. Set up reactions according to a template sheet plan. Determine how many wells are needed for each probe and then prepare a mixture of Gene Expression Assay probe + master mix. Prepare enough for one extra well to cover pipetting errors. For each well use 0.5 µL of 20× probe + 5 µL of 2× master mix.

2. Pipet 5.5 µL of this mixture into each well according to the template plan, then add 4.5 µL of diluted cDNA from the previous step (**Note 4.2.4**). Seal the plate and wrap it in aluminum foil until it is ready to be run.

3. The plate is inserted in the Applied Biosystems 7500 Fast Sequence Detector and the rates of accumulation of amplicons are measured. The output is then analyzed using the instrument software, which calculates the levels of the target sequence relative to the endogenous control in the various culture conditions being compared (**Note 4.2.5**). Figure 2 shows application to TCDD induction of CYP1A1 and 1B1 in keratinocytes in culture and in the skin [29].

4 Notes

4.1 Luciferase/EGFP Reporter Gene Bioassays

4.1.1. The fetal bovine serum (FBS) used in these and other gene expression-based cell bioassays is a critical factor affecting the level of assay sensitivity and magnitude of AhR-dependent induction of reporter and endogenous gene expression. Thus, prescreening of small amounts of several different lots of FBS from one or more vendors for low background and high signal is strongly encouraged, in order to identify the optimal serum lot. A sufficient amount of the same lot of FBS will not only ensure optimal bioassay characteristics, but also consistent assays over an extended time period.

4.1.2. TCDD is one of the most potent AhR agonists and it is the most commonly used agonist in AhR bioassays and is typically dissolved in DMSO. Use of TCDD as the positive control AhR agonist in these assays can be problematic in some laboratories as it is considered a highly toxic carcinogen, even though it is used at extremely low concentrations in these assays, and disposal costs can be high. When working with TCDD take extreme care to avoid contaminating work areas and personnel and use it only in appropriately designated areas with all necessary precautions, including the use of laboratory coats, protective eyewear, disposable benchtop paper, gloves, plastic ware and glassware. Follow all chemical safety guidelines for handling and disposal of these materials. It is particularly important that when handling TCDD or any hazardous chemical dissolved in DMSO or other solvent that you use appropriate solvent resistant nitrile gloves as latex gloves provide little or no barrier to solvent penetration and subsequent chemical exposure. Given the hazards rating associated with TCDD, laboratory use of these chemicals usually requires prior permission of your institutional chemical safety office.

4.1.3. A number of alternative highly potent non-toxic AhR agonists are available, but they have some limitations. Omeprazole is a non-toxic AhR agonist that has been used successfully as positive controls in human AhR cell-based bioassays, but it is a poor AhR agonist in rodent cells [30, 31]. Other non-toxic AhR agonists that have been used as positive controls in rodent AhR bioassays include BNF and indirubin [3, 7, 30–32]. In contrast to TCDD, these non-toxic AhR agonists can be enzymatically degraded within the cells resulting in the reduction in agonist concentration and a decrease in the overall induction of reporter gene activity over time (i.e. induction responses with these non-toxic agonists are transient). However, since use of relatively high concentrations of these non-toxic ligands in the bioassay can result in maximal induction of AhR-dependent gene expression for extended incubation times, under these conditions they can be used judiciously as positive controls.

4.1.4. During re-suspension, cells must be dispersed so that no clumps of cells remain. This is critical to allow for accurate cell counts for dilution and distribution into microplate wells. Failure to disperse the cells adequately will result in variable numbers of cells in each well and will lead to significant differences in luciferase activity between replicate samples and thus high assay variability with inaccurate potency determinations.

4.1.5. Cells in each well of the microplate must be examined microscopically prior to and after chemical treatment to confirm that each well contains comparable numbers of cells. Not only does this ensure that cells were not lost during the washing and processing steps in the protocol, but it also allows assessment of whether the test chemical produced any cell toxicity. Since luciferase requires lysing of the cells, there can be no post-assay evaluation of cell numbers. Failure to inspect the cells in each well can contribute significantly to variations in the final reporter gene activity.

4.1.6. In most mammalian cell bioassays, cells are passaged and incubated with test chemicals at the standard temperature of 37 °C. However, for recombinant cell bioassays using luciferase or EGFP reporter genes, it has been observed that incubation of treated cells at 33 °C instead of 37 °C results in a dramatic increase in luciferase activity and EGFP fluorescence [28]. This increased activity appears to result from increased activity of the proteins themselves (perhaps due to more optimal folding at the lower temperature) and not from an increase in gene expression. Accordingly, if greater reporter gene activity is desired, the incubation of cells with test chemical should be carried out at 33 °C. However, cells should still be maintained and passaged at 37 °C since this temperature is optimal for cell growth.

4.1.7. The luciferase activity in these assays is based on the sum of the activity present in a defined number of cells within each well of the plate, and activity is not normalized to protein concentration in each well. This is primarily because the detergent present in the Promega lysis buffer interferes with most protein assays (even though some company brochures indicate that their assay is unaffected by detergent). This interference can lead to substantial variation in results and inaccurate determinations of overall luciferase activity.

4.1.8. Reporter gene (luciferase and EGFP) activity is typically expressed as a percent of the maximum induction observed

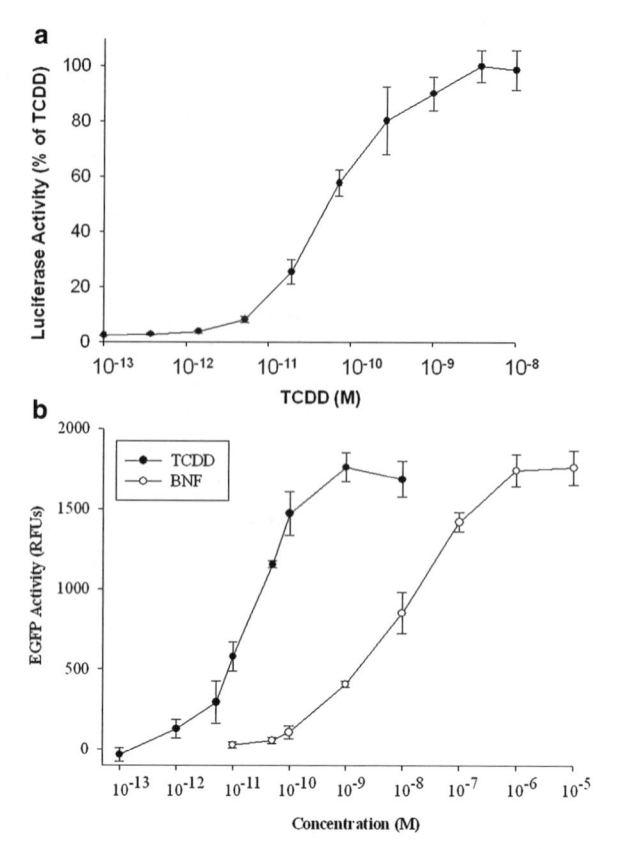

Fig. 3 Concentration-dependent induction of reporter gene activity in CALUX and CAFLUX cell bioassays. (**a**) Induction of luciferase activity by TCDD in mouse hepatoma (H1L6.1c2) CALUX cells. (**b**) Induction of EGFP fluorescence by TCDD and beta-naphthoflavone (BNF) in rat hepatoma (H4G1.1c3) CAFLUX cells

using a potent AhR agonist like TCDD or other AhR agonist, with the luciferase activity (relative light units (RLUs)) or EGFP fluorescence (relative fluorescent units (RFUs)) values representing the mean ± SD of at least triplicate determinations. Figure 3 illustrates the concentration-dependent induction of luciferase activity and EGFP fluorescence by TCDD in the H1L6.1c2 and H1G1.1c3 cell lines, respectively.

4.1.9. In order to allow normalization of EGFP results between experiments, the instrument fluorescence gain setting should be adjusted in each experiment so that the level of EGFP induction by 1 nM TCDD (or a maximal inducing concentration of another AhR agonist) produces a relative fluorescence of 9,000 relative fluorescence units (RFUs).

4.1.10. One major advantage of the EGFP bioassay over the luciferase bioassay is that EGFP fluorescence is measured in intact cells without having to remove the medium, and cells can be returned to the incubator after they have been read. Luciferase activity typically requires cell lysis. Accordingly, this allows repeated measurements of EGFP fluorescence from the same cells over time and provides an extremely easy avenue in which to examine the time course of induction by test chemicals.

4.1.11. The H1L6.1c2 luciferase cell bioassay protocol described here can readily be applied to other cell lines containing stably-transfected AhR-responsive luciferase reporter genes [19, 20, 25, 33, 34].

4.2 Quantitative Real Time PCR Bioassay

4.2.1. The experimentalist should become familiar with the valuable information included in "Guide to Performing Relative Quantitation of Gene Expression Using Real-Time Quantitative PCR" available from Applied Biosystems.

4.2.2. A fume hood is used when working with Trizol, phenol and chloroform. These solutions are discarded as hazardous waste.

4.2.3. In earlier protocols, contaminating genomic DNA is removed typically by pretreatment with DNase (e.g., DNA-Free Kit, Ambion, Austin, TX). This step guarded against forming products during real time PCR from contaminating genomic DNA. Design of primers to span introns so as to yield very large cDNA amplicons with genomic DNA has made this step moot. Even if the amplicon is within a single exon, removing genomic DNA may not be necessary in some cases where the mRNA being measured is abundant.

4.2.4. In real time PCR, a threshold level is set above the baseline but low enough to be within the exponential portion of an amplification curve. The cycle number at which the fluorescence signal arising from accumulation of an amplicon reaches the threshold is called the C_T. Dilutions of cDNA are most convenient when the number of PCR cycles to reach the C_T is in the range of 20–30. The difference between the C_T for CYP1A1 and the GUSB endogenous control gene is calculated for two of the conditions being compared. From the difference between these differences, the relative amount of CYP1A1 can be calculated as the antilog (base 2), called the $\Delta\Delta C_T$ method.

4.2.5. This protocol describes measurement of CYP mRNA levels in cultured cells relative to one or more endogenous housekeeping genes. In keratinocytes, two genes that change little in transcription with many treatment conditions are used for the present purpose for normalization, but this property must be verified for previously untried treatment conditions.

Acknowledgements

We thank Dr. Marjorie A. Phillips for expert advice regarding real time PCR. The recombinant cell lines and methods described here were developed as part of ongoing research supported by the National Institute of Environmental Health Sciences Superfund Research Program (ES004699).

References

1. Safe S (1990) Polychlorinated biphenyls (PCBs), dibenzo-p-dioxins (PCDDs), dibenzofurans (PCDFs), and related compounds: environmental and mechanistic considerations which support the development of toxic equivalency factors (TEFs). Crit Rev Toxicol 21:51–88
2. Hankinson O (1995) The aryl hydrocarbon receptor complex. Annu Rev Pharmacol Toxicol 35:307–340
3. Denison MS, Nagy SR (2003) Activation of the aryl hydrocarbon receptor by structurally diverse exogenous and endogenous chemicals. Annu Rev Pharmacol Toxicol 43:309–334
4. Beischlag TV, Morales JL, Hollingshead BD, Perdew GH (2008) The aryl hydrocarbon receptor complex in the control of gene expression. Crit Rev Eukaryot Gene Expr 18:207–250
5. White SS, Birnbaum LS (2009) An overview of the effects of dioxins and dioxin-like compounds on vertebrates, as documented in human and ecological epidemiology. J Environ Sci Health C Environ Carcinog Ecotoxicol Rev 27:197–211
6. Denison MS, Soshilov AA, He G, DeGroot DE, Zhao B (2011) Exactly the same but different: promiscuity and diversity in the molecular mechanisms of action of the aryl hydrocarbon (dioxin) receptor. Toxicol Sci 124:1–22
7. Denison MS, Seidel SD, Rogers WJ, Ziccardi M, Winter GM, Heath-Pagliuso S (1999) Natural and synthetic ligands for the Ah receptor. In: Puga A, Wallace KB (eds) Molecular biology approaches to toxicology. Taylor & Francis, Philadelphia, pp 393–410
8. Stanisić V, Lonard DM, O'Malley BW (2010) Modulation of steroid hormone receptor activity. Prog Brain Res 181:153–176
9. McMillan BJ, Bradfield CA (2007) The aryl hydrocarbon receptor sans xenobiotics: endogenous function in genetic model systems. Mol Pharmacol 72:487–498
10. Humblet O, Birnbaum L, Rimm E, Mittleman MA, Hauser R (2008) Dioxins and cardiovascular disease mortality. Environ Health Perspect 116:1443–1448
11. Linden J, Lensu S, Tuomisto J, Pohjanvirta R (2010) Dioxins, the aryl hydrocarbon receptor and the central regulation of energy balance. Front Neuroendocrinol 31:452–478
12. Vondracek J, Umannova L, Machala M (2011) Interactions of the aryl hydrocarbon receptor with inflammatory mediators: beyond CYP1A regulation. Curr Drug Metab 12:89–103
13. Montelone I, MacDonald TT, Pallone F, Montelenoe G (2012) The aryl hydrocarbon receptor in inflammatory bowel disease: linking the environment to disease pathogenesis. Curr Opin Gastroenterol 28:310–313
14. Bock KW (2013) The human Ah receptor: hints from dioxin toxicities to deregulated target genes and physiological functions. Biol Chem 394:729–739
15. Lawrence BP, Denison MS, Novak H, Vorderstrasse BA, Harrer N, Neruda W, Reichel C, Woisetschläger M (2008) Activation of the aryl hydrocarbon receptor is essential for mediating the anti-inflammatory effects of a novel low molecular weight compound. Blood 112:1158–1165
16. Zhang S, Lei P, Liu X, Walker K, Kotha L, Rowlands C, Safe S (2009) The aryl hydrocarbon receptor as a target for estrogen receptor-negative breast cancer chemotherapy. Endocr Relat Cancer 16:835–844
17. Jin U-H, Lee S-O, Safe S (2012) Aryl hydrocarbon receptor (AhR)-active pharmaceuticals are selective AhR modulators in MDA-MB-468 and BT474 breast cancer cells. J Pharmacol Exp Ther 343:333–341
18. Behnisch PA, Hosoe K, Sakai S (2001) Bioanalytical screening methods for dioxins and dioxin like compounds: a review of bioassay/biomarker technology. Environ Int 27:413–439

19. Hahn ME (2002) Biomarkers and bioassays for detecting dioxin-like compounds in the marine environment. Sci Total Environ 289:49–69
20. Denison MS, Zhao B, Baston DS, Clark GC, Murata H, Han D-H (2004) Recombinant cell related chemicals. Talanta 63:1123–1133
21. Han D-H, Nagy SR, Denison MS (2004) Comparison of recombinant cell bioassays for the detection of Ah receptor agonists. Biofactors 20:11–22
22. Whyte JJ, Schmitt CJ, Tillitt DE (2004) The H4IIE cell bioassay as an indicator of dioxin-like chemicals in wildlife and the environment. Crit Rev Toxicol 34:1–83
23. Garrison PM, Tullis K, Aarts JMMJG, Brouwer A, Giesy JP, Denison MS (1996) Species-specific recombinant cell lines as bioassay systems for the detection of 2,3,7,8-tetrachlorodibenzo-p-dioxin-like chemicals. Fundam Appl Toxicol 30:194–203
24. Nagy SR, Sanborn JR, Hammock BD, Denison MS (2002) Development of a green fluorescent protein based cell bioassay for the rapid and inexpensive detection and characterization of AhR agonists. Toxicol Sci 65:200–210
25. He G, Tsutsumi T, Zhao B, Baston DS, Zhao J, Heath-Pagliuso S, Denison MS (2011) Third generation Ah receptor-responsive luciferase reporter plasmids: amplification of dioxin responsive elements dramatically increases CALUX bioassay sensitivity and responsiveness. Toxicol Sci 123:511–522
26. Denison MS, Rogers JM, Rushing SR, Jones CL, Tetangco SC, Heath-Pagliuso S (2002) Analysis of the Ah receptor signal transduction pathway. In: Maines M, Costa LG, Reed DJ, Sassa S, Sipes IG (eds) Current protocols in toxicology. Wiley, New York, pp 4.8.1–4.8.45
27. Wall RJ, He G, Denison MS, Congiu C, Onnis V, Fernandes A, Bell DR, Rose M, Rowlands JC, Balboni G (2012) Novel 2-aminoisoflavones exhibit aryl hydrocarbon receptor agonist or antagonist activity in a species/cell-specific context. Toxicology 88:881–887
28. Zhao B, Baston DS, Khan E, Sorrentino C, Denison MS (2010) Enhancing the response of CALUX and CAFLUX cell bioassays for quantitative detection of dioxin-like compounds. Sci China Chem 53:1010–1016
29. Zhao B, Bohonowych JES, Timme-Laragy A, Jung D, Affatato AA, Rice RH, Di Giulio RT, Denison MS (2013) Common commercial and consumer products contain activators of the aryl hydrocarbon (dioxin) receptor. PLoS One 8:e56860
30. Denison MS, Pandini A, Nagy SR, Baldwin EP, Bonati L (2002) Ligand binding and activation of the Ah receptor. Chem Biol Interact 141:3–24
31. Dzeletovic N, McGuire J, Daujat M, Tholander J, Ema M, Fujii-Kuriyama Y, Bergman J, Maurel P, Poellinger L (1997) Regulation of dioxin receptor function by omeprazole. J Biol Chem 272:12705–12713
32. Knockaert M, Blondel M, Bach S, Leost M, Elbi C, Hager GL, Nagy SR, Han D, Denison MS, French M, Ryan XP, Magiatis P, Polychronopoulos P, Greengard P, Skaltsounis L, Meijer L (2004) Independent actions on cyclin-dependent kinases and aryl hydrocarbon receptor mediate the antiproliferative effects of indirubins. Oncogene 23:4400–4412
33. Chao HR, Wang YF, Wang YN, Lin DY, Gou YY, Chen CY, Chen KC, Wu WK, Chiang BA, Huang YT, Hsieh LT, Yeh KJ, Tsou TC (2012) An improved AhR reporter gene assay for analyzing dioxins in soil, sediment and fish. Bull Environ Contam Toxicol 89:739–743
34. Van der Linden SC, Heringa MB, Man HY, Sonneveld E, Puijker LM, Brouwer A, Van der Burg B (2008) Detection of multiple hormonal activities in wastewater effluents and surface water, using a panel of steroid receptor CALUX bioassays. Environ Sci Technol 42:5814–5820

Chapter 14

In Vitro CYP Induction Using Human Hepatocytes

Monica Singer, Carlo Sensenhauser, and Shannon Dallas

Abstract

Induction potential of compounds towards CYP1A2, 2B6 and 3A4 via the aryl hydrocarbon (AhR), constitutive androstane (CAR), and pregnane X (PXR) nuclear receptors (NRs), respectively, is routinely determined during small molecule drug development. Significant CYP induction can result in therapeutic failure from clinical exposure of a compound outside the therapeutic window, if the respective enzymes are responsible for a significant portion of the drugs overall metabolism and clearance. Co-medications can also be impacted in a similar manner. Additionally, if metabolism via the induced CYP enzyme results in toxic or pharmacologically active compounds being produced, exaggerated pharmacological effects may be seen resulting in direct and/or indirect toxicity. The following chapter will describe methodologies used for determining CYP induction using isolated human hepatocytes, the current gold standard for such *in vitro* assays. Where appropriate in the chapter recent guidelines will be highlighted by regulatory agencies, such as, the Food and Drug Administration (FDA) and the European Medicines Agency (EMA).

Key words CYP induction, Isolated human hepatocytes, CYP1A2, CYP2B6, CYP3A4, AhR, CAR, PXR, Nuclear receptors

1 Introduction

Preclinical drug-drug interaction (DDI) assessments are paramount to every small molecule drug development program and are a requirement of both the Food and Drug Administration (FDA) and European Medicines Agency (EMA) when submitting compound packages for approval [1, 2]. One important *in vitro* DDI assay included in most development programs is CYP induction assessment. Induction of CYPs at the transcriptional level is known to occur through activation of their respective nuclear receptors, including the aryl hydrocarbon (AhR), constitutive androstane (CAR), and pregnane X (PXR) nuclear receptors [3]. In a discovery setting, nuclear receptor luminescence assays such as PXR and AhR are therefore often used to determine CYP induction liabilities, since they are relatively easy to automate and can provide a first read-out on potential enzyme induction issues. These assays are described in more detail in this volume in Chaps. 11–14.

To understand whether results generated from nuclear receptor assays are clinically relevant, a more physiologically relevant model is required. Induction in multiple liver derived cell lines have been described by numerous investigators [4–8]. These systems have been criticized for their lack of some important physiologically relevant enzymes and/or uptake transporters [9]. Therefore, as with the nuclear receptor assays, these data are largely considered supplemental by the regulatory agencies. The gold standard for *in vitro* CYP induction assays remains isolated human hepatocytes [10, 11]. When cultured, hepatocytes are known to de-differentiate rapidly which results in a significant lowering of the basal levels of most CYPs important from a drug metabolism point of view, including CYP1A2, 2B6, multiple 2C's and 3A4. These models are therefore well suited to detect potential increases in CYP expression following induction. Here we describe general methods for using human hepatocytes (cryopreserved or freshly isolated) for CYP induction assessments *in vitro*.

2 Materials

2.1 Reagents and Consumables

All reagents and consumables can be obtained from commercial sources. Catalog numbers are given in Table 1 (**Note 1**).

2.2 Instrumentation

1. Laminar flow hood (The Baker Company, Sanford, ME)
2. Water bath (Thermo Scientific, West Palm Beach, FL, USA)
3. Bright field microscope (Olympus, Tokyo, Japan)
4. CO_2 incubator (Nuaire, Plymouth, MN)
5. Spectrophotometer capable of reading multiwell plates (Molecular Devices, Sunnyvale, CA)
6. Sciex API 4000™ triple quadrupole mass spectrometer (AB Sciex, Foster City, CA)
7. Centrifuge with rotor for 1.5 mL tubes (Brinkmann, Westbury, NJ)
8. Centrifuge with rotor for 96 well plates (Sigma, Harz, Germany)

2.3 Media and Buffer Preparation

2.3.1 Williams E. Medium (Complete)

Remove 59.3 mL of prepared Williams E Medium (1000 mL) and discard. To the remaining medium add 10 mL of penicillin/streptomycin, 10 mL of non-essential amino acids, 10 mL of insulin-transferrin-selenium, 29.3 mL of sodium bicarbonate, and 10 μL of 10 mM dexamethasone (final concentration of 100 nM). Adjust pH to 7.4 using 1.0 N sodium hydroxide and filter sterilize immediately (**Note 2**). Complete WEM should be stored at 4 °C until use and discarded after 1 month. Medium should be pre-warmed to 37 °C prior to use, and pH should be periodically checked.

Table 1
Chemicals, reagents and consumables

Chemicals/reagents/consumables	Catalog #	Vendor
Calcium chloride	C7902	Sigma (St. Louis, MO)
Dexamethasone	D4902	Sigma (St. Louis, MO)
DMSO	D2438	Sigma (St. Louis, MO)
Ethanol	E7023	Sigma (St. Louis, MO)
Glacial acetic acid	A-6283	Sigma (St. Louis, MO)
Hepatocyte plating medium	Z99029	Celsis In Vitro Technologies (Baltimore, MD)
Hepatocyte thawing medium	Z99019	Celsis In Vitro Technologies (Baltimore, MD)
Hepes	15630	Life Technologies (Grand Island, NY)
Human liver total RNA	7960	Ambion (Grand Island, NY)
Insulin-transferrin-selenium (100×)	25030	Life Technologies (Grand Island, NY)
Krebs henseleit buffer (KHB)	K3753	Sigma (St. Louis, MO)
2-Mercaptoethanol	M6250	Sigma (St. Louis, MO)
Methanol	MX0488-6	EM Science through VWR (Bridgeport. NJ)
Non-essential amino acids (100×)	11140	Life Technologies (Grand Island, NY)
Penicillin/streptomycin (10,000 U/mL)	15140	Life Technologies (Grand Island, NY)
Sodium bicarbonate (7.5 %)	25080	Life Technologies (Grand Island, NY)
Sodium hydroxide (1 N)	S2770	Sigma (St. Louis, MO)
Taqman one-step RT-PCR master mix	4309169	Applied Biosystems (Grand Island, NY)
Trypan blue	T8154	Sigma (St. Louis, MO)
Water, nuclease free	9930	Ambion (Grand Island, NY)
William's E Media	12551	Life Technologies (Grand Island, NY)
MicroAmp optical adhesive film	4311971	Applied Biosystems (Grand Island, NY)
Qiagen 96 RNeasy Kit	74181	Qiagen (Alameda, CA)
0.22 μm Filter bottle assembly (500 mL)	163-0020	Nalgene via VWR (Bridgeport, NJ)
Single and multichannel pipets	–	Eppendorf, Hamburg, Germany
Biocoat 96-well collagen coated plates	354649	BD Gentest, Woburn, MA, USA

2.3.2 Krebs Henseleit Buffer (KHB)

Dissolve 1 package of KHB powder in ~900 mL of ddH$_2$O. To this solution add 0.373 g of calcium chloride dehydrate, 12.5 mM Hepes and 28 mL of sodium bicarbonate (7.5 % w/v). Adjust pH to 7.4, bring solution to 1 L, and filter sterilize using a 0.22 μm bottle top filter apparatus immediately. KHB solution should be stored at 4 °C until use and discarded after 1 month. Solution should be pre-warmed to 37 °C and bubbled with oxygen prior to use.

2.3.3 Lysis Buffer

Add 2.2 mL of 2-mercaptoethanol to 250 mL of Buffer RLT from Qiagen RNeasy 96 Kit prior to use.

3 Methods

3.1 Procurement of Human Hepatocytes

Procurement of human liver tissue and isolation of freshly isolated human hepatocytes is performed by many reputable laboratories throughout the US and Europe routinely. These freshly isolated cells were the gold standard for CYP inductions studies just a decade ago [12]. The development of the cryopreservation technique revolutionized the field enabling scientists to procure many vials of the same donor once, which were then kept in stasis until use. Early cryopreserved cells were criticized for loss of activity, low viability and related issues, in relation to their freshly isolated counterparts. These days cryopreserved cells are readily available, of comparable quality to freshly isolated cells, and are a staple in many drug discovery and development laboratories [13, 14]. The experimental methods listed below are exactly the same whether cryopreserved or freshly isolated cells are used, with the exception of the initial set up procedures (Sects. 3.2 and 3.3).

3.2 Thawing and Plating Cryopreserved Hepatocytes

Cryopreserved hepatocytes will arrive in a pre charged cryocan and should be placed in liquid nitrogen as soon as possible, if not being used immediately. All steps should be performed in a sterile environment.

1. Dispense 48 mL of pre-warmed (37 °C) hepatocyte thawing medium into a 50 mL conical tube. One tube should be used for each donor thawed, and no more than three donors should be manipulated at once to ensure maximal viability.

2. Rapid thaw the hepatocyte vials in a 37 °C water bath for approximately 60–90 s. Once the pellet is thawed just enough to move within the vial, empty the entire contents into the conical tube containing the pre warmed thawing medium. The semi-frozen hepatocyte pellet will drop to the bottom of the conical tube.

3. Using a single use sterile pipette remove 1 mL of thawing medium from the top of the conical tube and add to the hepatocyte vial to recover all possible hepatocytes and produce the maximal yield from each vial. Cap and invert the unthawed hepatocytes in the 50 mL conical tube gently to dissolve the pellet fully and resuspend the cells homogeneously.

4. Centrifuge the hepatocytes at 700 rpm for 7 min in a centrifuge with a fixed rotor. Remove the supernatant carefully without disturbing the pellet and resuspend the cells in 2 mL of hepatocyte plating medium using gentle trituration. Once resuspended count the viable cells using a hemocytometer. During the counting procedure the capped conical tubes should be kept at 37 °C.

5. To count the cells, dilute 10 µL of cell suspension 10 fold (70 µL of Trypan Blue (0.4 % w/v) + 20 µL of pre warmed and oxygenated KHB). Add 10 µL of the diluted cell suspension to a hemocytometer and count cells (live and dead separately) under 10× magnification. Count all four quadrants on the hemocytometer and calculate total viable cell number according to (1):

$$\frac{\text{Total Viable Cells (in 4 quadrants)}}{4} \times 10 \times 10,000 = \text{Viable cells / ml} \quad (1)$$

where 10 is the dilution factor and 10,000 takes into account the total volume of the hemocytometer, i.e., 0.0001 mL.

Cell preparation viability (% viability) can be calculated according to (2) (**Note 3**).

$$\frac{\text{Viable Cells}}{\text{Total Cells}} \times 100 = \% \text{ Viability} \quad (2)$$

6. Once counted, adjust volume of each hepatocyte donor cell suspension to 0.5–0.6×10^6 million cells/mL using plating medium (**Note 4**). Dispense 100 µL of cell suspension to each well of a collagen coated 96 well plate using a multichannel pipette and allow 4–6 hours for cells to attach in a 37 °C CO_2 incubator. After 4 h of plating, replace plating medium in each well with 100 µL complete WEM (Day 0) (**Note 5**). Replenish cells with fresh WEM at 24 h (Day 1) post seeding.

3.3 Receiving and Preparing Commercially Purchased Freshly Isolated Hepatocytes

Purchased freshly isolated hepatocytes (with or without overlay) will arrive already plated (i.e., Day 2 cultures), and ready to be used for induction assays. All steps should be performed in a sterile environment.

1. Carefully remove cells from packaging and wipe plates with a lint free wipe soaked in 70 % ethanol.

2. Remove the transport buffer shipped with the cells, and wash monolayer twice with pre-warmed complete WEM.

3. Allow cells to equilibrate for 30 min in complete WEM at 37 °C before starting the hepatocyte incubations.

Table 2
List of positive control (PC) inducers for CYP induction

CYP	Inducer	Catalog number	Vendor
1A2	β-Naphthoflavone	N-3633	Sigma (St. Louis, MO)
2B6	Phenobarbital	P-1636	Sigma (St. Louis, MO)
2C9, 2C19, 3A4	Rifampicin	R-3501	Sigma (St. Louis, MO)

3.4 Hepatocyte Incubations

Hepatocyte incubations to determine CYP induction generally start 48 h post seeding (Day 2) (**Note 6**). Cultured hepatocytes de-differentiate rapidly and lose a considerable amount of CYP enzymes levels over the first 2 days. By waiting 48 h, the CYP levels have generally stabilized, making this system extremely useful to measure induction [11].

1. On day 2, prepare fresh stock solutions of the control inducers in DMSO (Table 2) (**Note 7**). Suggested concentrations are 50 mM β-naphthoflavone (CYP1A2) 1 M phenobarbital (CYP2B6) and 10 mM rifampicin (CYP2C9, 2C19, 3A4).

2. Further dilute the stock solutions of control inducers with complete WEM to achieve final incubation concentrations of 50 μM β-naphthoflavone, 1 mM phenobarbital and 10 μM rifampicin.

3. If solubility allows, prepare test compounds in DMSO at 1000× the final concentrations desired. The final concentration of DMSO in all of the treatments (control and test compound) will be 0.1 %. A vehicle control (0.1 % DMSO) should be included in every assay for each CYP isoform examined (**Note 8**).

4. Aspirate medium from 96 well plates and add 100 μL control inducers, test compounds or vehicle control (in triplicate for each isoform examined). A positive control (PC) for the cytotoxicity assay in triplicate (2 % Triton X-100) should also be included (see Sect. 3.4).

5. After 24 h of incubation remove 10 μL of medium from each well and transfer to a clean 96 well plate for cytotoxicity testing (LDH release). Aspirate remaining medium from wells and re-dose with freshly prepared control inducers, test compound or vehicle control.

3.5 Cytotoxicity

Determine potential cytotoxicity of compounds at 24 h post dosing by measuring, for example, LDH release from the cells (**Note 9**). Overall cell health and morphology should also be monitored throughout the assay.

3.5.1 LDH Assay

Using a commercially available lactate dehydrogenase (LDH) kit (Tox-7, Sigma), determine LDH release from cells.

1. Remove the LDH assay substrate solution, LDH dye solution and LDH assay cofactor from the kit and warm constituents to room temperature.
2. The LDH assay cofactor is initially lyophilized. Add 25 mL of ddH$_2$O to lyophilized cofactor container and gently invert to obtain a homogeneous solution (**Note 10**).
3. Prepare the LDH assay mixture by combining equal parts of the LDH assay substrate solution, LDH dye solution and LDH assay cofactor mixture.
4. Dilute each 10 μL sample (from Sect. 3.4, Step 5) by adding 30 μL ddH$_2$O.
5. Add a further 80 μL of LDH assay mixture (from step 3) to each well and incubate in the dark for 20 min at room temperature.
6. At the end of the incubation period, add 12 μL of 1.0 N HCl to each well and mix on a multimixer for 1 min (**Note 11**).
7. Measure absorbance in each well with a spectrophotometer at 490 nm using a 690 nm correction.
8. Compare absorbance (arbitrary units) of vehicle or negative control (NC; 0.1 % DMSO) to all treatments to determine if a statistically significant difference in LDH release is evident.

3.6 Activity Measurements (LC-MS/MS)

Substrates used for activity measurements should be freshly prepared using pre-warmed and oxygenated KHB on Day 4 at the end of the 48 h incubation period. Samples are analyzed by LC-MS/MS. CYP specific probe substrate metabolism is determined by monitoring metabolite formation in the presence of either test compound or known CYP inducers. The metabolites are quantified using a 12-point standard curve and five QC levels in triplicate on a Sciex API 4000™ triple quadrupole mass spectrometer in the Multiple Reaction Monitoring (MRM) scan mode using electrospray ionization (ESI).

1. Prepare stock solutions of phenacetin (CYP1A2), bupropion (CYP2B6), tolbutamide (CYP2C9), S-(+)-mephenytoin (CYP2C19) and testosterone (CYP3A4) in DMSO at 100 mM. Dilute stock substrate solutions 1,000× in KHB to achieve final concentrations of 200 μM (Table 3) (**Note 12**).
2. Aspirate medium from all wells and wash twice with 100 μL of KHB.
3. Incubate wells with appropriate CYP substrate at 37 °C for 45 min (**Note 13**). At the end of the incubation period remove 80 μL of medium using a manual pipette and transfer to a deep

Table 3
List of probe substrates

CYP	Substrate	Catalog number	Vendor
1A2	Phenacetin	A-2500	Sigma (St. Louis, MO)
2B6	Bupropion	B-102	Sigma (St. Louis, MO)
2C9	Tolbutamide	T-0891	Sigma (St. Louis, MO)
2C19	S-(+)-Mephenytoin	457053	BD Biosciences (San Jose, CA)
3A4	Testosterone	T-1500	Sigma (St. Louis, MO)

Table 4
List of metabolite reference standards

CYP	Metabolite/reference standard	Catalog number	Vendor
1A2	Acetaminophen	A-7085	Sigma (St. Louis, MO)
2B6	OH-Bupropion	451711	BD Biosciences (San Jose, CA)
2C9	OH-Tolbutamide	UC-160	Ultrafine (Woburn, MA)
2C19	(+/−)-4-OH-Mephenytoin	UC-126	Ultrafine (Woburn, MA)
3A4	6β-OH-Testosterone	T-1500	Sigma (St. Louis, MO)
1A2—I.S.	Acetaminophen-d4	P-909	Cerilliant (Round Rock, TX)
2B6—I.S.	OH-Bupropion-d6	451003	BD Biosciences (San Jose, CA)
2C9—I.S.	Oh-Tolbutamide-d9	SI-01099-002	Synfine (Ontario, Canada)
2C19—I.S.	(+/−)-4-OH-Mephenytoin-d3	SI-01099-001	Synfine (Ontario, Canada)
3A4—I.S.	6β-OH-Testosterone-d7	451009	BD Biosciences (San Jose, CA)

well 96 well-plate. Samples from the multiple CYPs can be pooled for cocktail analysis of the resulting metabolites (**Note 14**).

4. Aspirate remaining supernatant from all wells, wash the wells two times with 100 μL of KHB and add 100 μL of Lysis buffer to each well (Qiagen Kit). If mRNA measurements are not being performed immediately, cover plates and store at −80 °C until mRNA analysis (Sect. 3.6).

3.6.1 Analytical Assay

1. To the pooled CYP mediated post incubation medium samples (Sect. 3.6, step 3), add 150 μL of methanol and 100 μL of internal standard (a 1–1.5 μM mixture of the deuterated analogs in methanol). A vendor list for metabolites and reference standards is given in Table 4.

Table 5
Liquid chromatography parameters used in the LC/MS analysis of the CYP induction assay

Analytical column	Thermo Betasil Phenyl-Hexyl, 2.1 × 100 mm (Thermo Fisher Scientific, Bellefonte, PA)					
Flow rate	0.5 mL/min					
Run time (including equilibration)	6 min					
Column temperature	50 °C					
Mobile phase	A = 0.1 % acetic acid, B = methanol with 0.1 % acetic acid					
Time program	See below					
Time (min)	% B	Time (min)	% B	Time (min)	% B	
0	6	3	45	4.3	6	
0.4	6	3.4	80	6	Re-equilib.	
0.8	28	3.5	90			
2	38	4.2	90			

2. Evaporate the samples to dryness under a gentle stream of nitrogen, and reconstitute in 250 μL of mobile phase (1:1 methanol: water, 0.1 % acetic acid).

3. Prepare a 12-point standard curve and five QC levels in triplicate by spiking 240 μL of blank KHB with 100 μL of internal standard solution, as above, and 150 μL of the appropriate probe metabolite standard mixture in methanol.

4. Achieve chromatographic separation of the analytes using the parameters listed in Table 5. A representative chromatogram, obtained on a Shimadzu Nexera LC system with binary LC-30 AD pumps and a CTO-30A column oven, showing the separation of hepatocyte generated analytes is given in Fig. 1.

3.6.2 Data Analysis

Acquire and reduce data in Analyst 1.6.1 (Applied Biosystems/Sciex) and after weighted (1/x) linear regression analysis (coefficient of determination, R^2, >0.99) the standard curves are used to calculate the metabolic rates of formation. That is, the concentration of the analytes are converted to pmol, and divided by the time of incubation and number of cells per million, to yield a rate of formation expressed as pmol/min/million cells. Hepatocytes are shown to be inducible by demonstrating a fold increase in CYP activity resulting from incubation with classical inducer (PC) over non treated cells (vehicle control). The fold induction by the PC is defined as 100 % induction. The potential for drug candidates to

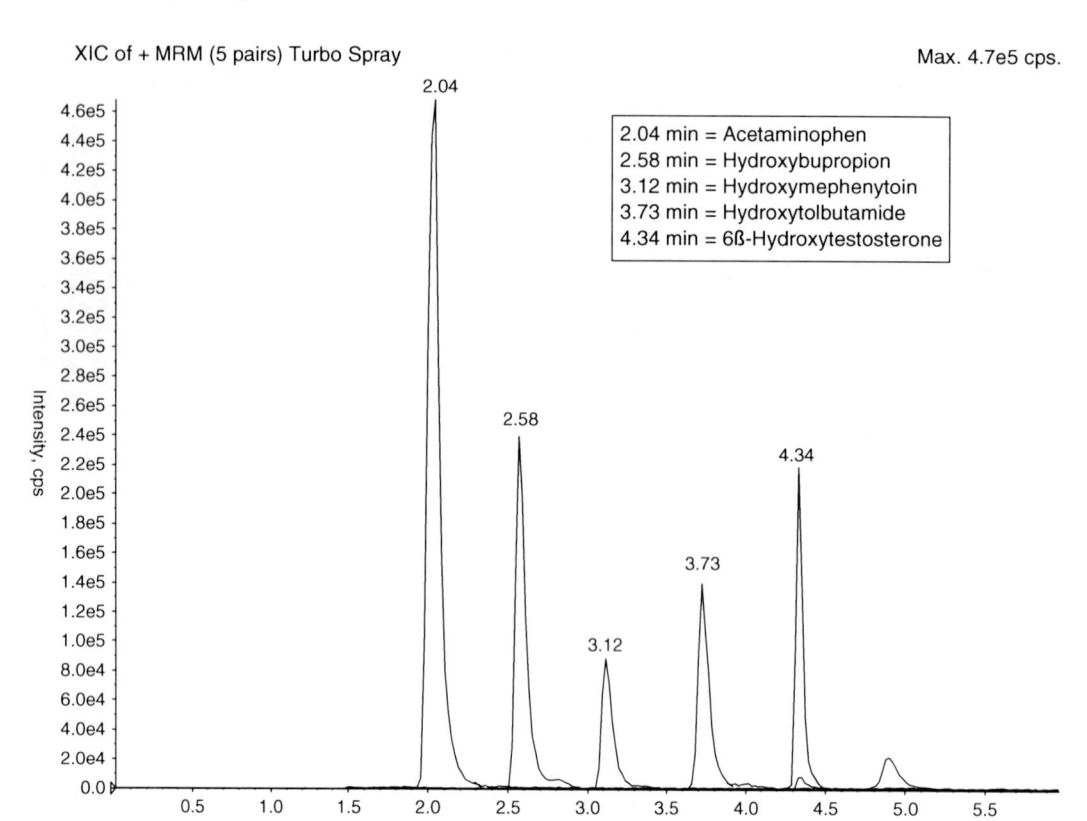

Fig. 1 Representative LC-MS/MS chromatogram depicting separation of hepatocyte generated probe substrate metabolites. This chromatogram was obtained on a Shimadzu Nexera LC System

cause CYP induction is described as a percentage of the induction observed with the classic PC inducer according to (3):

$$100 \times \frac{\text{Sample} - \text{NC}}{\text{PC} - \text{NC}} \quad (3)$$

The induction potential is also examined as the fold induction versus the NC according to (4).

$$\frac{\text{Rate of formation of sample}}{\text{Rate of formation of NC}} \quad (4)$$

3.7 RNA Measurements

3.7.1 RNA Preparation

If analysis of the RNA samples occurs immediately after the completion of the hepatocyte incubations (Sect. 3.6, Step 4), proceed immediately to Step 1. Otherwise, multiwell plate containing lysed cells should be removed from −80 °C and allowed to thaw at room temperature prior to RNA preparation.

1. Prepare 70 % ethanol using nuclease free water. Add 100 μL of prepared 70 % ethanol to each well, pipet up and down three times to mix.

2. Add entire sample to RNeasy 96 kit column. Centrifuge plate for 4 min at 5,100 rpm, room temperature and discard eluent.

3. Add 800 μL RW1 buffer to column (provided in kit), centrifuge for 4 min at 5,100 rpm, room temperature and discard eluent.

4. Add 800 μL RPE buffer column (provided in kit), centrifuge for 4 min at 5,100 rpm, room temperature and discard eluent. Repeat washing step twice.

5. Elute RNA using 50 μL H_2O (provided in kit) into microtube plate, centrifuge for 4 min at 5,100 rpm, room temperature and transfer eluent to a clean 96 well plate. Repeat elution step twice. Cover plate and store at −80 °C until PCR is performed.

3.7.2 PCR

1. Probes are purchased from Applied Biosystems—TaqMan Gene Expression Assays for use on a Viia7 Real Time PCR system instrument (Applied Biosystems, Grand Island, NY) or an equivalent PCR system. Human CYP probes are detailed in Table 6.

2. A TaqMan One-Step RT-PCR Master Mix is used for quantitative real time PCR in a 384 well format (**Note 15**). Per reaction, 2 μL RNA is added and OD values are not measured. That is, the sample is used directly from the RNA preparation step (**Note 16**).

3. For each reaction, mix together the following kit components in a tube: 25 μL Master Mix, 1.25 μL of 40× Multiscribe RNase inhibitor, 12.5 μL nuclease free water, and 1.25 μL Applied Biosystems TaqMan Gene Expression Assay probe (Table 6). Prepare enough volume for all samples to be

Table 6
Probes used for CYP induction

Assay ID	Catalog number	Gene symbol
Hs00167927_m1	4331182	CYP1A2
Hs04183483_g1	4331182	CYP2B6
Hs02383631_s1	4331182	CYP2C9
Hs00426380_m1	4331182	CYP2C19
Hs00604506_m1	4331182	CYP3A4
Hs99999901_s1	4331182	18S

measured. Add 8 μL of reaction mix into each well of a clean 384 well plate, containing 2 μL/per well of RNA sample.

4. Cover plate with MicroAmp optical adhesive film and centrifuge at 5,100 rpm for 2 min at room temperature.

5. Load plate into PCR machine. Run standard curve using the following parameters:

(a)	Stage 1	2 min	50 °C
(b)	Stage 2	10 min	95 °C
(c)	Stage 3 (40 cycles)	15 s	95 °C
		1 min	0 °C

3.7.3 Data Analysis

1. Each probe must have a standard curve generated for it (**Note 17**). RNA standards for CYPs are prepared by 1:2 dilutions of Human liver total RNA yielding a standard curve range from 0.1–200 ng. RNA standards for 18S are prepared by 1:5 dilutions of Human liver total RNA yielding a standard curve range from 0.0002–80 ng. All standard curve points are run in triplicate. Standard curve points are then plotted as mean cycle threshold (CT) versus log10 concentration and the equation of best fit line is generated. A standard curve is generated for each experiment.

2. The standard curve equations are used to calculate the relative ng of RNA for each sample, and for each probe.

3. Relative ng of 18S or CYP RNA are calculated using values generated from the 18S or CYP standard curves according to (5).

$$10^{(\text{mean CT} - y\text{-intercept}/\text{slope})} \quad (5)$$

4. Fold change (FC) induction for each sample compared to the NC is then calculated according to (6).

$$\frac{\text{individual ratio of CYP:18S of sample}}{\text{mean ratio of CYP:18S of NC}} \quad (6)$$

Overall FC is obtained by taking the mean of the individual values.

5. Percent of PC can also be calculated according to (7):

$$100 \times \frac{\text{individual treatment FC} - \text{mean NC FC}}{\text{mean PC FC} - \text{mean NC FC}} \quad (7)$$

6. Individual percent of PC fold changes can then be averaged for replicates in each group.

3.8 Data Interpretation

3.8.1 General Considerations and Follow-Up Studies

Data from both the activity and mRNA analyses are expressed as percent PC induction and FC from the NC, as detailed in Sects. 3.7.3 and 3.6.2. Both the FDA and EMA provide language around interpretation of assay results. However, the reader is encouraged to examine both documents carefully, as the requirements from the two regulatory agencies may be similar, but they are not identical. In general, and to be conservative, one can consider a two-fold increase from the NC, or a 20 % increase relative to the PC as a positive induction result if the changes are dose dependent. The concentration range chosen should bracket the known or suspected C_{max} of the compound in humans, and the highest concentration tested should be ideally 50× this C_{max}. As this concentration may be very high (>10 uM), prior testing of the compound for solubility and cytotoxicity is advised. Each donor should be presented separately, to capture potential donor differences in induction potential. The donor with the largest induction change by the test compound is considered as worst case. This method for interpreting the data is defined as a "basic" model. If only three concentrations were examined in the basic assay, follow on studies to generate EC_{50} and E_{max} values, using at least eight concentrations of compound can be undertaken, and compared with known CYP inducers. Several dynamic or static methods can then use this data for modeling of a potential DDI including the RIS correlation method [9] or more mechanistically intricate models such as Physiologically Based Pharmacokinetic (PBPK) modeling [15].

3.8.2 Special Cases I: Concurrent Inhibition

If a compound is not cytotoxic, does not show induction at the activity level, but significant induction is observed at the gene level, concurrent inhibition by the compound may be masking the activity results. If this information is unknown at the time of the assay, an inhibition control can be included in the assay as follows: 2 h prior to the end of the 48 h dosing period (Sect. 3.4), medium is removed from triplicate cells induced with the respective PC inducers, and is replaced with medium containing the respective probe substrates and incubated for an additional 2 h. After 2 h the activity measurements are undertaken as described previously. If a significant decrease in activity is observed between the pre-induced compound treated versus untreated cells, this may indicate a concurrent inhibition which is masking induction at the activity level. This hypothesis can be verified using the definitive assays for CYP inhibition, both reversible and mechanism based, which are described in detail in this volume in Chaps. 16–20.

3.8.3 Special Cases II: Non-specific Binding and Metabolism

Depending on the physicochemical properties of the compound, other issues related to non-specific binding or metabolism may complicate interpretation of generated CYP induction data. In both cases, the actual dose that the cells are exposed to will be overestimated, which might lead to false negatives, and inaccurate modeling conclusions. These issues have been highlighted in the recent FDA and EMA Guidances. In terms of non-specific binding, data generated from other assays including mass-balance in Caco-2 assays or non-specific binding assays used during protein binding assessments (using radiolabeled compound), as well as other non-specific binding assessments, should be given attention when considering the concentration range undertaken in the CYP induction assay. With respect to metabolism, the EMA has suggested taking medium samples on the last day of induction, which can be important for compounds that undergo significant lysosomal trapping, for example. Understanding metabolism of the compound in the assay format chosen could be undertaken using radiolabeled compound; however this will be challenging using the 96 well-format, for example, where the total amount of cells is extremely low.

4 Notes

1. The commercial reagents, consumables and equipment listed in this chapter are an immediate example of specific vendor source information that can be used for CYP induction assays. Multiple commercial sources exist for hepatocyte and culturing media, reagents and equipment listed in this chapter.

2. Studies are generally run under serum free/protein free conditions. If additional protein must be added to the assay, for example due to non-specific binding issues (Sect. 3.8.3) or to decrease toxicity (Sect. 3.5), the degree of protein binding will need to be considered when choosing the concentration range for the study.

3. If hepatocyte viability upon thawing of cryopreserved cells is found to be low (<75 %), or considerable debris is noted, the hepatocyte preparation can be cleaned up using a Percoll gradient. However, it is generally recommended to pre-validate cryopreserved hepatocytes donors prior to studies to ensure adequate viability, etc. without the need for a separate Percoll step.

4. Plating numbers should be optimized for the well format used (96, 24 or 6 well).

5. Prolonged absence of medium on the hepatocytes will result in cell death. Aspirate no more than three columns or rows of 96 wells at a time to ensure cells do not dry out during medium changes and compound dosing.

6. Culture and dosing times can vary depending on hepatocyte configuration used. When using Matrigel overlay, for example, cells can be dosed 24 h after plating, and the duration can be as long as 72 h. Culture times of greater than 5 days of simple

hepatocyte preparations are not recommended. General considerations on the dosing and culturing times used for CYP induction assays have recently been summarized by Hewitt et al. [11]. Information can also be found in the recent FDA [1], and EMA [2] Drug Interaction Guidances. Whichever method is chosen to ultimately undertake the studies, it is recommended to pre-validate cryopreserved donors for robust induction of the desired CYP isoforms using well characterized PC inducers prior to undertaking test compound studies.

7. Multiple suggested control inducers for CYP1A2, 2B6 and 3A4 are described in Drug Interaction Guidances by the FDA [1] and EMA [2]. Inducers for CYP2C9 and 2C19 are also included in this chapter for completeness, but these isoforms are generally not run routinely. Phenobarbital is a schedule IV controlled substance in the United States, use of this compound requires a license by the Drug Enforcement Agency.

8. If the test compound is not soluble in DMSO, dissolve in other organic solvents such as methanol or ethanol, ensuring the final concentration of the solvent does not exceed 0.1 %. In this case, separate vehicle controls for the control inducers and the test compounds will need to be included in the assay. Higher concentrations of organic solvents such as DMSO may inhibit the activity of the CYP enzymes and will, therefore, make interpretation of the data difficult.

9. LDH provides a relatively quick and easy measurement of potential cytotoxicity at 24 h. However, other measures of cytotoxicity that are more sensitive can be used, including ATP depletion, MTT, resorufin, etc. Commercial kits are available from multiple commercial sources.

10. Once reconstituted, the LDH Assay cofactor should be aliquoted (1 mL each) for future assays to avoid multiple freeze-thaw cycles.

11. Samples from the Triton-X 100 wells will be visibly pink after 20 min of incubation. If the PC is not visibly pink or the color is a deep red, the amount of sample used, the fold dilution of water, and/or the incubation time will need to be altered. The assay should be optimized ahead of time on the particular spectrometer that will ultimately be used for the assay.

12. Testosterone is a schedule III controlled substance in the United States. A controlled substance license is required to work with this material and must be obtained from the Drug Enforcement Agency.

13. Linearity of the CYP mediated reactions at the end of assay time point chosen, under laboratory specific experimental conditions, should be verified for all CYPs screened.

14. Analytical analysis of each single CYP isoform can be performed, but cocktail analysis will save considerable amount of time and resources.

15. A two-step protocol for mRNA expression can be performed by quantitative reverse transcription-polymerase chain reaction (qRT-PCR). First-strand cDNA can be synthesized using SuperScript® VILO™ MasterMix (Invitrogen cat# 11755), followed by analysis with Applied Biosystem's Viia7 Real Time PCR System (or equivalent PCR system) using Applied Biosystems TaqMan® Universal PCR Master Mix (cat#4304437).

16. When undertaking studies in a 96 well format, the concentration of RNA recovered is too low to obtain accurate 260/280 absorbance measurements. If using a 24 or 6 well plate format, measuring RNA concentrations directly will likely be more feasible.

17. The Comparative C_T method, also referred to as the $\Delta\Delta C_T$ method, is similar to the relative standard curve method described in Sect. 3.7.3., except it uses arithmetic formulas to achieve the result for relative quantitation. It is possible to eliminate the use of standard curves described in Sect. 3.7.3., and to use the $\Delta\Delta C_T$ Method for relative quantitation of RNA as long as the PCR efficiencies between the target(s) and endogenous control(s) are relatively equivalent. Details on this methodology can be found on the Applied Biosystems Website: http://www3.appliedbiosystems.com/cms/groups/mcb_support/documents/generaldocuments/cms_042380.pdf.

Acknowledgements

The views expressed here are solely those of the authors and do not reflect the opinions of Janssen Research & Development, LLC.

References

1. US Food and Drug Administration, US Department of Health and Human Services, Center for Drug Evaluation and Research, Center for Biologics Evaluation and Research (CBER) (2012) Guidance for industry: drug interaction studies—study design, data analysis, and implications for dosing and labeling. Silver Spring, MD. http://www.fda.gov/downloads/Drugs/GuidanceComplianceRegulatoryInformation/Guidances/UCM292362.pdf

2. European Medicines Agency (2012) Guideline on the investigation of drug interactions. http://www.ema.europa.eu/docs/en_GB/document_library/Scientific_guideline/2012/07/WC500129606.pdf

3. Amacher DE (2010) The effects of cytochrome P450 induction by xenobiotics on endobiotic metabolism in pre-clinical safety Studies. Toxicol Mech Methods 20(4):159–166

4. Gomez-Lechon MJ, Donato T, Jover R et al (2001) Expression and induction of a large set of drug-metabolizing enzymes by the highly differentiated human hepatoma cell line BC2. Eur J Biochem 268(5):1448–1459

5. Westerink WM, Schoonen WG (2007) Cytochrome P450 enzyme levels in HepG2 cells and cryopreserved primary human hepatocytes and their induction in HepG2 cells. Toxicol In Vitro 21(8):1581–1591

6. Hariparsad N, Carr BA, Evers R, Chu X (2008) Comparison of immortalized Fa2N-4 cells and human hepatocytes as in vitro models for cytochrome P450 induction. Drug Metab Dispos 36(6):1046–1055
7. McGinnity DF, Zhang G, Kenny JR et al (2009) Evaluation of multiple in vitro systems for assessment of CYP3A4 induction in drug discovery: human hepatocytes, pregnane X receptor reporter gene, and Fa2N-4 and HepaRG cells. Drug Metab Dispos 37(6):1259–1268
8. Gerets HH, Tilmant K, Gerin B et al (2012) Characterization of primary human hepatocytes, HepG2 cells, and HepaRG cells at the mRNA level and CYP activity in response to inducers and their predictivity for the detection of human hepatotoxins. Cell Biol Toxicol 28(2):69–87
9. Fahmi OA, Ripp SL (2010) Evaluation of models for predicting drug-drug interactions due to induction. Models for predicting drug-drug interactions due to induction. Expert Opin Drug Metab Toxicol 6(11):1399–1416
10. Chu V, Einolf HJ, Evers R et al (2009) In vitro and in vivo induction of cytochrome p450: a survey of the current practices and recommendations: a pharmaceutical research and manufacturers of America perspective. Drug Metab Dispos 37(7):1339–1354
11. Hewitt NJ, Lecluyse EL, Ferguson SS (2007) Induction of hepatic cytochrome P450 enzymes: methods, mechanisms, recommendations, and in vitro-in vivo correlations. Xenobiotica 37(10–11):1196–1224
12. Parkinson A, Mudra DR, Johnson C, Dwyer A, Carroll KM (2004) The effects of gender, age, ethnicity, and liver cirrhosis on cytochrome P450 enzyme activity in human liver microsomes and inducibility in cultured human hepatocytes. Toxicol Appl Pharmacol 199(3):193–209
13. Roymans D, Annaert P, Van Houdt J et al (2005) Expression and induction potential of cytochromes P450 in human cryopreserved hepatocytes. Drug Metab Dispos 33(7):1004–1016
14. Smith CM, Nolan CK, Edwards MA, Hatfield JB, Stewart TW, Ferguson SS, LeCluyse EL, Sahi J (2012) A comprehensive evaluation of metabolic activity and intrinsic clearance in suspensions and monolayer cultures of cryopreserved primary human hepatocytes. J Pharm Sci 101(10):3989–4002
15. Varma MV, Lin J, Rotter CJ, Fahmi OA, Lam J, El-Kattan AF, Goosen TC, Lai Y (2013) Quantitative prediction of repaglinide-rifampicin complex drug interactions using dynamic and static mechanistic models: delineating differential CYP3A4 induction and OATP1B1 inhibition potential of rifampicin. Drug Metab Dispos 41(5):966–974

Chapter 15

Assessment of CYP3A4 Time-Dependent Inhibition in Plated and Suspended Human Hepatocytes

J. George Zhang and David M. Stresser

Abstract

This chapter provides a step-by-step description of methodology used to assess time-dependent inhibition/inactivation (TDI) potential of cytochrome P450 3A4 (CYP3A4) by a test compound using human hepatocytes in a 96-well plate format. Human hepatocytes in suspension or plated cultures are pre-incubated with the test compound for different time periods and at different concentrations, prior to incubation with midazolam, a CYP3A4 probe substrate. The metabolite 1′-hydroxymidazolam is then quantified by LC/MS/MS. The TDI potential of the test compound may then be evaluated by determining the enzyme activity remaining after each condition. Methods to determine K_I and k_{inact} values from these data, as well as tips and considerations for robust assay outcomes are also provided.

Key words Time-dependent inhibition/inactivation, TDI, Cytochrome P450, CYP3A4, Human hepatocytes

1 Introduction

Time-dependent inhibition and/or inactivation (TDI) of cytochrome P450 enzymes by a drug candidate may be defined as the loss of catalytic activity occurring over time in incubation. Often inhibition is irreversible, occurring when reactive metabolites bind covalently to the protein and/or heme of the enzyme [1, 2]. Quasi-irreversible inhibition may result from tight, yet non-covalent binding of a metabolite rendering it catalytically inactive, yet functional activity can be restored under certain non-physiologic conditions. Time-dependent inhibition can also be reversible, as in a case where the metabolite exhibits much stronger inhibition of the enzyme than the parent drug, but without inactivation. TDI is a major concern during drug discovery and development as it can cause drug-drug interactions (DDI) [2]. Therefore, detection and elimination of TDI potential are usually critical steps in the process of bringing a drug to market.

Among all drug metabolizing enzymes, CYP3A4 is the most abundant, comprising about 30 % of the total hepatic P450 content [3]. It is also responsible for the metabolism of more than 50 % of commercially available drugs [4] and thus critical to evaluate as a mediator of DDI.

Comprehensive TDI assays are generally set up to evaluate the effect of multiple concentrations of test compound and multiple pre-incubation time points. Following the preincubation, the remaining enzyme activity (i.e. that which was not inactivated in the preincubation) is measured by determining the catalytic activity with an added probe substrate [1]. Importantly, any compound-independent activity loss should be controlled for and the potential for reversible inhibition from the parent drug should be minimized. If activity loss is significant, determination of kinetic parameters such as K_I, the inactivation rate constant or k_{inact}, the maximal rate of inactivation, may be possible. These parameters may be used for subsequent determination of DDI risk using various mathematical models [5, 6]. Simpler experimental designs with fewer (e.g. one or two) preincubation time points and concentrations are possible, although outputs tend to be more binary (yes/no). The IC_{50} "shift" assay, another abbreviated assay, is designed to measure the change in concentration achieving 50 % inhibition after preincubation (at least one time point), compared to suitable non-preincubated or non-metabolically competent control test system. Typically, if only one or two concentrations are tested (reducing confidence in calculation of an IC_{50}), additional preincubation time points are included, and percent inhibition over the time course evaluated. The endpoints in these simpler assays can sometimes serve as range-finding experiments to design the more complex assay needed to determine a K_I and k_{inact}.

Human liver microsomes (HLM) are a convenient test system and traditionally used to determine TDI. However, depending on the test compounds, this system may not always reproduce conditions *in vivo* and can result in over or under prediction of DDI potential [7, 8]. Human hepatocytes have been recently used for evaluation of P450 TDI and have demonstrated promise as an improved test system [7–17]. Unlike HLM, hepatocytes are intact cells with a complete complement of drug metabolizing enzymes and their necessary cofactors. In addition, hepatocytes enable factors that affect drug concentrations at the enzyme active site to be taken into account, such as active uptake and efflux transports, lysosomal trapping and binding to other intracellular components. A classic example illustrating the benefits of hepatocytes is gemfibrozil. In HLM, gemfibrozil is a weak inhibitor of CYP2C8 and inhibition is not time-dependent. However, in hepatocytes, gemfibrozil exhibits potent and time-dependent inhibition of CYP2C8, which occurs after conversion by glucuronosyl transferases within the cell to gemfibrozil-glucuronide [16]. At the same time, the value of HLM for mechanistic studies should not go underappreciated—for example the requirement for glucuronidation of gemfibrozil to elicit TDI may have gone undetected using hepatocytes alone.

TDI experiments in hepatocytes involve a pre-incubation with test compounds, followed by incubation with probe substrate, usually at saturating concentrations. Both plated and suspended hepatocytes can be used for TDI studies. Human hepatocytes in suspension are generally pooled from multiple donors. Typically, three methods have been used for TDI studies with hepatocytes in suspension, namely, the "wash method", the "add method" and the "dilution method" [7–10]. In the wash method, after incubation of hepatocytes with test compounds, the hepatocyte suspension (either entirely or in aliquot) is centrifuged, washed and resuspended in a buffer, followed by an addition of the probe substrate to assess remaining enzyme activity. The add method is conducted in a similar manner as the wash method, however, after the pre-incubation, a small volume of concentrated probe substrate, such as 10 % of the total incubation volume is added directly to the incubations [9]. The dilution method starts with a similar pre-incubation with test compounds as above, but with a higher cell density, such as 1 million cells/mL to permit adequate enzyme content prior to the dilution. After the pre-incubation, substrate is added in a larger volume, resulting in a 4- to 5-fold dilution [9]. In principle, the wash method would be optimal since it offers the possibility to completely remove extracellular test compound from pre-incubation, thus mitigating the direct/competitive inhibition in the subsequent incubation with the probe substrate. However, significant loss of enzyme activity in controls using the wash method has been reported, leading to difficulties in data interpretation [9]. In addition, the wash method is far more labor-intensive, is harder to control for ongoing TDI during the washing and centrifugation process, and has the potential for the introduction of more practical errors due to more sample handling when compared with either the dilution or add methods [9, 10]. Unfortunately, the addition and dilution methods take no or minimal steps in reducing the impact of inhibition due to direct/competitive inhibition, potentially confounding accurate assessment of kinetic parameters. Assays using plated hepatocytes in 24 or 96-wells have also been described for TDI studies [11–13]. Unlike suspension assays, the culture is typically prepared from a single donor. This may have the disadvantage of not incorporating potential interindividual variability in response. Another concern can be instability in the phenotype as most P450 isoforms tend to decline in abundance over time in culture. Nevertheless, this system offers unique advantages. Unlike hepatocytes in suspension, which generally survive for only a few hours, plated cells can be kept in culture for days. This is particularly useful for evaluating compounds that are slowly metabolized or require significant secondary or tertiary metabolism to elicit TDI [11]. Another major advantage in using plated hepatocytes is the ability for facile removal of test compounds after pre-incubation using simple wash steps without the need for centrifugation or excessive handling. Finally, hepatocyte monolayers in culture offer potential of a higher level of physiological relevance and by incorporating metabolism, transport, inhibition and induction and the net effect of all the processes. However, it should also be recognized that some attributes of the model (single donor, unstable phenotype) might make it more suitable for robust qualitative, rather than quantitative endpoints. In this chapter, the step-by-step procedures for both a suspension dilution assay and plated cell assay in a 96-well plate format are described.

2 Materials

2.1 Equipment

1. LC/MS/MS system, capable quantitative analysis in cell culture media (e.g. API 4000, Applied Biosystems, Foster City, CA)
2. Incubator with a humidified atmosphere of CO_2/air (5/95 %) at 37 °C (SL® SHEL Lab, Cornelius, OR)
3. Biosafety cabinet (SterilGard Hood, the Baker Company Inc., Sanford, ME)
4. Table top centrifuge capable of accommodating 96-well plates and 50 mL conical tubes (e.g. Model 5810 R, Eppendorf, Hauppauge, NY)
5. Phase contrast light microscope (Nikon Eclipse TS100, Nikon Instruments Inc., Melville, NY)
6. Precision water-bath capable of heating to 37 °C (Thermo Scientific, Tewksbury, MA)
7. PlateLoc thermal microplate sealer (Model Velocity 11, Agilent Technologies, Santa Clara, CA)
8. Suitable software to generate non-linear curve fits (e.g. XLFit™ software, IDBS, Guilford, Surrey, UK)

2.2 Consumables

1. Collagen I-coated 96 well plates (Catalog No. 354407) (Corning Life Sciences—Discovery Labware, Tewksbury, MA)
2. 96-well cell culture plates (Catalog No. 353075) (Corning Life Sciences—Discovery Labware, Tewksbury, MA)
3. 96-deep well polystyrene plates (2 mL) (Catalog No. 10011-942) (Corning Life Science—Axygen, Union City, CA)
4. Stop/Injection plates (e.g. Catalog No. 10011-228, Corning Life Sciences—Axygen, Union City, CA)
5. Falcon™ sterile serological pipettes (1, 2, 5, 10 and 25 mL) and polypropylene tubes (15 and 50 mL) (Corning Life Sciences—Discovery Labware, Tewksbury, MA)
6. Reagent reservoir (Catalog No. 89094-680) (VWR, Radnor, PA)
7. Heat seal (Catalog No. 24210-001) (Agilent Technologies, Santa Clara, CA)

2.3 Reagents and Solutions

1. Plateable cryopreserved human hepatocytes (PCHH) (e.g. Catalog No. 454543, Corning Life Sciences—Discovery Labware, Tewksbury, MA)

2. Cryopreserved human hepatocytes (CHH) (e.g. Catalog No. 454426, Corning Life Sciences—Discovery Labware, Tewksbury, MA)

3. High Viability Cryohepatocytes Recovery Kit (e.g. Catalog No. 454534, Corning Life Sciences—Discovery Labware, Tewksbury, MA)

4. William's Medium E (WME) (e.g. Sigma, Catalog No. W1878, St. Louis, MO)

5. Culture medium (CM): WME supplemented with 0.1 µM dexamethasone (Sigma, Catalog No. D4902), penicillin-streptomycin-glutamine (GIBCO 100×, Catalog No. 10378-016), insulin-transferrin-selenium (GIBCO 100×, Catalog No. 41400-045), 15 mM HEPPS (pH 7.4) (GIBCO, Catalog No. 15630-080)

6. 0.4 % Trypan blue (Sigma, Catalog No. T8154, St. Louis, MO)

7. Test compounds: ketoconazole (Catalog No. K1003), verapamil (Catalog No. V4629) and diltiazem (Catalog No. D2521, Sigma, St. Louis, MO)

8. CYP3A4 Substrate and internal standard: midazolam (Catalog No. 451028), 1'-hydroxymidazolam (Catalog No. 451038), 1'-hydroxymidazolam-[13C3] (Catalog No. 451010) (Corning Life Sciences—Discovery Labware, Tewksbury, MA)

9. HPLC mobile phases: A: 0.1 % formic acid in water; B: 0.1 % formic acid in acetonitrile (ACN)

10. Enzyme reaction-stop solution: ACN containing 0.1 µM 1'-hydroxymidazolam-[$^{13}C_3$] and 0.1 % formic acid

11. Substrate stock solution: Prepare by dissolving midazolam in dimethyl sulfoxide (DMSO) to achieve a 30 mM concentration.

12. Stock solutions of test compounds: Eight concentrations for each test compound are prepared by threefold serial dilution in DMSO, starting with upper concentrations of 15 mM for ketoconazole, 50 mM for verapamil and 30 mM for diltiazem. The upper concentration for each test compound is prepared by dissolving test compounds in DMSO

13. Fetal bovine serum (FBS) (Catalog No. 30-2020, ATCC, Manassas, VA)

3 Methods

The methods described below are intended to provide guidance for testing three compounds in a 96-well assay format.

3.1 Cell Suspension Dilution Method

3.1.1 Preparation of Working Solutions for Test Compounds

Prepare the working solutions of each compound by mixing 10 μL of stock solution with 1 mL WME in a 96-well deep plate. Warm up the plate in the incubator prior to the pre-incubation for at least 20 min but within 1 h.

3.1.2 Preparation of Working Solution for the Midazolam Probe Substrate

Prepare the working solution of 40 μM midazolam by diluting 30 mM midazolam stock solution with WME in a 50 mL polypropylene conical tube. Warm up the tube in a water-bath at 37 °C prior to the enzyme reaction. Scale volumes as necessary.

3.1.3 Preparation of Cryopreserved Human Hepatocytes (CHH) in Suspension

1. Rapidly thaw three vials of individual CHH lots in a water-bath at 37 °C and pour into a 50 mL tube containing the thawing medium provided in the kit. Mix the tube gently. Refer to manufacturer's instructions for more details of the thawing procedure.

2. Centrifuge tube at 100 g at room temperature for 10 min and aspirate the supernatant. Re-suspend cell pellet in 50 mL WME and centrifuge at 40 g for 3 min as a wash step to remove residual thawing medium.

3. Gently resuspend the cell pellet in approximately 5 mL pre-warmed WME and gently mix.

4. Take 20 μL of cell suspension and mix with 80 μL of trypan blue (0.4 %). Load 10 μL into a hemocytometer. Count both viable and non-viable cells. Calculate the density of viable cells in suspension.

5. Dilute cell density to 1.1 million viable cells/mL with pre-warmed WME.

3.1.4 Pre-Incubation with Test Compounds

Note: All incubations are conducted in duplicate with an initial equilibration at 37 °C for all reagents (see below).

1. Pour the cell suspension in a reagent reservoir and gently mix to archive homogeneity with a multi-channel pipette.

2. Add 45 μL cell suspensions in each well of a 96-well culture plate. Warm the plate in the incubator for 15 min. Note: It is recommended to gently mix cell suspension with a multi-channel pipette as above after every other load or so.

3. Using a multi-channel pipette, add 5 μL of pre-warmed working solution for each concentration of test compound.

4. Incubate plate at 37 °C for different time periods such as 0, 0.5, 1, 2 and 4 h.

3.1.5 Enzyme Reaction with Probe Substrate

1. After each pre-incubation time, add 150 µL midazolam working solution (final substrate concentration in the reaction is 30 µM).

2. Further incubate plate at 37 °C for 10 min. Stop the enzyme reaction by directly adding 50 µL of stop solution.

3. Seal the plate with the microplate sealer and store the plate at −20 °C until LC/MS/MS analysis.

3.2 Plated Cell Method

3.2.1 Thawing and Plating Plateable Cryopreserved Human Hepatocytes (PCHH)

Note: refer to manufacturer's instructions for more details of the thawing and plating procedures.

1. Rapidly thaw one lot of PCHH in a water-bath at 37 °C and pour into a 50 mL tube containing the thawing medium (usually obtained from the kit).

2. Mix the tube gently. Centrifuge tube at 100 g at room temperature for 10 min and aspirate the supernatants. Resuspend the cell pellet in approximately 2–5 mL pre-warmed plating medium. Note: the plating medium should be supplemented with 10 % fetal bovine serum prior to use.

3. Take 20 µL of cell suspension and mix with 80 µL of trypan blue (0.4 %).

4. Load 10 µL into a hemocytometer. Count both viable and non-viable cells. Calculate density of viable cells in suspension.

5. Dilute cell density to 0.6 million viable cells/mL with pre-warmed plating medium.

6. Pour the suspension in a reagent reservoir and gently mix with a multi-channel pipette.

7. Load 100 µL cell suspensions in each well of a 96-well collagen I-coated plate. Note: It is recommended to gently mix cell suspension with a multi-channel pipette as above after every other load or so.

8. Incubate plate for approximately 2–4 h until most cells are attached to the plate.

9. Aspirate medium in the plate and replace with pre-warmed culture medium. Maintain the plate in the incubator overnight.

3.2.2 Preparation of Working Solutions for Test Compounds

Prepare the working solutions of each compound by mixing 1.5 µL of stock solution with 1.5 mL WME in a 96-well deep plate. Warm up the plate in the incubator prior to the pre-incubation steps.

3.2.3 Preparation of Working Solution for the Probe Substrate Midazolam

Prepare the working solution of 30 µM midazolam by diluting 30 mM midazolam stock solution with WME (1,000-fold dilution) in a tube. Warm up the tube in a water-bath at 37 °C prior to the enzyme reaction.

3.2.4 Pre-Incubation with Test Compounds

Note: All incubations are conducted in duplicate with an initial equilibration at 37 °C for all reagents (see below).

1. After an overnight culture, aspirate culture media in the plate. Add 100 µL pre-warmed working solution for each concentration of test compound.
2. Incubate plate at 37 °C for different time periods, such as 0.25, 0.5, 1, 2 and 4 h.

3.2.5 Enzyme Reaction with Probe Substrate

1. After each pre-incubation time, aspirate medium in the plate. Wash the plate twice with 100 µL of pre-warmed WME.
2. Add 100 µL midazolam working solution (final substrate concentration in reaction is 30 µM) after the last wash step. Further incubate plate at 37 °C for 10 min. After incubation, remove 75 µL medium and dispense into a 96-well stop plate, preloaded 19 µL of stop solution.
3. Seal the plate with the microplate sealer and store the plate at −20 °C until LC/MS/MS analysis.

3.3 Preparation of Samples for LC/MS/MS Analysis

1. Thaw sample plates at room temperature and centrifuge the plates at 4,000 g for 20 min at room temperature.
2. Load 60 µL resulting supernatant samples into the injection plates.
3. Prepare 1-hydroxymidazolam standard curve in same matrix (one part of stop solution and four parts of WME) at concentrations of 2.53, 4.92, 9.56, 18.6, 36.1, 70.1, 136, 265, 515 and 1,000 nM. Centrifuge the plate at 3,200 g for 20 min. Load 100 µL samples into the injection plates.
4. Analyze samples using LC/MS/MS under the conditions as described in Tables 1 and 2. Calculate the metabolite formation for each unknown sample (µM).

4 Data Analysis

Determine IC$_{50}$ values, K$_I$ and k$_{inact}$, using XLfit (IDBS) or other curve-fitting software. Note: for time-dependent inhibitors, the IC$_{50}$ values should decrease with increasing pre-incubation time. Absence of TDI is indicated when IC$_{50}$ values remain constant (or show a modest increase).

4.1 Determination of the Percent Inhibition and Calculation of the IC$_{50}$ Values

1. Determine the percent remaining of enzyme catalytic activity caused by the test compound relative to that of solvent vehicle controls (DMSO) at each pre-incubation time (1).

$$\% \ remaining \ activity = \frac{Activity \ (\mu M) \ in \ inhibitor \ treated \ hepatocytes}{Activity \ (\mu M) \ DMSO \ vehicle \ control} \times 100 \qquad (1)$$

Table 1
LC/MS/MS conditions

Assay parameter	CYP3A4
Substrate	Midazolam
Analyte	1′-Hydroxymidazolam
Standard metabolite range (nM)	2.53–1,000
Internal standard	1′-Hydroxymidazolam-[$^{13}C_3$]
Mobile phase A	A: 0.1 % formic acid in water
Mobile phase B	B: 0.1 % formic acid in acetonitrile
Flow rate (µL/min)	750
Column	C18 2.1×50×5 µm
Ion source/mode	Electrospray/positive ion
Run time (min)	1.5
MRM (analyte)	342 → 203
MRM (internal standard)	347 → 208
Quantitation	Quadratic regression 1/X2 weighting

Table 2
HPLC gradient conditions

Time (min)	% Solvent A	% Solvent B
0	90	10
0.3	90	10
0.5	5	95
1.1	2	98
1.2	90	10
1.5	90	10

2. Enter the test concentrations as x values and the percent remaining of activity as y values in the XLfit software. Calculate IC_{50} values by using Michaelis-Menten Model fit (Model 250) (2), where f is the fraction of control activity, I_{max} is the maximal inhibitory effect (constrained to 100 %), I is the test compound concentration [µM], n is the slope factor or the Hill coefficient and IC_{50} is the test concentration causing a 50 % decrease in activity. Example IC_{50} values for model test compounds at each pre-incubation time point are shown in Table 3.

Table 3
Example of TDI parameters

Pre-incubation time (min)	IC$_{50}$ (µM) Ketoconazole (0.02–15 µM) Suspension	Plated cells	Verapamil (0.07–50 µM) Suspension	Plated cells	Diltiazem (0.04–30 µM) Suspension	Plated cells
0	0.37	Nd	>50	Nd	>30	Nd
15	Nd	0.34	Nd	19	Nd	29
30	0.46	0.21	27	5.5	>30	10
60	0.55	0.24	12	3.1	>30	3.7
120	1.4	0.20	11	1.2	15	2.3
240	0.72	0.37	2.4	1.0	4.5	1.1
K$_I$ (µM)	Na	Na	1.4	0.69	0.38	0.32
k$_{inact}$ (h^{-1})	Na	Na	0.29	0.41	0.18	0.25

Note: *Nd* not determined, *Na* not applicable

$$f = \frac{I_{max} \times I^n}{I^n + IC_{50}^n} \quad (2)$$

4.2 Determination of k$_{inact}$ and K$_I$

1. Determine the percent remaining of enzyme catalytic activity caused by the test compound relative to that of solvent vehicle controls (DMSO) at each pre-incubation time point.
2. Convert each percent remaining of enzyme activity at each pre-incubation time point to natural logarithm.
3. Plot the natural logarithm of percent remaining of enzyme activity versus pre-incubation time (Fig. 1, top plot). Visually inspect the curve to see if the curve is in a near linear range. Remove those points contributing to non-linearity of the fit as necessary.
4. Calculate slope of natural logarithm as function of pre-incubation time at each test compound concentration.
5. Enter the test concentrations as *x* values and the corresponding slopes as *y* values in the XLfit software. Calculate k$_{inact}$ and K$_I$ values by using Michaelis-Menten Model fit (Model 250), (3) where K_{obs} is the observed rate constant for inactivation (h^{-1}), k$_{inact}$ is the maximal rate of inactivation of enzyme activity (h^{-1}), I is the test compound concentration [µM], n is the

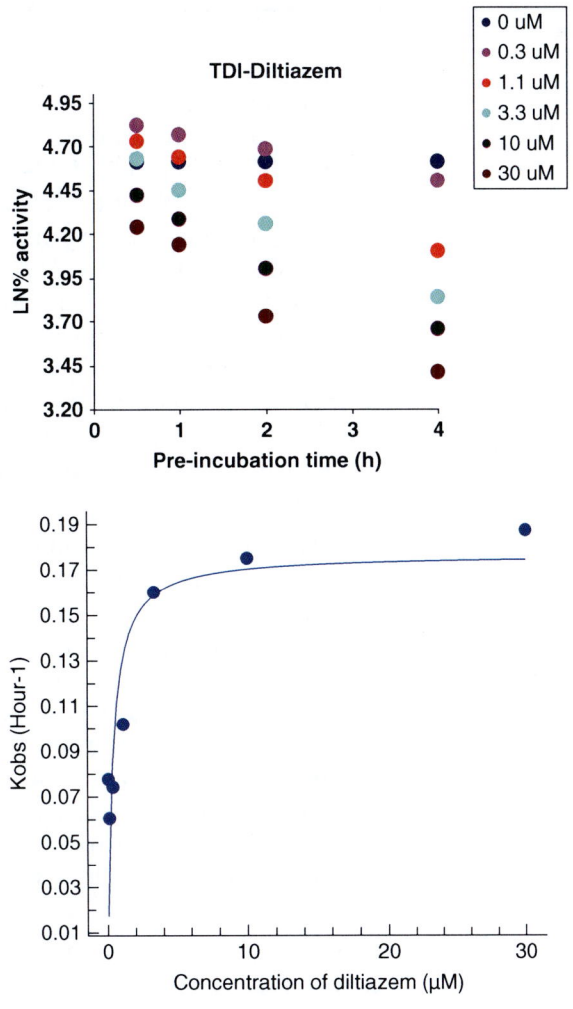

Fig. 1 Concentration and time-dependent CYP3A4 inhibition by diltiazem in suspended human hepatocytes (dilution method). *Top plot*—natural log of percentage remaining of CYP3A4 activity versus pre-incubation time in the presence of each test concentration. Values shown were corrected for compound-independent loss of activity due to the preincubation step. Therefore, vehicle only values are shown at 100 % for reference. *Bottom plot*—the rate of inhibition of CYP3A4 (k_{obs}) as a function of diltiazem concentrations. k_{obs} is the slope of the line for each test concentration on the *top plot*

slope factor or the Hill coefficient and K_I is the concentration of drug that yields 50 % of k_{inact}. The k_{inact} and K_I values for model test compounds are shown in Table 3.

$$k_{obs} = \frac{k_{inact} \times I^n}{I^n + K_I^n} \qquad (3)$$

5 Notes

1. Ketoconazole may be used as a negative control in the assay. This compound exhibits potent direct, but not time-dependent inhibition of CYP3A4 activity. Verapamil and diltiazem may be used as positive controls. As shown in Table 3, ketoconazole caused strong inhibition of CYP3A4 but no inhibition as a function of preincubation time in either the suspension or plated methods. Therefore K_I and k_{inact} could not be determined. Preincubation with verapamil and diltiazem resulted in TDI in both methods, as evidenced by the decrease in IC_{50} values with preincubation times. Note that within this limited data set, the plated cell method resulted in lower IC_{50} values but yielded similar K_I and k_{inact}, as compared with the suspension method.

2. The choices of probe substrate concentration and incubation time are important for robust outcomes. Using a saturating substrate concentration is preferable to limit confounding effects of competitive inhibition in a typical TDI assay. For the midazolam assay, we found that the K_m for both plated hepatocytes and cells in suspension was similar (2.4–2.8 μM). Metabolite formation was linear up to 20 min using 30 μM midazolam, for both suspension and plated hepatocyte assay. Short reaction times such as no longer than 10 min are recommended. This helps to limit the extent of TDI occurring during the incubation phase with probe substrate. The choice of preincubation times may require adjustment depending on the rate of inactivation exhibited by the test compound.

3. It should be noted that the viability of hepatocytes in suspension gradually decreases over the incubation time. For example, up to 30–40 % decrease in cell viability can be observed after a 4-h incubation in the absence of test compound. This may warrant caution in data interpretation. If a longer preincubation time is needed to elicit TDI, use of the plated cell method may be the optimal choice of model. In addition, the potential for cytotoxicity exhibited by the test compound may need to be considered in the suspension and plated assays.

4. As with any LC/MS bioanalytical method, the potential for matrix effects should be examined. During initial sample analysis, we found an approximate 70 % decrease in internal standard signal for unknown samples, compared with that in the standard curve. Further investigation showed that midazolam substrate present in incubation samples, but not present in the standard curve, accounted for the signal reduction. Since midazolam decreased analyte (1′-hydroxymidazolam) signal to the same extent, the overall impact in quantification of

metabolite formation in unknown samples was negligible. Nevertheless, this underscores the importance of matching standard curve matrices as closely as possible to the unknown samples.

5. The choice of solvents used to dissolve test compounds and their concentrations in incubation should be considered. DMSO is a commonly used solvent for cell based assays, but is well-known to inhibit cytochrome P450 catalytic activity. It has been shown that DMSO at 1 % can significantly attenuate the time-dependent inhibitory effects [13]. Using DMSO, even at levels of 0.2–0.5 % (v/v) in the incubation, may mask TDI properties [17]. In our hands, we found no difference in outcomes using 0.1 % DMSO compared to 0.5 % methanol on TDI parameters (IC_{50}, K_I and kinact) caused by verapamil and diltiazem (data not shown). Alternative solvents such as acetonitrile and methanol as the test compound delivery vehicle, at concentrations of 1 % or less in the preincubation are recommended in TDI studies with hepatocytes. If DMSO must be used, concentrations no higher than 0.1 % are recommended [13, 17].

6. Parameter estimates may be donor- or donor pool-dependent. Selection of donors with high CYP3A4 content may be useful to limit impact of non-CYP3A4 metabolism on CYP3A4 TDI.

7. Media selection for the pre-incubation phase among laboratories can vary. Complete culture media (WME, RPMI 1640 medium) and simple balanced physiological solutions such as KHB (Krebs-Henseleit buffer) have been used [8–10]. More recently, the use of plasma alone in the preincubation step has shown improved DDI predictions for some of CYP inhibitors [14, 15] compared to protein-free media. The choice of matrix and in particular, the potential effects of matrix components on the free fraction should be an important consideration in experimental design.

Acknowledgements

The authors thank Thuy Ho and Alanna Callendrello for their assistance in performing TDI experiments necessary for the development of these assays.

References

1. Grimm SW, Einolf HJ, Hall SD, He K, Lim HK, Ling KH, Lu C, Nomeir AA, Seibert E, Skordos KW et al (2009) The conduct of in vitro studies to address time-dependent inhibition of drug-metabolizing enzymes: a perspective of the pharmaceutical research and manufacturers of America. Drug Metab Dispos 37:1355–1370

2. Lin JH, Lu AY (1998) Inhibition and induction of cytochrome P450 and the clinical implications. Clin Pharmacokinet 35:361–390

3. Shimada T, Yamazaki H, Mimura M, Inui Y, Guengerich FP (1994) Interindividual variations in human liver cytochrome P-450 enzymes involved in the oxidation of drugs, carcinogens and toxic chemicals: studies with liver microsomes of 30 Japanese and 30 Caucasians. J Pharmacol Exp Ther 270:414–423
4. Lehmann JM, McKee DD, Watson MA, Willson TM, Moore JT, Kliewer SA (1998) The human orphan nuclear receptor PXR is activated by compounds that regulate CYP3A4 gene expression and cause drug interactions. J Clin Invest 102:1016–1023
5. FDA guidance (draft) (2012) Drug interaction studies—study design, data analysis, implications for dosing, and labeling recommendations http://www.fda.gov/downloads/Drugs/GuidanceComplianceRegulatoryInformation/Guidances/ucm292362.pdf
6. European Medicines Agency (2012) Guideline on the investigation of drug interactions (final) http://www.ema.europa.eu/docs/en_GB/document_library/Scientific_guideline/2012/07/WC500129606.pdf
7. Xu L, Chen Y, Pan Y, Skiles GL, Shou M (2009) Prediction of human drug-drug interactions from time-dependent inactivation of CYP3A4 in primary hepatocytes using a population-based simulator. Drug Metab Dispos 37:2330–2339
8. Zhao P, Kunze KL, Lee CA (2005) Evaluation of time-dependent inactivation of CYP3A in cryopreserved human hepatocytes. Drug Metab Dispos 33:853–861
9. Chen Y, Liu L, Monshouwer M, Fretland AJ (2011) Determination of time-dependent inactivation of CYP3A4 in cryopreserved human hepatocytes and assessment of human drug-drug interactions. Drug Metab Dispos 39: 2085–2092
10. Van LM, Swales J, Hammond C, Wilson C, Hargreaves JA, Rostami-Hodjegan A (2007) Kinetics of the time-dependent inactivation of CYP2D6 in cryopreserved human hepatocytes by methylenedioxymethamphetamine (MDMA). Eur J Pharm Sci 31:53–61
11. McGinnity DF, Berry AJ, Kenny JR, Grime K, Riley RJ (2006) Evaluation of time dependent cytochrome P450 inhibition using cultured human hepatocytes. Drug Metab Dispos 34: 1291–1300
12. Albaugh DR, Fullenwider CL, Fisher MB, Hutzler JM (2012) Time-dependent inhibition and estimation of CYP3A clinical pharmacokinetic drug-drug interactions using plated human cell systems. Drug Metab Dispos 40:1336–1344
13. Li AP, Doshi U (2011) Higher throughput human hepatocyte assays for the evaluation of time-dependent inhibition of CYP3A4. Drug Metab Lett 5:183–191
14. Mao J, Mohutsky MA, Harrelson JP, Wrighton SA, Hall SD (2011) Prediction of CYP3A mediated drug-drug interactions using human hepatocytes suspended in human plasma. Drug Metab Dispos 39:591–602
15. Mao J, Mohutsky MA, Harrelson JP, Wrighton SA, Hall SD (2012) Predictions of cytochrome P450-mediated drug-drug interactions using cryopreserved human hepatocytes: comparison of plasma and protein-free media incubation conditions. Drug Metab Dispos 40:706–716
16. Parkinson A, Kazmi F, Buckley DB, Yerino P, Ogilvie BW, Paris BL (2010) System-dependent outcomes during the evaluation of drug candidates as inhibitors of cytochrome P450 (CYP) and uridine diphosphate glucuronosyltransferase (UGT) enzymes: human hepatocytes versus liver microsomes versus recombinant enzymes. Drug Metab Pharmacokinet 25:16–27
17. Aasa J, Hu Y, Eklund G, Lindgren A, Baranczewski P, Malmquist J, Turek D, Bueters T (2013) Effect of solvents on the time-dependent inhibition of CYP3A4 and the biotransformation of AZD3839 in human liver microsomes and hepatocytes. Drug Metab Dispos 41:159–169

Chapter 16

Evaluation of Time-Dependent CYP3A4 Inhibition Using Human Hepatocytes

Yuan Chen and Adrian J. Fretland

Abstract

Time-dependent inhibition (TDI) is an important consideration in the drug development process. To date, methods to accurately predict the magnitude of a clinical interaction from pre-clinical TDI data have been lacking. Although more complex prediction algorithms have been developed, the accuracy has still improved little. This suggests alternate methods to collect input data may improve prediction robustness. Historically, human liver microsomes have been used to generate inhibition kinetic data used as inputs in the *in vivo* DDI predictions. Recently, it has been suggested that human hepatocytes and the kinetic data derived from this matrix may provide a better prediction for assessing clinical interactions related to TDI. This chapter reviews a detailed method to assess TDI related to CYP3A in human hepatocytes.

Key words Time-dependent inhibition (TDI), Hepatocyte, Drug-drug interaction (DDI)

1 Introduction

Assessment of drug-drug interaction (DDI) potential is an important component of the drug discovery process. Drug-drug interactions can result in serious consequences including hospitalization and occasionally death [1, 2]. As a result regulatory agencies globally have outlined their requirements in addressing DDIs both pre-clinically as well as clinically [3, 4]. Ideally, all clinical development drug candidates would have no measurable DDI potential from an *in vitro* perspective, but often the balance of other important drug-like properties, e.g. pharmacokinetics, potency, among other factors, results in compounds with some liability related to DDI. This results in the need for robust methods to fully quantify the risk of a clinically relevant DDI pre-clinically.

Drug-drug interactions can result from either the inhibition or induction of enzymes and/or transporters responsible for the clearance of a co-administered drug [5, 6]. This results in either a sub-therapeutic concentration of the victim drug (induction) or higher plasma concentrations (inhibition) which may lead to unwanted toxicities. The most common inhibition based DDIs result from the inhibition of cytochrome P450s (P450s), which are involved in the metabolism of the majority of administered drugs [7]. Because of the importance of P450 inhibition to patient safety, identifying P450 inhibition early in drug discovery is an important activity of drug metabolism scientists.

Inhibition-based DDIs can be the result of competitive, non-competitive, or time-dependent inhibition. Time-dependent inhibition (TDI) often is the result of covalent modification of the enzyme through metabolic activation [8]. Additionally, inhibitors that bind in a non-covalent manner, but irreversibly are also characterized as TDIs as *in vitro* their behavior is most similar to TDIs. The important distinction between competitive and TDI is the reversibility. Time dependent inhibition results in the destruction of the target enzyme, thus it take times to replace the degraded enzyme through the re-synthesize process which is part of normal cellular turnover. In patients the time required for enzyme re-synthesis can result in prolonged exposure to the victim drug, and leads to prolonged toxicity. Because of this screening for TDI is of utmost importance in drug development. Additionally, the ability to predict the magnitude of an interaction in the clinical setting would allow the progression of a potential new drug with a less than clinically relevant TDI signal but an *in vitro* TDI liability.

The primary manner in which DDIs are evaluated early in drug discovery is through the use of *in vitro* methods. These methods may utilize various matrices ranging from cellular fractions to more complex cellular systems. Due to their ready availability and ease of use, microsomal fractions from human liver (HLM) have been used to assess the inhibition of P450s historically. From a pure enzymatic perspective, these fractions may represent the true inhibition potential of a compound, but they fail to capture other important factors such as permeability, alternate pathways of metabolism, as well as other aspects that may impact intracellular concentration and distribution. Recently, the use of cellular systems, specifically, primary human hepatocytes have been evaluated for the assessment of inhibition of drug candidates. The use of more complex *in vitro* systems, such as cellular systems, may represent a more accurate method to assess TDI, thus a more accurate estimation of the *in vivo* inhibition.

Traditionally, the use of HLM has been the primary method for TDI evaluation. There have been several published methods, such as IC$_{50}$ shift, progress curve, and kinetic determination, to measure TDI in HLM [9]. These methods can easily be adapted to higher throughput methods that allow the screening of multitudes of compounds for rank ordering of TDI liability. They can also be used in estimating K_I and K_{inact} for more detailed characterization of TDI kinetics. However, a considerable drawback with HLM methods is the simplistic nature of the system in that it is only a subcellular fraction and fails to capture other important aspects that exist in a more physiologic system. Importantly, this drawback could impact the accuracy of *in vivo* DDI predictions and lead to the unnecessary attrition of viable drug candidates.

An integral part of risk assessment related to drug interactions, is the use of modeling and simulation to predict the potential magnitude of interaction that would be observed in a clinical study. Static models along with inhibition kinetic data using HLM have been used with varying degrees of success, but incorporation of the fraction metabolized of the victim drug along with other potential components of the interaction, e.g. competitive inhibition or induction, seem to be important [10]. However, in general gross over prediction of magnitude of interaction is the common consensus from these prediction methods. In attempt to increase the robustness of the predictions, more advanced dynamic models have been developed, such as software packages Simcyp and GastroPlus [11]. The dynamic models incorporate more parameters than just the kinetics for TDI, but also the pharmacokinetics of the perpetrator drug. Even using these more advanced models, over prediction of the magnitude of effect is common using the kinetic data derived from HLM assays [12]. Another potential source for the lack of prediction power of models is in system parameters. Specifically, k_{deg} (degradation rate of the enzyme inhibited) has been evaluated and found to affect DDI predictions in both static and dynamic models [12]. Nevertheless, the use of more complex systems, such as cellular systems for *in vitro* TDI determination may represent a mechanistic approach that can improve the accuracy of predictions.

Cellular systems provide a more comprehensive system for assessment of drug-drug interactions related to TDI. Recent publications have investigated the utility of primary human hepatocytes to measure the inhibition kinetics of known time dependent inhibitors, and incorporated these values into predictions of magnitude of DDI caused by TDI. The first report using hepatocytes to assess TDI was from Zhao *et al.* [13]. This work utilized cryopreserved human hepatocytes in suspension to assess the TDI of CYP3A of six known inhibitors. The authors measured the potency in hepatocytes and compared them with the IC_{50} values derived from human liver microsomes. The study found that the inactivation potency (IC_{50}) in hepatocytes is consistently lower for 5 out of 6 known inhibitors when compared to the inactivation potency predicted using HLM data. The next report by McGinity *et al.* [14] compared fresh plated human hepatocytes to two other enzyme sources, recombinant P450s and HLM. The purpose of the investigation of plated hepatocytes was to emphasize the importance of evaluating TDI in enzyme induction studies, and the potential impact they may have on interpretation of induction results. For this purpose, the incubations were performed for up to 48 h but TDI was assessed between 12 and 16 h using CL_{int} and estimating K_I and K_{inact} using these data. In general, there were some differences in inhibition kinetics between the different sources, but no bias for higher or lower parameters between the different enzymes sources was observed. In 2009, Xu et al. [15] published an additional manuscript on the use of cryopreserved human hepatocytes to assess TDI of CYP3A. This work used five well studied CYP3A TDIs and incorporated Simcyp for an assessment of magnitude of clinical interaction. The author's then compared accuracy of predictions when using either kinetic parameter's derived from HLM or those from human hepatocytes. The outcome of these studies showed that human hepatocytes do provide an advantage in accuracy of predictions when compared to TDI data from HLM. In 2011, our laboratory published a detailed characterization of different methods on the determination of TDI of CYP3A in human hepatocytes, and the outcome of predictions of magnitude of interaction using Simcyp [16]. This method evaluated three different methods for measuring CYP3A TDI in hepatocytes. The data suggest that two methods are superior in assessing CYP3A in hepatocytes, they were termed the add and dilution methods. One method, the wash method, was not suitable due to continued inactivation during processing of the cells. The inhibition parameters obtained from both the add and dilution methods were essentially indistinguishable from each other, and were used in subsequent clinical DDI predictions using Simcyp. As was shown in all previous publications, the accuracy of the predictions using TDI parameters derived from hepatocytes was superior to those resulting from HLM. The outcome of all of these studies is highly suggestive that human hepatocytes provide a better tool for the prediction of clinical DDIs, especially for CYP3A.

Drug discovery and development is a complex endeavor that must consider innumerable factors for the progression of a clinical drug candidate. In order to bring the right candidate forward, assessing potential liabilities, *e.g.* TDI, preclinically in an accurate and robust manner is of utmost importance. Obtaining TDI kinetic data from human hepatocytes is suggested to improve the accuracy of predictions of clinical DDIs. These data may lead to the progression of more viable clinical candidates with decreased potential for clinical DDIs related to TDI even with measurable TDI *in vitro*.

2 Materials

2.1 Chemical and Biological Reagents

1. Midazolam maleate salt, 1'-hydroxy midazolam, (Sigma-Aldrich, St. Louis, MO).
2. RPMI 1640 media, Percoll, trypan blue, NADPH (Sigma-Aldrich, St. Louis, MO).

3. Dimethyl sulfoxide (DMSO), formic acid, ammonium acetate, HPLC grade methanol and acetonitrile (Sigma-Aldrich, St. Louis, MO).

4. Cryopreserved human hepatocytes (10-door pool, mixed gender, 5 million cells/vial) (Celsis, Baltimore, MD).

5. Fetal bovine serum (FBS) and L-glutamine (Invitrogen, Carlsbad, CA).

6. Incubation medium: RPMI 1640 with 5 % FBS and 2 mM L-glutamine (Sigma-Aldrich, St. Louis, MO).

7. Percoll solution: 30 % Percoll in RPMI 1640 with 5 % FBS and 2 mM L-glutamine

8. HPLC mobile phase: (A) 0.1 % formic acid in water and (B) 0.1 % formic acid in methanol/acetonitrile 50:50.

2.2 Equipment

1. Shimadzu LC-10ADVP HPLC (Shimadzu Scientific Instruments, Columbia, MD)

2. BDS Hyperisil C_{18} 50×2.1 mm column (Fisher Scientific, Pittsburgh, PA)

3. AB Sciex API 4000 triple quadrupole mass spectrometer with an electrospray ionization source (AB Sciex, Foster City, CA)

4. Centrifuge (BECKMAN COULTER Allegra™ X-2, Batavia, IL)

5. BBD 6220 CO_2 Incubator (Thermo Scientific, Asheville, NC)

6. 96-well plate: 0.25 mL/well (for hepatocyte incubation) and 1 mL/well (for ACN precipitated samples)

3 Methods

3.1 Preparation of Stock Solutions

1. Midazolam and 1'-hydroxy midazolam, 10 mM in acetonitrile

2. Inhibitors to be tested: (a) stock solution, 10 mM preferably in ACN, however DMSO is acceptable if solubility is limiting; (b) working solution, 10X of the concentration in final incubation prepared in incubation medium. Total concentration of solvent in final incubations should be less than <1 % (v/v)

3.2 Preparation of Cryopreserved Human Hepatocytes

1. Thaw cryopreserved human hepatocytes quickly in a 37 °C water bath with gentle shaking ~1 min.

2. Rinse the thawed hepatocytes using 1:3 ratio Percoll solution (~30 % Percoll in RPMI 1640 with 5 % FBS and 2 mM L-glutamine), mix gently.

3. Centrifuge cell suspension at $100 \times g$ for 10 min, remove and discard supernatant.

4. Re-suspend the hepatocyte pellet in 2 mL of fresh incubation media (RPMI 1640 with 5 % FBS and 2 mM L-glutamine).

5. Determine the cell viability using the trypan blue method. In all experiments, the cell viability should be 90 % or higher. If viability is less than 90 %, seek a new batch of hepatocytes.

6. Prior to the initiation of TDI incubations, hepatocyte suspensions should be placed in a 37 °C incubator maintained at 5 % CO_2 and 95 % humidity for 30 min.

3.3 Hepatocyte Incubation

The TDI experiments can be carried out using either of two methods, the add method or dilution method. A comparison of major experiment parameters are presented in Table 1. All incubations are conducted in a 96-well plate with an initial equilibration at 37 °C for 30 min, as stated in Sect. 3.2.

3.3.1 Add Method

1. Place 90 µL of pre-warmed hepatocyte suspension in each well of a 96-well plate.

2. Add 10 µL of inhibitor (working solution, preparation see Sect. 3.1) to each well, the final cell density in each well is 0.3×10^6 viable cell/mL. For controls (without inhibitor) at each pre-incubation time points, add 10 µL of cell incubation media.

3. Place the 96-well plate in the incubator and start pre-incubation phase at 37 °C with 500 rpm of orbital shaking for 0–60 min.

Table 1
Experimental parameters

	Add method	Dilution method
Pre-incubation		
Volume (µL)	100	50
Cell density (cell/mL)	0.3×10^6	0.6×10^6
Inhibitor concentration[a] (µM)	0.1–100	0.1–100
Pre-incubation time[b] (min)	0–60	0–60
Incubation		
Volume (µL)	110	250
Cell density	0.245×10^6	0.125×10^6
Inhibitor concentration (µM)	1.25-fold dilution	fivefold dilution
Midazolam concentration (µM)	30	30
Incubation time (min)	10	10

[a] and [b] concentration and pre-incubation time vary depending on the inhibitor and experimental design

4. At each pre-incubation time point, 0, 10, 20, 30, 60 min, add 10 µL of incubation media containing 330 µM midazolam to each well, which results in a final midazolam concentration of 30 µM, at the start of the incubation phase.

5. Continue to incubate cell suspensions for an additional 10 min at 37 °C with orbital shaking. Terminate the reactions by the addition of 220 µL of ACN.

6. Centrifuge ACN quenched cell suspensions at 4,000 rpm for 10 min, collect and then transfer supernatants to a new plate (1 mL/well) for further analysis using LC/MS/MS.

3.3.2 Dilution Method

1. Place 45 µL of pre-warmed hepatocyte suspension in each well of a 96-well plate.

2. Add 5 µL of inhibitor (working solution, preparation see Sect. 3.1) to each well; the final cell density in each well should be 0.6×10^6 viable cell/mL. For controls (without inhibitor) at each pre-incubation time points, add 5 µL of cell incubation media.

3. Put the 96-well plate in incubator and start pre-incubation phase at 37 °C with 500 rpm of orbital shaking for 0–60 min.

4. At each pre-incubation time point, 0, 10, 20, 30, 60 min, add 200 µL of incubation media containing 37.5 µM midazolam to each well, resulting in a final midazolam concentration of 30 µM, and start incubation phase.

5. Continue to incubate cell suspensions for an additional 10 min at 37 °C, and terminate reactions by the addition of 220 µL of ACN.

6. Centrifuge ACN quenched cell suspensions at 4,000 rpm for 10 min. Collect the supernatants to a new plate for further analysis using LC/MS/MS.

3.4 HPLC-MS/MS Analysis

The CYP3A activity is determined using the formation of 1′-OH-midazolam (1′-OH-MDZ), the primary metabolite of midazolam. The analysis of 1′-OH-MDZ was carried out on a Shimadzu LC-10ADVP HPLC coupled to a Sciex API 4000 triple quadrupole mass spectrometer with an electrospray ionization source. A BDS Hyperisil C_{18} 50×2.1 mm column was utilized for separation. Mobile phase consisted of 5 mM ammonium acetate in 0.1 % formic acid in water (A) and 0.1 % formic acid in methanol/acetonitrile 50:50 (B). The initial condition was set at 5 % B for 1 min, increased to 95 % B over 1 min, and remained at 95 % B for 2 min, and finally brought back to the initial conditions in 0.1 min. The flow rate is at 0.45 mL/min, and total run time was 5 min. The retention time for 1′-OH-MDZ is at approximately 1.65 min. The tune parameters of the MS detector and the scan function parameters, including cone voltages and collision energies, should be optimized for detection of 1′-OH-MDZ (MS/MS transition monitored is 342.1/324.2).

3.5 Data Calculation

1. Data normalization: set the enzyme activity (1′-OH-MDZ formation) of control incubations (without inhibitors) to 100 % at each pre-incubation time point. The remaining enzyme activity in incubations with different inhibitor concentrations at each time point is determined by comparing 1′-OH-MDZ formation to the control incubations.

2. Calculation of remaining enzyme activity: For each test inhibitor concentration, the remaining enzyme activity after pre-incubation is calculated as percentage of control at corresponding concentration [I] without pre-incubation.

3. Determination of apparent rate of inactivation ($-K_{obs}$): Plot the natural log (ln) of the percentage remaining activity against the pre-incubation time. The slope ($-K_{obs}$) of each line is then calculated for the period of 0 min to the last linear time point of the pre-incubation phase. A negative value of k_{obs} is considered as zero.

4. Calculation of TDI kinetic parameters: The K_{inact} (maximum inactivation rate constant) and K_I (inhibitor concentration produces half-maximal rate of inactivation) can be obtained from the non-linear fitting of equation (1) to $-K_{obs}$ determined at different inhibitor concentration using Winonlin (Pharsight, Mountain View, CA) or other non-linear fitting data packages, *e.g.* GraphPad, Sigma Plot, *etc.*

$$K_{obs} = \frac{K_{inact} \cdot [I]}{K_I + [I]} \qquad (1)$$

4 Notes

4.1 Inhibitor Concentration

For the determination of TDI kinetics, inhibitor concentrations ranging 100-fold (e.g. 0.1–10 μM) with 5–7 concentration levels is typically used. However, for a test compound that is also a competitive inhibitor, the highest concentration included in the TDI experiment maybe of limited utility due to the profound inhibition of CYP3A activity at time zero. The significant loss (>20–30 %) of enzyme activity in control incubations (with inhibitor, but without pre-incubation) due to competitive inhibition could lead to inaccurate determination of TDI kinetic parameters.

4.2 Pre-Incubation Times

Pre-incubation times should be optimized to ensure adequate enzyme inactivation is observed while minimizing significant depletion of inactivator (preferred to be no more than 75 %) during pre-incubation period.

4.3 Midazolam (Probe Substrate) Concentration and Incubation Time

To prevent further inhibition (competitive and/or irreversible) during the incubation phase, a saturating level of substrate concentration is typically used. The common practice is to utilize a probe substrate at a concentration that is approximately tenfold higher than the K_m. Alternatively, the probe substrate concentration should be in a ratio that concentration/K_m exceeds that of concentration/K_I [10]. The optimal concentration of midazolam can be determined at which the maximum rate of 1'-OH-MDZ formation is obtained. To minimize the possible continued inhibition by the inactivator, the length of incubation with substrate should be relatively short compared with the pre-incubation time.

4.4 Effect of Cell Density

For the convenience of experimental design, different cell densities can be used with minimal effect on kinetic parameters. The effect of cell density at 0.1, 0.3, and 0.6×10^6 cells/mL on the measured CYP3A activity was tested in previous studies using verapamil at different concentrations [16]. It was found that formation of 1'-OH MDZ increased linearly with increased cell density (Fig. 1a), and following the same pattern in the absence and presence of inhibitor at different concentrations. However, the percentage inhibition of CYP3A activity (calculated based on formation of 1'-OH MDZ in the presence and absence of inhibitor) remained relatively consistent across difference cell densities for a given inhibitor concentration (Fig. 1b).

4.5 Effect of Dilution Factors, and Choosing Between the Add and Dilution Methods

The continued inactivation of enzyme by inhibitor during the incubation phase is a major concern in generating accurate *in vitro* TDI data. A 10- to 20-fold dilution is commonly used in TDI assays utilizing HLM to minimize further enzyme inactivation and/or competitive inhibition during the incubation phase; however, it is technically challenging to do this in hepatocyte-based assays. In a previous study, three methods (identified as wash, add, and dilution) with different approaches in dilution were compared [16]. Even though the wash method intends to remove the inhibitor from the pre-incubation mixture, it appeared that it did not

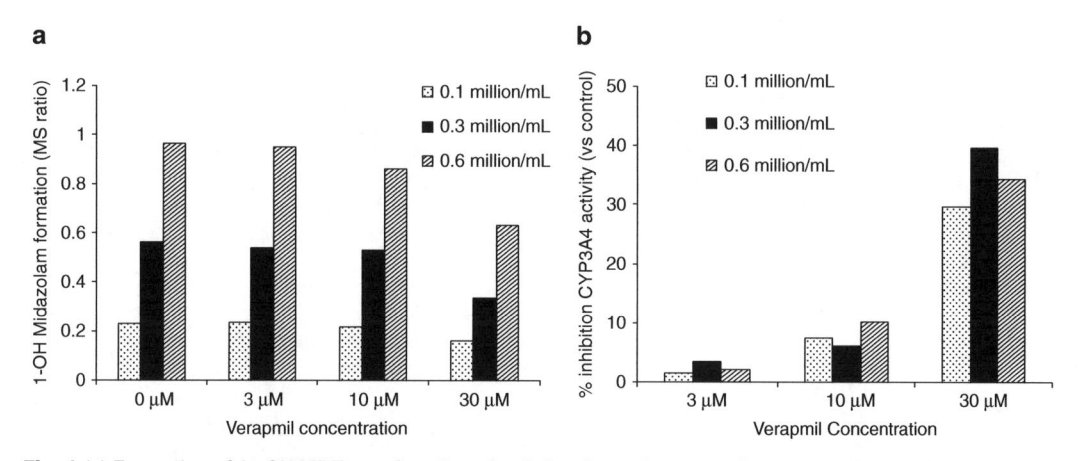

Fig. 1 (**a**) Formation of 1'-OH MDZ as a function of cell density and verapamil concentration. (**b**) Percent CYP3A inhibition of 1'-OH MDZ formation as a function of cell density and verapamil concentration

prevent ongoing inactivation during the washing and centrifugation process. The extra steps prolong the presence of the inhibitor in the cell incubation, which resulted in even more loss of CYP3A4 activity than that using either the add or dilution methods. In fact, both the add and dilution methods controlled for this loss of enzyme activity far better than the wash method. In addition, the wash method is far more labor intensive and has the potential for the introduction of more error due to greater sample handling when compared to either the dilution or add methods.

The effect of dilution factor on TDI kinetic parameters generated using the add method (1.25-fold dilution) and dilution method (5-fold dilution) for four known time-dependent inactivators were compared in previous studies in our laboratory [16]. The inhibition parameters from the dilution method tended to be slightly higher than those from the add method, but the difference in dilution factor appeared to have no significant impact. This is likely due to the differences in dilution efficiency between hepatocyte and HLM incubations because of the cell membrane present in hepatocytes. The add method (with a lower dilution factor) has less sample handling procedures which will reduce the chance of experimental error. However, a higher dilution may be beneficial for potent competitive P450 inhibitors which would require re-testing at higher concentrations in hepatocytes.

References

1. Juurlink DN, Mamdani M, Kopp A, Laupacis A, Redelmeier DA (2003) Drug-drug interactions among elderly patients hospitalized for drug toxicity. JAMA 289(13):1652–1658
2. Monahan BP, Ferguson CL, Killeavy ES, Lloyd BK, Troy J, Cantilena LR Jr (1990) Torsades de pointes occurring in association with terfenadine use. JAMA 264(21):2788–2790
3. http://www.fda.gov/downloads/Drugs/GuidanceComplianceRegulatoryInformation/Guidances/ucm292362.pdf
4. http://www.ema.europa.eu/docs/en_GB/document_library/Scientific_guideline/2012/07/WC500129606.pdf
5. Lin JH, Lu AY (1998) Inhibition and induction of cytochrome P450 and the clinical implications. Clin Pharmacokinet 35(5):361–390
6. Lin JH (2007) Transporter-mediated drug interactions: clinical implications and in vitro assessment. Expert Opin Drug Metab Toxicol 3(1):81–92
7. Lynch T, Price A (2007) The effect of cytochrome P450 metabolism on drug response, interactions, and adverse events. Am Fam Physician 76:391–396
8. Riley RJ, Grime K, Weaver R (2007) Time-dependent CYP inhibition. Expert Opin Drug Metab Toxicol 3(1):51–66
9. Grimm SW, Einolf HJ, Hall SD, He K, Lim HK, Ling KH, Lu C, Nomeir AA, Seibert E, Skordos KW, Tonn GR, Van Horn R, Wang RW, Wong YN, Yang TJ, Obach RS (2009) The conduct of in vitro studies to address time-dependent inhibition of drug-metabolizing enzymes: a perspective of the pharmaceutical research and manufacturers of America. Drug Metab Dispos 37:1355–1370
10. Fahmi OA, Hurst S, Plowchalk D, Cook J, Guo F, Youdim K, Dickins M, Phipps A, Darekar A, Hyland R, Obach RS (2009) Comparison of different algorithms for predicting clinical drug-drug interactions, based on the use of CYP3A4 in vitro data: predictions of compounds as precipitants of interaction. Drug Metab Dispos 37(8):1658–1666
11. Jamei M, Marciniak S, Feng K, Barnett A, Tucker G, Rostami-Hodjegan A (2009) The Simcyp population-based ADME simulator. Expert Opin Drug Metab Toxicol 5(2):211–223
12. Wang YH (2010) Confidence assessment of the Simcyp time-based approach and static

mathematical model in predicting clinical drug-drug interactions for mechanism-based CYP3A inhibitors. Drug Metab Dispos 38: 1094–1104
13. Zhao P, Kunze KL, Lee CA (2005) Evaluation of time-dependent inactivation of CYP3A in cryopreserved human hepatocytes. Drug Metab Dispos 33:853–861
14. McGinnity DF, Berry AJ, Kenny JR, Grime K, Riley RJ (2006) Evaluation of time-dependent cytochrome P450 inhibition using cultured human hepatocytes. Drug Metab Dispos 34: 1291–1300
15. Xu L, Chen Y, Pan Y, Skiles GL, Shou M (2009) Prediction of human drug-drug interactions from time-dependent inactivation of CYP3A4 in primary hepatocytes using a population-based simulator. Drug Metab Dispos 37:2330–2339
16. Chen Y, Liu L, Monshouwer M, Fretland AJ (2011) Determination of time-dependent inactivation of CYP3A4 in cryopreserved human hepatocytes and assessment of human drug-drug interactions. Drug Metab Dispos 39: 2085–2092

Chapter 17

Rapidly Distinguishing Reversible and Time-Dependent CYP450 Inhibition Using Human Liver Microsomes, Co-incubation, and Continuous Fluorometric Kinetic Analyses

Gary W. Caldwell and Zhengyin Yan

Abstract

In this chapter we have provided a step-by-step protocol of a 384-well plate fluorescence-based assay used for rapid identification of reversible and time-dependent CYP450 inhibition. This was accomplished by comparing the time-dependence pattern of IC_{50} values of potential test inhibitors using a co-incubation approach with continuous fluorometric kinetic measurements. Briefly, test compounds were mixed with NADPH and were serially diluted to eight different concentrations. The enzymatic reaction was initiated by adding a single recombinant CYP pre-mixed with its corresponding fluorescent substrate and a mixture of $NADP^+$, G6P and $MgCl_2$. The enzyme activity was measured every 2 min by fluorescence intensity (CYP product) over typically a 30-min time period. Inhibition percentages were calculated relative to controls that contained no inhibitors at each time point and IC_{50} values of inhibitors were calculated at different incubation time intervals. Plotting IC_{50} values vs. incubation time revealed three different patterns for test inhibitors that could be used to distinguish reversible and time-dependent inhibitors. IC_{50} values of reversible inhibitors either maintained within a narrow range, or increased with incubation time because of losing inhibitor as a result of metabolism or non-specific binding to the matrix. In contrast, IC_{50} values decreased with incubation time for time-dependent inhibitors because of irreversible reactions caused by progressive enzyme inactivation by reactive metabolite species generated during the incubation or other inactivation mechanisms. Results clearly suggest that this co-incubation *in vitro* continuous fluorometric kinetic assay using recombinant CYPs and fluorometric generating substrates is a valuable high-throughput assay for distinguishing reversible and time-dependent inhibitors for large compound collections.

Key words Metabolism, Reversible and irreversible kinetics, CYP450s, Fluorometric analyses, Time dependent inhibitions

1 Introduction

Drug discovery groups in the pharmaceutical industry have adopted over the years an assay tiered approach toward selecting potential new drug candidates with superior drug properties from large compound collections [1–4]. For example, consider Fig. 1 where each box in the diagram represents a particular assay for

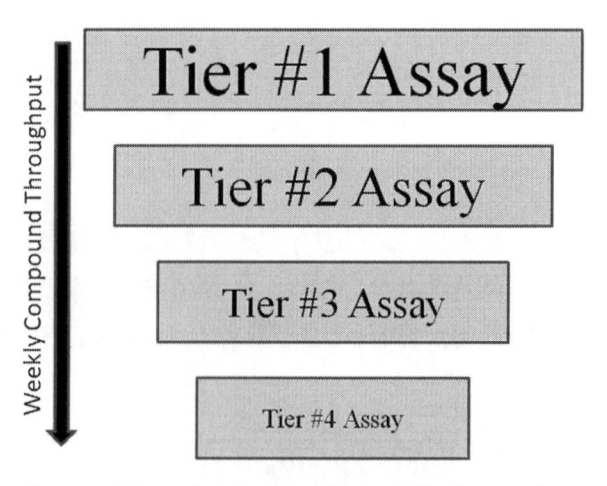

Fig. 1 Tiered approach toward evaluating drug properties for new drug candidates. Each *box* represents an assay for determining a particular drug property. The size of the *box* represents the compound throughput

determining a drug property with the size of the box representing the assay compound throughput. In this example, tier #1 and #2 assays represent early stage assays that can effectively evaluate large numbers of compounds while tier #3 and #4 are lower capacity assays allowing for more detailed mechanistic characterization of compounds. In this approach, thousands of compounds are funneled through a series of high-throughput capacity assays to lower capacity assays, which reveal more and more detailed information on a particular drug property. From this process, the selection of a few drug candidates can be obtained from hundreds of thousands of compounds.

A major assumption in this type of tiered approach is that the false-positive and false-negative characteristics (i.e., rate of occurrence) of the assays are understood well enough not to significantly interfere with compound selection. For example, assuming that a positive result in an assay would trigger the elimination of the compound from further consideration and a negative result would allow the compound to move to the next tier, an ideal assay would not produce any false-positive results. A false-negative result could be managed in a tiered approach since the next tier could potentially correct for this mistake. Based on our own experience, an acceptable assay in this approach should have a false-positive rate of approximately 5 % or less and a false-negative rate of approximately 30 % or less to be effective in a tiered approach.

This assay tiered approach has been successfully utilized to study metabolism properties of drug candidate compounds [5, 11]. For example, metabolism of xenobiotics (i.e., drug candidate compounds) by cytochrome P450 enzymes (CYPs) in the liver represents the major *in vivo* route of drug elimination in humans and animals.

CYP3A4, 2D6, 2C9, 2C19 and 1A2 are the primary CYP enzymes involved in P450-catalyzed drug biotransformation reactions with 3A4 being the most important since many therapeutic drugs are metabolized by it including immunosuppressant, antihypertensive, antiarrhytmic, analgesic, antibiotic, anticonvulsant, antidepressant, and antifungal drugs. Briefly, a general feature of P450 enzymes is that a single enzyme can metabolize multiple substrates; thus, if several substrates are taken *in vivo* at the same time (i.e., polypharmacy) a competition reaction between two substrates may occur for the same P450 enzyme and result in an undesirable elevation in systemic concentration of one of the substrates. Depending on the magnitude of the change in concentration in the substrate at the site of pharmacological action and the therapeutic index of the substrate, this could lead to a clinical drug-drug interaction (DDI). Thus, over the last decade, many tiered assay approaches have been developed to understand the mechanisms of CYP enzyme inhibition or inactivation that leads to DDIs for potential new drug candidates. The main approach has been to establish a series of *in vitro* assays to measure the potential of drug candidates to produce CYP enzyme reversibility and irreversibility leading to enzyme inhibition or inactivation. From these assays, drug candidates can be eliminated in early drug discovery programs before they enter costly drug development programs [5–11].

While it is understood, that the *in vivo* inhibition process is complicated, there are two drug binding properties that are typically examined to predict DDIs at early drug discovery stages; that is, reversible CYP inhibition reactions caused by competitive interactions from multiple substrates or inactivation of CYP caused by irreversible binding or quasi-irreversible inhibition which is sometimes referred to as a metabolite intermediate complex (MIC). All irreversible inhibition processes are time-dependent and thus, are commonly referred to as time dependent inhibition (TDI). If the TDI is caused by the reaction of intermediates formed during biotransformation, the time dependent inhibition is termed mechanism based inhibition (MBI). Both reversible inhibition and irreversible MBI are further illustrated in Schemes 1 and 2, respectively.

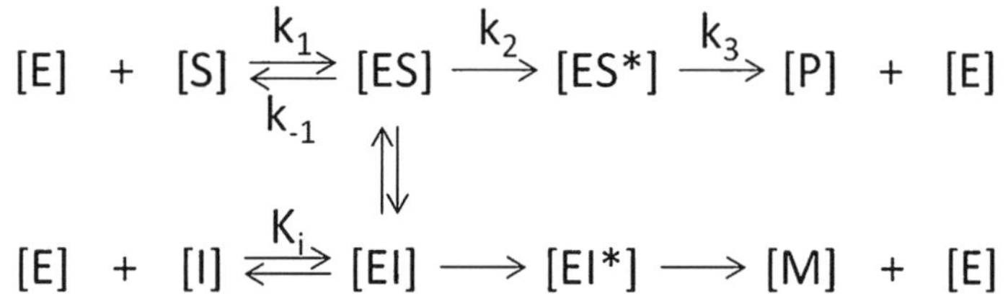

$$[E] + [S] \underset{k_{-1}}{\overset{k_1}{\rightleftarrows}} [ES] \overset{k_2}{\longrightarrow} [ES^*] \overset{k_3}{\longrightarrow} [P] + [E]$$

$$\Updownarrow$$

$$[E] + [I] \overset{K_i}{\rightleftarrows} [EI] \longrightarrow [EI^*] \longrightarrow [M] + [E]$$

Scheme 1 Hypothetical illustration of a competitive enzyme inhibition reaction [12]

$$[E] + [S] \underset{k_{-1}}{\overset{k_1}{\rightleftharpoons}} [ES] \overset{k_2}{\longrightarrow} [ES^*] \overset{k_3}{\longrightarrow} [P] + [E]$$

$$\Updownarrow$$

$$[E] + [I] \overset{K_i}{\rightleftharpoons} [EI] \longrightarrow [EI^*] \longrightarrow [M] + [E]$$

$[E\text{-}I^*]'$
Apoprotein
Covalent Binding
(Irreversible)

$[E\text{-}I^*]''$
Heme-prosthetic
Covalent Binding
(Irreversible)

$[E\cdot I^*]'''$
Complexation
Quasi-Irreversible

Scheme 2 Hypothetical illustration of mechanism-based inactivation (MBI) [12]

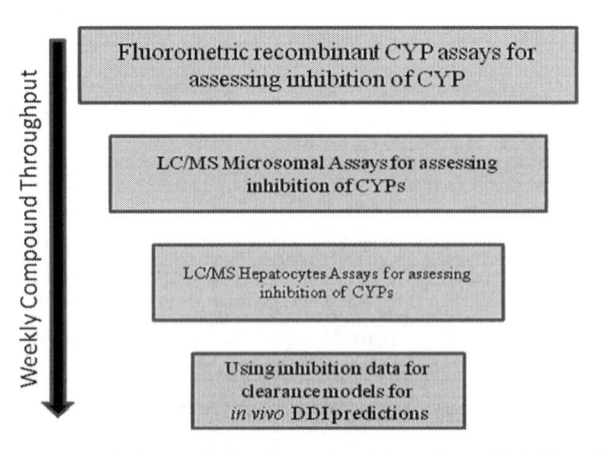

Fig. 2 Tiered approach for predicting drug-drug interactions (DDIs) based upon *in vitro* CYP inhibition assays [13–18]

The CYP reversible inhibition kinetics are illustrated in Scheme 1 where [E] denotes the concentration of the CYP enzyme, [S] denotes the substrate concentration (i.e., probe compound), [I] denotes the inhibitor concentration (i.e., test compound), [ES] and [EI] denotes bonding complexes at the active site of the enzyme, [ES*] and [EI*] denotes bonding complexes that are activated, [P] is the concentration of the metabolite generated from [S] and [M] is the concentration of the metabolite generated from [I]. Assuming that [S] is a biological active drug, then [I] could attenuate the physiological response of [S] via inhibition of [E] which is involved in its drug elimination via [P]. Note that in Scheme 1 the enzyme concentration [E] is not affected by the presence of [S] or [I]; thus, compounds that behave in this manner are referred to as competitive reversible inhibitors. This can be understood by noting that while [S] and [I] directly compete for the active site of [E] the enzymatic activity can be fully restored since the inhibitor [I] is depleted and removed from the system via its biotransformation to metabolite [M]. The inhibition potency of [I] is usually measured by the concentration of [I] where half maximal inhibitory occurs (IC_{50}) or the inhibition constant (K_i) that reflects the binding affinity of inhibitor to the CYP enzyme [EI] [12].

The tiered approach used to evaluate compounds for competitive reversible inhibitors (i.e., CYP inhibition potency) that might lead to clinical DDIs is illustrated in Fig. 2. *In vitro* CYP450 inhibition screening assays have been designed over the years using Scheme 1 to rank compounds for their inhibitory potential [5–8]. These protocols typically use a single substrate concentration [S] near the apparent K_m (i.e., $K_m = (k_{-1} + k_2 + k_3)/k_1$) and multiple concentrations of the test compound [I] where the concentration of [P] is measured over a time period under the experimental constraint that the concentration of [S] does not change by more than 10 % relative to its initial concentration. By measuring the initial velocity of the reaction in the presence of an increasing concentration of [I], the inhibitory potency (i.e., IC_{50}) of compounds relative to the substrate can be measured [12]. Several different types of high throughput assays that determine enzyme activity in a microtiter plate format by using recombinant CYPs (rCYPs) and fluorescent substrates have been used to measure CYP inhibition potency for large compound collections [7, 13, 15, 16]. These types of assays can measure thousands of compounds daily. During the past decade, a rapid increase in throughput of LC/MS analysis has led to a gradual shift of inhibition assays from rCYPs to liver microsomes [18]. When a smaller number of compounds need to be studied or compounds from a previous tiered assay have been selected a lower throughput method can be used to generate more detailed mechanistic information. Typically, these protocols involve using substrates that are only metabolized by a specific CYP in human liver microsomes [8, 17] or hepatocytes [14] using liquid chromatography interfaced with tandem mass spectrometry (LC-MS/MS). These types of assays are commonly used to elucidate inhibitory structural motifs within a chemical series and to dial out these undesirable properties. In addition, these types of assays are used to measure IC_{50} or inhibition constants K_i that can be used in *in vitro–in vivo* extrapolation (IVIVE) models for estimating the hepatic metabolic clearance (CL) of drugs [19].

The irreversible MBI enzyme kinetics of CYPs via reactive intermediates formed during biotransformation is illustrated in Scheme 2 where the notation used in Scheme 1 is again used in Scheme 2. In this case, the reactive intermediate complexed to the enzyme [EI*] now reacts via three pathways. The reactive intermediate irreversibly bonds to the CYP apoprotein to form the covalent specie [E-I*]′; it irreversibly bonds to the CYP prosthetic heme group to form the covalent specie [E-I*]″. Finally, the reactive intermediate [EI*] forms a quasi-irreversible complexation [E·I*]‴; that is, a complexation where enzyme activity is regained upon dialysis or gel filtration to deplete the inhibitor. Note that in Scheme 2 that the reactive intermediate I* does not leave the complex [EI*] but is immediately transformed to inactive the enzyme. This situation implies that the presence of exogenous scavenger nucleophiles such as glutathione should have no effect on the rate of enzyme inactivation. There are several other types of reactions that can inactivate the enzyme concentration and that can show TDI that are not shown in Scheme 2 such as, autoinactivation caused by production of superoxide and/or hydrogen peroxide formed during the reduction CYP cycle, the metabolite [M] could form irreversible complexes with the enzyme, tight binding of [ES] or [EI], and non-specific binding of [S] or [I] (e.g., to microsomal membranes). All of these mechanisms result in the permanent or apparent loss of enzymatic activity and thus, TDI. Unlike reversible inhibition, irreversible inactivation of CYP450s causes long-term effects on some drug pharmacokinetics, as the inactivated enzyme must be replaced by newly synthesized CYP protein [20].

The first assays in a tiered approach should distinguish reversible inhibitors and TDI followed by more thorough inactivation assays that would identify the mechanisms for TDI. Over the past decade, several different methods have been developed to identify and evaluate TDI potential using either a single rCYP enzyme derived from baculovirus infected insect cells [21–29] or human liver microsomes (HLM) containing a cocktail of human CYP enzymes [30–34] or hepatocytes [35]. In addition, several different experimental protocols have been developed for prediction of DDIs based on IC_{50} values or the kinetic parameters K_I (i.e., inhibitor concentration that support half the maximal rate of inactivation) and k_{inact} (i.e., the maximal rate of inactivation). Details of these approaches and their usefulness in DDI predictions can be found in the literature [11, 33–37]. In this chapter, we will only discuss the top tiered rCYPs with fluorometric substrate probes assays.

There are several different experimental designs to study TDI for potential test inhibitors. In Scheme 3 are shown the pre-incubation and the co-incubation experimental designs. The general method to obtain IC_{50} values or kinetic parameters following pre-incubation is to incubate the enzyme system [E] (i.e., rCYP or HLM), the inhibitor (i.e., test compound), and cofactors (CF) together for a set time period. Note that the enzymatic reaction starts at this step. This pre-incubation period can be a single time period such as 30 min to generate an IC_{50} value or multiple time periods to generate kinetic parameters such as K_I and k_{inact}.

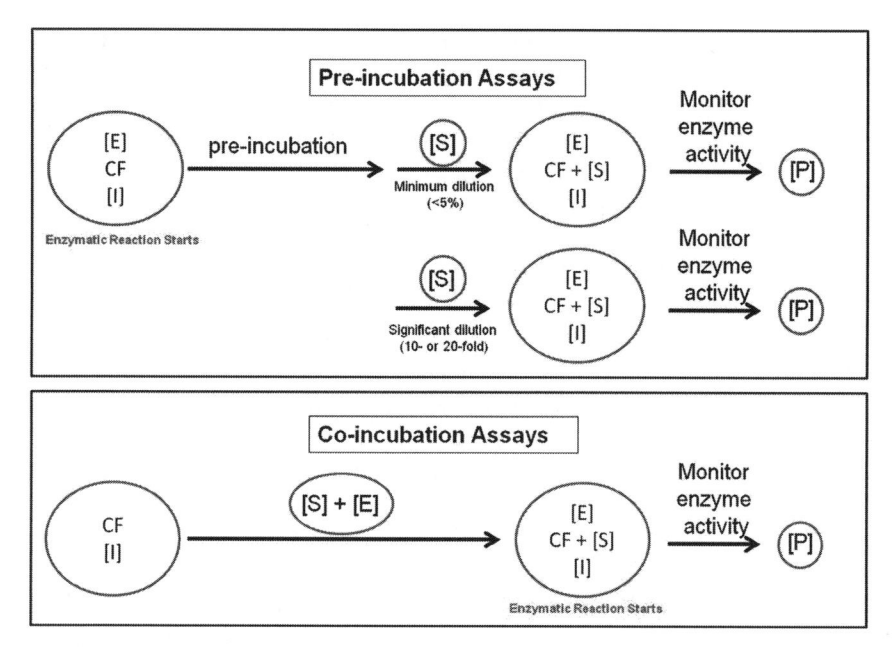

Scheme 3 Experimental designs for studying TDI

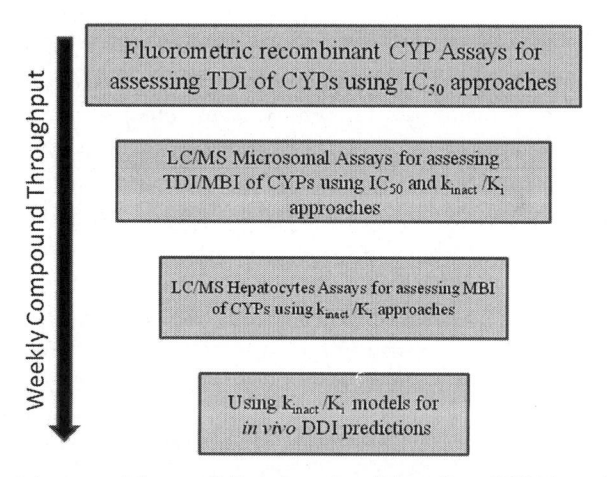

Fig. 3 Tiered approach for predicting drug-drug interactions (DDIs) based upon *in vitro* CYP assays [13–18]

To monitor enzyme activity, a substrate probe [S] is added either with minimum dilution of the enzyme system, the inhibitor and the cofactor concentration or with a 10- to 20-fold dilution of these reagents. The 10- to 20-fold dilution step allows the substrate [S] to compete more effectively for the CYP active site by diluting out the [I] concentration. For the pre-incubation approach, a control without pre-incubation is required to compare IC_{50} shift or K_I and k_{inact} determinations. The co-incubation assays are much simpler than the pre-incubation assays. Here, the cofactors and the inhibitor [I] are premixed and the enzymatic reaction is initiated by addition of the enzyme [E] premixed with the substrate [S]. In pre- and co-incubation assays, the appearance of metabolite [P] (Scheme 2), which is generated from [S], is used as a measure of enzyme activity. More details on these approaches can be found in the literature [11].

The tiered approach used to evaluate compounds for irreversible inhibition that potentially could lead to clinical DDIs is illustrated in Fig. 3. The majority of DDIs reported in the clinic are due to reversible inhibitions, which are difficult to predict from only *in vitro* data. However, irreversible inhibitors, such as MBIs, are in many cases involved in either toxicity or drug-drug interactions in the pre-clinical development stages and can be detected using *in vitro* assays. Thus, distinguishing reversible inhibitors and TDIs, using high throughput assays in the early drug discovery stage, is the major focus of the CYP inhibition tiered screening approach in Fig. 3. It should be emphasized that that these early high throughput assays are designed to detect TDI. The lower capacity assays are used to determine the mechanism or mechanisms for TDI; that is, MBI, MCI, or other inactivation mechanisms. The overall tiered *in vitro* approach can only suggest potential DDI inhibitors. However, the results, from these *in vitro* assays, can be helpful in designing appropriate *in vivo* animal studies and human clinical trials to study the impact of inhibitors on DDIs.

In the present chapter, we outline a 384-well plate TDI co-incubation high throughput continuous fluorometric kinetic assay using rCYP enzymes and fluorometric generating substrate probes to rapidly distinguish reversible CYP inhibitors and TDI. This work is an updated version of an earlier co-incubation TDI screening assay developed in our lab [21]. Briefly, a putative TDI inhibitor (a drug candidate) is first mixed with a NADPH that is serially diluted to different inhibitor concentrations at a constant NADPH concentration. The enzymatic reaction is initiated by adding a single recombinant CYP pre-mixed with a fluorescent substrate with NAPH, G6P, and $MgCl_2$. During the co-incubation, enzyme activity is measured every 2 min by fluorescence intensity (CYP product) in a continuous mode over a period of 30-min. Inhibition percentages are calculated in relative to the control (containing no inhibitor), and the IC_{50} values of the inhibitor are obtained at different incubation time intervals. Plotting IC_{50} values vs. incubation time can easily differentiate reversible and TDI. IC_{50} values of reversible inhibitors either maintain within a narrow range, or increase with incubation time because of losing inhibitor as a result of metabolism or non-specific binding to matrix. In contrast, for an TDI, the IC_{50} values decreased with incubation time because of progressive inactivation of the corresponding enzyme by reactive metabolite species generated during the incubation. Because all measurements are taken on the same sample plate without sample transferring or preparations, IC$_{50}$ values at different time intervals are very consistent and easy to be compared, and thus the pattern for reversible inhibitors and TDI is readily recognized.

2 Materials

2.1 Equipment

1. Assay Plates: polypropylene 96-well plates (0.5 mL) and Black wall Costar 384-deep well plates (Corning Incorporated, Corning, NY)
2. Multi-channel (8) automatic and pipettor (1–1,200 μL)
3. Variable Temperature Water bath set at 37 °C
4. pH meter
5. FL600 microplate fluorescence reader set at 37 °C (Biotek Instruments, Inc., Winooski, VT) (Note 1)

2.2 Reagents, Buffers and Organic Solvents

1. β-Nicotinamide adeninedinucleotide phosphate sodium salt hydrate (0.3 H$_2$O/mol) (NADP$^+$; Sigma: Cat# N0505). This compound has an anhydrous FW 765.39 g/mol.
2. D-Glucose 6-phosphate disodium salt hydrate (G-6-P-Na$_2$; Sigma: Cat# G7250). This compound C$_6$H$_{11}$Na$_2$O$_9$P. xH$_2$O is hydrated 2–4 H$_2$O/mol with an anhydrous FW 304.10 g/mol. Assuming 4 H$_2$O/mol, use the batch MW 376.10 g/mol.
3. Magnesium chloride (MgCl$_2$·6H$_2$O; Sigma: Cat# M2670) Batch MW 203.30 g/mol (Note 2)
4. Glucose-6-Phosphate Dehydrogenase 1KU (G6PDH; Sigma: Cat# G8404) Protein is suspended in 3.2 M (NH$_4$)$_2$SO$_4$ containing 50 mM Tris and 1 mM MgCl$_2$ at pH 7.5. Use the information on the bottle to calculate Units/mL. For our case, this number should be close to 3,000 U/mL.
5. Phosphate buffer 0.5 M @ pH 7.4 (BD Biosciences: Cat# 451201).
6. 0.5 M Tris base (2-amino-2-(hydroxymethyl)-1,3-propanediol); 121.14 g/mol; Sigma-Aldrich 77-86-1: Weigh 60.55 g solid Tris base and bring to 1,000 mL with deionized water. Stir to dissolve. Do not adjust pH.
7. Water (J.T. Baker-HPLC Solvent Grade; Cat# 4218-03).
8. Dimethyl Sulfoxide (DMSO) and acetonitrile (ACN) (Sigma-Aldrich).

2.3 CYPs

1. Human CYP1A2 + CYP reductase (1A2); BD Bioscience: 456203 (0.5 nmol of P450 in 0.5 mL): store at −80 °C
2. Human CYP2C9*1 (Arg144) + CYP reductase + b5 (2C9); BD Bioscience: 456258 (0.5 nmol of P450 in 0.5 mL): store at −80 °C

3. Human CYP2C19 + CYP reductase + b5 (2C19); BD Bioscience: 456259 (0.5 nmol of P450 in 0.5 mL): store at −80 °C

4. Human CYP2D6*1 (Val374) + CYP reductase (2D6); BD Bioscience: 456217 (0.5 nmol of P450 in 0.5 mL): store at −80 °C

5. Human CYP3A4 + CYP reductase + b5 (3A4); BD Bioscience: 456202 (0.5 nmol of P450 in 0.5 mL): store at −80 °C

2.4 Substrate Probes and Solution

1. 3-Cyano 7-ethoxycoumarin (CEC); BD Bioscience: 451014; (1A2) and (2C19): 20 mM CEC; 215.20 g/mol: Add 8.61 mg of CEC to 2.0 mL acetonitrile. Invert to dissolve (Notes 1 and 3).

2. 7-Methoxy-4-trifluoromethylcoumarin (MFC); BD Bioscience: 451740; (2C9): 25 mM MFC; 244.17 g/mol: Add 12.21 mg of MFC to 2.0 mL acetonitrile. Invert to dissolve and store at −20 °C (Note 1).

3. 3-[2-(N,N-diethyl-N-methylamino)ethyl]-7-methoxy-4-methylcoumarin iodide (AMMC); BD Bioscience: 451700 (451705); (2D6): 10 mM AMMC; 431.31 g/mol: Add 4.32 mg of AMMC to 1.0 mL acetonitrile. Invert to dissolve and store at −20 °C (Note 1).

4. 7-Benzyloxy-4-(trifluoromethyl)-coumarin (BFC); BD Bioscience: 451730; (3A4): 50 mM BFC; 320.26 g/mol: Add 16 mg BFC to 1.0 mL acetonitrile. Invert to dissolve and store at −20 °C (Note 1).

2.5 Positive Control Inhibitors and Solution

1. Furafylline: (10 mM; 260.25 g/mol), Sigma-Aldrich: F124-5MG. Add 2.6 mg of furafylline to 1 mL DMSO (Note 4).

2. Sulfaphenazole: (10 mM; 314.36 g/mol), Sigma-Aldrich: S0758-1MG. Add 3.14 mg of sulfaphenazole to 1 mL DMSO (Note 4).

3. Omeprazole: (10 mM; 345.42 g/mol), Sigma-Aldrich: O104-100MG. Add 3.46 mg of omeprazole to 1 mL DMSO (Note 4).

4. Quinidine: (0.1 mM; 324.42 g/mol; 0.032 mg/mL), Sigma-Aldrich: Q3635-5G. Add 3.2 mg of quinidine to 100 mL DMSO (Note 4).

5. (±) Verapamil HCl: (10 mM; 491.06 g/mol), Sigma-Aldrich: V4629-1G. Add 4.9 mg of verapamil to 1 mL DMSO (Note 4).

2.6 Solution A: NADPH Regenerating System for all CYP Except 2D6

A 50 mL solution containing 26.1 mM $NADP^+$, 66.0 mM G-6-P-Na2, and 66.0 mM $MgCl_2$ is required. This solution provides the necessary co-factors to catalyze a CYP450 enzyme reaction. To prepare this solution (Note 5):

1. Dissolve 998.9 mg of $NADP^+$ (FW 765.39 g/mol) in 10 mL of water (131 mM)

2. Dissolve 1241.1 mg of G-6-P-Na$_2$ (Batch MW 376.10 g/mol) in 10 mL of water (330 mM)

3. Dissolve 670.9 mg of MgCl$_2$·6H$_2$O (Batch MW 203.30 g/mol) in 10 mL of water (330 mM)

4. Combine the three 10 mL solutions and add 20 mL of water.

2.7 Solution B: NADPH Regenerating System for 2D6

A 50 mL solution containing 1.3 mM NADP$^+$, 66.0 mM G-6-P-Na$_2$, and 66.0 mM MgCl$_2$ is required. This solution provides the necessary co-factors to catalyze a CYP450 enzyme reaction. To prepare this solution (Note 5):

1. Dissolve 49.75 mg of NADP$^+$ (FW 765.39 g/mol) in 10 mL of water (6.5 mM)

2. Dissolve 1241.1 mg of G-6-P-Na$_2$ (Batch MW 376.10 g/mol) in 10 mL of water (330 mM)

3. Dissolve 670.9 mg of MgCl$_2$·6H$_2$O (Batch MW 203.30 g/mol) in 10 mL of water (330 mM)

4. Combine the three 10 mL solutions and add 20 mL of water.

2.8 Stop Solution

Prepare 500 mL by adding the components in the following order:

1. 400 mL ACN
2. 100 mL 0.5 M Tris Base

2.9 Serial Dilution Buffer

Prepare 150 mL by adding the components in the following order:

1. 135 mL deionized water
2. 15 mL 0.5 M KPO$_4$, pH 7.4
3. 0.05 mL of Glucose-6-Phosphate Dehydrogenase 1KU (G6PDH; 3000 U/mL) or 1 U/mL

2.10 Inhibitor Test Compounds

Prepare 0.5 mM (0.120 mL) by adding the components in the following order:

1. 0.114 mL ACN
2. 0.006 mL of a 10 mM DMSO Stock Solutions (Note 6)
3. Solution contains 5 % DMSO and 95 % ACN

2.11 Substrate and Enzyme Mixture

2.11.1 CEC and 1A2

Prepare 15 mL by adding the components in the following order:

1. 8.92 mL deionized water
2. 4.50 mL 0.5 M KPO$_4$, pH 7.4
3. 1.5 mL of Solution A; Total Concentration: 2.6 mM NADP$^+$, 6.6 mM G-6-P-Na$_2$, and 6.6 mM MgCl$_2$
4. 0.075 mL 1A2 (1 nmol/mL); Total amount: (75 pmol or 5 pmol/mL)
5. 0.008 mL substrate (20 mM CEC); Total Concentration: 10.7 µM (ACN 0.05 %)

2.11.2 MFC and 2C9

Prepare 15 mL by adding the components in the following order:

1. 13.26 mL deionized water
2. 1.5 mL of <u>Solution A</u>; Total Concentration: 2.6 mM NADP$^+$, 6.6 mM G-6-P-Na$_2$, and 6.6 mM MgCl$_2$
3. 0.15 mL 2C9 (1 nmol/mL); Total amount: (150 pmol or 10 pmol/mL)
4. 0.09 mL substrate (25 mM MFC); Total Concentration: 150 µM (ACN 0.05 %)

2.11.3 CEC and 2C19

Prepare 15 mL by adding the components in the following order:

1. 11.82 mL deionized water
2. 1.50 mL 0.5 M KPO$_4$, pH 7.4
3. 1.5 mL of <u>Solution A</u>; Total Concentration: 2.6 mM NADP$^+$, 6.6 mM G-6-P-Na$_2$, and 6.6 mM MgCl$_2$
4. 0.15 mL 2C19 (1 nmol/mL); Total amount: (150 pmol or 10 pmol/mL)
5. 0.03 mL substrate (20 mM CEC); Total Concentration: 40 µM (ACN 0.05 %)

2.11.4 AMMC and 2D6

Prepare 15 mL by adding the components in the following order:

1. 10.05 mL deionized water
2. 4.50 mL 0.5 M KPO$_4$, pH 7.4
3. 0.23 mL of <u>Solution B</u>; Total Concentration: 0.13 mM NADP$^+$, 6.6 mM G-6-P-Na$_2$, and 6.6 mM MgCl$_2$
4. 0.225 mL 2D6 (1 nmol/mL); Total amount: (225 pmol or 15 pmol/mL)
5. 0.0045 mL substrate (10 mM AMMC); Total Concentration: 3 µM (ACN 0.05 %)

2.11.5 BFC and 3A4

Prepare 15 mL by adding the components in the following order:

1. 2.82 mL deionized water
2. 10.5 mL 0.5 M KPO$_4$, pH 7.4
3. 1.5 mL of <u>Solution A</u>; Total Concentration: 2.6 mM NADP$^+$, 6.6 mM G-6-P-Na$_2$, and 6.6 mM MgCl$_2$
4. 0.15 mL 1A2 (1 nmol/mL); Total amount: (150 pmol or 10 pmol/mL)
5. 0.03 mL substrate (50 mM BFC); Total Concentration: 100 µM (ACN 0.05 %)

3 Methods

The fluorometric CYP inhibition assays to determine reversible/TDI kinetics were performed in black wall Costar 384-well plates using microsomal CYP enzymes that were obtained from baculovirus infected insect cells. The inhibitor test compounds were serially diluted (1:3 dilution) in 50 mM phosphate buffer (pH 7.4, pre-warmed at 37 °C) containing a constant concentration of glucose-6-phosphate dehydrogenase. Serial dilutions gave seven concentrations in the assay for calculation of IC_{50} values. The enzymatic reaction was initiated by adding individual CYP enzymes pre-mixed with fluorescent substrates in 50 mM phosphate buffer (pH 7.4) containing glucose-6-phosphate, $NADP^+$ and $MgCl_2$. Substrate concentrations were chosen to be close to the Km values (Note 4). A blank 384-well plate was prepared, which contained the inhibitor test, compounds serially diluted (1:3 dilution) in 50 mM phosphate buffer containing glucose-6-phosphate dehydrogenase. rCYP activity and blanks were continually measured for fluorescence at the excitation and emission wavelengths for the assay at intervals of 2 min on an FL600 microplate fluorescence reader for 30 min (Note 1). The plate holder of the microplate reader was heated at 37 °C before the reaction was started, and the temperature was held until the end of measurements.

The fluorometric CYP inhibition assay can be conducted manually or robotically, depending upon the throughput requirement. The current method is a manual version, but it can be easily modified to run the assay robotically using a liquid handler. The current experimental design allow for 48 test compounds can be measured per 384- well plate. Typically, the experiment is conducted in triplet.

3.1 The Main Steps in TDI Protocol

1. The serial dilution buffer, solution A and B, stop solution, and the inhibitor test compound stock solutions are prepared by adding the regents per instruction.

2. Adjust the 150 mL of the serial dilution buffer by adding 0.269 mL of DMSO and 5.2 mL of ACN. The serial dilution buffer now contains 0.97 U/mL G6PDH, 3.3 % ACN, and 0.2 % DMSO (Note 7).

3. All assay solutions, except rCYPs are warmed to 37 °C before the assay is started.

4. In a 96 well plate, add 432 μL of the adjusted Serial Dilution Buffer to the first column of six plates (Fig. 4). The wells now contain 0.97 U/mL G6PDH, 3.3 % ACN, and 0.2 % DMSO.

5. Add 300 μL of the adjusted Serial Dilution Buffer to the next seven columns of six plates (Fig. 4). Each well contains 0.97 U/mL G6PDH, 3.3 % ACN and 0.2 % DMSO.

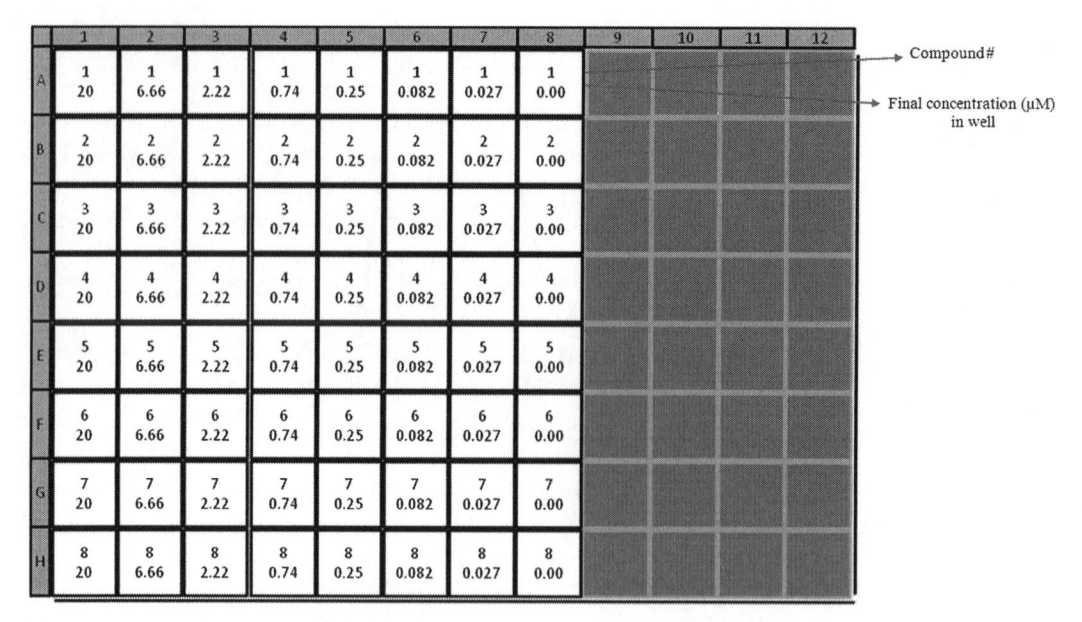

	1	2	3	4	5	6	7	8	9	10	11	12
A	1 20	1 6.66	1 2.22	1 0.74	1 0.25	1 0.082	1 0.027	1 0.00				
B	2 20	2 6.66	2 2.22	2 0.74	2 0.25	2 0.082	2 0.027	2 0.00				
C	3 20	3 6.66	3 2.22	3 0.74	3 0.25	3 0.082	3 0.027	3 0.00				
D	4 20	4 6.66	4 2.22	4 0.74	4 0.25	4 0.082	4 0.027	4 0.00				
E	5 20	5 6.66	5 2.22	5 0.74	5 0.25	5 0.082	5 0.027	5 0.00				
F	6 20	6 6.66	6 2.22	6 0.74	6 0.25	6 0.082	6 0.027	6 0.00				
G	7 20	7 6.66	7 2.22	7 0.74	7 0.25	7 0.082	7 0.027	7 0.00				
H	8 20	8 6.66	8 2.22	8 0.74	8 0.25	8 0.082	8 0.027	8 0.00				

→ Compound#
→ Final concentration (µM) in well

Fig. 4 First of six 96-well serial dilution plate layout

6. To the first column, add 18 µL of the 0.5 mM Inhibitor Test Compounds to the plate (note there are six plates). The concentration of the inhibitor test compound is 20 µM at 0.4 % DMSO and 7.0 % ACN and 0.95 U/mL G6PDH (Note 8).

7. Take 150 µL from the first column and add it to the second column, which contains 300 µL of adjusted serial dilution buffer. The first well now has a volume of 300 µL and the second well has a volume of 450 µL. The concentration of the inhibitor test compound in the second well is now 6.7 µM (1:3 dilutions) at 0.3 % DMSO, 4.5 % ACN and 0.96 U/mL G6PDH.

8. Repeat this 1:3 dilution process for the second column to the third column and so on. The concentration of the inhibitor test compounds in the third column (2.22 µM), fourth (0.74 µM), fifth (0.25 µM), sixth (0.08 µM), and seventh (0.027 µM) are produced For the eighth column, transfer 150 µL of the adjusted Serial Dilution Buffer (i.e., blank column). The average concentration in each well is 0.2 % DMSO, 3.3 % ACN and 0.97 U/mL G6PDH.

9. Distribute 25 µL of the each dilution of the compounds from step 8 (column 1–7) to five 384-well plates (i.e., one plate for each rCYP) and the eighth (0.00 µM) column solution to a blank plate (Fig. 5).

10. Lanes 22, 23, and 24 can be used for any type of negative or positive control experiments desired.

11. All the 384-well plates are kept at 37 °C at all times.

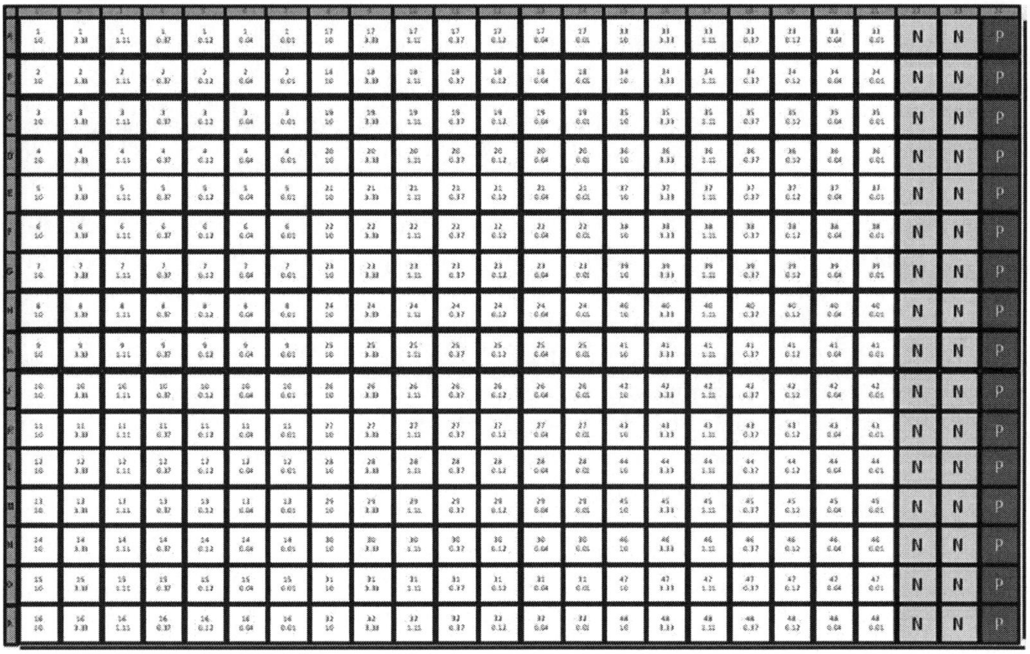

Fig. 5 First of five 384-well assay plate rCYPs layout and one blank 384-well assay plate. Lanes 22–24 can be used for negative (*N*) and positive (*P*) controls

12. Add 25 μL of the Stop Solution to each well to the blank 384-well plate.

13. The reaction is initiated by the addition of 25 μL of appropriate Enzyme and Substrate Mixture solution (i.e., 3A4, 2D6, 2C9, 2C19 and 1A2). The rCYP microsomes should be thawed and added just before the reaction is started and mixed gently by inverting the tube a few times. Add 25 μL of the adjusted serial dilution buffer to the blank plate instead of the E/S solution.

14. The concentration of the inhibitor test compounds in the plate is: first (10 μM), second (3.35 μM), third column (1.11 μM), fourth (0.37 μM), fifth (0.12 μM), sixth (0.04 μM), and seventh (0.0135 μM). Each well using Solution A contains 0.1 % DMSO and 1.8 % ACN, 1.3 mM NADP$^+$, 3.3 mM G-6-P-Na$_2$, 3.3 mM MgCl$_2$ and 0.48 U/mL G6PDH.

15. The reaction is initiated for each rCYP plate separately.

16. The 384-well assay plate along with blank plate is read using different excitation and emission wavelengths depending on the fluorescent substrate for 30 min (Note 1).

17. After subtracting background fluorescence from the blank 384-well plate, the plate reader readout is uploaded into a database for calculating IC$_{50}$ value for each rCYP.

18. The experiment is repeated to generate a control containing no inhibitor. In this experiment, buffer replaces the test inhibitor in step 6.

19. The experiment can be repeated depending on the number of replicates desired (Notes 9 and 10).

3.2 Calculation of IC50s and Formulas

The raw data obtained from fluorometric kinetic assays were transferred to a spread sheet (i.e., EXCEL; Microsoft). Spread sheet functions were created in house to automatically calculate IC_{50} values and their standard deviations, and generate time dependence curves of IC_{50} values [12]. Briefly, inhibition percentages at each time internal were calculated relative to a control that contained no inhibitor. IC_{50} values of inhibitions were calculated by linear extrapolation, using the equation shown below (Technical Manual, Gentest Corp.):

$$IC50 = \frac{(50\% - \text{low percentage}) * (\text{high conc.} - \text{low conc.})}{(\text{high percentage} - \text{low percentage})} + (\text{low conc.})$$

where,

- *low percentage* = highest percent inhibition less than 50%;
- *high percentage* = lowest percent inhibition greater than 50%;
- *low concentration* = concentration of test compound corresponding to the low percentage inhibition;
- *high concentration* = concentration of test compound corresponding to the high percentage inhibition.

Since all inhibition percentages are calculated from data taken on the same sample plate (i.e., no sample transferring or preparations), IC_{50} values at different time intervals were very consistent.

Because the level of fluorescent metabolite [P] (see Scheme 2) was relatively low in the early stage of incubation, larger error bars in the IC_{50} values were seen in data collected prior to 4 min. Thus, the initial IC_{50} value—when the MBI was still minimal—was set at 4 min and used as 100 %. Using this initial IC_{50} as 100 %, different plots could be generated which showed changes of relative IC_{50} vs. incubation time. The time dependent behaviors of IC_{50} values were readily recognized.

3.3 Time Dependent Behaviors of IC_{50} Values

Plotting IC_{50} values (relative to the initial IC_{50} value) vs. incubation time can easily differentiate reversible and irreversible. Using rCYP3A4 data from our lab [21], as shown in Fig. 6, IC_{50} values of reversible inhibitors either maintain within a narrow range (clotrimazole), or increase with incubation time (miconazole) presumably due to losing inhibitor as a result of metabolism or non-specific binding to matrix. In contrast, for an irreversible inhibitor (verapamil),

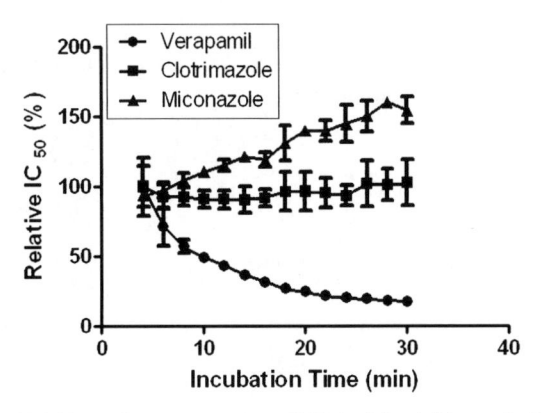

Fig. 6 The initial IC$_{50}$ values are verapamil (6.4 μM), clotrimazole (11 nM), and miconazole (0.11 μM). From reference [21]

the IC$_{50}$ values decreased with incubation time because of progressive inactivation of the corresponding enzyme by reactive metabolite species generated during the incubation.

Another way to examine the data is to compare the initial IC$_{50}$ value calculated at 4 min incubation time to the final IC$_{50}$ value calculated at 30 min incubation time for each test inhibitor. The percent ratios of the final IC$_{50}$ value to the initial IC$_{50}$ value can be used classify inhibitors as either reversible or irreversible (Table 1). In this case, when the percent ratio is less than 100 %, the test inhibitor is irreversible, and when the percent ratio is greater than or equal to 100 %, the test inhibitor is reversible.

4 Conclusion

The co-incubation *in vitro* continuous fluorometric kinetic assay using recombinant CYPs and fluorometric generating substrates is a valuable high-throughput assay for distinguishing reversible and time-dependent inhibitors for large compound collections.

Recently, 56 drugs were evaluated using an rCYP3A4 preincubation fluorometric assay and a human liver microsomes (HLM) LC/MS/MS assay [28]. In this rCYP3A4 assay, a preincubation step was added to allow for detection of enzyme inactivation and the enzymatic reaction was initiated by adding the DEF fluorometric substrate. The HLM assay used midazolam as the substrate probe. The prediction of reversible and irreversible inhibitors was found to be comparable between the two assays. The HLM assay predicted reversible inhibition for 31 compounds while the rCYP3A4 assay predicted 29 compounds. The HLM assay predicted irreversible inhibition for 25 compounds while the rCYP3A4 assay predicted 15 compounds. Thus, the rCYP3A4 assay had a false-positive rate of only 2 % and a false-negative rate of 18 %. When we compare our CYP3A4 results using a co-incubation

Table 1
Time-dependence pattern of IC$_{50}$ values from continuous fluorometric kinetic measurements using a co-incubation approach [21]

Compound (rCYP)	IC$_{50}$ (μM) "initial" (4 min)	IC$_{50}$ (μM) "final" (30 min)	Ratio Final/initial (%)	Classification
Clarithromycin (3A4)	19.1	5.6	29.3	Irreversible
Diltiazem (3A4)	10.2	7.9	77.5	Irreversible
Erythromycin (3A4)	16.2	6.1	37.7	Irreversible
Ethyinylestradiol (3A4)	43.2	0.6	1.4	Irreversible
Furafylline (1A2)	4	1.2 (20 min)	30.0	Irreversible
Midazolam (3A4)	3.4	1.6	47.1	Irreversible
Oleandomycin (3A4)	61.7	14.0	22.7	Irreversible
Troleandomycin (3A4)	1.1	0.3	27.2	Irreversible
Verapamil (3A4)	6.4	1.2	18.7	Irreversible
Clotrimazole	11.0	11.3	102.7	Reversible
Impramine (3A4)	80.7	90.7	112.4	Reversible
Ketoconazole (3A4)	0.023	0.044	191.3	Reversible
Miconazole (3A4)	0.11	0.18	163.6	Reversible
a-Naphthoflavone (1A2)	1.2	1.6	133.3	Reversible
Orphenadrine (3A4)	38.9	66.8	171.7	Reversible
Terfenadine (3A4)	0.31	0.96	309.7	Reversible

rCYP3A4 assay [21] to their pre-incubation assay [28] we find that the results for classifying reversible and irreversible inhibitors were identical. These results suggest that TDI fluorometric assays can be used as first tiered assays for distinguishing reversible inhibitors and TDIs for large compound collections.

Due to the "artificial" nature of expressed enzymes, it should be clearly understood that the fluorescence-based rCYP TDI assays are subject to at least three major limitations. First, the rCYP model system does not reflect the complexity of the *in vivo* scenario in which multiple enzymes co-exist and function sequentially or in parallel. Second, the system lacks phase II cytosolic enzymes which may be important in attenuating MBI and third, the system lacks drug transport across membranes. These limitations seriously limit the *in vitro–in vivo* extrapolation potential of rCYP model systems. In addition, the approach is also limited if the test compounds have a significant amount of fluorescent.

5 Notes

1. The FL600 plate reader requires systematic maintenance to guarantee its performance, which included regularly changing its air filter, and replacing its lamp. The plate was set at 37 °C and was shaken at a medium setting of 17 Hz (0.9 mm travel). Substrate probes are compounds that are selectively metabolized to a defined fluorescent metabolite by the rCYP isoform under investigation. The substrate probes, metabolite formed, the excitation, and emission bandwidths used in this assay are shown below. The sensitivity setting varied as required and the data collected was from the bottom.

rCYP enzyme	Substrate probe	Metabolite formed	Excitation (bandwidth)	Emission (bandwidth)
1A2	3-cyano 7-ethoxycoumarin (CEC)	3-cyano-7-hydroxycoumarin (CHC)	410 nm (±20 nm)	460 nm (±40 nm)
2C9	7-methoxy 4-trifluoromethyl-coumarin (MFC)	7-hydroxy-4-(trifluoromethyl) coumarin (HFC)	410 nm (±20 nm)	538 nm (±25 nm)
2C19	3-cyano 7-ethoxycoumarin (CEC)	3-cyano-7-hydroxycoumarin (CHC)	410 nm (±20 nm)	460 nm (±40 nm)
2D6	3-[2-(N,N-diethyl-N-methylammonium)ethyl]-7-methoxy-4-methylcoumarin (AMMC)	(3-[2-(N,N-diethylamino) ethyl]-7-hydroxy-4-methylcoumarin hydrochloride) (AHMC)	390 nm (±20 nm)	460 nm (±40 nm)
3A4	7-benzyloxy-4-(trifluoromethyl)-coumarin (BFC)	7-hydroxy-4-(trifluoromethyl) coumarin (HFC)	410 nm (±20 nm)	538 nm (±25 nm)

2. Magnesium chloride is extremely hydroscopic. Even new bottles of $MgCl_2 \cdot 6H_2O$ that are not opened may be hydrated more than the label indicates. Because water adds weight, it is impossible to obtain an accurate concentration of magnesium chloride by weighing it out. One trick is to prepare the entire lot of new bottles of $MgCl_2 \cdot 6H_2O$ as a 100× stock solution. Of course, we are assuming that the manufacturer known quantity of the material is correct.

3. Solution may precipitate upon storage at –20 °C but will redissolve when warmed to room temperature.

4. The substrate probe, positive control inhibitor, concentration, incubation time, kinetic parameters, and the IC_{50} values obtained from our lab for each rCYP isoform is shown below.

The numbers in the parentheses are the final amount or concentration in the assay well. The experimental conditions are such that the final concentrations of the cofactors are: 1.3 mM NADP$^+$, 3.3 mM G-6-P-Na$_2$, 0.48 U/mL G6PDH, 3.3 mM MgCl$_2$, 0.1 % DMSO and 2 % ACN.

rCYP enzyme (pmol)	Substrate probe (μM)	Positive control inhibitors (μM)	Incubation time (min)	Km (μM)	Vmax (min^{-1})	IC$_{50}$ (μM)
1A2 (0.125)	CEC (5.4)	Furafylline (10–0.0137)	20	3.5	3.4	2.0 ± 0.8
2C9 (0.25)	MFC (75)	Sulfaphenazole (10–0.0137)	45	78	2.1	0.32 ± 0.18
2C19 (0.125)	CEC (20)	Omeprazole (10–0.0137)	45	29	0.016	8.8 ± 0.5
2D6 (0.375)	AMMC (1.5)	Quinidine (0.10–0.000137)	30	1	1	0.015 ± 0.008
3A4 (0.25)	BFC (50)	Verapamil (10–0.0137)	30	>200	1.5 @ 40 μM	0.34 ± 0.16

5. This solution maybe frozen (–20 °C) in 1.5 mL aliquots and is typically used within 2–3 weeks.

6. It should be noted that dissolved test compounds from library collects have inaccurate concentrations. From our own experience, library stock solution of 10 mM in DMSO were measured and discovered to be closer to 6–7 mM in many cases.

7. This step ensures that there is a constant % of organic solvent in every well. The final % organic in each well will be approximately 0.1 % DMSO and 2 % ACN

8. The serial dilution range is set to 20–0.009 μM, which is diluted 1:2 in the final assay plate. To increase the range, increase the amount of inhibitor test compound; to decrease the range, decrease the amount of inhibitor test compound. For example to increase the inhibitor test compound concentration range to 40–0.018 μM. In step 4, add 396 μL of the adjusted Serial Dilution Buffer to the first column of six plates and in step 5, add 36 μL of the 0.5 mM Inhibitor Test Compounds to the plate.

9. Instead of inhibitor test compounds, one could add known MBI compounds as positive controls and known completive reversible inhibitors as negative controls. For example, here is a list of suggestions [36]:

CYP	Inhibitor	Mechanism of inhibition
3A4	Gestodene	MBI
	Ketoconazole	Reversible
	Erythromycin	MIC
2D6	Paroxetine	MBI
	Cisapride	Reversible
2C9	Tienilic acid	MBI
	Sulfaphenazole	Reversible
2C19	Ticlopidine	MBI
	Fluconazole	Reversible
1A2	Furafylline	MBI
	Fluvoxamine	Reversible

10. It should be remembered that some enzymatic reactions are markedly impaired by even small changes in the pH and the ionic strength of the solution. Therefore, in most *in vitro* enzymatic assays, it is necessary to add a buffer to the medium to stabilize the pH and inorganic salts to stabilize the ionic strength of the solution. If these conditions are changed during the experiment, the results may also be altered. The buffer conditions used in our assay do have effects on enzyme activity. Other factors to keep in mind are that the pH may also vary with temperature and the water in which the buffer substances are dissolved should be of the highest quality.

Acknowledgements

The views expressed here are solely those of the author and do not reflect the opinions of Janssen Research & Development, LLC.

References

1. Caldwell GW (2000) Compound optimization in early- and late-phase drug discovery: acceptable pharmacokinetic properties utilizing combined physicochemical, in vitro and in vivo screens. Curr Opin Drug Discov Devel 3: 30–41
2. Caldwell GW, Ritchie DM, Masucci JA, Hageman W, Yan Z (2001) The new pre-preclinical paradigm: compound optimization in early and late phase drug discovery. Curr Top Med Chem 1(5):353–366
3. Bjornsson TD, Callaghan JT, Einolf HJ, Fischer V, Gan L, Grimm S, John Kao S, King P, Miwa G, Ni L, Kumar G, McLeod J, Obach SR, Roberts S, Roe A, Shah A, Snikeris F, Sullivan JT, Tweedie D, Vega JM, Walsh J, Wrighton SA (2003) The conduct of in vitro and in vivo drug-drug interaction studies: a Pharmaceutical Research and Manufacturers of America (PhRMA). J Clin Pharmacol 43(5): 443–469
4. Caldwell GW, Yan Z, Tang W, Dasgupta M, Hasting B (2009) ADME optimization and toxicity assessment in early- and late-phase drug discovery. Curr Top Med Chem 9(11): 965–980

5. Lin JH, Lu AYH (1998) Inhibition and induction of cytochrome P450 and the clinical implications. Clin Pharmacokinet 5(5):361–390
6. Yan Z, Caldwell GW (2001) Metabolism profiling, and cytochrome P450 inhibition & induction in drug discovery. Curr Top Med Chem 1(5):403–425
7. Stresser DM (2004) High-through screening of human cytochrome P450 inhibitors using fluorometric substrates: methodology for 25 enzyme/substrate pairs. In: Yan Z, Caldwell GW (eds) Optimization in drug discovery: in vitro methods, Method in pharmacology and toxicology. Humana, Totowa, NJ
8. Yan Z, Caldwell GW (2004) Evaluation of cytochrome P450 inhibition in human liver microsome. In: Yan Z, Caldwell GW (eds) Optimization in drug discovery: in vitro methods, Method in pharmacology and toxicology. Humana, Totowa, NJ
9. Riley RJ, Grime K, Weaver R (2007) Time-dependent CYP inhibition. Expert Opin Drug Metab Toxicol 3(1):51–66
10. Polasek T, Miner JO (2007) In vitro approaches to investigate mechanism-based inactivation of CYP enzymes. Expert Opin Drug Metab Toxicol 3(1):321–329
11. Yan Z, Caldwell GW (2012) The current status of time dependent CYP inhibition assay and in silico drug-drug interaction predictions. Curr Top Med Chem 12(11):1291–1297
12. Caldwell GW, Yan Z, Lang W, Masucci J (2012) The IC_{50} concept revisited. Curr Top Med Chem 12(11):1282–1290
13. Moody GC, Griffin SJ, Mather AN, McGinnity DF, Riley RJ (1998) Fully automated analysis of activities catalysed by the major human liver cytochrome P450 (CYP) enzymes: assessment of human CYP inhibition potential. Xenobiotica 29(1):53–75
14. Mano Y, Usui T, Kamimura H (2007) Comparison of inhibition potentials of drugs against zidovudine glucuronidation in rat hepatocytes and liver microsomes. Drug Metab Dispos 35(4):602–606
15. Crespi CL, Stresser DM (2001) Fluorometric screening for metabolism-based drug-drug interactions. J Pharmacol Toxicol Methods 44(1):325–331
16. Crespi CL, Miller VP, Penman BW (1997) Microtiter plate assays for inhibition of human, drug-metabolizing cytochromes P450. Anal Biochem 248(1):188–190
17. Chu I, Favreau L, Soares T, Lin CC, Nomeir AA (2000) Validation of higher-throughput high-performance liquid chromatography/ atmospheric pressure chemical ionization tandem mass spectrometry assays to conduct cytochrome P450s CYP2D6 and CYP3A4 enzyme inhibition studies in human liver microsomes. Rapid Commun Mass Spectrom 14(4):207–214
18. Bell L, Bickford S, Nguyen PH, Wang J, He T, Zhang B, Friche Y, Zimmerlin A, Urban L, Bojanic D (2008) Evaluation of fluorescence- and mass spectrometry-based CYP inhibition assays for use in drug discovery. J Biomol Screen 13(5):343–353
19. Poulin P, Kenny JR, Hop CECA, Haddad S (2012) In vitro-in vivo extrapolation of clearance: modeling hepatic metabolic clearance of highly bound drugs and comparative assessment with existing calculation methods. J Pharm Sci 101(2):838–851
20. Imai H, Kotegawa T, Ohashi K (2011) Duration of drug interactions: putative time courses after mechanism-based inhibition or induction of CYPs. Expert Rev Clin Pharmacol 4(4):409–411
21. Yan Z, Rafferty B, Caldwell GW, Masucci JA (2002) Rapidly distinguishing reversible and irreversible CYP450 inhibitors by using fluorometric kinetic analyses. Eur J Drug Metab Pharmacokinet 27(4):281–287
22. Yamamoto T, Suzuki A, Kohno Y (2002) Application of microtiter plate assay to evaluate inhibitory effects of various compounds on nine cytochrome P450 isoforms and to estimate their inhibition patterns. Drug Metab Pharmacokinet 17(5):437–448
23. Yamamoto T, Suzuki A, Kohno Y (2004) High-throughput screening for the assessment of time-dependent inhibitions of new drug candidates on recombinant CYP2D6 and CYP3A4 using a single concentration method. Xenobiotica 34(1):87–101
24. Naritomi Y, Teramura Y, Terashita S, Kagayama A (2004) Utility of microtiter plate assays for human cytochrome P450 inhibition studies in drug discovery: application of simple method for detecting quasi-irreversible and irreversible inhibitors. Drug Metab Pharmacokinet 19(1):55–61
25. Turpeinen M, Korhonen LE, Tolonen A, Uusitalo J, Juvonen R, Raunio H, Pelkonen O (2006) Cytochrome P450 (CYP) inhibition screening: comparison of three tests. Eur J Pharm Sci 29(2):130–138
26. Krippendorff BF, Neuhaus R, Lienau P, Reichel A, Huisinga W (2009) Mechanism-based inhibition: deriving KI and kinact directly from time-dependent IC50 values. J Biomol Screen 14(8):913–923

27. Sekiguchi N, Higashida A, Kato M, Nabuchi Y, Mitsui T, Takanashi K, Aso Y, Ishigai M (2009) Prediction of drug-drug interactions based on time-dependent inhibition from high throughput screening of cytochrome P450 3A4 inhibition. Drug Metab Pharmacokinet 24(6): 500–510
28. Kajbaf M, Palmieri E, Longhi R, Fontana S (2010) Identifying a higher throughput assay for metabolism dependent inhibition (MDI). Drug Metab Lett 4(2):104–113
29. Salminen KA, Leppanen J, Venalainen JI, Pasanen M, Auriola S, Juvonen RO, Raunio H (2011) Simple, direct, and informative method for the assessment of CYP2C19 enzyme inactivation kinetics. Drug Metab Dispos 39(3): 412–418
30. Watanabe A, Nakamura K, Okudaira N, Okazaki O, Sudo K (2007) Risk assessment for drug-drug interaction caused by metabolism-based inhibition of CYP3A using automated in vitro assay systems and its application in the early drug discovery process. Drug Metab Dispos 35(7):1232–1238
31. Berry LM, Zhao Z (2008) An examination of IC50 and IC50-shift experiments in assessing time-dependent inhibition of CYP3A4, CYP2D6 and CYP2C9 in human liver microsomes. Drug Metab Lett 2(1):51–59
32. Burt HJ, Galetin A, Houston JB (2010) IC50-based approaches as an alternative method for assessment of time-dependent inhibition of CYP3A4. Xenobiotica 40(5):331–343
33. Burt HJ, Pertinez H, Sall C, Collins C, Hyland R, Houston JB, Galetin A (2012) Progress curve mechanistic modeling approach for assessing time-dependent inhibition of CYP3A4. Drug Metab Dispos 40(9): 1658–1667
34. Kenny JR, Mukadam S, Zhang C, Tay S, Collins C, Galetin A, Khojasteh SC (2012) Drug-drug interaction potential of marketed oncology drugs: in vitro assessment of time-dependent cytochrome P450 inhibition, reactive metabolite formation and drug-drug interaction prediction. Pharm Res 29(7): 1960–1976
35. Chen Y, Liu L, Monshouwer M, Fretland AJ (2011) Determination of time-dependent inactivation of CYP3A4 in cryopreserved human hepatocytes and assessment of human drug-drug interactions. Drug Metab Dispos 39(11): 2085–2092
36. Maeng HJ, Chow ECY, Fan J, Pang KS (2012) Physiologically based pharmacokinetic (PBPK) modeling: usefulness and applications. In: Lyubimov AV, Rodrigues AD, Sinz MA (eds) Encyclopedia of drug metabolism and interactions, vol 2. Wiley, New York, pp 637–684
37. Donato MT, Gomez-Lechon MJ (2006) Inhibition of P450 enzymes: an *in vitro* approach. Curr Enzym Inhib 2:281–304

Chapter 18

Identification of Time-Dependent CYP Inhibitors Using Human Liver Microsomes (HLM)

Kevin J. Coe, Judith Skaptason, and Tatiana Koudriakova

Abstract

Cytochrome P450s (CYPs) are responsible for the metabolism of a majority of marketed drugs and, as a consequence, alteration in CYP activity can result in clinically relevant drug-drug interactions (DDIs). Drugs that are time dependent inhibitors (TDIs) of CYPs have been reported to cause severe DDIs, leading to prescription adjustments and, in certain cases, have been withdrawn from the market (Zhou et al., Ther Drug Monit 29:687–710, 2007). Oftentimes, TDI is the result of mechanism-based inactivation (MBI), where CYPs catalyze the formation of reactive metabolites that irreversibly or quasi-irreversibly inhibit its own activity. In order to restore basal CYP activity lost as a result of MBI *de novo* enzyme synthesis is required, and therefore, MBI can have greater clinical consequences than reversible CYP inhibition. Methodologies capable of identifying MBIs in drug discovery are warranted to address this potential liability. The most commonly employed assay to identify MBIs is by measuring the IC_{50} for CYP enzymes with and without pre-incubation of discovery compounds with human liver microsomes. An IC_{50} shift assay for CYP3A will be described in greater detail given the enzyme's prominent role in drug metabolism and association with severe clinical DDIs resulting from MBI; however, the overall assay design can easily be adopted for other CYPs. While the IC_{50} shift assay can be used to build SAR to mitigate this liability in discovery, clinical risk assessment of an MBI requires the determination of the kinetic parameters K_I and k_{inact}.

Key words Drug–drug interactions, Time dependent inhibition, Mechanism-based inactivation, Irreversible and quasi-irreversible inhibition, IC_{50} shift

1 Introduction

1.1 Characteristics of Mechanism-Based Inactivators

Since the majority of marketed drugs are metabolized by CYPs, drugs that alter CYP activity can cause clinically relevant DDIs. For example, drugs that inhibit CYPs may reduce the metabolism and clearance of other co-administered drugs and possibly impact their efficacy and safety profiles. CYP inhibition can be reversible (competitive or non-competitive), quasi-irreversible or irreversible. Quasi-irreversible and irreversible inhibitors are known as MBIs. CYPs metabolize MBIs into reactive intermediates that bind to the enzyme active site and prevent catalysis. When incubated in liver microsomes supplemented with NADPH, MBIs demonstrate time dependent inhibition of CYP activity that cannot be restored after gel filtration or dialysis [2].

MBIs can cause CYP inactivation through multiple routes. Quasi-irreversible CYP inhibitors produce reactive metabolites (typically nitroso and carbene species) that form a tight, non-covalent interaction, often called the metabolite-intermediate complex (MIC), with the heme iron. The MIC transitions the heme iron from the ferrous to ferric state thereby preventing oxygen binding required for catalysis. CYP activity for quasi-irreversible inhibitors can be restored in vitro by the use of chelating agents such as potassium ferricyanide, which is capable of displacing nitroso MICs; however, enzymatic activity cannot be restored in vivo for quasi-irreversible inhibitors. Irreversible CYP inhibitors form reactive metabolites that bind covalently either to the prosthetic heme or to an amino acid residue of the apoprotein to result in enzyme inactivation.

1.2 Clinical Implications of CYP Mechanism-Based Inactivation

MBIs can cause severe clinical DDIs. By irreversibly inhibiting CYP, MBIs deplete the pool of active CYP present for drug metabolism, leading to a substantial reduction in metabolic capacity after repeat administration. The concentration of a co-administered drug, whose clearance relies upon this CYP, may be significantly increased after prolonged dosing with an MBI. Furthermore, after irreversible or quasi-irreversible CYP inhibition, de novo enzyme synthesis is required in order to restore basal CYP activity. This time lag may require multiple days after an MBI administration, illustrating the long lasting consequences of MBI. In addition, MBIs have been associated with idiosyncratic hepatotoxicity [3], possibly due to the generation of autoantibodies against protein adducts. A number of CYP MBIs, including dihydralazine (CYP1A2), halothane (CYP2E1), and tienilic acid (CYP2C9), induce the formation of autoantibodies against the CYPs they inactivate.

Considering that CYP3A is the most abundantly expressed CYP in the liver and responsible for the metabolism of ~50 % of marketed drugs [4], CYP3A MBIs can impact a wide spectrum of drug classes and result in severe DDIs [1]. For example, the drug terfenadine has a narrow therapeutic index for cardiotoxicity and is metabolized extensively by CYP3A into its active and less toxic acid metabolite. Patients co-administered terfenadine with the CYP3A MBI erythromycin had increased levels of terfenadine and were at greater risk for developing torsades de pointes [5]. In the case of the anti-hypertensive mibefradil, a CYP3A MBI, its co-administration with β-blockers resulted in four incidences of cardiogenic shock and one death [6]. Such severe cases of CYP3A MBI led to prescription adjustments either in dose or drug selection and prompted the withdrawal of certain drugs from the market, including terfenadine and mibefradil. Only in rare instances (as in the case of ritonavir) has CYP3A MBI been used beneficially to "boost" the exposure of anti-retrovirals otherwise subject to high clearance and low exposure [7]. In most instances, the progression of a CYP3A MBI is considered a liability that may require costly DDI studies before marketing and narrow the patient population for intended drug therapy. As result, the identification of MBIs, particularly of CYP3A, is a critical activity in drug discovery.

1.3 Methodologies to Identify MBIs—The IC50 Shift Method

One of the common assays utilized in drug discovery to identify MBIs is the IC_{50} shift method, where the CYP IC_{50} value of a test compound is compared with and without a pre-incubation step in NADPH supplemented human liver microsomes (HLMs) [8]. In this assay design, multiple concentrations of a test compound are either pre-incubated in HLMs fortified with NADPH (the pre-incubate mix) or added to the mix just prior to the addition of a CYP-selective probe. CYP activity is measured, and the CYP inhibitory potency is compared between with and without pre-incubation. Since MBIs demonstrate time dependent inhibition, greater CYP inhibition is observed after preincubation with a MBI (*see* Note 1), leading to a leftward-shift in a test compound's inhibitory potency relative to its inhibition profile without pre-incubation. Figure 1 illustrates the CYP3A inhibition by troleandomycin, erythromycin, and ketoconazole in HLMs with and without a pre-incubation step.

Troleandomycin and erythromycin are both CYP3A MBIs and demonstrate a leftward shift in their inhibitory potency with a pre-incubation step. In addition, troleandomycin exhibits a significant degree of reversible CYP inhibition relative to erythromycin, which shows minimal CYP inhibition without a pre-incubation step. Ketoconazole is a reversible CYP3A inhibitor and demonstrates similar inhibitory potency with and without a pre-incubation step (*see* **Notes 2** and **3**).

There are a few assay variables that should be taken into consideration before implementation, such as test compound concentration range, selection and concentration of the CYP-selective

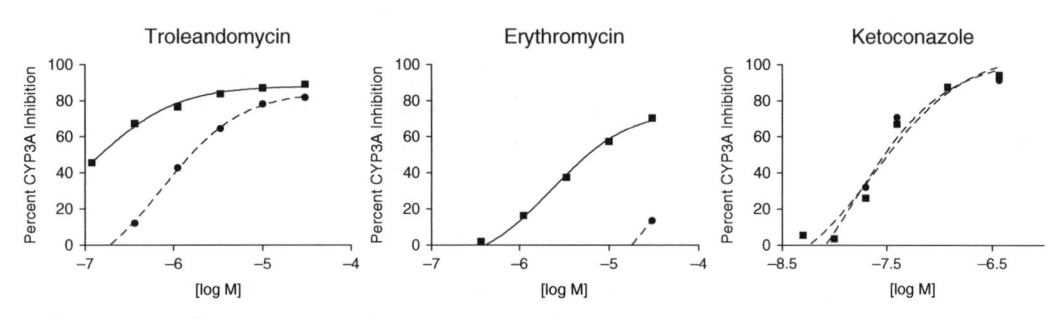

Fig. 1 The inhibition curves are determined by measuring the formation of the CYP3A-mediated metabolite 6β-hydroxy-testosterone in the presence of different concentration of inhibitor without pre-incubation (*dashed circle*) and with pre-incubation (*solid square*) for 30 min at 37 °C in HLMs supplemented with NADPH. CYP activity is normalized to vehicle control incubations

probe substrate, the amount of HLMs employed, the duration of the pre-incubation and probe reaction steps, and the selection of assay reference compounds.

Test Compound Concentration Range: In order to adequately derive a concentration-response, at least six concentrations of inhibitor are necessary. Test concentrations can be limited by compound solubility, but top concentrations of 10 or 30 µM are often achievable for most compounds. Although acetonitrile and methanol show less inhibition of CYPs than DMSO, DMSO is often employed as the solvent of choice for preparing drug discovery compounds stock solutions. In order to avoid solvent inhibition of CYPs, it is advisable to minimize the organic solvent content in the incubations, particularly of DMSO to ≤0.1 % [9].

CYP-Probe Selection: A selective probe substrate should be chosen for the CYP isoenzyme to be explored through TDI. For assays assessing CYP3A TDI, midazolam or testosterone serve as selective substrates to gauge inhibition of CYP3A activity through the measurement of 1-hydroxy-midazolam or 6β-hydroxy-testosterone, respectively. During the probe incubation, the probe substrate should be added at its estimated K_M concentration to allow for sufficient metabolite formation to measure inhibition relative to control incubations without test compound.

HLM Content: The amount of HLM protein should balance the requirements of permitting sufficient turnover of compound and probe substrate but minimize the possibility of extensive turnover of the probe substrate, where substrate depletion may reduce levels of the monitored metabolite.

Pre-incubation and Probe Reaction Time Points: A pre-incubation of 30 min is often employed for the IC_{50} shift method. Since incubations of HLMs with NADPH can result in generation of reactive oxygen species that inactivate CYPs, longer pre-incubations (>60 min) may attenuate CYP activity compromising sufficient formation of the probe metabolite. Alternatively, shorter pre-incubations may not allow for sufficient CYP inactivation to elicit a shift in the IC_{50} plots relative to the no-preincubation arm thereby compromising data interpretation. For the probe reaction, the incubation time used should maximize the formation of the probe metabolite but not carried on further, where sequential metabolism may minimize assay robustness.

Assay Reference Controls: In order to gauge the success of the assay, TDI positive and negative assay controls should be included for each test study. These controls should include at least one known MBI and one reversible inhibitor of the CYP enzyme evaluated. The IC_{50} values observed for these controls (along with their observed IC$_{50}$ shift values) ensure that both the pre-incubation and probe reaction steps yield values within a pre-defined assay range to validate each study.

2 Materials

The methods described herein are for assessing TDIs of CYP3A, given its clinical bases for severe DDI. Nonetheless, the foundation of the assay methodology described herein can be adapted to other CYP TDI assays [8, 10, 11].

2.1 CYP3A IC$_{50}$ Shift Method

2.1.1 Preparation of Test Compound

1. Test compound of known molecular weight
2. Dimethyl sulfoxide (DMSO), anhydrous (ARCO)
3. HPLC grade acetonitrile (JT Baker, Phillipsburg, NH)
4. HPLC grade water (JT Baker, Phillipsburg, NH)

2.1.2 Reagents for Incubations

1. Potassium phosphate buffer, 0.5 M, pH 7.4 (BD Gentest, San Jose, CA)
2. β-Nicotinamide adenine dinucleotide phosphate, reduced form (NADPH) (Sigma, St. Louis, MO)
3. Human liver microsomes (HLMs), 20 mg/mL (BD Gentest, San Jose, CA)
4. Ketoconazole (Sigma, St. Louis, MO)
5. Troleandomycin (Sigma, St. Louis, MO)
6. Testosterone (Sigma, St. Louis, MO)
7. 6β-Hydroxy-testosterone (Sigma, St. Louis, MO) for LC/MS/MS method development
8. HPLC grade acetonitrile (JT Baker, Phillipsburg, NH) for sample extraction
9. HPLC grade methanol (JT Baker, Phillipsburg, NH) for sample extraction

2.1.3 Materials for Incubation

1. 1.2-mL 96-Well polypropylene reaction plates and seals
2. Plate mixing apparatus
3. Shaking water bath set to 37 °C
4. Ice tray and crushed ice
5. Automated platform, e.g. Beckman FX and Tecan instruments (*optional*)

2.1.4 Preparation of Samples for LC/MS/MS Analyses

1. Selected internal standard (e.g. phenytoin) diluted in 50:50 acetonitrile:methanol
2. Refrigerated centrifuge capable of $10,000 \times g$
3. 96-Head transfer pipette, e.g., Apricot Designs i-Pipette 96 head personal transfer pipette (*optional*)

2.2 Materials for Data Analyses from Samples Generated from Incubations

2.2.1 LC/MS/MS Materials

1. API 4000 LC/MS/MS system (Applied Biosystems, Concord, Ontario, Canada)
2. Agilent 1100 Series or Shimadzu HPLC (Santa Clara, CA)
3. Autosampler (LEAP PAL, Carrboro, NC)
4. Zorbax™ SB-Phenyl column, 2×50 mm, 5 µm (Agilent, Santa Clara, CA)
5. HPLC grade acetonitrile (JT Baker, Phillipsburg, NH)
6. HPLC grade water (JT Baker, Phillipsburg, NH)
7. Formic Acid (MP Biomedicals, Inc., Solon, OH)

3 Method

3.1 YP3A IC50 Shift Method

3.1.1 Preparation of CYP3A Substrate, Assay Reference Controls, and Test Compound Stock Solutions

1. Testosterone is dissolved in HPLC-grade acetonitrile as a 10 mM stock solution.
2. Ketoconazole (negative CYP3A TDI control) is dissolved in DMSO as a 0.6 mM stock.
3. Troleandomycin (positive CYP3A TDI control) is dissolved in DMSO as a 10 mM stock.
4. Dissolve test compound in DMSO at a stock concentration of 10 mM stock.

3.1.2 Serial Dilution of Assay Reference Controls and Test Compound

1. Transfer assay reference controls or test compound DMSO stock solutions (3 µL) to 20 % acetonitrile:water (147 µL) in a 1.2 mL polypropylene 96-well plate to prepare a secondary stock. This results in an organic solvent concentration of 2 % DMSO and ~20 % acetonitrile to ensure the DMSO contest does not exceed 0.1 % (v/v) during the pre-incubation step. Mix thoroughly by pipetting up and down repeatedly.
2. Use the same organic solvent mixture, 2 % DMSO and 96 % (20 % acetonitrile: 80 % water), for 1:2 serial dilutions of the secondary stock to prepare at least six test concentrations.
3. Use the above solvent composition without test compound to serve as the assay vehicle control.

3.1.3 Assay Reagent Preparation

1. Potassium phosphate, 0.5 M, pH 7.4, stock buffer is diluted to 0.1 M in HPLC-grade water.
2. HLMs (20 mg/mL) are thawed and diluted in 0.1 M potassium phosphate buffer to 0.6 mg/mL (e.g. 0.3 mL HLMs added to 9.7 mL buffer).
3. NADPH is prepared as a 2 mM stock concentration in 0.1 M potassium phosphate buffer. An aliquot of this NADPH solution is spiked with testosterone to a concentration of (50 µM) for the probe reaction step.

4. All HLM and NADPH solutions are pre-warmed in a 37 °C water bath for 5 min prior to initiation of the assay.

5. A 50:50 organic mixture of acetonitrile:methanol spiked with an appropriate internal standard (e.g. 0.05 μg/mL phenytoin) is prepared and kept ice-cold to facilitate reaction termination and protein precipitation.

3.1.4 Pre-incubation Mix Preparation

1. HLMs are mixed with NADPH in a 1:1 ratio (v/v) to yield a pre-incubation mix with final HLM and NADPH concentrations of 0.3 mg/mL protein and 1 mM, respectively, in 0.1 M potassium phosphate buffer.

3.1.5 Probe Reaction Procedure

1. Dispense the pre-warmed pre-incubation mix (100 μL) into two 96-well, 1.2-mL deep polypropylene plates. Plate 1 and Plate 2 represent the "No pre-incubation" and "Pre-incubation" assay arms, respectively.

2. Add test compound (5 μL) to Plate 2 and mix quickly, where the top assay concentration will be 10 μM when starting from a 10 mM compound stock solution. This results in an organic composition of 0.2 % DMSO and ~1 % acetonitrile.

3. Pre-incubate both plates for 30 min in a 37 °C shaking water-bath.

4. After pre-incubation, add test compound (5 μL) to Plate 1.

5. Follow with the addition of 100 μL NADPH containing 50 μM testosterone to both plates for the probe reaction. Mix quickly and place plates into a 37 °C shaking water-bath for 10 min. This results in a probe reaction condition of 0.15 mg/mL HLMs, 1 mM NADPH, and 25 μM testosterone.

6. Terminate the probe reaction by adding the ice-cold organic solvent mixture containing internal standard.

7. Seal plates, vortex vigorously and precipitate proteins for 15 min on ice.

8. Centrifuge plates at 14,000 g for 10 min at 4 °C.

9. Remove supernatant (200 μL) and mix with water (100 μL) for subsequent LC/MS/MS analyses.

3.2 LC/MS/MS Analyses of CYP3A4 IC$_{50}$ Shift Method

Samples are loaded onto 2 × 50 mm SB-Phenyl columns using a reverse-phase HPLC method at 0.8 mL/min flow rate to elute 6β-hydroxy-testosterone and the internal standard (phenytoin). Both HPLC reagents water (A) and acetonitrile (B) are spiked with 0.1 % formic acid and separation is achieved during a 1.0 min gradient from 10 to 95 % organic. The 6β-hydoxy-testosterone is quantified on a 4000 triple quadruple MS/MS instrument by single ion monitoring (m/z 305 > 269), where mass spectral counts are normalized to the response of the internal standard.

3.2.1 Determination of Percent Remaining and IC$_{50}$ Values

The MS response of 6β-hydroxy-testosterone formed in vehicle-control reactions is set to 100 % activity (or 0 % inhibition). The ratio of 6β-hydroxy-testosterone formed in test compound incubations over the vehicle-control reactions is converted to a percentage of CYP inhibition. Using graphing software such as Prism, the percent CYP inhibition over time is plotted to obtain an IC$_{50}$ value for the test compound. The IC$_{50}$ shift value is calculated by dividing the IC$_{50}$ observed with no pre-incubation by the observed IC$_{50}$ with pre-incubation (*see* Note 4).

The IC$_{50}$ shift value is used as means of determining whether a drug discovery compound demonstrates TDI and is more qualitative than quantitative (as described in Sect. 3.3). Use of established, commercially available CYP MBIs should be employed beforehand to help calibrate the assay. In general, a leftward IC$_{50}$ shift of twofold is considered a "positive" for TDI (*see* Note 5).

3.3 Scope and Limitations of the IC$_{50}$ Shift Method

The IC$_{50}$ shift assay is useful in the discovery setting to identify compounds that demonstrate TDI and may be MBIs. The assay provides sufficient throughput to build SAR in order for medicinal chemists to modify or remove sub-structures responsible for TDI [12]. Nonetheless, the assay has limitations. First, the IC$_{50}$ shift method does not elucidate the mechanism of TDI, which may be through MBI or through alternative mechanisms (e.g. formation of metabolites with more potent reversible CYP inhibition than parent compound). Second, given the high concentrations of the discovery compound present in the probe reaction, the assay may be inadequate to resolve an MBI with a strong reversible inhibitory component. Third, the IC$_{50}$ shift value itself is not an indicator of potency for an MBI. To better gauge the potency of an MBI and, importantly, its clinical consequences, the kinetic parameters, K$_I$ and k$_{inact}$, must be obtained. K$_I$ is the half-maximal inactivation concentration, and k$_{inact}$ is the inactivation rate constant. More resource-intensive studies, often beyond the scope of drug discovery, are required to obtain these kinetic parameters [11].

3.4 Conclusions

Progressing drug discovery compounds that are MBIs, particularly of CYP3A, can result in severe clinical DDIs, sometimes life-threatening, that have prompted the removal of MBIs from the market as illustrated by mibefradil. A pragmatic initial approach in drug discovery is to set-up throughput assays to identify this risk. One such methodology is the IC$_{50}$ shift method to identify TDIs, which are most often the result of MBI. Although the assay does not definitively subscribe TDI to MBI nor define an MBI's potency, it does offer an important first read in order to identify and SAR away from this liability early in the drug discovery phase. In addition, results from the IC$_{50}$ shift method may prompt further mechanistic work, such as trapping reactive metabolites and obtaining the kinetic parameters K$_I$ and k$_{inact}$ to better gauge clinical risks.

4 Notes

1. MBIs represent a subset of CYP time dependent inhibitors as other mechanisms may result in TDI. For example, the formation of a metabolite with increased reversible CYP inhibition than the parent compound may result in TDI. Dialysis should restore CYP activity and identify these metabolism-based inhibitors. In addition, certain CYP substrates may promote uncoupling of the CYP catalytic cycle to form reactive oxygen species (e.g. superoxide anion, hydrogen peroxide) that potentially may increase CYP inhibition with time. Pre-incubations conducted in the presence of free radical scavengers may mitigate this type of TDI.

2. For compounds that are both strong reversible inhibitors as well as MBIs, clear separation of the inhibitory potency may not be possible. Increased pre-incubation time-periods (e.g. 60 min) may improve this resolution but not in all instances (as is the case for ritonavir).

3. For some reversible CYP inhibitors, a rightward shift in inhibitory potency may be observed. This is most often caused by extensive metabolism of the inhibitor during the pre-incubation step to less inhibitory metabolites thereby depleting its overall inhibitory potency.

4. For some compounds, where there is little to no evidence of inhibition in the no-pre-incubation arm (as in the instance of erythromycin in Fig. 1), the IC50 shift can be reported by dividing the top concentration used in the assay by the observed IC50 in the pre-incubation assay arm. This IC50 shift value can be reported as greater than calculated value.

5. Discovery compounds that exhibit an IC50 shift value of <2 may still be TDIs but are beyond the assay's resolution given the inherent assay variability when comparing inhibition between the two assay arms.

Acknowledgements

The views expressed here are solely those of the author and do not reflect the opinions of Janssen Research & Development, LLC.

References

1. Zhou SF et al (2007) Clinically important drug interactions potentially involving mechanism-based inhibition of cytochrome P450 3A4 and the role of therapeutic drug monitoring. Ther Drug Monit 29(6):687–710
2. Silverman R (1998) Mechanisim-based enzyme inactivation: chemistry and enzymology. CRC Press, Boca Raton, FL
3. Tucker GT, Houston JB, Huang SM (2001) Optimizing drug development: strategies to

assess drug metabolism/transporter interaction potential—towards a consensus. Br J Clin Pharmacol 52(1):107–117
4. Plant N (2007) The human cytochrome P450 sub-family: transcriptional regulation, interindividual variation and interaction networks. Biochim Biophys Acta 1770(3):478–488
5. Dresser GK, Spence JD, Bailey DG (2000) Pharmacokinetic-pharmacodynamic consequences and clinical relevance of cytochrome P450 3A4 inhibition. Clin Pharmacokinet 38(1):41–57
6. Mullins ME et al (1998) Life-threatening interaction of mibefradil and beta-blockers with dihydropyridine calcium channel blockers. JAMA 280(2):157–158
7. Moyle GJ, Back D (2001) Principles and practice of HIV-protease inhibitor pharmacoenhancement. HIV Med 2(2):105–113
8. Obach RS, Walsky RL, Venkatakrishnan K (2007) Mechanism-based inactivation of human cytochrome p450 enzymes and the prediction of drug-drug interactions. Drug Metab Dispos 35(2):246–255
9. Chauret N, Gauthier A, Nicoll-Griffith DA (1998) Effect of common organic solvents on in vitro cytochrome P450-mediated metabolic activities in human liver microsomes. Drug Metab Dispos 26(1):1–4
10. Atkinson A, Kenny JR, Grime K (2005) Automated assessment of time-dependent inhibition of human cytochrome P450 enzymes using liquid chromatography-tandem mass spectrometry analysis. Drug Metab Dispos 33(11):1637–1647
11. Grime KH et al (2009) Mechanism-based inhibition of cytochrome P450 enzymes: an evaluation of early decision making in vitro approaches and drug-drug interaction prediction methods. Eur J Pharm Sci 36(2–3):175–191
12. Orr ST et al (2012) Mechanism-based inactivation (MBI) of cytochrome P450 enzymes: structure-activity relationships and discovery strategies to mitigate drug-drug interaction risks. J Med Chem 55(11):4896–4933

Chapter 19

CYP Time-Dependent Inhibition (TDI) Using an IC$_{50}$ Shift Assay with Stable Isotopic Labeled Substrate Probes to Facilitate Liquid Chromatography/Mass Spectrometry Analyses

Gary W. Caldwell and Zhengyin Yan

Abstract

In this chapter we have provided a step-by-step protocol for a time-dependent inhibition (TDI) IC$_{50}$ shift assay using stable isotopic labeled probe substrates. The assay is performed in a 96-well format and can be fully automated and extended to a 384-well format if desired. Since the IC$_{50}$ shift assay requires parallel paired incubations to obtain two inhibition curves for comparison, the use of stable isotopic labeled probe substrates and non-labeled probe substrates allows the two sets of incubation samples to be combined and then simultaneously analyzed by LC/MS/MS in the same batch run. Compared to the traditional method, this sample pooling approach in combination with a short LC/MS/MS sample analysis time significantly enhances the throughput of the TDI screening assay by reducing sample analysis time and thus can be easily implemented in drug discovery to evaluate a large number of compounds without adding additional resources.

Key words Metabolism, Reversible and irreversible kinetics, CYP450s, LC/MS/MS analyses, Time dependent inhibition, IC$_{50}$ shift

1 Introduction

Due to the common practice of poly-therapy drug regiments for treating diseases in patients, the systemic concentration of the individual drugs may be significantly altered as compared to a mono-therapy treatment. Depending on the magnitude of the change in concentration in the drugs at the site of pharmacological action and the therapeutic index of the drugs, the co-administered drugs may be lowered to non-efficacious levels or elevated to toxic levels. Thus, poly-therapy drug regiments, in many cases, lead to clinical drug-drug interactions (DDIs), which are particularly prevalent in elderly patients attending public primary health care systems and taking more than six prescribed drugs at the same time [1, 2].

These types of DDIs represent a significant percentage of all adverse drug reactions (ADRs) reported by the general population [3].

The mechanism of DDIs is frequently associated with the competition of drugs for the same active site of cytochrome P450 (CYP) enzymes by creating inhibitor effects [4]. Metabolism of xenobiotics (i.e., drugs) by CYPs via biotransformation reactions in the small intestine and the liver represents the major *in vivo* route for drug elimination in humans and animals. CYP1A2, CYP2B6, CYP2C8, CYP2C9, CYP2C19, CYP2D6, CYP3A4 and CYP3A5 are the primary CYP enzymes involved in catalyzed drug biotransformation reactions with CYP3A4 being the most important for marketed drugs [5]. A general feature of CYPs is that a single enzyme can metabolize multiple drugs. If several drugs are taken *in vivo* at the same time, a competition reaction between two drugs or more may occur for the same CYP. This situation may result in an undesirable elevation or lowering in the systemic concentration of one or all of the drugs. In other words, DDIs cause an inhibition of the CYP's normal basal function of eliminating drugs from the body. This CYP inhibitory effect can be produced by either or both reversible and irreversible CYP kinetics. Reversible CYP kinetics is referred to as reversible inhibition while irreversible kinetics is referred to as time-dependent inhibition (TDI). The consequences of reversible inhibition are considered to be less serious than irreversible inhibition. Reversible inhibition brought about by DDIs, as the name implies, is a situation where the enzymatic function of the CYPs are returned to basal activity once the drugs are eliminated from the body. Here the inhibition of the CYPs is related to the elimination half-life of the drugs. TDI is the opposite situation since the CYPs are inactivated by the formation of tight or covalent bonds between the drugs and the CYPs. If the TDI is caused by the reaction of drug intermediates formed during biotransformation, the inhibition is termed mechanism based inhibition (MBI). All irreversible inhibition processes are time-dependent and can persist even after the drug has been eliminated from the body because CYP activity is only restored to basal levels by de novo CYP enzyme synthesis which typically requires a few days in humans [6].

Over the last decade, many *in vivo* and *in vitro* CYP screening approaches for potential new drug candidates have been developed to predict and understand the mechanisms of CYP enzyme inhibition that leads to DDIs [6–39]. Since *in vivo* CYP inhibition processes are complicated, the main approach in pharmaceutical drug discovery and development groups has been to establish a series of *in vitro* CYP kinetic assays to measure the potential of new drug candidates to produce CYP reversible inhibition and TDI [6, 9–39]. Due to the large number of new drug candidates that are synthesized in drug discovery groups, drug candidate TDI screening are typically funneled through a series of tiered assays where successively more detailed and revealing mechanistic data is collected with each assay. In many pharmaceutical companies [16, 29], reversible CYP inhibition assays over a limit range of inhibitor concentrations [9–17] and CYP inactivation rate assays at a single inhibitor concentration [18, 39] are used as early TDI assays while CYP IC_{50} shift [28, 30, 33] and CYP kinetic inactivation parameters [29] are used to further evaluate the TDI mechanism. The CYP kinetic parameters measured in these inactivation rate assays are K_I (i.e., inhibitor concentration that support half the maximal rate of inactivation) and k_{inact} (i.e., the maximal rate of inactivation) [8].

Reversible CYP reactions are caused by the rapid association (k_1) and dissociation (k_{-1}) between a CYP enzyme [E] and a drug [S] to form enzyme complexes [ES] and [ES*] leading to the formation of a metabolite [P] (Scheme 1). It is termed reversible since the overall enzyme [E] concentration remains constant during the reaction. When another drug [I] is present in the reaction mixture a competitive interaction for the CYP enzyme binding site may occur causing a potential inhibition for either drug. If the dissociation rate of the enzyme [E] and drug [I] is slow, then this situation could lead to reversible inhibition of the drug [S] which would decrease the formation of metabolite [P] as compared to metabolite [M]. This reaction is illustrated in Scheme 1 by the size of the black arrows. The opposite situation could occur if the dissociation rate of the enzyme [E] and drug [S] is slow, then this situation could lead to reversible inhibition of the drug [I] which would decrease the formation of metabolite [M] metabolite as compared

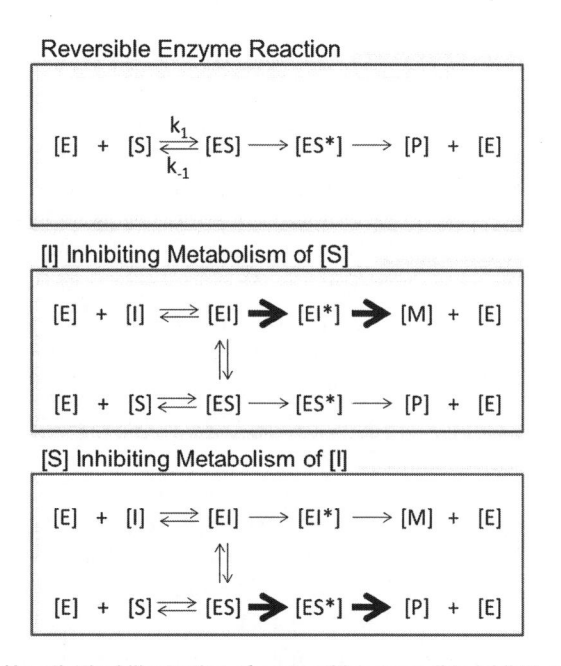

Reversible Enzyme Reaction

$$[E] + [S] \underset{k_{-1}}{\overset{k_1}{\rightleftharpoons}} [ES] \longrightarrow [ES^*] \longrightarrow [P] + [E]$$

[I] Inhibiting Metabolism of [S]

$$[E] + [I] \rightleftharpoons [EI] \Rightarrow [EI^*] \Rightarrow [M] + [E]$$
$$\updownarrow$$
$$[E] + [S] \rightleftharpoons [ES] \longrightarrow [ES^*] \longrightarrow [P] + [E]$$

[S] Inhibiting Metabolism of [I]

$$[E] + [I] \rightleftharpoons [EI] \longrightarrow [EI^*] \longrightarrow [M] + [E]$$
$$\updownarrow$$
$$[E] + [S] \rightleftharpoons [ES] \Rightarrow [ES^*] \Rightarrow [P] + [E]$$

Scheme 1 Hypothetical illustration of competitive reversible inhibition

to metabolite [P]. In this case the black arrows would be switched. Thus, we see that [I] could attenuate the physiological response of [S] via inhibition of [E] which is involved in its drug elimination via [P] or *visa-versa*. It should be clear that while [S] and [I] directly compete for the active site of [E] the enzymatic activity can be fully restored to basal activity since both [S] and [I] are depleted in time by being removed from the system via its biotransformation to metabolite [P] and [M], respectively.

There are many different types of *in vitro* CYP kinetic assays that have been designed to measure drug inhibition based upon Scheme 1 [9–17]. The most common way is to select CYP isoform-specific substrates [S] that have well defined kinetic parameters (V_{Max}, and K_m) and one major metabolite [P] that is structurally known. In this case, the CYP isoform-specific substrates [S] are incubated individually with human liver microsomes (HLM) [E] and a range of drug candidate concentrations [I]. For each of the drug candidate compound concentrations, the incubation is terminated and the concentration of metabolite [P] is measured by liquid chromatography tandem mass spectrometry (LC/MS/MS). A decrease in the concentration of the metabolite [P] compared to vehicle control is used to calculate an IC_{50} value for the test drug candidate compound [I]. The IC_{50} values are typically classified as:

- $IC_{50} < 1$ μM (Potent Reversible Inhibition)
- $IC_{50} < 10$ μM and >1 μM (Moderate Reversible Inhibition)
- $IC_{50} > 10$ μM (Weak Reversible Inhibition)

Irreversible CYP reactions are caused by the drug [I] covalently bonding to the enzyme [E-I*] or forming strong complexes [E·I*] (Scheme 2). It is termed irreversible since the overall enzyme concentration does not remain constants during the reaction. Note that in Scheme 2 that the reactive intermediate I* does not leave the complex [EI*] but is immediately transformed to inactive the enzyme. In this case, the reactive intermediate complexed to the enzyme [EI*] now reacts via three pathways. The reactive intermediate irreversibly bonds to the CYP apoprotein to form the covalent specie [E-I*]′, it irreversibly bonds to the CYP prosthetic heme group to form the covalent specie [E-I*]″ and finally, the reactive intermediate [EI*] forms a quasi-irreversible complexation [E·I*]‴. Quasi-irreversible inhibition is a tight binding complex between the drug [I] and the heme prosthetic group of the CYP. This complex is sometimes referred to as a metabolite intermediate complex (MIC). The formation of these types of complexes formed during biotransformation of [I] are referred to as MBI. There are several other types of reactions that can inactivate the enzyme that are not shown in Scheme 2 such as, autoinactivation caused by production of superoxide and/or hydrogen peroxide formed during the reduction CYP cycle, the metabolite [M] could form irreversible complexes

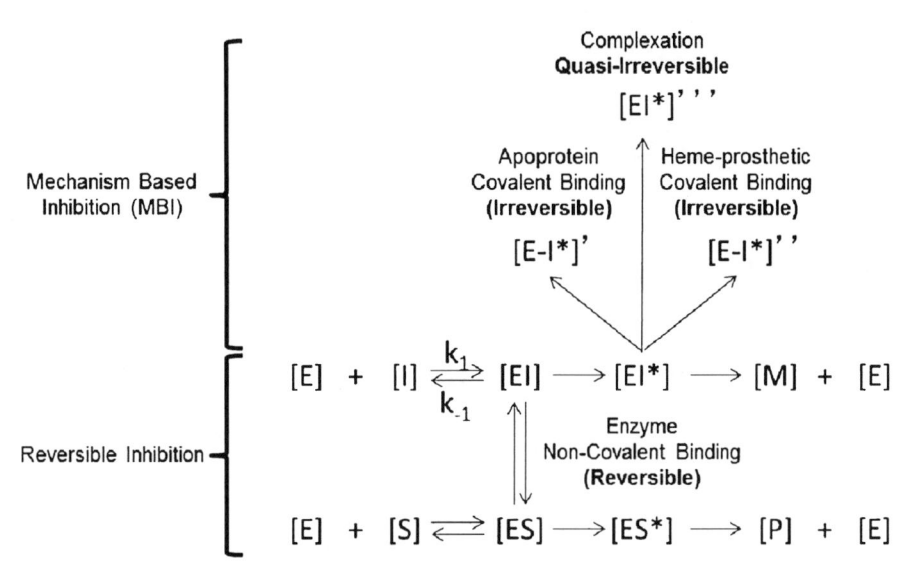

Scheme 2 Hypothetical illustration of drug-drug interactions (DDI)

with the enzyme, tight binding of [ES] or [EI], and non-specific binding of [S] or [I] (e.g., to microsomal membranes) [37]. All of these mechanisms result in the permanent or apparent loss of enzymatic activity. Unlike reversible inhibition, irreversible inactivation of CYP450s causes long-term effects on some drug pharmacokinetics, as the inactivated enzyme must be replaced by newly synthesized CYP protein.

There are several *in vitro* CYP kinetic assays used to predict TDI for potential new drug candidates [6, 14–39]. These *in vitro* TDI assays can be classified as co-incubation [14, 15] or as pre-incubation assays [16, 18, 28–30, 33, 39]. In this chapter, we will only discuss the pre-incubation IC_{50} shift assay [28, 30, 33] which measures the inhibitor concentration which results in 50 % inhibition of enzyme activity (Scheme 3). Thus, TDI is defined in the IC_{50} shift assay as an interaction where there is an enhanced inhibition if the test inhibitor [I] is pre-incubated with the metabolizing system prior to the addition of the probe substrate [S]. Typically, three assays are required to discriminate between drug candidates which cause reversible inhibition, TDI, or both reversible inhibition and TDI. These include the direct inhibition assay (i.e., pre-incubation 0 min), the TDI assay with non-labeled [S] (i.e., pre-incubation 30 min –NADPH) and the TDI assay with labeled [S*] (i.e., pre-incubation 30 min +NADPH).

For the direct inhibition assay, the incubation mixture contains human liver microsomes (HLM) [E], a probe substrate [S], a test drug candidate [I], and an NADPH-regenerating system (NADPH). The enzymatic reaction is initiated by the addition of NADPH and at a desired incubation time the enzymatic reaction is stopped. The concentration of [P] is measured typically using a

Direct Inhibition Assay

[Diagram: [E] [S]+[I] → Pre-incubation 0 minutes → + NADPH → [E] [S]+[I] NADPH (Enzymatic Reaction Starts) → Monitor enzyme activity 10-30 minutes → [P] (Enzymatic Reaction Stopped)]

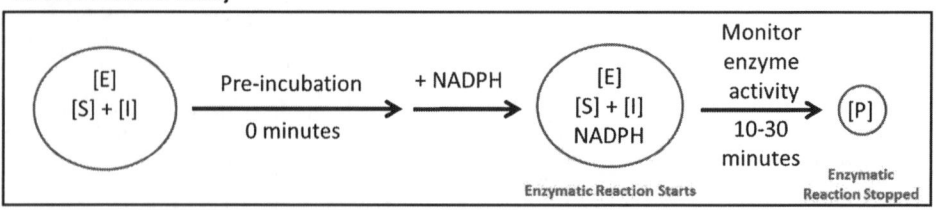

IC$_{50}$ Shift Assay: TDI with Non-Labeled [S] (-NADPH)

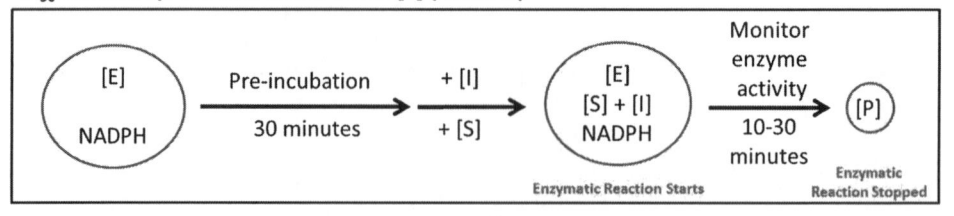

[Diagram: [E] NADPH → Pre-incubation 30 minutes → + [I] + [S] → [E] [S]+[I] NADPH (Enzymatic Reaction Starts) → Monitor enzyme activity 10-30 minutes → [P] (Enzymatic Reaction Stopped)]

IC$_{50}$ Shift Assay: TDI with Isotope-Labeled [S*] (+NADPH)

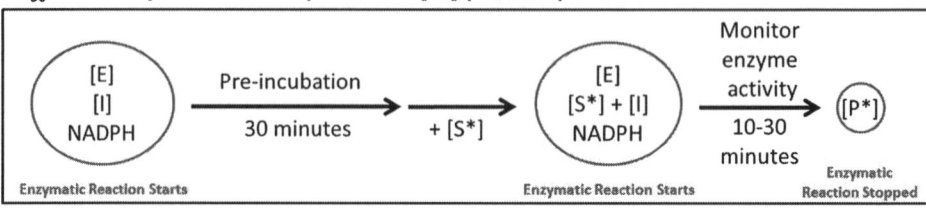

[Diagram: [E] [I] NADPH (Enzymatic Reaction Starts) → Pre-incubation 30 minutes → + [S*] → [E] [S*]+[I] NADPH (Enzymatic Reaction Starts) → Monitor enzyme activity 10-30 minutes → [P*] (Enzymatic Reaction Stopped)]

Scheme 3 Illustration of reversible and TDI IC$_{50}$ shift assays

LC/MS/MS method and an IC$_{50}$ value is calculated. In many cases, the direct inhibition assay is run prior to running the IC$_{50}$ shift assay such that the most appropriate concentration range for the IC$_{50}$ shift assay can be determined and utilized.

For the IC$_{50}$ shift assay with non-labeled substrate [S], the mixture containing HLM [E] and NADPH is incubated for 30 min. Since the test drug candidate [I] is not present in this mixture, this assay is sometimes referred to as the −NADPH assay. Following the pre-incubation, a probe substrate [S], and a test drug candidate [I] are added to the mixture without significant dilution of the reaction solution. The enzymatic reaction is initiated at this step. After a desired incubation time, the enzymatic reaction is stopped, the concentration of [P] is measured and an IC$_{50}$ value is calculated. The IC$_{50}$ shift assay with non-labeled substrate [S] is used as a control and should give the same IC$_{50}$ value as the direct inhibition assay. If shifts in the IC$_{50}$ values are observed, this may indicate potential non-NADPH mediated metabolisms. For the IC$_{50}$ shift assay with isotope-labeled substrate [S*], the mixture containing HLM [E], NADPH and a test drug candidate [I] is incubated for 30 min. Since the test drug candidate [I] is present in this mixture, this assay is sometimes referred to as the +NADPH assay. Following the pre-incubation, an isotope-labeled probe substrate [S*] is added to the mixture without significant dilution of the reaction solution. The second enzymatic reaction is initiated at this step. After a desired incubation time, the enzymatic reaction is stopped, the concentration of [P] (or [P*]) is measured and an IC_{50} value is calculated. If the test drug candidate [I] is only a reversible inhibitor, no increase in potency in the IC_{50} values will be observed between the −NADPH and +NADPH assays. If the test drug candidate [I] is a reversible inhibitor and a TDI or solely a TDI, an increase in potency in the IC_{50} values will be observed between the −NADPH and +NADPH assays. The ratio of these two values gives the IC_{50} shift.

- IC_{50} fold shift ≤1 (Reversible Inhibition)
- IC_{50} fold shift >1 and <2 (Moderate TDI or TDI and Reversible Inhibition)
- IC_{50} fold shift ≥2 (Significant TDI or TDI and Reversible Inhibition)

A 96-well format step-by-step protocol for a time-dependent inhibition (TDI) IC_{50} shift assay using stable isotopic labeled substrate probes is provided in this chapter. Since the IC_{50} shift assay requires parallel paired incubations to obtain two inhibition curves for comparison, the use of stable isotopic labeled substrate probes and non- labeled substrate probes allows the two sets of incubation samples be combined and then simultaneously analyzed by LC/MS/MS in the same batch run to reduce the run time. The assay can be fully automated or converted to a 384-well format [40]. This approach of combining sample pooling and short LC/MS/MS gradient significantly enhances the throughput of TDI screening and thus can be easily implemented in drug discovery to evaluate a large number of compounds without adding additional resources. This work is an updated version of an earlier pre-incubation TDI screening assay developed in our lab [41].

2 Materials

2.1 Equipment

Some specific equipment used in this assay is described; however, any model of comparable capability can be easily substituted.

1. Assay Plates: 96-deep well plates (1.2 mL/well) (Corning Incorporated, Corning, NY)
2. PlateLoc thermal microplate sealer (e.g. Model Velocity 11, Agilent Technologies, Santa Clara, CA)
3. Liquid handling and compound dilutions: BioTek Precision XL and CyBio liquid handler with a 96 channel dispenser system (Woburn, MA)
4. Precision water-bath capable of heating to 37 °C (e.g. Thermo Scientific, Tewksbury, MA)
5. Table top centrifuge capable of accommodating 96-well plates (e.g. Model 5810 R, Eppendorf, Hauppauge, NY)

6. pH meter (VWR, Bridgeport, NJ)
7. LC/MS/MS system, capable quantitative analysis: ABI Sciex 4000 QTRAP triple quadrupole mass spectrometer interfaced to a HPLC system (MDS, Toronto, Canada).
8. HPLC system: Shimadzu 20A coupled with a CTC LEAP autosampler interfaced to the electrospray apparatus of a triple quadrupole mass spectrometer (Canby, OH)
9. HPLC column: Princeton SPHER-100 C18 column (2.0×50 mm, 5 µm) was used for the chromatographic separations.

2.2 Chemical, Reagents, Buffers and Organic Solvents

1. (S)-Mephenytoin (451032), Furafylline (451037), and Tienilic acid (451000) (BD Biosciences Discovery Labware, Woburn, MA)
2. Dextromethorphan (D2531), Erythromycin (E0774), Itraconazole (I6657), Mifepristone (M8046), Nicardipine (N7510), Paroxetine (P9623), Phenacetin (77440), Propranolol (P8688), Testosterone (T1500), Tolbutamide (T0891), Ticlopidine (T6654), Terfenadine (T9652), Troglitazone (T2573), and Verapamil (V4629), (Sigma-Aldrich, St. Louis, MO)
3. Troleandomycin (BML-E1249-0050), (Enzo Life Sciences, Inc., Farmingdale, New York)
4. Tolbutamide-d9 (D6169) (C/D/N Isotopes, Quebec, Canada)
5. Phenacetin-d3 (SI-01359-001) (Synfine Research Inc., Richmond Hills, ON, Canada)
6. Mephenytoin-d3 (M225002) and Dextromethorphan-d3 (KIT0597) (Toronto Research Chemicals, Inc., North York, ON, Canada)
7. Testosterone-d2 (DLM683) (Cambridge Isotope Laboratories, Inc., Andover, MA)
8. NADPH regenerating solution (15265) (AAT Bioquest, Sunnyvale, CA) (*see* **Notes 1** and **2**)
9. Phosphate buffer 100 mM @ pH 7.4 (451201) (BD Biosciences, San Jose, CA: Cat# 451201)
10. Water (J.T. Baker-HPLC Solvent Grade; Cat# 4218–03)
11. Dimethyl sulfoxide (DMSO) and acetonitrile (ACN) (Sigma-Aldrich, St. Louis, MO)

2.3 Test Compound Stock Solutions

Prepare 0.5 mM (500 µL) by adding the components in the following order:

1. 475 µL ACN
2. 25 µL of a 10 mM test compound DMSO stock solutions (**Note 3**)
3. Solution contains 5 % DMSO and 95 % ACN

2.4 Microsomes

1. All human (20 mg/mL) microsomes were purchased from BD Biosciences Discovery Labware (Woburn, MA, U.S.A.). Human liver microsomes were prepared from livers with mixed gender pool of 20 donors that was fully characterized for CYP activity.

2.5 Stop Solution

Metabolic reactions were terminated by mixing enzymatic solution with approximately one-third volume of stop solution. To prepare 500 mL of acetonitrile (ACN) containing 1 μg/mL of propranolol (259.34 g/mol) as an internal standard, use the following procedure:

1. Dissolve 10 mg of propranolol in 10 mL ACN (1 mg/mL or 3.86 mM)
2. Add 500 μL of the 3.86 mM propranolol stock solution to 499.5 mL ACN

3 Methods

Three assays are required to discriminate between drug candidates that our potential reversible inhibitors, TDI, or both reversible inhibitors and TDI (see Scheme 3). These include the direct co-incubation inhibition assay and two pre-incubation TDI IC_{50} shift assays. The direct inhibition assay is run prior to running the TDI IC_{50} shift assays such that the correct concentration range is used for the TDI IC_{50} shift assay (Sect. 3.1). The TDI IC_{50} shift assays are described in Sect. 3.2.

3.1 The Main Steps in Direct Inhibition Protocol

This assay is an updated version of an earlier direct inhibition screening assay developed in our lab [20]. To make the protocol easier to understand, the current method is a manual version, but it can be easily modified to run the assay robotically using a liquid handler. Briefly, the incubation mixture contains a probe substrate, a test drug candidate, and HLM in 100 mM potassium buffer (pH 7.4). Reactions were initiated by the addition of an NADPH-regenerating system (NADPH) to the mixture and at a desired incubation time the enzymatic reaction was stopped. These conditions are listed in Table 1. The concentration of the metabolite of the probe substrate is measured using a LC/MS/MS method (Sect. 3.4) and an IC_{50} value is calculated (Sect. 3.5).

3.1.1 Probe Substrate Stock Solutions

1. Prepare 10 mL of a 50 mM phenacetin (179.22 g/mol) solution by adding 89.6 mg in 10 mL of ACN.
2. Prepare 10 mL of a 300 mM of tolbutamide (270.35 g/mol) solution by adding 811.0 mg in 10 mL of ACN.
3. Prepare 10 mL of a 40 mM of (S)-mephenytoin (218.25 g/mol) solution by adding 84.3 mg in 10 mL of ACN.

Table 1
Microsomal incubation conditions [41]

Enzymes	Probe substrates [S] (molecular weight)	HLM conc. (mg/mL)	Probe substrates [S] conc. (µM)	Incubation time (min)
CYP1A2	Phenacetin (179.22 g/mol)	0.20	80	15
	Phenacetin-d3 (182.23 g/mol)	0.20	80	15
CYP2C9	Tolbutamide (270.35 g/mol)	0.15	100	10
	Tolbutamide-d9 (279.39 g/mol)	0.15	100	10
CYP2C19	(S)-Mephenytoin (218.25 g/mol)	0.20	30	30
	Mephenytoin-d3 (221.26 g/mol)	0.20	60	30
CYP2D6	Dextromethorphan (271.40 g/mol)	0.20	3	15
	Dextromethorphan-d3 (274.41 g/mol)	0.20	3	15
CYP3A4	Testosterone (288.42 g/mol)	0.15	25	10
	Testosterone-d2 (290.43 g/mol)	0.15	25	10

4. Prepare 10 mL of a 20 mM of dextromethorphan (271.40 g/mol) solution by adding 54.3 mg in 10 mL of ACN.

5. Prepare 10 mL of a 20 mM of testosterone (288.42 g/mol) solution by adding 57.7 mg in 10 mL of ACN.

3.1.2 Preparation of Microsome-Substrate Mixtures

Microsome (0.2 mg/mL) Substrate Mixtures for Phenacetin, (S)-Mephenytoin and Dextromethorphan

1. To make 13 mL of a 0.4 mg/mL microsome solution containing various substrate concentrations of phenacetin, (S)-mephenytoin and dextromethorphan, add 2.6 mL of 100 mM potassium phosphate (pH 7.4) to three 15 mL tubes (i.e., one tube for each substrate).

2. Add 260 µL of HLM (20 mg/mL).

3. Add an appropriate amount of probe substrate stock (Sect. 3.1.1) to generate substrate concentrations at twice the concentrations listed in Table 1 (*see* **Note 4**).

4. Add enough deionized water to bring to the final volume (13 mL).

5. Invert the tube repeatedly to mix and keep on ice.

6. Repeat the procedure for all probe substrates.

Microsome (0.15 mg/mL) Substrate Mixtures for Tolbutamide and Testosterone

1. To make 13 mL of a 0.3 mg/mL microsome solution containing various substrate concentrations of tolbutamide and testosterone, add 2.6 mL of 100 mM potassium phosphate (pH 7.4) to two 15 mL tubes (i.e., one tube for each substrate).

2. Add 195 µL of HLM (20 mg/mL).

3. Add an appropriate amount of probe substrate stock (Sect. 3.1.1) to generate substrate concentrations at twice the concentrations listed in Table 1 (*see* **Note 4**).

4. Add enough deionized water to bring to the final volume (13 mL).
5. Invert the tube repeatedly to mix and keep on ice.
6. Repeat the procedure for all probe substrates.

3.1.3 Preparation of Test Compounds Dilution Plate

1. Dilute 0.5 mM test compound stock solutions (Sect. 2.3) to an 80 µM stock solution (6.25-fold dilution) by adding 240 µL of the 0.5 mM stock solution to 1.5 mL of ACN.
2. In a 96-well plate, add 750 µL of 100 mM potassium phosphate (pH 7.4) to each well.
3. Transfer 250 µL of the 80 µM test compound stock solution to the first well. The concentration of the test compound is 20 µM.
4. Transfer 250 µL of the 20 µM test compound stock solution in the first well to the second well. The concentration of the test compound is 5 µM.
5. Repeat this fourfold dilution three more times to create five concentrations (i.e., 20, 5, 1.25, 0.31, and 0.078 µM). Solution contains 0.2 % DMSO and 3.8 % ACN.
6. In the sixth well, transfer 250 µL of ACN which can be used as a control.
7. Repeat this procedure depending on the number of replicates desired. That is, a single 96-well plate will contain a maximum of 16 different compounds at six concentrations, eight different compounds for duplicate studies, and four different compounds for studies done it triplicate.

3.1.4 Microsomal Incubation Procedure

1. In five 96-well plates, dispense 125 µL of the HLM-substrate mixtures to each well using a multichannel pipet or liquid handler (*see* **Note 5**). Use the HLM-substrate mixture in Sect. 3.1.2.1 for probe substrates phenacetin, (S)-mephenytoin and dextromethorphan and the HLM-substrate mixture in Sect. 3.1.2.2 for tolbutamide and testosterone.
2. Transfer 2.5 µL of test compounds in the dilution plate (Sect. 3.1.3) to the microsomal incubation plate. Solution contains ≤0.1 % of DMSO and ACN.
3. Pre-warm the plate at 37 °C for 5 min.
4. Add 122.5 µL of the NADPH generating solution (*see* **Note 6**).
5. Incubate 96-well plate at 37 °C for the specific substrate incubation time as indicated in Table 1 (*see* **Note 7**).
6. Add 80 µL of ice-cold stop solution to all well to stop the reaction (Sect. 2.5).
7. Repeat the procedure for the remaining four substrates (Sect. 3.1.1).

8. Centrifuge the five 96-well plates at 4 °C for 30 min at 5,000 ×g to pellet down proteins.

9. Pool supernatant for all five substrates corresponding to each concentration of the test compound in a 96-deep well plate. In duplicate, transfer 200 μL of this pool substrate mixture to 96-well plates.

10. Evaporate solvents off and seal plate for LC/MS/MS analysis (Sect. 3.4).

3.2 The Main Steps in TDI IC50 Shift Protocol

The TDI IC$_{50}$ shift assay with non-labeled substrate [S] in which there is no test compound in the pre-incubation step is used as a control (Scheme 4). This assay is similar to the direct inhibition assay (Sect. 3.1) except that a mixture containing only HLM and NADPH is incubated first for 30 min. The control IC$_{50}$ shift assay with non-labeled substrate [S] should give the same IC$_{50}$ value as the direct inhibition assay (*see* **Note 8**). Following this pre-incubation step, a test drug compound and an non-labeled probe substrate containing NAPDH are added to the mixture without significant dilution of the reaction solution. The second incubation is initiated at this step. After a desired incubation time, the enzymatic reaction is stopped and centrifuged. In parallel with the

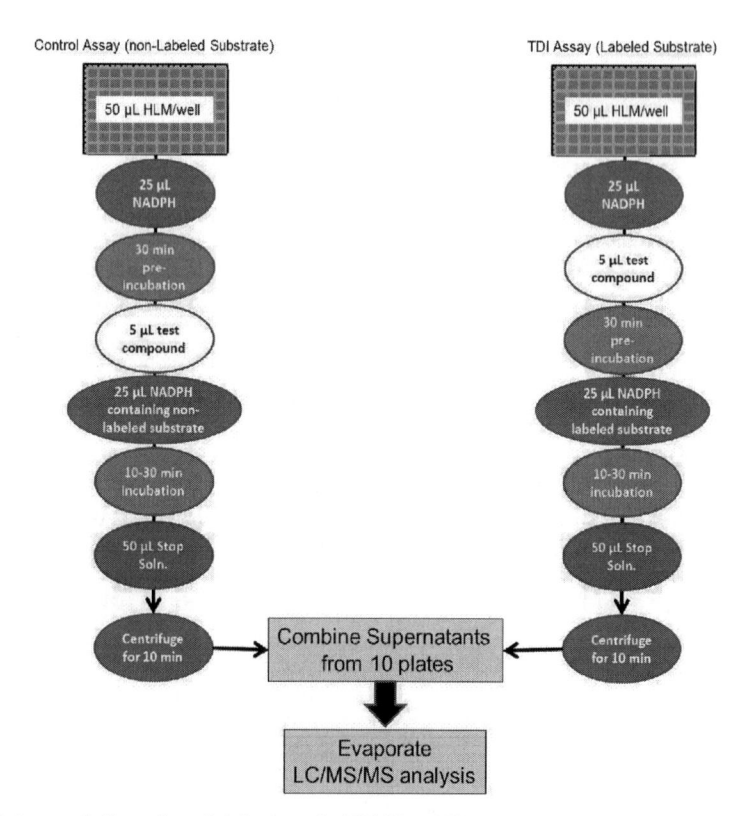

Scheme 4 Experimental designs for TDI IC$_{50}$ shift assay

control experiment, the TDI IC$_{50}$ shift assay in which there is test compound in the pre-incubation step is used to determine the present of TDI. Following this pre-incubation step, a test drug compound and a deuterium labeled probe substrate containing NAPDH are added to the mixture without significant dilution of the reaction solution. After a desired incubation time, this enzymatic reaction is stopped and centrifuged. The supernatants of the two assays are combined, the concentration of [P] is measured using a LC/MS/MS method (Sect. 3.3) and an IC$_{50}$ value is calculated (Sect. 3.5).

3.2.1 Preparation of Microsome Solution

Microsome (0.2 mg/mL) Solution

1. To make 13 mL of a 0.42 mg/mL microsome solution, add 2.6 mL of 100 mM potassium phosphate (pH 7.4) to a 15 mL tubes.
2. Add 273 μL of HLM (20 mg/mL).
3. Add enough deionized water to bring to the final volume (13 mL).
4. Invert the tube repeatedly to mix and keep on ice.

Microsome (0.15 mg/mL) Solution

1. To make 13 mL of a 0.315 mg/mL microsome solution, add 2.6 mL of 100 mM potassium phosphate (pH 7.4) to a 15 mL tubes.
2. Add 204.8 μL of HLM (20 mg/mL).
3. Add enough deionized water to bring to the final volume (13 mL).
4. Invert the tube repeatedly to mix and keep on ice.

3.2.2 Microsomal Incubation Procedure: Control Assay (Non-Labeled Substrates)

1. In five 96-well plates, dispense 50 μL of the HLM mixture 3.2.1.1 for the probe substrates phenacetin, (S)-mephenytoin and dextromethorphan and 50 μL of the HLM mixture 3.2.1.2 for the probe substrates tolbutamide and testosterone to each well using a multichannel pipet or liquid handler (*see* **Note 5**).
2. Add 25 μL of the NADPH generating solution (*see* **Note 9**).
3. Pre-incubate the plate at 37 °C for 30 min.
4. Add 5 μL of the diluted test compounds (*see* **Note 10**)
5. Add 25 μL of NADPH containing specific substrate concentrations as indicated in Table 1 (*see* **Note 11**).
6. Incubate 96-well plate at 37 °C for the specific substrate incubation time as indicated in Table 1 (*see* **Note 7**).
7. Add 50 μL of ice-cold stop solution to all well to stop the reaction (Sect. 2.5).
8. Centrifuge the five 96-well plates at 4 °C for 30 min at 5,000 g to pellet down proteins.
9. Pool supernatants from five 96-well plates with supernatant from Sect. 3.2.3 **Microsomal Incubation Procedure.**

3.2.3 Microsomal Incubation Procedure: TDI Assay (Deuterium-Labeled Substrates)

1. In five 96-well plates, dispense 50 µL of the HLM mixture 3.2.1.1 for the probe substrates phenacetin, (S)-mephenytoin and dextromethorphan and 50 µL of the HLM mixture 3.2.1.2 for the probe substrates tolbutamide and testosterone to each well using a multichannel pipet or liquid handler (*see* **Note 5**).
2. Add 25 µL of the NADPH generating solution (*see* **Note 9**).
3. Add 5 µL of the diluted test compounds (*see* **Note 10**)
4. Pre-incubate the plate at 37 °C for 30 min.
5. Add 25 µL of NADPH containing specific deuterium labeled substrate concentrations as indicated in Table 1 (*see* **Note 11**).
6. Incubate 96-well plate at 37 °C for the specific substrate incubation time as indicated in Table 1 (*see* **Note 7**).
7. Add 50 µL of ice-cold stop solution to all well to stop the reaction (Sect. 2.5).
8. Centrifuge the five 96-well plates at 4 °C for 30 min at 5,000 g to pellet down proteins.
9. Pool supernatant for all ten substrates corresponding to each concentration of the test compound and the corresponding control in a 96-deep well plate. For sample pooling, 90 µL CYP2C9 incubation samples were combined with 25 µL of incubation samples of the other CYP incubations. In duplicate, transfer 200 µL of this pool non-labeled/deuterium labeled substrate mixture to 96-well plates.
10. Evaporate solvents off and seal plates for LC/MS/MS analysis (Sect. 3.4).

3.3 LC/MS/MS Protocol

A LC/MS/MS assay was designed to elute and simultaneous detect the ten metabolites derived from both non-labeled and deuterium-labeled probe substrates in a single injection. Basically, multiple reaction monitoring (MRM) channels of deuterium-labeled metabolites (Table 2) were used to scan non-labeled metabolites in samples generated from HLM incubations with non-labeled probes whereas MRM transitions of non-labeled metabolites were utilized to monitor deuterium labeled metabolites in incubation samples generated with deuterium labeled probes.

The entire run time was 2.5 min per sample which corresponded to 15 min per test compound. The LC/MS/MS was operated in the ESI+ mode using the following conditions: ion spray voltage 5,500 V, turbo gas temperature 450 °C, entrance potential 10 V, nebulizing gas 30, and turbo gas 30. A dwell time of 50 and 8 ms inter-channel delay were set for each analyte. The MS analytical parameters for each analyte are listed in Table 2. Aliquots of 30 µL were injected onto a Princeton SPHER-100 C18 column with a mobile phase of 1 % acetic acid in water and acetonitrile at a flow rate of 0.4 mL/min. The metabolites were

Table 2
LC/MS analysis parameters [41]

Enzymes	Probe substrates [S]	Probe metabolite products [P]	MRM transition (Da)	CE/DP (eV)
CYP1A2	Phenacetin	Acetamidophenol	152 → 110	28/65
	Phenacetin-d3	Acetamidophenol-d3	155 → 110	28/65
CYP2C9	Tolbutamide	4-Hydroxytolbutamide	287 → 107	38/65
	Tolbutamide-d9	4-Hydroxytolbutamide-d9	296 → 107	38/65
CYP2C19	(S)-Mephenytoin	4′-Hydroxymephenytoin	235 → 150	28/70
	Mephenytoin-d3	4′-Hydroxymephenytoin-d3	238 → 150	28/70
CYP2D6	Dextromethorphan	Dextrorphan	258 → 133	38/65
	Dextromethorphan-d3	Dextrorphan-d3	261 → 133	38/65
CYP3A4	Testosterone	6β-Hydroxytestosterone	305 → 269	24/55
	Testosterone-d2	6β-Hydroxytestosterone-d2	307 → 271	24/55

eluted using a single gradient from 95 % aqueous to 80 % acetonitrile over 1.8 min, and then the column was flushed with 95 % acetonitrile for 12 s before re-equilibration at the initial condition. During the run, the divert valve was activated to direct the HPLC eluant to the waste line for the first 30 s of elution and then switched to the mass spectrometer for analysis.

3.4 Calculation of IC$_{50}$s

The raw data obtained from the LC/MS/S assays were processed by Analyst 1.4.2 (ABI Sciex) to obtain peak areas that were normalized relative to the propranolol internal standard. Inhibition percentages were calculated in relative to the control incubation in which test compound was absent. IC$_{50}$ values of inhibitions were calculated by linear extrapolation, using the equation shown below (Technical Manual, Gentest Corp.):

$$IC50 = \frac{(50\% - \text{low percentage}) * (\text{high conc. low conc.})}{(\text{high percentage low percentage})} + (\text{low conc.})$$

Where

- *low percentage = highest percent inhibition less than 50 %;*
- *high percentage = lowest percent inhibition greater than 50 %;*
- *low concentration = concentration of test compound corresponding to the low percentage inhibition;*
- *high concentration = concentration of test compound corresponding to the high percentage inhibition.*

Plotting control %inhibition (−NADPH) and %TDI (+NADPH) data, the time dependent behaviors of drugs were readily recognized.

Table 3
IC$_{50}$ shift assay results for selected test compounds [41]

CYP	Test compound	IC$_{50}$ (µM) (−NADPH)	IC$_{50}$ (µM) (+NADPH)	Ratio[a] (−/+)	Mechanism
CYP1A2	Furafylline	4.4	0.70	6.29	TDI
CYP2C9	Tilenic acid	1.34	0.15	8.93	TDI
CYP2C19	Ticlopidine	1.75	0.72	2.43	TDI
CYP2D6	Paroxetine	1.04	0.18	5.78	TDI
3A4	Itraconazole	0.005	0.006	0.83	Non-TDI
3A4	Nicardipine	0.11	0.19	0.58	Non-TDI
3A4	Terfenadine	1.67	1.64	1.02	Non-TDI
3A4	Troglitazone	2.62	0.76	3.45	TDI
3A4	Erythromycin	28.0	5.2	5.38	TDI
3A4	Mifepristone	>1	0.52	>1.92	TDI
3A4	Troleandomycin	>10	0.19	>52.63	TDI
3A4	Verapamil	26.67	2.16	12.35	TDI

[a]This ratio is calculated using IC$_{50}$ values; that is, IC$_{50}$ (−NADPH)/IC$_{50}$ (+NADPH) with a greater than twofold criteria for defining TDI

3.5 Time Dependent Behaviors of Drugs

A number of test compound inhibitors specific to individual CYP isoforms were selected to illustrate the assay (Table 3). For example, terfenadine a non-TDI test compound was selected and, as expected, did not exhibit significant shifting in its inhibition curves obtained with testosterone-d2 (with NADPH in pre-incubation) and testosterone (without NADPH in pre-incubation) (Fig. 1). The solid control curve (non-labeled probe substrate assay) had an IC$_{50}$ = 1.67 µM while the dashed curve (deuterium labeled probe substrate assay) had an IC$_{50}$ = 1.64 µM. The IC$_{50}$ ratio was approximately 1; thus, indicating that terfenadine was not a TDI of CYP3A4.

For potent TDI inhibitors such as furafylline, the curve shifting is more pronounced and their IC$_{50}$ values changed more dramatically (Fig. 2). The solid control curve (non-labeled probe substrate assay) had an IC$_{50}$ = 4.4 µM while the dashed curve (deuterium labeled probe substrate assay) had an IC$_{50}$ = 0.70 µM. The IC$_{50}$ ratio was 6.3; thus, indicating that furafylline was a potent TDI of CYP1A2. In Table 3, are listed several other test compound inhibitors specific to individual CYP isoforms. The results demonstrated that this method is reliable to differentiate TDI inhibitors from reversible inhibitors.

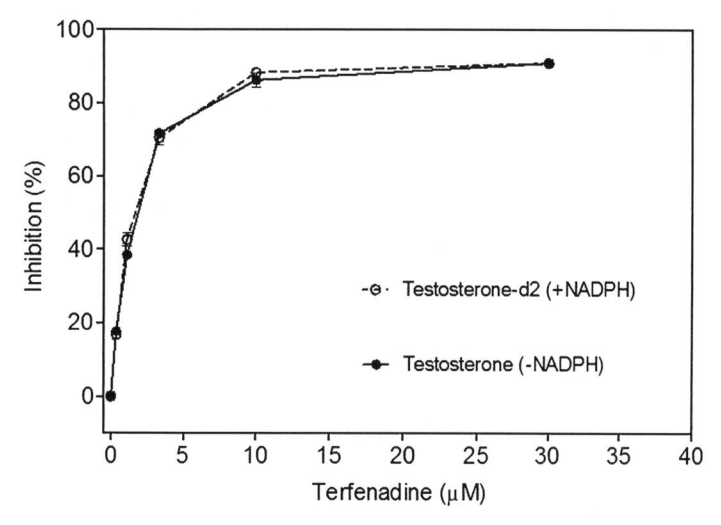

Fig. 1 Inhibition curve shift of terfenadine a known reversible inhibitor of CYP3A4. CYP activity is normalized to vehicle control incubations

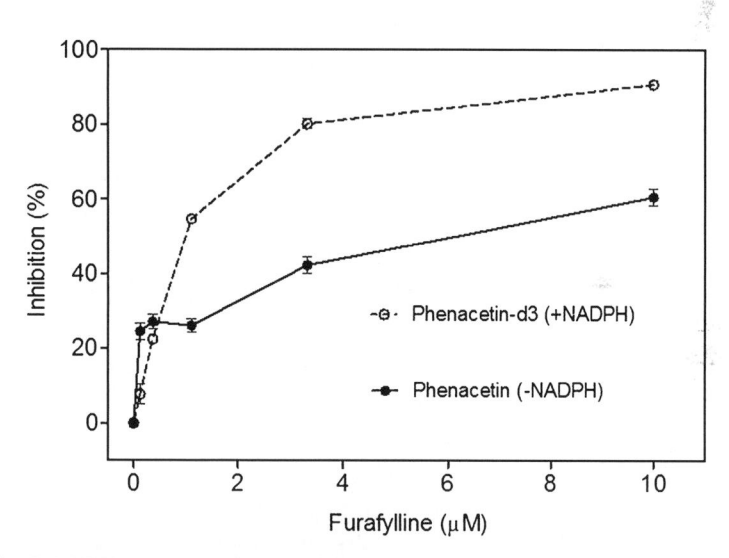

Fig. 2 Inhibition curve shift of furafylline a known TDI of CYP1A2. CYP activity is normalized to vehicle control incubations

It should be noted that the IC_{50} shift assay has limitations. The assay described here can only indicate the present of TDI for a test compound. It cannot elucidate if the mechanism of TDI is caused by MBI, a strong reversible inhibitory or through alternative mechanisms. To get a better understanding of the TDI, its mechanism, and its possible clinical consequences, a variety of other assays are required [29].

4 Notes

1. AAT Bioquest's RediUse™ NADPH Regenerating Kit provides three ready-to-use solutions including Component A Buffer I (25 mL), Component B Buffer II (25 mL), and Component C containing 400 units/mL of 500× Glucose-6-phosphate dehydrogenase (100 μL) to regenerate NADPH by a simple mixing these solutions together. About 300–500 enzyme assays can be performed using this kit depending on the experimental design. To make a 2× NADPH Regenerating Solution, thaw Component A, Component B, and Component C at room temperature then add the whole content of Component B and Component C into Component A and mix well. Solution can be aliquoted and stored at ≤ -20 °C.

2. NADPH is a required cofactor to activate CYPs within microsomes. The NADPH supplies electrons to the CYP enzyme reaction through NADPH-CYP reductase where NADPH is generated from NADP$^+$ with the use of glucose-6-phosphate dehydrogenase. While the readymade NADPH generating solution is convenient, it can be prepared in the following manner if needed. To prepare a 4 mL solution containing 1.3 mM NADP$^+$, 3.3 mM G-6-P-Na$_2$, 3.3 mM MgCl$_2$ and 0.4 U/mL of glucose-6-phosphate dehydrogenase, the following procedure can be used. This solution provides the necessary co-factors to catalyze a CYP450 enzyme reaction and should be used immediately. To prepare this solution:

   (a) Dissolve 39.8 mg of NADP$^+$ in 1 mL of 100 mM potassium phosphate buffer using β-nicotinamide adeninedinucleotide phosphate sodium salt hydrate (0.3 H$_2$O/mol) (NADP$^+$; Sigma: Cat# N0505). This compound has an anhydrous FW 765.39 g/mol.

   (b) Dissolve 37.2 mg of G-6-P-Na$_2$ in 1 mL of 100 mM potassium phosphate buffer using d-glucose 6-phosphate disodium salt hydrate (G-6-P-Na$_2$; Sigma: Cat# G7250). This compound C$_6$H$_{11}$Na$_2$O$_9$P·x H$_2$O is hydrated 2–4 H$_2$O/mol with an anhydrous FW 304.10 g/mol. Assuming 4 H$_2$O/mol, use the batch MW 376.10 g/mol.

   (c) Dissolve 26.8 mg of MgCl$_2$·6H$_2$O in 1 mL of water using magnesium chloride (MgCl$_2$·6H$_2$O; Sigma: Cat# M2670). Batch MW 203.30 g/mol.

   (d) Dissolve 11.5 U of glucose-6-phosphate Dehydrogenase 1KU (G6PDH; 3,000 U/mL) into 1 mL of 100 mM potassium phosphate buffer. Glucose-6-phosphate dehydrogenase 1KU (G6PDH; Sigma: Cat# G8404) was used where the protein was suspended in 3.2 M (NH$_4$)$_2$SO$_4$ containing 50 mM Tris and 1 mM MgCl$_2$ at pH 7.5.

Table 4
Probe substrate volumes

Enzymes	Probe substrates [S] (molecular weight)	Probe substrates stock conc. (mM)	Probe substrates stock volume (μL)	Probe substrates conc. (μM)
CYP1A2	Phenacetin (179.22 g/mol)	50	41.6	160
CYP2C9	Tolbutamide (270.35 g/mol)	300	8.7	200
CYP2C19	(S)-Mephenytoin (218.25 g/mol)	40	19.5	60
CYP2D6	Dextromethorphan (271.40 g/mol)	20	3.9	6
CYP3A4	Testosterone (288.42 g/mol)	20	32.5	50

Use the information on the bottle to calculate Units/mL. For our case, the number was 3,000 U/mL.

(e) Combine the four 1 mL solutions.

3. It should be noted that dissolved test compounds from library collects have inaccurate concentrations. From our own experience, library stock solution of 10 mM in DMSO were measured and discovered to be closer to 6–7 mM in many cases. DMSO solvent content in the incubations should be ≤0.1 % in order to avoid solvent inhibition of CYPs.

4. The probe substrate volumes found in Table 4 can be used.

5. The assay can be fully automated using a liquid handler such as a TECAN [21, 28].

6. This NADPH generating solution contains 2.6 mM NADP$^+$, 6.7 mM G-6-P-Na$_2$, 0.8 U/mL G6PDH, 6.6 mM MgCl$_2$.

7. It should be remembered that some enzymatic reactions are markedly impaired by even small changes in the pH and the ionic strength of the solution. Therefore, in most *in vitro* enzymatic assays, it is necessary to add a buffer to the medium to stabilize the pH and inorganic salts to stabilize the ionic strength of the solution. If these conditions are changed during the experiment, the results may also be altered. The buffer conditions used in our assay do have effects on enzyme activity. Other factors to keep in mind are that the pH may also vary with temperature and the water in which the buffer substances are dissolved should be of the highest quality.

8. If the IC$_{50}$ values between the direct assay and this assay are different, this may indicate potential non-NADPH mediated

metabolisms. For examples of these non-NADPH reactions, see reference [12].

9. This NADPH generating solution contains 5.46 mM NADP$^+$, 13.86 mM G-6-P-Na$_2$, 1.68 U/mL G6PDH, 13.86 mM MgCl$_2$.

10. The concentration range that derives a TDI concentration-response curve should be somewhere between 30 and 0.01 µM in most cases. Using the direct inhibition assay results, the concentration can be correctly chosen to improve the accuracy of the experiment. It should be noted that test compound concentrations can be limited by compound solubility.

Acknowledgements

We thank Dr. Weimin Tang and Malini Dasgupta for expert advice regarding assay method development.

The views expressed here are solely those of the author and do not reflect the opinions of Janssen Research & Development, LLC.

References

1. Obreli Neto PR, Nobili A, Marusic S, Pilger D, Guidoni CM, Baldoni Ade O, Cruciol-Souza JM, Da Cruz AN, Gaeti WP, Cuman RK (2012) Prevalence and predictors of potential drug-drug interactions in the elderly: a cross-sectional study in the Brazilian primary public health system. J Pharm Pharm Sci 15(2):344–354
2. Lin C-F, Wang C-Y, Bai C-H (2011) Polypharmacy, aging and potential drug-drug interactions in outpatients in Taiwan: a retrospective computerized screening study. Drugs Aging 28(3):219–225
3. FDA, Adverse event reporting system. http://www.fda.gov/Safety/MedWatch/HowToReport/ucm053074.htm. Accessed May 2013
4. Doan J, Zakrzewski-Jakubiak H, Roy J, Turgeon J, Tannenbaum C (2013) Prevalence and risk of potential cytochrome P450-mediated drug-drug interactions in older hospitalized patients with polypharmacy. Ann Pharmacother 47(3):324–332
5. McGraw J, Waller D (2012) Cytochrome P450 variations in different ethnic populations. Expert Opin Drug Metab Toxicol 8(3):371–382
6. Imai H, Kotegawa T, Ohashi K (2011) Duration of drug interactions: putative time courses after mechanism-based inhibition or induction of CYPs. Expert Rev Clin Pharmacol 4(4):409–411
7. Tang C, Prueksaritanont T (2010) Use of in vivo animal models to assess pharmacokinetic drug-drug interactions. Pharm Res 27(9):1772–1787
8. Yan Z, Caldwell GW (2012) The current status of time dependent CYP inhibition assay and in silico drug-drug interaction predictions. Curr Top Med Chem 12(11):1291–1297
9. Crespi CL, Miller VP, Penman BW (1997) Microtiter plate assays for inhibition of human, drug-metabolizing cytochromes P450. Anal Biochem 248(1):188–190
10. Moody GC, Griffin SJ, Mather AN, McGinnity DF, Riley RJ (1998) Fully automated analysis of activities catalysed by the major human liver cytochrome P450 (CYP) enzymes: assessment of human CYP inhibition potential. Xenobiotica 29(1):53–75
11. Lin JH, Lu AYH (1998) Inhibition and induction of cytochrome P450 and the clinical implications. Clin Pharmacokinet 5(5):361–390
12. Yan Z, Caldwell GW (2001) Metabolism profiling, and cytochrome P450 inhibition & induction in drug discovery. Curr Top Med Chem 1(5):403–425
13. Crespi CL, Stresser DM (2001) Fluorometric screening for metabolism-based drug-drug

14. Yan Z, Rafferty B, Caldwell GW, Masucci JA (2002) Rapidly distinguishing reversible and irreversible CYP450 inhibitors by using fluorometric kinetic analyses. Eur J Drug Metab Pharmacokinet 27(4):281–287
15. Yamamoto T, Suzuki A, Kohno Y (2002) Application of microtiter plate assay to evaluate inhibitory effects of various compounds on nine cytochrome P450 isoforms and to estimate their inhibition patterns. Drug Metab Pharmacokinet 17(5):437–448
16. Bjornsson TD, Callaghan JT, Einolf HJ, Fischer V, Gan L, Grimm S, John Kao S, King P, Miwa G, Ni L, Kumar G, McLeod J, Obach SR, Roberts S, Roe A, Shah A, Snikeris F, Sullivan JT, Tweedie D, Vega JM, Walsh J, Wrighton SA (2003) The conduct of in vitro and in vivo drug-drug interaction studies: a pharmaceutical research and manufacturers of America (PhRMA). J Clin Pharmacol 43(5):443–469
17. Stresser DM (2004) High-through screening of human cytochrome P450 inhibitors using fluorometric substrates: methodology for 25 enzyme/substrate pairs. In: Yan Z, Caldwell GW (eds) Optimization in drug discovery: in vitro methods. Method in pharmacology and toxicology. Humana Press Inc, Totowa, NJ
18. Yamamoto T, Suzuki A, Kohno Y (2004) High-throughput screening for the assessment of time-dependent inhibitions of new drug candidates on recombinant CYP2D6 and CYP3A4 using a single concentration method. Xenobiotica 34(1):87–101
19. Naritomi Y, Teramura Y, Terashita S, Kagayama A (2004) Utility of microtiter plate assays for human cytochrome P450 inhibition studies in drug discovery: application of simple method for detecting quasi-irreversible and irreversible inhibitors. Drug Metab Pharmacokinet 19(1):55–61
20. Yan Z, Caldwell GW (2004) Evaluation of cytochrome P450 inhibition in human liver microsome. In: Yan Z, Caldwell GW (eds) Optimization in drug discovery: in vitro methods. Method in pharmacology and toxicology. Humana Press Inc, Totowa, NJ
21. Atkinson A, Kenny JR, Grime K (2005) Automated assessment of time-dependent inhibition of human cytochrome P450 enzymes using liquid chromatography-tandem mass spectrometry analysis. Drug Metab Dispos 33(11):1637–1647
22. Turpeinen M, Korhonen LE, Tolonen A, Uusitalo J, Juvonen R, Raunio H, Pelkonen O (2006) Cytochrome P450 (CYP) inhibition screening: comparison of three tests. Eur J Pharm Sci 29(2):130–138
23. Venkatakrishnan K, Obach RS, Rostami-Hodjegan A (2007) Mechanism-based inactivation of human cytochrome P450 enzymes: strategies for diagnosis and drug-drug interaction risk assessment. Xenobiotica 37(10/11):1225–1256
24. Watanabe A, Nakamura K, Okudaira N, Okazaki O, Sudo K (2007) Risk assessment for drug-drug interaction caused by metabolism-based inhibition of CYP3A using automated in vitro assay systems and its application in the early drug discovery process. Drug Metab Dispos 35(7):1232–1238
25. Riley RJ, Grime K, Weaver R (2007) Time-dependent CYP inhibition. Expert Opin Drug Metab Toxicol 3(1):51–66
26. Polasek TM, Miners JO (2007) In vitro approaches to investigate mechanism-based inactivation of CYP enzymes. Expert Opin Drug Metab Toxicol 3(1):321–329
27. Mano Y, Usui T, Kamimura H (2007) Comparison of inhibition potentials of drugs against zidovudine glucuronidation in rat hepatocytes and liver microsomes. Drug Metab Dispos 35(4):602–606
28. Berry LM, Zhao Z (2008) An examination of IC50 and IC50-shift experiments in assessing time-dependent inhibition of CYP3A4, CYP2D6 and CYP2C9 in human liver microsomes. Drug Metab Lett 2(1):51–59
29. Grimm SW, Einolf HJ, Hall SD, He K, Lim H-K, Ling K-HJ, Lu C, Nomeir AA, Seibert E, Skordos KW et al (2009) The conduct of in vitro studies to address time-dependent inhibition of drug-metabolizing enzymes: a perspective of the pharmaceutical research and manufacturers of America. Drug Metab Dispos 37(7):1355–1370
30. Krippendorff B-F, Neuhaus R, Lienau P, Reichel A, Huisinga W (2009) Mechanism-based inhibition: deriving KI and kinact directly from time-dependent IC50 values. J Biomol Screen 14(8):913–923
31. Sekiguchi N, Higashida A, Kato M, Nabuchi Y, Mitsui T, Takanashi K, Aso Y, Ishigai M (2009) Prediction of drug-drug interactions based on time-dependent inhibition from high throughput screening of cytochrome P450 3A4 inhibition. Drug Metab Pharmacokinet 24(6):500–510
32. Kajbaf M, Palmieri E, Longhi R, Fontana S (2010) Identifying a higher throughput assay for metabolism dependent inhibition (MDI). Drug Metab Lett 4(2):104–113

33. Burt HJ, Galetin A, Houston JB (2010) IC50-based approaches as an alternative method for assessment of time-dependent inhibition of CYP3A4. Xenobiotica 40(5):331–343
34. Chen Y, Liu L, Monshouwer M, Fretland AJ (2011) Determination of time-dependent inactivation of CYP3A4 in cryopreserved human hepatocytes and assessment of human drug-drug interactions. Drug Metab Dispos 39(11):2085–2092
35. Salminen KA, Leppanen J, Venalainen JI, Pasanen M, Auriola S, Juvonen RO, Raunio H (2011) Simple, direct, and informative method for the assessment of CYP2C19 enzyme inactivation kinetics. Drug Metab Dispos 39(3):412–418
36. Burt HJ, Pertinez H, Sall C, Collins C, Hyland R, Houston JB, Galetin A (2012) Progress curve mechanistic modeling approach for assessing time-dependent inhibition of CYP3A4. Drug Metab Dispos 40(9):1658–1667
37. Kenny JR, Mukadam S, Zhang C, Tay S, Collins C, Galetin A, Khojasteh SC (2012) Drug-drug interaction potential of marketed oncology drugs: in vitro assessment of time-dependent cytochrome P450 inhibition, reactive metabolite formation and drug-drug interaction prediction. Pharm Res 29(7):1960–1976
38. Murray BP (2009) Mechanism-based inhibition of CYP3A4 and other cytochromes P450. Annu Rep Med Chem 44:535–553
39. Obach RS, Walsky RL, Venkatakrishnan K (2007) Mechanism-based inactivation of human cytochrome P450 enzymes and the prediction of drug-drug interactions. Drug Metab Dispos 35(2):246–255
40. Mukadam S, Tay S, Tran D, Wang L, Delarosa EM, Khojasteh SC, Halladay JS, Kenny JR (2012) Evaluation of time-dependent cytochrome P450 inhibition in a high-throughput, automated assay: introducing a novel area under the curve shift approach. Drug Metab Lett 6(1):43–53
41. Dasgupta M, Tang W, Caldwell GW, Yan Z (2010) Use of stable isotope labeled probes to facilitate liquid chromatography/mass spectrometry based high-throughput screening of time-dependent CYP inhibitors. Rapid Commun Mass Spectrom 24(15):2177–2185

Chapter 20

Screening for P-Glycoprotein (Pgp) Substrates and Inhibitors

Qing Wang and Tina M. Sauerwald

Abstract

P-glycoprotein (P-gp), the product of the human ABCB1 gene and often called MDR1, is the best understood membrane protein known to be involved in the active transport of drugs across biological membranes. In addition to mediating or limiting the absorption, distribution, excretion, and toxicity of many drugs, P-gp is the potential locus of a number of pharmacokinetic drug-drug interactions when two drugs, one a substrate and the other a substrate or inhibitor of the transporter, are co-administered. This last point is the reason for the interest of regulatory authorities around the world, several of which (most notably the U.S. FDA and the EMA) now require that all new molecular entities (NMEs) be evaluated as P-gp substrates and inhibitors. This chapter will cover model test systems, including *in vitro* assays for human P-gp such as cell-based (over-expressing and knockdown cells) and subcellular (membrane vesicle) approaches, as well as *in vivo* animal models. The chapter will conclude with examples of two cell-based systems, MDR1-MDCK (over-expressing) and Caco-2 (parental and P-gp knockdown cell lines).

Key words P-gp, MDR1, ABCB1, Transporter, Caco-2, MDCK, MDR1-MDCK, ADME, DDI

1 Introduction

Membrane transporters play a crucial role in drug efficacy and safety because they dictate a drug's pharmacokinetics, referred to as ADME (absorption, distribution, metabolism, excretion), as well as drug-drug interactions (DDIs) with co-administered drugs. The ATP-binding cassette (ABC) superfamily is comprised of ATP-driven pumps responsible for the efflux of a plethora of therapeutic drugs, peptides, and lipid-like compounds across biological membranes. One such transporter, P-glycoprotein (P-gp/ABCB1), is expressed widely, interacts with diverse classes of substrates and inhibitors, and is over-expressed in many cancers. As a consequence, P-gp has become the best understood transporter in the ABC superfamily. The necessity for P-gp testing continues to be dictated by regulatory authorities (in particular, the U.S. Food and

Drug Administration (FDA) [1] and the European Medicines Agency (EMA) [2]) for any new NME in development.

P-gp was discovered in 1976, when Chinese hamster ovary cell mutants were selected for resistance to colchicine, an inhibitor of microtubule polymerization and, ultimately, mitosis. The expression of this cell surface protein (later called MDR1, for multidrug resistance) was correlated with resistance to a wide variety of cytotoxic drugs [3].

Tissue distribution of P-gp is widespread, including the apical surfaces of epithelial cells in the liver, intestine, kidney, adrenal gland, and pancreas [4]. Its significance in terms of pharmacokinetics is well established, as it is implicated in the excretion of drugs and other xenobiotics (and their metabolites) into urine, bile, and the lumen of the gastrointestinal tract. Additionally, P-gp is localized to the blood-brain barrier and other blood-tissue barriers such as the testes, skin, and placenta [5, 6]. Many human cancers, particularly carcinomas, over-express P-gp, such that it imparts resistance to chemotherapy [7].

P-gp is a 170 kDa monomeric surface glycoprotein with two-fold pseudo-symmetry comprised of two transmembrane domains containing six transmembrane α-helices each. This results in twelve hydrophilic extracellular loops, with the first loop containing three N-linked glycosylation sites. The resultant cylindrical conformation allows for a large aqueous pore in the membrane to be open to the cell surface and closed at the cytoplasmic face [8]. The aqueous pore, referred to as the drug binding pocket, contains numerous binding sites for substrates and inhibitors. The bottom of the pocket contains more charged and polar residues, whereas the top of the pocket has primarily aromatic and other hydrophobic residues [9]. Additionally, two nucleotide binding domains are present on the cytoplasmic side of the transporter. Upon ATP binding, the α-helices rotate, causing a reorganization of the transmembrane domains, exposing residues of drug binding sites to the aqueous and lipid phases [10]. This conformational change opens a central pore along the length of the transporter, mediating the efflux of substrates from the lipid bilayer to the aqueous pore [11]. Consequently, ATP binding is the impetus for transport as a result of signal transduction within the transmembrane domains [10], followed by ATP hydrolysis, which reverts P-gp to its unbound initial state, reinitiating the transport cycle [12].

P-gp substrates are structurally diverse, lipid-soluble compounds, many of which are organic cations or neutral compounds with aromatic rings. P-gp substrates (reviewed in [13]) include analgesics (e.g., morphine), antibiotics (e.g., erythromycin), anticancer drugs (e.g., doxorubicin, paclitaxel, and vinblastine), Ca^{2+} channel blockers (e.g., verapamil), cardiac glycosides (e.g., digoxin), cholesterol lowering agents (e.g., lovastatin and simvastatin), fluorescent dyes (e.g., Hoechst 33342), HIV protease inhibitors (e.g., ritonavir), linear peptides (e.g., leupeptin), natural products (e.g., actinomycin D), and steroids (e.g., dexamethasone).

Similarly, P-gp inhibitors vary both structurally and in their mechanism of action. Some bind to the large, flexible binding pocket, blocking access to it by substrates; others function as competitive, non-transported "substrates" in that they bind and activate ATP hydrolysis but apparently engage in futile cycling by the transporter, thereby inhibiting the transport of other substrates; finally, a small group interferes with ATP hydrolysis. Inhibitors of P-gp (reviewed in [16]) include cyclosporine, tariquidar, and verapamil; note that verapamil is also a substrate, which makes the point that any substrate of P-gp can potentially act as a competitive inhibitor of any other substrate.

Numerous test systems, both *in vitro* and *in vivo*, have been developed for the identification and characterization of P-gp substrates and inhibitors. The most widely known *in vitro* test system involves the study of vectorial (directional) transport across monolayers of polarized mammalian epithelial cells such as the human colon adenocarcinoma cell line Caco-2, the porcine kidney cell line LLC-PK$_1$, and the Madin-Darby canine kidney cell line MDCK. The last two (LLC-PK$_1$ and MDCK) have been engineered to over-express P-gp [14–16]. In some cases, a basolateral uptake transporter and an apical efflux transporter (such as P-gp) have been over-expressed in the same cell to enable the study of vectorial trans-cellular transport of compounds requiring transporters on both poles of the cell [17]. P-gp has also been over-expressed in the unpolarized human embryonic kidney (HEK) cell line by electroporation [18]. In contrast, the opposite approach has been taken with Caco-2 cells, which natively express human P-gp: they have been engineered to partially silence (knock down) P-gp expression and function [19] or to completely knock out transporter expression and function [20]. All of the aforementioned adherent cell lines offer the benefit of greater consistency, lower expense, and ease of scalability compared with primary hepatocytes (see below); one advantage of HEK and Caco-2 cells is that they are of human origin.

An alternative test system involves the use of primary cells such as human hepatocytes. For example, sandwich-cultured primary hepatocytes can be cultured in a sandwich configuration between gelled collagen (below) and Matrigel (above), in which the cells retain many organotypic functions, including polarized expression of membrane transporters [21]. However, a key disadvantage to primary cells is donor-to-donor variability.

Finally, another *in vitro* P-gp test system involves the use of inside-out plasma membrane vesicles from either mammalian or insect cells over-expressing human P-gp [22, 23]. After preparing plasma membrane vesicles, those with inside-out orientation

(ATP- and substrate-binding sites of the transporter on the outer side facing the buffer) are the only vesicles that will be detected in the assay although they comprise only about half of the total vesicles. Substrates are taken up in an ATP-dependent manner. It has been reported that modifying the membrane composition by adding cholesterol to vesicles from insect cells normalizes transporter function to that seen in mammalian cells [24].

P-gp can be tested *in vivo* through the use of P-gp knockout mice [25, 26]. Tested in parallel with wild-type mice, these genetically engineered animals serve as a model system for the study of the involvement of P-gp in the pharmacokinetics of drugs. For a drug developer, a P-gp knockout mouse is a preclinical model of a human subject genetically lacking P-gp expression or currently taking a strong P-gp inhibitor.

Note that even with the plethora of test systems available, none may precisely model the actions of transporters in humans because of the biological complexity of the numerous transporters and the way they interact with each other.

2 Materials

2.1 Equipment

Descriptions of specific equipment and reagents used are given; however, equivalent substitutions may be made.

1. FLUOstar OPTIMA fluorescence microplate reader equipped with MARS software (BMG Labtech, Durham, NC, USA) for the monolayer integrity test.
2. Humidified cell culture incubator capable of maintaining 5 % CO_2 and 37 °C.
3. Microplate shaker.
4. Class II biological safety cabinet for aseptic operations.
5. Electronic pipettor in the range of 200–1,000 μL.

2.2 Reagents and Solutions

1. Dulbecco's Modified Eagle Medium (DMEM; Life Technologies, Carlsbad, CA, USA): cell culture base medium.
2. Fetal bovine serum (FBS; Life Technologies): cell culture medium supplement.
3. MEM non-essential amino acids (NEAA; Life Technologies): cell culture medium supplement.
4. MEM sodium pyruvate solution (Life Technologies, Cat): cell culture medium supplement.
5. Penicillin-streptomycin solution (Life Technologies): cell culture medium supplement.

6. Colchicine (Sigma, St. Louis, MO, USA): cell culture medium supplement; selection agent for MDR1-MDCK cells.
7. Puromycin dihydrochloride (Life Technologies): selection agent for CPT-P1 (P-gp knockdown) cells.
8. Dulbecco's phosphate-buffered saline (D-PBS; Life Technologies) for rinsing cells.
9. Ethylenediaminetetraacetic acid (EDTA; Life Technologies)
10. Trypsin-EDTA (Life Technologies) for dissociating cells from a plate.
11. T150 cm^2 cell culture flask (Becton Dickinson, Franklin Lakes, NJ, USA): culture vessel for stock cell cultures.
12. 12-well Costar Transwell® plates (1.13 cm^2 insert area, 0.4 µm pore size), Corning: dual-chamber plates in which cell monolayers are cultured in inserts for transport experiments.
13. Rat-tail collagen (BD Biosciences, Worcester, MA, USA) for coating Transwell culture inserts.
14. Hanks' Balanced Salt Solution (HBSS, Life Technologies) for assay buffer.
15. 4-(2-Hydroxyethyl)-1-piperazineethanesulfonic acid (HEPES; Life Technologies) for assay buffer.
16. D-Glucose (Sigma) for assay buffer.
17. HBSSg, pH 7.4 (assay buffer): HBSS with 10 mM HEPES and 15 mM D-glucose.
18. Digoxin (Sigma): probe substrate for P-gp inhibition assay.
19. Atenolol (Sigma): low passive permeability probe for batch QC assay.
20. Estrone-3-sulfate (E3S; Sigma): BCRP substrate for batch QC assay.
21. Propranolol (Sigma): high passive permeability probe for batch QC assay.
22. Lucifer yellow (LY; Life Technologies): monolayer integrity probe.
23. Cyclosporine A (CsA; Calbiochem, La Jolla, CA, USA): P-gp inhibitor.

2.3 Cell Lines

1. Caco-2 cells were from American Type Culture Collection (Clone #C2BBe1, Rockville, MD, USA).
2. MDR1-MDCK cells were from the National Institutes of Health (Bethesda, MD, USA).
3. CPT-P1 is a proprietary cell line created by Absorption Systems, with reduced expression level of human P-gp compared to the parental Caco-2 cells.

3 Methods

3.1 Cell Culture

Perform all cell culture operations in a biological safety cabinet under sterile conditions and with proper aseptic technique.

1. Maintain stock cultures of each cell line in its optimal culture medium in a humidified incubator. Details of the cell culture media are listed in Table 1. Change the culture medium three times weekly, and observe cell growth by light microscopy.

2. When the stock cultures become confluent, harvest the cells by trypsinization and seed at a density of 60,000 cells/cm^2 on 12-well Costar Transwell plates containing collagen-coated, microporous (0.4 μm pore size), polycarbonate filter membranes for experimental monolayers (**Note 1**). Add culture medium (1.5 mL) to each bottom well, and cell suspension (0.5 mL) to each insert.

3. Grow cell monolayers to confluence on the filter membranes in a humidified incubator. Change the culture medium every other day until use. A schematic illustration of a cell monolayer is shown in Fig. 1.

Table 1
Culture media

Cell types	Caco-2	CPT-P1	MDR1-MDCK	MDCK
Base medium	DMEM with 10 % FBS, 1 % NEAA, 1 mM sodium pyruvate, 100 IU/mL penicillin, and 100 μg/mL streptomycin			
Selection reagent	None	Puromycin (10 μg/mL)	Colchicine (80 ng/mL)	None

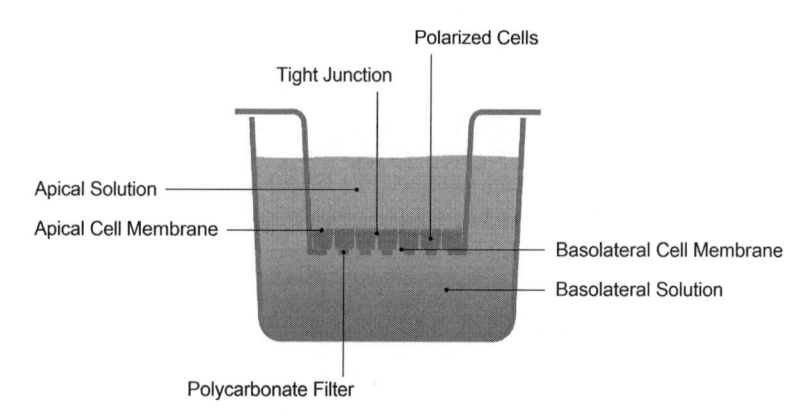

Fig. 1 Illustration of a polarized cell monolayer for a bidirectional permeability experiment

Table 2
Batch QC acceptance criteria for Caco-2 cell monolayers

Parameter	Acceptance criteria
TEER (Ω cm^2)	450–650
Lucifer yellow P_{app} (10^6 cm/s)	≤ 0.4
Atenolol P_{app} (10^6 cm/s)	≤ 0.5
Propranolol P_{app} (10^6 cm/s)	10–30
Digoxin efflux ratio	≥ 10
Estrone-3-sulfate efflux ratio	≥ 25

3.2 Batch Quality Control Assay

To confirm the barrier properties and polarized expression of apical efflux transporters, and to ensure consistent performance of the model over many years, it is recommended that a rigorous quality control (QC) assessment be run with several (at least six) randomly selected monolayers from each batch of cell monolayers. As an example, the batch QC acceptance criteria established in our laboratory for Caco-2 cell monolayers are shown in Table 2.

3.3 Description of the Model

In polarized cell monolayers expressing P-gp, the efflux transporter resides in the apical plasma membrane. The monolayers are used in a bidirectional assay format, with samples taken from the basolateral (BL) and apical (AP) compartments after dosing a test compound into the opposite compartment (separate monolayers in each direction). The ratio of the BL-to-AP apparent permeability (P_{app}) to the AP-to-BL P_{app} (the "efflux ratio") is a measure of the contribution of P-gp to the transport of a compound across the monolayer. In general, a test compound is scored as a substrate of P-gp if its efflux ratio is greater than or equal to 2; non-substrates cross the monolayer at approximately the same rate in either direction, resulting in an efflux ratio of ~1. The efflux ratio of a P-gp substrate is, in part, a function of the level of expression of P-gp in the test system; in the case of MDR1-over-expressing MDCK cells, a strong P-gp substrate could have an efflux ratio over 100 (there is no theoretical upper limit). Classification of a test compound as a P-gp substrate can be confirmed by challenge with a P-gp inhibitor. Because the absolute value of the efflux ratio depends on the test system, a positive control compound should be incorporated into the study design (refer to [27] for review).

The FDA draft guidance on drug interaction studies [1] lists bidirectional assays in Caco-2 cells or P-gp-over-expressing cell lines (the most common being MDR1-transfected MDCK cells) as the preferred method for *in vitro* evaluation of the P-gp substrate potential of test compounds. In the case of over-expressing cells, the non-transfected parental cells should be run in parallel as a control; the "ratio of efflux ratios" (transfected vs. non-transfected cells) indicates the magnitude of efflux specifically attributable to the transfected human P-gp. A more elegant approach, mentioned in the EMA guideline on the investigation of drug interactions [2], is to compare bidirectional transport in P-gp knockdown cells with that in control cells [19]. In this case, the relative efflux ratio (control vs. knockdown cells) is a very specific marker of the degree of P-gp-mediated efflux of a test compound. One advantage of the latter approach is that it reduces the reliance on pharmacologic inhibitors, which in most cases are not specific for a single transporter.

The bidirectional assay format is also used to assess the P-gp inhibitor potential of test compounds; in that case, the efflux ratio (or simply the B-to-A P_{app}) of digoxin, a well-characterized P-gp probe substrate, is measured in the presence and absence of one or more concentrations of a test compound.

3.4 P-gp Substrate Assessment

The plates containing the experimental monolayers are kept in a humidified incubator for the duration of the assay. The assay buffer is HBSS with 15 mM D-glucose and 10 mM HEPES, pH 7.4 ("HBSSg"). Experimental conditions and a representative sampling profile for a test compound (TC) are summarized in Table 3; the positive control P-gp substrate digoxin is run in parallel. A confirmatory experiment with the P-gp inhibitor CsA (following a 30-min pre-incubation), summarized in Table 4, is performed either in parallel (for all TCs) or subsequently (both with and without CsA, only for positives from the initial screen; **Note 2**).

Table 3
P-gp substrate assessment (condition 1)

Pre-incubation with test compound dosing solution	Matrix composition		Sampling volume (µL)		Sampling time points (min)	
	AP	BL	AP	BL	AP	BL
No[a]	10 µM TC[b] or Digoxin + 200 µM LY	HBSSg	50	200	5 and 120	120
No[a]	HBSSg	10 µM TC[b] or Digoxin + 200 µM LY	200	50	120	5 and 120

[a]**Note 3**
[b]TC concentration should be low enough to avoid saturating the transporter while still allowing sufficient analytical sensitivity; a concentration in the range of 1–10 µM is generally used

Table 4
P-gp substrate assessment (condition 2)

30-min Pre-incubation with P-gp inhibitor[a]		AP-to-BL directional permeability		BL-to-AP directional permeability	
AP	BL	AP	BL	AP	BL
5 µM CsA	5 µM CsA	10 µM TC or Digoxin + 200 µM LY + CsA	HBSSg + CsA	HBSSg + CsA	10 µM TC or Digoxin + 200 µM LY + CsA

[a]Note 4

For receiver samples, aliquots (200 µL) are taken at one or more pre-selected time points (a single 120 min receiver time point is shown in Table 3; note that linearity with time is required for accurate calculation of transport rate, and the transport characteristics of TCs are generally unknown prior to an experiment, which may necessitate collection of receiver samples at multiple time points). When receiver samples are taken at multiple time points, they are replaced by an equal volume of fresh HBSSg at each time point except the final one and the calculation of cumulative receiver concentration must take into account the multiple sample/replace steps [28]. A portion of each receiver sample is used for analysis of the TC (generally by LC-MS/MS), and another portion for measurement of co-dosed lucifer yellow (LY, a monolayer integrity marker). Aliquots (50 µL) are also taken from the donor compartment without replacement (at 5 and 120 min in the experimental design illustrated in Table 3). Figure 2 is a decision tree from the FDA draft guidance on drug interaction studies [1], illustrating the consequences of the possible outcomes of an *in vitro* P-gp substrate assessment.

3.5 P-gp Inhibitor Assessment

For assessment of P-gp inhibition, the bidirectional transport of the P-gp probe substrate digoxin is measured in the absence and presence of a TC and, in parallel, a positive control (PC) P-gp inhibitor such as cyclosporine A (CsA). The assay conditions are summarized in Table 5. The permeability assay is preceded by a 30-min pre-incubation with TC or PC on both sides of the monolayers (**Note 4**). Following the pre-incubation, for AP-to-BL transport 0.5 mL dosing solution (digoxin with or without TC or PC, depending on the pre-incubation condition for a given monolayer) is added to the AP side, and 1.5 mL of HBSSg (with or without TC or PC) is added to the BL side. For BL-to-AP transport, 1.5 mL dosing solution (digoxin with or without TC or PC) is added to the BL side, and 0.5 mL of HBSSg (with or without TC or PC) is added to the AP side. The monolayers are

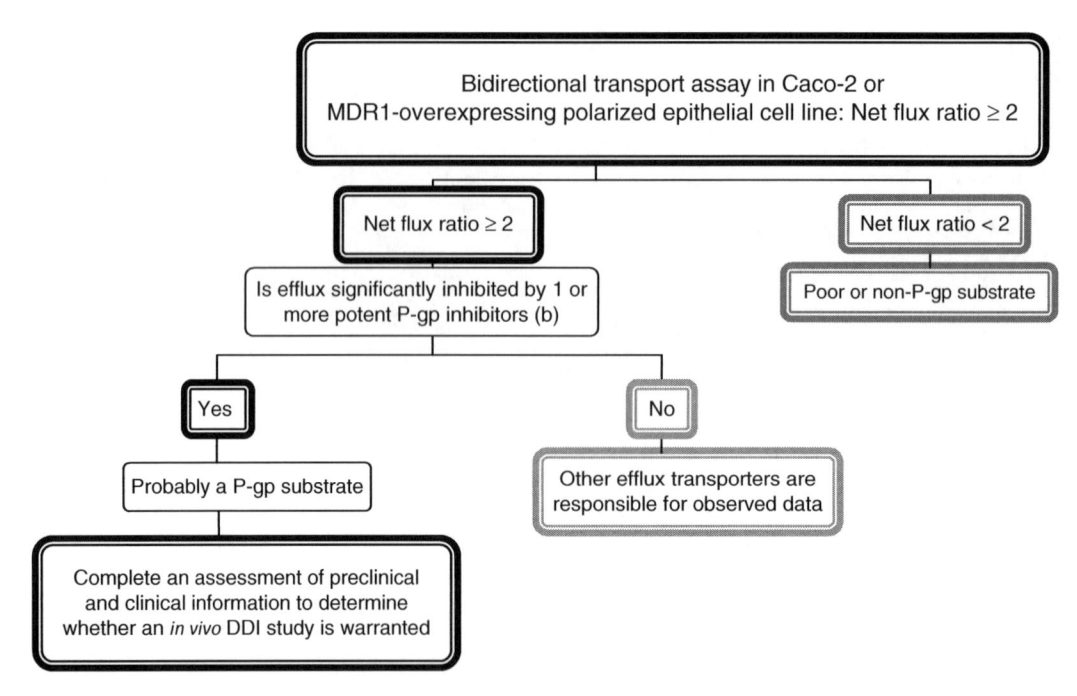

Fig. 2 Decision tree to assess P-gp substrate potential, adapted from [1]. "Net flux ratio" is identical to the efflux ratio defined above (Sect. 3.3) and in (2), Sect. 3.6. The only difference in the corresponding EMA guideline [2] is that net flux ratio >2 (as opposed to ≥) triggers further investigation

Table 5
P-gp inhibitor assessment conditions

30-min Pre-incubation[a]	Matrix composition AP	BL	Sampling volume (µL) AP	BL	Sampling time points (minutes) AP	BL
HBSSg with or without TC[b] or PC	Digoxin (10 µM) with or without TC[b] or PC	HBSSg with or without TC[b] or PC	50	200	5 and 120	120
HBSSg with or without TC[b] or PC	HBSSg with or without TC[b] or PC	Digoxin (10 µM) with or without TC[b] or PC	200	50	120	5 and 120

[a]**Note 4**
[b]*See* **Note 5** for selection of TC concentration

incubated in a humidified incubator for 120 min; note that the rate of transport of the probe substrate digoxin is being measured, and it is a compound whose transport characteristics (e.g., linearity for at least 120 min) are well characterized ahead of time. For receiver samples, aliquots (200 µL) are taken at 120 min, and 50 µL aliquots

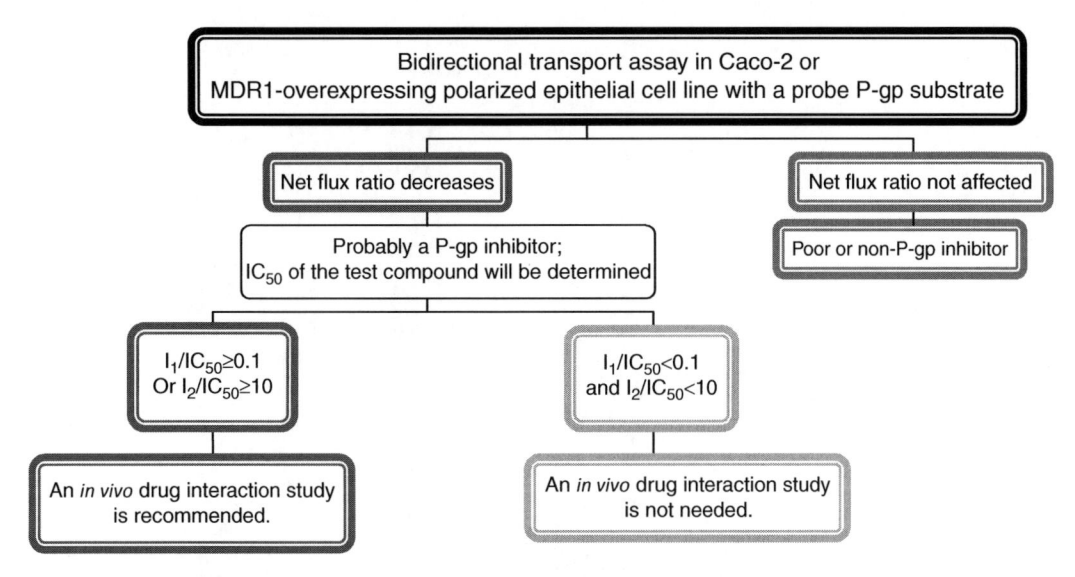

Fig. 3 Decision tree to assess P-gp inhibitor potential, adapted from [1]. I_1 is the total systemic C_{max} and I_2 is an estimate of the maximal gut concentration: the highest dose strength dissolved in 250 mL. The EMA decision tree is very similar, the one difference being a higher safety margin for I_1, which the EMA guideline [2] defines as the <u>unbound</u> systemic C_{max}: I_1/K_i (or IC_{50}) ≥ 0.02 triggers an *in vivo* drug interaction study

are taken from the donor compartment at 5 and 120 min. Figure 3 is a decision tree from the FDA draft guidance on drug interaction studies [1], illustrating the consequences of the possible outcomes of an *in vitro* P-gp inhibitor assessment.

3.6 Calculations

$$P_{app} = (C_R/120) \times V_R / (A \times C_{D5} \times 60) \quad (1)$$

$$\text{Efflux ratio} = \frac{P_{app} \text{BL-to-AP}}{P_{app} \text{AP-to-BL}} \quad (2)$$

$$\text{Relative efflux ratio (RER)} = \frac{\text{efflux ratio}_{Caco-2}}{\text{efflux ratio}_{KD}} \quad (3)$$

$$\text{Corrected Efflux Ratio} = \text{Efflux ratio} - 1 \quad (4)$$

$$\text{Percentage efflux remaining} = \frac{\text{corrected efflux ratio}_{TC}}{\text{corrected efflux ratio}_{No\,TC}} \times 100 \quad (5)$$

$$Y = Y_L + (Y_H - Y_L) / (1 + 10 \wedge ((\text{LogIC}_{50} - X) * \text{Hillslope})) \quad (6)$$

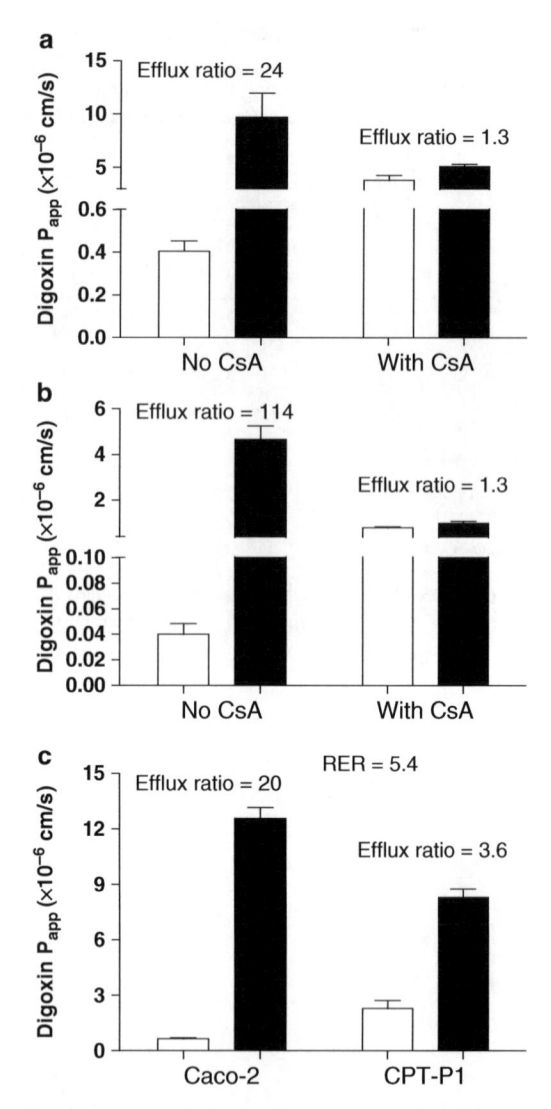

Fig. 4 Bidirectional transport of the positive "test compound" digoxin (10 μM) in (**a**) Caco-2, (**b**) MDR1-MDCK, and (**c**) Caco-2 and CPT-P1 cell monolayers. *Open bar*: AP-to-BL; *Filled bar*: BL-to-AP. Efflux ratio: (2), Sect. 3.6. Relative efflux ratio: (3), Sect. 3.6

C_R: Concentration of test compound in the receiver compartment (μM);

V_R: Volume of the receiver compartment (BL: 1.5 cm^3; AP: 0.5 cm^3);

A: Cell monolayer area (1.13 cm^2 for a 12-well Transwell);

C_{D5}: Donor concentration at 5 min (μM);

KD: CPT-P1 (P-gp knockdown) cell line

The value "120" in (1) represents the duration of the permeability assessment (120 min), and the value "60" is a conversion factor (minutes to seconds). Values of remaining efflux activity are used for IC$_{50}$ determination by nonlinear regression using GraphPad Prism (version 5.0), with the minimum value constrained to 0. For (6), X is the logarithm of the nominal concentration of inhibitor, Y is the value of the remaining efflux activity of the transporter at a given concentration, Y$_L$ represents the lowest response (at the highest concentration of inhibitor), and Y$_H$ represents the highest response (vehicle control). A value of 1 % of the lowest inhibitor concentration is entered instead of zero.

3.7 Results

Representative substrate assessment results are shown in Fig. 4 for digoxin as a positive "test compound" and CsA as a confirmatory inhibitor. In the absence of CsA, the efflux ratio ((2), Sect. 3.6) of digoxin was 24 in Caco-2 cell monolayers, and CsA decreased it to 1.3 (Fig. 4a); the corresponding values were 114 and 1.3 in MDR1-MDCK cell monolayers (Fig. 4b). When Caco-2 cells were run in parallel with CPT-P1 cells (derivatives of Caco-2 in which P-gp expression is knocked down), the efflux ratios of digoxin were 20 and 3.6, respectively, for a relative efflux ratio (RER; parental vs. knockdown) ((3), Sect. 3.6) of 5.4 (Fig. 4c).

In the P-gp inhibitor assessment, a test compound is first tested for inhibition of P-gp at a single concentration (**Note 5**); if it reduces the efflux ratio of digoxin by at least 50 %, the potency is assessed by determining the IC$_{50}$, as illustrated for the potent P-gp inhibitor CsA (IC$_{50}$ of 0.311 µM) in Fig. 5.

4 Notes

1. Monolayers can also be cultured on inserts without collagen coating and/or with a different pore size, but in any case appropriate ranges for cell batch QC and assay performance must be established under each laboratory's set of conditions. The conditions given are those used in our laboratory.

2. If Caco-2 or MDR1-MDCK cells are used for P-gp substrate assessment, the assays are conducted with a two-step approach; the two steps can be run in parallel or sequentially. In the first step, the efflux ratio of the test compound is determined. If the efflux ratio is ≥2.0, the bidirectional permeability of the test compound will be challenged with a P-gp inhibitor such as CsA in the second step. If the efflux ratio is reduced more than 50 % by CsA, the test compound is classified as a P-gp substrate (Fig. 4a, b). If, on the other hand, Caco-2 and CPT-P1 cells are used, classification is based on the efflux ratio of the test compound in both cell lines, followed by calculation of the

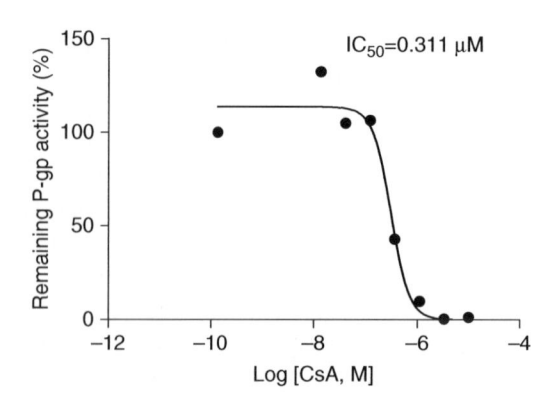

Fig. 5 Effect of CsA on the transport of the P-gp probe substrate digoxin (10 μM) across Caco-2 cell monolayers

relative efflux ratio (RER); if the RER is ≥2.0, the test compound is classified as a P-gp substrate (Fig. 4c).

3. Depending on the physicochemical properties of the test compound (i.e., if it is hydrophobic and tends to bind nonspecifically to plastic), a brief (5- or 10-min) pre-incubation with the test compound may be needed to reduce non-specific binding to the experimental device.

4. For assessment of P-gp inhibition, pre-incubation with test compound was recommended in the 2006 FDA draft guidance on drug interaction studies, and a 30-min pre-incubation has been used in our laboratory since then.

5. For compounds intended for oral administration, the test concentration in the inhibitor assay is typically either 10 μM or 0.1× the concentration equal to the highest intended clinical dose strength dissolved in 250 mL ($[I]_2$ [1]). For non-orally administered compounds, the test concentration in the inhibitor assay is typically either 10 μM or 10× the clinical total (bound plus unbound) systemic C_{max} ($[I]_1$ [1]). It also may depend on the results of suitability assessments such as solubility of the test compound and tolerability of the cells.

References

1. US Department of Health and Human Services, Food and Drug Administration, Center for Drug Evaluation and Research (CDER) (2012) Draft guidance for industry. Drug interaction studies—study design, data analysis, implications for dosing, and labeling recommendations. http://www.fda.gov/downloads/Drugs/GuidanceComplianceRegulatoryInformation/Guidances/UCM292362.pdf
2. European Medicines Agency, Committee for Human Medicinal Products (CHMP) (2012) Guideline on the investigation of drug interactions. http://www.ema.europa.eu/docs/en_GB/document_library/Scientific_guideline/2012/07/WC500129606.pdf
3. Juliano RL, Ling V (1976) A surface glycoprotein modulating drug permeability in Chinese hamster ovary cell mutants. Biochim Biophys Acta 455(1):152–162
4. Thiebaut F, Tsuruo T, Hamada H, Gottesman MM, Pastan I, Willingham MC (1987) Cellular localization of the multidrug-resistance gene product P-glycoprotein in normal human

tissues. Proc Natl Acad Sci USA 84: 7735–7738

5. Cordon-Cardo C, O'Brien JP, Casals D, Rittman-Grauer L, Biedler JL, Melamed MR, Bertino JR (1989) Multidrug-resistance gene (P-glycoprotein) is expressed by endothelial cells at blood-brain barrier sites. Proc Natl Acad Sci USA 86:695–698

6. Cordon-Cardo C, O'Brien JP, Boccia J, Casals D, Bertino JR, Melamed MR (1990) Expression of the multidrug resistant gene product (P-glycoprotein) in human normal and tumor tissues. J Histochem Cytochem 38(9):11277–11287

7. Goldstein LJ, Galski H, Fojo A, Willingham M, Lai SL, Gazdar A, Pirker R, Green A, Crist W, Brodeur GM et al (1989) Expression of a multidrug resistance gene in human cancers. J Natl Cancer Inst 81(2):116–124

8. Rosenberg MF, Callaghan R, Ford RC, Higgins CF (1997) Structure of the multidrug resistance P-glycoprotein to 2.5 nm resolution determined by electron microscopy and image analysis. J Biol Chem 272(16):10685–10694

9. Aller SG, Yu J, Ward A, Weng Y, Chittaboina S, Zhuo R, Harrell PM, Trinh YT, Zhang Q, Urbatsch IL, Chang G (2009) Structure of P-glycoprotein reveals a molecular basis for poly-specific drug binding. Science 323(5922):1718–1722

10. Rosenberg MF, Velarde G, Ford RC, Martin C, Berridge G, Kerr ID, Callaghan R, Schmidlin A, Wooding C, Linton KJ, Higgins CF (2001) Repacking of the transmembrane domains of P-glycoprotein during the transport ATPase cycle. EMBO J 20(20):5615–5625

11. Rosenberg MF, Kamis AB, Callaghan R, Higgins CF, Ford RC (2003) Three-dimensional structures of the mammalian multidrug resistance P-glycoprotein demonstrate major conformational changes in the transmembrane domains upon nucleotide binding. J Biol Chem 278(10):8294–8299

12. Tombline G, Muharemagic A, Bartholomew White L, Senior AE (2005) Involvement of the "occluded nucleotide conformation" of P-glycoprotein in the catalytic pathway. Biochemistry 44:12879–12886

13. Sharom FJ (2011) The P-glycoprotein multidrug transporter. Essays Biochem 50:161–178

14. Pastan I, Gottesman MM, Ueda K, Lovelace E, Rutherford AV, Willingham MC (1988) A retrovirus carrying an MDR1 cDNA confers multidrug resistance and polarized expression of P-glycoprotein in MDCK cells. Proc Natl Acad Sci USA 85:4486–4490

15. Tanigawara Y, Okamura N, Hirai M, Yasuhara M, Ueda K, Kioka N, Komano T, Hori R (1992) Transport of digoxin by human P-glycoprotein expressed in a porcine kidney epithelial cell line (LLC-PK1). J Pharmacol Exp Ther 263:840–845

16. Ueda K, Okamura N, Hirai M, Tanigawara Y, Saeki T, Kioka N, Komano T, Hori R (1992) Human P-glycoprotein transports cortisol, aldosterone, and dexamethasone, but not progesterone. J Biol Chem 267:24248

17. Matsushima S, Maeda K, Kondo C, Hirano M, Sasaki M, Suzuki H, Sugiyama Y (2005) Identification of the hepatic efflux transporters of organic anions using double-transfected Madin-Darby canine kidney II cells expressing human organic anion-transporting polypeptide 1B1 (OATP1B1)/multidrug resistance-associated protein 2, OATP1B1/multidrug resistance 1, and OATP1B1/breast cancer resistance protein. J Pharmacol Exp Ther 314(3):1059–1067

18. Crouthamel MH, Wu D, Yang Z, Ho RJY (2006) A novel MDR1 G1199T variant alters drug resistance and efflux transport activity of P-glycoprotein in recombinant Hek cells. J Pharm Sci 95:2767–2777

19. Darnell M, Karlsson JE, Owen A, Hidalgo IJ, Li J, Zhang W, Andersson TB (2010) Investigation of the involvement of P-glycoprotein and multidrug resistance-associated protein 2 I the efflux of ximelagatran and its metabolites by using short hairpin RNA knockdown in Caco-2 cells. Drug Metab Dispos 38:491–497

20. Pratt J, Venkatraman N, Brinker A, Xiao Y, Blasberg J, Thompson DC, Bourner M (2012) Use of zinc finger nuclease technology to knock out efflux transporters in C2BBe1 cells. Curr Protoc Toxicol Chapter 23:Unit 23.2

21. Hoffmaster KA, Turncliff RZ, LeCluyse EL, Kim RB, Meier PJ, Brouwer KLR (2004) P-glycoprotein expression, localization, and function in sandwich-cultured primary rat and human hepatocytes: relevance to the hepatobiliary disposition of a model opioid peptide. Pharm Res 21(7):1294–1302

22. Karlsson JE, Heddle C, Rozkov A, Rotticci-Mulder J, Tuvesson O, Hilgendorf C, Andersson TB (2010) High-activity P-glycoprotein, multidrug resistance protein 2, and breast cancer resistance protein

membrane vesicles prepared from transiently transfected human embryonic kidney 293-Epstein-Barr virus nuclear antigen cells. Drug Metab Dispos 38(4):705–714

23. Kodon A, Shibata H, Matsumoto T, Terakado K, Sakiyama K, Matsuo M, Ueda K, Kato H (2009) Improved expression and purification of human multidrug resistance protein MDR1 from baculovirus-infected cells. Protein Expr Purif 66:7–14

24. Pál Á, Méhn D, Molnar É, Gedey S, Mészáros P, Nagy T, Glavinas H, Janáky T, von Richter O, Báthori G, Szente L, Krajcsi P (2007) Cholesterol potentiates ABCG2 activity in a heterologous expression system: improved *in vitro* model to study function of human ABCG2. J Pharmacol Exp Ther 321(3):1085–1094

25. Schinkel AH, Smit JJM, van Tellingen O, Beijnen JH, Wagenaar E, van Deemter L, Mol CAAM, van der Valk MA, Robanus-Maandag EC, te Riele HPJ, Berns AJM, Borst P (1994) Disruption of the mouse mdr1a P-glycoprotein gene leads to a deficiency in the blood-brain barrier and to increased sensitivity to drugs. Cell 77:491–502

26. Schinkel AH, Mayer U, Wagenaar E, Mol CAAM, van Deemter L, Smit JJ, van der Valk MA, Voordouw AC, Spits H, van Tellingen O, Zijlmand JMJM, Fibbe WE, Borst P (1997) Normal viability and altered pharmacokinetics in mice lacking mdr1-type (drug-transporting) P-glycoproteins. Proc Natl Acad Sci USA 94:4028–4033

27. Giacomini KM, Huang SM, Tweedie DJ, Benet LZ, Brouwer KL, Chu X, Dahlin A, Evers R, Fischer V, Hillgren KM, Hoffmaster KA, Ishikawa T, Keppler D, Kim RB, Lee CA, Niemi M, Polli JW, Sugiyama Y, Swaan PW, Ware JA, Wright SH, Yee SW, Zamek-Gliszczynski MJ, Zhang L (2010) Membrane transporters in drug development. Nat Rev Drug Discov 9:215–236

28. Hidalgo IJ (1996) Cultured intestinal epithelial cell models. In: Borchardt RT et al (eds) Models for assessing drug absorption and metabolism. Plenum Press, New York, pp 35–50

Chapter 21

In Vitro Characterization of Intestinal Transporter, Breast Cancer Resistance Protein (BCRP)

Chris Bode and Li-Bin Li

Abstract

The organ distribution and substrate specificity of breast cancer resistance protein (BCRP), the product of the human ABCG2 gene, overlaps considerably with that of P-glycoprotein. Both are up-regulated in some cancers, leading to drug resistance, and can mediate drug-drug interactions when two drugs, one a substrate and the other an inhibitor or substrate of the same transporter, are co-administered. Thus, the U.S. FDA and the EMA now require that all NCEs be evaluated as substrates and inhibitors of BCRP. This chapter will cover *in vitro* assays for human BCRP, including cell-based (over-expressing and knockdown cells) and subcellular (membrane vesicle) approaches; the advantages of intact cells for discerning the complex interplay among the various uptake and efflux transporters will be discussed. Only recently have cell lines become available in which human BCRP is over-expressed; the advantages and limitations of this model will be illustrated. The uses and limitations of existing pharmacologic reagents, and the importance of pairing a given probe substrate with the appropriate biological model, will be discussed and illustrated as well.

Key words BCRP, ABCG2, Transporter, Caco-2, MDCK, BCRP-MDCK, ADME, DDI

1 Introduction

Breast cancer resistance protein (BCRP), the product of the *ABCG2* gene, is one of several members of the ABC (ATP-binding cassette) superfamily of transporters that play particularly important roles in the disposition and safety of drugs [1]. BCRP is expressed on the apical side of many polarized cells, including the intestinal epithelium, where it limits the absorption of xenobiotics, including many drugs in clinical use. It is also expressed in the liver (in the bile canalicular membrane of hepatocytes) and the kidney (in the brush border membrane of renal epithelial cells), where it is involved in the excretion of drugs into the bile and urine, respectively. Along with P-gp, it is located in the apical membrane of brain capillary endothelial cells, acting as a barrier to the entry of compounds into the brain across the blood-brain barrier.

BCRP (originally called MXR) was discovered due to its up-regulation in drug-resistant breast cancer cells [2–4], hence the name.

Because it is expressed at so many pharmacokinetically important interfaces, BCRP is a potential locus of drug-drug interactions (DDIs), when one drug that is a substrate of the transporter is co-administered with another drug that is a substrate or an inhibitor of the same transporter. Depending on the site at which the interaction is most significant, the pharmacokinetic result can be either an increase or a decrease in systemic (plasma) exposure of the affected drug, e.g., due to reduced renal clearance or intestinal absorption, respectively. In addition, one should always keep in mind that transporter-mediated DDIs can be insidious, in that they can lead to elevated tissue concentrations with little or no change in plasma levels [5]. Due to its potential involvement in DDIs (and the consequent drug safety implications), both the U.S. Food and Drug Administration (FDA) and the European Medicines Agency (EMA) require that all new drugs be tested *in vitro* for interactions as both a substrate and an inhibitor of BCRP [6, 7]. One DDI in which BCRP may be involved is that between methotrexate (substrate) and benzimidazole proton pump inhibitors [8].

BCRP plays a role in the excretion of clinically important drugs such as topotecan [9] and appears in the FDA-approved label for both lapatinib [10] and topotecan [11]. It transports both neutral and negatively charged molecules, sulfated conjugates of drugs and hormones, and environmental and dietary toxins; many substrates of BCRP are also transported by P-gp. There is evidence for the involvement of BCRP in the maintenance of cellular folate homeostasis [12], and recent work suggests that it is involved in the extra-renal (intestinal) elimination of uric acid [13].

Compared with P-gp, fewer preclinical test systems have been developed for BCRP. In the most commonly used assay format, bidirectional transport is monitored across monolayers of polarized mammalian epithelial cell lines such as Caco-2 or, only recently, BCRP-transfected MDCK cells [14]. In some cases, a basolateral uptake transporter and an apical efflux transporter (such as BCRP) have been over-expressed in the same polarized cell to enable the study of vectorial trans-cellular transport of compounds requiring transporters on both poles of the cell [15]. The opposite approach has also been taken with Caco-2 cells, which natively express human BCRP: they have been engineered to substantially silence (knock down) BCRP expression and function [16, 17]. Finally, BCRP function has been studied via uptake into inside-out plasma membrane vesicles prepared from BCRP-transfected insect cells [18] or mammalian cells [19].

Note that it is a challenge to design an *in vitro* system to model accurately the complexity of *in vivo* transporter functions and the interplay among them. In general, intact cell models may achieve the closest approximation, although data interpretation can be complicated. Simpler systems such as inside-out membrane vesicles, on the other hand, are susceptible to both false positives (e.g., an observed result that could never occur *in vivo* because the compound never gets into the cell) and false negatives (highly permeable compounds that are not retained within the vesicles after being pumped in).

This chapter will cover bidirectional transport assays across monolayers of polarized mammalian epithelial cells.

2 Materials

2.1 Equipment

Descriptions of specific equipment and reagents are provided; however, equivalent substitutions may be made.

1. Fluorescence microplate reader for the monolayer integrity test.
2. Humidified cell culture incubator capable of maintaining 5 % CO_2 and 37 °C.
3. Microplate shaker.
4. Class II biological safety cabinet for aseptic operations.
5. Electronic pipettor in the range of 200–1,000 μL.

2.2 Reagents and Solutions

1. Dulbecco's Modified Eagle Medium (DMEM; Life Technologies, Carlsbad, CA, USA): cell culture base medium.
2. Fetal bovine serum (FBS; Life Technologies): cell culture medium supplement.
3. MEM Non-essential amino acids (NEAA; Life Technologies): cell culture medium supplement.
4. MEM Sodium pyruvate solution (Life Technologies, Cat): cell culture medium supplement.
5. Penicillin-streptomycin solution (Life Technologies): cell culture medium supplement.
6. Neomycin (Life Technologies): cell culture medium supplement; selection agent for BCRP-MDCK cells.
7. Puromycin dihydrochloride (Life Technologies): selection agent for CPT-B1 (BCRP knockdown) cells.
8. Dulbecco's Phosphate-buffered Saline (D-PBS; Life Technologies) for rinsing cells.
9. Ethylenediaminetetraacetic acid (EDTA; Life Technologies)
10. Trypsin-EDTA (Life Technologies) for dissociating cells.
11. T150 cm^2 cell culture flask (Becton Dickinson, Franklin Lakes, NJ, USA): culture vessel for stock cell cultures.

12. 12-well Costar Transwell® plates (1.13 cm² insert area, 0.4 μm pore size), Corning: dual-chamber plates in which cell monolayers are cultured in inserts for transport experiments.

13. Rat-tail collagen (BD Biosciences, Worcester, MA, USA) for coating Transwell culture inserts.

14. Hanks' Balanced Salt Solution (HBSS, Life Technologies) for assay buffer.

15. 4-(2-Hydroxyethyl)-1-piperazineethanesulfonic acid (HEPES; Life Technologies) for assay buffer.

16. D-Glucose (Sigma-Aldrich) for assay buffer.

17. HBSSg, pH 7.4 (assay buffer): HBSS with 10 mM HEPES and 15 mM D-glucose.

18. Atenolol and propranolol (Sigma-Aldrich): low and high passive permeability probes, respectively, for batch QC assay.

19. Estrone-3-sulfate (E3S; Sigma-Aldrich): BCRP substrate for batch QC assay and positive control in BCRP substrate assessment.

20. Digoxin (Sigma-Aldrich): P-gp substrate for batch QC assay.

21. Cladribine (Sigma-Aldrich): probe substrate for BCRP inhibition assay.

22. Lucifer yellow (LY; Life Technologies): monolayer integrity probe.

23. Ko143 and fumitremorgin C (FTC) (Sigma-Aldrich): BCRP inhibitors.

2.3 Cell Lines

1. Caco-2 cells are from ATCC (Clone #C2BBe1, Rockville, MD, USA).

2. MDCK (Madin-Darby Canine Kidney) cells are from the National Institutes of Health (Bethesda, MD, USA).

3. BCRP-MDCK is a proprietary cell line created by Absorption Systems, stably transfected to over-express human BCRP.

4. CPT-B1 is a proprietary cell line created by Absorption Systems, with stably reduced expression of human BCRP compared to the parental Caco-2 cells.

5. CPT-P1 is a proprietary cell line created by Absorption Systems, with stably reduced expression of human P-gp compared to the parental Caco-2 cells.

3 Methods

3.1 Cell Culture

Perform all cell culture operations in a biological safety cabinet under sterile conditions and with proper aseptic technique.

Table 1
Culture media

Cell line	Caco-2	CPT-B1	BCRP-MDCK	MDCK
Base medium	DMEM with 10 % FBS, 1 % NEAA, 1 mM sodium pyruvate, 100 IU/mL penicillin, and 100 μg/mL streptomycin			
Selection reagent	None	Puromycin (10 μg/mL)	Neomycin (800 μg/mL)	None

1. Maintain stock cultures of each cell line in its optimal culture medium in a humidified incubator. Details of the cell culture media are listed in Table 1. Change the culture medium three times per week, and observe cell growth by light microscopy.

2. When the stock cultures become confluent, harvest the cells by trypsinization and seed at a density of 60,000 cells/cm² on 12-well Costar Transwell plates containing collagen-coated, microporous (0.4 μm pore size), polycarbonate filter membranes for experimental monolayers (**Note 1**). Add culture medium (1.5 mL) to each bottom well and cell suspension (0.5 mL) to each insert.

3. Grow cell monolayers to confluence on the filter membranes in a humidified incubator, changing the culture medium every other day. A schematic illustration of a cell monolayer is shown in Fig. 1.

3.2 Batch Quality Control Assay

To confirm the barrier properties and polarized expression of apical efflux transporters and ensure consistent performance of any cell-based model over many years, it is recommended that a rigorous quality control (QC) program be implemented by testing several (at least six) randomly selected monolayers from each batch of cell monolayers and archiving the results for later review. As an example, the batch QC acceptance criteria established in our laboratory for Caco-2 cell monolayers are shown in Table 2.

3.3 Description of the Model

In polarized cell monolayers expressing BCRP, the efflux transporter resides in the apical plasma membrane. The monolayers are used in a bidirectional assay format, with samples taken from the basolateral (BL) and apical (AP) compartments after dosing a test compound into the opposite compartment (separate monolayers in each direction). The ratio of the BL-to-AP apparent permeability (P_{app}) to the AP-to-BL P_{app} (the "efflux ratio") is a measure of the contribution of BCRP to the transport of a compound across the monolayer. In general, a test compound is scored as a substrate of BCRP if its efflux ratio is greater than or equal to 2; non-substrates cross the monolayer at approximately the same rate in either

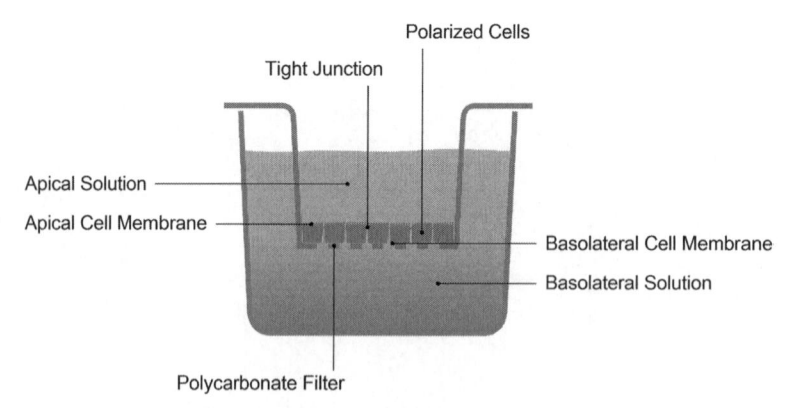

Fig. 1 Illustration of a polarized cell monolayer for a bidirectional permeability experiment

Table 2
Batch QC acceptance criteria for Caco-2 cell monolayers

Parameter	Acceptance Criteria
TEER (Ω cm^2)	450–650
Lucifer yellow P_{app} (10^6 cm/s)	≤0.4
Atenolol P_{app} (10^6 cm/s)	≤0.5
Propranolol P_{app} (10^6 cm/s)	10–30
Digoxin efflux ratio	≥10
Estrone-3-sulfate efflux ratio	≥25

direction, resulting in an efflux ratio of ~1. The efflux ratio of a BCRP substrate is, in part, a function of the level of expression of BCRP in the test system. In the case of BCRP-over-expressing MDCK cells, a strong BCRP substrate could have an efflux ratio over 100; there is no theoretical upper limit. Initial classification of a test compound as a BCRP substrate can be confirmed by a subsequent or parallel challenge with a BCRP inhibitor. Because the absolute value of the efflux ratio depends on the test system, a positive control substrate should be incorporated into the study design.

The FDA draft guidance on drug interaction studies [6] lists bidirectional assays in Caco-2 cells or BCRP-over-expressing cell lines as the preferred method for *in vitro* evaluation of the BCRP substrate potential of test compounds. In the case of over-expressing cells, the non-transfected parental cells should be run in parallel as a control; the "ratio of efflux ratios" (transfected vs.

non-transfected cells) indicates the magnitude of efflux specifically attributable to the transfected human BCRP. The opposite approach, mentioned in the EMA guideline on the investigation of drug interactions [7], is to compare bidirectional transport in BCRP knockdown cells with that in control cells. In this case, the relative efflux ratio (control vs. knockdown cells) is a very specific marker of the degree of BCRP-mediated efflux of a test compound. One advantage of the latter approach is that it reduces the reliance on pharmacologic inhibitors, which in most cases are not specific for a single transporter.

The bidirectional assay format is also used to test for BCRP inhibition; in that case, the efflux ratio (or simply the BL-to-A P_{app}) of cladribine, a BCRP probe substrate, is measured in the presence and absence of one or more concentrations of a test compound. Although cladribine is primarily a BCRP substrate, it does interact (much less) with P-gp (data not shown); in order to minimize the possibility of P-gp interference with the assessment of BCRP inhibition by a test compound, it is recommended to use cladribine with a cell line (1) specifically over-expressing BCRP (e.g., BCRP-MDCK cells) or (2) with reduced expression of P-gp (e.g., CPT-P1, engineered from Caco-2 cells with substantially silenced P-gp expression) [20].

3.4 BCRP Substrate Assessment

The plates containing the experimental monolayers are kept in a humidified incubator for the duration of the assay. The assay buffer is HBSS with 15 mM D-glucose and 10 mM HEPES, pH 7.4 ("HBSSg"). Experimental conditions and a representative sampling profile for a test compound (TC) are summarized in Table 3; the positive control BCRP substrate E3S is run in parallel. A confirmatory experiment with one or more BCRP inhibitors (following a 30-min pre-incubation), summarized in Table 4, is performed either in parallel (for all test compounds) or subsequently (both with and without inhibitor(s), only for positives from the initial screen; **Note 2**). The receiver compartment is sampled (aliquots of 200 μL) at one or more pre-selected time points (a single 120 min receiver time point is shown in the experimental design summarized in Table 3). The transport characteristics of TCs are generally unknown prior to an experiment, which may necessitate collection of receiver samples at multiple time points to ensure linearity with time (required for accurate calculation of transport rate). When receiver samples are taken at multiple time points, they are replaced by an equal volume of fresh HBSSg at each time point except the final one and the calculation of cumulative receiver concentration must take into account the multiple sample/replace steps [21]. A portion of each receiver sample is used for quantification of the TC (generally by LC-MS/MS), and another portion for measurement of co-dosed LY (monolayer integrity marker). Samples (50 μL aliquots) are also

Table 3
BCRP substrate assessment (condition 1)

Pre-incubation with test compound dosing solution	Matrix composition		Sampling volume (µL)		Sampling time points (minutes)	
	AP	BL	AP	BL	AP	BL
No[a]	5 µM TC[b] or E3S + 200 µM LY	HBSSg	50	200	5 and 120	120
No[a]	HBSSg	5 µM TC[b] or E3S + 200 µM LY	200	50	120	5 and 120

[a]Note 3
[b]TC concentration should be low enough to avoid saturating the transporter while still allowing sufficient analytical sensitivity; a concentration in the range of 1–10 µM is generally used

Table 4
BCRP substrate assessment (condition 2)

30-min Pre-incubation with BCRP inhibitor[a]		AP-to-BL directional permeability		BL-to-AP directional permeability	
AP	BL	AP	BL	AP	BL
10 µM Ko143 or FTC	10 µM Ko143 or FTC	5 µM TC or E3S + Ko143 or FTC + 200 µM LY	HBSSg + Ko143 or FTC	HBSSg + Ko143 or FTC	5 µM TC or E3S + Ko143 or FTC + 200 µM LY

[a]Note 4

taken from the donor compartment without replacement (at 5 and 120 min in Table 3).

Representative results with Caco-2 cells are shown in Table 5 for cladribine as a positive "test compound" and both Ko143 and FTC as confirmatory BCRP inhibitors.

To illustrate the use of a BCRP-over-expressing cell line for substrate assessment, results for cladribine in BCRP-MDCK cells (efflux ratio of 136) are shown in Table 6, confirmed by comparison with bidirectional permeability in non-transfected MDCK cells (efflux ratio of 0.84) and in the presence of the BCRP inhibitor Ko143 (efflux ratio of 0.63).

An alternative approach is to knock down the expression of BCRP in a cell line, such as Caco-2, in which it is expressed natively. Bidirectional assays with parental (Caco-2) cells and knockdown (CPT-B1) cells yield efflux ratios ((2), Sect. 3.6), from which a

Table 5
Permeability of the BCRP substrate cladribine across Caco-2 cell monolayers

Treatment	Direction	P_{app} (10^6 cm/s) R1	R2	R3	Mean ± SD	Efflux ratio[a]
Cladribine only (10 μM)	AP-to-BL	1.89	1.05	2.17	1.71 ± 0.582	14.9
	BL-to-AP	23.7	25.4	27.1	25.4 ± 1.72	
Cladribine + FTC (10 μM)	AP-to-BL	2.78	3.73	2.71	3.07 ± 0.570	0.7
	BL-to-AP	1.84	2.26	2.21	2.10 ± 0.230	
Cladribine + Ko143 (10 μM)	AP-to-BL	2.92	1.38	3.24	2.51 ± 0.993	1.3
	BL-to-AP	2.84	3.43	3.56	3.28 ± 0.385	

[a]Equation (2), Sect. 3.6

Table 6
Transport of the BCRP substrate cladribine across BCRP-MDCK cell monolayers

Treatment	Cell line	P_{app} (10^{-6} cm/s) AP-to-BL	BL-to-AP	Efflux ratio[a]
None	BCRP-MDCK	0.15 ± 0.05	20.16 ± 1.84	136
	MDCK	0.50 ± 0.14	0.42 ± 0.02	0.84
10 μM Ko143	BCRP-MDCK	0.67 ± 0.13	0.42 ± 0.03	0.63

The concentration of cladribine was 10 μM
[a]Equation (2), Sect. 3.6

relative efflux ratio ((3), Sect. 3.6) is calculated; compounds with a relative efflux ratio ≥2 are scored as BCRP substrates, as shown in Table 7.

Figure 2 is a decision tree from the FDA draft guidance on drug interaction studies [6], illustrating the consequences of the possible outcomes of an *in vitro* BCRP substrate assessment.

3.5 BCRP Inhibitor Assessment

For assessment of BCRP inhibition, the bidirectional transport of the BCRP probe substrate cladribine is measured in the absence and presence of a TC and, in parallel, a positive control (PC) BCRP inhibitor such as Ko143. The assay conditions are summarized in Table 8. The permeability assay is preceded by a 30-min pre-incubation with TC or PC on both sides of the monolayers (**Note 4**). Following the pre-incubation, for AP-to-BL transport 0.5 mL dosing solution (cladribine with or without TC or PC, depending on the pre-incubation condition for a given monolayer) is added to the AP side, and 1.5 mL of HBSSg (with or without TC or PC) is added to the BL side. For BL-to-AP transport, 1.5 mL

Table 7
BCRP substrate assessment via BCRP knockdown

Compound	Caco-2 efflux ratio[a]	CPT-B1 efflux ratio[a]	Relative efflux ratio[b]	BCRP substrate
Mitoxantrone	54.7	3.20	17.1	Yes
Sulfasalazine	124	9.35	13.3	Yes
Doxorubicin	14.3	1.33	10.8	Yes
Estrone-3-sulfate	32.4	3.30	9.82	Yes
Etoposide	43.9	8.15	5.39	Yes
Irinotecan	40.8	10.5	3.89	Yes
Rosuvastatin	62.3	16.7	3.73	Yes
Daunorubicin	82.0	21.6	3.80	Yes
SN-38	69.6	19.6	3.55	Yes
Topotecan	17.2	5.27	3.26	Yes
Cladribine	21.7	6.90	3.14	Yes
Quinidine	12.0	9.75	1.23	No
Lamivudine	1.93	1.60	1.21	No
Tamoxifen	2.66	2.62	1.02	No
Cerivastatin	1.16	1.23	0.94	No

All compounds were tested at 3 μM
[a]Equation (2), Sect. 3.6
[b]Equation (3), Sect. 3.6

dosing solution (cladribine with or without TC or PC) is added to the BL side, and 0.5 mL of HBSSg (with or without TC or PC) is added to the AP side. The monolayers are incubated in a humidified incubator for 120 min; note that it is the rate of transport of the probe substrate cladribine (a compound whose transport characteristics are well characterized ahead of time) that is being measured, and it is known to be linear for at least 120 min. Receiver samples (200 μL aliquots) are taken at 120 min, and 50 μL aliquots are taken from the donor compartment at 5 and 120 min.

A test compound is first tested for inhibition of BCRP at a single concentration (**Note 5**); if it reduces the efflux ratio of cladribine by at least 50 %, the potency is assessed by determining the IC_{50}, as illustrated for the potent BCRP inhibitor Ko143 (IC_{50} of 0.073 μM) in Fig. 3. Figure 4 is a decision tree from the FDA draft guidance on drug interaction studies [6], illustrating the consequences of the possible outcomes of an *in vitro* BCRP inhibitor assessment.

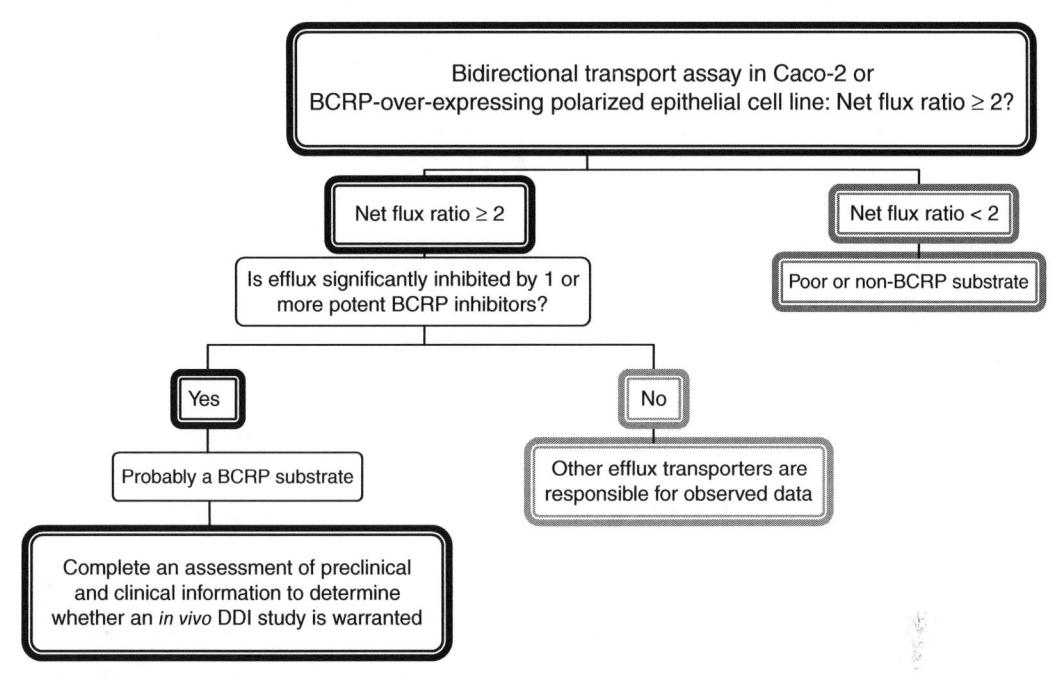

Fig. 2 Decision tree to assess BCRP substrate potential, adapted from [6]. "Net flux ratio" is identical to the efflux ratio defined above (Sect. 3.3) and in (2), Sect. 3.6. The only difference in the corresponding EMA guideline [7] is that net flux ratio >2 (as opposed to ≥) triggers further investigation

Table 8
BCRP inhibitor assessment conditions

30-min Pre-incubation[a]	Matrix composition AP	BL	Sampling volume (µL) AP	BL	Sampling time points (min) AP	BL
HBSSg with or without TC[b] or PC	Cladribine (10 µM) with or without TC[b] or PC	HBSSg with or without TC[b] or PC	50	200	5 and 120	120
HBSSg with or without TC[b] or PC	HBSSg with or without TC[b] or PC	Cladribine (10 µM) with or without TC[b] or PC	200	50	120	5 and 120

[a]**Note 4**
[b]*See* **Note 5** for selection of TC concentration

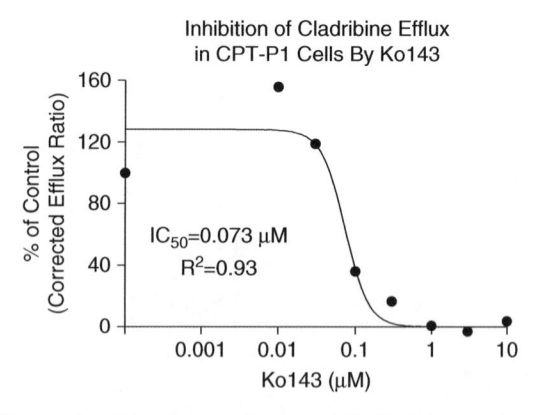

Fig. 3 Inhibition of cladribine transport across CPT-P1 cell monolayers by Ko143

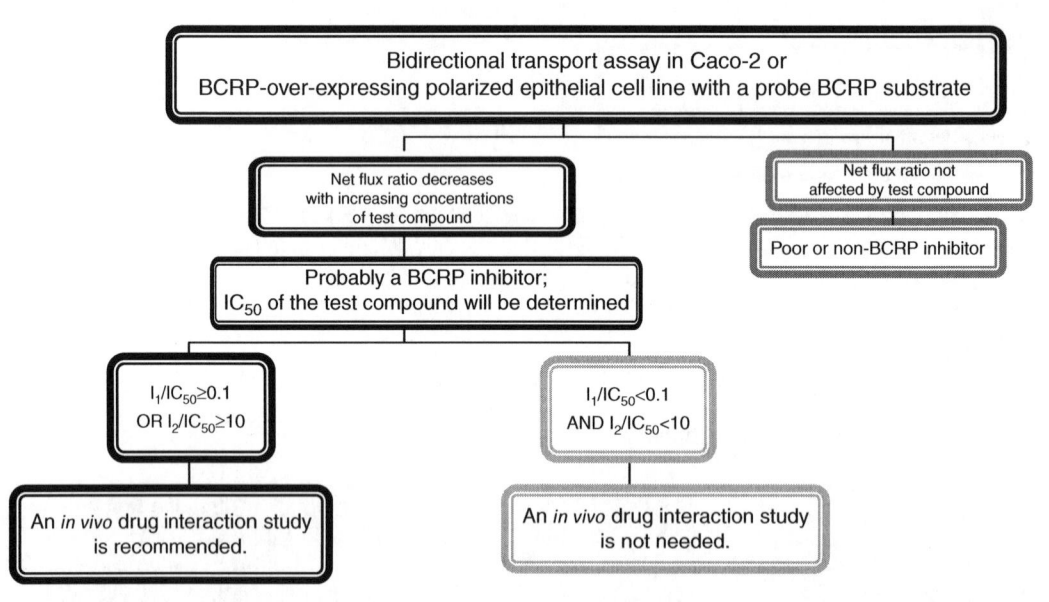

Fig. 4 Decision tree to assess BCRP inhibitor potential, adapted from [6]. I_1 is the total systemic C_{max} and I_2 is an estimate of the maximal gut concentration: the highest dose strength dissolved in 250 mL. The EMA decision tree is very similar, the one difference being a higher safety margin for I_1, which the EMA guideline [7] defines as the <u>unbound</u> systemic C_{max}: I_1/K_i (or IC_{50}) ≥ 0.02 triggers an *in vivo* drug interaction study

3.6 Calculations

$$P_{app} = (C_R / 120) \times V_R / (A \times C_{D5} \times 60) \quad (1)$$

$$\text{Efflux ratio} = \frac{P_{app} BL - to - AP}{P_{app} AP - to - BL} \quad (2)$$

$$\text{Relative efflux ratio (RER)} = \frac{\text{efflux ratio}_{Caco-2}}{\text{efflux ratio}_{KD}} \quad (3)$$

$$\text{Corrected Efflux Ratio} = \text{Efflux ratio} - 1 \qquad (4)$$

$$\text{Percentage efflux remaining} = \frac{\text{corrected efflux ratio}_{TC}}{\text{corrected efflux ratio}_{No\,TC}} \times 100 \qquad (5)$$

$$Y = Y_L + (Y_H - Y_L) / (1 + 10^{\wedge}((\text{LogIC}_{50} - X) * \text{Hillslope})) \qquad (6)$$

C_R: Concentration of test compound in the receiver compartment (μM);

V_R: Volume of the receiver compartment (BL: 1.5 cm^3; AP: 0.5 cm^3);

A: Cell monolayer area (1.13 cm^2 for a 12-well Transwell);

C_{D5}: Donor concentration at 5 min (μM);

KD: CPT-B1 (BCRP knockdown) cell line

The value "120" in (1) represents the duration of the permeability assessment (120 min), and the value "60" is a conversion factor (minutes to seconds). Values of remaining efflux activity were used for IC$_{50}$ determination by nonlinear regression using GraphPad Prism (version 5.0), with the minimum value constrained to 0. For (6), X is the logarithm of the nominal concentration of inhibitor, Y is the value of the remaining efflux activity of the transporter at a given concentration, Y_L represents the lowest response (at the highest concentration of inhibitor), and Y_H represents the highest response (vehicle control). A value of 1 % of the lowest inhibitor concentration was entered instead of zero.

4 Notes

1. Monolayers can also be cultured on inserts without collagen coating and/or with a different pore size, but in any case appropriate ranges for cell batch QC and assay performance must be established under each laboratory's set of conditions. The conditions given are those used in our laboratory.

2. If Caco-2 or BCRP-MDCK cells are used for BCRP substrate assessment, the assays are conducted with a two-step approach; the two steps can be run in parallel or sequentially. In the first step, the efflux ratio of the test compound is determined. If the efflux ratio is ≥2.0, the bidirectional permeability of the test compound will be challenged with a BCRP inhibitor such as Ko143 and/or FTC in the second step. If the efflux ratio is reduced more than 50 % by the inhibitor(s), the test compound is classified as a BCRP substrate (Tables 5 and 6). If, on the other hand, Caco-2 and CPT-B1 cells are used, classification is based on the efflux ratio of the test compound in both cell lines, followed by calculation of the relative efflux ratio

(RER); if the RER is ≥2.0, the test compound is classified as a BCRP substrate (Table 7).

3. Depending on the physicochemical properties of the test compound (i.e., if it is hydrophobic and tends to bind non-specifically to plastic), a brief (5- or 10-min) pre-incubation with the test compound may be needed to reduce non-specific binding to the experimental device.

4. For assessment of BCRP inhibition, pre-incubation with test compound was recommended in the 2006 FDA draft guidance on drug interaction studies, and a 30-min pre-incubation has been used in our laboratory since then.

5. For compounds intended for oral administration, the test concentration in the inhibitor assay is typically either 10 μM or 0.1× the concentration equal to the highest intended clinical dose strength dissolved in 250 mL ($[I]_2$ [6]). For non-orally administered compounds, the test concentration in the inhibitor assay is typically either 10 μM or 10× the clinical total (bound plus unbound) systemic C_{max} ($[I]_1$ [6]). It also may depend on the results of suitability assessments such as solubility of the test compound and tolerability of the cells.

References

1. The International Transporter Consortium, Giacomini KM et al (2010) Membrane transporters in drug development. Nature Rev Drug Discov 9(3):215–236
2. Doyle LA, Yang W, Abruzzo LV, Krogmann T, Gao Y, Rishi AK, Ross DD (1998) A multidrug resistance transporter from human MCF-7 breast cancer cells. Proc Natl Acad Sci USA 95:15665–15670
3. Miyake K, Mickley L, Litman T, Zhan Z, Robey R, Cristensen B, Brangi M, Greenberger L, Dean M, Fojo T, Bates SE (1999) Molecular cloning of cDNAs which are highly overexpressed in mitoxantrone-resistant cells: demonstration of homology to ABC transport genes. Cancer Res 59(1):8–13
4. Litman T, Brangi M, Hudson E, Fetsch P, Abati A, Ross DD, Miyake K, Resau JH, Bates SE (2000) The multidrug-resistant phenotype associated with overexpression of the new ABC half-transporter, MXR (ABCG2). J Cell Sci 113(11):2011–2021
5. Polli JW, Olson KL, Chism JP, St. John-Williams L, Yeager RL, Woodard SM, Otto V, Castellino S, Demby VE (2009) An unexpected synergist role of P-glycoprotein and breast cancer resistance protein on the central nervous system penetration of the tyrosine kinase inhibitor lapatinib (N-{3-chloro-4-[(3-fluorobenzyl)oxy]phenyl}-6-[5-({[2-(methylsulfonyl)ethyl]amino}methyl)-2-furyl]-4-quinazolinamine; GW572016). Drug Metab Dispos 37(2):439–442
6. US Department of Health and Human Services, Food and Drug Administration, Center for Drug Evaluation and Research (CDER) (2012) Draft guidance for industry. Drug interaction studies—study design, data analysis, implications for dosing, and labeling recommendations. http://www.fda.gov/downloads/Drugs/GuidanceComplianceRegulatoryInformation/Guidances/UCM292362.pdf
7. European Medicines Agency, Committee for Human Medicinal Products (CHMP) (2012) Guideline on the investigation of drug interactions. http://www.ema.europa.eu/docs/en_GB/document_library/Scientific_guideline/2012/07/WC500129606.pdf
8. Breedveld P, Zelcer N, Pluim D, Sonmezer O, Tibben MM, Beijnen JH, Schinkel AH, van Tellingen O, Borst P, Schellens JH (2004) Mechanism of the pharmacokinetic interaction between methotrexate and benzimidazoles: potential role for breast cancer resistance pro-

tein in clinical drug–drug interactions. Cancer Res 64(16):5804–5811
9. Jonker JW, Smit JW, Brinkhuis RF, Maliepaard M, Beijnen JH, Schellens JH, Schinkel AH (2000) Role of breast cancer resistance protein in the bioavailability and fetal penetration of topotecan. J Natl Cancer Inst 92(20):1651–1656
10. Highlights of prescribing information for TYKERB® (lapatinib). http://www.accessdata.fda.gov/drugsatfda_docs/label/2010/022059s007lbl.pdf
11. Highlights of prescribing information for HYCAMTIN® (topotecan). http://us.gsk.com/products/assets/us_hycamtin_capsules.pdf
12. Ifergan I, Shafran A, Jansen G, Hooijberg JH, Scheffer GL, Assaraf YG (2004) Folate deprivation results in the loss of breast cancer resistance protein (BCRP/ABCG2) expression. A role for BCRP in cellular folate homeostasis. J Biol Chem 279(24):25527–25534
13. Hosomi A, Nakanishi T, Fujita T, Tamai I (2012) Extra-renal elimination of uric acid via intestinal efflux transporter BCRP/ABCG2. PLoS One 7(2):e30456
14. Li J, Wang Y, Zhang W, Huang Y, Hein K, Hidalgo IJ (2012) The role of a basolateral transporter in rosuvastatin transport and its interplay with apical breast cancer resistance protein in polarized cell monolayer systems. Drug Metab Dispos 40(11):2102–2108
15. Kitamura S, Maeda K, Wang Y, Sugiyama Y (2008) Involvement of multiple transporters in the hepatobiliary transport of rosuvastatin. Drug Metab Dispos 36(10):2014–2023
16. Zhang W, Li J, Allen SM, Weiskircher EA, Huang Y, George RA, Fong RG, Owen A, Hidalgo IJ (2009) Silencing the breast cancer resistance protein expression and function in Caco-2 cells using lentiviral vector-based short hairpin RNA. Drug Metab Dispos 37(4):737–744
17. Li J, Volpe DA, Wang Y, Zhang W, Bode C, Owen A, Hidalgo IJ (2011) Use of transporter knockdown Caco-2 cells to investigate the in vitro efflux of statin drugs. Drug Metab Dispos 39(7):1196–1202
18. Huang L, Wang Y, Grimm S (2006) ATP-dependent transport of rosuvastatin in membrane vesicles expressing breast cancer resistance protein. Drug Metab Dispos 34(5):738–742
19. Karlsson JE, Heddle C, Rozkov A, Rotticci-Mulder J, Tuvesson O, Hilgendorf C, Andersson TB (2010) High-activity P-glycoprotein, multidrug resistance protein 2, and breast cancer resistance protein membrane vesicles prepared from transiently transfected human embryonic kidney 293-Epstein-Barr virus nuclear antigen cells. Drug Metab Dispos 38(4):705–714
20. Darnell M, Karlsson JE, Owen A, Hidalgo IJ, Li J, Zhang W, Andersson TB (2010) Investigation of the involvement of P-glycoprotein and multidrug resistance-associated protein 2 I the efflux of ximelagatran and its metabolites by using short hairpin RNA knockdown in Caco-2 cells. Drug Metab Dispos 38:491–497
21. Hidalgo IJ (1996) Cultured intestinal epithelial cell models. In: Borchardt RT et al (eds) Models for assessing drug absorption and metabolism. Plenum Press, New York, pp 35–50

Chapter 22

In Vitro Characterization of Intestinal and Hepatic Transporters: MRP2

Ravindra Varma Alluri, Peter Ward, Jeevan R. Kunta, Brian C. Ferslew, Dhiren R. Thakker, and Shannon Dallas

Abstract

The transporter field has grown extensively over the past decade. Analogous to drug metabolizing enzymes such as the cytochrome P450s, transporters play a major role in defining pharmacokinetic, safety and efficacy profiles of drugs. Multidrug resistance-associated protein 2 (MRP2, *ABCC2*) is an ATP-dependent efflux pump that belongs to the ATP binding cassette (ABC) superfamily of transporters, and is localized at the apical membrane of polarized cells from a variety of human tissues including enterocytes, hepatocytes and renal proximal tubules. It is highly expressed in liver, intestine and kidney, with lesser expression in other tissues. MRP2 primarily transports organic anions and large bulky conjugated compounds and shares some overlapping substrate specificity with other ABC family members including P-glycoprotein (P-gp) and the breast cancer resistance protein (BCRP).

Understanding whether investigational compounds are potential MRP2 substrates or inhibitors during drug discovery and development may potentially help to explain why drug candidates show poor bioavailability or are rapidly cleared by hepatic efflux. This chapter outlines various *in vitro* techniques that can be used to examine whether compounds are substrates and/or inhibitors of MRP2 (ATPase assays, vesicular transport assays and/or MDCKII-MRP2 overexpressing cells) and assess the role of MRP2 in attenuating intestinal absorption of drugs (wild-type and MRP2 knockdown Caco-2 cells) or in mediating their hepatobiliary excretion (wild-type and MRP2 knockdown human hepatocytes cultured in sandwich configuration). The primary aim of the chapter is to provide a range of assay options. However, the strategy around when/if/why/ or how a specific assay(s) should be used will depend on a number of factors such as physiochemical properties, drug target, overall distribution, etc, and is therefore ultimately left to the reader.

Key words ABCC2, Drug-drug interactions, Efflux, Intestinal MRP2, Hepatic MRP2, ATPase assays, Vesicular transport assays, MDCKII, Caco-2, Sandwich-cultured hepatocytes, MRP2 knockdown

1 Introduction

The ATP binding cassette (ABC) superfamily of transporters play a major role in influencing the absorption and disposition of drugs [1]. The ABCC subfamily contains nine transporters, which differ in structures, substrate specificities and intracellular localization [2, 3].

In this chapter, *in vitro* methods for one member of this subfamily, namely the multidrug resistance-associated protein 2 (MRP2; *ABCC2*), will be discussed in detail, with emphasis on its potential role in the intestinal absorption and hepatobiliary excretion of drug candidates. In the intestine, MRP2 is localized at the apical (AP) membrane of enterocytes. Its expression is highest in the proximal duodenum, with progressively lower expression in the jejunum and ileum [4]. MRP2 is also expressed on the canalicular membrane of hepatocytes [5] where it plays an important role in detoxification by transporting large bulky anions such as endogenous glucuronide and glutathione conjugates into bile. Therefore, MRP2 substrates are structurally diverse and include conjugates of lipophilic compounds with glutathione, glucuronate and sulfate [6, 7]. MRP2 also transports some unconjugated compounds such as vincristine through co-transport with glutathione [8]. Kinetic studies with different substrates have shown evidence for at least two similar but non-identical ligand binding sites on MRP2, resulting in complex inhibition and stimulation patterns [9, 10]. Despite not being emphasized in recent regulatory agency guidelines on drug-drug interactions [11, 12], the possible role of MRP2 in limiting the intestinal absorption of drugs and mediating canalicular efflux of many compounds suggests the need to understand if compounds are potential substrates and/or inhibitors of MRP2 [13–17].

Several *in vitro* test systems can be used to identify whether compounds are potential substrates or inhibitors of MRP2. It is essential that these *in vitro* systems are well characterized with known substrates and inhibitors to verify the validity of the test results [18]. Also, the robustness of the experiments should be demonstrated by determining the intra-plate and inter-day variability of the results. The systems, discussed below, can be used in a tiered approach from simple to complex (from the perspective of time, resources, impact (risk/benefit ratio) and clinical relevance) to gain a broader mechanistic understanding on the potential role of MRP2 in influencing the pharmacokinetic behavior of drug candidates. The choice of experiment can also be driven by the (apparent) permeability and/or lipophilicity of the test compounds. Data obtained from these *in vitro* systems, along with other relevant data on compounds (chemical structure, dose, exposure, co-medications in target population, etc.) can help to guide the relevance of MRP2 in the overall drug disposition of a compound, potential back-up compound strategies to discharge potential MRP2 liabilities if desired, and evaluate the need to conduct dedicated clinical drug-drug interaction studies later in drug development.

2 Membrane Based ATPase Assay to Screen MRP2 Substrates and Inhibitors

The ATPase assay is a relatively inexpensive and high throughput assay that can be readily automated for screening test compounds that interact with MRP2, both as inhibitors and substrates. The principle of ATPase assay is based on the utilization of ATP by ABC transporters as the energy source to translocate substrates against a concentration gradient. The inorganic phosphate (P_i) that is released following hydrolysis of ATP in the presence of interacting compounds is used as a measure of MRP2 activity. This assay is performed with commercially available membrane preparations from recombinant baculovirus-infected *Spodoptera frugiperda* ovarian (Sf9) cells that overexpress MRP2. These membranes exhibit ATPase activity that can be inhibited by sodium orthovanadate (NaOV), which is a well characterized inhibitor of ATPase. However, the ABC transporter membranes also exhibit ATPase activity that is insensitive to inhibition by NaOV. To obtain inhibitor sensitive ATPase activity, the assays are undertaken in the presence and absence of NaOV. To determine whether a compound is a substrate, the "activation" mode of the assay is performed by incubating test compounds with cell membranes and ATP in the presence and absence of NaOV, and the release of P_i is measured colorimetrically [19–23]. An appropriate positive control stimulator (MRP2 substrate) such as probenecid is included in each assay. The data are reported as fold-stimulation in the presence of test compounds relative to basal ATPase activity in the presence of a vehicle control. To determine whether a compound is an inhibitor, the "inhibition" mode of the assay is performed by incubating a known MRP2 substrate (e.g. probenecid) with cell membranes in the presence and absence of multiple concentrations of test compounds and modulation of the probe substrate stimulated ATPase activity is measured [24]. A positive control inhibitor such as the general MRP inhibitor MK-571 is also included and IC_{50} values can be calculated. It is however worth pointing out that the interpretation of data generated in this assay needs to be weighed against some of the drawbacks of this technique such as (1) inconsistency between ATPase activity and the transport rate of some substrates and inhibitors, (2) high incidence of false positives and negatives and (3) requirement of high substrate concentrations [18].

2.1 Materials

Chemicals, reagents and equipment that can be used for multiple MRP2 methodologies listed in this chapter are given in Table 1 (**Note 1**). Chemicals and reagents needed specifically for the ATPase assay are detailed in Table 2.

Table 1
Chemicals, reagents and equipment used for multiple MRP2 methodologies

Chemical/reagents/membranes/equipment	Catalog #	Vendor
(Hydroxymethyl)aminomethane (Tris)	252859	Sigma, St. Louis, MO, USA
Potassium chloride (KCl)	P9541	Sigma, St. Louis, MO, USA
Adenosine 5′-triphosphate magnesium salt (MgATP)	A9187	Sigma, St. Louis, MO, USA
(E)-3-(((3-(2-(7-chloro-2-quinolinyl) ethenyl)phenyl)((3-(dimethylamino)-3-oxopropyl)thio)methyl)thio)-propanoic acid, sodium salt (MK-571 sodium salt)	70720	Cayman Chemical Co. Ann Arbor, Michigan, USA
Dimethyl sulfoxide (DMSO)	D2438	Sigma, St. Louis, MO, USA
Methanol	34860	Sigma, St. Louis, MO, USA
Fetal bovine serum (FBS)	S12450	Atlanta Biologicals, Lawrenceville, GA, USA
Minimum essential medium nonessential amino acids (MEM NEAA)	11140-050	Life Technologies, Grand Island, NY, USA
Antiobiotic-Antimycotic (ABX)	15240-062	Life Technologies, Grand Island, NY, USA
Hank's Balanced Salt Solution (1×) with calcium and magnesium (HBSS)	21-023-CV	Cellgro, Manassas, VA, USA
HEPES buffer	15630-106	Life Technologies, Grand Island, NY, USA
Glucose	G7528	Sigma, St. Louis, MO, USA
Sulfobromophthalein disodium salt hydrate (BSP)	S0252	Sigma Aldrich, St. Louis, MO, USA
Mannitol	M4125	Sigma Aldrich, St. Louis, MO, USA
(^{3}H/^{14}C) Mannitol	NET101001MC/ NEC314050UC	Perkin Elmer, Waltham, MA, USA
Epithelial Voltohmmeter (EVOM2)	EVOM2	World Precision Instruments, Sarasota, FL, USA
Chopstick electrode	STX2	World Precision Instruments, Sarasota, FL, USA
Forma Series II Water-Jacketed CO_2 Incubators	3110	Thermo Scientific, West Palm Beach, FL, USA
Biological safety cabinet (Type II)	NU-427	NUAIRE, Plymouth, MN, USA
Inverted microscope	TS100	Nikon Eclipse, Melville, NY, USA
Transwell tissue culture plates (6-well, 12-well, 24-well polycarbonate membrane)	3412, 3401, 3397	Corning, NY, USA
T-75 Flasks	430641	Corning, NY, USA

Table 2
Chemicals and reagents required for conducting MRP2 ATPase assays

Chemical/reagents/membranes/equipment	Catalog #	Vendor
Human MRP2 expressing Sf9 membranes (5 mg/mL)	453332 or SB-MRP2-sf9-ATPase	BD Gentest, Woburn, MA, USA or Solvo Biotechnology, Boston, MA, USA
2-(N-Morpholino)ethanesulfonic acid hydrate (MES hydrate)	69890	Sigma, St. Louis, MO, USA
Dithiothreitol (DTT)	D9779	Sigma, St. Louis, MO, USA
Ethylene glycol-bis (2-aminoethylether)-N,N,N′,N′-tetraacetic acid (EGTA)	E3889	Sigma, St. Louis, MO, USA
Sodium azide	71289	Sigma, St. Louis, MO, USA
Sodium dihydrogen phosphate (NaH_2PO_4)	S3139	Sigma, St. Louis, MO, USA
Probenecid	P8761	Sigma, St. Louis, MO, USA
Sodium orthovanadate (NaOV)	450243	Aldrich, St. Louis, MO, USA
Sodium dodecyl sulfate (SDS)	L4390	Sigma, St. Louis, MO, USA
Ammonium molybdate	277908	Aldrich, St. Louis, MO, USA
Zinc acetate	383317	Aldrich, St. Louis, MO, USA
Ascorbic acid	A0278	Sigma, St. Louis, MO, USA
Absorbance Microplate Reader	SpectraMax Plus384	Molecular Devices, Sunnyvale, CA, USA
Single and multichannel pipets	–	Eppendorf, Hamburg, Germany
Flat clear 96-well plates	353075	BD Gentest, Woburn, MA, USA
Microplate shaker	12620-926	VWR, Radnor, PA, USA
Water bath	2864	Thermo Scientific, West Palm Beach, FL, USA

2.2 Reagent Composition

Store all reagents at −20 °C. Stop solution can be stored at room temperature or at 4 °C. Prepare all reagents in phosphate free water and organic solvents. Perform work in a phosphate free environment (**Notes 2** and **3**).

1. Assay buffer (50 mM Tris-MES, pH 6.8 containing 50 mM KCl, 2 mM DTT, 2 mM EGTA, 5 mM Sodium Azide)
2. 50 mM MgATP in water
3. 10 mM NaH_2PO_4 (*phosphate standard*)
4. 100 mM probenecid in DMSO (general *MRP substrate—positive control*)
5. MK-571 (*general MRP inhibitor*)
6. 10 mM NaOV in water (*ATPase inhibitor*)

7. 10 % SDS in water (*stop solution*)

8. 70 mM ammonium molybdate pH 5.0 and 30 mM zinc acetate pH 5.0 (*colorimetric reagents*)

9. 10 % ascorbic acid pH 5.0 (*reducing agent*) (**Note 4**)

2.3 Assay Preparation

Remove all reagents from −20 °C. Thaw MgATP on ice. The remaining reagents can be rapidly thawed in a 37 °C water bath. After thawing, place all reagents on wet ice. Assay buffer, 10 % SDS and probenecid solutions can be kept at room temperature.

1. Phosphate standards: Prepare blank and seven different standards (0, 3, 9, 30, 60, 90, 120 and 150 nmol) by diluting 10 mM NaH_2PO_4 in assay buffer (**Note 5**).

2. Stock solutions for screening compounds as substrates of MRP2: Prepare stock solutions of test compound and probenecid in a suitable solvent at ≥50× the final concentration. The organic content in the final incubations should be less than 2 % when DMSO is used as a solvent (**Note 6**). Dilute the stock solution of test compounds and probenecid separately in assay buffer to prepare 3× working stocks of 60 µM and 3 mM, respectively. To assess baseline ATPase activity, prepare a vehicle control by adding the appropriate volume of vehicle to the assay buffer (**Note 7**).

3. Stock solutions for screening compounds as inhibitors of MRP2: Prepare seven stock solutions (≥100×) of test compound and MK-571 ranging between 0.01 and 10 mM in a suitable solvent. Dilute probenecid in assay buffer to a concentration of 3 mM (3×). Split solution into eight aliquots. To the first aliquot, add an appropriate volume of vehicle control. To the remaining seven aliquots, add an appropriate volume of different concentrations of test compounds or MK-571 to prepare 3× working stock solutions ranging between 0.3 and 300 µM. To assess baseline ATPase activity, prepare a vehicle control by adding the appropriate volume of vehicle to the assay buffer. The final concentration of organic (<2 % total) should be equivalent between all the samples.

4. Dilution of Membranes: Dilute MRP2 membranes (5 mg/mL) to a concentration of 1 mg/mL (3×) in assay buffer, divide into two aliquots and add an appropriate volume of 10 mM NaOV to one of the aliquots to prepare a 3× working stock solution of 1.2 mM. Add an equal volume of water to the second sample (**Note 8**).

5. Preparation of MgATP Solution: Dilute 50 mM MgATP in assay buffer to prepare 3× working stock of 12 mM.

6. Preparation of Colorimetric Solution: Add 2.5 mL of 70 mM ammonium molybdate pH 5.0 and 2.5 mL of 30 mM zinc acetate pH 5.0 to 20 mL of freshly prepared 10 % ascorbic acid pH 5.0.

2.4 Assay Procedure

1. With a multichannel pipette, add 60 µL of phosphate standards in duplicate to the first two columns of a 96-well plate. Transfer 20 µL of cell membranes with and without NaOV in triplicate into separate wells of the plate (**Note 9**). The final concentration of membranes is 0.02 mg/well.

2. When screening substrates, add 20 µL of the test compound, probenecid (positive control) or assay buffer containing vehicle to the respective wells with the cell membranes and incubate at 37 °C for 5 min on a microplate shaker (~100 rpm). The final concentration of test compound and probenecid will be 20 µM and 1 mM, respectively (**Note 10**).

3. When screening inhibitors, add 20 µL of vehicle control, probenecid, probenecid + different concentrations of the test compounds or MK-571 to the appropriate wells and incubate at 37 °C for 5 min on a microplate shaker. The final concentration of test compounds and MK-571 typically range between 0.1 and 100 µM (**Note 11**).

4. Initiate the reaction by adding 20 µL of 12 mM MgATP solution to wells containing membranes (4 mM MgATP final concentration). Incubate the plate at 37 °C for 40 min on a microplate shaker (~100 rpm).

5. Stop the reaction by adding 30 µL of 10 % SDS to all wells, including the phosphate standards (**Note 12**).

6. Add 200 µL of colorimetric solution to all wells, including standards, and incubate for 20 min at 37 °C with gentle shaking (**Note 13**).

7. Determine absorbance at 800 nm using a SpectraMax M2 or similar spectrophotometer.

2.5 Data Analysis

1. Generate a standard curve for inorganic phosphate by plotting nmoles of phosphate on the X-axis versus the corresponding optical density (OD) values on the y-axis and run linear regression analysis to determine the slope, r^2 and intercept values. Alternatively, some spectrophotometers are already programmed to calculate these values.

2. Determine the amount of inorganic phosphate formed in the test, probenecid, probenecid + test compound or MK-571 and vehicle control samples with the linear regression values obtained from the phosphate standard curve. Calculate mean values of inorganic phosphate from triplicate data.

3. Subtract (mean data with NaOV) from (mean data without NaOV) to obtain the mean nmoles of NaOV sensitive inorganic phosphate generated.

4. Calculate NaOV sensitive ATPase activity (nmol/min/mg protein) with (1) shown below [22]:

$$\text{ATPase activity (nmol/min/mg protein)} = \text{(generated inorganic phosphate (nmol))} / \text{(reaction time (min))} / \text{(mg protein)} \quad (1)$$

where: Protein (1 mg/mL) × 20 μL/well = 20 μg/well (0.02 mg/well) and Reaction time = 40 min.

5. Calculate probenecid stimulated ATPase activity in absence and presence of different concentrations of test compounds or MK-571 to determine relative IC_{50} values (**Note 14**). Calculate percent inhibition by dividing ATPase activity at different concentrations of test compounds by the vehicle control ATPase activity. Determine IC_{50} values by nonlinear regression using (2).

$$y = Bottom + \frac{(Top - Bottom)}{(1 + 10^{\wedge}((LogIC_{50} - X) * Hillslope))} \quad (2)$$

where X: Log of concentration, Y: Response, decreases as X increases, Top and Bottom: Plateaus in same units as Y, and Hill Slope: Slope factor, unit less

2.6 Data Reporting and Interpretation

2.6.1 Substrate Screening

Report test compound and positive control stimulated ATPase activity as fold-stimulation relative to baseline ATPase activity in the absence of drug (vehicle control).

Calculation:

- Probenecid stimulated ATPase activity: X nmoles phosphate/min/mg protein
- Basal activity (vehicle control): Y nmoles phosphate/min/mg protein
- Fold-stimulation = Probenecid stimulated ATPase activity/Basal ATP activity (vehicle control)

A compound is classified as a substrate if the fold-stimulation is greater than two-fold over vehicle control (**Note 15**). Validation studies with a known set of weak and strong substrates will further help in classifying the test compounds as weak or strong substrates of MRP2 (*see* Fig. 1).

2.6.2 Inhibitor Screening

Report $IC_{50} \pm$ S.E. values as calculated. Test compounds can be classified as weak (>10 μM), moderate (1–10 μM) or strong (<1 μM) inhibitors by benchmarking against positive controls

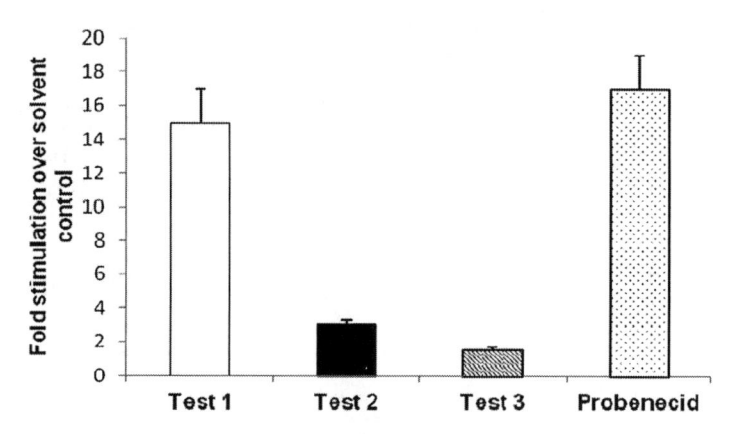

Fig. 1 Representative data showing fold stimulation of MRP2 ATPase activity by three different test and positive control compounds. Greater than twofold stimulation over the solvent control classifies test compounds 1, 2 and the positive control probenecid as MRP2 substrates, whereas test compound 3 (Ratio <2) is not a MRP2 substrate with this assay

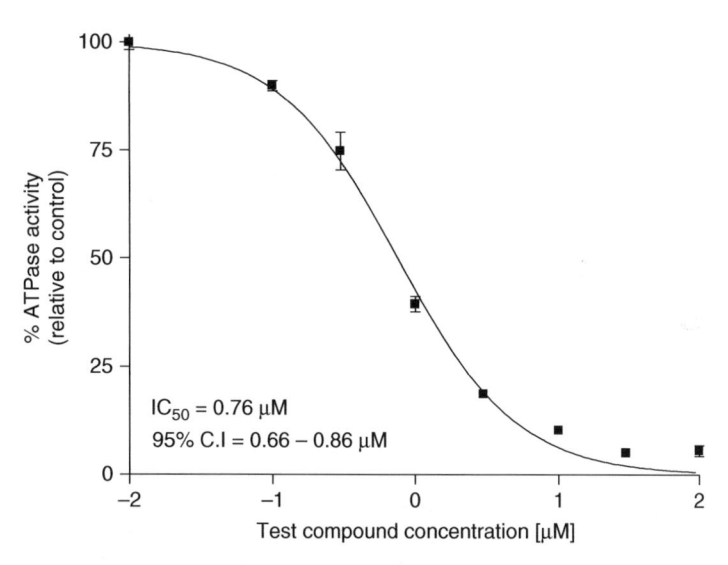

Fig. 2 Representative data showing the dose-response curves for inhibition of probenecid stimulated ATPase activity in the presence of different concentrations of test compound

(*see* Fig. 2). The interpretation of the degree of inhibition (weak to strong) may vary based on internal validation data sets in any given laboratory.

Since ATPase assays are not functional assays, additional systems such as inside-out oriented vesicles can be used as alternative approaches for screening substrates and inhibitors of MRP2.

3 Vesicular Transport Assays

Vesicular transport assays are used to measure the actual disposition of the test compounds across cell membranes. These assays are performed with inside-out-oriented vesicles prepared from cell membranes of different sources (e.g. insect cells (Sf9) or mammalian cell lines over expressing MRP2) [24–26]. These vesicles contain an ATP binding site and a substrate binding site of the transporter facing the buffer side, which is particularly useful for compounds that would otherwise not be permeable in a cellular system such as conjugated compounds. Additionally, issues around potential compound metabolism during the assay are mitigated since the background cells are generally non-liver derived and have limited metabolic capabilities. For screening substrates of MRP2, a 'direct mode' of assay is performed in which the translocation of test compounds into the vesicles is determined in the presence and absence of ATP. Studies with a known positive control substrate such as tritium labeled leukotriene C4 ((^{3}H)-LTC4) or estradiol-17-β-D-glucuronide ((^{3}H)-E$_2$ 17βG) are also included. The data generated from these studies are represented as ATP-dependent uptake activity. The direct mode of assay is sensitive to the passive permeability of test compounds and is more suited for low permeability compounds [27]. Compounds with medium to high passive permeability will not be retained inside the vesicles making the transport measurements difficult to perform and interpret. For screening compounds as inhibitors, an 'indirect mode' of the assay is performed where inhibition of a known probe substrate (e.g. fluorescence compound: 5(6)-Carboxy-2',7'-Dichlorofluorescein (CDCF) or radiolabeled compound ((^{3}H)-LTC4 or (^{3}H)-E$_2$ 17βG) uptake into vesicles is determined in the presence of single or multiple concentrations of test compounds and a positive control inhibitor such as MK-571 [28–30]. The assay is performed in the presence and absence of ATP at all concentrations and the data can be reported as IC$_{50}$ values. This assay format is not sensitive to the passive permeability of test compounds, or to inhibition of a probe substrate by a very low affinity investigational compound; interpretation of the data should be performed carefully [24, 25, 28, 31–33].

3.1 Materials

See Table 3.

3.2 Reagent Composition

Store assay buffer and wash buffer at 4 °C. Rest of the chemicals/reagents should be stored at –20 °C. Stock solution of LTC4 should be stored at –80 °C (**Note 16**).

1. Assay buffer (50 mM MOPS-Tris, 65 mM KCl, 7.5 mM MgCl$_2$, pH 7.0)
2. Wash Buffer (50 mM MOPS-Tris, 70 mM KCl, pH 7.0)
3. 200 mM MgATP in water (*Cofactor*)

Table 3
Chemicals, reagents and equipment required for conducting MRP2 vesicular assays

Chemical/reagents/membranes/equipment	Catalog #	Vendor
Human MRP2 vesicles (5 mg/mL)	453450 or SB MRP2 Sf9 VT Membrane	BD Gentest, Woburn, MA, USA or Solvo Biotechnology, Boston, MA, USA
3-(N-morpholino)propanesulfonic acid (MOPS)	M9381	Sigma, St. Louis, MO, USA
Magnesium chloride (MgCl$_2$)	M8266	Sigma, St. Louis, MO, USA
Glutathione (GSH)	G4251	Sigma, St. Louis, MO, USA
LTC4	L4886	Sigma, St. Louis, MO, USA
(^{3}H)-LTC4	–	Perkin Elmer, Waltham, MA, USA
E$_2$ 17βG	E1127	Sigma, St. Louis, MO, USA
(^{3}H)-E$_2$ 17βG	–	Perkin Elmer, Waltham, MA, USA
5(6)-Carboxy-2′,7′-Dichlorofluorescein (CDCF)	21884	Sigma, St. Louis, MO, USA
Sodium hydroxide	S8045	Sigma, St. Louis, MO, USA
Betaplate scintillation fluid	1205-440	Perkin Elmer, Waltham, MA, USA
Glass-fiber filter plates	6005177 or MAFBN0B	Perkin Elmer, Waltham, MA, USA or Millipore, Billerica, MA, USA
Wallac 1450 MicroBeta Trilux,	1450	Perkin Elmer, Waltham, MA, USA
Cell harvester or vacuum manifold	C961960 or MSVMHTS00	Perkin Elmer, Waltham, MA, USA Millipore, Billerica, MA, USA
Water bath	2864	Thermo Scientific, West Palm Beach, FL, USA

4. 300 mM GSH in water (*Cofactor*)
5. 1 mM CDCF in DMSO (*Fluorescence substrate for MRP2*)
6. 100 µM LTC4 in DMSO (*General MRP2 substrate*)
7. (^{3}H) LTC4 in ethanol (**Note 17**)
8. 100 µM E$_2$ 17βG in assay buffer (*General MRP2 substrate*)
9. (^{3}H) E$_2$ 17βG in ethanol (**Note 17**)

3.3 Assay Preparation [25, 32, 33] Warm assay buffer to 37 °C in a water bath. Remove rest of the reagents/chemicals from −20 or −80 °C and thaw at room temperature. After thawing, place all reagents on wet ice. Assay buffer and probe substrates can be kept at room temperature. Store wash buffer at 4 °C or on wet ice before use.

1. <u>Vesicular/substrate and cofactor mix for screening compounds as substrates of MRP2:</u> Prepare stock solutions of test compound, LTC4 or E_2 17βG in a suitable solvent at ≥50× the final concentration (**Note 6**). Dilute the stock solution of test compounds in assay buffer to prepare 1.5× working stock of 1.5 μM (final concentration 1 μM) (**Note 18**). Similarly, prepare 0.15 μM (1.5×) working stock solutions of LTC4 (final concentration 0.1 μM) or 1.5 μM (1.5×) E_2 17βG (final concentration 1.0 μM) by adding appropriate volumes of labeled and/or unlabeled LTC4 or E_2 17βG to the assay buffer (**Note 19**). Add an appropriate volume of 300 mM GSH to these working stock solutions to get a concentration of 3 mM (1.5×) (final concentration 2.0 mM). Dilute MRP2 vesicles (5 mg/mL) in the above working stock solutions to a concentration of 0.833 mg/mL. The final vesicular concentration will be 0.05 mg/well. The total organic content in the final incubations should not exceed 2 %.

2. <u>Vesicular/substrate and cofactor mix for screening compounds as inhibitors of MRP2:</u> Prepare seven stock solutions (≥100×) of test compound and MK-571 ranging between 0.01 and 10 mM in a suitable solvent (final concentrations of test compounds are 0.1 and 100 μM) (**Note 11**). Dilute CDCF or labeled and/or unlabeled LTC4 or E_2 17βG in assay buffer to a concentration of 7.5 μM (1.5×, final concentration 5 μM), 0.15 μM (1.5×, final concentration 0.1 μM) and 1.5 μM (1.5×, final concentration 1.0 μM), respectively (**Notes 19, 20**). Add an appropriate volume of 300 mM GSH to these working stock solutions to get a concentration of 3 mM (1.5×, final concentration 2.0 mM). Dilute MRP2 vesicles (5 mg/mL) in the above working stock solutions to a concentration of 0.845 mg/mL. The final vesicular concentration will be 0.05 mg/well. The total organic solvent content in the final incubations should not exceed 2 %.

3. <u>Preparation of MgATP:</u> Dilute 200 mM MgATP in assay buffer to prepare 5× working stock solution of 25 mM (final concentration 5 mM).

3.4 Assay Procedure

3.4.1 Substrate Screening: Reaction Initiation

1. With a multichannel pipette, add 60 μL of the above vesicular/substrate and cofactor mix (prepared in Sect. 3.3, step 1) in triplicate to the appropriate wells of 96 well plates.

2. Prepare 25 mM ATP in assay buffer as described above (Sect. 3.3, 3).

3. Preincubate the plates at 37 °C for 5 min on a microplate shaker (~100 rpm). Preincubate the diluted ATP (25 mM) at 37 °C for the same amount of time.

4. With a multichannel pipette, initiate the reaction by adding 15 μL of 25 mM ATP or 15 μL of assay buffer (control) to the appropriate wells and incubate at 37 °C for 4 min (**Note 21**).

3.4.2 Inhibitor Screening with Fluorescence or Unlabeled and/or Radiolabeled Substrates: Reaction Initiation

1. With a multichannel pipette, add 57.25 μL of the above mix (prepared in Sect. 3.3, 2) in duplicates to the appropriate wells of 96 well plates.

2. Add 0.75 μL of serially diluted stock solutions of test compounds or MK-571 to the appropriate wells.

3. Prepare 25 mM ATP in assay buffer as described above (Sect. 3.3, 3).

4. Preincubate the plates at 37 °C for 5 min on a microplate shaker (~100 rpm). Preincubate the diluted ATP at 37 °C for the same amount of time.

5. With a multichannel pipette, initiate the reaction by adding 15 μL of 25 mM ATP or 15 μL of assay buffer (control) to the appropriate wells. Incubate the wells containing CDCF as probe substrate for 15 min. Similarly, incubate the wells containing LTC4 or E$_2$ 17βG as probe substrates for 4 or 10 min, respectively.

3.4.3 Reaction Termination

This step is common for both screening modes.

1. Stop the reaction by adding 200 μL of ice cold 1× wash buffer to all the wells with a multichannel pipette.

2. Transfer the entire contents in the wells to glass-fiber filter plates and place on a vacuum manifold. Apply vacuum. Wash the filter plate five times with 200 μL of wash buffer and allow to dry at room temperature for 2–3 h (**Note 22**).

3.4.4 Sample Preparation

1. Unlabeled substrates: Following drying, pass ~1 mL of methanol/water (80:20) or ethanol (100 %) through the appropriate wells and collect the filtrate. Evaporate the filtrate to dryness and reconstitute the compounds in a suitable solvent. Analyze samples with LC-MS/MS [34–36].

2. Radiolabeled substrates: Add 50 μL of Betaplate scintillation fluid to the wells and measure radioactivity in the wells with MicroBeta Trilux or a similar scintillation counter.

3. Fluorescence substrates: Add 100 μL of 0.1 N NaOH to the wells and incubate for 10 min at room temperature. Apply vacuum and collect the filtrate into fresh 96 well plates. Measure fluorescence (Excitation 510 nm and Emission 535 nm) with SpectraMax M2 or a similar spectrophotometer.

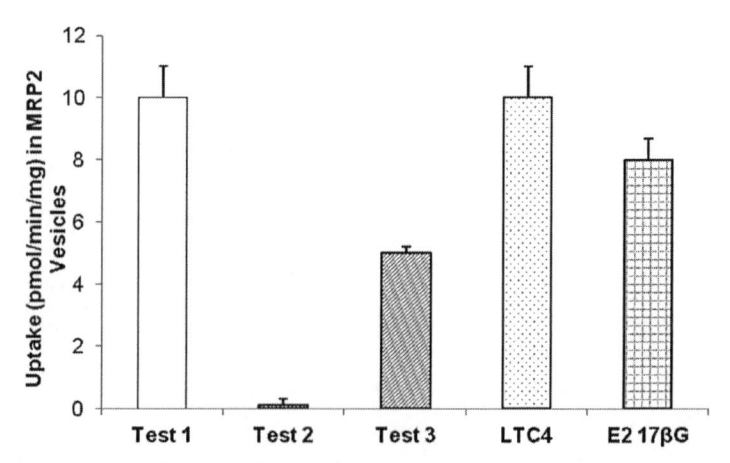

Fig. 3 **Representative** data showing ATP dependent uptake rate for three different test compounds and positive controls in MRP2 vesicles. With this assay, test compounds 1, 3 and positive controls were identified as substrates while compound 2 is not a substrate

3.5 Data Analysis

3.5.1 Substrate Screening

Calculate the rate of uptake (pmol/min/mg) in the presence and absence of ATP by determining the amount (e.g. pmol) of test compound or positive control present in each well and dividing it by incubation time (minutes) and the amount of protein per well (mg).

Calculate ATP dependent uptake activity using (3) shown below.

$$\text{ATP dependent uptake activity} = \text{Uptake activity with ATP (pmol/min/mg)} - \text{Uptake activity in absence of ATP (pmol/min/mg)} \quad (3)$$

Additionally, apparent transport kinetics (K_m and V_{max}) can also be determined.

3.5.2 Inhibitor Screening (Indirect Assay)

IC_{50} determination: Calculate the ATP dependent uptake (pmol) of probe substrate by subtracting the measured values obtained in the absence of ATP from those obtained in the presence of ATP. Determine percent inhibition by dividing ATP dependent uptake observed for probe substrate at different concentrations of test compounds by the vehicle control uptake value. Calculate $IC_{50} \pm S.E.$ values using nonlinear regression.

3.6 Data Reporting and Interpretation

3.6.1 Substrate Screening

ATP dependent uptake of test compounds into vesicles indicates that the test compounds are substrates of MRP2 (**Note 23**) (*see* Fig. 3). Validation studies with a known set of weak and strong substrates may further help in classifying the test compounds as weak, moderate or strong substrates of MRP2.

3.6.2 Inhibitor Screening

Report $IC_{50} \pm S.E.$ values as calculated. Test compounds can be classified as weak, moderate or strong inhibitors by benchmarking against positive controls.

4 Cell Based Assay (Madin-Darby Canine Kidney II Cells Overexpressing MRP2 (MDCKII-MRP2))

Dog-derived MDCKII cells are known to form tight junctions and spontaneously differentiate into polarized cell monolayers with well-defined AP and basolateral (BL) membranes, when grown to confluence on porous membranes. In addition, MDCKII cells can be readily transfected with cDNA that encodes for transporters, and that subsequently route to the appropriate membrane, making them a valuable tool to investigate the role of AP or BL transporters in altering the flux of test compounds [13, 37, 38]. MDCKII-MRP2 cells grown on porous membranes (6, 12 or 24-well plates) can be used for both substrate and inhibitor studies of MRP2. Although amenable to a higher throughput format, these assays are more expensive and laborious compared to ATPase and vesicular assays. The most significant advantage with MDCKII-MRP2 cells is that they reach confluence and achieve differentiation within 4–6 days, which is significantly shorter than the time required by other cell lines such as classical Caco-2 (~21 days) [39]. The assay to identify substrates for MRP2 can be performed by measuring unidirectional (BL to AP) or bidirectional (AP to BL and BL to AP) permeability of test compounds in MDCKII wild-type (MDCKII) and MDCKII-MRP2 cells. A positive control substrate such as bromosulfophthalein (BSP) or topotecan (or other appropriate MRP2 substrate) is included in each experiment to qualify the batch of cells and verify robustness of the assay. A twofold increase in BL to AP apparent permeability (P_{app}) in MDCKII-MRP2 cells compared to that in MDCKII cells or an efflux ratio ($P_{appBL-AP}/P_{appAP-BL}$) of ≥2 in MDCKII-MRP2 cells indicates that the test compound is a substrate for MRP2 [11, 12, 18, 40, 41]. To identify potential inhibitors, efflux ratio ($P_{appBL-AP}/P_{appAP-BL}$) or BL to AP permeability of a known MRP2 substrate (e.g. BSP) can be tested in MDCKII-MRP2 cells, in the absence and presence of different concentrations of test compounds or known positive control inhibitors. While conducting unidirectional permeability assays, AP to BL permeability can also be used to identify substrates and inhibitors of MRP2. Due to low level or lack of expression of uptake transporters in MDCKII cells, this cell system may give false negative results for compounds that depend on transporters to enter the cell, or are normally formed intracellularly, such as compound derived metabolites. Moreover, the compounds with very high and poor permeability may not be identified as substrates or inhibitors of MRP2.

4.1 Materials

See Table 4.

Table 4
Reagents and cell lines required for conducting MDCK assays

Chemical/reagents/cells/equipment	Catalog #	Vendor
Dulbecco's Modified Eagles Medium (DMEM)	11965-092	Life Technologies, Grand Island, NY, USA
MDCKII cells (passage # 30–45)	CRL-2936	ATCC, Manassas, VA, USA
MDCKII-MRP2 overexpressing cells	–	Can be developed in-house, acquired from an academic group or purchased commercially

Table 5
Plate formats for seeding MDCKII-MRP2 cells

Plate format	Cell density (cells/cm^2)	Volume in AP/BL chambers (mL)
6 well	1×10^5	1.5/2.6
12 well		0.5/1.5
24 well		0.1/0.6

4.2 Reagent Composition

1. Growth medium: DMEM supplemented with 10 % FBS, 1 % MEM NEAA and 1 % ABX.
2. Transport buffer: HBSS supplemented with 10 mM HEPES buffer and 25 mM Glucose, adjusted to pH 7.4 (**Note 24**).

4.3 Seeding and Maintenance of MDCKII/MDCKII-MRP2 Cells in Transwells

1. Seed and maintain MDCKII and MDCKII-MRP2 cells in T-75 tissue culture flasks as recommended by the vendor [42] (**Note 25**).
2. Trypsinize cells from T-75 flask at 80–90 % confluence. Count viable cells with a hemocytometer and adjust the cell number to the appropriate density depending on the plate format used for the assay (Table 5) (**Note 26**).
3. Add an appropriate volume of cell suspension to the AP chamber, and growth medium without cells to the BL chamber of the transwell plates (Table 5) (**Note 27**).
4. Transfer the plates to an incubator maintained at 37 °C, 5 % CO_2, and 95 % relative humidity.
5. Change the culture medium every other day.
6. Use the plates for transport studies after 4–6 days post-seeding. Measure transepithelial electrical resistance (TEER) values for each transwell with an EVOM2 and chopstick electrode to verify the formation of tight junctions (**Note 28**). TEER values should be in the range of 150–350 Ω/cm^2 with little variation [43, 44].

4.4 Assay Preparation

1. <u>Stock solutions for screening substrates of MRP2:</u> Prepare 2 mM stock solutions of test compound and BSP (*general substrate for MRP2*) in a suitable solvent. Dilute the stock solutions 200-fold in transport buffer to prepare a working solution of 10 μM. The final concentration of the organic solvent should be ≤1 %.

2. <u>Stock solutions for screening test compounds as inhibitors of MRP2:</u> Prepare seven different stock solutions of test compounds ranging between 0.02 and 20 mM in a suitable solvent. Dilute 2 mM stock solution of BSP 200-fold in transport buffer to prepare a working solution of 10 μM, and divide into eight aliquots. For the first aliquot, add an appropriate volume of vehicle control. To the remaining seven aliquots, add appropriate volumes of the stock solutions of test compound to obtain final working concentrations ranging between 0.1 and 100 μM, respectively (Note 11). Prepare working solutions of 0.1–100 μM test compounds, separately without BSP. Use these solutions to incubate cells with test compounds during pre-incubation and transport studies.

4.5 Assay Procedure

The entire procedure can be performed outside a biological safety cabinet.

4.5.1 Screening of MRP2 Substrates

1. Aspirate cell culture medium from the transwells containing MDCKII and MDCKII-MRP2 cells and wash each well 2× with pre-warmed transport buffer (the volume of buffer used for washing is similar to the volumes mentioned above for respective plate formats). Following washing, pre-incubate the transwells on both sides with drug free transport buffer for 30–60 min.

2. Measure and record TEER values for each transwell well. It is recommended to omit wells that have TEER values <150 Ω/cm². An alternative method to ensure monolayer integrity is to include (^{14}C) mannitol or other appropriate low permeable compounds in each well (**Note 29**). This step is common for screening substrates or inhibitors.

3. Remove transport medium from transwells. For measuring AP-BL permeability, add an appropriate volume (based on plate format) of dose solution containing test compound or positive control substrate into the AP compartment. Add the respective volume of fresh 37 °C drug free (vehicle control) transport buffer into the BL compartment. Similarly, for measuring BL-AP permeability, add dose solutions to BL chamber and drug free transport buffer to AP compartment.

4. Collect samples (~50 μL for 24 well and ~100 μL for 6 and 12 well) from acceptor compartments at 5, 15, 30, 60, 90 and

120 min. Replace the volume withdrawn at each time point with an equal volume of drug free transport buffer to maintain sink conditions (**Note 30**).

5. Measure and record TEER values at the end of the experiment to ensure that monolayer integrity has not been compromised during the experiment. Final TEER values <150 Ω/cm² should be rejected as an outlier [43, 44]. Alternatively, determine the P$_{app}$ values for (^{14}C) mannitol to identify wells that may have been compromised during the experiment. Analyze all samples with scintillation counting, HPLC or LC-MS/MS, as appropriate.

4.5.2 Screening of MRP2 Inhibitors

1. Following washing, pre-incubate the transwells containing MDCKII-MRP2 cells on both sides with transport buffer containing vehicle control or different concentrations of positive control inhibitor or test compound for 30–60 min.

2. Remove transport medium from transwells and add an appropriate volume of fresh 37 °C transport buffer containing BSP (10 μM) in the absence or presence of either the positive control inhibitor or test compound(s) at different concentrations into the AP compartment.

3. Collect samples (~50 μL for 24 well and ~100 μL for 6 and 12 well) from basolateral compartments at 5, 15, 30, 60, 90 and 120 min. Replace the volume withdrawn at each time point with an equal volume of transport buffer.

4. Measure and record TEER values at the end of the experiment to ensure that monolayer integrity has not been compromised during the experiment. Final TEER values <150 Ω/cm² should be rejected as potential outliers. Alternatively, determine the P$_{app}$ values for (^{14}C) mannitol or low permeable compound to identify wells that may have been compromised during the experiment. Analyze all samples with scintillation counting, HPLC or LC-MS/MS, as appropriate.

4.6 Data Analysis and Interpretation

Calculate P$_{app}$ from the linear plot of drug accumulated in the receiver side versus time using (4).

4.6.1 Substrate Studies

$$P_{app} = \frac{dQ/dt}{A * C_0} \quad (4)$$

1. Where dQ/dt (flux) is the cumulative amount of test compound appearing in the receiver compartment (Q) over time (t) during the experiment, A is the surface area of the membrane, and C$_0$ is the initial concentration (μM) of test compound on donor side.

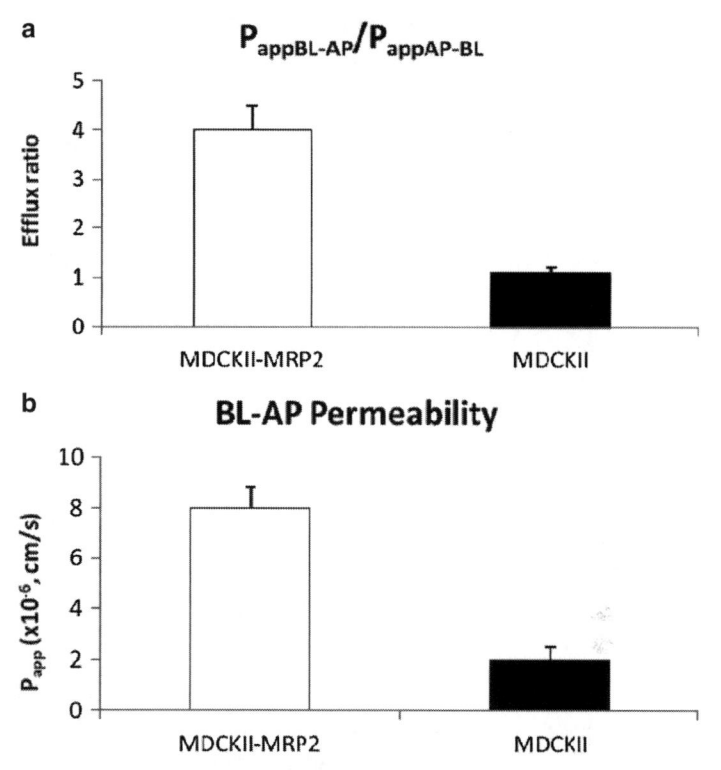

Fig. 4 Example showing (**a**) efflux ratio ($P_{appBL-AP}/P_{appAP-BL}$) and (**b**) BL to AP permeability of test compound in MDCKII-MRP2 and MDCKII cells. An efflux ratio and BL to AP permeability of >2 in MDCKII-MRP2 transfected cells indicate that the test compound is a substrate for MRP2

2. An efflux ratio ($P_{appBL-AP}/P_{appAP-BL}$) of >2 in MDCKII-MRP2 or BL-AP permeability ratio of (P_{app} in MDCKII-MRP2/P_{app} in MDCKII) greater than 2 indicates that the test compound is a substrate for MRP2 (*see* Fig. 4a). Since MDCKII-MRP2 cells tend to have higher expression of MRP2, the BL-AP permeability ratio would be expected to be much higher than 2 for strong substrates (*see* Fig. 4b). Benchmarking test compounds against internally validated positive control substrates (of varying affinities), will help in appropriately classifying the test compounds as low, medium or strong substrates [11, 12, 18]. Additionally, apparent transport kinetics (K_m and V_{max}) can also be determined.

3. The results from *in vitro* data need to be corroborated with the (i) physicochemical properties of the test compound (solubility, permeability), along with its metabolism data and (ii) the relative contribution of MRP2 to the overall clearance of the test compound, to determine if an inhibitor of MRP2 will have a major effect on the disposition of the test compound [18].

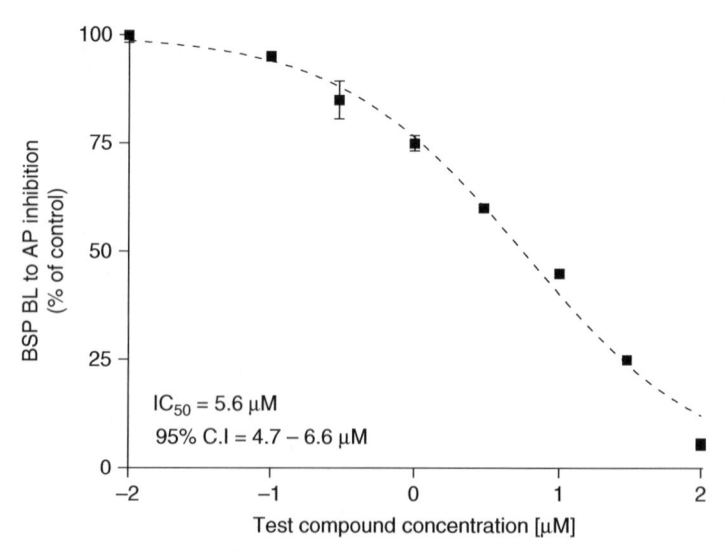

Fig. 5 Example showing dose-response curves for inhibition of BSP transport in BL to AP direction in the presence of different concentrations of test compound

4.6.2 Inhibitor Studies

Determine efflux ratio ($P_{appBL-AP}/P_{appAP-BL}$) or BL to AP P_{app} for BSP in absence and presence of different concentrations of test compound (*see* Fig. 5). Calculate the percent inhibition at different concentrations of test compound by dividing the efflux ratio or BL to AP P_{app} by the vehicle control efflux ratio or P_{app} value. Calculate $IC_{50} \pm S.E.$ values using nonlinear regression.

The test compound can be classified as a weak, moderate or strong inhibitor by benchmarking against positive control inhibitors (**Note 31**).

5 Caco-2 Cell Monolayers to Study the Role of Intestinal MRP2 in Limiting Oral Absorption

Caco-2 cells derived from human colorectal carcinoma are widely used to model drug behavior in the human intestine. This cell line when grown on porous membranes spontaneously differentiates into mature enterocytes that represent the lining of the small intestine [45, 46]. Upon long-term culture, Caco-2 cells become polarized and express many transporters, including MRP2, which is appropriately routed along with P-gp and BCRP on the AP membrane of the cells [47]. The mRNA levels of MRP2 in Caco-2 cells correlate well with the respective levels found in the human jejunum [48]. With the exception of BCRP, Caco-2 cells express the major uptake and efflux transporters to a comparable level as found in human jejunum [48]. Caco-2 cells grown on Transwell plates for ~21–28 days (to confluency) are therefore, a more

Table 6
Reagents and cell lines required for conducting Caco-2 assays

Chemical/reagents/cells/equipment	Catalogue #	Vendor
Minimum Essential Medium (MEM)	11095	Life Technologies, Grand Island, NY, USA
Caco-2 cells (passage # 27–35)	HTB-37	ATCC, Manassas, VA, USA
Caco-2-MRP2 KD	–	Can be developed in-house, acquired from academic groups or purchased through commercial vendors

physiologically relevant model (as compared to MDCKII) and can be used to investigate the overall absorption of a compound within the intestine. However, due to the abundance of transporters in the system, it becomes challenging to dissect out the contribution of one specific transporter, particularly when a compound is transported by multiple proteins since no MRP2 specific substrates or inhibitors have been identified to date. Therefore, MRP2 knockdown Caco-2 (Caco-2-MRP2 KD) models may be more informative on the specific contribution of MRP2 to overall test compound intestinal absorption. The methods for creating a transporter knockdown in Caco-2 cells are detailed elsewhere [49, 50]. The assay is performed by measuring unidirectional (AP to BL) permeability of test compounds in both Caco-2 and Caco-2-MRP2 KD cells. A positive control substrate such as BSP or topotecan is included in each experiment to qualify the batch of cells and verify robustness of the assay. A twofold increase in AP to BL P_{app} in Caco-2-MRP2 KD cells compared to that in the wild-type Caco-2 cells indicates that MRP2 plays a potential role in attenuating transport of test compounds.

5.1 Materials

See Table 6.

5.2 Reagent Composition

1. Growth medium: MEM supplemented with 10 % FBS, 1 % MEM NEAA and 1 % ABX.

5.3 Seeding and Maintenance of Caco-2 and Caco-2-MRP2 KD Cells in Transwell Plates

1. Seed and maintain Caco-2 and Caco-2-MRP2 KD cells in T-75 tissue culture flasks as recommended by the vendor [51] (**Note 25**).

2. Trypsinize cells from T-75 flask at 80–90 % confluence. Count viable cells using a hemocytometer and adjust the cell number to appropriate density depending on the plate format used for the assay (Table 7) (**Note 26**).

Table 7
Plate formats for Caco-2 and Caco-2 MRP2 KD cells

Plate format	Cell density (cells/cm²)	Volume in AP/BL chambers (mL)
6 well	0.6×10^5	1.5/2.6
12 well		0.5/1.5
24 well		0.1/0.6

3. Add an appropriate volume of cell suspension to the AP chamber, and growth medium without cells to the BL chamber of the transwell plates (Table 7) (**Note 27**).

4. Transfer the plates into an incubator maintained at 37 °C, 5 % CO_2, and 95 % relative humidity.

5. Change the culture medium every other day. Measure TEER values for each transwell starting from day 10 with an EVOM2 and chopstick electrode to verify the formation of tight junctions. Ensure that TEER measurement is performed in a sterile fashion. TEER values should be in the range of 230–750 Ω/cm² and should vary as little as possible [44].

6. Use the plates for transport studies between 21 and 28 days post-seeding.

5.4 Assay Preparation

1. Prepare stock solutions of test compound and BSP (2 mM) in a suitable solvent. Dilute the stock solution 200-fold in transport buffer to prepare working solutions of 10 µM. The final concentration of the organic solvent should be ≤1 %.

5.5 AP to BL Permeability With Caco-2 and Caco-2-MRP2 KD Cell Monolayers

1. Aspirate cell culture medium from the transwells and wash each well 2× with pre-warmed transport buffer (the volume of buffer used for washing is similar to the volumes mentioned above for respective plate formats). Following washing, pre-incubate the transwells on both sides with drug free transport buffer for 30–60 min.

2. Measure and record TEER values for each transwell well. It is recommended to omit wells that have TEER values <230 Ω/cm². An alternative way to ensure monolayer integrity is to include (^{14}C) mannitol or other appropriate low permeable compound in each well (**Note 29**).

3. Remove transport medium from transwells. Add appropriate volume (based on plate format) of dose solutions containing test compounds or BSP into the AP compartment. Add the respective volume of fresh 37 °C drug free (vehicle control) transport buffer into the BL compartment.

4. Collect samples (~50 µL for 24 well and ~100 µL for 6 and 12 well plates) from BL compartment at 5, 15, 30, 60, 90 and 120 min.

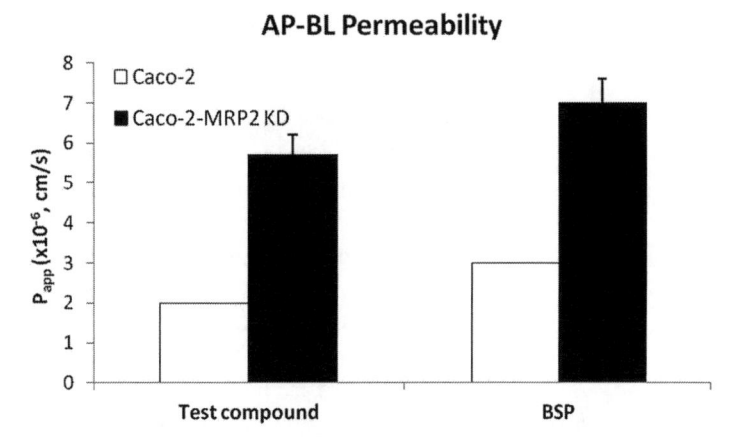

Fig. 6 Example showing AP to BL permeability of test compound and MRP2 substrate in Caco-2 and Caco-2-MRP2 KD cells. In this example, MRP2 plays a role in limiting the AP to BL permeability of the test compound

Replace the volume withdrawn at each time point with an equal volume of drug free transport buffer.

5. Measure and record TEER values at the end of the experiment to ensure that monolayer integrity has not been compromised during the experiment. Final TEER values <230 Ω/cm^2 should be rejected. Alternatively, determine the P$_{app}$ values for (^{14}C) mannitol or low permeable compounds to identify wells that may have been compromised during the experiment. Analyze all samples with scintillation counting, HPLC-UV or LC-MS/MS, as appropriate.

5.6 Data Analysis and Interpretation

Calculate the P$_{app}$ (AP-BL) obtained for test compounds in Caco-2 and Caco-2-MRP2 KD cells using equation (4), as described in Sect. 4.6.1.

A twofold increase in AP to BL permeability (P$_{app}$) in Caco-2-MRP KD cells indicates that MRP2 plays a potential role in limiting the intestinal absorption of the test compound (*see* Fig. 6). The results from Caco-2 studies can be corroborated with the physicochemical properties of the test compound (solubility, permeability) along with its metabolism data to determine if an inhibitor of MRP2 might lead to enhanced absorption of the test compound.

6 Sandwich-Cultured Human Hepatocytes (SCH) to Study the Role of MRP2 in Canalicular Efflux of Test Compounds

SCH are a commonly used *in vitro* tool for studying the hepatobiliary disposition of test compounds [52]. Hepatocytes when grown between two layers of collagen or matrigel (sandwich configuration) form functional "canalicular-like" networks and maintain the expression, localization and function of uptake and efflux transporters, relative to *in vivo* [52]. The canalicular networks formed in SCH are isolated from the media by tight junctions between the cells. Liu et al. [53, 54] demonstrated that the integrity of the tight junctions are maintained when cells are incubated with buffer containing calcium. However, in calcium free buffer, the tight junctions open and release contents from the canaliculi into the media. This calcium modulation technique (U.S. Pat. No. 6,780,580, Pat. No. 7,604,934 and International patents both issued and pending) enables quantification of cellular accumulation, and cell plus bile accumulation of substrate; thus, the AP efflux (biliary excretion) of substrate by transporters (MRP2, P-gp, BCRP etc.) can be assessed [55, 56]. SCH express all of the uptake and efflux transporters normally present in hepatocytes, making it difficult to dissect out the contribution of one specific transporter. Therefore, MRP2 knock-down SCH (MRP2 KD-SCH) models may be more informative on the specific contribution of MRP2 in the efflux of test compound's into bile, and overall importance of MRP2 to a compounds liver distribution. The methods for transient knockdown of transporters in SCH cells are detailed elsewhere (U.S. Pat. No. 7,601,494 and International patents both issued and pending) [57–60]. The assay is performed by incubating test compounds or positive controls such as 5-(and-6)-carboxy-2′,7′-dichlorofluorescein diacetate (CDFDA) (*positive control for MRP2*) and ^{3}H taurocholate (*positive control for overall function of SCH system*) with SCH and MRP2 KD-SCH with and without intact bile canaliculi. The accumulation of test compounds/positive control in the bile canaliculi are measured at the end of incubation (**Note 32**). The results are represented as accumulation in cells and in cells + bile (pmol/mg protein), biliary excretion index (BEI; %) and *in vitro* intrinsic biliary clearance (Cl′$_{biliary}$) (mL/min/kg) [61–63]. The data generated from this *in vitro* system can be used to predict *in vivo* hepatobiliary disposition and transporter-based hepatic drug-drug interactions [52].

6.1 Materials

See Table 8.

6.2 Reagent Composition

The reagent composition (plating and feeding medium) mentioned below is for growing freshly isolated human hepatocytes [64].

1. Seeding medium: DMEM containing 10 % FBS, 1 μM dexamethasone, 2 mM glutamine, 1 % MEM NEAA, 10 μM insulin, and 1 % penicillin/streptomycin.

2. Feeding medium: DMEM containing 0.1 μM dexamethasone, 2 mM glutamine, 1 % MEM NEAA, 1 % insulin/transferrin/selenium, and 1 % penicillin/streptomycin.

3. Triton X-100: 0.5 % in phosphate-buffered saline.

Table 8
Chemicals, reagents and equipment required for conducting SCH assays

Chemical/reagents/cells/equipment	Catalog #	Vendor
Freshly isolated (preferred) (Note 33) Transporter-Certified Human CryoHepatocytes B-CLEAR® Biliary Excretion/DDI Kit	–	Commercial Hepatocyte Vendor Qualyst Transporter Solutions, Durham, NC, USA
Cryopreserved hepatocyte recovery medium (CHRM) (Note 33) Cryopreserved Human Hepatocyte Thawing Medium (Note 33)	–	Life Technologies, Grand Island, NY, USA Triangle Research Laboratories, RTP, NC, USA
Dulbecco's modified Eagle's medium, high glucose, no glutamine, no phenol (DMEM)	31053	Life Technologies, Grand Island, NY, USA
FBS	26140-079	Life Technologies, Grand Island, NY, USA
Dexamethasone	D4902	Sigma-Aldrich, St. Louis, MO, USA
Glutamine	25-005-Cl	Cellgro, Manassas, VA, USA
Insulin	12585-014	Life Technologies, Grand Island, NY, USA
Penicillin G sodium & streptomycin sulfate	15140-122	Life Technologies, Grand Island, NY, USA
Insulin/transferrin/selenium (ITS+™ + Premix)	354352	BD Biosciences, San Jose, CA, USA
HBSS with $CaCl_2$	H-1387	Sigma-Aldrich, St. Louis, MO, USA
HBSS without $CaCl_2$	H-4891	Sigma-Aldrich, St. Louis, MO, USA
Triton X-100	X100	Sigma-Aldrich, St. Louis, MO, USA
Bicinchoninic acid (BCA) protein assay kit	23225	Pierce Chemical Co., Rockford, IL, USA
24-well BD BioCoat™ collagen I plates 6-well BD BioCoat™ collagen I plates	356408 354400	BD Biosciences, San Jose, CA, USA
Matrigel™ basement membrane matrix	354234	BD Biosciences, San Jose, CA, USA
5-(and-6)-carboxy-2′,7′-dichlorofluorescein diacetate (CDFDA)	C-369	Life Technologies, Grand Island, NY, USA
Sodium taurocholate	86339	Sigma-Aldrich, St. Louis, MO, USA
^{3}(H) taurocholate	–	Perkin Elmer, Waltham, MA, USA
Bio-Safe II	111196	Research Products International Corp., Mount Prospect, IL, USA
Sonic dismembrator	100	Fisher Scientific, Pittsburgh, PA, USA
Liquid scintillation spectrometry (Packard Tri-Carb scintillation counter)	3110TR	PerkinElmer Life and Analytical Sciences

Table 9
Plate format for sandwich culture studies

Plate format	Cell density (cells/mL)	Volume per well (mL)
6 well	1.17×10^6	1.5
24 well	0.7×10^6	0.5

6.3 Seeding and Maintenance of Human Hepatocytes

1. Thaw transporter-certified human cryohepatocytes in hepatocyte recovery/thawing medium with manufacturer's protocol. The protocol for seeding and maintaining cryopreserved hepatocytes in SCH configuration are detailed elsewhere [58, 65–67]. Once the cultures are established, accumulation studies can be undertaken with the procedure mentioned below (Sect. 6.4).

2. The protocol detailed below is for seeding and maintaining freshly isolated human hepatocytes in SCH configuration. Suspend hepatocytes in plating medium and adjust the cell number to appropriate density based on the plate format used for the assay (Table 9) [52].

3. Add appropriate volume of cell suspension to 6 well or 24 well BioCoat™ collagen I plates, swirl gently (using a figure-8-motion and north/south and east/west motions) to distribute cells evenly throughout well. After filling each plate, gently shake plates back and forth and side to side before transferring the plates into an incubator (**Note 34**). Add 1.5/0.5 mL of plating medium without hepatocytes to few wells (two or three wells per compound) for determining nonspecific binding. Incubate the plate in an incubator at 37 °C, 5 % CO_2 and 95 % relative humidity for 2–4 h (**Note 35**).

4. After the initial attachment period, swirl plates vigorously to dislodge any loose cells and aspirate medium with a suction device. Add 1.5 mL/0.5 mL of fresh pre-warmed plating medium to all wells. At this point, add small interfering RNA (siRNA) or small hairpin RNA (shRNA) targeting MRP2 to one set of hepatocytes and incubate for additional 20–22 h to generate MRP2 KD-SCH. Similarly, add negative control siRNA or shRNA to control cells (Day 0) (**Note 36**). The methods for transient knockdown of transporters in SCH cells are detailed elsewhere (U.S. Pat. No. 7,601,494 and International patents both issued and pending) [57–60]. The MRP2 KD-SCH model should be validated for substrate screening studies (**Note 37**). If MRP2 KD cannot be achieved with these approaches, chemical inhibition approach can be used as an alternative method [61] (**Note 38**).

5. Matrigel Overlay: Thaw Matrigel overnight in fridge (~4 °C). Keep Matrigel and feeding medium in wet ice bucket (**Note 39**). Prepare 0.25 mg/mL Matrigel solution by adding appropriate volume of Matrigel to the feeding medium. Aspirate medium from the wells and add 2 mL (6 well) or 0.5 mL (24 well) of Matrigel solution. Leave the plates in the incubator overnight (Day 1).

6. Gently aspirate the medium from the plates without disturbing the Matrigel matrix overlay and add 1.5 mL (6 well) or 0.5 mL (24 well) of fresh pre-warmed feeding medium (Day 2). Change medium every 24 h. Based on the formation of the canalicular network and functional data on MRP2 KD, accumulation studies can be performed between days 5 and 7.

6.4 Assay Preparation

Prepare stock solutions of test compounds (2 mM) and CDFDA (0.5 mM) in a suitable solvent. Dilute the stock solutions separately in standard HBSS (HBSS containing Ca^{2+}) to prepare working solutions of 10 and 2 μM, respectively (**Note 40**). Prepare vehicle control by adding an appropriate volume of vehicle to standard HBSS.

6.5 Assay Procedure

6.5.1 Fluorescent Microscopy Study to Test the Formation and Integrity of Bile Canaliculi

Before initiating studies with SCH and MRP2 KD-SCH, the formation of bile canaliculi and functional activity of MRP2 must be tested by visualizing the retention of 5(and 6)-carboxy-2′,7′dichlorofluorescein (CDF) in bile canaliculi [68] (**Note 41**). Since MRP2 at the apical membrane mediates efflux of CDF into bile, a reduction in the accumulation of CDF can be used as a surrogate to determine the extent of MRP2 KD in siRNA treated SCH. Alternatively; a reduction in BEI of CDF can be used to quantitatively measure the extent of MRP2 KD.

Procedure

Rinse hepatocytes with 2 mL (6 well)/0.6 mL (24 well) of standard buffer and add 1.5 mL (6 well)/0.5 mL (24 well) of CDFDA (2 μM) in standard buffer to the cells. Incubate for 10 min, remove the buffer and wash with 1.5 mL/0.5 mL of standard HBSS. Image the cells and bile canaliculi using an inverted fluorescent microscope [65, 68].

6.5.2 Accumulation Study to Determine the Role of MRP2 in the Canalicular Efflux of Test Compounds

To ensure the functional activity of bile canaliculi in MRP2 KD-SCH, a general substrate (e.g. (^{3}H) taurocholate) for other efflux transporters like the bile salt export pump should be included in the assay. For consistency, the same positive control should also be tested in SCH.

Procedure

1. Aspirate medium and rinse SCH and MRP2 KD-SCH 2× with 2 mL (6 well)/0.6 mL (24 well) of standard HBSS (cells + bile) or Ca^{2+} free HBSS (cells). Add 1.5 mL/0.5 mL of standard HBSS or Ca^{2+} free HBSS and incubate for 10 min to maintain or disrupt the tight junctions sealing the bile canalicular networks, respectively.

2. Aspirate buffer and add 1.5 mL/0.5 mL of test compound (10 μM) or (^{3}H) taurocholate (1 μM) diluted in standard HBSS to separate wells and incubate for 10 min.

3. To determine non-specific binding, add 1.5 mL/0.5 mL of test compound (10 μM) or (^{3}H) taurocholate (1 μM) diluted in standard HBSS to wells without hepatocytes.

4. Remove dose solution, save for analysis and wash all the wells 3 times with 1.5 mL/0.5 mL of ice-cold HBSS.

5. Lyse the cells containing (^{3}H) taurocholate by adding 1 mL/ 0.5 mL of Triton X-100 and sonicate for 20 s with a sonic dismembrator. Place on a rotating shaker for 20 min before collecting cell lysates and before analyzing the samples sonicate for 20 sec with a sonic dismembrator.

6. Transfer the lysed samples to 7.5 mL of Bio-Safe II scintillation cocktail in glass vials and analyze the samples by liquid scintillation spectroscopy.

7. If the test compounds are not radiolabeled, lyse the hepatocytes with ~1 mL (6 well)/~0.5 mL (24 well) of 70 % (v/v) ice-cold methanol/water, scrape the cells off the plates and sonicate for 20 s with a sonic dismembrator (**Note 42**). Analyze the samples by HPLC-UV or LC-MS/MS.

8. Correct the accumulation of test compound in cells for non-specific binding.

9. Determine the protein concentration in each well using a BCA protein assay kit as instructed by the manufacturer and normalize the amount of test compound in each well to the protein concentration.

6.6 Data Analysis and Interpretation

Determine the accumulation (pmol/mg of protein/min) of test compounds in cells + bile (standard HBSS) and cells (Ca$^+$-free HBSS) in SCH and MRP2 KD-SCH. Calculate the biliary excretion index (BEI %) (5) (*see* Fig. 7) and unbound intrinsic biliary clearance (Cl'$_{biliary}$, mL/min/kg) (6) in SCH and MRP2 KD-SCH [53, 56].

$$\text{BEI} = \frac{Accumulation_{cell+Bile} - Accumulation_{cell}}{Accumulation_{cell+Bile}} \times 100 \quad (5)$$

$$\text{Intrinsic Cl'biliary} = \frac{Accumulation_{cell+Bile} - Accumulation_{cell}}{AUC_{Media} X f_u} \times 100 \quad (6)$$

where f_u is the unbound fraction of test compound or probe substrate in the incubation media, which is equal to 1 in this example (**Note 43**). AUC$_{media}$ represents the area under the substrate concentration-time curve. This is determined by dividing the sum of the substrate concentration in the incubation medium at the beginning and end of the incubation period by 2 and multiplying by the

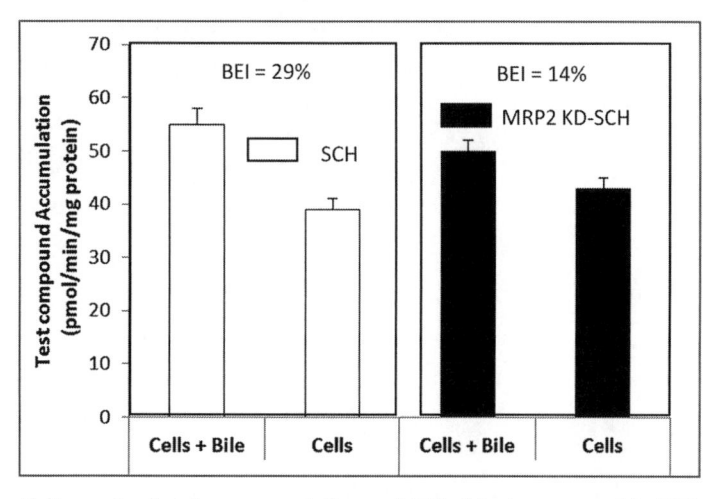

Fig. 7 Example showing accumulation and BEI of test compound in SCH and MRP2 KD-SCH. A decrease in the accumulation of test compound in cells + bile and a reduction in BEI of test compound in MRP2 KD-SCH indicate that the test compound excretion into bile is mediated by MRP2

incubation time (10 min). Cl'$_{biliary}$ can be converted to milliliter per minute per kilogram based on scaling factors [52].

A significant reduction of BEI in MRP2 KD-SCH compared to SCH indicates that MRP2 likely plays a significant role in the canalicular efflux of test compound.

Since, it is now well established that if there are multiple transporters that can excrete the substrate into bile, loss-of-function of one transporter may be totally compensated for by another transporter. If no change is observed in biliary clearance or BEI between MRP2 KD-SCH and SCH then MRP2 likely does not have significant relevance in the liver for the compound being studied.

7 Notes

1. The catalog numbers and vendor information provided serve as immediate references. Equivalent versions of all chemicals/reagents/equipment can be used.

2. Contamination of reagents/assay with phosphate can lead to inconsistent results. Use phosphate-free glass vials, tubes and reservoirs. Decontaminate work area with bleach prior to use.

3. Reagents can be prepared and stored at recommended temperatures ahead of time. According to instructions from BD Gentest, the performance of reagents will not be affected until eight freeze-thaw cycles [22].

4. Make fresh 10 % ascorbic acid solution prior to the preparation of colorimetric reagent.

5. The range and concentrations of phosphate standards may vary depending upon the sensitivity and linearity range of the spectrophotometer. Prior studies to determine the linear range of the respective spectrometer should be undertaken. An R^2 value of >0.98 is generally recommended.

6. Based on solubility, test compounds can be dissolved in water or other solvents like DMSO, ethanol, methanol or acetonitrile. When ethanol, methanol or acetonitrile are used as solvents ensure that the final percentage of organic solvent in the incubation is ≤1 % [23].

7. Include appropriate vehicle controls for test and positive control compounds to determine the effect of vehicle on basal ATPase activity.

8. Mix the cell membranes gently using a pipet or by inversion shaking. Do not vortex.

9. Do not add cell membranes to wells containing phosphate standards.

10. For screening purposes, 20 μM is recommended as the final concentration of test compound. This is based on literature that most compounds give robust signals at this concentration in ATPase assays [69]. However, a clinically relevant concentration can be chosen based on the theoretical maximal gastrointestinal concentration after oral administration calculated as the highest clinical dose (mg) in a volume of 250 mL and mean steady state C_{max} of unbound drug [18].

11. The suggested concentrations are generally used to determine IC_{50} values of test compounds. The range may vary based on the solubility of the test compounds in assay buffer. Alternatively, a clinically relevant concentration can be chosen based on the theoretical maximal gastrointestinal concentration after oral administration calculated as the highest clinical dose (mg) in a volume of 250 mL and mean steady state C_{max} of unbound drug [18].

12. Care should be taken to minimize the formation of bubbles while adding 10 % SDS, as bubbles may interfere with readout on spectrophotometer. To avoid bubble formation, antifoam A can be added to the 10 % SDS solution [69].

13. Prepare the colorimetric solution 10 min prior to the end of incubation.

14. Due to the presence of multiple binding sites on MRP2, some test compounds can further enhance probenecid stimulated ATPase activity rather than inhibit it. This could probably result from the test compounds altering the affinity of the binding sites for probenecid.

15. Slowly transported substrates may not be detected in the activation mode of the assay. To minimize the number of false negatives, the inhibition mode of ATPase assays should be performed to identify any possible interaction [24].

16. Reagents can be prepared and stored at recommended temperatures ahead of time. According to instructions from BD Gentest, the performance of reagents will not be affected until six freeze-thaw cycles [25].

17. Bring the radiolabeled substrates to room temperature, 30 min prior to running the assay.

18. Due to size limitation, the amount of compound that gets into the vesicles is usually in the range of picomoles. When using non-radiolabelled compounds, LC-MS/MS methods must be optimized to achieve enough sensitivity.

19. The concentrations of MRP2 substrates chosen are based on their reported K_m values [29, 70].

20. It has been shown that E2 17βG uptake can be inhibited or stimulated several fold in the presence of interacting compounds [10, 71]. If such a situation is encountered, it can be considered to increase the concentration of E_2-17βG to ~50 μM, to minimize the effect of transport stimulation [70].

21. Alternatively, assay buffer containing 25 mM AMP can be used as a control.

22. Ensure that the plates are completely dried. Incomplete drying can lead to low signal-to-noise ratio [25].

23. The physicochemical properties of the test compounds should be taken into consideration. For example, the transport of highly lipophilic ('sticky') compounds may be masked and incorrectly identified as a 'non-substrate' using this assay.

24. Transport buffer is good for 2 months when stored at 4 °C.

25. It is best to grow cells without antibiotics. Growing cells in presence of antibiotics may lead to selection for cells that transport out the antibiotic and may cover up a starting infection.

26. It is good practice to count cells at each passage. If cells suddenly start to grow faster/slower it may indicate that something may be wrong.

27. This is a critical step: Cell suspension must be used quickly; swirling of the cells in the well should be avoided at all cost and ensure that cell suspensions are homogenous. The flow cabinet and incubator should be free of vibrations as it can cause cells going to the center of the well.

28. The electrodes used for measuring TEER should be decontaminated with 70 % ethanol.

29. Where plausible, radiolabeled test compound and positive control substrates (^{3}H/^{14}C) are recommended to decrease analytical resources required during this assay. Here, leakiness of the membranes can be assessed at every time point and in every well by inclusion of radiolabeled (^{3}H/^{14}C) mannitol or another appropriate low permeability compound. As an example, if the test compound is tritiated, (^{14}C) mannitol would be used. Additionally, issues around non-specific binding can be addressed only using radiolabeled compound and determining overall compound mass balance. If radiolabeled studies are not possible (e.g. in early discovery setting) the assays can also be performed using unlabeled compound; however significant analytical resources (e.g. LC-MS/MS) would be required to determine permeability values.

30. Sink conditions refers to the situation in which back diffusion of drug in the receiver side to cell monolayer is negligible. Sink condition would be satisfied when the sampling is conducted within the time interval that drug concentration in the receiver side remains <10 % of the loading concentration or when drug in the serosal side is removed rapidly and irreversibly, leaving no chance for the drug to return back to the enterocyte [72, 73].

31. Due to intra-laboratory variability in IC$_{50}$ values reported for known inhibitors, each laboratory should establish its own set of cutoff values based on historical data to qualify and interpret the data obtained from this cell system.

32. The method described below do not reflect current methods used in the commercially available technology.

33. Multiple commercial companies sell transporter qualified cryohepatocytes, along with their respective optimized mediums. Laboratories should undertake their own validation using the positive controls to see what cell and medium combination works best.

34. Shaking the plates helps to prevent the formation of cell clumps.

35. This step is performed to allow hepatocytes to attach to the plates.

36. The feasibility of knocking down transporters in rat and human SCH using siRNA techniques has been demonstrated in few studies (e.g. Mrp2, Mrp3 and Bcrp KD in rat and Organic Anion Transporting Polypeptide (OATP) in human) [57–60]. However, to date successful knockdown of MRP2 in human SCH has not been reported.

37. The up regulation of other relevant efflux transporters should be evaluated in MRP2 KD-SCH. This will help to rule out the

possibility of other efflux transporters compensating for the knockdown of MRP2. This becomes particularly relevant for substrates that are effluxed by multiple efflux transporters.

38. Chemical inhibitors are fairly non-specific. The data obtained from these studies should be interpreted with caution.

39. Keep feeding media, Matrigel, sterile bottle/conical tubes on wet ice. Keep tips, boxes, and serological pipets in a freezer and remove them as needed. Mix Matrigel between each use to avoid it from settling.

40. Alternatively, compounds can be tested at a clinically relevant concentration (steady state plasma C_{max}) based on the availability of data.

41. CDFDA is rapidly hydrolyzed in hepatocytes to 5 (and 6)-carboxy-2′,7′dichlorofluorescein (CDF) and is excreted into bile canaliculi by AP MRP2 [68].

42. Use appropriate solvents for lysis, based on the nature of the test compound and the analytical procedures employed.

43. Since the free fraction of the test compounds or probe substrates in the incubation is not measured, f_u is considered as 1 for intrinsic clearance calculations.

Acknowledgements

The authors would like to thank Kim L. Brouwer, Ruth S Everett, Maarten Huisman, and Ian Templeton for helpful comments made during the preparation of the manuscript.

The views expressed here are solely those of the authors and do not reflect the opinions of Janssen Research & Development, LLC.

References

1. DeGorter MK et al (2012) Drug transporters in drug efficacy and toxicity. Annu Rev Pharmacol Toxicol 52:249–273
2. Jedlitschky G, Hoffmann U, Kroemer HK (2006) Structure and function of the MRP2 (ABCC2) protein and its role in drug disposition. Expert Opin Drug Metab Toxicol 2(3): 351–366
3. Dallas S, Miller DS, Bendayan R (2006) Multidrug resistance-associated proteins: expression and function in the central nervous system. Pharmacol Rev 58(2):140–161
4. Zimmermann C et al (2005) Mapping of multidrug resistance gene 1 and multidrug resistance-associated protein isoform 1 to 5 mRNA expression along the human intestinal tract. Drug Metab Dispos 33(2):219–224
5. Keppler D, Konig J (1997) Hepatic canalicular membrane 5: expression and localization of the conjugate export pump encoded by the MRP2 (cMRP/cMOAT) gene in liver. FASEB J 11(7):509–516
6. Homolya L, Varadi A, Sarkadi B (2003) Multidrug resistance-associated proteins: export pumps for conjugates with glutathione, glucuronate or sulfate. Biofactors 17(1–4): 103–114
7. Kruh GD, Belinsky MG (2003) The MRP family of drug efflux pumps. Oncogene 22(47): 7537–7552

8. Borst P et al (2000) A family of drug transporters: the multidrug resistance-associated proteins. J Natl Cancer Inst 92(16):1295–1302
9. Gerk PM et al (2007) Human multidrug resistance protein 2 transports the therapeutic bile salt tauroursodeoxycholate. J Pharmacol Exp Ther 320(2):893–899
10. Zelcer N et al (2003) Evidence for two interacting ligand binding sites in human multidrug resistance protein 2 (ATP binding cassette C2). J Biol Chem 278(26):23538–23544
11. US Food and Drug Administration, US Department of Health and Human Services, Center for Drug Evaluation and Research, Center for Biologics Evaluation and Research (CBER) (2012) Guidance for industry: drug interaction studies—study design, data analysis, and implications for dosing and labeling. Silver Spring, MD. http://www.fda.gov/downloads/Drugs/GuidanceComplianceRegulatoryInformation/Guidances/UCM292362.pdf
12. European Medicines Agency (2010) Guideline on the investigation of drug interactions. http://www.ema.europa.eu/docs/en_GB/document_library/Scientific_guideline/2012/07/WC500129606.pdf
13. Ming X, Knight BM, Thakker DR (2011) Vectorial transport of fexofenadine across Caco-2 cells: involvement of apical uptake and basolateral efflux transporters. Mol Pharm 8(5):1677–1686
14. Dahan A, Sabit H, Amidon GL (2009) Multiple efflux pumps are involved in the transepithelial transport of colchicine: combined effect of p-glycoprotein and multidrug resistance-associated protein 2 leads to decreased intestinal absorption throughout the entire small intestine. Drug Metab Dispos 37(10):2028–2036
15. Yoshida K, Maeda K, Sugiyama Y (2013) Hepatic and intestinal drug transporters: prediction of pharmacokinetic effects caused by drug-drug interactions and genetic polymorphisms. Annu Rev Pharmacol Toxicol 53:581–612
16. Gerk PM, Vore M (2002) Regulation of expression of the multidrug resistance-associated protein 2 (MRP2) and its role in drug disposition. J Pharmacol Exp Ther 302(2):407–415
17. Sugie M et al (2004) Possible involvement of the drug transporters P glycoprotein and multidrug resistance-associated protein Mrp2 in disposition of azithromycin. Antimicrob Agents Chemother 48(3):809–814
18. Giacomini KM et al (2010) Membrane transporters in drug development. Nat Rev Drug Discov 9(3):215–236
19. Sarkadi B et al (1992) Expression of the human multidrug resistance cDNA in insect cells generates a high activity drug-stimulated membrane ATPase. J Biol Chem 267(7):4854–4858
20. Drueckes P, Schinzel R, Palm D (1995) Photometric microtiter assay of inorganic phosphate in the presence of acid-labile organic phosphates. Anal Biochem 230(1):173–177
21. Mazur CS et al (2012) Human and rat ABC transporter efflux of bisphenol a and bisphenol a glucuronide: interspecies comparison and implications for pharmacokinetic assessment. Toxicol Sci 128(2):317–325
22. BD Gentest, ATPase Assay Kit Data Sheet, BD Biosciences, Woburn, MA. http://www.bdj.co.jp/gentest/1f3pro00000vv95q-att/Cat459006-lot05056.pdf
23. ATPase Assay Protocol, GenoMembrane, Kanagawa, Japan. http://www.google.com/url?sa=t&rct=j&q=genomembrane%20atpase%20protocol&source=web&cd=1&sqi=2&ved=0CC8QFjAA&url=http%3A%2F%2Fwww.genomembrane.com%2FATPase_assay_Ver.6.5.n.pdf&ei=sKc4UeUjg8z2BImNgPAH4UeUjg8z2BImNgPAH&usg=AFQjCNEV3fbT5XiVZkeS1XLRRcCq5kQybg&bvm=bv.43287494,d.eWU&cad=rja
24. Glavinas H et al (2008) Utilization of membrane vesicle preparations to study drug-ABC transporter interactions. Expert Opin Drug Metab Toxicol 4(6):721–732
25. BD Gentest MRP/BCRP Vesicle Assay Kit Data Sheet, BD Biosciences, Woburn, MA. http://www.bdj.co.jp/gentest/1f3pro00000vv95q-att/Cat459010-lot79163.pdf
26. Karlsson JE et al (2010) High-activity p-glycoprotein, multidrug resistance protein 2, and breast cancer resistance protein membrane vesicles prepared from transiently transfected human embryonic kidney 293-epstein-barr virus nuclear antigen cells. Drug Metab Dispos 38(4):705–714
27. Glavinas H et al (2007) Passive permeability is a crucial parameter in choosing the right assay to detect the interaction of compounds with ABCB1 (Pgp) and ABCG2 (BCRP). International society for the study of xenobiotics annual meeting (October 9th – 12th), Sendai, Japan (Poster ID # 7808). http://issx.confex.com/issx/intl8/webprogram/Paper7808.html
28. Heredi-Szabo K et al (2009) Multidrug resistance protein 2-mediated estradiol-17beta-D-glucuronide transport potentiation: in vitro-in vivo correlation and species specificity. Drug Metab Dispos 37(4):794–801

29. Heredi-Szabo K et al (2008) Characterization of 5(6)-carboxy-2,7-dichlorofluorescein transport by MRP2 and utilization of this substrate as a fluorescent surrogate for LTC4. J Biomol Screen 13(4):295–301
30. Kidron H et al (2012) Impact of probe compound in MRP2 vesicular transport assays. Eur J Pharm Sci 46(1–2):100–105
31. Szeremy P et al (2011) Comparison of 3 assay systems using a common probe substrate, calcein AM, for studying P-gp using a selected set of compounds. J Biomol Screen 16(1):112–119
32. Vesicular Transport Assay Protocol (RI labeled compounds as substrates), GenoMembrane, Kanagawa, Japan. http://www.google.com/url?sa=t&rct=j&q=genomembrane%20vesicular%20protocol&source=web&cd=1&sqi=2&ved=0CC8QFjAA&url=http%3A%2F%2Fwww.genomembrane.com%2FVT_assay_protocol_Ver.7.3_for_RI.pdf&ei=KLc4UcWkG4OC9QSYuIG4Cw&usg=AFQjCNG66ew6EFfqon48ToyfnOb5o-9DzA&bvm=bv.43287494,d.eWU&cad=rja
33. Vesicular Transport Assay Protocol (Inhibition study using fluorescent substrates) GenoMembrane, Kanagawa, Japan. http://www.google.com/url?sa=t&rct=j&q=genomembrane%20vesicular%20protocol&source=web&cd=2&sqi=2&ved=0CDMQFjAB&url=http%3A%2F%2Fwww.genomembrane.com%2FVT_assay_protocol_Ver.7.3_for_fluorescent_probe.pdf&ei=KLc4UcWkG4OC9QSYuIG4Cw&usg=AFQjCNGO6TcaMJf7_tD6M0GdFRyckT6Y1A&bvm=bv.43287494,d.eWU&cad=rja
34. Krumpochova P et al (2012) Transportomics: screening for substrates of ABC transporters in body fluids using vesicular transport assays. FASEB J 26(2):738–747
35. Kato Y et al (2008) Involvement of multidrug resistance-associated protein 2 (Abcc2) in molecular weight-dependent biliary excretion of beta-lactam antibiotics. Drug Metab Dispos 36(6):1088–1096
36. Yamaguchi K et al (2009) Measurement of the transport activities of bile salt export pump using LC-MS. Anal Sci 25(9):1155–1158
37. Ming X et al (2009) Transport of dicationic drugs pentamidine and furamidine by human organic cation transporters. Drug Metab Dispos 37(2):424–430
38. Matsushima S et al (2005) Identification of the hepatic efflux transporters of organic anions using double-transfected Madin-Darby canine kidney II cells expressing human organic anion-transporting polypeptide 1B1 (OATP1B1)/multidrug resistance-associated protein 2, OATP1B1/multidrug resistance 1, and OATP1B1/breast cancer resistance protein. J Pharmacol Exp Ther 314(3):1059–1067
39. Cho MJ et al (1989) The Madin Darby canine kidney (MDCK) epithelial cell monolayer as a model cellular transport barrier. Pharm Res 6(1):71–77
40. Agarwal S, Pal D, Mitra AK (2007) Both P-gp and MRP2 mediate transport of Lopinavir, a protease inhibitor. Int J Pharm 339(1–2):139–147
41. Evers R et al (2000) Vinblastine and sulfinpyrazone export by the multidrug resistance protein MRP2 is associated with glutathione export. Br J Cancer 83(3):375–383
42. MDCK.2, ATCC, Manassas, VA, USA. http://www.atcc.org/ATCCAdvancedCatalogSearch/ProductDetails/tabid/452/Default.aspx?ATCCNum=CRL-2936&Template=cellBiology
43. Ward PD et al (2002) Phospholipase C-gamma modulates epithelial tight junction permeability through hyperphosphorylation of tight junction proteins. J Biol Chem 277(38):35760–35765
44. Ward PD, Tippin TK, Thakker DR (2000) Enhancing paracellular permeability by modulating epithelial tight junctions. Pharm Sci Technol Today 3(10):346–358
45. Hidalgo IJ, Raub TJ, Borchardt RT (1989) Characterization of the human colon carcinoma cell line (Caco-2) as a model system for intestinal epithelial permeability. Gastroenterology 96(3):736–749
46. Gan LSL, Thakker DR (1997) Applications of the Caco-2 model in the design and development of orally active drugs: elucidation of biochemical and physical barriers posed by the intestinal epithelium. Adv Drug Deliv Rev 23(1):77–98
47. Schinkel AH, Jonker JW (2003) Mammalian drug efflux transporters of the ATP binding cassette (ABC) family: an overview. Adv Drug Deliv Rev 55(1):3–29
48. Hilgendorf C et al (2007) Expression of thirty-six drug transporter genes in human intestine, liver, kidney, and organotypic cell lines. Drug Metab Dispos 35(8):1333–1340
49. Ming X, Thakker DR (2010) Role of basolateral efflux transporter MRP4 in the intestinal absorption of the antiviral drug adefovir dipivoxil. Biochem Pharmacol 79(3):455–462
50. Li J et al (2011) Use of transporter knockdown caco-2 cells to investigate the in vitro efflux of statin drugs. Drug Metab Dispos 39(7):1196–1202
51. Caco-2, ATCC, Manassas, VA, USA. http://www.atcc.org/ATCCAdvancedCatalogSearch/ProductDetails/tabid/452/Default.aspx?ATCCNum=htb-37&Template=cellBiology

52. Swift B, Pfeifer ND, Brouwer KL (2010) Sandwich-cultured hepatocytes: an in vitro model to evaluate hepatobiliary transporter-based drug interactions and hepatotoxicity. Drug Metab Rev 42(3):446–471

53. Liu X et al (1999) Biliary excretion in primary rat hepatocytes cultured in a collagen-sandwich configuration. Am J Physiol 277(1 pt 1):G12–G21

54. Liu X et al (1999) Use of Ca2+ modulation to evaluate biliary excretion in sandwich-cultured rat hepatocytes. J Pharmacol Exp Ther 289(3): 1592–1599

55. LeCluyse EL, Brouwer KL, Liu X (2004) Method of screening candidate compounds for susceptibility to biliary excretion, US 6,780,580 B2, 24 Aug 2004. The University of North Carolina at Chapel Hill, Chapel Hill, NC

56. LeCluyse EL, Kim RB, Liu X (2009) Method of screening candidate compounds for susceptibility to biliary excretion by endogenous transport systems, US 7,604,934 B2, 20 Oct 2009. The University of North Carolina at Chapel Hill, Chapel Hill, NC

57. Tian X et al (2004) Modulation of multidrug resistance-associated protein 2 (Mrp2) and Mrp3 expression and function with small interfering RNA in sandwich-cultured rat hepatocytes. Mol Pharmacol 66(4):1004–1010

58. Liao M et al (2010) Inhibition of hepatic organic anion-transporting polypeptide by RNA interference in sandwich-cultured human hepatocytes: an in vitro model to assess transporter-mediated drug-drug interactions. Drug Metab Dispos 38(9): 1612–1622

59. Yue W, Abe K, Brouwer KL (2009) Knocking down breast cancer resistance protein (Bcrp) by adenoviral vector-mediated RNA interference (RNAi) in sandwich-cultured rat hepatocytes: a novel tool to assess the contribution of Bcrp to drug biliary excretion. Mol Pharm 6(1):134–143

60. Tian X, Zhang P, Brouwer KL (2009) Method of screening candidate compounds for susceptibility to biliary excretion, US 7,601,494 B2, 13 Oct 2009. The University of North Carolina at Chapel Hill, Chapel Hill, NC

61. Swift B, Yue W, Brouwer KL (2010) Evaluation of (99m)technetium-mebrofenin and (99m)technetium-sestamibi as specific probes for hepatic transport protein function in rat and human hepatocytes. Pharm Res 27(9):1987–1998

62. Liu X et al (1999) Correlation of biliary excretion in sandwich-cultured rat hepatocytes and in vivo in rats. Drug Metab Dispos 27(6): 637–644

63. Abe K, Bridges AS, Brouwer KL (2009) Use of sandwich-cultured human hepatocytes to predict biliary clearance of angiotensin II receptor blockers and HMG-CoA reductase inhibitors. Drug Metab Dispos 37(3):447–452

64. Ghibellini G et al (2007) In vitro-in vivo correlation of hepatobiliary drug clearance in humans. Clin Pharmacol Ther 81(3): 406–413

65. Bi YA, Kazolias D, Duignan DB (2006) Use of cryopreserved human hepatocytes in sandwich culture to measure hepatobiliary transport. Drug Metab Dispos 34(9):1658–1665

66. Kimoto E et al (2011) Characterization of digoxin uptake in sandwich-cultured human hepatocytes. Drug Metab Dispos 39(1):47–53

67. Kotani N et al (2011) Culture period-dependent changes in the uptake of transporter substrates in sandwich-cultured rat and human hepatocytes. Drug Metab Dispos 39(9): 1503–1510

68. Wolf KK et al (2008) Effect of albumin on the biliary clearance of compounds in sandwich-cultured rat hepatocytes. Drug Metab Dispos 36(10):2086–2092

69. Polli JW et al (2001) Rational use of in vitro P-glycoprotein assays in drug discovery. J Pharmacol Exp Ther 299(2):620–628

70. Pedersen JM et al (2008) Prediction and identification of drug interactions with the human ATP-binding cassette transporter multidrug-resistance associated protein 2 (MRP2; ABCC2). J Med Chem 51(11):3275–3287

71. Bodo A et al (2003) Differential modulation of the human liver conjugate transporters MRP2 and MRP3 by bile acids and organic anions. J Biol Chem 278(26):23529–23537

72. Sun H, Pang KS (2008) Permeability, transport, and metabolism of solutes in Caco-2 cell monolayers: a theoretical study. Drug Metab Dispos 36(1):102–123

73. Troutman MD, Thakker DR (2003) Rhodamine 123 requires carrier-mediated influx for its activity as a P-glycoprotein substrate in Caco-2 cells. Pharm Res 20(8):1192–1199

Chapter 23

In Vitro Characterization of Hepatic Transporters OATP1B1 and OATP1B3

Blair Miezeiewski and Allison McLaughlin

Abstract

The hepatic transporters OATP1B1 and OATP1B3 contribute (to varying degrees, depending on the drug) to the uptake of many anionic drugs, including several of the widely used statins. Because statins are prescribed for many patients and the consequences of pharmacokinetic interactions with uptake inhibitors can be severe (even fatal), the U.S. Food and Drug Administration (FDA) and the European Medicines Agency (EMA) require that all NCEs be evaluated as inhibitors of OATP1B1 and OATP1B3. In addition, if hepatic clearance is expected to be a major pathway of elimination of an NCE, it must also be evaluated as a substrate of both transporters. Cell-based assays with over-expressing cell lines are useful for screening both substrates (based on uptake of the test compound) and inhibitors (based on interference with the uptake of a probe substrate by the test compound). The approach will be illustrated with real data, and subtle but important technical details will be discussed.

Key words OATP, OATP1B1, OATP1B3, Transporter, NCE, FDA, EMA

1 Introduction

The organic anion transporting polypeptides (OATPs) are members of the solute carrier (SLC) family of transporters and are responsible for the uptake of a wide range of substrates. OATP1B1 and OATP1B3 are predominantly expressed at the sinusoidal membrane of hepatocytes and play an important role in the uptake of numerous compounds prior to metabolism and/or biliary excretion [1]. Pharmacologic inhibition of OATPs, or expression of variants with reduced activity, can lead to marked elevations in the circulating concentrations of OATP substrates [2–4]. In fact, this mechanism resulted in dozens of fatal drug-drug interactions (DDIs) involving cerivastatin (Baycol®), leading to its withdrawal from the market in 2001 [5]. As a result, drug regulatory agencies, including the FDA and the EMA, recommend evaluation of the potential of all new chemical entities (NCEs) as inhibitors of

OATP1B1 and OATP1B3, and as substrates if they display significant hepatic/biliary clearance [6, 7]. Thus, it is important to investigate these transporters early in drug development.

Evaluation of the inhibitor and substrate potential of NCEs with OATP1B1 and OATP1B3 includes both clinical and non-clinical methods. Clinical methods focus on comparative PK in patients expressing different polymorphic variants, while non-clinical methods include knock-out animals and in vitro, cell-based assays. As highlighted in the current FDA draft guidance on drug interaction studies [6], in vitro studies are part of an integrated approach to (1) assess a new drug's safety and effectiveness, (2) design appropriate clinical trials, and (3) avoid performing unnecessary ones: "Along with clinical pharmacokinetic data, results from in vitro studies can serve as a screening mechanism to rule out the need for additional in vivo studies, or provide a mechanistic basis for proper design of clinical studies using a modeling and simulation approach."

This chapter will focus on the use of cell-based assays to evaluate inhibitor and substrate potential with the OATP1B1 and OATP1B3 transporters. Robust and reliable in vitro assay systems are required to identify substrates and inhibitors of OATP1B1 and OATP1B3, either of which could pose a risk of clinical DDIs with co-medications. Stably transfected human cell lines such as HEK293 are ideal models in many ways: as human cell lines, they exhibit no interference from animal transporters (unlike other cell-based models); as stable transfectants, they are well characterized and provide reproducible results. They also show little to no expression of other uptake transporters, ensuring that transporter-mediated uptake is a result of the transporter of interest.

2 Materials

2.1 Equipment

Descriptions of specific equipment and materials used are given; however, any suitable equivalent may be substituted.

1. FLUOstar OPTIMA fluorescence microplate reader equipped with MARS software (BMG Labtech, Durham, NC, USA) for QC assays and protein determination.
2. Forma Scientific Steri-Cult 2000 CO_2 incubator, Thermo Stericycle CO_2 incubator for cell culture and assay incubations.
3. Microplate shaker.
4. Biological safety cabinet: NuAire Class II Type A/B3, Labconco Purifier Class II for aseptic operations.
5. Electronic pipettor.

2.2 Reagents and Solutions

1. Dulbecco's modified Eagle medium (DMEM; Life Technologies, Carlsbad, CA, USA): cell culture base medium.
2. Fetal bovine serum (FBS; Life Technologies): cell culture medium supplement.
3. L-glutamine (Life Technologies): cell culture medium supplement.
4. Penicillin-streptomycin (Life Technologies): cell culture medium supplement.
5. G418 (Life Technologies): cell culture medium supplement; selection agent for transfected cells.
6. Dulbecco's phosphate-buffered saline (DPBS; Life Technologies) for rinsing cells in a plate.
7. Trypsin-EDTA (Life Technologies) for dissociating cells from a plate.
8. Cellbind T150 cm^2 cell culture flask (Corning, Corning, NY, USA): culture vessel for stock cell cultures.
9. 24-well poly-D-lysine Biocoat® plates (Becton Dickinson, Franklin Lakes, NJ, USA): vessels in which cells are cultured for uptake experiments.
10. Hanks' balanced salt solution (HBSS, Life Technologies) for assay buffer.
11. 4-(2-Hydroxyethyl)-1-piperazineethanesulfonic acid (HEPES; Life Technologies) for assay buffer.
12. D-Glucose (Sigma, St. Louis, MO, USA) for assay buffer.
13. HBSSg, pH 7.4 (assay buffer): HBSS with 10 mM HEPES and 15 mM D-glucose.
14. Acetonitrile (EMD Millipore, Gibbstown, NJ, USA) for lysing cells.
15. BCA kit (Thermo Pierce, Pittsburgh, PA, USA) for protein determination.
16. Atorvastatin-d5, (Toronto Research Chemicals): analytical internal standard.
17. Fluorescein-methotrexate (FMTX; Life Technologies): QC assay substrate.
18. Radioimmunoprecipitation assay (RIPA) buffer (Santa Cruz Biotechnology, Santa Cruz, CA, USA) for lysing cells at the end of the uptake assay.
19. Opaque, black 96-well plates for fluorometric detection in the QC assay.

3 Methods

3.1 Cell Culture

The cell-based assay uses stably transfected HEK-OATP1B1, HEK-OATP1B3, and HEK-vector control (VC) cells created by Absorption Systems (Exton, PA, USA). Perform all cell culture operations in a biological safety cabinet under sterile conditions and with proper aseptic technique.

1. HEK-OATP1B1, HEK-OATP1B3, and HEK-VC cells are cultured in Cellbind T150 flasks and maintained in DMEM supplemented with 10 % FBS, 1 % L-glutamine, 100 IU/mL penicillin, and 100 µg/mL streptomycin. In addition, selection pressure is maintained using G418 at 800 µg/mL.

2. To subculture the cells, wash once with DPBS and dissociate the cells with trypsin-EDTA. Seed the cells at a density of 2.5×10^5 cells/well on 24-well poly-D-lysine-coated plates using 1 mL of cell suspension per well.

3. Maintain the cultured cells in a humidified incubator at 37 °C with 5 % CO_2 (**Note 1**).

3.2 Batch Quality Control

In order to confirm adequate transporter activity in each batch of cells (**Note 2**), it is recommended that a quality control (QC) assessment be run prior to each uptake experiment. Typically, a fluorescent substrate is used because results can be obtained immediately, unlike with LC-MS/MS. QC must be performed on both transporter-transfected cells and the corresponding VC cells. Representative batch QC data are shown in Table 1.

1. Prepare QC dosing solution of the probe substrate FMTX at 1 µM in HBSSg, pH 7.4 and a standard curve of FMTX in RIPA buffer (**Note 3**).

2. Gently aspirate the culture medium, taking care not to disturb the cells, and add 0.5 mL QC dosing solution into triplicate wells (n = 3) for each cell line (transporter-transfected and VC).

3. Incubate the cells at 37 °C for 10 min in a humidified incubator with 5 % CO_2.

Table 1
Representative QC data

Cell line	Average FMTX influx rate (pmol/mg/min)[a]	SD	Influx rate ratio[b]
HEK-VC	0.88	0.09	–
HEK-OATP1B1	55.42	1.43	63.25
HEK-OATP1B3	85.42	1.89	97.48

[a]Equation (1), Sect. 3.5
[b]Transporter-transfected cells vs. VC; (2), Sect. 3.5

4. At 10 min, aspirate the dosing solution (taking care not to disturb the cells) and wash twice with ice-cold HBSSg, pH 7.4 (**Note 4**).

5. Lyse the cells with 0.15 mL ice-cold RIPA buffer and shake at 4 °C for 15 min. Transfer 100 μL of lysate from each replicate to an opaque, black 96-well plate. Read the fluorescence in a fluorescence microplate reader with 485 nm excitation and 520 nm emission wavelengths. Calculate the concentrations of FMTX taken up into cells by comparison with the standard curve.

6. Save an aliquot of the lysate at −20 °C for protein determination using a BCA kit, following the manufacturer's protocol. The average protein concentration for each cell line is used to normalize the uptake rate for a given batch of cells.

3.3 Substrate Assay

In order to evaluate potential OATP substrates, a matrix protocol is recommended, with multiple dosing concentrations and multiple time points. Representative data, for the OATP substrate atorvastatin as a "test compound," is shown in Table 2 for HEK-OATP1B1 and in Table 3 for HEK-OATP1B3, using three concentrations and four time points for each cell line (transporter-transfected and VC). The criterion for scoring a test compound as a substrate is influx rate ratio (IRR; (2), Sect. 3.5) ≥2 under at least one set of conditions; otherwise, it is classified as a non-substrate. Atorvastatin (Table 2) is positive for both transporters under all test conditions, but that is not the case for many other substrates.

1. Prepare dosing solutions by diluting the substrate stock in HBSSg, pH 7.4 (**Note 5**).

2. Gently aspirate the culture medium, taking care not to disturb the cells, and add 0.5 mL dosing solution (each test concentration) into triplicate wells (n = 3) per time point for each cell line. Incubate at 37 °C in a humidified incubator with 5 % CO_2.

Table 2
HEK-OATP1B1 substrate matrix for atorvastatin

		Influx rate ratio[a]			
		Duration (min)			
	Conc. (μM)	2	5	10	20
Atorvastatin	0.1	13.62	21.31	35.40	29.22
	0.5	12.70	17.44	19.47	13.61
	2	6.79	6.99	7.31	5.07

[a]Transporter-transfected cells vs. VC; (2), Sect. 3.5

Table 3
HEK-OATP1B3 substrate matrix for atorvastatin

	Conc. (µM)	Influx rate ratio[a] Duration (min) 2	5	10	20
Atorvastatin	1	9.44	16.86	28.92	24.00
	5	10.26	16.21	22.78	16.57
	20	7.74	10.26	14.17	9.44

[a]Transporter-transfected cells vs. VC; (2), Sect. 3.5

3. At each time point, terminate the assay by aspirating the dosing solution and washing two times with ice-cold HBSSg.

4. With the assay plate on ice (**Note 6**), lyse the cells by adding 0.4 mL per well ice-cold 75 % acetonitrile:25 % H_2O (v:v) containing an internal standard (deuterated version of the test compound, if available). Lyse on ice for at least 2 min and collect 300 µL of lysate into a 96-well deepwell plate for LC-MS/MS analysis.

It is recommended that positive results from the substrate test matrix be confirmed by challenging uptake of the test compound with a known inhibitor of the transporter(s) of interest. Rifamycin SV (50 µM) is suitable for both OATP1B1 and OATP1B3. In general, use the lowest test compound concentration and the shortest assay duration that tested positive in the substrate matrix. To ensure that the test compound concentration in the inhibitor challenge is below the K_m (and therefore susceptible to inhibition in that step), an optional intermediate step is to determine the kinetic parameters (K_m and V_{max}) of the test compound (**Note 7**).

3.4 Inhibitor Assay

This assay is used to evaluate the inhibitor potential of a test compound at OATP1B1 and OATP1B3, based on reduction of the rate of uptake of the probe substrate atorvastatin into transporter-transfected cells. The time course of atorvastatin uptake is shown in Table 4 and Fig. 1 for HEK-OATP1B1 cells and in Table 5 and Fig. 2 for HEK-OATP1B3 cells. At a concentration of 0.15 µM, uptake is linear for approximately 5 min with HEK-OATP1B1 cells and perhaps longer with HEK-OATP1B3 cells; inhibition assays should be conducted under linear conditions. The same concentration (0.15 µM) of the probe substrate atorvastatin, which is well below its K_m (data not shown) is used in the inhibitor assay. The results of screening a panel of reported OATP inhibitors at two concentrations (*see* **Note 8** for selection of test compound concentration) for inhibition of atorvastatin uptake into OATP1B1-HEK

Table 4
Atorvastatin uptake time course in HEK-OATP1B1 cells

Duration (min)	Atorvastatin uptake (pmol/mg) HEK-VC	HEK-OATP1B1	Net uptake (pmol/mg)[a]
2	2.12	20.04	17.92
5	3.03	44.39	41.36
10	4.19	57.30	53.11
20	5.61	72.74	67.13

[a]Difference between HEK-OATP1B1 and HEK-VC; (3), Sect. 3.5

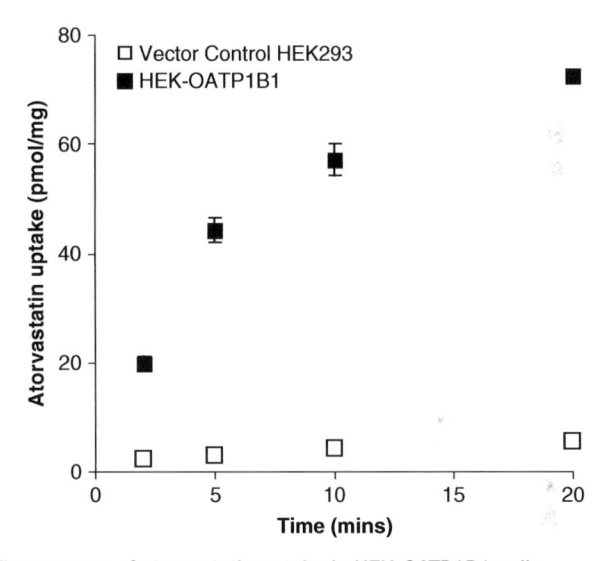

Fig. 1 Time course of atorvastatin uptake in HEK-OATP1B1 cells

Table 5
Atorvastatin uptake time course in HEK-OATP1B3 cells

Duration (min)	Atorvastatin uptake (pmol/mg) HEK-VC	HEK-OATP1B3	Net uptake (pmol/mg)[a]
2	2.12	13.27	11.15
5	3.03	34.44	31.41
10	4.19	51.21	47.02
20	5.61	66.03	60.42

[a]Difference between HEK-OATP1B3 and HEK-VC; (3), Sect. 3.5

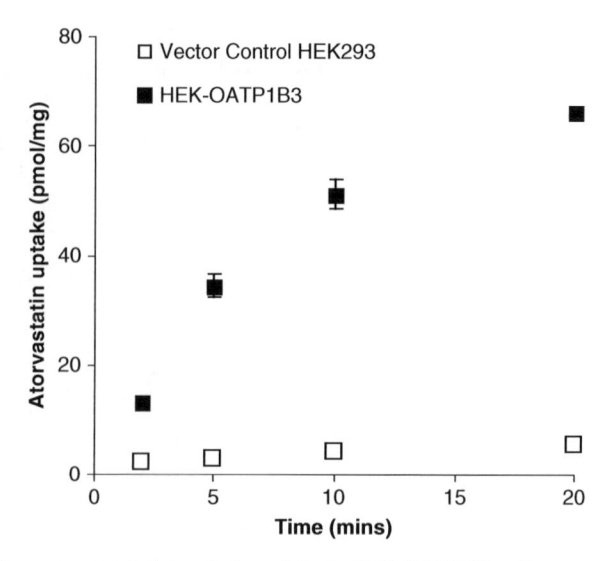

Fig. 2 Time course of atorvastatin uptake in HEK-OATP1B3 cells

cells are shown in Table 6. Positive scores (generally defined as compounds that inhibit uptake of the probe substrate by at least 50 % at the initial test concentration) are followed up with an IC$_{50}$ determination, by co-incubating the probe substrate and several concentrations of the test compound. IC$_{50}$ determination for the representative "test compound" rifamycin SV with OATP1B1 is shown in Table 7 and Fig. 3.

1. Prepare dosing solutions (probe substrate alone and with one or more concentrations of test compound) by diluting the stocks in HBSSg, pH 7.4. *See* **Note 8** for selection of test compound concentration.

2. Gently aspirate the culture medium, taking care not to disturb the cells, and add 0.5 mL dosing solution (probe substrate with or without test compound) into triplicate wells (n = 3) for HEK-OATP1B1 and/or HEK-OATP1B3 cells. Incubate for 5 min at 37 °C in a humidified incubator with 5 % CO$_2$.

3. Terminate the assay by aspirating the dosing solution and washing two times with ice-cold HBSSg.

4. With the assay plate on ice, lyse the cells by adding 0.4 mL per well ice-cold 75 % acetonitrile: 25 % H$_2$O (v:v) containing deuterated atorvastatin (atorvastatin-d5) as an analytical internal standard. Lyse on ice for at least 2 min and collect 300 μL of lysate into a 96-well deepwell plate for LC-MS/MS analysis.

3.5 Data Analysis

Uptake data is processed using the following equations:

$$\text{Influx Rate (IR)} = \frac{C_S \times V_S}{(C_P \times V_P) \times T} \quad (1)$$

Table 6
Effect of test compounds on atorvastatin uptake in HEK-OATP1B1 cells

Compound	Compound at 10 μM Avg. atorvastatin influx rate (pmol/mg/min)[a] HEK-VC	HEK-OATP1B1	% Activity remaining[b]	Compound at 100 μM Avg. atorvastatin influx rate (pmol/mg/min)[a] HEK-VC	HEK-OATP1B1	% Activity remaining[b]
Control	0.20	6.91	100.00	0.20	6.91	
Atenolol	0.11	7.32	107.35	0.09	7.27	106.92
CsA[c]	0.24	2.22	29.52	0.39	0.85	7.39
Digoxin	0.13	5.83	84.88	0.18	3.68	52.14
E3S	0.13	1.84	25.35	0.10	1.59	22.15
Estradiol	0.13	4.10	59.13	0.08	1.48	20.76
Gemfibrozil	0.11	6.44	94.33	0.15	2.60	36.39
Ketoconazole	0.17	5.24	75.45	0.20	0.68	7.22
Propranolol	0.13	6.14	89.55	0.12	5.06	73.68
Rifampicin	0.32	1.37	15.69	0.47	0.58	1.63
Rifamycin SV	0.36	0.75	5.92	0.50	0.62	1.76
Ritonavir	0.25	1.60	20.02	0.30	0.60	4.40
Rosuvastatin	0.12	4.33	62.73	0.09	1.73	24.55
Taurocholic acid	0.26	5.19	73.40	0.33	2.15	27.20
Tolbutamide	0.31	7.33	104.62	0.21	7.78	112.65
Verapamil	0.23	5.68	81.28	0.27	2.73	36.61

[a]Equation (1), Sect. 3.5
[b]Equation (7), Sect. 3.5
[c]CsA was dosed at 1 and 10 μM rather than 10 and 100 μM

$$\text{Influx Rate Ratio (IRR)} = \frac{\text{IR}_{\text{Transfectd cells}}}{\text{IR}_{\text{VC}}} \quad (2)$$

$$\text{Net Influx} = \text{Influx}_{\text{Transfected cells}} - \text{Influx}_{\text{VC}} \quad (3)$$

$$\text{Net Influx Rate (NIR)} = \text{IR}_{\text{Transfected cells}} - \text{IR}_{\text{VC}} \quad (4)$$

$$v = \frac{V_{\max} \times [S]}{K_m + [S]} \quad (5)$$

$$\% \text{ of Inhibition} = \left[1 - \left(\frac{\text{NIR}_{\text{TC}}}{\text{NIR}_{\text{w/oTC}}}\right)\right] \times 100 \quad (6)$$

Table 7
OATP1B1 IC₅₀ of rifamycin SV

Rifamycin SV conc. (μM)	Avg. atorvastatin influx rate (pmol/mg/min)[a] HEK-VC	HEK-OATP1B1	% Activity remaining[b]	IC₅₀ (μM)[c]
0	0.30	7.98	100.00	0.30
0.03	0.26	6.19	77.11	
0.1	0.30	6.37	79.12	
0.3	0.27	4.01	48.66	
1	0.28	1.92	21.42	
3	0.38	1.24	11.27	
10	0.42	1.09	8.83	
30	0.48	1.06	7.58	

The concentration of the probe substrate atorvastatin was 0.15 μM and the assay duration was 5 min
[a]Equation (1), Sect. 3.5
[b]Equation (7), Sect. 3.5
[c]Equation (8), Sect. 3.5

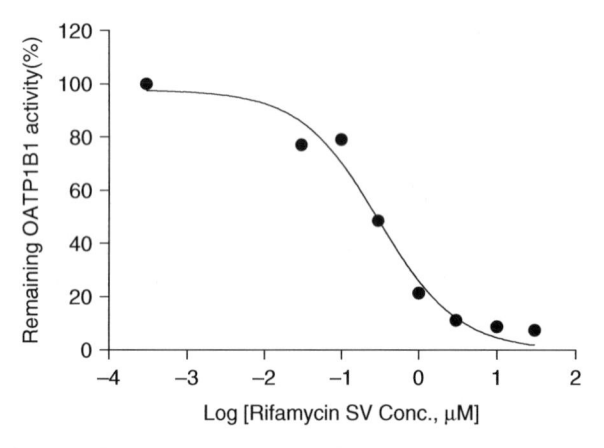

Fig. 3 Rifamycin SV inhibition curve for OATP1B1. The concentration of the probe substrate atorvastatin was 0.15 μM and the assay duration was 5 min

$$\% \text{ of Activity Remaining} = \left(\frac{\text{NIR}_{\text{TC}}}{\text{NIR}_{\text{w/oTC}}}\right) \times 100 \quad (7)$$

$$Y = Y_L + \frac{(Y_H 0 Y_L)}{1 + 10^{((\text{Log IC}_{50} 0X) \times \text{Hillslope})}} \quad (8)$$

C_s: Concentration of test compound in the cell lysate (µM);

V_s: Cell lysate volume in substrate assessment (mL);

C_p: Protein concentration in the cell lysate (mg/mL);

V_p: Cell lysate volume in protein determination (mL);

T : Time (min) ;

VC: Vector control cells;

v: Uptake velocity (pmol/mg/min);

V_{max}: Maximal velocity (pmol/mg/min);

[S]: Substrate concentration (µM);

K_m: Substrate concentration at half-maximal velocity;

TC: Test compound;

Y_L: % of control activity at the highest concentration of inhibitor;

Y_H: % of control activity at the lowest concentration of inhibitor;

X: $\log_{10}$ of inhibitor concentration.

K_m, V_{max}, and IC_{50} are determined by nonlinear regression analysis using GraphPad Prism software.

4 Notes

1. In our settings, the cells are best used within 2 days of seeding on 24-well plates. It may be helpful to run a shelf life evaluation at 1, 2, 3, 4, and 5 days after seeding.

2. We recommend an acceptance criterion of FMTX uptake rate at least three times greater in OATP-transfected cells than in vector control cells.

3. The FMTX standard curve is prepared in RIPA buffer beginning with 1 µM and diluted with twofold serial dilutions to 0.0078125 µM.

4. The addition of ice-cold buffer terminates the uptake of the substrate and removes residual dosing solution. Due to the brevity of the assay, precise and accurate timing of the initial wash step is critical.

5. We recommend pre-warming the dosing solution to 37 °C in the incubator to minimize cell shock.

6. Keep all components chilled on ice, especially during the lysis step, in order to avoid evaporation. The addition of a deuterated internal standard (in this case, 50 nM atorvastatin-d5) aids in monitoring solvent evaporation.

7. Active, transporter-mediated uptake is a saturable process; a plot of uptake rate vs. concentration will reach a plateau approaching the V_{max}; the concentration at $0.5 \times V_{max}$ is the K_m.

Use at least eight concentrations of the test compound, spanning at least two log units, and the shortest assay duration that tested positive in the substrate matrix. Calculate active, transporter-mediated uptake (net influx rate; (4), Sect. 3.5) at each concentration of test compound as the difference in uptake between transporter-transfected and VC cells. Fit the data to the Michaelis-Menten equation ((5), Sect. 3.5) with a software package such as GraphPad® Prism®.

8. The test compound concentration in the initial inhibitor assay screen is typically either 10 μM or, if clinical PK data is available, 10× the mean (unbound, if available) steady-state C_{max} at the highest dose ($[I]_1$ [6, 7]). It also depends on the results of suitability assessments such as solubility of the test compound and tolerability of the cells.

References

1. Niemi M (2007) Role of OATP transporters in the disposition of drugs. Pharmacogenomics 8(7):787–802
2. Shitara Y, Hirano M, Sato H, Sugiyama Y (2004) Gemfibrozil and its glucuronide inhibit the organic anion transporting polypeptide 2 (OATP2/OATP1B1:*SLC21A6*)-mediated hepatic uptake and CYP2C8-mediated metabolism of cerivastatin: analysis of the mechanism of the clinically relevant drug-drug interaction between cerivastatin and gemfibrozil. J Pharmacol Exp Ther 311(1):228–236
3. Niemi M, Backman JT, Kajosaari LI, Leathart JB, Neuvonen M, Daly AK, Eichelbaum M, Kivisto KT, Neuvonen PJ (2005) Polymorphic organic anion transporting polypeptide 1B1 is a major determinant of repaglinide pharmacokinetics. Clin Pharmacol Ther 77:468–478
4. Kalliokoski A, Niemi M (2009) Impact of OATP transporters on pharmacokinetics. Br J Pharmacol 158(3):693–705
5. Staffa JA, Chang J, Green G (2002) Cerivastatin and reports of fatal rhabdomyolysis. N Engl J Med 346:539–540
6. US Department of Health and Human Services, Food and Drug Administration, Center for Drug Evaluation and Research (CDER) (2012) Draft guidance for industry. Drug interaction studies—study design, data analysis, implications for dosing, and labeling recommendations. http://www.fda.gov/downloads/Drugs/GuidanceCompliance RegulatoryInformation/Guidances/UCM 292362.pdf
7. European Medicines Agency, Committee for Human Medicinal Products (CHMP) (2012) Guideline on the investigation of drug interactions. http://www.ema.europa.eu/docs/en_GB/document_library/Scientific_guideline/2012/07/WC500129606.pdf

Chapter 24

In Vitro Characterization of Renal Transporters OAT1, OAT3, and OCT2

Ying Wang and Nicole Behler

Abstract

The human transporters organic anion transporter 1 (OAT1), organic anion transporter 3 (OAT3) and organic cation transporter 2 (OCT2) are membrane proteins involved in the renal clearance of substances from the body. While the original purpose of these clearance pathways was likely the removal of naturally occurring metabolism byproducts, they are also responsible for secretion of many drugs. Various assay systems have been developed to study the interactions of drugs with these transporters as either substrates or inhibitors, since a reduction in drug clearance (due to inhibition of transport of one drug by a co-administered drug) could result in elevated exposure and toxicity. This chapter will provide a brief background on the transporters and how they function, highlight the importance of using *in vitro* test systems to evaluate the potential for drug-drug interactions (DDIs), and provide a detailed procedure for an assay using transfected human embryonic kidney (HEK293) cells.

Key words Transporter, OAT1, OAT3, OCT2, HEK293, Drug-drug interaction, DDI, Substrate, Inhibitor, NCE, FDA, EMA

1 Introduction

OAT1 (SLC22A6), OAT3 (SLC22A8) and OCT2 (SLC22A2) are uptake transporters belonging to the solute carrier (SLC) superfamily of transporters. While present in other tissues, they are primarily expressed in the proximal tubule epithelial cells of the kidney, where they are localized to the basolateral membrane separating the cytosol from the blood-facing interstitial fluid [1]. Within the kidney, these transporters play an important role in the clearance of endogenous substances, drugs, and drug metabolites by means of vectorial transport: uptake of the solute from the blood and efflux across the apical membrane into the urine [2]. Renal uptake transporters are responsible for the first half of the process. For OAT1 and OAT3, uptake involves exchange of two substrates in opposite directions: organic anions in the blood are exchanged for endogenous, cellular α-ketoglutarate, driven by the flow of the latter down its electrochemical potential gradient [3]. OCT2, while also capable of exchange, typically transports substrates across the basolateral membrane via facilitated diffusion. Cations are pumped into the cell, theoretically until the electrochemical potential (inside negative) reaches equilibrium [2]; in reality, equilibrium is never reached because the electrochemical potential is maintained by numerous processes, including efflux of OCT2's cationic substrates across the opposite pole of the cell into the urine. In instances where OCT2 is acting as an antiporter (i.e., exchanger), the extra- and intracellular substrates are cations.

Drugs that are cleared primarily by the kidney have an increased likelihood of interacting with at least one of these renal uptake transporters, either as a substrate or as an inhibitor, and interfering with their function can result in accumulation of the substance, either in the kidney or systemically. The result of the former can be renal toxicity; the latter can be detected as an increase in the circulating concentration of the substance in the blood and possibly as systemic side effects. Since so many patients take multiple medications, which can interact with each other or dietary constituents or herbal supplements, it is important in terms of drug safety to determine if a drug has the potential to be either a victim or a perpetrator of a drug-drug interaction (DDI). Fatalities have resulted from the use of the OAT1/OAT3 substrate methotrexate, a widely used chemotherapy agent, when it is co-administered with certain non-steroidal anti-inflammatory drugs (NSAIDS), likely due to the ability of the latter to inhibit OATs [4]. The prescribing information for the antiviral drug cidofovir specifically warns of the danger of nephrotoxicity if co-administered with the OAT inhibitor probenecid [1]. Case studies such as these are why both the U.S. Food and Drug Administration (FDA) and the European Medicines Agency (EMA) now require that interactions of new chemical entities (NCEs) with these transporters be evaluated during drug development [5, 6].

Cell-based *in vitro* test systems for renal uptake transporters employ intact cells, either in a directional assay (across monolayers of polarized epithelial cells in a dual-chamber apparatus) or in an uptake assay. Due to a lack of existing cell lines with robust expression of the renal uptake transporters, cell lines are generally modified to over-express the transporter(s) of interest through transfection. Examples of cell lines used for transfection include Chinese hamster ovary (CHO), Madin-Darby canine kidney (MDCK), and human embryonic kidney (HEK293) [3]. In addition, *Xenopus* oocytes can be injected with complementary RNA coding for an uptake transporter, which will then be expressed on the surface. Of these, only MDCK cells can be used in the directional assay format; after over-expression of both a basolateral uptake transporter and an apical efflux transporter, transport is monitored in the basolateral-to-apical direction and compared to transport in a control cell line (non-transfected, or transfected only with the efflux transporter). In non-polarized cells such as CHO or HEK, a cellular uptake assay is performed in which a test compound (or probe substrate with and without test compound) is incubated with cells, and the amount of test compound (or probe substrate) inside the cells is measured.

This chapter will focus on uptake assays using HEK293 cells that have been transfected with a single uptake transporter.

2 Materials

2.1 Chemicals

1. *p*-Aminohippuric acid (PAH; Sigma, St. Louis, MO, USA): probe substrate of OAT1
2. Furosemide (Sigma): probe substrate of OAT3
3. 1-Methyl-4-phenylpyridinium iodide (MPP+; Sigma): probe substrate of OCT2
4. Imipramine (Sigma): inhibitor of OCT2
5. 6-Carboxyfluorescein (6-CF; Toronto Research Chemicals [TRC]): QC probe substrate for OAT1
6. 5-Carboxyfluorescein (5-CF; TRC): QC probe substrate for OAT3
7. 4-(4-(Dimethylamino)styryl)-N-methylpyridinium iodide (ASP; Life Technologies, Carlsbad, CA, USA): QC probe substrate for OCT2

2.2 Cell Culture

1. Cell lines: stably transfected OAT1-, OAT3-, OCT2-, and vector control (VC)-transfected HEK293 cells were created by Absorption Systems (Exton, PA, USA).
2. Culture medium: Dulbecco's modified Eagle's medium (DMEM) supplemented with 10 % fetal bovine serum, 1 % L-glutamine, 100 IU/mL penicillin, and 100 µg/mL streptomycin (all from Life Technologies, Carlsbad, CA, USA), and 800 µg/mL G418 (Cellgro, Corning, NY, USA).
3. Dulbecco's phosphate-buffered saline (Life Technologies) for rinsing cells.
4. Trypsin (Life Technologies) for subculturing cells.

2.3 Uptake Assay

1. Assay buffer: HBSSg, pH 7.4, consisting of Hanks' balanced salt solution (HBSS; Life Technologies) with 10 mM HEPES (4-(2-hydroxyethyl)piperazine-1-ethanesulfonic acid; Life Technologies) and 15 mM D-glucose (Sigma)

2. Culture/assay plates: 24-well poly-D-lysine-coated Biocoat® plates (Becton Dickinson, Franklin Lakes, NJ USA)

3. BCA assay reagent (Pierce, Rockford, IL, USA) for protein determination

4. RIPA buffer (Santa Cruz Biotechnology, Santa Cruz, CA, USA): cell lysis buffer for protein samples

5. Acetonitrile (EMD Chemicals, Gibbstown, NJ, USA): cell lysis reagent for test compound/probe substrate samples

3 Methods

3.1 Cell Culture

Perform all cell culture operations in a biological safety cabinet under sterile conditions and with proper aseptic technique.

1. HEK-OAT1, HEK-OAT3, HEK-OCT2, and HEK-VC cells are cultured in Cellbind T150 cm² cell culture flasks (Corning, Corning, NY, USA) and maintained in a humidified incubator at 37 °C with 5 % CO_2.

2. To subculture the cells, wash once with DPBS and dissociate the cells with trypsin-EDTA. Seed the cells at a density of 2.5×10^5 cells/well on 24-well poly-D-lysine-coated plates using 1 mL of cell suspension per well.

3. Counting the day of plating as Day 0, use the cells for uptake experiments on Day 2.

3.2 Batch Acceptance (Quality Control)

In order to confirm adequate transporter activity in each batch of cells, a quality control (QC) assay with a fluorescent probe substrate should be performed with both transporter-transfected cells and the corresponding vector control-transfected cells prior to each uptake experiment. Representative batch QC data are shown in Table 1.

1. Prepare QC dosing solutions of each probe substrate (at the concentrations indicated in Table 1) in HBSSg, pH 7.4, and standard curves in RIPA buffer.

2. Gently aspirate the culture medium, taking care not to disturb the cells (**Note 1**), and add 0.5 mL of the appropriate QC dosing solution (Table 1) into triplicate wells (n = 3) for each cell line.

3. Incubate the cells at 37 °C for 5 min (HEK-OAT1 and HEK-OCT2) or 10 min (HEK-OAT3), in a humidified incubator with 5 % CO_2.

4. Aspirate the dosing solution (taking care not to disturb the cells) and wash twice with ice-cold HBSSg, pH 7.4.

Table 1
Representative QC data

Cell line	QC probe substrate	Concentration (μM)	Average influx rate (pmol/mg/min)[a]	Influx rate ratio[b]
HEK-OAT1/HEK-VC	6-CF[c]	5	15.8/0.918	17.2
HEK-OAT3/HEK-VC	5-CF[c]	20	17.2/1.69	10.2
HEK-OCT2/HEK-VC	ASP[c]	1	102/6.71	15.2

[a]Equation (1), Sect. 3.5
[b]Transporter-transfected cells vs. VC; Equation (2), Sect. 3.5
[c]6-CF 6-carboxyfluorescein, 5-CF 5-carboxyfluorescein, ASP 4-(4-(dimethylamino)styryl)-N-methylpyridinium iodide

5. Lyse the cells with 0.15 mL ice-cold RIPA buffer and shake at 4 °C for 15 min. Transfer 100 μL of lysate from each replicate to an opaque, black 96-well plate. Read the fluorescence in a fluorescence microplate reader with excitation and emission wavelengths of 485 and 520 nm (6-CF and 5-CF) or 492 and 620 nm (ASP). Calculate the concentrations taken up into cells by comparison with the standard curve.

6. Save an aliquot of the lysate at −20 °C for protein determination using a BCA kit (Thermo Pierce, Pittsburgh, PA, USA), following the manufacturer's protocol. The average protein concentration per well for each cell line is used to normalize the uptake rate for a given batch of cells.

3.3 Substrate Assay

The objectives of substrate assessment are to investigate the potential of a test compound as a substrate of a transporter, confirm an initial positive result by challenge with an inhibitor of the same transporter, and/or further characterize its uptake profile by determining the K_m and V_{max}.

3.3.1 Step 1

In the first step, substrate potential is assessed using a matrix uptake assay protocol (**Note 2**), with three dosing concentrations and four time points for each cell line (**Note 3**). The criterion for scoring a test compound as a substrate is influx rate ratio (IRR; (2), Sect. 3.5) ≥2 under at least one set of conditions; otherwise, it is classified as a non-substrate. To illustrate the approach, representative data, for the OAT1 substrate PAH as a "test compound," is shown in Table 2.

1. Conduct assays in triplicate (n = 3).
2. Prepare dosing solutions by diluting stock solutions of test compounds in HBSSg, pH 7.4 (**Note 4**).

Table 2
Uptake of PAH by OAT1

	Test compound		IRR			
			Duration (min)			
Cell line	Name	Conc. (µM)	2	5	10	20
HEK-OAT1	PAH	10	49.0	98.9	193.5	215.1
		40	61.4	143.0	410.4	766.3
		100	24.9	52.4	156.7	368.3

3. Gently aspirate the medium from each well, taking care not to disturb the cells.
4. Add pre-warmed (**Note 5**) dosing solution (0.5 mL) to each well (**Note 6**).
5. Incubate for 2, 5, 10, and 20 min at 37 °C in a humidified incubator with 5 % CO_2.
6. Terminate the uptake assay by washing twice with ice-cold HBSSg.
7. With the assay plate on ice, lyse the cells by adding 0.4 mL per well ice-cold 75 % acetonitrile:25 % H_2O (v:v) containing an internal standard (deuterated version of the test compound, if available). Lyse on ice for 2 min, then transfer 0.3 mL of cell lysate to a 96-well block for analysis by LC-MS/MS.
8. In a parallel set of wells (n = 3), aspirate the medium gently, wash twice with ice-cold HBSSg, and lyse the cells with 0.15 mL RIPA buffer on a shaker for 15 min at 4 °C. These lysates will be used for protein concentration determination with a BCA kit.
9. Normalize the uptake in each well to the average measured protein concentration per well, and express uptake in units of pmol/mg protein or uptake rate in units of pmol/mg/min.

3.3.2 Step 2 (Optional)

The second step of substrate assessment, which is recommended but optional, is to determine the kinetic parameters (K_m and V_{max}) of test compounds that were positive in Step 1. Active, transporter-mediated uptake is a saturable process; a plot of uptake rate vs. concentration will reach a plateau approaching the V_{max}. Representative data is shown in Fig. 1 for the "test compounds" PAH (OAT1), furosemide (OAT3), and MPP+ (OCT2). The concentration at $0.5 \times V_{max}$ is the K_m; one advantage to performing this step now is to ensure that in Step 3 of substrate assessment the test

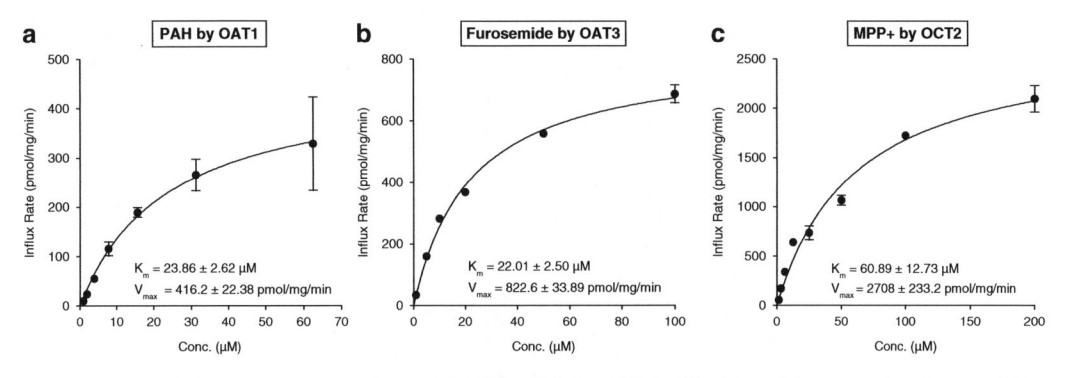

Fig. 1 Kinetics of OAT1-mediated uptake of PAH (**a**), OAT3-mediated uptake of furosemide (**b**), and OCT2-mediated uptake of MPP⁺ (**c**). In each case, the *solid line* is the fit of the kinetic data to the Michaelis-Menten equation. Each point represents the mean ± SD of three wells

compound concentration will be below the K_m, ensuring that it will be susceptible to inhibition in that step.

1. Prepare dosing solutions of the test compound, with at least eight concentrations.

2. Follow the procedure above (**Step 1 of substrate assessment**). Select the assay duration for **Step 2** based on the results of Step 1:

   - If the test compound scores positive at only one time point in **Step 1**, use that time point in **Step 2**.
   - If it scores positive at multiple time points in **Step 1**, select the shortest positive time point for **Step 2** or plot uptake vs. time at the lowest positive concentration and select a time point for **Step 2** that is within the linear range.

3. Calculate active, transporter-mediated uptake (net influx rate; (4), Sect. 3.5) at each concentration of test compound as the difference in uptake between transfected and vector control cells. Analyze the results using the Michaelis-Menten equation ((5), Sect. 3.5) with a software package such as GraphPad® Prism®.

3.3.3 Step 3

In the third step of substrate assessment, uptake of the test compound is challenged with a known inhibitor of the transporter(s) of interest. Representative results, for the "test compounds" PAH (OAT1), furosemide (OAT3), and MPP⁺ (OCT2), are shown in Table 3.

1. Prepare dosing solutions of the test compound with and without a known inhibitor of each transporter of interest. Select the test compound concentration as (1) the lowest positive concentration in **Step 1** (if **Step 2** was not done), or (2) ≤K_m determined in **Step 2**. The concentration and identity of the inhibitor used for each transporter are shown in Table 3.

Table 3
Uptake of PAH by HEK-OAT1 cells, furosemide by HEK-OAT3 cells, and MPP+ by HEK-OCT2 cells

Cell line	Test compound (conc., μM)	Inhibitor (conc., μM)	Time (min)	Uptake rate (pmol/mg/min) Transfected	VC	Net influx rate[a]	% Inhibition[b]
HEK-OAT1[c]	PAH (10)	None	5	74.9	14.0	60.9	0
		Probenecid (200)		1.64	0.848	0.80	98.7
HEK-OAT3[d]	Furosemide (5)	None	5	173	1.37	172	0
		Probenecid (200)		4.36	1.29	3.07	98.2
HEK-OCT2[c]	MPP+ (5)	None	5	98.1	4.04	94.1	0
		Imipramine (300)		4.79	0.854	3.93	95.8

[a]Equation (4), Sect. 3.5
[b]Equation (6), Sect. 3.5
[c]No pre-incubation with inhibitor
[d]Cells pre-incubated ±inhibitor for 25 min prior to initiating uptake assay by replacing the pre-incubation solution with dosing solution (test compound ±inhibitor)

2. Follow the procedure above (**Step 1 of substrate assessment**). Select the assay duration as described above (**Step 2, No. 2**). The only difference is that HEK-OAT3 cells are pre-incubated with inhibitor (three wells) or buffer alone (three wells) for 25 min, after which the medium is removed and the uptake assay initiated by adding dosing solution with inhibitor (to the cells that were pre-incubated with inhibitor) or without inhibitor (to the cells that were pre-incubated without inhibitor).

3. Calculate net influx rate ((4), Sect. 3.5) and percent inhibition ((6), Sect. 3.5).

3.4 Inhibitor Assay

The objectives of inhibitor assessment are to investigate the potential of a test compound as an inhibitor of a transporter and to determine its inhibitory potency (IC_{50}), by monitoring its effect on uptake of a probe substrate of the transporter of interest.

The probe substrate for the OAT1 inhibitor assay is PAH (10 μM), furosemide (5 μM) for OAT3, and MPP+ (5 μM) for OCT2. In each case, the probe substrate concentration is well below the K_m (Fig. 1). Uptake of each probe substrate by HEK cells transfected with the respective transporter is robust and time-dependent (Fig. 2). Uptake of PAH by OAT1 and furosemide by OAT3 is linear up to 10 min, while uptake of MPP+ by OCT2 reaches a plateau rapidly, within 5 min. Uptake by the vector control cells is negligible. An early time point, 5 min, is used for inhibitor assays in each case, within the linear range for HEK-OAT1 and HEK-OAT3, and as close as feasible to the linear range for HEK-OCT2.

3.4.1 Step 1

In the first step of inhibitor assessment, uptake of a probe substrate of each transporter is determined in the presence and absence of a single concentration of test compound. Representative results, for a known inhibitor of each transporter as "test compounds," are shown in Table 4.

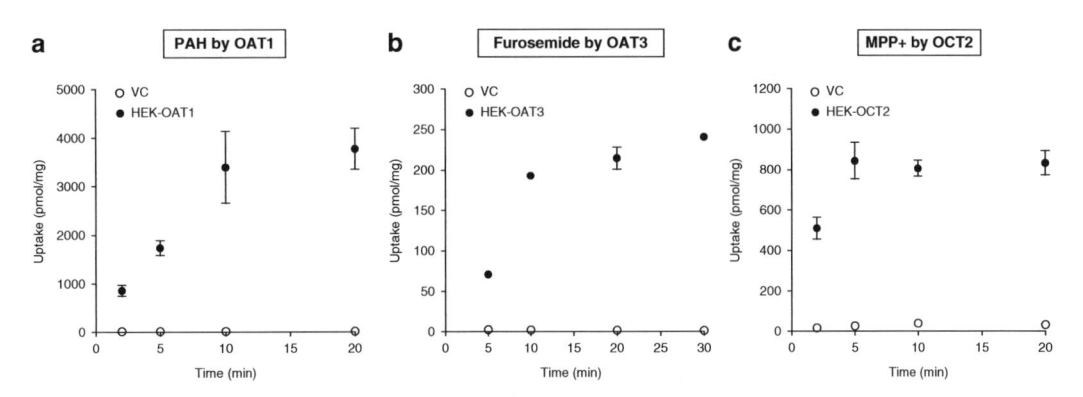

Fig. 2 Time-dependent uptake of PAH (**a**) by HEK-OAT1 cells, furosemide (**b**) by HEK-OAT3 cells, and MPP+ (**c**) by HEK-OCT2 cells. Transfected cells are depicted by *closed circles* and vector control cells by *open circles*. Each point represents the mean ± SD of three wells

Table 4
Effect of probenecid on OAT1 and OAT3, and imipramine on OCT2

Cell line	Probe substrate (conc., µM)	Test compound (conc., µM)	Time (min)	Uptake rate (pmol/mg/min) Transfected	VC	Net influx rate[a]	% Inhibition[b]
HEK-OAT1c	PAH (10)	None	5	123	1.86	121	0
		Probenecid (10)		38.5	1.05	37.5	69.0
HEK-OAT3d	Furosemide (5)	None	5	136	1.28	135	0
		Probenecid (10)		53.4	1.75	51.6	61.8
HEK-OCT2c	MPP+ (5)	None	5	127	4.91	122	0
		Imipramine (10)		73.9	5.43	68.5	43.9

[a]Equation (4), Sect. 3.5
[b]Equation (6), Sect. 3.5
[c]No pre-incubation with inhibitor
[d]Cells pre-incubated ±inhibitor for 25 min prior to initiating uptake assay by replacing the pre-incubation solution with dosing solution (test compound ±inhibitor)

1. Prepare dosing solutions, containing the probe substrate of the transporter of interest (PAH for OAT1, furosemide for OAT3, MPP+ for OCT2), with and without a test compound (*see* **Note 7** for selection of test compound concentration).

2. Follow the procedure outlined in **Step 1 of substrate assessment**. The assay duration is 5 min in each case. The only difference is that HEK-OAT3 cells are pre-incubated with test compound (three wells per test compound) or buffer alone (three wells) for 25 min, after which the medium is removed and the uptake assay initiated by adding dosing solution with test compound (to the cells that were pre-incubated with test compound) or without test compound (to the cells that were pre-incubated without test compound).

3. Calculate net influx rate ((4), Sect. 3.5) and percent inhibition ((6), Sect. 3.5).

3.4.2 Step 2

In the second step of inhibitor assessment, the potency of inhibition by the test compound is assessed by determining the IC_{50} for inhibition of uptake of a probe substrate of the transporter of interest. Representative results, for the "test compounds" probenecid (OAT1 and OAT3) and imipramine (OCT2), are shown in Fig. 3.

1. Prepare dosing solutions, containing the probe substrate of the transporter of interest (PAH for OAT1, furosemide for OAT3, MPP+ for OCT2), with and without a test compound (at least eight concentrations spanning at least two log units).

2. Follow the procedure outlined in **Step 1 of substrate assessment**. The assay duration is 5 min in each case. The only difference is that, HEK-OAT3 cells are pre-incubated with test compound (three wells per test compound) or buffer alone (three wells) for 25 min, after which the medium is removed

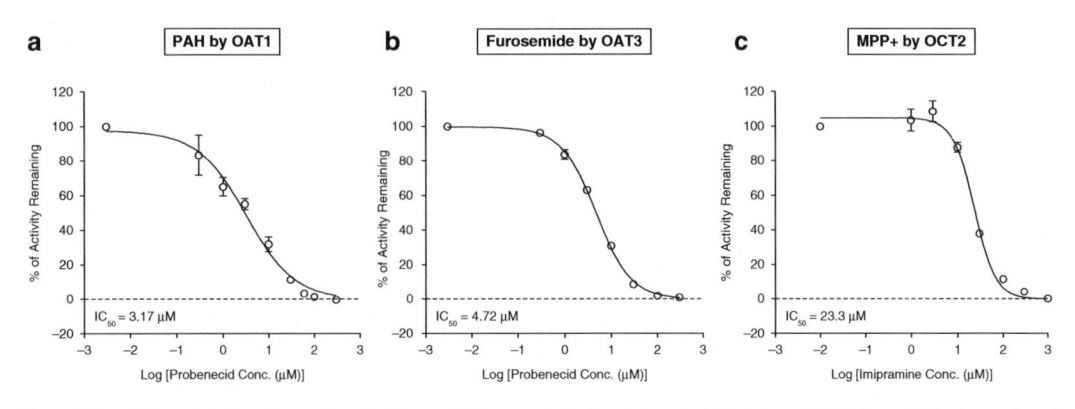

Fig. 3 IC_{50} determination for probenecid vs. OAT1-mediated uptake of PAH (**a**) and OAT3-mediated uptake of furosemide (**b**), and for imipramine vs. OCT2-mediated uptake of MPP+ (**c**). IC_{50} values are shown in the graph. Each point represents the mean ± SD of three wells

and the uptake assay is initiated by adding dosing solution with test compound (to the cells that were pre-incubated with test compound) or without test compound (to the cells that were pre-incubated without test compound).

3. Calculate net influx rate ((4), Sect. 3.5) and percent activity remaining ((7), Sect. 3.5). Analyze the results using (8), Sect. 3.5, with a software package such as GraphPad Prism.

3.5 Data Analysis

Uptake data is processed using the following equations:

$$\text{Influx Rate (IR)} = \frac{C_S \times V_S}{(C_P \times V_P) \times T} \quad (1)$$

$$\text{Influx Rate Ratio (IRR)} = \frac{IR_{\text{Transfectd cells}}}{IR_{VC}} \quad (2)$$

$$\text{Net Influx} = \text{Influx}_{\text{Transfected cells}} - \text{Influx}_{VC} \quad (3)$$

$$\text{Net Influx Rate (NIR)} = IR_{\text{Transfected cells}} - IR_{VC} \quad (4)$$

$$v = \frac{V_{\max} \times [S]}{K_m + [S]} \quad (5)$$

$$\% \text{ of Inhibition} = \left[1 - \left(\frac{NIR_{TC}}{NIR_{w/o\ TC}}\right)\right] \times 100 \quad (6)$$

$$\% \text{ of Activity Remaining} = \left(\frac{NIR_{TC}}{NIR_{w/o\ TC}}\right) \times 100 \quad (7)$$

$$Y = Y_L + \frac{(Y_H - Y_L)}{1 + 10^{((\text{Log IC}_{50} - X) \times \text{Hillslope})}} \quad (8)$$

C_s: Concentration of test compound in the cell lysate (µM);
V_s: Cell lysate volume in uptake assay (mL);
C_p: Protein concentration in the cell lysate (mg/mL);
V_p: Cell lysate volume in protein determination (mL);
T: Time (min);
VC: Vector control cells;
v: Uptake velocity (pmol/mg/min);
$V_{\max}$: Maximal velocity (pmol/mg/min);
[S]: Substrate concentration (µM);
K_m: Substrate concentration at half-maximal velocity;
TC: Test compound;
Y_L: % of control activity at the highest concentration of inhibitor;
Y_H: % of control activity at the lowest concentration of inhibitor;
X: $\log_{10}$ of inhibitor concentration;
K_m, $V_{\max}$, and IC_{50} are determined by nonlinear regression analysis using GraphPad Prism software.

4 Notes

1. All assay steps, including aspirating, dosing, washing, lysing, and sample collection, should be done with caution so as not to disturb the cells.

2. The substrate assessment protocol uses a matrix design with multiple dosing concentrations and multiple time points to avoid potential misclassification of test compounds due to incorrect assumptions about the optimal test concentration and uptake interval.

3. The test compound concentrations in the substrate matrix vary from one test compound to another, but are typically 0.5, 2, and 10 µM. If clinical PK data is available, the test concentrations are in the range of the mean unbound steady-state C_{max} at the highest dose ($[I]_1$ [5, 6]), taking into account the results of suitability assessments such as solubility of the test compound and tolerability of the cells as well.

4. If DMSO is the vehicle for a stock solution, the final DMSO concentration in the dosing solution should be ≤ 0.8 %.

5. Pre-warm dosing solutions to 37 °C to avoid cell shock.

6. Multichannel pipettes are widely used in cell culture and assay procedures such as seeding cells, dosing test compounds, collecting samples, etc. to ensure dispensing and collection of consistent volumes from well to well.

7. The test compound concentration in Step 1 of the inhibitor assay is typically either 10 µM or, if clinical PK data is available, 10× the mean (unbound, if available) steady-state C_{max} at the highest dose ($[I]_1$ [5, 6]). It also depends on the results of suitability assessments such as solubility of the test compound and tolerability of the cells.

References

1. Giacomini KM, Huang SM, Tweedie DJ et al (2010) Membrane transporters in drug development. Nat Rev Drug Discov 9(3):215–236
2. Giacomini KM, Sugiyama Y (2005) Membrane transporters and drug response. In: Brunton L et al (eds) Goodman and Gilman's The pharmacological basis of therapeutics. McGraw-Hill, New York, pp 41–70
3. Lepist EI, Ray AS (2012) Renal drug-drug interactions: what we have learned and where we are going. Expert Opin Drug Metab Toxicol 8(4):433–448
4. Mizuno N et al (2003) Impact of drug transporter studies on drug discovery and development. Pharmacol Rev 55(3):425–461
5. US Department of Health and Human Services, Food and Drug Administration, Center for Drug Evaluation and Research (CDER) (2012) Guidance for industry. Drug interaction studies—study design, data analysis, implications for dosing, and labeling recommendations. http://www.fda.gov/downloads/Drugs/GuidanceComplianceRegulatoryInformation/Guidances/UCM292362.pdf
6. European Medicines Agency, Committee for Human Medicinal Products (CHMP) (2012) Guideline on the investigation of drug interactions. http://www.ema.europa.eu/docs/en_GB/document_library/Scientific_guideline/2012/07/WC500129606.pdf

Chapter 25

General Guidelines for Setting Up an In Vitro LC/MS/MS Assay

John A. Masucci and Gary W. Caldwell

Abstract

In this chapter, we will discuss the choice of proper chromatographic and mass spectrometric conditions that are critical for establishing sensitive and robust *in vitro and in vivo* assays. Liquid chromatography combined with mass spectrometry (LC/MS) has been a primary tool for quantitation of analytes for *in vitro* and *in vivo* assays. This is due to the sensitivity and selectivity of modern LC/MS instrumentation and materials which routinely allow detection to the low ng/mL range for many analytes. The preparation of *in vitro* and *in vivo* generated samples requires that analytes are efficiently extracted with minimal chemical or enzymatic degradation both before and during LC/MS analysis. Modern chromatographic and MS ionization methods allow quantitation of analytes from small, polar organic molecules to large proteins and peptides over a wide dynamic range. Tandem MS methods via multiple reaction monitoring (MRM) has extended quantitation to low abundance analytes previously requiring radiometric detection. Quantitation software is integral to all MS operating software and will accurately determine concentrations of unknown analyte solutions based on calibration curves generated from known standard solutions.

Key words In vitro ADME, Liquid chromatography/mass spectrometry, LC/MS, Quantitation, In vitro assay, Ex vivo assay, In vivo assay

1 Introduction

For almost 20 years, liquid chromatography coupled with mass spectrometry (LC/MS) has been a major analytical method for quantification of both small organic molecules and larger biomolecule analytes supporting *in vitro* methods [1, 2]. This is due mainly to the introduction of LC- compatible ionization/interface methods such as electrospray ionization (ESI), which allowed coupling of mass spectrometers to chromatographs operating under essentially standard LC conditions for flow and mode of separation. Prior to this time, gas chromatography coupled with mass spectrometry (GC/MS) was a primary tool for mixture analysis and quantitation for those analytes that are thermally stable and sufficiently volatile [3]. This limitation of thermal stability and volatility was a major obstacle for analysis of many endogenous metabolites, drugs and particularly biomolecules, as these compounds are frequently polar and thermally unstable and require LC for separation from components typically found in biological matrices such as cell extracts or plasma. It is interesting to note that in more recent years there has been a resurgence in the use of GC/MS as a complement to LC/MS in the field of metabolomics as many small molecular mass metabolites are important in the biochemical processes being monitored and GC/MS is the method of choice for many of these compounds [4].

In addition to LC-compatible ionization methods, the development of tandem mass spectrometry (MS/MS) [5] allows the required MS selectivity to prevent interference from co-eluting sample components that would otherwise prevent detection of low concentration species. The ability to select multiple species (e.g. analyte and internal standard) for quantitation via MS/MS is called multiple reaction monitoring (MRM) and is the standard method now utilized for high sensitivity applications. The combination of these two developments has enabled routine detection and quantitation to low ng/mL levels and below for many analytes related to *in vitro* and *in vivo* applications.

In vitro applications of LC/MS have included such areas as determination of cytochrome P450 activities [6–9], Caco-2 permeability studies [10, 11], determination of reactive metabolites [12], identification of *in vitro* glucuronidation [13], determination of enzyme pathways [14], and high throughput methods for various *in vitro* assays [15–23].

2 Sample Preparation

2.1 In Vitro Sample Preparation

In vitro samples typically include enzymatic incubates and cell lysates. Incubates are initially quenched by either adding organic solvent, changing pH, or by freezing to stop enzymatic activity. Organic solvent and pH quenched incubates can be analyzed directly following a brief centrifugation or filtering step to remove particulates. Frozen incubates can either be thawed and maintained at 5 °C prior to LC/MS analysis to prevent additional enzymatic conversion, or can subsequently be treated with organic solvents or acid/base to quench activity prior to analysis.

As required, cellular samples can be lysed using a number of approaches including: organic solvent disruption of membranes; the use of MS-compatible detergents, such as sodium deoxycholate; hypotonically using deionized water to rupture the membrane via osmotic pressure; by ultrasonication. Cell samples can also be frozen to disrupt the membranes by expansion of intracellular water upon freezing. These methods are usually followed by centrifugation to remove cellular debris prior to analysis. Quenching of enzyme activity is usually also performed once the cells are lysed using one of the approaches described above.

2.2 Ex-Vivo/In-Vivo Sample Preparation

For MS analysis, whole blood samples should be collected in tubes which are treated with anticoagulant. These can include heparin, EDTA or citrate. It is possible for the anticoagulant to sometimes interfere with quantitation of a specific analyte, but this is a rare occurrence. Blood is normally centrifuged to precipitate red blood cells and the plasma supernatant is removed for analysis. If whole blood analysis is required then lysis of the precipitated red blood cells via sonication or a related method can be employed to access the intracellular fluid for analysis.

In vivo samples such as plasma can usually be prepared using several methods. One of the most popular is protein precipitation via organic solvent treatment with a miscible solvent such as acetonitrile. This will cause many of the abundant solubilized proteins to precipitate and are then separated by centrifugation to form a pellet. The supernatant can then be removed for direct LC/MS analysis. For analytes that require a more aggressive approach, liquid/liquid extraction can also be utilized using a non-miscible solvent such as hexane, chloroform or methylene chloride. The use of protein precipitation followed by liquid/liquid extraction can also be useful when drugs or metabolites are particularly unstable [24]. More recent formats include the use of microtiter-plate based devices for precipitation/filtration steps to increase throughput.

For the analysis of urine samples several preparatory approaches can be used. The simplest approach is to inject the urine directly without any treatment except for perhaps centrifugation or filtration to remove particulates that may clog the LC system, etc. When analyte concentrations are high enough, a simple dilution step (so called "dilute and shoot") can be utilized to reduce the introduction of salts and other possible interferences. And since urine is an aqueous sample, it is also a good candidate for solid phase extraction for removal of salts and possible enrichment of analytes, when required. And liquid/liquid extraction can also be applied to urine using the previously mentioned solvents.

Tissue specimens are normally prepared by extraction with an appropriate solvent along with an aggressive mechanical grinding step. This will result in tissue homogenates that can either be treated the same as plasma samples or directly centrifuged to pellet cellular and other tissue debris and generate a clarified supernatant that can be directly analyzed by LC/MS. For those analytes located within the cell, multiple grinding steps are sometimes employed along with sonication to increase efficiency or freezing with liquid nitrogen can also be used [25].

2.3 Sample Preparation for Protein and Peptide Quantitation

The analysis of biomolecules such as proteins and larger peptides frequently require enrichment/purification techniques to allow quantitation from biological matrices [26]. Offline methods such as ion chromatography (IC), gel permeation chromatography (GPC), polyacrylamide gel electrophoresis (PAGE) or immunoaffinity capture (IAC), are routinely utilized prior to LC/MS. Enrichment factors of 5,000 fold or more are frequently required for detection of low level biomolecules in complex matrices and even this approach is frequently insufficient for detection of very low level or otherwise poorly behaved analytes.

Mass spectral analysis of protein samples frequently requires digestion to generate characteristic peptides which can be use for both qualitative identification purposes or for quantitation via signature peptides. Digestion protocols will depend on the nature of the sample (e.g. in-gel digestion vs. solution). The general approach to protein digestion consists of reducing disulfide bonds, alkylation of the resulting reactive sulfhydryls and proteolytic cleavage of the protein, in that order. Additionally, it is also frequently necessary to sufficiently denature the protein prior to treatment to ensure that all residues can be exposed to chemical and enzymatic transformations [25]. This can be achieved with 6 M urea treatment, or more frequently for MS purposes, a proprietary detergent such as Rapigest™. Reduction of disulfides is normally achieved by treating the protein with dithiothreitol (DTT) or other reducing reagent. Alkylation is then performed using a reagent such as iodoacetamide. Finally digestion is performed using trypsin or an alternative protease, depending on application and the nature of the protein.

A general procedure for protein digestion is show in Fig. 1 for the use of Rapigest™.

3 Quantitation By LC/MS

Analysis of *in vitro* samples with the intent of quantitating drug, enzymatic substrates or products or other species is a multistep process. Initially this requires the development of an LC/MS method in which the analyte(s) is separated from interfering components via chromatographic separation and a sensitive, specific MS method is established by which the analyte(s) of interest can be determined. Finally, the quantitation is performed by calibration of the mass spectrometer by properly fitting the instrument response via creation of the calibration curve over the analyte range of interest followed by sample analysis.

3.1 Chromatography

Separation of analytes is typically achieved by using an MS-compatible chromatographic separation. Often a reversed-phase gradient LC method is utilized in which analytes separated based on hydrophobicities with less hydrophobic compounds

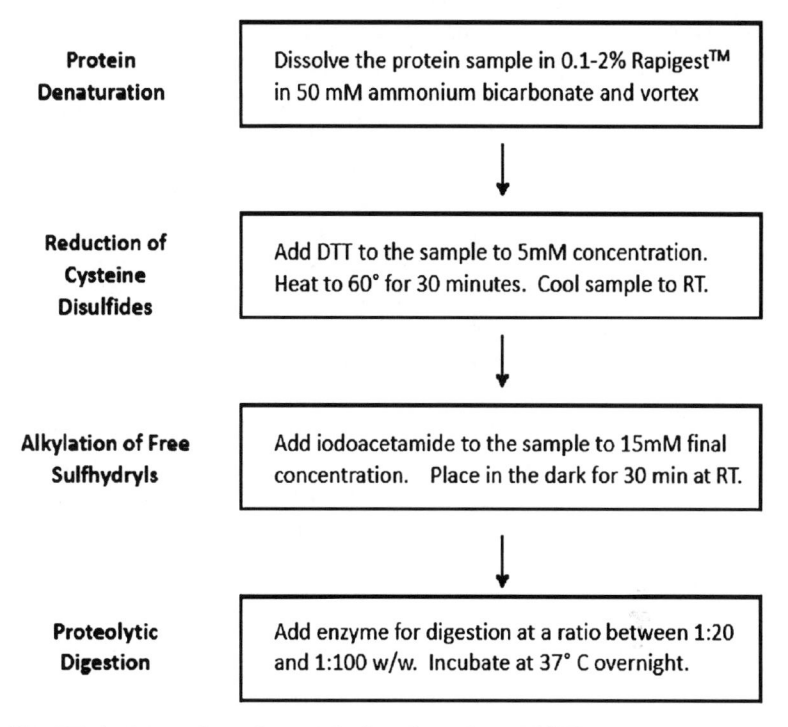

Protein Denaturation	Dissolve the protein sample in 0.1-2% Rapigest™ in 50 mM ammonium bicarbonate and vortex
Reduction of Cysteine Disulfides	Add DTT to the sample to 5mM concentration. Heat to 60° for 30 minutes. Cool sample to RT.
Alkylation of Free Sulfhydryls	Add iodoacetamide to the sample to 15mM final concentration. Place in the dark for 30 min at RT.
Proteolytic Digestion	Add enzyme for digestion at a ratio between 1:20 and 1:100 w/w. Incubate at 37° C overnight.

Fig. 1 Typical procedures for protein digestion prior to LC/MS analysis are shown. All reagents, including buffers, are MS-compatible

eluting first. Reversed-phase methods utilize non-polar stationary phases such as C4, C8, C18, Phenyl, Cyano and others. Large pore versions (300 Å) are used for proteins and peptides to allow for analyte partitioning while packings with only a porous outer shell and fused core are used to decrease peak width due to shorter diffusional distances of analyte molecules, thus replicating a smaller fully porous particle size [27].

There has also been a growing interest in the use of hydrophilic interaction liquid chromatography (HILIC) in which analytes elute in order of increasing polarity with mobile phase gradients starting at higher organic and going to higher aqueous over the course of the separation [28]. HILIC is especially suited for small polar organics, which are otherwise poorly retained via reversed-phase methods and are therefore poorly separated from inorganic salts and other potential MS-interfering species. Stationary phases utilized in HILIC include unbonded silica, cationic, anionic or zwitterionic bonded phases. Size exclusion and ion chromatography are also possible with MS detection. Size exclusion would be applicable for separation of larger sample components, such as proteins from smaller species, but it is limited by lack of chromatographic efficiencies normally required for in-vitro assays [29]. Ion chromatography can also be useful for analysis of ionized species of interest, but the need for column rejuvenation via high concentration of salts makes it less suited for MS use [30].

3.2 Mass Spectrometry

Quantitative mass spectrometry is most often performed using a triple quadrupole mass spectrometer. These spectrometers are capable of very high sensitivities and have a good dynamic range for quantitation of up to five orders of magnitude reported for some systems [31]. For quantitation, a triple quadrupole is usually operated in MRM mode, in which multiple analytes can be quantitated in a single chromatographic run. Often this comprises the analyte of interest and an appropriate internal standard. Internal standards are frequently isotopically labeled versions of the analyte(s) or a structurally related compound which exhibits similar chromatographic and mass spectral response. The purpose of the internal standard is to correct for run to run variations and/or to compensate for extraction efficiencies or other possible losses.

Most modern MS instruments utilize an atmospheric ionization method such as electrospray (ESI) or atmospheric pressure chemical ionization (APCI). ESI is the most popular ionization technique due to its applicability to a wide range of analyte types [32]. ESI is best for nitrogen containing analytes as a primary mechanism for ionization is protonation of nitrogen atoms within the molecule. This is particularly useful for proteins in which multiple charge states are observed, which can be utilized for molecular mass verification. Any of the observed charge states for proteins and peptides could be used as precursors for MRM quantitation (typically peptides) or for direct quantitation via selected ion monitoring (SIM) (small proteins or peptides). APCI is useful for simple hydrocarbon analytes which do not contain a nitrogen atom. The mechanism for APCI is a chemical-based ionization that results from the interaction of discharge-generated reactive cluster ions from water and organic solvents contained in the LC eluent, with the analyte molecules in the MS ion source [33]. This generates a protonated analyte molecule in positive ionization mode which can be detected directly or acts as a precursor for the MRM analysis. The choice of using positive or negative ionization is normally based on the nature of the analyte to be detected. Compounds containing basic nitrogen atoms, such as many drugs, are easily protonated and are usually analyzed under positive ionization conditions. Molecules containing acidic functionalities are often run in negative ionization mode as these typically lose a proton to form the anion.

When developing an MS method, a standard solution of the analyte of interest is infused into the MS ion source via syringe pump. It is best if this standard solution is tee'd into the LC eluent operating at the normal LC flow rate for analysis. In this way, the spectrometer will be optimized under actual conditions and not the reduced flow typical of syringe pump infusion. To establish the correct transition for MRM analysis, an appropriate precursor mass (often the molecular ion), is chosen and fragmented via collisionally induced dissociation (CID), in the collision cell of the mass spectrometer. One or more of the major resulting fragments is then selected as the specific product ion(s) for quantitation. This process is then repeated for all other analytes to be quantitated in the same MS run.

With the advent of automated software utilities which will tune MS systems and automatically establish MRM transitions for quantitation of analytes, it is common for some MS system users to be less familiar with manual optimization procedures. This can be a problem in cases in which analytes are not well behaved and manual optimization could improve overall MS signal to noise intensity. This is particularly true when operating very close to limits of quantitation where small improvements can be the difference between having an assay which meets the requirements and not being able to accurately determine analytes of interest. For this reason, it is very important that those scientists who are involved in MS method development are familiar with manual optimization methods for MRM and do not exclusively rely on automated tuning procedures.

3.3 Calibration Standards Preparations

The calibration concentration range is normally chosen to bracket the expected analyte values from a particular in vitro or in vivo source. For many assays, the lowest concentration standard is chosen at the limit of quantitation, which is defined as the level that generates a 3 to 1 signal to noise ratio. When possible, standards should be prepared in the identical matrix as the samples to be analyzed. This could be incubation media, blank cell lysate solution, blank plasma, etc. When it is not possible to obtain a matrix which does not contain the analyte as a background constituent, then a choice can be made to use a synthetic version of the matrix, such as 4 % albumin in PBS buffer as a substitute for plasma. Alternately, the actual matrix can be used and the background level subtracted or accounted for as the intercept in the calibration curve. Calibration standards can be prepared in neat buffer solutions, but extraction efficiency corrections should then be applied to the actual sample results to correct for losses. When relative concentrations are sufficient for diagnostic purposes, neat buffer standards can be used without correction as the relative results can be used directly. Calibration standards should be prepared using the same procedure as the unknown samples to account for any losses during preparation. This includes liquid or solid phase extraction steps, derivatizations, filtrations, centrifugations and any dilutions or transfers. When possible, it is also best to add the internal standard to the sample at the earliest step to correct for any analyte changes due to degradation, evaporation, adsorption to container walls or other surfaces, protein binding, etc.

It is normally advised to choose a fixed internal standard concentration at one half the upper calibration standard concentration values. This is usually sufficient for correction across the quantitation range. When using a structurally related compound as the internal standard, you can adjust the concentration to a level that generates the desired MS signal, as it is typical for analytes to vary in MS response factors.

3.4 Data Processing and Quantitation

LC/MS data is normally displayed as a plot of signal intensity vs. time for each MRM transition being monitored for a specific assay. As analytes elute from the chromatographic system and enter into the MS ion source, they will generate a characteristic peak. The area under these generated peaks is used for quantitation as representative of the analyte concentration. When internal standards are used, peak area ratios are calculated as the ratio of analyte peak area to internal standard peak area. These peak area ratios are then used in place of analyte areas for establishing calibration curves and quantitation of unknown samples.

Modern MS instruments will usually contain a quantitation module as part of their system software. These typically include features for defining retention times of analytes and internal standards, calibration utilities which can create a calibration curve using either linear or quadratic fits of the standard data and appropriate weightings and the ability to account for sample dilutions, etc.

When calibrating the MS response for a particular analyte, a linear relationship is first assumed for response vs. concentration via (1) shown below:

$$\mathbf{y = mC + b} \qquad (1)$$

Where, y is analyte response in area counts or area/ratio values, m is the slope of the calibration line, C is the concentration of the analyte and b is the intercept value or blank value of the analysis [34]. Frequently data will fit this relationship very well over a specific linear operating range. When analyte concentrations increase, resulting in larger signals close to the saturation value, then this relationship will typically deviate from linearity and become more hyperbolic in shape as shown in Fig. 2 as the region from C_1 to C_2.

Curvature of the calibration plot can be accommodated by current software, but more concentration values will be required to properly define the non-linear portion of the calibration curve. Correction of the non-linearity is frequently accomplished by applying a curve weighting factor such as $1/x$ or $1/x^2$, which decreases the influence of the higher concentrations which deviate most from linearity.

When quantitation software fits data to a linear relationship using linear regression analysis, it also determines the quality of fit as defined by the correlation coefficient (R). For a perfect linear fit, R=1 and it is not unusual for R values to be 0.998 or better for many MS generated calibration curves over the measurement range. If R drops to below 0.9, caution should be taken when quantitating

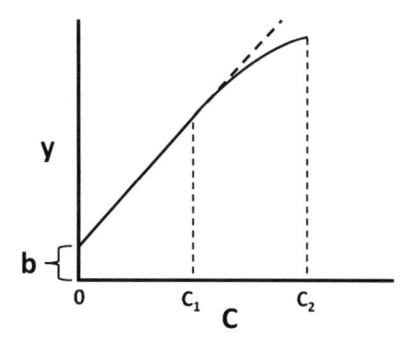

Fig. 2 Typical calibration curve for MS quantitation is shown. Terms are defined in text

analytes with this calibration. It is better to reanalyze fresh calibration standards or apply a more appropriate fit in this case.

When an isotopically labeled version of the analyte is used, it is possible to perform the quantitation using a single point calibration using the internal standard concentration value and measured area ratios. Concentration is then calculated using the formula shown below in (2).

$$C_U = (A_U / A_{IS})C_{IS} \quad (2)$$

where C_U is the concentration of unknown sample, A_U is the area count for the unknown sample, A_{IS} is the area count for internal standard and C_{IS} is the concentration of internal standard in the sample. This approach works best when the analyte concentration is in the linear response range and close to the internal standard value.

4 Experimental Tips

4.1 Sample Analysis Order

The order for running samples will vary depending on the nature of the study but it usually best to analyze from lowest to highest expected concentration of analyte to minimize any possible LC carryover effects. For example, with *in vivo* studies in which compounds are dosed intravenously, it is usually best to analyze the samples from last to first time point, as concentration is usually at a maximum at the earliest time points and decreases according to the compound elimination half life. However for oral dosing of slowly absorbed compounds, it is usually best to analyze from earliest to latest time points as the concentrations will initially be low.

With *in vitro* studies, it will depend on the nature of the assay. When monitoring the disappearance of substrate you should usually analyze from last time point to first, while monitoring of enzymatic product would proceed in the opposite manner. Subsequently, when analyzing inhibition via dose response, it is usually best to analyze from highest inhibitor concentration to lowest.

4.2 LC Buffers

Trifluoroacetic acid (TFA) is commonly used as an LC buffer component/ion pairing agent for coupling with positive ionization electrospray MS. It is very useful in that mode at low concentrations (0.1 %) but if used under negative ionization conditions, it typically will quench ionization and it will also create a strong background ion at m/z 113. Once it is in a system, it is difficult to get rid of and it is particularly easy to contaminate membrane-based in-line degassers which are commonly used by various HPLC vendors. There have been reports of some investigators attempting to exchange TFA with a weaker acid, such as acetic or propionic acid, post column to recover some of the positive ionization signal intensity [35]. If possible, it is best to dedicate a system to TFA use in positive ionization and have another LC for use in negative ionization. For similar reasons, quaternary ammonium salts should also be avoided in positive ionization mode.

In addition any non-volatile buffers should generally be avoided as they will contaminate the MS source and possibly the first analyzer region which can affect instrument performance. This includes the common LC buffers such as phosphate, citrate, borate, and nonvolatile versions of formate, acetate, carbonate and hydroxide (usually sodium or potassium salts). However, small quantities of non-volatile buffers can be tolerated in the samples themselves but diversion of the initial salt peak from the chromatographic eluent (reversed-phase separation) to waste is recommended to minimize instrument contamination.

4.3 Chromatographic Peak Shapes

Chromatographic peak shape can be quite diagnostic when optimizing LC conditions. Normal peak shape is Gaussian as the random movement of molecules in a chromatographic packing will result in a distribution of distances traveled and the off rate for partitioning is rapid. When peaks show significant tailing that indicates that more than one retention mechanism is in place. Although you may think that mostly hydrophobic interactions account for the retention mechanism in reversed-phase chromatography using packing such as C_{18}, there are also a large number of free silanol groups in the chromatographic packing which can cause tailing for many basic analytes. There is usually some tailing in most chromatography, but when it prohibits proper integration of peaks for quantitation or affects separation of closely eluting components then it should be corrected. This can be achieved in several ways including using a modifier which will mask the effect of silanol groups such as triethylamine; lowering the pH of the mobile phase to reduce silanol ionization; using an end capped packing with fewer free silanols; using a non-silica column such as organic polymer based. Peak tailing can also be minimized by using a solvent gradient or increasing the current gradient. Peak fronting is rarely observed in chromatographic separations. The most common cause of this phenomenon is column overload in which analyte molecules are not able to sufficiently interact due to saturation of the packing. This situation is simply corrected by injecting a lower volume or less concentrated sample. You will also observe peak fronting when the column is failing due to loss of liquid phase or general contamination and it cannot properly partition the analyte.

4.4 Correct Mass-Wrong Retention Time

Occasionally it is possible for the correct molecular mass to be observed at a different retention time than the reference standard solution for an analyte of interest. There can be several possible explanations for this observation. If utilizing reversed phase separations, peak splitting of the eluted analyte is sometimes possible when the sample solution contains too high a percentage of organic solvent. The organic solvent portion of the sample can actually cause a portion of the analyte to be unretained by the column and elute prior to the expected retention time. Frequently this results in two peaks observed with the same molecular mass. This can be corrected by decreasing the organic proportion of the sample solution. Another possibility for detection of the correct mass at the wrong retention time can occur when conjugates of a drug are generated. These conjugates are typically more polar than the analytes and can elute much earlier under reversed-phase HPLC conditions. When a conjugate enters the MS ion source it can be fragmented to the free drug via in-source dissociation processes. When this occurs, the MS-generated free drug is essentially the same species that is then detected via the MRM process, but at the wrong retention time. Analysts should take particular care not to misassign a peak as the analyte/drug of interest in ADME studies as this could invalidate an entire series of studies.

5 Notes

1. As a general rule, all solutions should be prepared for LC/MS analysis using 18 MΩ water and total organic content of less than three parts per billion. Final water polishing cartridges can be used with purification systems to minimize last traces of organics.

2. For data subject to FDA regulatory submissions, apply the 4-6-20 rule in which 4 of 6 calibration standards fall within 20 % of theoretical values on the calibration curve [36].

3. When methods are developed for high/low ranking purposes, a 20–30 % variation may be sufficient for diagnostic use. Biological variation is almost always greater than analytical precision of measurement for equivalent samples.

4. It is best to generally start with conventional LC conditions such as 2.1 mm ID columns and reversed-phased packings unless you have prior knowledge based on experience or literature.

5. Increasing column temperature up to 60 °C may improve your chromatographic performance when attempts to resolve closely eluting components are otherwise unsuccessful.

6. When attempting to diagnose a clog in your chromatographic system always start at the connection furthest downstream and work your way back to the pumps, looking for major decreases in system pressure.

Acknowledgements

The views expressed here are solely those of the author and do not reflect the opinions of Janssen Research & Development, LLC.

References

1. Chu I, Nomeir AA (2006) Utility of mass spectrometry for in-vitro ADME assays. Curr Drug Metab 7:467–477
2. Wan H, Holmén AG (2009) High throughput screening of physicochemical properties and in-vitro ADME profiling in drug discovery. Comb Chem High Throughput Screen 12:315–329
3. Hirtz J (1986) Importance of analytical methods in pharmacokinetic and drug metabolism studies. Biopharm Drug Dispos 7(4):315–326
4. Han J, Datla R, Chan S, Borchers CH (2009) Mass spectrometry-based technologies for high-throughput metabolomics. Bioanalysis 1(9):1665–1684
5. Schwartz JC, Cooks GR (1988) Recent developments in tandem mass spectrometry. Spectroscopy 5(1–3):49–63
6. Dixit V, Hariparsad N, Desai P, Unadkat JD (2007) In-vitro LC-MS cocktail assays to simultaneously determine human cytochrome P40 activities. Biopharm Drug Dispos 28:257–262
7. Lin T, Pan K, Mordenti J, Pan L (2007) In-vitro assessment of cytochrome P450 inhibition: strategies for increasing LC/MS based assay throughput using a one-point IC_{50} method and multiplexing high performance liquid chromatography. J Pharm Sci 96(9):2485–2493
8. Tolonen A, Petsalo A, Turpeinen M, Uusitalo J, Pelkonen O (2007) In-vitro interaction cocktail assay for nine major cytochrome P450 enzymes with 13 probe reactions and a single LC/MSMS run: analytical validation and testing with monoclonal anti-CYP antibodies. J Mass Spectrom 42:960–966
9. Smalley J, Marino AM, Xin B, Olah T, Balimane PV (2007) Development of a quantitative LC-MS/MS analytical method coupled with turbulent flow chromatography for digoxin for the *in vitro* P-gp inhibition assay. J Chromatogr B Analyt Technol Biomed Life Sci 845:260–267
10. Wang Z, Hop CECA, Leung KH, Pang J (2000) Determination of *in vitro* permeability of drug candidates through caco-2 cell monolayer by liquid chromatography/tandem mass spectrometry. J Mass Spectrom 35:71–76
11. Caldwell GW, Easlick SM, Gunnet J, Masucci JA, Demarest K (1998) In vitro permeability of eight β-blockers through caco-2 monolayers utilizing liquid chromatography/electrospray ionization mass spectrometry. J Mass Spectrom 33:607–614
12. Soglia JR, Contillo LG, Kalgutkar AS, Zhao S, Hop CE, Boyd JG, Cole MJ (2006) A semi-quantitative method for the determination of reactive metabolite conjugate levels in vitro utilizing liquid chromatography-tandem mass spectrometry and novel quaternary ammonium glutathione analogues. Chem Res Toxicol 19:480–490
13. Chen M, Howe D, Leduc B, Kerr S, Williams DA (2007) Identification and characterization of two chloramphenicol glucuronides from the in vitro glucuronidation of chloramphenicol in human liver microsomes. Xenobiotica 37(9):954–971

14. Bradshaw HB, Rimmerman N, Hu SSJ, Benton VM, Stuart JM, Masuda K, Cravatt BF, O'Dell DK, Walker JM (2009) The endocannabinoids anandamide is a precursor for the signaling lipid N-arachidonoyl glycine by two distinct pathways. BMC Biochem 10:14
15. Lindqvist A, Hilke S, Skoglund E (2004) Generic three-column parallel LC-MS/MS system for high-throughput in vitro screens. J Chromatogr A 1058:121–126
16. Castro-Perez J, Plumb R, Liang L, Yang E (2005) A high-throughput liquid chromatography/tandem mass spectrometry method for screening glutathione conjugates using exact mass neutral loss acquisition. Rapid Commun Mass Spectrom 19:798–804
17. Luippold AH, Arnhold T, Jörg W, Krüger B, Süssmuth RD (2011) Application of a rapid and integrated analysis system (RIAS) as a high throughput processing tool for in vitro ADME samples by liquid chromatography/tandem mass spectrometry. J Biomol Screen 16:370–377
18. Luippold AH, Arnhold T, Jörg W, Süssmuth RD (2010) An integrated platform for fully automated high-throughput LC-MS/MS analysis of in vitro metabolic stability assay samples. Int J Mass Spectrom 296:1–9
19. Drexler DM, Edinger KJ, Mongillo JJ (2007) Improvements to the sample manipulation design of a LEAP CTC HTS PAL autosampler used for high-throughput qualitative and quantitative liquid chromatography-mass spectrometry assays. J Assoc Lab Autom 12:152–156
20. Drexler D, Barlow DJ, Falk P, Cantone J, Hernandez D, Ranasinghe A, Sanders M, Warrack B, McPhee F (2006) Development of an on-line automated sample clean-up method and liquid chromatography-tandem mass spectrometry analysis: application in an in vitro proteolytic assay. Anal Bioanal Chem 384:1145–1154
21. Shin S, Fung H (2011) Evaluation of an LC-MS/MS assay for ^{15}N-nitrite for cellular studies of L-arginine action. J Pharm Biomed Anal 56:1127–1131
22. Cox JM, Troutt JS, Knierman MD, Siegel RW, Qian Y, Ackerman BL, Konrad RJ (2012) Determination of cathepsin S abundance and activity in human plasma and implications for clinical investigation. Anal Biochem 430:130–137
23. Otten JN, Hingorani GP, Hartley SD, Kragerud SD, Franklin RB (2011) An in vitro, high throughput, seven CYP cocktail inhibition assay for the evaluation of new chemical entities using LC-MS/MS. Drug Metab Lett 5:17–24
24. Grosse CM, Davis IM, Arrendale RF, Jersey J, Amin J (1994) Determination of remifentanil in human blood by liquid-liquid extraction and capillary GC-HRMS-SIM using a deuterated internal standard. J Pharm Biomed Anal 12(2):195–203
25. Cañas B, Piñeiro C, Calvo E, López-Ferrer D, Gallardo JM (2007) Trends in sample preparation for classical and second generation proteomics. J Chromatogr A 1153:235–258
26. Shi T, Su D, Liu T, Tang K, Camp DG, Qian W, Smith RD (2012) Advancing the sensitivity of selected reaction monitoring-based targeted quantitative proteomics. Proteomics 12:1074–1092
27. Badman ER, Beardsley RL, Liang Z, Bansal S (2010) Accelerating high quality bioanalytical LC/MS/MS assays using fused-core columns. J Chromatogr B Analyt Technol Biomed Life Sci 878:2303–2313
28. Gama MR, Silva R, Collins CH, Bottoli CBG (2012) Hydrophilic interaction chromatography. Trends Anal Chem 37:44–60
29. Meira GR, Vega JR (2010) Band broadening in GPC/SEC. In: Cazes J (ed) Encyclopedia of chromatography, vol 1. CRC Press, Boca Raton, FL, pp 147–156
30. Burgess K, Creek D, Dewsbury P, Cook K, Barrett MP (2011) Semi-targeted analysis of metabolites using capillary-flow ion chromatography coupled to high resolution mass spectrometry. Rapid Commun Mass Spectrom 25:3447–3452
31. Kane MA, Folias AE, Wang C, Napoli JL (2008) Quantitative profiling of endogenous retinoic acid in vivo and in vitro by tandem mass spectrometry. Anal Chem 80:1702–1708
32. Strege MA (1999) High-performance liquid chromatographic-electrospray ionization mass spectrometric analyses for the integration of natural products with modern high-throughput screening. J Chromatogr B Biomed Sci Appl 725(1):67–78
33. Lee H (2005) Pharmaceutical applications of liquid chromatography coupled with mass spectrometry. J Liq Chrom Relat Tech 28:1161–1202
34. Woodget BW, Cooper D (1987) Analytical standards and calibration curves from samples and standards. Wiley, London, pp 109–145
35. Shou WZ, Weng N (2005) Simple means to alleviate sensitivity loss by trifluoroacetic acid (TFA) mobile phases in the hydrophilic interaction chromatography-electrospray tandem mass spectrometric (HILIC-ESI/MS/MS) bioanalysis of basic compounds. J Chromatogr B Analyt Technol Biomed Life Sci 825(2):186–192
36. Hoffman D (2009) Statistical considerations for assessment of bioanalytical incurred sample reproducibility. AAPS J 11(3):570–580

Chapter 26

Metabolite Identification in Drug Discovery

Wing W. Lam, Jie Chen, Rongfang Fran Xu, Jose Silva, and Heng-Keang Lim

Abstract

Early knowledge on the structures of metabolites from *in vitro* and *in vivo* metabolism studies is very useful for improving the biopharmaceutical, efficacy and safety properties of lead candidates in drug discovery. The recognition of the value in what metabolite identification brings to drug discovery led to its inclusion in ADMET toolbox and recent trend to multiplex assessment of metabolic stability and metabolite identification. In this chapter, we will cover the *in vitro* and *in vivo* systems typically used for metabolite identification. Fast LC-MS/MS with capability of data-dependent multiple-stage mass analysis is the instrument of choice and the workhorse for multiplexing assessment of metabolic stability and metabolite identification from a single analysis. Therefore, various LC-MS/MS instruments and techniques used for metabolite identification including software to speed up data-mining along with estimating major metabolites based on UV methods will be discussed. In general, the exact site of biotransformation is difficult to obtain based solely on MS data. As a result, microchemistry will also be discussed to help narrowing down site of modification in metabolites.

Key words LC-MS/MS, Hepatocytes, Liver microsomes, Liver S9, Plasma, Metabolite quantitation and identification, Derivatization

1 Introduction

Lead candidates, from drug discovery hit-to-lead stages, often do not possess desirable biopharmaceutical properties for development [1]. One of the developability issues commonly encountered early in discovery is that the lead candidates often have high systemic clearance due to low metabolic stability. Early identification of the major metabolites helps the medicinal chemists to optimize the chemical structure of the lead candidate leading to metabolically more stable compound that eventually translated into having better systemic exposure and with reasonable half-life for once-a-day oral dosing. In addition, early screening for reactive intermediate formation through trapping experiments will provide guidance to medicinal chemists to modify the chemical structures to minimize bioactivation to reactive metabolites. This is because minimization bioactivation is the pragmatic approach taken across pharmaceutical industry to reduce drug attrition from idiosyncratic drug-induced toxicities due to poor predictivity of preclinical animal models [2]. Another application of metabolite identification is to provide mechanistic understanding to resolve disconnects between pharmacokinetics and pharmacodynamics of unchanged drug, and provide rationalization for involvement of active metabolite.

Therefore, knowing the structure of metabolites can be critical for the optimization of lead candidates in discovery [3]. In this chapter, we will cover the following topics including *in vitro* and *in vivo* systems, LC-MS/MS data dependent mass analysis, quantitation of metabolites, metabolite identification, and softwares to aid metabolite identification.

2 Materials

2.1 Hepatocyte Suspension Incubations

1. Krebs-Henseleit buffer can be purchased from Sigma-Aldrich (St. Louis, MO).
2. Cryopreserved or freshly prepared hepatocytes (e.g. male rat, female rat, dog, and human) are available from different vendors (e.g. BD (Woburn, MA), IVT (Woburn, MA); CellDirect, (Durham, NC)).

2.2 Microsomal and S9 Incubations

1. Monobasic (1.0 M) and dibasic (1.0 M) potassium phosphate solutions (Sigma, St. Louis, MO).
2. Ethylenediaminetetraacetic acid disodium salt solution (500 mM, Sigma, St. Louis, MO).
3. Magnesium chloride solution (1.0 M, Sigma, St. Louis, MO).
4. Oxygen gas (O_2) (Air gas, Piscataway, NJ).
5. NADPH regenerating system (NRS, Solution A, cat# 451220, consisting of 26.0 mM $NADP^+$, 66 mM glucose-6-phosphate, and 66 mM $MgCl_2$ in H_2O and Solution B, cat# 451200, consisting of 40 U/mL glucose-6-phosphate dehydrogenase in 5 mM sodium citrate) from BD Gentest™ (Woburn, MA).
6. Glutathione, reduced (GSH, Sigma-Aldrich, St. Louis, MO)
7. GSH, γ-glutamylcysteinylglycine-$^{13}C_2$-^{15}N (Cambridge Isotope laboratories, Andover, MA).
8. Uridine 5′-diphosphoglucuronic acid ammonium salt (UDPGA, Sigma-Aldrich, St. Louis, MO) and uridine 5′-diphospho-N-acetylglucosamine (UDPAG, Sigma-Aldrich, St. Louis, MO).
9. Hepatic and non-hepatic microsomes and S9 (e.g. 20 mg/mL) can be obtained from BD (Woburn, MA) or other vendors.

2.3 Equipments for Incubations, Sample Work Up and Profiling

1. Dubnoff Reciprocal Shaking Water Baths (Thermo Scientific, Chicago, IL).
2. Centrifuge CT422 (Jouan, Winchester, VA).
3. Eppendorf 5417R microcentrifuge.
4. HPLC grade acetonitrile and water (EMD Chemicals, Inc, Gibbstown, NJ).

2.4 Derivatizing Reagents

1. Methylation of carboxy and phenolic groups: Trimethylsilyl diazomethane (2.0 M in diethylether) (Sigma Aldrich, St. Louis, MO)
2. N-Oxide reduction: titanium trichloride (Sigma Aldrich, St. Louis, MO)

2.5 LC Chromatography

1. Columns: AquaStar C18 column and its guard column accessories (150×2.1 mm ID) (Thermo Scientific, Bellefonte, PA). Others columns can also be used. For example, Waters XBridge C18 column (150×2.1 mm ID, 3.5 μm) and guard cartridges (Waters Technologies Corp., Milford, MA) is also appropriate for sample analysis
2. Solvents: Common HPLC grade solvents are acetonitrile, methanol, water (EMD Chemicals, Inc, Gibbstown, NJ)

2.6 LC-MS/MS

There are a number of MS instruments can be used for metabolite identification. Below are just few that are widely used for this purpose.

1. Thermo LTQ/Orbitrap (Thermo Scientific, Inc, Bremen, Germany)
2. Waters Synapt G2 MS (Waters Corporation, Milford, MA)
3. AB Sciex API5500, 4000 QTRAP (AB Sciex, San Jose, CA)
4. Agilent 6530 Accurate-Mass Q-TOF LC/MS System (Agilent Technologies, Santa Clara, CA)

3 Methods

3.1 Hepatocyte, Microsomal, and S9 Incubations

3.1.1 Hepatocyte Suspended Culture Incubation

Below is a procedure for conducting hepatocyte incubations (*see* **Note 1**)

1. Cryopreserved cells removed from liquid nitrogen are quickly placed in the water bath (37 °C) for 1–2 min.
2. Transfer and suspend cells in warm thawing media and centrifuge at 100×g for 7 min at room temperature.
3. The supernatant is discarded and replaced with 2 mL of Kreb-Henseleit media containing 12.5 mM HEPES pH 7.4 (KHB).

4. Prepare sample to determine cell viability [4] (http://www.celsisivt.com/working-with-hepatocytes-in-suspension) by mixing e.g. 300 µL of KHB, 80 µL of 0.4 % Trypan Blue and 20 µL of diluted cells.

5. Count cells under microscope with viable cells clear and dead cells blue

6. Cell viability is determined by Trypan Blue exclusion method: Viable cells/Total cells × 100 %.

7. Cell viability (typically >85 % for all species).

8. Back calculate the number of viable cells in suspension and dilute to 1×10^6 cells/mL of suspension.

9. The incubations are carried out at a cell concentration of 1×10^6 cells/mL (in KHB) in a total incubation volume of 0.50 mL in a 24-well plate at 37 °C under 5 % CO_2 in a humidified incubator with constant mixing.

10. Cells are typically incubated for 2 h with drug at 10 µM concentration or lower due to solubility issue.

11. The cell incubates are transferred to eppendorf tubes and sonicated for 5 min, then quench with six volumes of ice-cold acetonitrile + 0.02 % formic acid.

12. Precipitated protein is then pelleted by centrifugation at 2,359 g for 10 min.

13. The supernatant is transferred to a glass test tube and evaporated to dryness under a gentle stream of nitrogen at room temperature.

14. After drying, the resulting residue from the incubate is reconstituted in 250 µL of a mixture of nine parts 0.1 % formic acid, and one part acetonitrile (or the initial mobile phase for sample profiling).

15. Prior to analysis, this sample is filtered by centrifugation at room temperature through a 0.45 µm Nylon membrane at 14,000 rpm for 2 min.

16. The filtrate is transferred into a 96-well plate for liquid chromatography/tandem mass spectrometry (LC-MS/MS) analysis.

17. Controls hepatocyte incubates: diclofenac can be included as a positive control and without drug and 0 min samples will be the negative controls.

3.1.2 Liver S9 or Microsomal Incubations

Below is a procedure for conducting Liver S9 or microsomal incubations (**see** Note 2).

Phosphate Buffer Preparation

Phosphate buffer (pH 7.40, 100 mM) is typically used for microsomal and S9 incubations. In order to prepare 1,000 mL of the buffer solution, the following method can be used.

1. Combine 77.8 mL of dibasic phosphate solution (1.0 M) and 22.4 mL of monobasic phosphate (1.0 M) solution in a 1 L Volumetric Flask.
2. Add 2.0 mL of ethylenediaminetetraacetic acid disodium salt solution (EDTA, 500 mM) to get a 1 mM final concentration after dilution.
3. Add 5.0 mL of magnesium chloride solution (1.0 M) to get a 5 mM final concentration after dilution.
4. Dilute the mixture to 990 mL using HPLC grade water to get close to 100 mM concentration.
5. To adjust the pH, add monobasic to lower the pH or dibasic to raise the pH and add water to 1,000 mL to give exactly 100 mM. The solution is mixed well and re-checked to confirm its pH.

Stock Solution Preparation

1. For example, compound A has a molecular weight of 359. Weigh out accurately to e.g. 3.59 mg and dissolve in 1 mL of organic solvent (e.g. Acetonitrile:DMSO/4:1) to give a 10 mM concentration stock solution.
2. Dilute the 10 mM stock solution tenfold with the phosphate buffer to give a 1 mM working solution.

Reagent Preparations

1. NADPH regenerating system (NRS): Mix Solution A and Solution B in 5:1 ratio.
2. UDPGA/UDPAG solution: Uridine 5′-diphosphoglucuronic acid ammonium salt (UDPGA, Sigma-Aldrich, St. Louis, MO) dissolve in phosphate buffer (500 mM), uridine 5′-diphospho-N-acetylglucosamine was prepared similarly to give 100 mM. Mix 1:1 ratio to give a stock solution of UDPGA/UDPAG (250:50 mM).
3. Prepare a mixture of glutathione (GSH, γ-glutamylcysteinylglycine) and the stable-isotope labeled compound (GSH, γ-glutamylcysteinylglycine-$^{13}C_2$-^{15}N)/1:1 [5, 6] 250 mM solution in phosphate buffer.

3.1.3 In Vitro Incubations: General Procedure for Incubation with S9 or Microsomes (see **Note 3**)

1. Bubble oxygen gas into approximately 100 mL of phosphate buffer (100 mM, pH 7.40) fortified with magnesium chloride (5 mM) and EDTA (1 mM) for 10 min.
   (a) Add the following reagents in sequence below: 840 μL phosphate buffer
   (b) 50 μL of 20 mg/mL microsome (final 1 mg/mL protein concentration). For S9, use 2 mg/mL protein and adjust volume accordingly with phosphate buffer for total volume
   (c) 10 μL of 1 mM test compound solution

2. Pre-incubate the mixture in the Dubnoff Reciprocal Shaking Water Bath at 37 °C for 3 min

3. Add appropriate co-factor(s) to initiate reaction

4. Control sample during microsomal/S9 incubation, typically without the use of the NADPH regenerating system or other co-factors. The volumes of cofactors are replaced with phosphate buffer

5. Incubate for 60 min

6. Add 5.0 mL of acetone:acetonitrile (1:1) fortified with 0.1 % formic acid to quench the reaction and efficiently precipitating the proteins

7. Vortex mix the mixture

8. Centrifuge the mixture at 2,359 g at 4 °C for 10 min

9. Decant supernatant into a tube for drying under nitrogen

3.1.4 Plasma Samples

1. Remove 1.0 mL plasma

2. Add 5.0 mL of pre-chilled acetone:acetonitrile (1:1) fortified with 0.1 % formic acid to precipitate proteins

3. Vortex the mixture to ensure complete protein precipitation mix the mixture

4. Centrifuge the mixture at 2,359 g at 4 °C for 10 min

5. Decant supernatant into a tube for drying under nitrogen

3.1.5 Sample Preparation for Metabolic Profiling

1. Dissolve the nitrogen dried residues from hepatocytes, microsomes, S9 or plasma using 250–300 µL of initial LC mobile phase gradient.

2. Filter through a 0.45 µm Nylon filter using a microcentrifuge at 14,000 g for 2 min prior to LC-MS/MS analysis.

3.2 LC-MS/MS Metabolic Profiling of Incubates

The procedures for both quantitative and qualitative metabolite profiling of incubates will be discussed (see Note 4)

3.2.1 Quantitative

1. *In Vitro* Samples

    (a) Run microsomal incubation of parent drug as described above at 50 µM (final concentration)

    (b) Spike parent drug standard in microsome similarly (50 µM final concentration)

    (c) Quench samples and work up as described above

    (d) Run LC-UV-MS/MS at $UV_{220\,nm}$

    (e) Compare UV peak areas in the chromatograms from incubate (a) and spike sample (b) above and get the concentration of parent drug and each metabolites in incubate assuming that the extinction coefficients at this wavelength are very similar for all metabolites.

2. **Plasma Samples (e.g. from PK study):**
   Quantitation of metabolites can be challenging for plasma samples because of low level of drug related components present in high amount of complex mixture of endogenous compounds. However, its metabolite levels can be estimated through the in-vitro samples.

   (a) Use the high concentration microsomal incubate (50 µM) sample described in previous section with known concentration of parent drug and its metabolites and dilute 20-fold with blank plasma from the same PK study (1 mL total volume).

   (b) Use 1 mL PK plasma samples

   (c) Quench and work up samples similarly

   (d) Run LC-UV-MS/MS analysis of the plasma diluted *in-vitro* sample as well as plasma samples

   (e) Use the plasma diluted *in-vitro* sample as a single point calibration standard for calculating the concentration of the metabolites in the PK samples based on the MS peak area ratio.

3.2.2 Qualitative

Typically, the qualitative analysis starts with an optimization of the instrument followed by a full scan mass analysis. Run LC-MS/MS experiments with data dependent acquisition using any of the commercially available instruments. The data dependent scan experiments commonly used are selective ion monitoring (SIM), precursor ion (PI) and neutral loss (NL) scanning. These scans can be set up as follows (**see** Note 5).

1. *Optimization by tee-infusion*
   The mass spectrometer is first optimized by infusion of 10 ng/µL unchanged drug in mobile phases A and B (50:50, v/v) using the anticipated flow rate (e.g. 400 µL/min). The collision energy should be set to a value with ≥80 % attenuation of the precursor ion.

2. *Full scan mass analysis*
   Set up full scan mass analysis over an appropriate mass range covering both phase I and II metabolites including its dimer using shortest scan speed or dwell time without impacting on its sensitivity.

3. *Data-dependent full scan mass analysis*
   There are many ways to triggered data-dependent full scan mass analysis and this procedure is dependent on the tandem mass spectrometer used for metabolite profiling as shown below (**see** Note 6):

(a) Tandem-in-time mass spectrometer (LTQ and LTQ/Orbitrap): mass list or isotopic pattern triggered data-dependent scan

Select mass-list triggered data-dependent scan if compound has no unique intrinsic isotopic pattern or if there is no possibility of creating an extrinsic isotopic pattern like mixing unlabeled and radio or stable-isotopically labeled in predefined ratio. The mass-list essentially consists of m/z of unchanged drug and its postulated metabolites, which can be generated using vendor's software. However, the isotopic pattern triggered data-dependent scan is preferred for compound with intrinsic or extrinsic isotopic pattern. The success of isotopic pattern triggered data-dependent scan is heavily dependent on the tolerance set for the isotopic ratio.

(b) Tandem-in-space mass spectrometer (TSQ, QTrap and QTOF): SIM, MS^2 (precursor ion scan, and neutral loss scan). Below is a summary of their uses and how a data dependent experiment can be set up using TSQ Quantum as an example. Tuning is the same as above.

1. Selective ion monitoring (SIM): this is a highly sensitive method and can detect low level of expected metabolites
   (a) First scan event: Full scan
   (b) Second event: enter the list of expected ions
   (c) Third event: MS^2 for data dependent product ion acquisition if metabolites detected from the SIM list to shed light on the site for modification based on fragmentations of the metabolite of interest.

2. Precursor ion scanning (PI): to include fragments based on parent drug and it's expected mass modifications for the fragments.
   (a) First scan event: Full scan
   (b) Second event: precursor fragments. For example, to add the parent fragment mass and its 16 amu to the particular fragment. This readily detects metabolites with monooxygenation occurring within the fragment.

3. Neutral loss (NL) experiments:
   (a) First scan event: Full scan
   (b) Second event: neutral loss for detections of e.g. 176 for glucuronide, 80 for sulfate, 129 for glutathione adduct
   (c) Third event: triggers MS^2 for data dependent product ion acquisition only for metabolites detected with neutral loss to shed light on the region of the molecule for conjugation or structure elucidation of the aglycone.

3.3 Structure Elucidation of Metabolites

3.3.1 Data Mining

Software from mass spectrometry vendors usually is capable of aiding the identification of metabolites. For examples, Thermo Xcalibur software can be used to process and identify metabolites. To elucidate metabolite structures are usually the most time-consuming part of the work, therefore, additional software to speed up this process becomes very desirable. Recently software packages are available to speed up metabolite identification and will be discussed in the following sections.

3.3.2 Software to Aid Metabolite Identification

Softwares from mass spectrometry vendors such as MetWorks (Thermo Scientific), Metabolynx (Waters), LightSight and MetabolitePilot (Applied Biosystem) and MassHunter Metabolite ID (Agilent) are capable of speeding up data-mining and aiding the identification of metabolites. All the above softwares have capability of improving throughput of data-mining by applying mass defect filtering of accurate mass data.

Mass Defect Filtering (MDF)

Mass defect filtering (MDF) is a useful way to remove ions of endogenous compounds in biological matrices and leave the metabolites of interest in the mass chromatogram (see Note 7). A typical procedure based on Thermo (Metworks) is described below.

1. Launch the Metwork program
2. Open the metabolism control and sample files (to include background subtraction)
3. Define the parent drug of interest
4. Add list of possible modifications
5. Set range for mass defects (e.g. 50 mDa)
6. Run the MDF program
7. Mass chromatogram will be much cleaner for easier metabolite identification than the non-filtered chromatogram (see Note 8)

Mass Meta-Site

Mass-MetaSite [7] is a computer assisted method for the interpretation of LC–MS/MS data that combines prediction of a compound's Site of Metabolism (SoM) with the processing of MS spectra and rationalization based on fragment analysis. The procedure consists of three steps: (a) automatic detection of the chromatographic peaks related to the parent compound and its metabolites; (b) structure elucidation by proposing a potential metabolite structure based on the fragmentation pattern for each peak detected in the previous step and (c) for all the potential metabolite structures compatible with the extracted fragment information, a ranking is performed using the MetaSite SoM prediction algorithm.

The procedure for structure elucidation is summarized as below.

1. Open the Mass-MetaSite application
2. Input the structure of parent drug
3. Import three data files, which include
   (a) A blank matrix file that is used to monitor the noise and to distinguish signal from noise by comparison with the other files;
   (b) A substrate file that is used to analyze the fragmentation pattern of the substrate and
   (c) an incubation file (or any metabolism file) that contains metabolites.
4. Run the program
5. Summary results are provided within the program and can be exported to spreadsheet for analysis or report.

MsMetrix

Recently another software program, MsXelerator RM, is available to aid the identification of metabolites [8]. The program is powerful and has multiple functions. It can do differential analysis, mass defect filtering, isotope pattern and neutral loss automatically to facilitate the data analysis. With these functions, it helps to capture relevant metabolites and filter out false positive peaks. Details set up will not be discussed here.

Others

Other software, ACD and Pallas, are also available to aid structure elucidation but will not be discussed here.

3.3.3 Determination of Metabolite Functional Groups

Microchemistry (including H/D exchange) is often employed to supplement structural elucidation by mass spectrometry in cases where diagnostic neutral loss is not readily apparent from CID as described below.

Carboxylic Acid and Phenolic Groups by Methylation

Metabolite with carboxylic acid or phenolic functionality can be readily identified by mass spectrometry following conversion to its methyl ester or methyl ether, respectively, by trimethylsilyldiazomethane. This is illustrated by the identification of O-methyl-(−)-epicatechin-O-sulfate metabolites [9] and a carboxylic acid metabolite [10, 11] after methylation using (TMSD). The method can be generalized as below.

1. Remove the solvent from isolated sample in a vial
2. Dissolve samples in 0.5 mL diethylether:methanol/1:1 (v/v) in a vial
3. Add excess trimethylsilyl diazomethane (TMSD, 2.0 M in diethylether) until a yellow solution is formed, and then cap the vial. The yellow color indicates an excess of TMSD.

4. Stir for 1–18 h and evaporate solvent under nitrogen. An hour stirring is sufficient in forming the methyl-carboxylate. Excess TMSD can also be removed under vacuum [10]

5. Dissolve residue using initial LC solvent system for LC-MS analysis

6. Filter the sample through a Nylon filter (0.45 µm) prior to LC-MS analysis

N-Oxide and Sulfoxide

There are many published procedures for identification of N-oxide metabolite which included reduction with titanium trichloride back to amine [12], absence of active hydrogen by H/D exchange to determine if it is an N-oxide/sulfoxide or hydroxyl metabolite [13–15] and thermal-induced deoxygenation during APCI analysis. Peiris et al., [16] have demonstrated using APCI to effectively deoxygenate the oxygen from N-oxide to its amine product.

1. **Reduction with titanium trichloride**

   A general method for the reduction of N-oxides to amines using titanium trichloride (ca. 10 wt% solution in 20–30 wt% hydrochloric acid) is described by [12]. If it is a hydroxyl metabolite, no reduction will occur under this condition. A brief description is summarized as follows:

   (a) Rat urine (250 µL) and 7.5 µL of $TiCl_3$ are added to 250 µg sample of substrate in 250 µL of methanol at 5 °C and allowed to stir for 2 h.

   (b) After 2 h at 5 °C, an aliquot from the reaction mixture is diluted 25 times with a 3:1 mixture of 0.2 % aqueous formic acid/acetonitrile and centrifuged.

   (c) Supernatant is analyzed using LC-MS/MS.

2. **Deuterium Exchange Methods**

   Deuterium exchange methods are often used to differentiate an N-oxide or sulfoxide metabolite from a hydroxyl metabolite [13–15]. If it is a hydroxyl metabolite, an additional 1 amu will be observed from the H/D exchange experiment due to the presence of an active hydrogen atom from the hydroxy group. A general procedure to conduct the deuterium exchange experiment is described below.

   (a) Prepare deuterated mobile phases A and B. Include deuterated rinsing solvents for needle and injector

   (b) Prepare the parent standard and the (e.g. incubate) sample in deuterated solvents (initial mobile phases prepared using deuterated solvents).

   (c) Equilibrate the entire LC-MS system using deuterated mobile phases

   (d) Run the parent standard first to ensure the experiment comes out as expected

(e) Run the sample and check if the oxygenated metabolite contains an active hydrogen for exchange

3. **Thermal Deoxygenation of N-Oxide**
   A brief description is summarized as follows:

   (a) Run LC-MS/MS sample using the electrospray ionization (ESI)

   (b) Switch probe to atmospheric pressure chemical ionization (APCI) and run the same sample

   (c) Compare the MS spectra obtained from ESI and APCI and check if the oxygen from N-oxide metabolite has been deoxygenated from the APCI but not the ESI experiments.

4 Notes

1. Investigation of *in vitro* metabolism using suspended or plated hepatocytes provides an *in vitro* metabolite profile that is both qualitative and quantitative closer to the *in vivo* metabolite profile because in theory, hepatocytes contain all the drug metabolizing enzymes [17]. However, hepatocytes still have limitation in providing an *in vitro* metabolite profile resembling that *in vivo* for slowly turnover compounds [17]. Also, it is more complicated in automation of hepatocyte incubation for high throughput screening of metabolite stability and identification.

2. In general, the liver S9 contains more drug metabolizing enzymes than liver microsomes, and therefore, *in vitro* metabolism conducted with S9 would provide a greater coverage of both phase I and II metabolite pathways than from microsomal incubation. However, both *in vitro* systems require fortification with the appropriate co-factor requires by the drug metabolizing enzyme.

3. The following generic incubation procedure using S9 or microsomes is for investigation of *in vitro* metabolism by the cytochrome P450 enzymes. Co-factors for other drug metabolizing enzymes need to be included. The following incubation has a final test compound concentration of 10 μM, 1 mM EDTA, 5 mM $MgCl_2$, 1 mg/mL of microsomes or 2 mg/mL S9, 1.3 mM NADP+, 3.3 mM Glucose-6-Phosphate, and 0.4 units of glucose-6-phosphate dehydrogenase in 1 mL total incubation volume in a test tube. Other co-factors: NAD+ may be included (3.3 mM) for S9 incubation, UDPGA/UDPAG (5 and 1 mM final concentrations) for glucuronide formation and glutathione (5 mM final concentrations, GSH mixture) for trapping reactive intermediates.

4. In general, the reference standard of metabolite is not available. One approach in quantification of metabolite used unchanged drug as reference standard at a universal response UV wavelength of 220 nm by itself or in combination with MS as in UV$_{220\ nm}$ calibrated MS responses [18, 19]. For metabolite identification, multiple-stage data-dependent MS methods is preferred to improve throughput. In essence, both quantitative and qualitative aspects of the metabolites can be achieved using LC-MS/MS methods in a single run.

5. Liquid chromatography-tandem mass spectrometry (LC-MS/MS) analysis has well established attributes of selectivity, sensitivity and speed coupled with the ability for data-dependent scan that has contributed to routine use of LC-MS/MS for metabolite ID in Discovery [20]. Common techniques used to detect metabolites in complex biological matrices include selective ion monitoring (SIM), precursor ion (PI) and neutral loss (NL) scanning [21–24] using hybrid linear ion trap or triple quadrupole mass spectrometers [25–27].

6. It is recommended to test out the data-dependent scan function from injection on column of ≤1 ng of the test compound prior to using for profiling. Any ions detected from full scan mass analysis which either matched the ions listed in the mass-list or has the correct isotopic pattern will be selected as precursor ion for collision-induced dissociation to generate product ions prior to full scan mass analysis.

7. The concept of mass defect filtering for identifying drug metabolite ions was first reported by Zhang et al. [28–30]. Typically, the mass defects of phase I and phase II metabolite ions fall within 50 mDa relative to that of the parent drug. Thus, the MDF program will filter out all compounds outside the set range of interest (e.g. 50 mDa) of the parent drug.

8. Additional MS detection methods along with MDF data mining techniques have been reported to reduce or eliminate false positive peaks displayed in ion chromatograms. For examples, Mortishire-Smith et al. [31] successfully utilized MDF to remove false positive peaks. Lim et al. [32] used MDF along with background subtraction and isotope pattern filtering for detecting reactive metabolites. Cuyckens et al. [33] applied a combination of MDF, neutral loss and isotope pattern filtering to improve the detection selectivity of fecal metabolites.

Acknowledgements

The views expressed here are solely those of the author and do not reflect the opinions of Janssen Research & Development, LLC.

References

1. Cheng KC, Korfmacher WA, White RE, Njoroge FG (2007) Lead optimization in discovery drug metabolism and pharmacokinetics/case study: the hepatitis C virus (HCV) protease inhibitor SCH 503034. Perspect Medicin Chem 1:1–9
2. Kalgutkar AS, Dalvie D, Obach RS, Smith DA (2012) Role of reactive metabolites in drug-induced toxicity—the tale of acetaminophen, halothane, hydralazine, and tienilic acid. In: Reactive drug metabolites. Eds. Mannhold R, Kubinyi H, and Folkers G. Wiley-VCH Verlag GmbH & Co. KGaA, Weinheim
3. Zhang Z, Zhu M, Tang W (2009) Metabolite identification and profiling in drug design: current practice and future directions. Curr Pharm Des 15:2220–2235
4. Working with hepatocytes in suspension (Video tutorials included 1) Thawing cryosuspension hepatocytes; 2) Cell counting using the Trypan Blue exclusion method and 3) Reconstitution of cryosuspension hepatocytes. http://www.celsisivt.com/working-with-hepatocytes-in-suspension from Celsis/In Vitro Technology
5. Yan Z, Caldwell GW, Maher N (2008) Unbiased high-throughput screening of reactive metabolites on the linear ion trap mass spectrometer using polarity switch and mass tag triggered data-dependent acquisition. Anal Chem 80:6410–6422
6. Mutlib A, Lam W, Atherton J, Chen H, Galatsis P, Stolle W (2005) Application of stable isotope labeled glutathione and rapid scanning mass spectrometers in detecting and characterizing reactive metabolites. Rapid Commun Mass Spectrom 19:3482–3492
7. Bonn B, Leandersson C, Fontaine F, Zamora I (2010) Enhanced metabolite identification with MS(E) and a semi-automated software for structural elucidation. Rapid Commun Mass Spectrom 24:3127–3138
8. Ruijken MMA (2010) MsXelerator RM: a software platform for reactive metabolite detection using low and high resolution mass spectrometry data. The 58th American Society for Mass Spectrometry (ASMS). Salt Lake City, UT, 23–27 May
9. Actis-Goretta L, Lévèques A, Giuffrida F, Destaillats F, Nagy K (2012) Identification of O-methyl-(−)-epicatechin-O-sulphate metabolites by mass-spectrometry after O-methylation with trimethylsilyldiazomethane. J Chromatogr A 1245:150–157
10. Nagy K, Redeuil K, Williamson G, Rezzi S, Dionisi F, Longet K, Destaillats F, Renouf M (2011) First identification of dimethoxycinnamic acids in human plasma after coffee intake by liquid chromatography-mass spectrometry. J Chromatogr A 1218(2011):491
11. Lamoureux G, Aguero C (2009) A comparison of several modern alkylating agents. ARKIVOC i:251–264
12. Kulanthaivel P, Barbuch RJ, Davidson RS, Yi P, Rener GA, Mattiuz EL, Hadden CE, Goodwin LA, Ehlhardt WJ (2004) Selective reduction of N-oxides to amines: application to drug metabolism. Drug Metab Dispos 32:966–972
13. Chen G, Daaro I, Pramanik BN, Piwinski JJ (2009) Structural characterization of in vitro rat liver microsomal metabolites of antihistamine desloratadine using LTQ-Orbitrap hybrid mass spectrometer in combination with online hydrogen/deuterium exchange HR-LC/MS. J Mass Spectrom 44:203–213
14. Lam W, Ramanathan R (2002) In electrospray ionization source hydrogen/deuterium exchange LC–MS and LC–MS/MS for characterization of metabolites. J Am Soc Mass Spectrom 13:345–353
15. Tolonen A, Turpeinen M, Uusitalo J, Pelkonen O (2005) A simple method for differentiation of monoisotopic drug metabolites with hydrogen–deuterium exchange liquid chromatography/electrospray mass spectrometry. Eur J Pharm Sci 25:155–162
16. Peiris DM, Lam W, Michael S, Ramanathan R (2004) Distinguishing N-oxide and hydroxyl compounds: impact of heated capillary/heated ion transfer tube in inducing atmospheric pressure ionization source decompositions. J Mass Spectrom 39:600–606
17. Wang WW, Khetani SR, Krzyzewski S, David Duignan D, Obach RS (2010) Assessment of micropatterned hepatocyte co-culture system to generate metabolites. Drug Metab Dispos 38:1900–1905
18. Josephs JL, Sanders M (2005) Chapter 13: an integrated LC-MS strategy for preclinical candidate optimization. In: Lee MS (ed) Integrated strategies for drug discovery using mass spectrometry. Wiley, New York, p 379
19. Yang Y, Grubb MF, Luk CE, Humphreys WG, Josephs JL (2011) Quantitative estimation of circulating metabolites without synthetic standards by ultra-high-performance liquid chromatography/high resolution accurate mass spectrometry in combination with UV correction. Rapid Commun Mass Spectrom 25:3245–3251
20. Sanders M, Ruzicka J, McHale K, Shipkova P. Thermo application notes # 476: accurate and

sensitive all-ions quantitation using ultra-high resolution LC/MS http://planetorbitrap.com/data/uploads/ZFS1327867271789_AN476.pdf

21. Baillie TA, Davis MR (1993) Mass spectrometry in the analysis of glutathione conjugates. Biol Mass Spectrom 22:319–325

22. Lafaye A, Junot C, Gall BR, Fritsch P, Ezan E, Tabet J-C (2004) Profiling of sulfoconjugates in urine using precursor ion and neutral loss scans in tandem mass spectrometry. Application to the investigation of heavy metal toxicity in rats. J Mass Spectrom 39:655–664

23. Xia Y-Q, Miller JD, Bakhtiar R, Franklin RB, Liu DQ (2003) Use of a quadrupole linear ion trap mass spectrometer in metabolite identification and bioanalysis. Rapid Commun Mass Spectrom 17:1137–1145

24. Liu DQ, Hop CECA (2005) Strategies for characterization of drug metabolites using liquid chromatography-tandem mass spectrometry in conjunction with chemical derivatization and on-line H/D exchange approaches. J Pharm Biomed Anal 37:1–18

25. Jackson PJ, Brownsill RD, Taylor AR, Walther B (1995) Use of electrospray ionization and neutral loss liquid chromatography/tandem mass spectrometry in drug metabolism studies. J Mass Spectrom 30:446–451

26. Clarke NJ, Rindgen D, Korfmacher WA, Cox KA (2001) Systematic LC/MS metabolite identification in drug discovery. Anal Chem 73:430A–439A

27. Kostiainen R, Kotiaho T, Kuuranne T, Auriola S (2003) Liquid chromatography/atmospheric pressure ionization-mass spectrometry in drug metabolism studies. J Mass Spectrom 38:357–372

28. Zhang H, Zhang D, Ray K (2003) A software filter to remove interference ions from drug metabolites in accurate mass liquid chromatography/mass spectrometric analyses. J Mass Spectrom 38:1110–1112

29. Zhang H, Zhang D, Ray K, Zhu M (2009) Mass defect filter technique and its applications to drug metabolite identification by high-resolution mass spectrometry. J Mass Spectrom 44:999–1016

30. Zhu M, Zhang H, Yao M, Zhang D, Ray K, Skiles GL (2004) Detection of metabolites in plasma and urine using a high resolution LC/MS-based mass defect filter approach: comparison with precursor ion and neutral loss scan analyses. Drug Metab Rev 36(Suppl 1):43

31. Mortishire-Smith RJ, O'Connor D, Castro-Perez JM, Kirby J (2005) Accelerated throughput metabolic route screening in early drug discovery using high-resolution liquid chromatography/quadrupole time-of-flight mass spectrometry and automated data analysis. Rapid Commun Mass Spectrom 19:2659–2670

32. Lim HK, Chen J, Cook K, Sensenhauser C, Silva J, Evans DC (2008) A generic method to detect electrophilic intermediates using isotopic pattern triggered data-dependent high-resolution accurate mass spectrometry. Rapid Commun Mass Spectrom 22:1295–1311

33. Cuyckens F, Hurkmans R, Castro-Perez JM, Leclercq L, Mortishire-Smith RJ (2009) Extracting metabolite ions out of a matrix background by combined mass defect, neutral loss and isotope filtration. Rapid Commun Mass Spectrom 23:327–332

Chapter 27

Drug, Lipid, and Acylcarnitine Profiling Using Dried Blood Spot (DBS) Technology in Drug Discovery

Wensheng Lang, Jenson Qi, and Gary W. Caldwell

Abstract

We provide here step-by-step protocols for the quantification of drugs/endogenous metabolites on dried blood spots (DBS) cards. DBS is a micro-volume blood collection technique in which aliquots of whole blood are deposited on specially manufactured filter paper, dried at ambient temperature, extracted, and then analyzed. We have developed liquid chromatography tandem mass spectrometry (LC/MS/MS) based drug and metabolite profiling assays for assessing triglycerides synthesis (energy storage) and acylcarnitine profiling for evaluation of fatty acid oxidation after drug exposure. Briefly, blood samples are spotted on dried blood filter paper. A metal punch with a 3-mm or 6-mm diameter is used to accurately cut a certain size disc. For extraction, an appropriate volume of extraction solvent containing internal standards is applied to extract analyte(s) from the loaded blood disc. The supernatant is separated and transferred to a new set of 96- or 384-well plates for LC/MS/MS analysis. There are two types of DBS cards now commercially available including chemical impregnated and no chemical treated. Chemical impregnated DBS cards offer instant blood cells and bacteria lysis, viral deactivation and enzymatic inhibition, generally leading to improved drugs and metabolites stability during sample collection, storage and transport process. The major pitfall of chemical impregnated DBS cards is the increased matrix effect which may affect the assay precision and accuracy. Therefore, a preliminary evaluation is recommended prior to application to select a specific type of cards with the best recovery and minimal matrix effects. The change in blood endogenous metabolite levels in response to drug treatment, using these assays, can serve as biomarkers for studies on pharmacokinetic (PK) and pharmacodynamic (PD) correlations.

Key words Dried blood spots (DBS), Liquid chromatography tandem mass spectrometry (LC/MS/MS), Hydrophilic interaction liquid chromatography (HILIC), Lipid profiling, Acylcarnitine profiling

1 Introduction

The dried blood spot (DBS) technique was first described in 1913 by Dr. Ivar Bang for estimation of sugar and lipid levels using small quantities of blood absorbed in filter paper [1–3]. The technique later became established in 1963 by Dr. Guthrie's work on quantification of blood phenylalanine levels for diagnostics of newborns' phenylketonuria. This assay was ideal for newborns since babies can only provide a limited amount of blood for testing [4].

Gary W. Caldwell and Zhengyin Yan (eds.), *Optimization in Drug Discovery: In Vitro Methods*, Methods in Pharmacology and Toxicology, DOI 10.1007/978-1-62703-742-6_27, © Springer Science+Business Media New York 2014

Recent years, the advancement in detection technology, particularly in the area of liquid chromatography (LC) and tandem mass spectrometry (MS/MS) has made newborn screening possible to identify over 50 inherited metabolic disorders with a single dried blood spot.

The DBS technique has been utilized in the pharmaceutical industry. In clinical drug development programs, the application and implementation of DBS to pharmacokinetic (PK) and toxicokinetic (TK) studies have become increasingly practical [5, 6]. Clinical PK and TK DBS micro-volume blood collection techniques have many advantages including significant reduction in blood sample volume producing less invasive procedures for patients, and cost reductions of shipping/handling of blood samples. For human PK and TK studies, the hematocrit value for a patient needs to be determined to understand the precision and accuracy of the DBS assay. For example, there are many factors in patients, such as gender, age, and disease stage, which cause wide variations in the percentage of red blood cells in a particular blood volume. The blood samples with a lower hematocrit value tend to spread a larger area after being spotted on DBS cards, resulting in a lower measured concentration than those with a higher hematocrit value, since the quantification of drug levels in DBS are based on the punched disc size. In addition, extensive blood and plasma (B/P ratio) partition validation studies are required to relate PK data derived from blood to PK data derived from plasma. The applications of DBS techniques in drug discovery have also gained broad interests particularly in the situation involving use of small animal (e.g., mouse or juvenile rat) PK/TK studies [7, 8]. The implementation of DBS techniques can significantly reduce sample volume, therefore avoid composite study design and reduce the total number of animals used. More importantly, the data quality is improved by construction of PK profiles and parameters (i.e., concentration–time curve, AUC, clearance, and etc.) from individual animals, not combined data from different animals. In the drug discovery stage, the impact of hematocrit of small animals on the assay results for PK/TK studies may not be of primary concern because the use of laboratory animals is under well-controlled conditions including gender, age, diet, temperature, humidity and environmental light/dark cycles [9]. However, in some cases in which the blood removal exceeds 15 % of circulating blood volume within 24 h, hematocrit correction is needed [10]. Additionally, the blank blood used for preparation of calibration standards and quality control samples should be from the same sources. For chronic efficacy studies, evaluation of the effect of test compounds on hematocrit is recommended, and hematocrit correction may be necessary [11].

DBS techniques are superior to conventional plasma or blood analysis when the analytes are labile, for example, the application of DBS to analysis of prodrugs, which may undergo plasma/blood enzyme-mediated degradation [12, 13]. In such an application, DBS offers improved metabolic stability to avoid potential degradation during samples handling and storage. There are two types of DBS cards commercially available including chemical impregnated and no chemical-treated DBS cards. Chemical impregnated DBS cards offer instant blood cells and bacteria lysis, viral deactivation and enzymatic inhibition, generally leading to improve drugs and metabolites stability during sample collection, storage and transport process. A pitfall of using the chemical impregnated DBS cards is high matrix-effect, which adversely affects assay sensitivity, precision and accuracy.

The DBS technique is not only suitable for determination of drug and its metabolites in PK and TK studies, but also for measuring body response to the treatment in terms of endogenous metabolite profiling for assessment of PK/PD correlation [14–17]. In this chapter, we describe (a) LC/MS-based lipid profiling method on 3-mm DBS discs for assessing triglycerides synthesis, (b) HILIC/MS/MS acylcarnitine profiling on the DBS discs for evaluation of fatty acid oxidation, and (c) use of rimonabant as an example for small molecule drug applications. The detailed procedure and limitation of these methods will be described and discussed.

2 Materials

2.1 DBS Cards

1. GE Healthcare Life Sciences (Piscataway, NJ) offers three types of cellulose-based FTA DMPK cards A, B and C. FTA DMPK-A card coded with a red strip on the left front panel and FTA DMPK-B card coded with a black strip are chemical-impregnated (Fig. 1). Each card has four printed circles with 1 cm in diameter for sample loading. A maximum of approximately seven discs can be punched from one fully loaded spot with a 3-mm punch. Use of FTA DMPK-A & -B cards commonly achieves better sample stability for enzyme-susceptive drugs due to instant termination of blood enzymatic activities. FTA DMPK-C card color-coded with a blue strip is made from filter paper without chemical treatment. This DBS card offers minimum matrix effect, better sensitivity, precision and accuracy in general. The FTA DMPK cards are not just suitable for collection of whole blood sample, but also for other biological fluids, plasma, serum, urine, synovial fluid or cerebrospinal fluid. It is difficult to visualize the spotted light color or colorless biological matrices particularly after dried. In order to clearly identify the spotted areas on DBS cards, GE

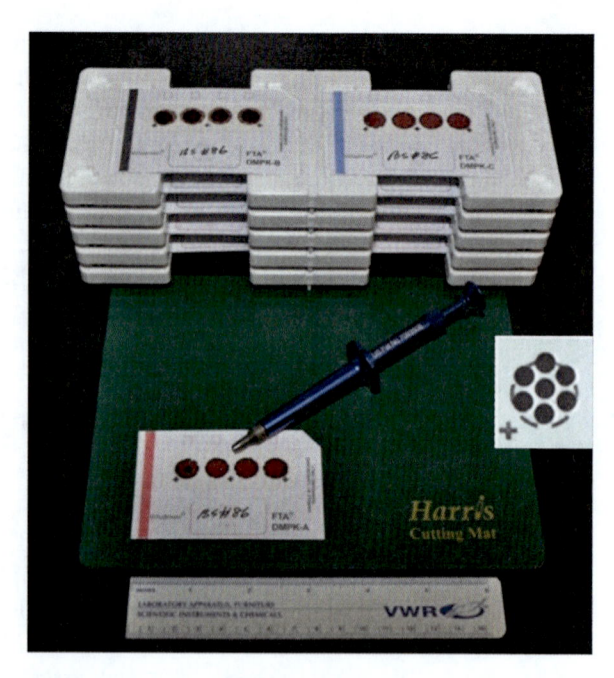

Fig.1 FTA DMPK cards, punch, cutting mat and drying racks from GE Healthcare Corp

Healthcare Biosciences has developed three corresponding indicating cards called as FTA DMPK A, B and C IND. The coated blue background on these DBS cards will turn to white after sample loading, and makes the loaded sample spots easily visualized and identified.

2. Agilent Technologies (Santa Clara, CA) also provides dried matrix spotting (DMS) cards in the similar format as GE Healthcare Biosciences. Agilent DMS cards are made from non-cellulose fiber materials, which are highly homogenous, non-hydroscopic and allowing blood evenly spread radially. The cellulose-free matrix DMS cards without chemical treatment reduce non-specific binding and independent blood hematocrit levels. Currently, Agilent offers a fully integrated system of Agilent Automated Card Extraction (AACE) and LC/M analysis.

3. Tomtec (Hamden, CT) has adopted a different polystyrene encased slide format with three grades of cotton fiber media DMPK200, DMPK300 and DMPK400. The difference in these media is the thickness, DMPK200 in 0.016 in., DMPK300 in 0.026 in. and DMPK400 in 0.032 in. The same size of disc on thicker media carries more analyte than that on the thin media. BSD Robotics provides Semi-automated punch system BSD600 Duet and high-throughput BSD700 Series.

2.2 DBS Cards, Chemicals and Equipment

Specific equipment used in the assays is described; however any model of comparable capability can be easily substituted.

1. FTA DMPK cards (GE Healthcare Life Sciences, Piscataway, NJ).
2. Harris punch, cutting mat and card drying racks (GE Healthcare Life Sciences, Piscataway, NJ).
3. Ziplock plastic bags for storage of dried blood spot cards and desiccant packets (GE Healthcare Life Sciences, Piscataway, NJ).
4. Triglyceride reference standard 1,3-di-heptadecanoyl-2-(10Z-heptadecenoyl)-glycerol-d5 (17:0–17:1–17:0) (Avanti Polar Lipids Inc., Alabaster, AL).
5. Stable isotope labeled carnitine and acylcarnitines reference standards (Cambridge Isotope Laboratories; Milford, MA).
6. Rimonabant (Cayman Chemical Co., Ann Arbor, MI).
7. Acetonitrile, isopropyl alcohol and ammonium formate (Sigma-Aldrich, St. Louis, MO).
8. Male Sprague Dawley rat whole blood with Na^+-EDTA (Biological Specialty Corp., Colmar, PA).
9. 96-wellplates with glass inserts (350 µL) (Agilent Technologies; Santa Clara, CA).
10. Red Rotor PR70 rotating shaker (Hoefer Pharmacia Biotech, Inc., San Francisco, CA).
11. Beckman Allegra 6 Centrifuge (Beckman Coulter, Inc.; Fullerton, CA).
12. Volumetric flasks and 8-channel pipette (VWR, Bridgeport, NJ).
13. Agilent 1100 Liquid Chromatographic system (Agilent Technologies; Santa Clara, CA).
14. Micromass triple-quadrupole *Quattro Ultima/Quattro Micro* mass spectrophotometer (Waters Corp., Milford, MA).
15. HPLC columns: Zorbax Extend-C18 column (2.1×50 mm, particle size=5 µm) and Zorbax Eclipse XDB-C8 column (2.1×50 mm, particle size=3.5 µm) (Agilent Technologies; Santa Clara, CA); ZIC HILIC column (2.1×50 mm, particle size=3.5 µm), (The Nest Group, Inc. Southborough, MA).

3 Method

3.1 Blood Collection on DBS Cards

Clearly label each DBS card with appropriate identity and specificity prior to applying blood on the card. Pipette 10–40 µL of whole blood with a pipette and gently press to dispense a drop of blood on the printed circle area of a DBS card without touching the surface by the pipette tip, or deposit multiple blood drops that touch the same spot before each blood drop completely soak through. Do not layer successive drops of blood or apply blood more than once in the same printed circle. Check the back side of the DBS card to make sure the blood drop(s) completely soaked through and appeared big enough for the appropriate punch size to be used (*see* **Note 1**). Do not try to make up the blood size by adding additional blood on the back side. Put the loaded DBS card on a drying rack and maintain horizontally at room temperature for at least 2 h. If possible, keep the loaded DBS cards in dark to avoid intensive light exposure (*see* **Notes 2** and **3**).

3.1.1 Package and Storage of DBS Cards

After the spotted DBS cards completely dried at ambient temperature, check both sides of each card to confirm a valid blood sample collection. Put one or two desiccant packets into a 4"×6" Ziplock plastic bag first, and then insert one DBS card (*see* **Note 4**). Remove air from the bag and seal it. Put the bags in a cardboard box to avoid direct light exposure and store the cardboard box at ambient temperature or refrigerator (4 °C) or freezer (−20 or −80 °C) depending on the stability features of the analytes interested. Do not staple the plastic bags together, which will cause air leaking and moisturizing samples. In a humid environment, the moisture can promote bacterial growth and facilitates enzyme-mediated degradation during storage.

3.1.2 Manual Punch of DBS Cards

Select the appropriate size punch (3-mm or 6-mm in diameter) to be used. To avoid the potential carry over between samples, clean the punching tool by cutting a waste disc from an unused part of the card or stacked cleaning tissues. Gently press to remove the area and release the disc to a glass insert or a well on a 96-wellplate (*see* **Note 5**).

3.2 Extraction from DBS Cards

3.2.1 Drug Molecules

1. Prepare an extraction solvent by mixing acetonitrile/methanol with water at a ratio of 2:1–4:1, which is depending on the analyte solubility. Transfer the appropriate amount of the mixed solvent into a volumetric flask. Add the appropriate volume of an internal standard stock solution. Make up the volume to graduate with the mixed extraction solvent.

2. For 3-mm disc samples, transfer aliquots of 100 μL of extraction solvent containing the internal standard into the wells with glass inserts (350 μL) on a 96-wellplate with an 8-channel pipette. Place the 96-wellplate on a shaker and gently shake at ambient temperature for 45–60 min. Load the 96-wellplates on Beckman Allegra 6 Centrifuge and centrifuge at 3,000 rpm for 10 min. Transfer the supernatant into a new set of 96-wellplate with an 8-channel pipette for LC/MS/MS analysis.

3. Prepare calibration standards: Use fresh whole blood for preparation of spiked blood samples with a test compound reference in organic solvent or mixed solvent. The content of

organic solvent in the initial spiked blood sample should be controlled at 1 % or less of the blood volume used to minimize the potential hemolysis. Prepare a series of dilutions from the initial spiked sample with the blank whole blood. Spot aliquots of 10–40 μL of the spiked blood samples at each concentration level on FTA-DMPK cards with a pipette and dry at ambient temperature as described above (*see* **Notes 6** and **7**).

3.2.2 Lipid Profiling

1. Prepare an extraction solvent by mixing isopropyl alcohol (IPA) with water at a ratio of 9:1. Transfer the appropriate amount of the mixed solvent into 100 mL volumetric flask. Add 100 μL of 1.0 mM 1,3-di-heptadecanoyl-2-(10Z-heptadecenoyl)-glycerol-d5 (TG 17:0–17:1–17:0, internal standard) in IPA stock solution. Make up the volume with the mixed extraction solvent, giving the final concentration of 1 μM internal standard.

2. For 3-mm disc samples, transfer aliquots of 100 μL of extraction solvent containing the internal standards into the glass inserts on a 96-wellplate. Place the 96-wellplate on Red Rotor PR70 shaker and gently shake at ambient temperature for 45–60 min. Transfer the supernatant into a new set of 96-well plate for LC/MS/MS analysis.

3.2.3 Acylcarnitines Profiling

1. Prepare an extraction solvent by mixing acetonitrile with water at a ratio of 9:1.

2. Add 1.0 mL of a mixed solvent of acetonitrile–water (1:1) into the vial containing stable isotope labeled carnitine and acylcarnitine reference standards set (Cambridge Isotope Laboratories cat# NSK-B). Transfer 250 μL of the working solution into a 50 mL volumetric flask. Make up the volume with the mixed extraction solvent.

3. For 3-mm disc samples, transfer aliquots of 100 μL of extraction solvent containing the stable isotope labeled carnitine and acylcarnitine reference standards set into the glass inserts/wells on a 96-wellplate. Place the 96-wellplate on Red Rotor PR70 shaker and gently shake at ambient temperature for 45–60 min. Transfer the supernatant into a new set of 96-well plate for LC/MS/MS analysis.

3.3 LC/MS/MS

3.3.1 Determination of Drug Molecules by LC/MS/MS

Agilent 1100 Liquid Chromatographic system interfaced with a Micromass triple-quadrupole *Quattro Ultima* mass spectrometer through a Z-spray electrospray ion source was used for determination of blood levels of rimonabant on DBS. Separation of rimonabant was performed on a Agilent Zorbax Extend-C18 column (2.1 × 50 mm, particle size = 5 μm) eluted with 50–95 %B in 2 min, hold 95 %B for 3.5 min and return to 50 %B in 0.15 min. The mobile phase A was 0.1 % formic acid in water, and B was

0.1 % formic acid in acetonitrile. The flow rate was 0.3 mL/min. Selective MRM detection of rimonabant and its acidic hydrolyzed metabolite was conducted in the positive ion mode. The MRM mass transition for rimonabant was m/z 463.0 > 362.9 at a collision energy of 25 eV and m/z 380.9 > 362.9 for its acidic hydrolyzed metabolite (ce 25 eV). The MS parameters were as follows: capillary voltage, 3.2 kV; cone voltage, 25 V; extractor, 2 V; RF lens, 0.1 V; source temperature, 120 °C; desolvation temperature, 300 °C; LM1, HM1, LM2 and HM2 resolutions 12.5; ion energy1, 1; entrance, 15; exit, 15; cone gas flow, 50 L/h; and desolvation gas flow, 700 L/h. Masslynx software version 4.0 was used for system control and data processing.

3.3.2 LC/MS-Based Lipid Profiling

Agilent 1100 Liquid Chromatographic system interfaced with a Micromass triple-quadrupole *Quattro Micro* mass spectrometer through a Z-spray electrospray ion source was used for lipid profiling. Separation of triglycerides was performed on an Agilent Zorbax Eclipse XDB-C8 column (2.1 × 50 mm, particle size = 3.5 μm) eluted with 30–80 %B in 10 min, hold 80 %B for 2 min and return to 30 %B in 0.1 min. The mobile phase A was 5 mM ammonium formate in acetonitrile–water (95:5), and B was 5 mM ammonium formate in isopropyl alcohol–water (95:5). The flow rate was 0.3 mL/min. The mass spectrometer was operated in the positive ion mode. The MS parameters were as follows: capillary voltage, 3.2 kV; cone voltage, 25 V; extractor, 2 V; RF lens, 0.1 V; source temperature, 120 °C; desolvation temperature, 300 °C; LM1, HM1, LM2 and HM2 resolutions 14; ion energy1, 1; entrance, 15; collision, 1; exit, 15; cone gas flow, 50 L/h; and desolvation gas flow, 700 L/h; mass scan range of 300–1,100 amu in a second. Masslynx software version 4.0 was used for system control and data processing.

A representative LC/MS map showing lipid profile of rat blood in FTA DMPK card is given in Fig. 2. Triacylglycerols as energy storage contain a glycerol backbone and 3 lipid acyl sidechains. The acyl sidechain contents vary in carbon chain length and double bond number. This class of compounds is highly diverse and lipophilic. Additionally, triacylglycerols are hardly protonated in positive ion electrospray ionization, therefore, they were converted to corresponding ammonium adduct ions for MS detection in the presence of ammonium formate salt in the mobile phases. Chromatographic separation of triacylglycerols was performed on a reversed phase C8 column eluted with the increased ratio of isopropyl alcohol to acetonitrile. The triacylglycerols with the same acyl sidechain length, but different double bond number were separated under the LC conditions (Fig. 3). Quantitation of triglycerides was done by integration of each extracted ion chromatographic peak against that of the deuterated internal standard. It has

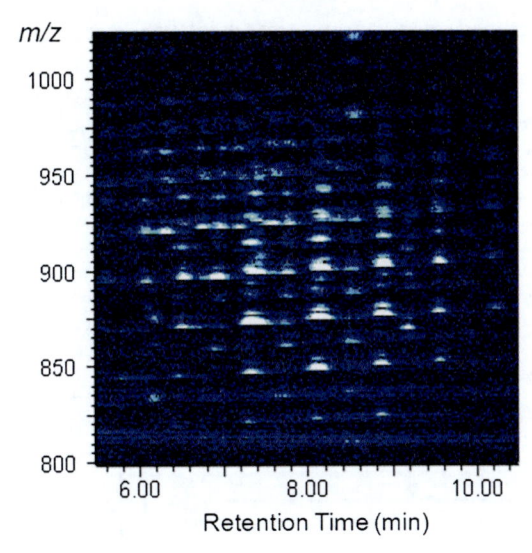

Fig. 2 Lipid profile of rat blood in FTA-DMPK card

been reported that the acyl sidechain content of plasma triacylglycerols can be used as a biomarker for prediction of diabetes risk [18].

3.3.3 Acylcarnitine Profiling by HILIC/LC/MS

The free carnitine and acycarnitines can be directly analyzed using HILIC/MS/MS without chemical derivatization. In this regard, an Agilent 1100 Liquid Chromatographic system was interfaced with a Micromass triple-quadrupole *Quattro Ultima* mass spectrometer through a Z-spray electrospray ion source. Separation of free carnitine and acylcarnitines was performed on a ZIC HILIC column (2.1×50 mm, particle size = 3.5 μm). Mobile phases A. 50 mM ammonium formate & 0.1 % fromic acid in water; and B. acetonitrile. A gradient elution was conducted at a flow rate of 0.3 mL/min with 5–55 %A within 5 min, hold 55 %A for 3 min, return to 5 %A in 0.1 min. The injection volume was 10 μL. The mass spectrometer was operated in the positive ion mode. Parent scan of m/z 85 from 150 to 500 amu within 0.8 s at collision energy of 20 eV was conducted for selective detection of carnitine and acylcarnitines (Figs. 4 and 5). The MS parameters were as follows: capillary voltage, 3.2 kV; cone voltage, 25 V; extractor, 2 V; RF lens, 0.1 V; source temperature, 120 °C; desolvation temperature, 300 °C; cone gas flow, 50 L/h; and desolvation gas flow, 700 L/h; LM1 = HM1 resolution = 14.5; LM2 = HM2 resolutions = 14.5; ion energy1, 1.0; entrance, -8; exit, 8; multiplier, 650; and interscan time 0.05 s. Masslynx software version 4.0 was used for system control and data processing.

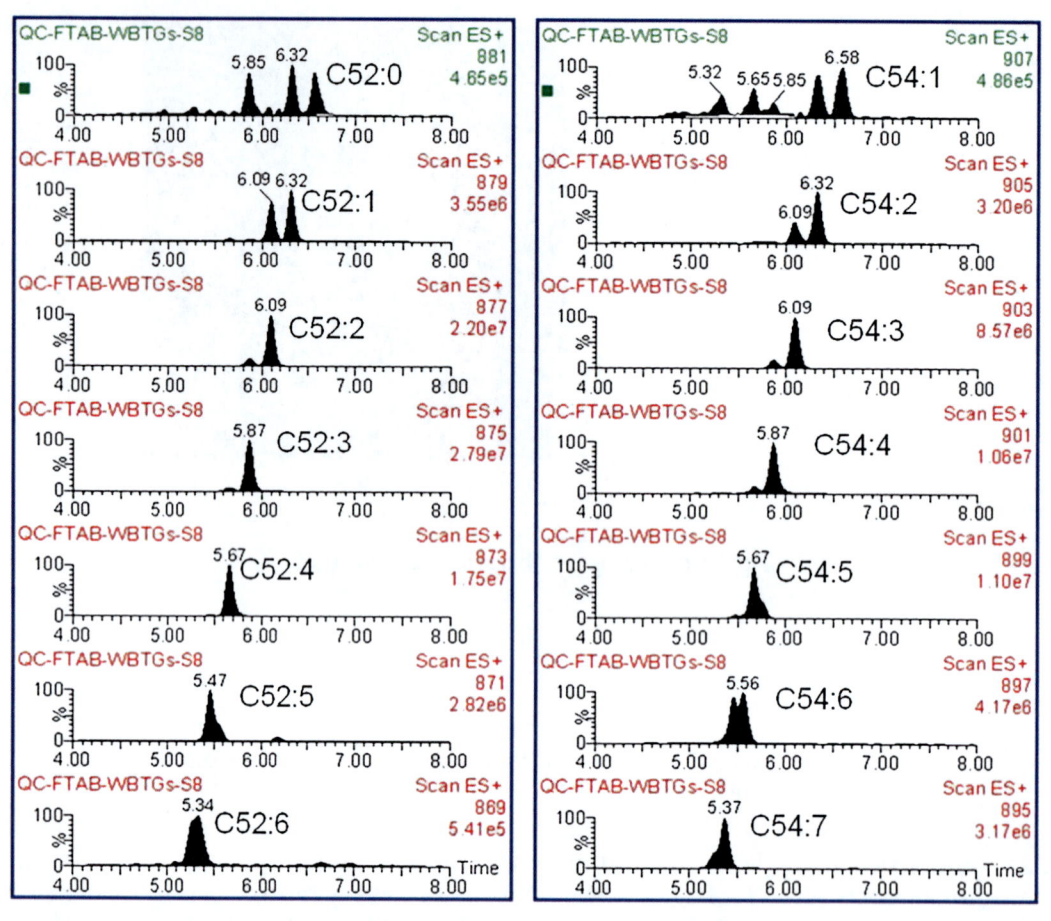

Fig. 3 Extracted ammonium adduct ion chromatograms of blood triacylglycerols (acyl carbon number:double bond number)

Fig. 4 Collision induced dissociation of acylcarnitine, generating product ion m/z 85

In electrospray ionization positive ion mode, carnitine and acylcarnitines undergo CID-fragmentation and generates a strong product ion of m/z 85 (Fig. 4). Precursor ion scan of m/z 85 was done over a mass range of 150–500 amu for selectively detection

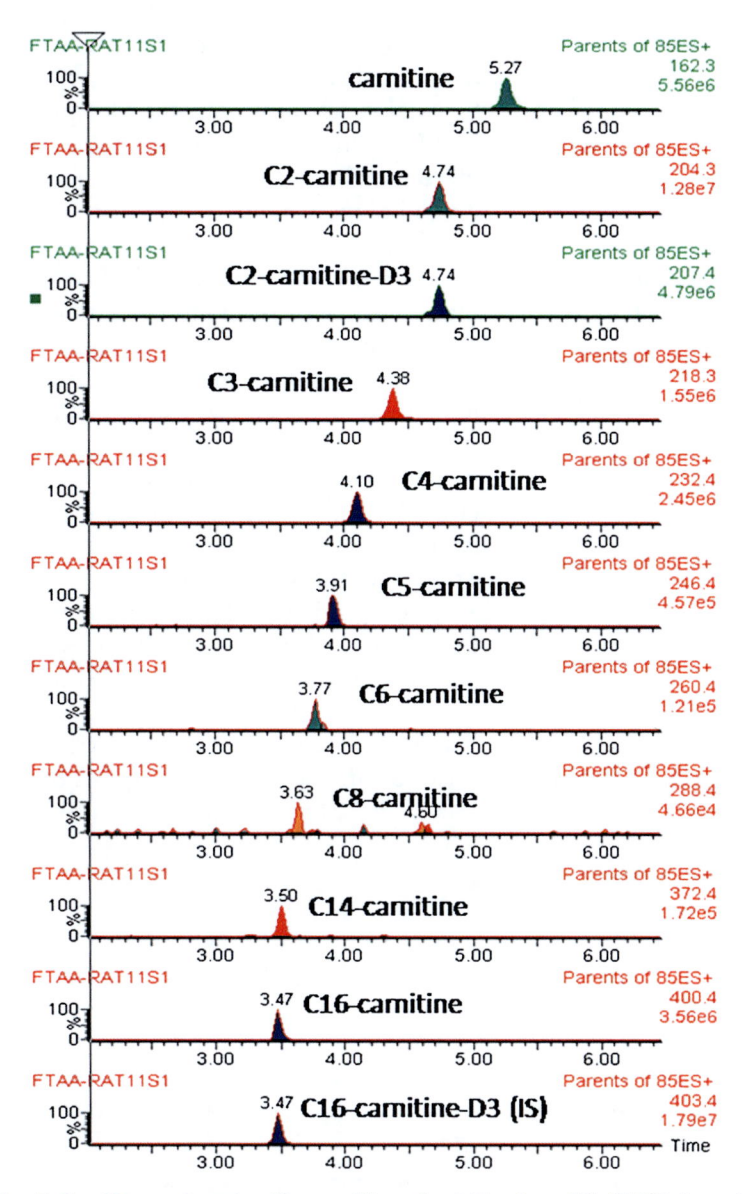

Fig. 5 Carnitine and acylarnitine profiling of rat blood on FTA DMPK-A cards, extracted ion chromatograms of parent ion scan of *m/z* 85

of free carnitine and acylcarnitines in DBS [19–21]. Quantification of blood levels of carnitine and acylcarnitines was conducted using isotope dilution method with a set of 8 deuterated carnitine and acylcarnitines ($C_2, C_3, C_4, C_5, C_8, C_{14}$ and C_{16}) reference standards labeled (D_3- or D_9-) at the trimethylamine moiety (Cambridge Isotope Lab. MA). The same product ion of m/z 85 was obtained for these deuterated carnitine and acylcarnitine reference standards. The relative levels of endogenous carnitine and acylcarnitines were calculated based on the added concentrations of corresponding deuterated reference standards or the closest acylcarnitine analog. Separation of carnitine and acylcarnitine was conducted on a ZIC HILIC column (2.1×50 mm, 3.5 μm). The extracted ion chromatograms at each precursor mass for rat blood in DBS are given in Fig. 5.

3.4 Stability of Rimonabant

3.4.1 Stability of Rimonabant in Rat Blood

To illustrate the effect of DBS cards on the stability of small molecules, cannabinoid receptor (CB1) inversed agonist rimonabant was used as an example for the comparative study of its stability in rat blood and in DBS cards. The stability of rimonabant in rat whole blood was first evaluated at 37 °C and ambient temperature (~20 °C). An aliquot of 1.5 μL of 200 μM rimonabant in acetonitrile stock solution was spiked in 300 μL of rat blood (EDTA) and mixed in triplicate. The rat blood samples were incubated at 37 °C or at ambient temperature. Aliquots of 10 μL of the incubates were withdraw at 0, 10, 20, 30, 40, 60, 90 and 120 min and added into glass inserts containing 200 μL of extraction solvent (acetonitrile–water, 4:1). The results of percentage remaining–time curves are given in Fig. 6.

3.4.2 Stability of Rimonabant on DBS Cards

Stability of rimonabant in FT DMPK cards during drying process was evaluated by spotting aliquots of 7 μL of a 1 μM rimonabant spiked rat blood sample onto FTA-DMPK cards. The entire

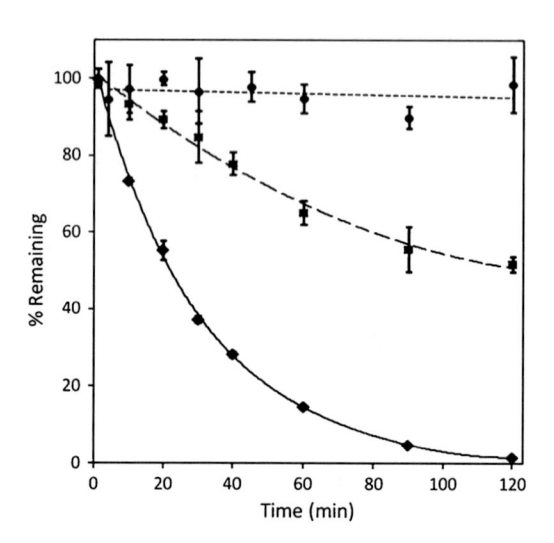

Fig. 6 Stability of small drug molecule rimonabant in rat whole blood and in spotted FTA DMPK-A card, *closed diamond*, % remaining of rimonabant in rat blood at 37 °C at various times for 2 h; *closed square*, % remaining of rimonabant in rat blood at ambient temperature for 2 h; *closed circle*, % remaining of rimonabant in rat blood sample spotted on FTA-DMPK-A card at ambient temperature during drying process for 2 h. The data represent averages of three independent measurements. The experimental conditions are given in Sects. 2 and 3

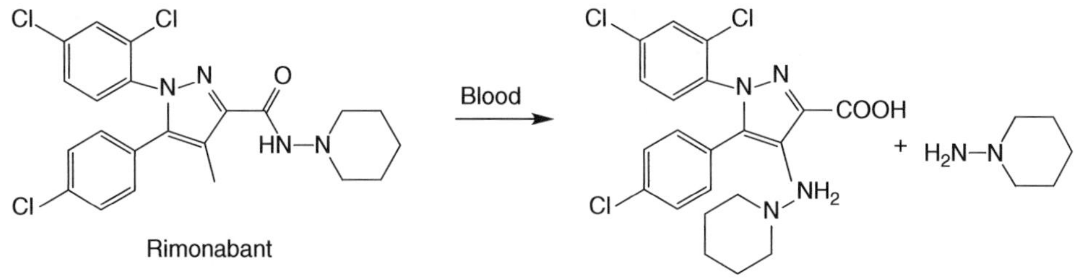

Fig. 7 Degradation pathway of rimonabant in rat blood

Table 1
Stability of rimonabant in FTA DMPK-A cards under different storage conditions

	Room temp (~20 °C)	Refrigerator (4 °C)	Freezer (−20 °C)
Day 1	10.46 ± 0.46	10.81 ± 0.63	9.28 ± 0.49
1 Week	10.39 ± 0.29	9.97 ± 0.83	9.63 ± 0.39
2 Weeks	8.72 ± 0.14	9.29 ± 0.24	9.13 ± 0.41

blood spot was punched with a 6-mm punch at 0, 10, 20, 30, 45, 60, 90 and 120 min in triplicate. Each 6-mm disc was transferred into a glass insert containing 200 µL of extraction solvent (acetonitrile–water, 4:1) on a 96-wellplate. The samples were centrifuged at 3,000 rpm at ambient temperature for 10 min on a Beckman Allegra 6 centrifuge. The supernatant was transferred into a new set of 96-wellplate with an 8-channel pipette for LC/MS/MS analysis. A comparative stability plot of rimonabant in liquid whole blood and in DBS cards is given in Fig. 6. The proposed degradation pathway of rimonabant thought enzyme-mediated hydrolysis of the amide bond is given in Fig. 7. The stability of rimonabant in DBS cards under different storage conditions were also evaluated and summarized in Table 1. The labile molecule in DBS cards showed good stability at ambient temperature for 1 week.

4 Notes

1. Generally load a minimum of 10 µL of fresh whole blood or 5 µL of plasma using a capillary/pipette on a DBS card for a 3 mm punch or a minimum of 15 µL of whole blood for 6 mm punch. You can load multiple blood drops on the same spot as long as the filter paper spot remains wet.

2. In the in-life phase of animal handling, avoid excessive squeezing tissues during blood collection, e.g., tail vein bleeding, which causes hemolysis and tissue fluid contamination.
3. Collect total blood less than 15 % of total circulation blood volume in animals within a 24 h period.
4. Pack the DBS cards individually with Ziploc plastic bags containing desiccant packets. Check indicator color "blue" through the desiccant pack window before use. Do not use desiccant packets when the content turns to pink.
5. To avoid volcano effect from FTA DMPK cards, punch location of DBS cards (central or close to the edge) should be consistent among samples, calibration standards and quality control samples.
6. The loading volume or blood spot size for preparation of calibrations standards and QCs on DBS cards should be close to that for samples.
7. Always use fresh whole blood containing anticoagulant for preparation of calibration standards for each test compound to avoid hemolysis causing decrease in hematocrit values and increase in spot size after an extensive storage period.

Acknowledgements

The views expressed here are solely those of the author and do not reflect the opinions of Janssen Research & Development, LLC.

References

1. Bang I (1913) A method for microdetermination of blood components. Biochem Z 49: 19–39
2. Bang I (1918) The microdetermination of glucose. II. Biochem Z 92:344–346
3. Bang I (1918) Microdetermination of blood lipoids. Biochem Z 91:235–256
4. Guthrie R, Suzi A (1963) A simple phenylalanine method for detecting phenyketonuria in large populations of newborn infants. Pediatrics 32:338–343
5. Xu Y, Woolf EJ, Agrawal NGB, Kothare P, Pucci V, Bateman KP (2013) Merck's perspective on the implementation of dried blood spot technology in clinical drug development—why, when and how. Bioanalysis 5:341–350
6. Kapur S, Kapur S, Zava D (2008) Cardiometabolic risk factors assessed by a finger stick dried blood spot method. J Diabetes Sci Technol 2:236–241
7. Dainty TC, Richmond ES, Davies I, Blackwell MP (2012) Dried blood spot bioanalysis: an evaluation of techniques and opportunities for reduction and refinement in mouse and juvenile rat toxicokinetic studies. Int J Toxicol 31:4–13
8. Turpin PE, Burnett JE, Goodwin L, Foster A, Barfield M (2010) Application of the DBS methodology to a toxicokinetic study in rats and transferability of analysis between bioanalytical laboratories. Bioanalysis 2:1489–1499
9. Ghanta VK, Hiramoto NS, Soong SJ, Hiramoto RN (1991) Survey of thymic hor-

mone effects on physical and immunological parameters in C57BL/6NNia mice of different ages. Ann N Y Acad Sci 621:239–255
10. Nahas K, Provost JP (2002) Blood sampling in the rat: current practices and limitations. Comp Clin Pathol 11:14–37
11. O'Mara M, Hudson-Curtis B, Olson K, Yueh Y, Dunn J, Spooner N (2011) The effect of hematocrit and punch location on assay bias during quantitative bioanalysis of dried blood spot samples. Bioanalysis 3:2335–2347
12. Alfazil AA, Anderson RA (2008) Stability of benzodiazepines and cocaine in blood spots stored on filter paper. J Anal Toxicol 32:511–515
13. D'Arienzo CJ, Ji QC, Discenza L, Cornelius G, Hynes J, Cornelius L, Santella JB, Olah T (2010) DBS sampling can be used to stabilize prodrugs in drug discovery rodent studies without the addition of esterase inhibitors. Bioanalysis 2:1415–1422
14. Clavijo CF, Hoffman KL, Thomas JJ, Carvalho B, Chu LF, Drover DR, Hammer GB, Christians U, Galinkin JL (2011) A sensitive assay for the quantification of morphine and its active metabolites in human plasma and dried blood spots using high-performance liquid chromatography-tandem mass spectrometry. Anal Bioanal Chem 400:715–728
15. Michopoulos F, Theodoridis G, Smith CJ, Wilson ID (2010) Metabolite profiles from dried biofluid spots for metabonomic studies using UPLC combined with oaToF-MS. J Proteome Res 9:3328–3334
16. Kong ST, Lin H, Ching J, Ho PC (2011) Evaluation of dried blood spots as sample matrix for gas chromatography/mass spectrometry based metabolomic profiling. Anal Chem 83:4314–4318
17. Wilson I, Mereside AP (2011) Global metabolic profiling (metabonomics/metabolomics) using dried blood spots: advantages and pitfalls. Bioanalysis 3:2255–2257
18. Rhee EP, Cheng S, Larson MG, Walford GA, Lewis GD, McCabe E, Yang E, Farrell L, Fox CS, O'Donnell CJ, Carr SA, Vasan RS, Florez JC, Clish CB, Wang TJ, Gerszten RE (2011) Lipid profiling identifies a triacylglycerol signature of insulin resistance and improves diabetes predction in humans. J Clin Invest 121:1402–1411
19. Fingerhut R, Roschinger W, Muntau AC, Dame T, Kreischer J, Arnecke R, Superti-Furga A, Troxler H, Liebl B, Olgemoller B, Roscher AA (2001) Hepatic carnitine palmitoyltransferase I deficiency: acylcarnitine profiles in blood spots are highly specific. Clin Chem 47:1763–1768
20. Paglia G, D'Apolito O, Corso G (2008) Precursor ion scan profiles of acylcarnitines by atmospheric pressure thermal desorption chemical ionization tandem mass spectrometry. Rapid Commun Mass Spectrom 22:3809–3815
21. Fingerhut R, Regina E, Roschinger W, Ro W, Arnecke R, Olgemöller B, Roscher AA (2009) Stability of acylcarnitines and free carnitine in dried blood samples: implications for retrospective diagnosis of inborn errors of metabolism and neonatal screening for carnitine transporter deficiency. Anal Chem 81:3571–3575

Chapter 28

In Vitro Trapping and Screening of Reactive Metabolites Using Liquid Chromatography-Mass Spectrometry

Zhengyin Yan and Gary W. Caldwell

Abstract

Metabolism catalyzed by the cytochrome P450 enzymes (CYPs) represents the most important clearance pathways for most drugs in humans. However, CYP-mediated metabolism can also lead to drug bioactivation resulting to formation of reactive metabolites that can potentially induce idiosyncratic toxicity by covalently binding to endogenous proteins and nucleic acids. Therefore, it has become imperative to implement strategies for screening and identifying bioactivation liability of drug candidates as an integrated approach to reduce the attrition rate in drug discovery and development. This chapter describes a detailed protocol for the *in-vitro* stable isotopic trapping and screening for reactive metabolites using common LC-MS/MS methodologies such as neutral loss scan and precursor ion scan.

Key words Cytochrome P450 mediated bioactivation, Screening of reactive metabolites

1 Introduction

Drugs are commonly metabolized by a variety of oxidative enzymes predominantly as cytochrome P450s (CYPs) to form stable and more polar metabolites that can be readily eliminated from human body, and thus oxidative metabolism is usually recognized as a detoxification process. However, for some drugs, CYP-mediated metabolism can also lead to bioactivation which results in formation of chemically reactive species that can covalently modify endogenous proteins and nucleic acids. Such covalent modifications of endogenous components resulting from drug bioactivation are proposed to play an important role in drug-induced idiosyncratic toxicity, although exact toxicological mechanisms can be highly drug-specific, and largely remain to be elucidated [1].

In general, drug-induced idiosyncratic toxicities are very difficult to predict primarily for two reasons: a high degree of individual susceptibility and lacking simple dose responses [2]. Thus, idiosyncratic drug reactions are not detected and reported until a large population of patients has been studied after approval.

Because severe idiosyncratic drug reactions can be life-threatening and lead to restricted use and even withdrawal from the market, a drug candidate undergoing bioactivation is less favorable for further development, despite a clear correlation between idiosyncratic drug reactions and bioactivation largely remains to be established. As a major effort to reduce the attrition rate in drug development, screening and structural characterization of chemically reactive metabolites has widely been implemented in the lead optimization process of drug discovery, since such information can be very helpful for medicinal chemists to optimize lead compounds at an early stage of drug discovery [3, 4].

1.1 Classification of Reactive Metabolites

Drug bioactivation can lead to formation of a wide variety of reactive metabolites [5]. Those chemically reactive species can be grossly classified into "soft" and "hard" reactive metabolites, based on their chemical reactivity. "Soft" reactive metabolites constitute of a majority of electrophilic metabolites which include quinones, quinone imines, iminoquinone methides, epoxides, arene oxides and nitrenium ions, and they can readily react with "soft" electrophiles such as the sulfhydryl group in cysteine. In contrast, "hard" reactive metabolites, most commonly seen as aldehydes, preferentially react to "hard" electrophiles such as amines of lysine, arginine and nucleic acids [6]. Because of their instability, direct detection and characterization of reactive metabolites is not technically feasible. A commonly utilized approach is to trap reactive metabolites in microsomal incubations with a proper capture molecule, resulting in formation of stable adducts that are subsequently characterized by tandem mass spectrometry.

1.2 In-Vitro Trapping Agents

For "soft" reactive electrophilic metabolites, glutathione (GSH) is the most commonly used agent to trap a vast majority of reactive metabolites formed in microsomal incubations [5]. Resulting GSH adducts are analyzed and structurally characterized by liquid chromatography-tandem mass spectrometry (LC-MS/MS), and structural information of reactive metabolites can be elucidated. However, the same strategy cannot be applied to the detection of "hard" electrophiles, largely due to low trapping efficiency of GSH. Alternative trapping agents such as semicarbazide, methoxylamine and α-acetyllysine have been used to mimic lysine residue of proteins for capturing aldehyde metabolites. Recently, a bi-functional peptide has been introduced for simultaneously trapping and rapidly screening of both "hard" and "soft" reactive metabolites [6]. Basically, the glycine of GSH is replaced by a lysine residue leading to a dual functional peptide (Υ-glutamyl-cystein-lysine, GSK) that can capture both "hard" and "soft" reactive metabolites formed in microsomal incubations by conjugation to either the sulfhydryl group of cysteine or the amine group of lysine, and resulting GSK adducts are subsequently analyzed by LC-MS/MS [6].

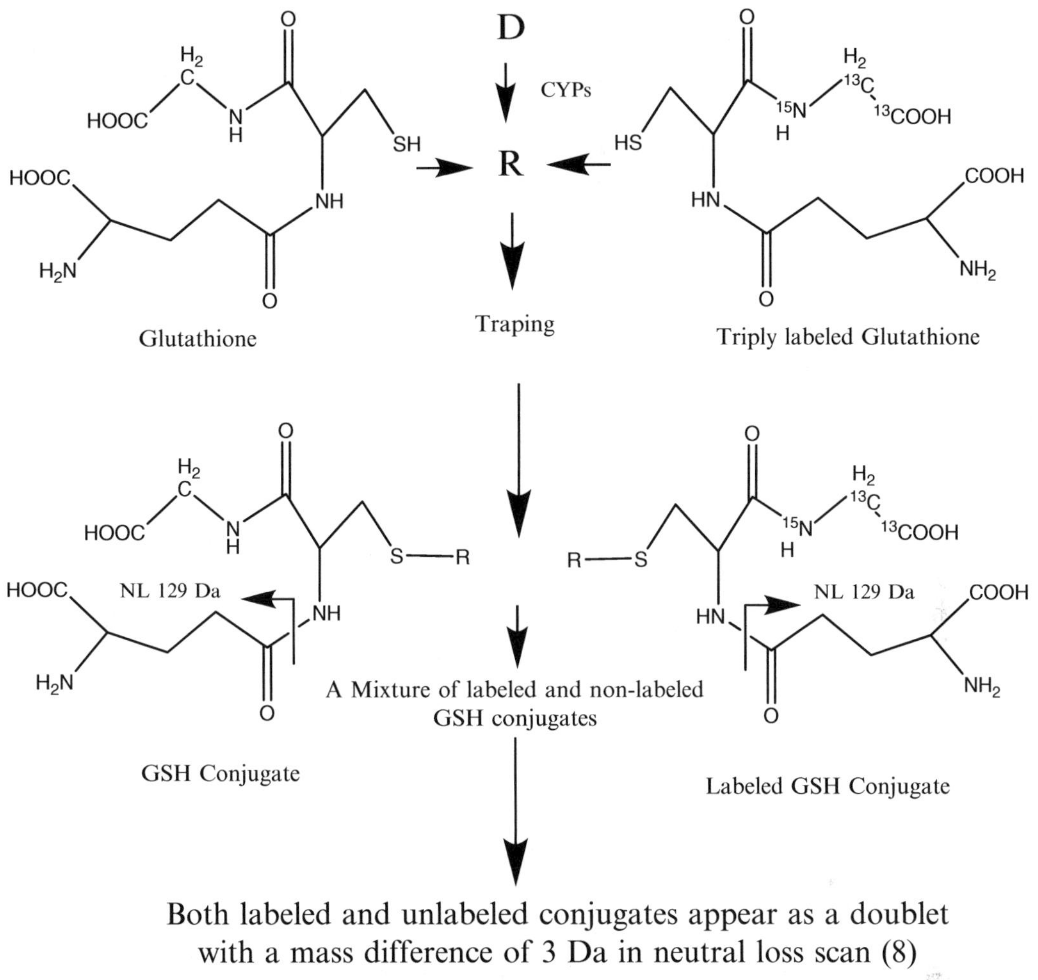

Both labeled and unlabeled conjugates appear as a doublet
with a mass difference of 3 Da in neutral loss scan (8)

Fig. 1 Stable isotopic trapping of reactive metabolites using glutathione as the capturing agent [7]. D and R denote drug molecules and corresponding reactive metabolites formed in microsomal incubations

1.3 MS Methodologies for Screening Reactive Metabolites

Previous structural characterization by mass spectrometry demonstrated that nearly all GSH adducts undergo a common neutral loss of 129 Da (the ϒ-glutamyl moiety) under the collision-induced dissociation (CID) in the positive mode, and thus CID-induced neutral loss scanning has been widely used as a generic method to rapidly detect GSH adducts formed in microsomal incubations [7, 8]. It has also been reported that, in the negative mode, all GSH adducts give rise of a product ion at m/z 272 corresponding to deprotonated γ-glutamyldehydroalanyl-glycine moiety [9]. Therefore, an alternative MS method is CID-induced precursor ion scan of m/z 272 in the negative mode to screen for GSH conjugates.

This chapter describes a detailed protocol for *in-vitro* isotopic trapping and rapid screening of reactive metabolites. As depicted in Fig. 1, a mixture of stable isotope labeled and non-labeled glutathione is used in microsomal incubation to capture reactive metabolites to form a pair of isotopic GSH conjugates that are easily detected by LC-MS/MS either using neutral loss scan [7, 8, 10] or isotope pattern-triggered MS/MS data acquisition [10–12]. This protocol can be easily modified to trap and detect structurally different reactive metabolites [6, 13, 14].

2 Material

2.1 Buffers, Cofactors and Stop Solution

All reagents were obtained from Sigma-Aldrich (St. Louis, MO) except for those specified.

1. 0.5 M Potassium phosphate buffer (pH 7.4) is prepared as the following:

    (a) 0.5 M Potassium phosphate, KH_2PO_4, monobasic. Dissolve 34 g KH_2PO_4 in 450 mL deionized water, and then bring the final volume to 500 mM with deionized water;

    (b) 0.5 M Potassium phosphate, K_2HPO_4, dibasic. Dissolve 57 g $K_2HPO_4 \cdot 3H_2O$ in 450 mL deionized water, and then bring the final volume to 500 mM with deionized water;

    (c) Mix 60 mL 0.5 M KH_2PO_4 with 280 mL 0.5 M K_2HPO_4, and check with a pH meter for a pH value of 7.4. If necessary, adjust pH with either KH_2PO_4 or K_2HPO_4.

2. Sodium citrate (5 mM), tribasic. Dissolve 14.7 mg sodium citrate in 100 mL deionized water, and store at 4 °C.

3. Co-factors: Dissolve 400 mg nicotinamide adenine dinucleotide phosphate (NADP+), 400 mg glucose-6-phosphate, and 266 mg $MgCl_2 \cdot 6H_2O$ in 18 mL deionized water, and then adjust the final volume to 20 mL with deionized water. Aliquot and store at −20 °C;

4. Glucose-6-phosphate dehydrogenase (G6PDH): 40 U/mL, prepared in 5 mM sodium citrate. Aliquot and store at −20 °C;

5. Stop solution: 45 % trichloroacetic acid (TCA, w/v).

6. Acetonitrile and methanol (EMD Chemicals, Gibbstown, NJ).

2.2 Trapping Agents

1. Glutathione (GSH): dissolved in deionized water to make a fresh solution of 40 mM (Note 1);

2. Stable isotope labeled glutathione (GSX, ϒ-glutamyl-cystein-glycin-$^{13}C2$-^{15}N, Cambridge Isotope Laboratories, Andover, MA);

3. Other trapping agents (optional): potassium cyanide and stable isotope labeled cyanide [13, 14] (**Note 2**);

4. Pooled human liver microsomes: HLM (20 mg/mL) prepared from 150 donors was obtained from BD Biosciences (Woburn, MA) and stored at −80 °C;

5. SEP-PAK 100 mg C18 packaged cartridges (Waters Corp., Milford, MA) or equivalent C18 packaged cartridges.

2.3 LC-MS Instrumentation

A triple quadrupole mass spectrometer such as ABI/MDS Sciex 4000 QTRAP mass spectrometer (Toronto, Canada) or a comparable MS interfaced with an auto-sampler and HPLC system such as Shimadzu 20A (Canby, OH). Alternatively, a Thermo Fisher LCT ion trap mass spectrometer (San Jose, CA) can be used to replace the triple quadrupole mass spectrometer if isotopic pattern-triggered MS/MS data acquisition is utilized [10].

3 Methods

3.1 In-Vitro Trapping of Reactive Metabolites

1. Testing compound working solution: Dry compounds are dissolved in acetonitrile or another proper solvent such as methanol and DMSO to make a working solution of 5 mM;

2. Positive control: Acetaminophen can be used as a positive control for the assay, and it can be dissolved in acetonitrile to make a working solution of 5 mM;

3. Three pairs of 2-mL micro-centrifuge tubes are required for every test compound and the positive control. For each compound, label 2-mL micro-centrifuge tubes separately as compound name plus "0 min" and "60 min" (**Note 2**);

4. Glutathione working solution: A total volume of 60 μL GSH working solution is required for each test compound. Weigh a proper amount of GSH powder based on the number of compounds, and dissolve in deionized water to make a fresh solution of 40 mM (**Note 3**);

5. Stable isotope labeled glutathione (GSX) working solution: A total volume of 60 μL GSX is required for each test compound. Weigh a proper amount of GSX powder based on the number of compounds, and dissolve GSX powder in deionized water to make a fresh solution of 40 mM (Note 3).

6. Mix 1.0 μL of both labeled (GSX) and unlabeled GSH solution in a clean 2-mL microcentrifuge tube and dilute to 2 mL with deionized water in order to check their relative ratio in **Step 7**;

7. Directly infuse diluted GSH-GSX mixture with a syringe to a mass spectrometer that is in operational conditions (ESI, positive mode) to check for the relative intensity of corresponding molecular ions at m/z 308 and 311 (**Note 4**);

8. Based on the ion intensity ratio estimated in Step 7, mix both labeled (GSX) and unlabeled GSH working solution

proportionally to ensure that the ratio of two isotopic ions is in the range of 1:0.80 to 1:0.6 approximately (**Note 5**);

9. For each test compound or positive control, 1 mL HLM working solution (2 mg protein/mL) is required. Calculate the total volume of HLM working solution needed for the assay based on the total number of compounds to be tested;

10. Prepare HLM working solution (2 mg protein/mL) according to the following proportion for each individual compound (**Note 6**):
    (a) 200 μL 0.5 M phosphate buffer;
    (b) 100 μL human liver microsomes (20 mg/mL protein);
    (c) 700 μL deionized water;
    (d) invert the tube repeatedly to mix well;

11. Dispense 500 μL HLM working solution (2 mg protein/mL) to individual labeled microcentrifuge tubes;

12. Add 2 μL test compound or control (5 mM) to each pair of labeled microcentrifuge tubes (0, 60 min) containing 500 μL HLM working solution (2 mg protein/mL);

13. Add 100 μL mixed GSH-GSX working solution to every labeled microcentrifuge tube containing drug-HLM mixture;

14. For each test compound or positive control, 1.0 mL NADPH working solution is needed. Calculate the total volume of NADPH working solution needed for the assay based on the total number of compounds to be tested;

15. Prepare NADPH regenerating solution in a clean 50 mL tube according to the following proportion for each individual compound (Note 7):
    (a) 200 μL 0.5 M phosphate buffer;
    (b) 680 μL deionized water;
    (c) 100 μL NADP$^+$ cofactor mixture;
    (d) Supply 20 μL 40 U/mL G6PDH to the mixture
    (e) Vortex briefly;

16. Dispense 500 μL NADPH regenerating solution to every "60-min" labeled microcentrifuge tube containing HLM-drug mixture, and invert tubes repeatedly; "0-min" labeled tubes received no NADPH regenerating solution (negative control);

17. Put all microcentrifuge tubes containing incubation mixture in a pre-heated water bath, and incubate at 37 °C for 60 min;

18. Dispense 500 μL NADPH regenerating solution to every "0-min" labeled microcentrifuge tube containing HLM-drug mixture at the end of incubation;

19. Immediately add 300 μL TCA stop solution to each microcentrifuge tube to terminate the reaction (**Note 8**). Alternatively, acetonitrile can be used to precipitate microsomal protein if no solid phase extraction is performed (Note 9);
20. Transfer all microcentrifuge tubes into a bench-top centrifuge, and centrifuge for 10 min at the highest speed (12,000 rpm) to precipitate microsomal protein;
21. Collect all supernatants from each microcentrifuge tube, and transfer to a new set of labeled microcentrifuge tubes correspondingly for further sample cleaning by solid phase extraction (SPE) as described below;

3.2 Solid Phase Extraction for Sample Clean

1. Each incubation sample requires one SEP-PAK cartridge packed with 100 mg of sorbent C_{18}; Label cartridges in the same order of incubation samples, and put on a proper cartridge holder with a waste container (**Note 2**);
2. All SEP-PAK cartridges are first activated and conditioned by flushing with 1 mL methanol, and repeat the flushing for four more times with 1 mL methanol each;
3. Before cartridges are completely dried, wash with 1 mL of deionized water to equilibrate cartridges, and repeat the washing for additional four times (**Note 10**);
4. As water is nearly depleted from the cartridges, all supernatants resulting from protein precipitation/centrifugation are individually loaded into the corresponding cartridges;
5. As aqueous is nearly depleted by gravity from the cartridges, every cartridge is washed with 1 mL of water to remove salts and residual proteins, and repeat the washing for additional three times;
6. Remove the waste container from the cartridge holder, and replace with a new set of labeled microcentrifuge tubes to collect eluted samples;
7. Add 1 mL of methanol to each individual cartridge to elute components of interest;
8. Eluted components are dried on a SpeedVac dryer.
9. The dried samples are reconstituted in 150 μL of water-acetonitrile (95:5), and then are ready for MS analyses.

3.3 LC-MS Analysis of Reactive Metabolites

3.3.1 Detecting GSH Conjugates Using a Triple-Quadrupole Mass Spectrometer [7]

1. Check LC-MS system to ensure that all components are interfaced properly;
2. Purge the LC system to remove any potential air bulbs;
3. Warm up the mass spectrometer for at least 60-min;
4. Tune MS instrument in the ESI+ mode by directly infusing 0.1 μM GSH solution or another alternative compound according to the manufacturer's instruction manual, and then save the MS parameters;

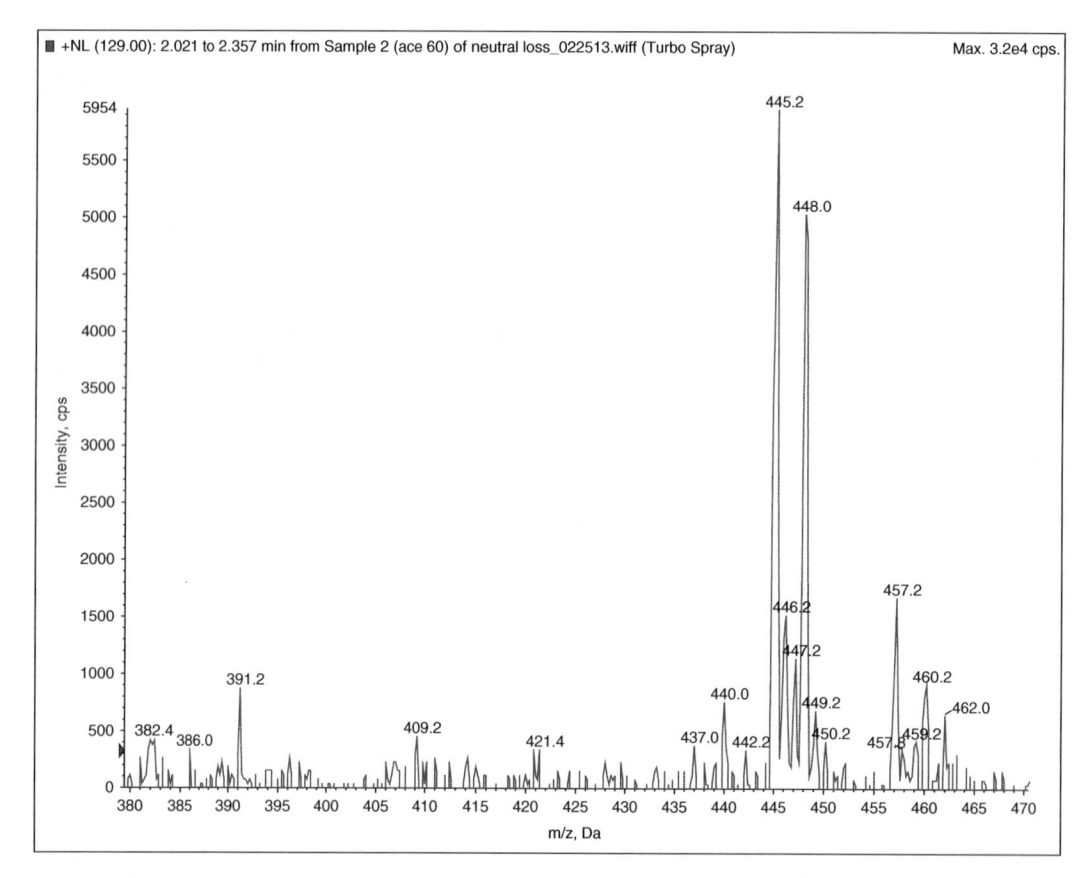

Fig. 2 MS² spectrum of a GSH conjugate detected by neutral loss scan using a triple-quadrupole mass spectrometer. A pair of isotopic ions appeared at m/z 445 and 448 respectively. A mass difference of 3 Da and an expected ratio suggested formation of a GSH conjugate in microsomal incubation

5. Create a LC-MS/MS method using neutral loss scan for 129 Da at low collision energy [7] over a proper mass range estimated by molecular weight of the test compound (**Note 11**).

6. Generic LC mobile phases and gradient profile can be used for LC-MS/MS analysis, and a representative one is given below: An Agilent Zorbax SB C18 column (2.1×50 mm) was used for chromatographic separations. The starting mobile phase consisted of 95 % water (0.5 % acetic acid), and metabolites were eluted using a single gradient of 95 % water to 95 % acetonitrile over 7 min at a flow rate of 0.3 mL/min. At 7 min, the column was flushed with 95 % acetonitrile for 2 min before re-equilibration at initial conditions (**Note 12**);

7. For each sample, an aliquot of 10 μL is injected for LC-MS/MS analysis;

8. After data acquisition is completed, examine total ion chromatogram (TIC) and look for appearance of a doublet with a mass difference of 3 Da and an intensity ratio of 1.0:0.8 approximately such as shown in Fig. 2 [7].

9. If a positive signal is found, MS² spectra are subsequently acquired for both isotopic ions under the same MS conditions;

10. Inspect MS/MS spectra of both doublet ions and look for characteristic ions resulting from CID to confirm formation of reactive metabolites (**Note 13**).

3.3.2 Detecting GSH Conjugates Using Thermo Fisher LCT Ion Trap Mass Spectrometer (San Jose, CA) [10]

1. Check LC-MS system to ensure that all components are interfaced properly;

2. Purge the LC system to remove any potential air bulbs;

3. Warm up the MS at least for 60-min;

4. Tune MS instrument in the ESI+ mode by directly infusing 0.1 μM GSH solution or other alternative compound according to the manufacturer's instruction manual, and then save the MS parameters;

5. Create a LC-MS/MS method using the isotopic MS pattern to selectively trigger data dependent MS² scans of both labeled and non-labeled GSH conjugates under low collision energy [10] over a proper mass range estimated by molecular weight of the test compound (**Note 11**).

6. Generic LC mobile phases and gradient profile can be used for LC-MS/MS analysis, and a representative one is given below: An Agilent Zorbax SB C18 column (2.1 × 50 mm) was used for chromatographic separations. The starting mobile phase consisted of 95 % water (0.5 % acetic acid), and metabolites were eluted using a single gradient of 95 % water to 95 % acetonitrile over 43 min at a flow rate of 0.3 mL/min. At 43 min, the column was flushed with 95 % acetonitrile for 2 min before re-equilibration at initial conditions (**Note 14**);

7. For each sample, an aliquot of 10 μL is injected for LC-MS/MS analysis;

8. After data acquisition is completed, examine MS² spectra of both isotopic molecular ions to look for appearance of characteristic product ions to confirm the identity of GSH conjugates (**Note 13**);

4 Notes

1. Other trapping agents such as semicarbazide, methoxylamine, α-acetyllysine and potassium cyanide can be used to replace glutathione in incubations to trap a wide variety of reactive metabolites.

2. A 96-well plate can be used to run a high volume of compounds simultaneously. As a result, sample preparation will need to be modified accordingly;

3. It is important to make both labeled and non-labeled glutathione fresh to prevent formation of dimers via S-S linkage.

4. Molecular ions of glutathione appear at m/z 308, and the stable isotopic ions appear at m/z 311;

5. It is essential to maintain the ratio of two ions in the range of 1:0.80 to 1:0.6 approximately. Otherwise, GSH conjugates will not be detected by the ion trap mass spectrometer using isotope pattern-triggered MS/MS data acquisition [10]. This requirement is not critical if triple-quadrupole mass spectrometer is used in constant neutral loss mode [7]; Other MS methodologies such as precursor ion scan can also be utilized for detecting GSH conjugates [12].

6. It is highly recommended that some extra HLM solution (e.g. 0.5 mL) is prepared to avoid pipetting errors;

7. It is highly recommended that some extra NADPH solution (e.g. 0.5 mL) is prepared to avoid pipetting errors;

8. It is important to add enough TAC to precipitate microsomal proteins. Incomplete protein removal of microsomal proteins can lead to a loss of GSH conjugates in solid phase extraction.

9. An alternative approach is to precipitate microsomal protein with cold acetonitrile followed by centrifugation and vacuum dry.

10. It is important to avoid air-drying of cartridges for too long which can lead to low recovery of GSH conjugates in SPE cleanup;

11. The mass range is dependent on the molecular weight of the test compound. The low end mass must be greater than 350 Da, and the high end mass can be roughly estimated as compound molecular weight plus 400.

12. The LC-MS method has been described in detail in literature [10]. It is highly recommended that one completely understand the methodology before practicing.

13. Both isotopic labeled and non-labeled GSH conjugate display a characteristic ion resulting from a neutral loss of 129 Da in collision-induced dissociation (Fig. 3). However, some may also exhibit a second characteristic ion resulting from neutral loss of 75 and 78 Da for unlabeled and labeled GSH conjugates, respectively;

14. The isotopic pattern triggered data dependent MS^2 scan has been described in detail in literature [11]. It is highly recommended that one completely understand the methodology before practicing. One must note that ion trap mass spectrometers from other vendors may not have this feature (the isotopic pattern triggered data dependent MS^2 scan).

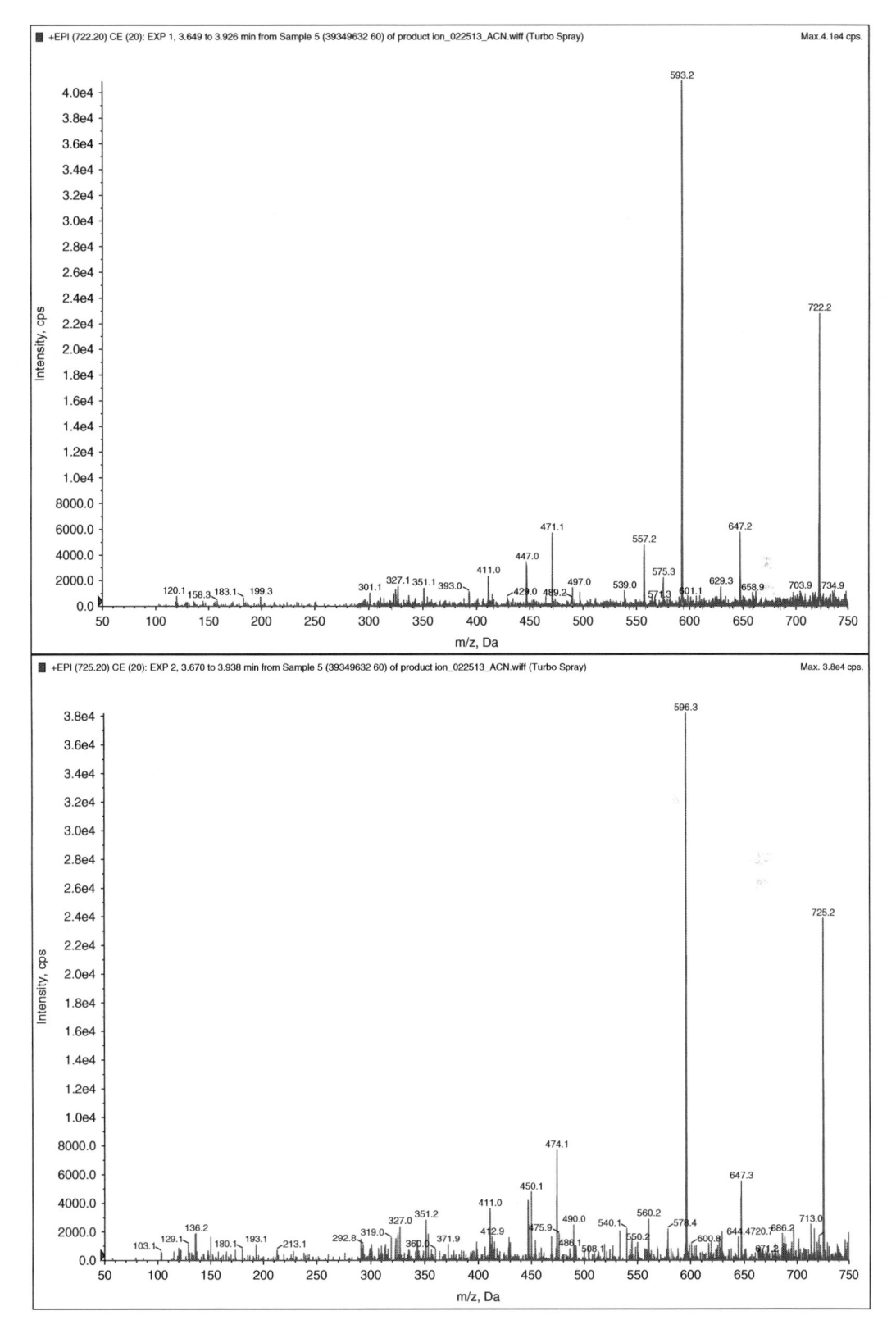

Fig. 3 MS² spectra of both non-labeled and labeled GSH conjugate. For non-labeled conjugate (top, m/z 722), neutral losses of 129 and 75 gave rise of characteristic product ions at m/z 593 and 647, respectively; for labeled GSH conjugate (bottom, m/z 725), neutral losses of 129 and 78 gave rise of characteristic product ions at m/z 596 and 647, respectively

Acknowledgements

The views expressed here are solely those of the author and do not reflect the opinions of Janssen Research & Development, LLC.

References

1. Ulrich RG (2007) Idiosyncratic toxicity: a convergence of risk factors. Annu Rev Med 58:17–34
2. Boelsterli UA (2003) Idiosyncratic drug hepatotoxicity revisited: new insights from mechanistic toxicology. Toxicol Mech Methods 13:3–20
3. Caldwell GW, Yan Z (2006) Screening for reactive intermediates and toxicity assessment in drug discovery. Curr Opin Drug Discov Devel 9:47–60
4. Prakash C, Sharma R, Gleave M, Nedderman A (2008) In vitro screening techniques for reactive metabolites for minimizing bioactivation potential in drug discovery. Curr Drug Metab 9:952–964
5. Kalgutkar AS, Gardner I, Obach RS, Shaffer CL, Callegari E, Henne KR, Mutlib AE, Dalvie DK, Lee JS, Nakai Y et al (2005) A comprehensive listing of bioactivation pathways of organic functional groups. Curr Drug Metab 6:161–225
6. Yan Z, Maher N, Torres R, Huebert N (2007) Use of a trapping agent for simultaneous capturing and high-throughput screening of both "soft" and "hard" reactive metabolites. Anal Chem 79:4206–4214
7. Yan Z, Caldwell GW (2004) Stable-isotope trapping and high-throughput screenings of reactive metabolites using the isotope MS signature. Anal Chem 76:6835–6847
8. Yan Z, Maher N, Torres R, Caldwell GW, Huebert N (2005) Rapid detection and characterization of minor reactive metabolites using stable-isotope trapping in combination with tandem mass spectrometry. Rapid Commun Mass Spectrom 19:3322–3330
9. Wen B, Ma L, Nelson SD, Zhu M (2008) High throughput screening and characterization of reactive metabolites using polarity switching of hybrid triple quadrupole linear ion trap mass spectrometry. Anal Chem 80:1788–1799
10. Yan Z, Caldwell GW, Maher N (2008) Unbiased high-throughput screening of reactive metabolites on the linear ion trap mass spectrometer using polarity switch and mass tag triggered data-dependent acquisition. Anal Chem 80:6410–6422
11. Ma L, Wen B, Ruan Q, Zhu M (2008) Rapid screening of glutathione-trapped reactive metabolites by linear ion trap mass spectrometry with isotope pattern-dependent scanning and postacquisition data mining. Chem Res Toxicol 21:1477–1483
12. Liao S, Ewing NP, Boucher B, Materne O, Brummel CL (2012) High-throughput screening for glutathione conjugates using stable-isotope labeling and negative electrospray ionization precursor-ion mass spectrometry. Rapid Commun Mass Spectrom 26:659–669
13. Rousu T, Pelkonen O, Tolonen A (2009) Rapid detection and characterization of reactive drug metabolites in vitro using several isotope-labeled trapping agents and ultra-performance liquid chromatography/time-of-flight mass spectrometry. Rapid Commun Mass Spectrom 23:843–855
14. Jian W, Liu H-F, Zhao W, Jones E, Zhu M (2012) Simultaneous screening of glutathione and cyanide adducts using precursor ion and neutral loss scans-dependent product ion spectral acquisition and data mining tools. J Am Soc Mass Spectrom 23:964–976

Chapter 29

Quantitative Assessment of Reactive Metabolites

Jie Chen, Rongfang Fran Xu, Wing W. Lam, Jose Silva, and Heng-Keang Lim

Abstract

Quantitation of reactive intermediates from bioactivation of drug via covalent protein binding using radiolabeled drug is the gold standard for quantitation of reactive metabolite formation in the absence of synthetic standard. However, radiolabeling many compounds during lead optimization can be resource intensive and expensive, which led to development of alternative method for quantitation of reactive metabolites using trapping agents as a surrogate to covalent protein binding. Quantitation of reactive metabolite formation using trapping agents can be broadly divided into two categories: (1) Radiolabeled trapping agents such as [^{35}S]-cysteine, [^{35}S]-glutathione, [^{3}H]-glutathione and [^{14}C]-cyanide. In general, the concentration of the radiolabeled adducts of reactive intermediates can be calculated from the peak area in dpm and the specific activity. (2) Non-radiolabeled trapping agents such as dansyl glutathione and quaternary ammonium glutathione analogs. Here quantitation is based on the intrinsic spectroscopic property of the chemical tag, for example, the dansyl glutathione adduct is quantitated using the emission and excitation wavelength of the fluorescence dansyl moiety. On the other hand, quantitation of quaternary ammonium glutathione conjugate is dependent on the similarity in MS responses of the precursor molecular ion and similar efficiency in formation of the product ion containing the quarternary ammonium moiety of analyte and internal standard. In this chapter, we will describe the experimental procedures for quantitation of reactive metabolite formation using these trapping agents.

Key words Reactive metabolite, Quantitative assessment, Trapping agent

1 Introduction

Drug-induced toxicity is the leading cause for failure of drugs in clinics and withdrawal of marketed drugs [1, 2]. In particular, idiosyncratic drug toxicity is not predicted during drug development because the preclinical animal models used are not predictive of human toxicity [3–5]. In general, drugs are typically biotransformed to inert metabolites by phase I and II drug metabolic enzymes. However, these enzymes may occasionally mediate bioactivation of drugs by catalyzing the formation of chemically reactive metabolites, which have potential for irreversibly binding to essential cellular macromolecules such as proteins and DNA [6–9].

Gary W. Caldwell and Zhengyin Yan (eds.), *Optimization in Drug Discovery: In Vitro Methods*, Methods in Pharmacology and Toxicology, DOI 10.1007/978-1-62703-742-6_29, © Springer Science+Business Media New York 2014

Although a direct link between toxicity and bioactivation has not been demonstrated in most drug-induced toxicities, however, it was observed that a majority of the withdrawn drugs were bioactivated to reactive metabolites. As a result, *in vitro* screening for reactive metabolite formation has been widely adopted across the pharmaceutical industry to minimize bioactivation as a strategy to reduce drug failure due to toxicity. The structure elucidation of reactive metabolite is useful for establishment of structure-activity-relationship of bioactivation for the purpose of designing out reactive metabolite formation. However, this qualitative information alone is not adequate for assessment of risk for development of toxicity, which requires quantitation of reactive metabolites.

Amongst the drug metabolizing enzymes, the cytochrome P450 enzymes are frequently associated with bioactivation of drugs and this is not surprising since cytochrome P450 enzymes use reactive activated oxygen species as oxidant. Therefore, this chapter will focus on quantitation of reactive metabolites produced by cytochrome P450 enzymes using liver microsomes. Conventional covalent protein binding study using radiolabeled compound can provide information on the % of the drug bioactivated by quantitation of the irreversible incorporation of radioactivity into NADPH-fortified liver microsomal proteins [10, 11]. This well-established methodology has recently been automated using a cell harvester to collect precipitated liver microsomal proteins and provided adequate throughput for discovery support. The high cost associated with radiolabeling many lead compounds led to development of alternative way for quantitation using less expensive methods involving trapping agents. Historically, nucleophilic trapping agents like glutathione, cysteine and cyanide have been used to trap reactive electrophilic metabolites for structural elucidation to provide mechanistic insight into the bioactivation pathway by cytochrome P450 enzymes. Recently, these nucleophilic trapping agents were modified for quantitation of reactive metabolites and they can be broadly grouped as radiolabeled and non-radiolabeled trapping agents.

Quantitation of reactive metabolite formation using radiolabeled trapping agent has been reported using [^{35}S]-cysteine, [^{35}S]-glutathione, [^{3}H]-glutathione and [^{14}C]-cyanide [12–16]. Reactive electrophilic metabolites were trapped as radioactive adducts followed by quantitation using liquid chromatography-radioactivity detector-mass spectrometry. These radiolabeled nucleophiles have been used for quantitative assessment of bioactivation potential of lead candidates in discovery after validation with selected withdrawn hepatotoxic marketed drugs. Non-radiolabeled nucleophilic trapping agents are essentially glutathione modified with chemical tags for quantitation using the intrinsic spectroscopic property of the chemical tag. Quantitation using dansyl and quarternary ammonium glutathione analogs have been reported for semi-quantitation of reactive metabolites during optimization of lead candidates in discovery [17–19].

2 Materials

2.1 Equipment

Descriptions of specific equipment used in these experiments are given; however, any model of comparable capability can easily be substituted.

1. The LC-RAD-MS (liquid chromatography-radioactivity detector-mass spectrometry) system consisted of an HP-1100 solvent delivery pump, an HP-1100 membrane degasser, an HP-1100 autosampler (Agilent Technologies, Wilmington, DE), coupled with an on-line v. ARC radioactivity detector (AIM Research Company, Hockessin, DE) and an LTQ (Thermo Scientific, Inc., San Jose, CA).

2. LC-FD-MS (liquid chromatography-fluorescence detector-mass spectrometry) system comprised of a Shimadzu LC-10Avp LC, a diode array detector, a fluorescence detector (Shimadzu, Columbia, MD), and an API-4000 Q-trap mass spectrometer (Applied Biosystems, Foster City, CA).

3. Capillary liquid chromatography system (Dionex Corp., Sunnyvale, CA).

4. Quantum triple quadrupole mass spectrometer (Thermo-Finnigan Corp., San Jose, CA).

5. HPLC columns: Eclipse XDB-C18, 150×2.1 mm ID, 5 μm (Agilent Technologies, Wilmington, DE); Prodigy ODS2, 150×4.6 mm ID, 5 μm (Phenomenex, Torrance, CA); Symmetry Shield RP18 column, 250×4.6 mm ID, 5 μm (Waters Corp., Milford, MA) coupled with a guard cartridge, a Symmetry Shield RP18, 20×3.9 mm ID, 5 μm (Waters Corp., Milford, MA), Synergy HydroRP 250×4.6 mm ID, 4 μm, (Phenomenex, Torrance, CA).

6. Millipore A10 Water Purification System (Millipore Corporation, Bedford, MA).

7. LSC-6005364 Liquid scintillation counter (Perkin Elmer, Waltham, MA).

2.2 Reagents and Solutions

1. L-[^{35}S]-Cysteine hydrochloride (100 mCi/mmol) was purchased from GE Healthcare (Piscataway, NJ).

2. [^{14}C]-Potassium Cyanide (53 mCi/mmol) was purchased from American Radiolabeled Chemicals, Inc. (Saint Louis, MO).

3. [^{35}S]-Glutathione (125 mCi/mmol) was obtained from Perkin Elmer (Wellesley, MA).

4. [^{3}H]-Glutathione (20 Ci/mmol) was purchased from American Radiolabeled Chemicals, Inc. (Saint Louis, MO).

5. Quarternary ammonium glutathione analog was available from AnaSpec (Freemont, CA).

6. Magnesium chloride, L-cysteine, ammonium acetate, glutathione, potassium phosphate, NADPH, acetaminophen, caffeine, diclofenac, clozapine, furosemide, indomethacin, nefazodone, flutamide, sulfamethoxazole, zomepirac and carbamazepine were purchased from Sigma Chemical Co. (St. Louis, MO).

7. Pooled human liver microsomes (HLM, n = 50, mixed gender, 20 mg of protein/mL) were obtained from XenoTech (Kansas City, KS).

8. All HPLC grade solvents used (reagent purity was >97 % unless otherwise specified) were purchased from J. T. Baker (Mallinckrodt Baker Inc., Phillipsburg, NJ).

9. StopFlow™ AD scintillation cocktail (AIM Research Company, Hockessin, DE).

3 Methods

3.1 Incubations

3.1.1 In Vitro Liver Microsomal Incubations with Radiolabeled Trapping Agents

The method used for quantitation of radiolabeled adducts of reactive metabolites was adapted from reported procedure for quantitation using [^{35}S]-glutathione [14, 15], [^{3}H]-glutathione [13], L-[^{35}S]-cysteine [13, 16] and [^{14}C]-cyanide [16]. Stock solutions of 10 mM test compounds were prepared in acetonitrile: methanol (9:1 v/v) and an aliquot were spiked directly to keep final organic content less than 1 % (v/v). Typical incubation conducted in 16 mL glass test-tube and consisted of the following solutions added in the order listed below:

1. 780 µL 0.1 M potassium phosphate buffer (pH 7.4)

2. 20 µL of 200 mM magnesium chloride to give final concentration of 4 mM

3. 40 µL of 50 mM EDTA to give final concentration of 2 mM

4. 100 µL of human liver microsomes (20 mg/mL) to give final concentration of 1 mg/mL

5. 10 µL of 10 mM test compound to give final concentration of 100 µM

6. Trapping agent (0.2 mM [^{35}S]-glutathione; 25 µCi/tube, 0.2 mM [^{3}H]-glutathione; 52 µCi/tube, 0.2 mM L-[^{35}S]-cysteine hydrochloride; 20 µCi/tube or 0.1 mM [^{14}C]-potassium cyanide; 25 µCi/tube) (**Note 1**).

The amount of each radiolabeled trapping agent used is a balance between trapping efficiency and its contribution to background. The radioactivity of each trapping agent required for incubation was calculated from its specific activity and the targeted incubation concentration (**Note 2**). For example, a final concentration of 0.2 mM or 31.4 µg/mL L-[^{35}S]-cysteine hydrochloride (specific activity of 100 mCi/mmol or 100 µCi/157 µg) was required for incubation, and calculated to contain 20 µCi [(31.4/157)×100 µCi] L-[^{35}S]-cysteine hydrochloride. The tubes were preincubated at 37 °C for 5 min, and the reactions were initiated by addition of 50 µL of 100 mM NADPH to give a final concentration of 5 mM NADPH and a final volume of 1.0 mL in each tube. It has been observed that addition of [^{14}C]-cyanide at concentration higher than 0.1 mM inhibited human liver microsomal metabolism and therefore, the rationale for using this concentration in this assay.

Negative control corresponded to replacement of test compound with same volume of vehicle solution (acetonitrile:methanol, 9:1 v/v) or replacement of NADPH with same volume of 0.1 M potassium phosphate buffer, pH 7.4. The minus test compound was used as a negative control in calculation (**Note 3**). The reaction in each tube was terminated by the addition of 3 mL of ice-cold acetonitrile after incubation at 37 °C for 60 min. Then 1 mL of 1 M ammonium acetate solution was added to the resulting mixtures to improve extraction of radiolabeled adducts [16]. After centrifugation of the mixtures (room temperature; 10,000 g, 5 min), the supernatants were removed and transferred to another tube. The protein pellet was re-extracted one more time as described above and the supernatants were from same tube were pooled together prior to drying under a gentle nitrogen stream at room temperature. The residue was reconstituted in 200 µL of water:acetonitrile (9:1 v/v) followed by sonication and vortexing prior to filtration using a 0.45 µm nylon filter at 20,000 g at room temperature for 5 min. The filtrate was transferred to a 96-well plate for immediate LC-RAD-MS analysis to minimize degradation of reactive metabolites (**Note 4**).

3.1.2 In Vitro Liver Microsomal Incubations with Non-radiolabeled Trapping Agent

Dansyl Glutathione

The method used for quantitation of dansyl glutathione adducts of reactive metabolites following trapping with dansyl glutathione was adapted from published method [17, 18]. The dansyl glutathione is not commercially available and synthesis is briefly described here. Dansyl glutathione was synthesized by reaction of 0.25 mM oxidized glutathione (GSSG) with 0.5 mM dansyl chloride to form dansyl GSSG. The subsequent reduction of the disulfide bond of dansyl GSSG (0.26 mM) with dithiothreitol (0.5 mM) yielded dansylated glutathione for purification by HPLC before use [17].

Solutions of 10 mM test compounds were prepared as described in Sect. 3.1.1. Typical incubation conducted in 16-mL glass test-tube and consisted of the following solutions added in the order listed below:

1. 785 µL 0.1 M potassium phosphate buffer (pH 7.4)
2. 20 µL of 200 mM magnesium chloride to give final concentration of 4 mM

3. 40 μL of 50 mM EDTA to give final concentration of 2 mM

4. 100 μL of human liver microsomes (20 mg/mL) to give final concentration of 1 mg/mL

5. 5 μL of 10 mM test compound to give final concentration of 50 μM

The tubes were preincubated for at 37 °C for 5 min and the reactions were initiated by the addition of 50 μL of 20 mM dansyl glutathione (dGSH) to give a final concentration of 1 mM dGSH and a final volume of 1.0 mL in each tube. The lower concentration of dansyl glutathione used is to minimize interference from fluorescence background while maintaining trapping efficiency. Samples without substrate or dGSH were used as blanks or controls, respectively. The reaction in each tube was terminated by the addition of three volumes of ice-cold methanol containing 5 mM dithiothreitol at the end of incubation at 37 °C for 30 min (**Note 5**). Dithiothreitol was added to minimize oxidation of dansyl glutathione to dansyl GSSG to minimize interference by dansyl GSSG peak. The tubes were vortexed and centrifuged as described in Sect. 3.1.1, and the resulting supernatants were transferred to new tubes for overnight storage at −80 °C prior to analysis by LC-FD-MS (**Note 6**). The overnight storage of samples at −80 °C before LC-FD-MS analysis was performed to reduce interference from dansyl GSSG.

Quaternary Ammonium Glutathione

The method used for quantitation of quarternary ammonium glutathione adducts of reactive metabolites following trapping with quarternary ammonium glutathione was adapted from previous published method [19]. Unlike dansyl glutathione, the quarternary ammonium glutathione analog is commercially available from AnaSpec. Solutions of 10 mM test compounds were prepared in acetonitrile/methanol (9:1, v/v) as described in Sect. 3.1.1.

Typical incubation conducted in 16-mL glass test-tubes and consisted of the following solutions added in the order listed below:

1. 785 μL 0.1 M potassium phosphate buffer (pH 7.4)

2. 20 μL of 200 mM magnesium chloride to give final concentration of 4 mM

3. 40 μL of 50 mM EDTA to give final concentration of 2 mM

4. 100 μL of human liver microsomes (20 mg/mL) to give final concentration of 1 mg/mL

5. 5 μL of 10 mM test compound to give final concentration of 50 μM

6. 20 μL of 50 mM quarternary ammonium glutathione (QA-GSH) to give final concentration of 1 mM

The tubes were preincubated for at 37 °C for 5 min and the reactions were initiated by the addition of 50 µL of 100 mM NADPH to give a final concentration of 5 mM NADPH and a final volume of 1.0 mL in each tube. Samples without substrate or NADPH added were used as negative controls. After 45 min of incubation at 37 °C, 3 mL of acetonitrile containing 0.1 µM N-(1-pyrenyl)maleimide-QA-GSH conjugate as internal standard (IS) was added to the incubations, and then centrifuged as described in Sect. 3.1.1. The supernatant was transferred to another tube prior to evaporation to dryness under a gentle stream of nitrogen at room temperature. The residue was reconstituted in 200 µL of water:acetonitrile (9:1, v/v) with sonication and vortexing prior to filtration using a 0.45 µm nylon filter at 20,000 g at room temperature for 5 min. The filtrate was transferred to a 96-well plate for LC-ESI-MS analysis.

3.2 Analytical Procedure

3.2.1 Radioactivity Detection of Reactive Metabolites Trapped by Radiolabeled Trapping Agents

All analyses were carried out using the LC-RAD-MS system to simultaneously quantitate and identify the radiolabeled adducts. Chromatographic separations were performed on a Symmetry Shield RP18 column, 250×4.6 mm ID, 5 µm (Waters Corp., Milford, MA) coupled with a guard cartridge, a Symmetry Shield RP18, 20×3.9 mm ID, 5 µm (Waters Corp.) eluted with a gradient from 10 mM ammonium acetate/H_2O/acetonitrile = 10/85/5 (v/v/v, solvent A) to 10 mM ammonium acetate/acetonitrile = 1/9 (v/v, solvent B) at a flow rate of 1 mL/min over 85 min, the eluant from the LC column was split post-column with 200 µL/min of the flow into the MS and 800 µL/min of the flow into RAD. Detail of the LC gradient was described in Table 1.

The initial 6.5-min eluant from the HPLC was diverted to waste to eliminate residual unreacted [^{35}S]-glutathione, [^{3}H]-glutathione, [^{35}S]-cysteine and [^{14}C]-cyanide so as to maintain a low background baseline for the remaining run. It is important to countercheck radioactivity with MS data to minimize false negative due to similar retention time of conjugate as dGSH.

It is important to emphasize the need for baseline chromatographic separation of remaining radiolabeled trapping agent from radiolabeled adducts for successful quantitation using this method. Alternatively, removal of remaining radiolabeled trapping agent by SPE may be needed to improve quantitation. The RAD was operated in the homogenous liquid scintillation dynamic flow counting mode with the addition of 0.2 mL/min of StopFlow™ AD scintillation cocktail (AIM Research Company, Hockessin, DE) to the eluant and mixed prior to detection by RAD. Quantitation of the radiolabeled adducts were carried out by integration of the individual chromatographically separated radioactivity peaks using the ARC software ARC101 Evaluate. The formation rate of each radiolabeled adducts (^{35}S-glutathionyl RM, ^{3}H-glutathionyl RM, ^{35}S-cysteinyl RM, and ^{14}C-cyano RM, where RM refers to reactive

Table 1
LC gradient

Time (min)	% of Solvent B
0	0
10	0
40	30
60	100
75	100
75.1	0
85	0

metabolite) was calculated according to the following equations below using ^{35}S-glutathionyl RM [RM-GS] as an example:

$$RM\text{-}GS\ formation\,[pmol] = Area_{GS[counts]} / f_{[counts/dpm]} / SA_{^{35}S\text{-}GSH\,[dpm/pmol]} \quad (1)$$

where $SA_{^{35}S\text{-}GSH\,[dpm/pmol]}$: specific activity of ^{35}S-GSH used for incubation and expressed as dpm/pmol;

$$RM\text{-}GS\ formation\ rate\,[pmol/\min/mg\ protein] = Area_{GS[counts]} / Area_{total[counts]} \times 100\% \quad (2)$$

$$Area_{total} = E\!f\!fective\ radioactivity\ injected_{[dpm]} \times f_{[counts/dpm]} \quad (3)$$

$$E\!f\!fective\ radioactivity\ injected_{[dpm]} = total\ radioactivity\ injected_{[dpm]} \times K \quad (4)$$

$$K = Area_{all\ peak} - Area_{excess\ trapping\ agent\ peak} / Area_{all\ peak} \quad (5)$$

Area $_{GS\,[counts]}$: the sum of radioactivity counts of RM-GS peaks observed in the radio-chromatogram. Area $_{total\,[counts]}$: the total radioactivity counts corresponded to all radiolabeled adduct peaks in the radio-chromatogram excluding unreacted trapping agent peak. $f_{[counts/dpm]}$: correction constant was calculated from the total radioactivity counts observed on the radio-chromatogram (LC with no flow divert) of a 20-fold diluted sample compared to its radioactivity counts measured using a liquid scintillation counter. The unreacted trapping agent peak area was also measured and the K constant was calculated described above.

The radiolabeled peaks were analyzed using an ion trap mass spectrometer operated in the positive ion electrospray mode. The ion spray voltage, capillary voltage and tube lens offset were optimized for highest sensitivity of detection of drug-derived compounds by tee-infusion of 10 ng/μL of test compound into LC flow of 50:50 mobile phase A and B. The LC eluant was sprayed into the mass spectrometer at 5 kV spray voltage and with other parameters such as sheath, auxiliary and countercurrent gas set at 65, 15 and 5 arbitrary units, respectively, and heated capillary temperature kept at 275 °C. The adducts of reactive metabolites were detected using mass list or isotopic pattern triggered data-dependent multiple-stage mass analysis with isolation width of 2 Da, normalized collision energy of 25, 30, and 35 for MS^2, MS^3 and MS^4, respectively, an activation q of 0.25 and activation time of 30 ms. The normalized collision energy can be adjusted to give more than 80 % attenuation of precursor ion. Data acquisition and reduction was carried out using Xcalibur™ 2.0 (Thermo Fisher Scientific, Inc., San Jose, CA). The radiolabeled peaks were identified as adducts of reactive metabolites from diagnostic neutral mass losses like 75, 129 and 307 Da in product ion mass spectrum.

This method was validated using "problematic drugs" reported to undergo bioactivation to reactive metabolites such as acetaminophen, carbamazepine, flutamide, furosemide, indomethacin, nefazodone, procainamide, sulfamethoxazole, tienilic acid, and zomepirac [15]. "Safe drugs" such as levofloxacin, moxifloxacin and caffeine, which are not associated with bioactivation, were tested for comparison. The formation rates of RM-GS from drugs tested are summarized in Fig. 1. For the problematic drugs tested, the calculated formation rates of RM-GS of ticlopidine, nefazodone, clozapine, and acetaminophen were relatively high and corresponded to 47, 37, 20, and 18 pmol/min/mg protein,

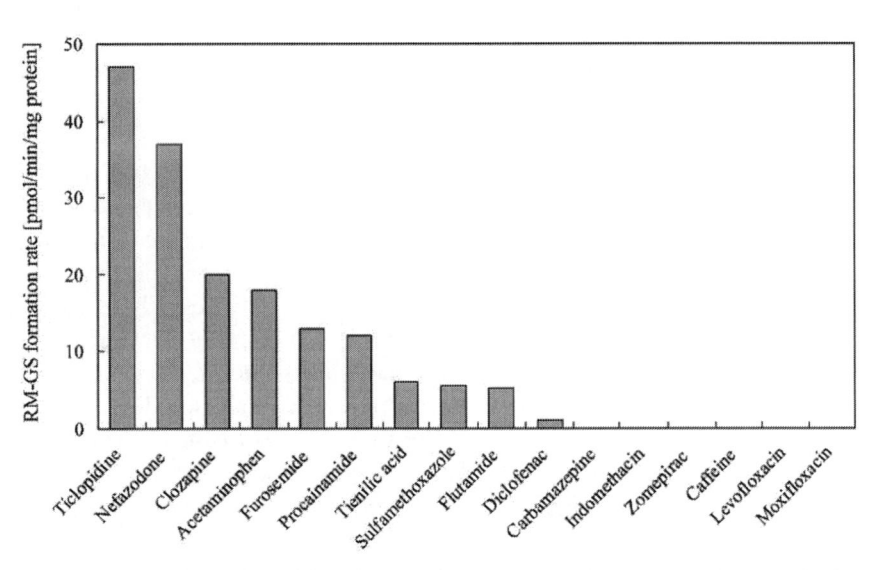

Fig. 1 The formation rates of RM-GS of marketed drugs reported to undergo metabolic bioactivation and compared to drugs known not to undergo metabolic bioactivation

Table 2
LC gradient

Time (min)	% of Solvent B
0	10
30	60
33	100
35	100
35.1	10
40	10

respectively. However, no RM-GS peak was detected in the radio-chromatograms for carbamazepine, indomethacin, and zomepirac, which are also known to form reactive metabolites. Importantly, the safe drugs such as levofloxacin, moxifloxacin and caffeine did not produce any RM-GS peak in their radio-chromatograms. Furthermore, the quantitation of radiolabeled glutathione adducts using [^{35}S]-glutathione gave reasonably good correlation with covalent protein binding using radiolabeled drug and suggests that indeed quantitation using radiolabeled trapping agent is a valid surrogate for covalent protein binding [15].

3.2.2 Fluorescent Detection of Reactive Metabolites Trapped by Dansyl Glutathione

All analyses were carried out using the LC-FD-MS system to simultaneously quantitate and identify the dansyl glutathione adducts. The chromatographic separation was carried out using a reverse phase LC column (Synergy HydroRP 250×4.6 mm ID, 4 μm, Phenomenex, Torrance, CA) using a 5 mM ammonium acetate buffer containing 0.02 % formic acid as mobile phase A and acetonitrile as mobile phase B at a flow rate of 1.0 mL/min (**Note 7**). The LC gradient was described in Table 2.

It is important that the LC gradient system provides baseline separation of remaining dansyl glutathione from dansyl glutathione adducts of reactive metabolites because the detection of dansyl glutathione adducts is based on the appearance of new peaks in the fluorescence chromatograms. Also, this is needed for accurate quantitation of adducts. The LC flow rate of 1 mL/min was split post-column with 0.75 and 0.25 mL/min diverted to the fluorescence detector and the mass spectrometer, respectively. The metabolic turnover of test compound was calculated by dividing its peak area at the UV λ_{max} from incubation in the presence of NADPH and dansyl glutathione over that of control incubation in the absence of NADPH and dansyl glutathione and multiplies by 100. The fluorescent detector was set at an excitation wavelength (λ_{ex}) of 340 nm and emission wavelength (λ_{em}) of 525 nm specifically to monitor for the dansyl moiety. The peaks detected by fluorescent detector are confirmed as dansyl glutathione adducts by mass spectrometry prior to inclusion for quantitation. The concentration of the detected dansyl glutathione adduct was calculated by comparing the peak area of the dansyl glutathione adduct against the external standard curve generated using dGSH. The external standard curve covering concentrations from 0, 1, 2, 4, 8, 16, 32 and 64 μM was prepared by spiking an aliquot of methanolic working solution of dGSH into methanol containing 5 mM DTT for termination of reaction of control incubate without test compound and processed as described previously.

The peaks detected by fluorescent detector are confirmed by mass spectrometry in both positive and negative electrospray ionization modes (**Note 8**). The deprotonated dansyl moiety (5-dimethylamino-1-naphthalenesulfinic acid at m/z 234) is used as the product ion for multiple-reaction monitoring (MRM) in negative ionization mode, and the product ions from neutral loss (NL) of dansyl glutamic acid (362 Da) were used for MRM in the positive ionization mode. The dansyl glutathione adducts detected by both MRM triggered data-dependent product ion scans (MRM-EPI) were used for structural elucidation of the dansyl glutathione adducts quantified by the fluorescence detector. In general, the negative MRM is more sensitive than the positive MRM for detection of most of the dansyl glutathione adducts tested, however, positive product ion scans gave structurally more informative fragmentation than negative.

3.2.3 ESI-MS Detection of Quaternary Ammonium Glutathione Conjugates

All chromatography was performed using a capillary liquid chromatography system with a modified autosampler for injecting samples from 96 well plates. The chromatography column consisted of a Waters Symmetry Shield, RP-18, 50 × 1 mm ID column packed with 3.5 μm particles. The mobile phase consisted of solvent system A, 5 mM ammonium formate containing 0.05 % (v/v) formic acid, and solvent system B, acetonitrile. The flow rate used was 50 μL/min. A limited dispersion injection technique was used, and the injection volume was 2 μL. The LC gradient used for chromatographic separation of quaternary ammonium glutathione adducts started out at 5 % B and ramped to 60 % B within 0.2 min and held isocratically for 10 min before ramped down to initial condition of 5 % B for equilibration for another 10 min before the next injection. This resulted in a total cycle time per analysis of 20 min. Detail of the LC gradient was described in Table 3.

MS analyses were performed using a Quantum triple quadrupole mass spectrometer equipped with an orthogonal electrospray ionization (ESI) source fitted with a 30 gauge stainless steel needle for efficient and stable ESI performance. The LC eluant was sprayed at 3.8 kV and nebulized with a nitrogen sheath gas at 8 psi and desolvation aided by source transfer capillary temperature at 250 °C. Collision-induced dissociation for tandem mass spectrometry

Table 3
LC gradient

Time (min)	% of Solvent B
0	5
0.2	60
10	60
10.2	5
20	5

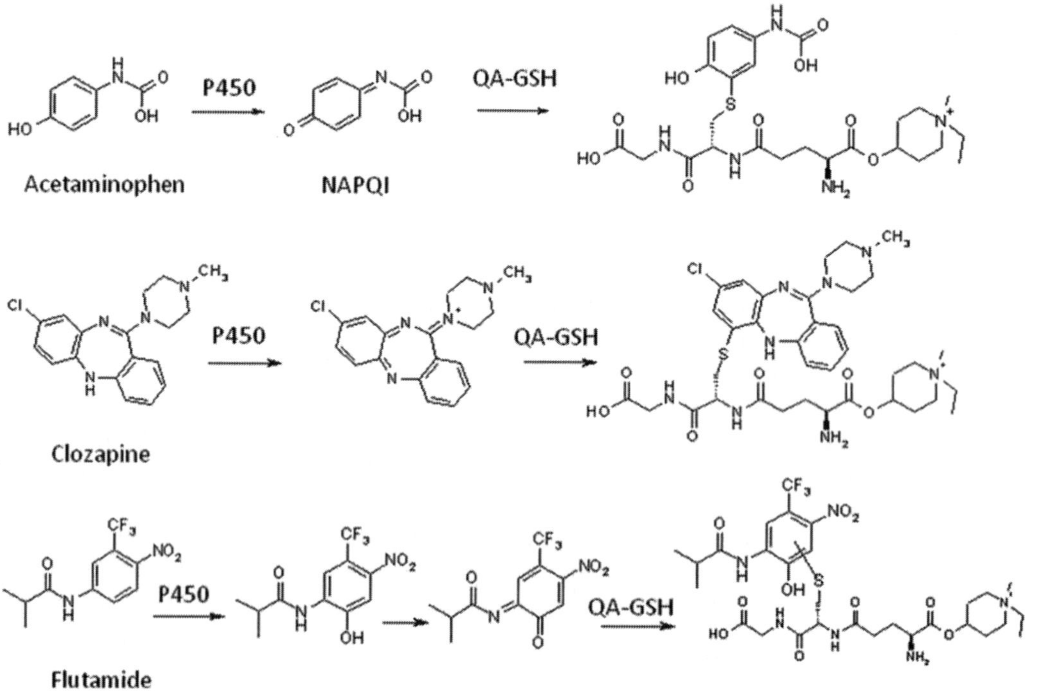

Scheme 1 Postulated bioactivation pathways of model compounds used in the analysis

analysis was carried out at Q2 offset voltage of 55 V and collision cell pressure of 2.0 mTorr. The resolution for Q1 and Q3 was set at 0.7 peak width half-height.

Three model drugs such as acetaminophen, clozapine and flutamide were chosen as positive controls based on their well-defined toxicity and bioactivation by human liver microsomes to reactive metabolites. The bioactivation pathways leading to the formation of potential quarternary ammonium glutathione conjugates with acetaminophen, clozapine and flutamide are shown in Scheme 1 based on published analogous reactions with glutathione [20–22].

The comparison of the MS responses of the molecular ion [M⁺] of quarternary ammonium glutathione adducts of several compounds gave only ~3-folds difference in MS responses, which led to conclusion that the MS response of each compound is based on the preformed charged of the quarternary ammonium nitrogen and little influence by the diverse structures of the parent drug or less susceptible to matrix suppression effect.

The quarternary ammonium glutathione adduct (QA-GS-adduct) was semi-quantitated based on IS approach using N-(1-pyrenyl)maleimide-QA-GSH as an internal standard. The precursor corresponding to M⁺ or MH⁺² ion of the QA-GS-adduct was determined experimentally using a precursor ion scan for formation of product ion at m/z 144 (4-hydroxy-N-methylethyl-piperidinium ion) during CID. Then a MRM method was set up to monitor MS transition corresponding to M⁺ or MH⁺² ion to m/z 144 for both analyte and IS. The amount of QA-GS-adduct formed was calculated as described below:

$$C_{RM\ QAGS} = PA_{RM\ QAGS} \times C_{IS} / PA_{IS} \qquad (6)$$

where $C_{RM\text{-}QAGS}$: concentration of quarternary ammonium glutathione adduct of reactive metabolite, $PA_{RM\text{-}QAGS}$: peak area of quarternary ammonium glutathione adduct of reactive metabolite, C_{IS}: concentration of internal standard and PA_{IS}: peak area of internal standard. This certainly overcomes the need to synthesize reference standard. Furthermore, this method does not have the drawback of high background from remaining trapping agent associated with using radiolabeled or dansyl trapping agents described above.

4 Conclusions

The quantitative assessments of reactive metabolite formation using radiolabeled and non-radiolabeled trapping agents are valid alternative method for quantitation of reactive metabolite by covalent protein binding using radiolabeled compound. These alternative methods are useful tool in discovery for minimizing the risk of drug-induced toxicities including assessment of development liability by calculation of body burden of reactive metabolites as follows:

$$D_{rm} = D \times f_a \times f_m \times f_{rm} \qquad (7)$$

where D_{rm} is the daily burden of reactive metabolites, D is the total daily dose (mg/day), f_a is fraction absorbed, f_m is fractional clearance via oxidative metabolism, and f_{rm} is the fraction of oxidative metabolism leading to adduct formation [18].

The body burden of reactive metabolites is one critical piece of drug metabolism sciences data requires for prioritization of lead candidate into development in the event that bioactivation to reactive metabolite cannot be design out in discovery. All methods discussed in this chapter have limitations and using data generated under correct context can be a viable strategy for mitigating bioactivation risk during lead optimization and lead candidate selection for development.

5 Notes

1. Higher concentration than 0.1 mM [^{14}C]cyanide could inhibit the metabolic reaction, the concentration 0.1 mM was used in this assay.

2. The radioactivity of the trapping agents used for experiments were determined by its specific activity and incubation concentration.

3. The (−) substrate incubation was adopted as a negative control not the (−) NADP$^+$ incubation.

4. Processed samples should keep in refrigerator and analysis them the same day as the incubation was conducted to prevent reactive metabolites degrade.

5. Dithiothreitol (DTT) was added along with the reaction quenching solution to prevent dansyl glutathione (dGSH) oxidation.

6. It was noted that during the incubation and subsequent analysis period, dGSH underwent oxidation resulting in a dansylated GSSG. Store the samples at −80 °C before LC/MS analysis will reduce dGSSG interference.

7. The HPLC gradient system used should have adequate separation of dGSH from dGSH trapped reactive metabolites. This is because the detection of adducts is based on the appearance of new peaks in the fluorescence chromatograms.

8. The drugs with positive signals were subsequently evaluated with control experiments without the addition of either NADPH or dGSH. The use of a step LC gradient also ensured that the organic solvent composition of the mobile phase remained constant so differences in MS detection responses resulting from ESI solvent effects were minimized.

Acknowledgements

The views expressed here are solely those of the author and do not reflect the opinions of Janssen Research & Development, LLC.

References

1. Liebler DC, Guengerich FP (2005) Elucidating mechanisms of drug-induced toxicity. Nat Rev Drug Discov 4:410–420
2. Walgren JL, Mitchell MD, Thompson DC (2005) Role of metabolism in drug-induced idiosyncratic hepatotoxicity. Crit Rev Toxicol 35:325–361
3. Baillie TA (2006) Future of toxicology-metabolic activation and drug design: challenges and opportunities in chemical toxicology. Chem Res Toxicol 19:889–893
4. Evans DC, Baillie TA (2005) Minimizing the potential for metabolic activation as an integral part of drug design. Curr Opin Drug Discov Devel 8:44–50
5. Amacher DE (2006) Reactive intermediates and the pathogenesis of adverse drug reactions: the toxicology perspective. Curr Drug Metab 7:219–229
6. Doss GA, Baillie TA (2006) Addressing metabolic activation as an integral component of drug design. Drug Metab Rev 38:641–649
7. Tang W (2007) Drug metabolite profiling and elucidation of drug-induced hepatotoxicity. Expert Opin Drug Metab Toxicol 3:407–420
8. Caldwell GW, Yan Z (2006) Screening for reactive intermediates and toxicity assessment in drug. Curr Opin Drug Discov Devel 9:47–60
9. Uetrecht J (2003) Screening for the potential of a drug candidate to cause idiosyncratic drug reactions. Drug Discov Today 8:832–837
10. Day SH, White R, Schulz-Utermoehl T, Miller R, Beconi MG (2005) A semi-automated method for measuring the potential for protein covalent binding in drug discovery. J Pharmacol Toxicol Methods 52:278–285
11. Evans DC, Watt AP, Nicoll-Griffith DA, Baillie TA (2004) Drug-protein adducts: an industry perspective on minimizing the potential for drug bioactivation in drug discovery and development. Chem Res Toxicol 17:3–16
12. Masubuchi N, Makino C, Murayama N (2007) Prediction of in vivo potential for metabolic activation of drugs into chemically reactive intermediates: correlation of in vitro and in vivo generation of reactive intermediates and in vitro glutathione conjugate formation in rats and human. Chem Res Toxicol 20:455–464
13. Mulder GJ, Le CT (1988) A rapid, simple in vitro screening test, using [^{3}H]glutathione and L-[^{35}S]cysteine as trapping agents, to detect reactive intermediates of xenobiotics. Toxicol In Vitro 2:225–230
14. Hartman NR, Cysyk RL, Bruneau-Wack C, Thenot JP, Parker RJ, Strong JM (2002) Production of intracellular ^{35}S-glutathione by rat and human hepatocytes for the quantification of xenobiotics reactive intermediates. Chem Biol Interact 142:43–55
15. Takakusa H, Masumoto H, Makino C, Okazaki O, Sudo K (2009) Quantitative assessment of reactive metabolite formation using ^{35}S-labeled glutathione. Drug Metab Pharmacokinet 24(1):100–107
16. Inoue K, Shibata Y, Ttakahashi H, Ohe T, Chiba M, Ishii Y (2009) A trapping method for semi-quantitative assessment of reactive metabolite formation using [^{35}S]Cysteine and [^{14}C]Cyanide. Drug Metab Pharmacokinet 24(3):245–254
17. Gan J, Harper TW, Hsueh MM, Qu Q, Humphreys WG (2005) Dansyl glutathione as a trapping agent for the quantitative estimation and identification of reactive metabolites. Chem Res Toxicol 18:896–903
18. Gan J, Ruan Q, He B, Zhu M, Shyu WC, Humphreys WG (2009) In vitro screening of 50 highly prescribed drugs for thiol adduct formations. Comparison of potential for drug-induced toxicity and extent of adduct formation. Chem Res Toxicol 22:690–698
19. Soglia RJ, Contillo GL, Kalgutkar SA, Zhao S, Hop ECAC, Boyd GJ, Cole JM (2006) A semi-quantitative method for the determination of reactive metabolite conjugate levels in vitro utilizing liquid chromatography-tandem mass spectrometry and novel quaternary ammonium glutathione analogues. Chem Res Toxicol 19:480–490
20. Hinson JA, Reid AB, McCullough SS, James LP (2004) Acetaminophen-induced hepatotoxicity: role of metabolic activation, reactive oxygen/nitrogen species, and mitochondrial

permeability transition. Drug Metab Rev 36:805–822

21. Maggs JL, Williams D, Pirmohamed M, Park BK (1995) The metabolic formation of reactive intermediates from clozapine, a drug associated with agranulocytosis in man. J Pharmacol Exp Ther 275:1463–1475

22. Berson A, Wolf C, Chachaty C, Fisch C, Fau D, Eugene D, Loeper J, Gauthier JC, Beaune P, Pompon D (1993) Metabolic activation of the nitroaromatic antiandrogen flutamide by rat and human cytochromes P-450, including forms belonging to the 3A and 1A subfamilies. J Pharmacol Exp Ther 265:366–372

Chapter 30

In Vitro Assessment of the Reactivity of Acyl Glucuronides

Rongfang Fran Xu, Wing W. Lam, Jie Chen, Michael McMillian, Jose Silva, and Heng-Keang Lim

Abstract

Methodology is described to evaluate the reactivity of acyl glucuronides of carboxylic acid-containing compounds. In this chapter, zomepirac is presented as an example where the reactivity is determined in two complimentary ways: (1) Acyl glucuronides are prepared with UDPGA-fortified microsomes and 1-O-β acyl glucuronide is established based on its sensitivity to hydrolysis by β-glucuronidase. The chemical degradation of 1-O-β isomer in 0.1 M phosphate buffer at pH 7.4 is monitored over time for acyl migration by LC-MS/MS analyses. Half-life of disappearance of 1-O-β acyl glucuronide is used as an index of reactivity. (2) Another index of reactivity of acyl glucuronide is from covalent binding to a protein. Acyl glucuronides are incubated with human serum albumin for 2 h. After pelleting and extensively washing the protein pellets, covalent binding is determined by quantitation of the released aglycone from alkaline hydrolysis.

Key words Acyl glucuronides, In vitro, Acyl migration, Reactivity, Covalent binding to protein

1 Introduction

Carboxylic acid-containing drugs contributed approximately 14 % of the prescribed drugs that were withdrawn worldwide from 1960 to 1999 because of safety issues [1]. The carboxylic acid functionality has been reported to form reactive coenzyme A thioesters and acyl glucuronides, which have been postulated to be the bioactivation pathways for carboxylate drugs [2]. Acyl glucuronidation is the major metabolic conjugation reaction of most carboxylic acid-containing drugs in mammals and it is responsible for their elimination from the body via both biliary and urinary excretions. Although glucuronidation is generally considered a detoxification route of drug metabolism, the chemical reactivity of acyl glucuronides has been linked with the toxic properties of drugs that contain carboxylic acid moieties. Some of these acyl glucuronides circulate systemically instead of being eliminated and may exceed the threshold required for safety testing as mentioned in the regulatory guidance for safety testing of metabolites (MIST and ICH/M3 guidances) [3, 4]. To better characterize the reactivity of acyl glucuronides, it was necessary to investigate the in vitro chemical degradation of 1-O-β glucuronide of carboxylic acid drugs and quantitate the extent of covalent binding of acyl glucuronide to proteins.

Zomepirac is a good example in this group of compounds, as its acyl glucuronide is known to be reactive, and this drug was withdrawn from the market due to anaphylactic reactions. In this chapter, we describe experimental procedures for in vitro assessment of reactivity of acyl glucuronides using zomepirac acyl glucuronide as a typical example.

2 Materials

2.1 In Vitro Biosynthesis

1. Ammonium formate, formic acid, 1 M phosphate buffer, DMSO (dimethyl sulfoxide), zomepirac and clozapine were purchased from Sigma-Aldrich (St. Louis, MO).
2. EDTA: Ethylenediaminetetraacetic acid disodium salt, E-7889, Sigma-Aldrich (St. Louis, MO).
3. $MgCl_2$: Magnesium chloride solution, for molecular biology, 1.00 M, M1028, Sigma-Aldrich (St. Louis, MO).
4. UDPGA: Uridine 5′-diphosphate glucuronic acid, triammonium salt, U5625, Sigma-Aldrich (St. Louis, MO).
5. UDPAG: Uridine 5′-diphospho-N-acetylglucosamine sodium salt, U4375, Sigma-Aldrich (St. Louis, MO).
6. HLM: Human liver microsome, BD science (Bedford, MA).
7. HSA: Human Serum Albumin, Sigma-Aldrich (St. Louis, MO).

2.2 Analytical Method

1. Ammonium acetate, KOH, HCl and trifluroacetic acids were purchased from Sigma-Aldrich (St. Louis, MO).
2. HPLC grade methanol, acetonitrile and water were obtained from EMD Chemicals, Inc (Gibbstown, NJ).
3. Kinetex PFP (150 × 2.1 mm ID, 2.6 µm, 100 Å) (Phenomenex, Torrance, CA), inline filter and guard cartridge were purchased from Thermo Fisher Scientific, Inc., (Bellefonte, PA).
4. SPE (solid phase extraction) cartridge: Strata X (Phenomenex, Torrance, CA).
5. LC- UV-MS: LTQ (Thermo Scientific, Inc., San Jose, CA).

3 Methods

3.1 LC-MS Analysis

All LC-MS analyses were carried out using an Accela LC (Thermo Scientific, Inc., San Jose, CA) coupled to an LTQ (Thermo Scientific, Inc., San Jose, CA). All mass spectrometers were operated in the positive electrospray ionization mode. An aliquot of 10–50 µL of reconstituted sample was injected for chromatographic separation of zomepirac and its acyl glucuronides by a gradient elution with 10 mM ammonium acetate and acetonitrile containing 0.02 % trifluoroacetic acid. The gradient elution was carried out at a flow rate of 0.2 mL/min. Details of the LC gradient are described in Table 1.

Chromatographic separation of the unchanged drug and its acyl glucuronide metabolites was achieved using a Kinetex PFP column (150 × 2.1 mm ID, 2.6 µm, 100 A) kept at 50 °C. It is important to emphasize the need for baseline chromatographic separation to present cross-talk from upfront collision induced dissociation (CID). The eluant was nebulized with sheath and auxillary gas set at 70 and 20 arbitrary units, respectively, and with a counter current sweep gas at 5 arbitrary units to minimize contamination of the atmospheric pressure interface of the mass spectrometer. Desolvation of the solvent droplets was aided by heated capillary temperature of 275 °C. The mass spectrometer was optimized by infusion of 10 ng/mL of zomepirac in 10:90 acetonitrile:water directly into a 50:50 mobile phase (A and B) at 0.2 mL/min. The unchanged drug and its acyl glucuronide metabolites were detected using product ion mass analysis using an isolation width of 2 Da, normalized collision energy of 20 % for MS^2, respectively, activation q of 0.25, and an activation time of 30 msec. Ions were detected with electron multiplier 1 and 2 set at 850 and 840 V, respectively. Data acquisition and reduction was carried out using Xcalibur 2.0.7 (San Jose, CA).

Table 1
LC gradient (mobile phase A: 10 mM ammonium acetate containing 0.02 % trifluoroacetic acid; and B: acetonitrile) for chromatographic separation of zomepirac and its acyl glucuronide

Time (min)	B%	Time (min)	B%	Time (min)	B%
0	5	14	22	26	68
2	6	16	27	28	80
4	7	18	34	30	95
6	8	20	40	35	95
10	14	22	48	35.1	5
12	18	24	57	45	5

3.2 In Vitro Biosynthesis of Acyl Glucuronides

The in vitro biosynthesis of acyl glucuronide was modified from previous published procedures [5]. Stock solutions of zomepirac were prepared at a concentration of 40 mM in DMSO. Biosynthesis of acyl glucuronide was performed in 16 mL glass test-tubes, in 1 mL total volume, and consisted of the addition of the following reagents in the order listed below:

1. Add 797.5 µL of 0.1 M phosphate buffer, pH 7.4 to tube.
2. Add 20 µL of 50 mM EDTA to tube.
3. Add 50 µL of 100 mM MgCl$_2$ to tube.
4. Add 100 µL of 20 mg/mL of human liver microsomes to tube.
5. Add 12.5 µL of 40 mM zomepirac to tube.
6. The tubes contents were gently mixed and pre-incubated for 5 min at 37 °C in a water bath.
7. Add 20 µL of a mixture of UDPGA:UDPAG (5:1) (i.e., 250 mM UDPGA mixed with 50 mM UDPAG) to start the biochemical reactions.
8. The final concentrations are: 1.0 mM EDTA, 5 mM MgCl$_2$, 2 mg/mL human liver microsomes, 500 µM zomepirac, 5 mM uridine 5′-diphosphoglucuronic acid triammonium salt (UDPGA) and 1.0 mM uridine 5′-diphospho-N-acetylglucosamine sodium salt (UDPAG).

Incubations were carried out in duplicate for 2 h at 37 °C and then 1 mL of the reaction mixtures were transferred to 1 mL ultracentrifuge tubes from Beckman (Palo Alto, CA). The reaction was terminated by ultracentrifugation at 434,513 × g for 20 min to pellet microsomal proteins using a Beckman Coulter benchtop Optimatm MAX-XP ultracentrifuge (Brea, CA) fitted with a rotor TLA 120.1. The amount of zomepirac acyl glucuronide biosynthesized was quickly estimated by LC-UV-MS analysis at UV 220 nm using a known amount of zomepirac injected on the column (**Note 1**).

3.3 Enzymatic Hydrolysis by β-Glucuronidase

The procedure for enzymatic hydrolysis of acyl glucuronide was modified from a previous published procedure [6]. An aliquot of 1 mL of biosynthesized acyl glucuronide product from Sect. 3.2 was adjusted to pH 5.0 with 0.1 M ammonium acetate buffer, pH 5. A 250 µL aliquot was removed following incubation at 37 °C for 0 or 2 h with 5,000 units/mL of β-glucuronidase from *Helix pomatia*. A control sample without β-glucuronidase was also incubated under identical conditions. The hydrolysis was terminated by adding 6 volumes of acetonitrile containing 0.02 % formic acid (**Note 2**) prior to vortexing and centrifugation in a Jouan Centrifuge CT-422 (Thermo Scientific, Inc., San Jose, CA) at 200 × g for 10 min at 5 °C.

There was hydrolysis of the direct glucuronide conjugate of zomepirac as indicated by approximately 90 % decrease in peak

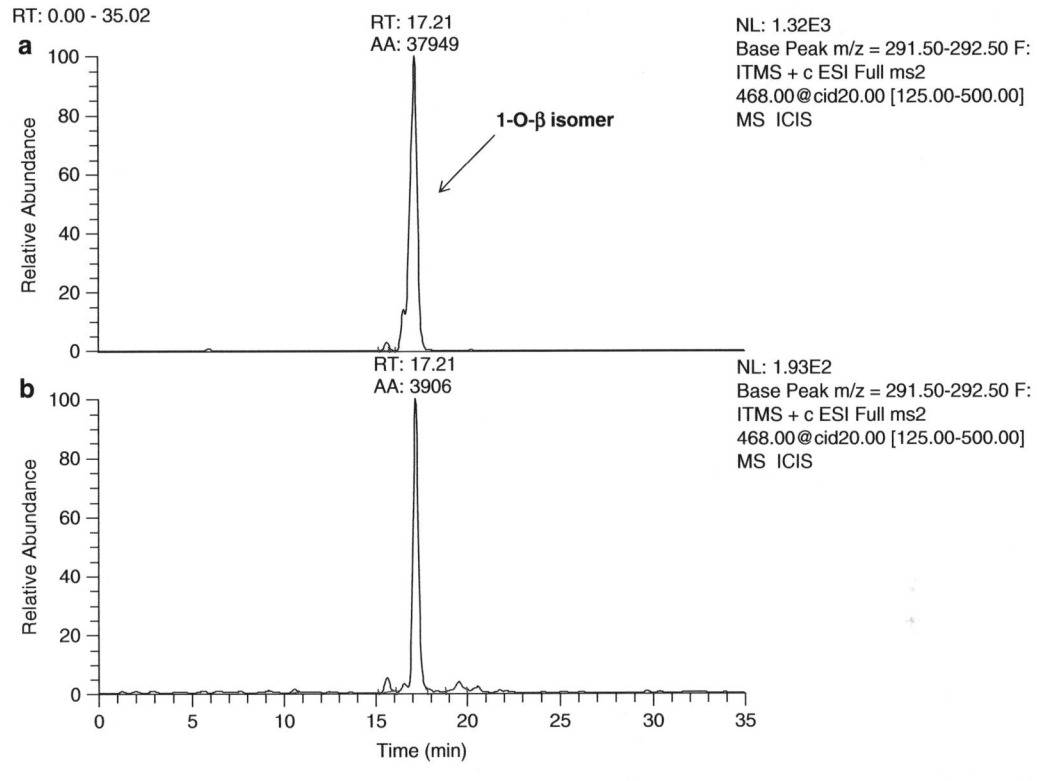

Fig. 1 Reconstructed ion chromatograms (RICs) corresponding to zomepirac and its 1-O-β acyl glucuronide from LC-MS/MS analysis of incubation of crude microsomal generated 1-O-β acyl glucuronide of zomepirac without β-glucuronidase at 37 °C for 2 h (**a**) and with β-glucuronidase hydrolysis at 37 °C for 2 h (**b**)

area of the glucuronide peak at 17.21 min following incubation with β-glucuronidase from *Helix pomatia* in 0.1 M ammonium acetate buffer (pH 5.0) at 37 °C for 2 h (see Fig. 1). This together with detection of the aglycone ion at m/z 292, from the diagnostic neutral mass loss of 176 Da from CID of the glucuronide conjugate at m/z 468 (see Fig. 2), and the observation of multiple acyl migrated products with incubation time (see Fig. 3) allowed the assignment of the peak at approximately 17 min as a 1-O-β isomer of acyl glucuronide of zomepirac.

3.4 Chemical Degradation Half-Life

The method used for assessment of chemical degradation was adapted from the methods described by Ebner et al. [7] and Sawamura et al. [8] (**Note 4**). Each incubation (final volume of 0.2 mL) consisted of approximately 2 μg/mL 1-O-β acyl glucuronide of zomepirac (estimated from UV signal) in 0.1 M phosphate buffer (pH 7.4). The incubations were carried out at 37 °C and the reaction was stopped at 0, 10, 20, 30, 60 and 120 min by adding equal volume of 0.02 % formic acid (pH 3.0) (**Note 2**) in acetonitrile, followed by addition of 10 μL of 0.2 μM

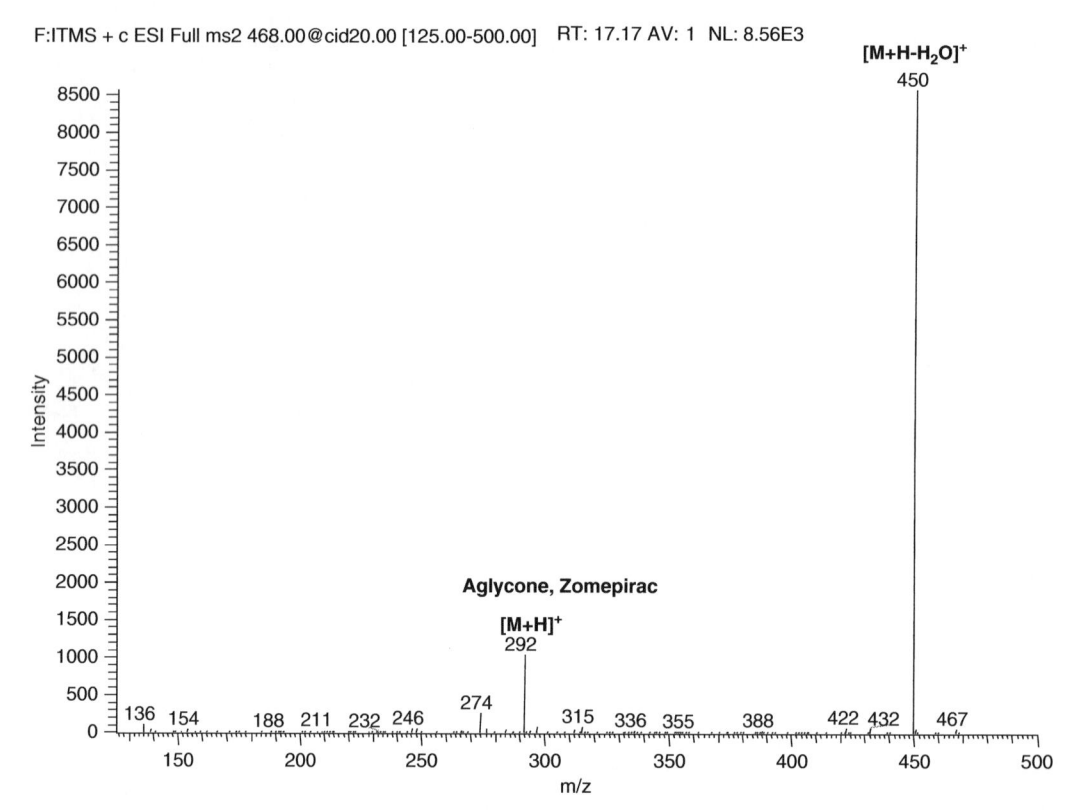

Fig. 2 Product ion mass spectrum of 1-O-β acyl glucuronide of zomepirac from CID of [M+H]⁺ at *m/z* 468 from LC-MS/MS analysis of incubation of crude microsomal generated 1-O-β acyl glucuronide of zomepirac

clozapine as internal standard. Each incubation was conducted in duplicate and was staggered to reduce delay time after each injection to minimize continuing chemical degradation. The sample was transferred to a 96-well plate for LC-MS analysis and the typical RICs corresponding to 1-O-β acyl glucuronide from its chemical degradation in buffer over time is illustrated in Fig. 3. The RIC corresponding to acyl glucuronide from the product ion scan at *m/z* 468 of the 0.5 h sample indicated that the decrease in peak area of 1-O-β isomer is accompanied by the appearances of several peaks from acyl migration (**Note 3**).

The ratio of the peak area of 1-O-β acyl glucuronide of zomepirac to that of an added internal standard clozapine was determined at each time point and normalized to the peak area ratio from time 0 min, which was set to 100 %. An example of data from this chemical degradation experiment is tabulated in Table 2. The data in Table 2 was graphed as a semi-log plot of % 1-O-β isomer remaining as a function of chemical degradation time in minutes. The half-life of the 1-O-β acyl glucuronide of zomepirac was determined with the assumption that the chemical degradation followed zero order kinetics (Fig. 4) (**Note 5**).

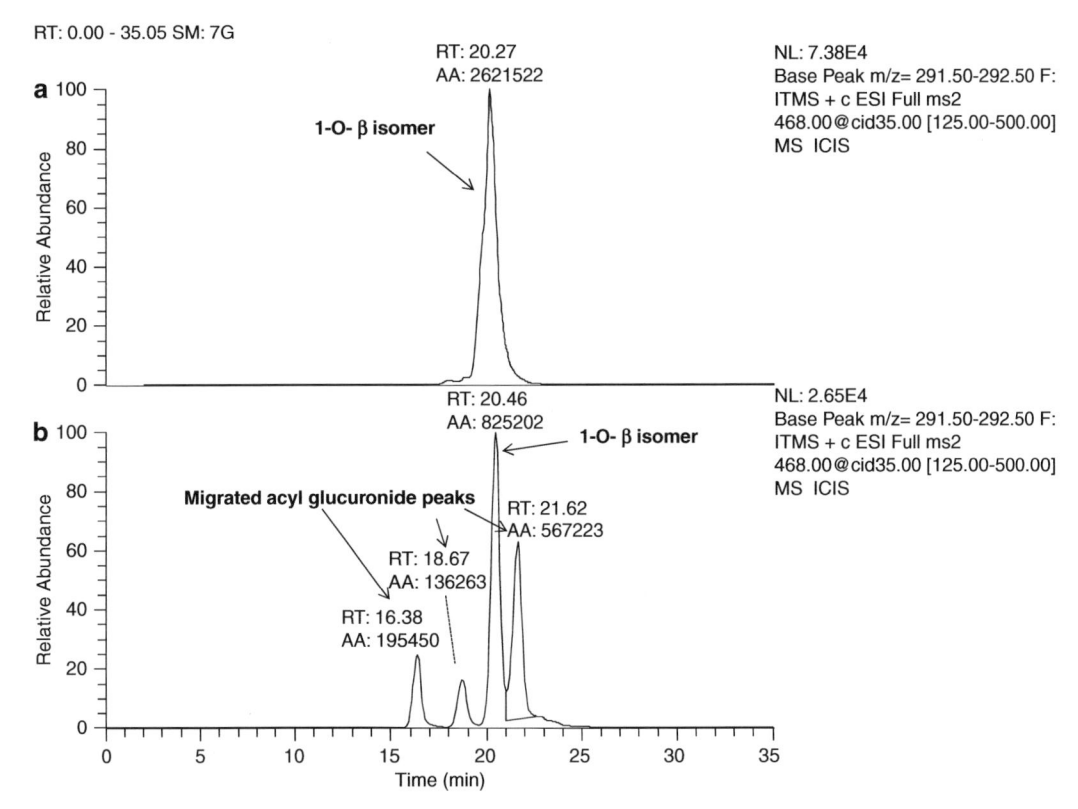

Fig. 3 RICs corresponding to acyl glucuronides from the chemical degradation of 1-O-β acyl glucuronide of zomepirac in 0.1 M phosphate buffer (pH 7.4) at 0 (**a**) and 0.5 (**b**) h

Table 2
The chemical degradation time in minutes, peak area ratios of 1-O-β isomer to clozapine and % remaining from normalization to time 0 min were summarized from the chemical degradation experiment

Time (min)	Peak area ratio	Remaining (%)
0	0.639	100.0
10	0.461	72.2
20	0.385	60.2
30	0.247	38.6
60	0.122	19.1
120	0.043	6.7

The chemical degradation half-life of 1-O-β acyl glucuronide was calculated from the equation $T_{1/2} = 0.693/k$, where k was the zero order kinetic degradation constant of 0.024 determined from Fig. 4. The $T_{1/2}$ was calculated to be 28.9 min and verified

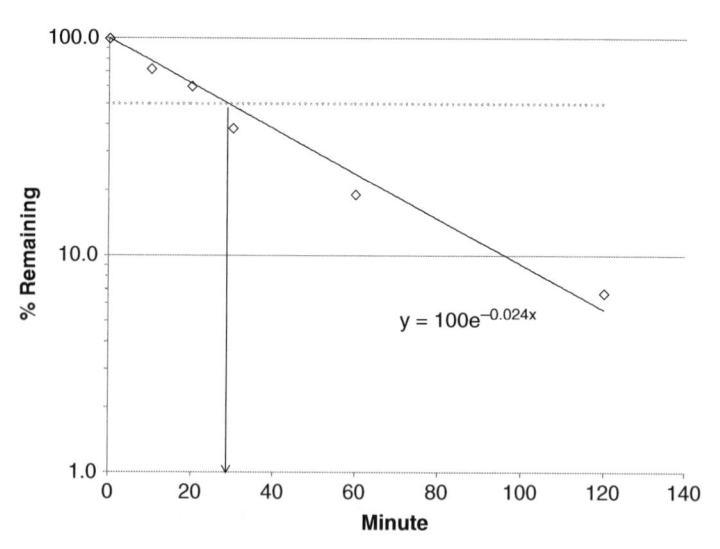

Fig. 4 Calculation of half-life from chemical degradation of 1-O-β acyl glucuronide of zomepirac in 0.1 M phosphate buffer (pH 7.4)

graphically as illustrated from the plot in Fig. 4. The average chemical degradation half-life from 3 different analyses was 30.2 ± 6.99 min for the 1-O-β acyl glucuronide of zomepirac is in agreement with findings of Ebner et al. [7] and Sawamura et al. [8] and shorter than the reported half-lives of other tested acyl glucuronides [8].

3.5 Quantitation of Covalent Binding of Acyl Glucuronides to Protein

The method used to quantitate covalent binding of acyl glucuronides to protein was adapted from the method described by Bolze et al. [5]. In a 16 mL glass test tube the following reagents were added in the order listed below:

1. Add 500 μL of 0.5 mM human serum albumin (HSA) to the tube: Weight out 16.5 mg of HSA powder and dissolved in 500 μL of 0.1 M phosphate buffer, pH 7.4.

2. Add 500 μL of biosynthesized zomepirac acyl glucuronide as outlined in Sect. 3.2 (0.1 M phosphate buffer, pH 7.4) to the tube.

3. The tubes were incubated in duplicate at 37 °C and a 0.5 mL aliquot was withdrawn at time points corresponding to 0 and 2 h.

4. The reaction was terminated by protein precipitation after addition of 3 mL of a 1:1 mixture of acetonitrile:acetone containing 4 % (v/v) trifluroacetic acid, and vortexed for 10 min

prior to centrifugation at 3,500×*g* with a Beckman Coulter Allegra™ X-22R Centrifuge (Brea, CA) at 5 °C for 10 min.

5. The pellets were washed with 3 mL of 4 % TFA, and centrifuged as above, and followed by another wash with 3 mL methanol and centrifuged as above. The wash step was repeated two more times.

6. The washed pellets were hydrolyzed using 1 mL of 1 N KOH by heating at 80 °C for 3 h. The alkaline hydrolysis of the ester bond of all covalently bound acyl glucuronides to amino groups of amino acids in HSA led to release of zomepirac.

7. The solution was neutralized with 1 mL of 1 N HCl prior to loading onto an SPE cartridge, which was conditioned by washing with 1 mL methanol followed by three washes with 1 mL water.

8. The loaded SPE cartridge was washed by 1 mL water three times, and eluted with 3 mL acetonitrile.

9. The eluent was evaporated to dryness under a gentle steam of nitrogen.

10. After reconstitution in a 9:1 acetonitrile:water mixture, the samples were ready for quantitation by LC-MS using a calibration curve.

A series of calibration standards corresponding to 0, 100, 200, 300, 750, 1,000, and 4,000 ng/mL of zomepirac (**Note 7**) were prepared for quantification of the above samples from the covalent binding of acyl glucuronides study. Aliquots of 200 μL of calibrators and the above prepared samples were spiked with 10 μL of 0.2 μM of clozapine (internal standard) prior to protein precipitation with 6 volumes of acetonitrile containing 0.02 % formic acid. This was followed by vortex-mixing and then centrifugation at 200×*g* at 5 °C for 10 min. The supernatants were evaporated to dryness under a gentle stream of nitrogen at room temperature and the residues were reconstituted in 250 μL of water:acetonitrile (9:1) containing 0.02 % formic acid. This was followed by vortexing to aid solubilization of drug-derived materials prior to filtration using 0.45 μm nylon filter at 3,500×*g* at room temperature for 10 min. The filtrate was transferred to a 96-well plate for LC-UV-MS analysis.

The quantity of aglycone (zomepirac) recovered from the HSA samples after the 2 h incubation with zomepirac acyl glucuronide is taken to quantitatively reflect covalent binding to HSA as illustrated in Fig. 5. The 0 h incubation of zomepirac acyl glucuronide with HSA serves as a background control. This method for determination of covalent binding of acyl glucuronide only works if the aglycone is stable during alkaline hydrolysis (**Note 6**).

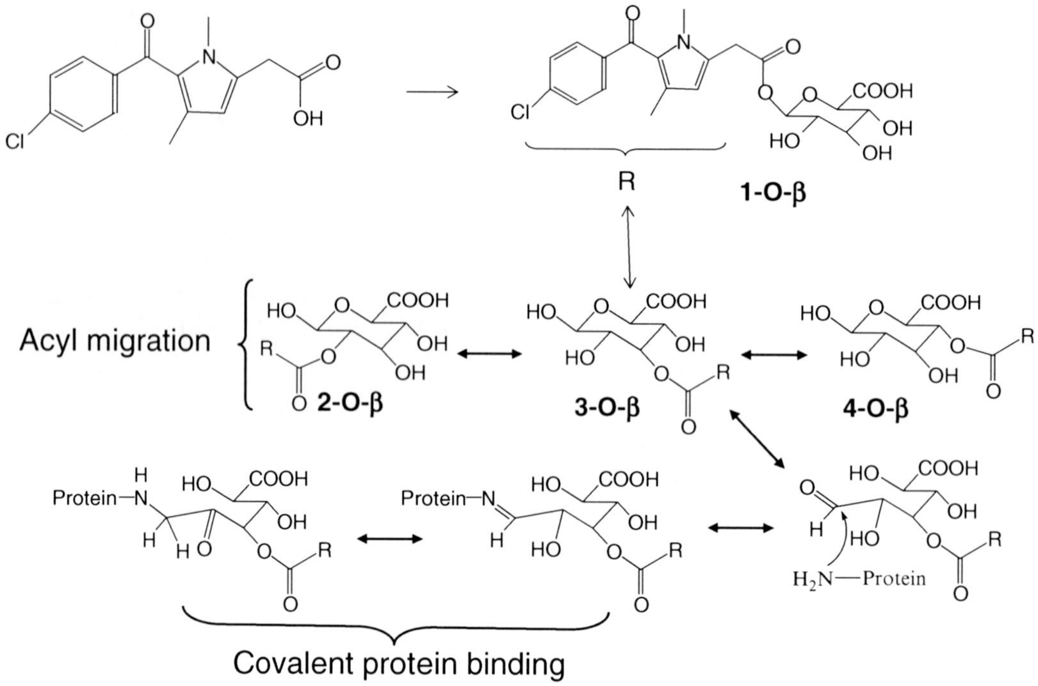

Fig. 5 Schematic illustrating formation of acyl glucuronide of zomepirac and products from acyl migration

4 Conclusions

Two in vitro methods for assessment of reactivity of acyl glucuronides have been described in this chapter: chemical degradation half-life of 1-O-β acyl glucuronide of carboxylic acid drugs and quantitation of covalent binding of acyl glucuronides to proteins. The availability of the chemical degradation half-life together with covalent binding data for estimation of body burden of bound reactive acyl glucuronide metabolites should facilitate risk assessment of carboxylate drugs. Zomepirac, a typical carboxylate drug was used as an example to illustrate the methods.

5 Notes

1. It is important to estimate the amount of biosynthesized acyl glucuornide by UV at λ_{max} of 220 nm so that similar amount of acyl glucuronide is used for further experiments. This information is useful for comparison of formation of acyl glucronides from different compounds.
2. There is a need to stabilize the samples to approximately pH 3 since acyl glucronide is more stable at this pH.

3. There should be good chromatographic separation of acyl glucuronides for correct identification of 1-O-β isomer, which is crucial in correct determination of chemical degradation half-life.

4. It is recommended to conduct a scouting experiment to crudely assess the chemical degradation half-life for decision making on the total run time needed for bioanalytical method, whether to stagger incubation and analysis to accurately determine the chemical degradation half-life, etc.

5. It is desirable to use semi-log plot to allow visual corroboration of calculated chemical degradation half-life.

6. It is recommended to conduct chemical stability of the parent drug in 1 N KOH at 80 °C for 3 h to decide whether quantitation of covalent protein by acyl glucuronide using base hydrolysis will work for this acyl glucuornide. This method only works if the parent compound is stable during base hydrolysis or if completely further converted to a single and stable product.

7. The dynamic range of the calibration curve used should include the incubated acyl glucuronide concentration as mid-point in the curve for better accuracy.

Acknowledgements

The views expressed here are solely those of the author and do not reflect the opinions of Janssen Research & Development, LLC.

References

1. Fung F, Thornton A, Mybeck M, Hornbuckle K, Muniz E (2001) Evaluation of the characteristics of safety withdrawal of prescription drugs from worldwide pharmaceutical markets—1960–1999. Drug Inf J 35:293–317
2. Regan SL, Maggs JL, Hammond TG, Lambert C, Williams DP, Park BK (2010) Acyl glucuronides: the good, the bad and the ugly. Biopharm Drug Dispos 31:367–395
3. FDA (2008) Guidance for industry: safety testing of drug metabolites. http://www.fda.gov/cder/guidance/index.htm
4. ICH (2010) Guidance for Industry: M3(R2) Nonclinical Safety Studies for the Conduct of Human Clinical Trials and Marketing Authorization for Pharmaceuticals, http://www.fda.gov/downloads/Drugs/GuidanceComplianceRegulatoryInformation/Guidances/ucm073246.pdf; Guidance for Industry: M3(R2) Nonclinical Safety Studies for the Conduct of Human Clinical Trials and Marketing Authorization for Pharmaceuticals Questions and Answers(R2), http://www.fda.gov/downloads/Drugs/GuidanceComplianceRegulatoryInformation/Guidances/UCM292340.pdf
5. Bolze S, Bromet N, Gay-Feutry C, Massiere F, Boulieu R, Hulot T (2002) Development of an in vitro screening model for the biosynthesis of acyl glucuronide metabolites and the assessment of their reactivity toward human serum albumin. Drug Metab Dispos 30(4):404–413
6. Comble J, Blake JW, Nugent TE, Tobin T (1982) Morphine glucuronide hydrolysis: superiority of β-glucuronidase from patella vulgate. Clin Chem 28(1):83–86
7. Ebner T, Heinzel G, Prox A, Beschke K, Wachsmuth H (1999) Disposition and chemical stability of telmisartan 1-O-acyl glucuronide. Drug Metab Dispos 27(10):1143–1149

8. Sawamura R, Okudaira N, Watanabe K, Murai T, Kobayashi Y, Tachibana M, Ohnuki T, Masuda K, Honma H, Kurihara A, Okazaki O (2010) Predictability of idiosyncratic drug toxicity risk for carboxylic acid-containing drugs based on the chemical stability of acyl glucuronide. Drug Metab Dispos 38(10): 1857–1864

Chapter 31

In Vitro Comet Assay for Testing Genotoxicity of Chemicals

Haixia Lin, Nan Mei, and Mugimane G. Manjanatha

Abstract

The Comet assay, also known as the single cell gel electrophoresis (SCGE) assay, is widely applied as one of the standard methods to assess DNA damage caused by a range of DNA damaging agents. The unique aspect of this method is its ability to detect DNA damage in individual cells. During the last 2 decades, the Comet assay has been used in a broad variety of applications, including genotoxicity testing, human biomonitoring, ecological monitoring, clinical studies, and as a tool for detection of DNA damage in different cell types. Comet assay protocols have been adopted and optimized by many laboratories around the world. In this chapter, the authors provide an example of in vitro Comet assay (neutral and alkaline) application with detailed procedures used in their laboratory for the analysis and interpretation of Comet assay data.

Key words Comet assay, Single cell gel electrophoresis, In vitro, DNA damage, Genotoxicity

1 Introduction

In 1984, two scientists in Sweden developed a new method that used a microgel and electrophoresis to study radiation-induced DNA damage in individual mouse lymphoma cells (L5178Y-S) and Chinese hamster fibroblast cells (Cl-1) [1]. In this method, cells were lysed in a neutral detergent solution, electrophoresed in a weak electric field, stained with acridine orange, and viewed under a fluorescence microscope. DNA double-strand breaks (DSB) induced by radiation introduced a relaxation of the supercoiling DNA that migrated further than the nucleus towards the anode [1]. Four years later in 1988, Singh and his colleagues modified the microgel electrophoresis technique under alkaline conditions (pH > 13) in human lymphocytes; a modification could allow the detection of DNA single-strand breaks (SSB) and alkali-labile sites, in addition to DSB [2]. Since the microgel electrophoresis technique has the ability to detect DNA damage at the level of the individual cells, it has been named the single cell gel electrophoresis (SCGE) assay. Because the resulting image of a cell with broken DNA resembles a "comet" with a brightly fluorescent head and a tail region, the SCGE assay is also called the Comet assay [3]. The head of the comet shape is formed by the intact DNA which is not pulled during electrophoresis, while the tail consists of the damaged or broken DNA which is migrated out of the nuclei in the electrophoresis process. The extent of DNA damage can be expressed by the percentage of DNA in the tail, i.e., higher the DNA intensity in the tail (usually larger tail), greater the extent of DNA damage.

The Comet assay is a rapid, simple, sensitive, and quantitative method for measuring DNA damage [4]. Although it also has some disadvantages, such as low throughput, relatively poor reproducibility (between slides, users, and laboratories), time-consuming and potentially biased image analysis [5], this assay has been accepted as a valuable tool for investigating DNA damage and repair by many laboratories around the world. Over the decades, the Comet assay has become a basic tool to be used for in vitro, ex vivo, and in vivo systems and is increasingly being employed in a broad range of scientific fields, including the genotoxicity testing [6], human biomonitoring [7, 8], clinical studies [9], lifestyle studies (e.g., smoking) [10], genetic ecotoxicology [11], and fundamental research in DNA damage and repair using different cell types [12].

Since its development, protocols for conducting in vitro and in vivo Comet assays have been published by different expert panels, such as the International Workshop on Genotoxicity Testing [13, 14] and the International Comet Assay Workshop [15], and by different laboratories [16–18]. Scientific committees and regulatory agencies have been recommending the use of the in vivo Comet assay to develop a weight of evidence based approach in genetic toxicology testing. An international collaborative trial on in vivo Comet assay in 14 laboratories from Europe, Japan, and USA evaluated 15 chemicals in rat liver, and the results supported the notion that the liver Comet assay is a reasonable alternative to liver unscheduled DNA synthesis assay to detect genotoxicity of chemicals in liver [19]. Recently, the International Conference on Harmonization (ICH) has included a second in vivo assay in Option 2 of the standard test battery for genotoxicity and recommended that the Comet assay in liver could be one of the second in vivo assays [20].

During the early screening of drug candidates for genotoxicity, the mutagenicity testing in bacteria and cultured mammalian cells, and clastogenicity study in cultured cells are typically conducted. It has been suggested that the in vitro Comet assay is useful as an alternative to cytogenetic assays in early genotoxicity or photogenotoxicity screening of drug candidates [6]. For this purpose, the in vitro Comet assay is most commonly applied to mammalian cells, including cultured cells from different cell lines or

6. Countess cell counting chamber slides, Cat. #C10228 (Life Technologies).
7. Horizontal electrophoresis apparatus, CometAssay ES, Cat. #4245-040-ESK (Trevigen Inc.; Gaithersburg, MD).
8. Electrophoresis power supply, VWR power source 300 V, Cat. #93000-744 (VWR International; West Chester, PA).
9. Coplin jars (opaque), Cat. #08-815-10 (Fisher Scientific Inc.; Waltham, MA).
10. Microscope slides, Cat. #12-550-123 (Fisher).
11. Cover slips (22×50 mm), Cat. #12-548-5E (Fisher).
12. Phosphate buffered saline (PBS) without Ca^{2+} and Mg^{2+} (1×), Cat. #SH3025601 (Fisher).
13. Dimethyl sulfoxide (DMSO), Cat. #D8418 (Sigma-Aldrich Co.; St. Louis, MO).
14. Microscope slide tray (aluminum), Cat: #2870 (Eberbach Corporation; Ann Arbor, MI).
15. Nikon 501 fluorescent microscope (Nikon Instruments Inc.; Melville, NY).
16. Image analysis software, Comet Assay IV (Perceptive Instruments Ltd.; Bury St. Edmunds, UK).

2.2 Cell Lines

Theoretically, any eukaryotic cell can be used in the in vitro Comet assay. Below is the list of some cell lines often used in the in vitro Comet assay.

1. Human lymphoblast cells (TK6).
2. Human acute lymphocytic leukemia cells (Jurkat).
3. Human alveolar epithelial cells (A549).
4. Human cervical cancer cells (HeLa).
5. Human hepatocellular carcinoma cells (HepG2).
6. Mouse lymphoma cells (L5178Y, subclone 3.7.2.c).
7. Chinese hamster ovary cells (CHO).
8. Chinese hamster lung cells (V79).

2.3 Cell Culture

1. Eagle's Minimum Essential Medium (EMEM) is used for A549 cell line. Dulbecco's Modified Eagle's Medium (DMEM) is used for HeLa and HepG2 cells. Fischer's medium for leukemic cells of mice was developed for culturing L5178Y cells. RPMI 1640 will also support the growth of many cell lines in suspension (L5178Y, TK6, Jurkat) or as monolayer (V79, CHO).
2. Animal sera are commonly used to support the growth of cells in culture. Except L5178Y cell line which usually requires

Table 1
Preparation of S9 mix for each sample[a]

Constituents	Formula 1 Conc. in S9 mix (mM)	Formula 1 Volume (ml)	Formula 2 Conc. in S9 mix (mM)	Formula 2 Volume (ml)	Formula 3 Conc. in S9 mix (mM)	Formula 3 Volume (ml)
S9		0.1–0.2		0.1–0.2		0.1–0.2
Sodium phosphate buffer, pH 8.0	50	0.8				
NADP	4	1	6.5	0.3	75	0.4
KCl	30		30			
G-6-P	5		128			
CaCl$_2$	10					
MgCl$_2$·6H$_2$O	10					
C$_6$H$_5$O$_7$Na$_3$					375	0.4
Total[b]		2		0.5		1
Cell culture		8		9.5		9
Final S9		1–2 %		1–2 %		1–2 %

[a]Multiply these numbers in the table by the number of samples plus 1 (n+1) for each experiment. Prepare the cofactors in distilled water (dH$_2$O) or serum-free medium, adjust pH to 7.0 with 1 N NaOH, and filter sterilize before use. Keep all ingredients chilled during preparation, and add the S9 fraction last
[b]When using less than the volume of S9, add dH$_2$O or medium to make up the final volume

horse serum, most cell lines are cultured with media supplemented with fetal bovine serum. The complete growth media for different cell lines may require the addition of other growth components and antibiotic agents to prevent contamination.

3. When subculturing the monolayer cells (e.g., A549, HeLa, HepG2, CHO, and V79), trypsin-EDTA (0.05 % trypsin-EDTA, Cat. #25300054; Life Technologies) is used to break both intercellular and intracellular surface bonds (**Note 1**).

2.4 Metabolism Activation (S9)

Some chemicals should be tested for their genotoxicity in the presence of S9. Table 1 shows different S9 cofactor formulae containing an aroclor 1254-induced male Sprague-Dawley rat liver post-mitochondrial fraction (S9) (Cat. #11-101, Molecular Toxicology Inc.; Boone, NC) mixed with nicotinamide adenine dinucleotide phosphate (NADP)-generating system. The constituents of the S9 mix formula may depend on the property of a test chemical.

2.5 Positive Controls

A positive control should be used when performing the in vitro Comet assay. Table 2 lists some chemicals which are widely used as positive control in the genotoxicity studies.

Table 2
Positive controls with or without S9 used in the in vitro Comet assay

Chemical[a]	CAS	S9	Company	Cat No.
4-nitroquinoline N-oxide (4-NQO)	56-57-5	−	Sigma-Aldrich	N8141
Ethyl methanesulfonate (EMS)	62-50-0	−		64292
Hydrogen peroxide solution (H$_2$O$_2$)	7722-84-1	−		31642
Methyl methanesulfonate (MMS)	66-27-3	−		129925
Benzo[a]pyrene (BP)	50-32-8	+		B1760
Cyclophosphamide (CPA)	6055-19-2	+		C0768

[a]The concentration of these chemicals used in this assay is different for different cell lines

2.6 Cytotoxicity Determination

Trypan blue stain (0.4 %, Cat. #T10282; Life Technologies) is used to examine the cell viability. Other equivalent stains or similar methods can also be used in this assay.

2.7 Agarose (Note 2)

1. 0.5 % low melting point agarose (LMPA), Cat. #BP165-25 (Fisher).
2. 1.0 % normal melting point agarose (NMPA), Cat. #16500500 (Ultrapure agarose; Life Technologies).

2.8 Lysis Solutions (Note 3)

1. Lysis solution A for detection of DSB DNA (neutral Comet assay): 30 mM Na$_2$EDTA and 0.5 % SDS, pH 8.0. This solution can be stored for several weeks at room temperature.
2. Lysis solution B (1 l, stock solution) for detection of SSB DNA and base damage (alkaline Comet assay): add 2.5 M NaCl, 100 mM EDTA, 10 mM Tris to about 700 ml dH$_2$O and begin stirring the mixture (adjust pH to 10.0 with approximately 7.5 g of solid NaOH). The stock lysis solution B can be stored at room temperature. The working lysis solution (100 ml) is freshly prepared for each experiment: add 1 % (v/v, 1 ml) of Triton X-100, and 10 % (v/v, 10 ml) DMSO to 89 ml of stock lysis solution (**Note 4**). Usually, 50 ml of this lysis solution in the Coplin jar is used for up to ten slides.

2.9 Electrophoresis Solutions

1. Electrophoresis solution A (TBE buffer) for neutral Comet assay: 90 mM Tris, 90 mM boric acid, 2 mM Na$_2$EDTA, pH 8.5.
2. Electrophoresis solution B for alkaline Comet assay (also can be used as unwinding solution): 300 mM NaOH and 1 mM EDTA, pH > 13, and mix well in cold room for at least 30 min prior to use. The stock solutions: 10 M NaOH (200 g in 500 ml dH$_2$O) and 200 mM EDTA (22.33 g in 300 ml dH$_2$O), pH 10. The fresh solution should be made before each electrophoresis run (for 1× buffer: add 30 ml 10 M NaOH and 5 ml 200 mM EDTA, quantity sufficient (q.s.) to 1,000 ml, mix well). The total volume of the buffer used depends on the ge

chamber capacity. The pH of the buffer should be measured to ensure >13 prior to use. In addition, the electrophoresis solution can also be prepared with solid NaOH pellets (12 g/l dH$_2$O) and 0.5 M EDTA (2.0 ml/l dH$_2$O).

2.10 Neutralization Buffer

For alkaline Comet assay, add 0.4 M Tris (48.5 g) to 800 ml dH$_2$O, and adjust pH to 7.5 with HCl, and q.s. to 1,000 ml with dH$_2$O.

2.11 DNA Staining and Image Analysis

1. SYBR Gold, Cat. #S11494 (Life Technologies) or SYBR Green, Cat. #4250-050-05 (Trevigen) dissolved in TE buffer (10 mM Tris-HCl, 1 mM EDTA, pH 7.5) or TBE to prepare a 1:10,000 dilution (**Note 5**).

2. For analysis, it is preferable to capture and analyze the images in real time, not on stored images. However, it is not absolutely necessary to use real-time image analysis for scoring. In our laboratory, the COMET IV analysis system by Perceptive Instruments is used.

3 Methods

Table 3 shows the procedure for conducting the in vitro Comet assay. The basic steps include cell treatment with a chemical, preparation of a single cell suspension, slide preparation, lysis of cells to liberate DNA, exposure to neutral solution (pH 8.5) or alkaline solution (pH > 13) to obtain DSB and/or SSB DNA, electrophoresis, neutralization, DNA staining, and comet scoring. In this section, the authors describe the procedures to conduct the in vitro Comet assay using a mouse lymphoma L5178Y cell line as an example.

3.1 Cell Treatment and Single Cell Suspension

1. The in vitro Comet assay for a chemical should be conducted both in the presence and absence of S9 (Table 1). The culture cells (e.g., L5178Y cells) are treated with serial dilution of the chemical in 50 ml tubes containing 10 ml Fischer's medium (supplemented with 10 % heat-inactivated horse serum), and then incubated at 37 °C in a humidified incubator for 3–6 h or 24 h at 37 °C in a humidified incubator with 5 % (v/v) CO$_2$ in air depending on the design of the experiment (**Note 6**). For most cell lines, 4-h treatment is often used for the in vitro Comet assay.

2. Generally, at least four concentrations of the chemical are used, together with at least one negative control and one positive control (Table 2). The negative control can be solvents, such as culture media, water, or DMSO (**Note 7**).

3. After chemical exposure, remove the culture media and the chemical by centrifugation at 200 g for 10 min, wash the cells

Table 3
The procedure for conducting the in vitro Comet assay

Step	In vitro Comet assay	
	Neutral	**Alkaline**
1. Cell treatment	Cells treated with different chemicals	
2. Single cell suspension	Suspending the cells in cold PBS	
3. Slide preparation	Embedding cells in agarose on microscope slides (1.0 % NMPA and 0.5 % LMPA)	
4. Lysis of the cells	Under neutral lysis solution (pH 8.0)	Under alkaline lysis solution (pH 10.0)
5. DNA unwinding or denaturation	Neutral buffer or TBE buffer (pH 8.5)	Alkaline solution (pH > 13)
6. Electrophoresis	Neutral electrophoresis solution or TBE buffer	Alkaline electrophoresis solution
7. Neutralization	—[a]	Neutralization buffer
8. Slide staining	Comet visualization using stain or dye (SYBR Gold or SYBR Green)	
9. Image analysis	Comet scoring using a fluorescence microscope connected with an image analysis system	

[a]It is generally used to rinse the slides using distilled water after electrophoresis

once with cold PBS, and resuspend the cells in cold PBS as single cell suspension (**Note 8**). Handle the samples under dimmed or yellow light to prevent DNA damage from ultraviolet light. Perform the Comet assay immediately.

3.2 Slide Preparation   Any microscope slide with single frosted end is appropriate for use in the Comet assay for most applications. However, to avoid the agarose detachment, fully frosted microscope slides (Fisher) are generally used to keep the agarose firmly attached to the slide.

1. Prepare 1 % NMPA in dH$_2$O (1 g in 100 ml) in a beaker: microwave (or heat in boiling water bath) until the agarose dissolves and then place the beaker at 55–60 °C to keep the agarose liquid before making comet slides.

2. When NMPA agarose is hot, dip the clean slides (without dust or machine oil) into the molten NMPA (about one-third of the frosted area), and then gently remove or spread the slides with 50–150 μl molten agarose.

3. Wipe underside of slide to remove agarose and lay the slide on a microscope slide tray (or a flat surface) to allow the gel to dry at room temperature or 4 °C and form a thin film (**Note 9**). The agarose should be spread evenly on the slides.

Table 4
Cell viability for each sample

Study title or No.: _____ Chemical: _____ Date: _____					
Dose	Viability (%)	Cell density (cells/ml)	Total volume of cell suspension (1×10^6 cells/ml)	Cell density adjusted to 1×10^6 cells/ml	
				Cell culture (ml)	PBS (ml)
NC					
PC					
Dose 1					
Dose 2					
Dose 3					
Dose 4					
↓					
Dose N					

NC negative control, *PC* positive control

4. After drying slides, place them into a box to store at room temperature until needed. Before use, the slides should be labeled with chemical concentration, slide ID, preparation date, etc.

5. Usually, pre-coated slides are prepared at least 1 day before use. Alternatively, commercially-available pre-coated slides can be used (**Note 2**).

3.3 Cell Viability

After preparing the single cell suspensions for each sample, the cell viability (Table 4) should be examined (**Note 10**). Trypan blue (0.4 %) dye exclusion test is commonly used to count cells and determine the cell viability.

1. Mix 10 μl of cell suspension with 10 μl of trypan blue in a microtube, and then pipet 10 μl mixture into a Countess cell counting chamber slide.

2. Insert slide into the Countess automated cell counter to score cell number and determine the cell viability (Table 4).

3.4 Mixing Cells with Agarose

1. The cell concentration for each sample is adjusted to 1×10^6 cells/ml (Table 4) using cold PBS before making the slides, and optimal results are usually obtained with $5–100 \times 10^3$ cells per slide based on the size of slides.

2. Prepare 0.5 % LMPA in PBS. Microwave (or heat in boiling water bath) until the agarose dissolves (**Note 11**). Once dissolved, keep the agarose in a water bath (or a heat blocker) at 37 °C to cool and stabilize the temperature.

3. Mix 50 μl of 1×10^6/ml cell suspensions with 450 μl of molten LMPA (37 °C) at a ratio of 1:9 (v/v) in a 1.5 ml microtube for each sample, rapidly transfer 150 μl of this mixture onto the surface of the pre-coated slide and cover with a cover slip (**Note 12**).

4. Lay the slides on the microscope slide tray (or a flat surface), place them either at room temperature or at 4 °C (or in a cold room) for at least 20 min until the gel is set, and then gently remove the cover slips from slides. Duplicate slides should be used in this assay for each sample including the treatment, negative control, and positive control.

3.5 Lysis for Detection of DSB

1. Immerse slides in a Coplin jar with pre-chilled neutral lysis solution A (see Sect. 2.8). The lysis solution should be pre-chilled at 4 °C or placed in a cold room for at least 30 min prior to use in order to inhibit endogenous damage and repair in the unfixed cells during sample preparation.

2. After lysis at 4 °C for 30 min (if needed, the time can be prolonged to overnight), remove slides from the lysis buffer, drain excess buffer from slides and gently place slides in a container with 1× TBE buffer (identical with the electrophoresis solution A) for 30 min at 4 °C or in a cold room.

3.6 Lysis for Detection of SSB or DSB and Unwinding

1. Immerse slides in a Coplin jar with the freshly made alkaline lysis solution B that has been chilled at 4 °C in a refrigerator or cold room (see Sect. 2.8). Lyse cells for at least 30 min (if needed, the time can be prolonged to overnight) at 4 °C (or in a cold room). Re-use of this solution beyond the day of preparation is not recommended, except to rinse before electrophoresis. Since fluorescent lights during the alkaline treatment can cause DNA breaks, it is important to protect slides from exposure to light during alkaline lysis, rinse, and electrophoresis.

2. After lysis, remove slides from the Coplin jar, drain excess solution from slide and carefully immerse in a container with cold neutralization buffer for 5 min to remove residual detergent and salts prior to alkali-unwinding step.

3. After rinsing, place slides in a container with freshly prepared alkaline unwinding solution (identical to the electrophoresis solution B) at 4 °C in the dark for at least 30 min (**Note 13**). Discard rinse buffer after use.

4. In case of detection of oxidative DNA damage, the assay can be combined with different enzymes (such as Endo III, FPG, and hOGG1) by adding several extra steps. Firstly, prepare

enzyme reaction buffer (**Note 14**): wash slides three times (5 min each time) using enzyme reaction buffer (including 40 mM HEPES, 0.1 M KCl, 0.5 mM EDTA, 0.2 mg/ml bovine serum albumin, pH 8.0) in a Coplin jar at 4 °C (or in a cold room); remove slides from the Coplin jar, and tap off excess liquid from the slides. Then pipet 50–150 µl of enzyme buffer (including Endo III, FPG, or hOGG1) onto the gels of corresponding slides, cover with cover slips, and put slides in a moist box (prevent from desiccation). Finally incubate at 37 °C for 45 min for Endo III or 30 min for FPG or hOGG1. After incubation, gently remove the cover slips, and lay the slides in the gel electrophoresis chamber at 4 °C (or in a cold room).

3.7 Electrophoresis and DNA Staining

1. After lysis and unwinding, place the slides randomly in the gel electrophoresis chamber (CometAssay ES) and fill the chamber with 950 ml electrophoresis solution A (neutral electrophoresis) or freshly made electrophoresis solution B (alkaline electrophoresis) (**Note 15**).

2. Adjust the level of electrophoresis buffer about 1–3 mm (no more than 5 mm) above the agarose on the slide. Before electrophoresis, make sure the electrophoresis tank is leveled by using a bubble leveling device.

3. Adjust the voltage to 21 V (or 1 V/cm) in the neutral Comet assay or no more than 300 mA (or 1 V/cm) in the alkaline Comet assay (**Note 16**), and then run electrophoresis for 30 min (**Note 17**). It is critical to avoid exposure to fluorescent lights in the room during electrophoresis.

4. After electrophoresis, gently drain excess electrophoresis buffer, rinse slides by submersing in a large volume of distilled water (neutral Comet assay) or neutralization buffer (alkaline Comet assay) at 4 °C or in a cold room for 5 min. If needed, repeat rinsing two more times.

5. After rinsing, immerse the slides in 70 % ethanol for 5 min, and then dry the slides at room temperature or at 45 °C for 10 min (quick drying). The slides can be stored in a covered box at room temperature until staining.

6. Place slides in a microscope slide tray (or metal or cardboard tray), add 50–200 µl of SYBR Gold or SYBR Green (1:10,000) working solution onto each slide for 10 min and keep the slides from exposing to light before scoring. The staining solution can be used within 24 h of preparation if stored in the dark at 2–8 °C.

7. If scoring immediately, gently blot away excess staining solution from each slide.

8. After staining, the slides can be scored immediately or placed in a humidified cover box and stored in a cold room for about a month [29]. When ready for scoring, the slides are re-hydrated in water.

Comet Assay 529

3.8 Image Analysis

1. Examine the comet images with blue light excitation if stained with SYBR Gold or SYBR Green. Before scoring slides, the slides should be randomly coded so they can be scored blind in order to avoid bias associated with scoring.

2. The overall slide condition should be checked before scoring the slides for comets (**Note 18**).

3. Any image analysis system may be used for the visualization and quantitation of comets. The comets are measured via a digital video camera linked to an image analyzer system using a fluorescence microscope at magnification of 200×. In our lab, the Perceptive Instruments Comet Assay IV image analysis system with a digital camera attached to Nikon 501 fluorescence microscope is used to quantify the length of DNA migration and the percentage of migrated DNA (for example, just by clicking the image of each cell nucleus) (**Note 19**).

3.9 Evaluation Criteria

1. At least 50 comets are randomly scored per slide and 100 comets or nuclei are scored for each duplicate sample.

2. The most important index (best parameter) collected for the evaluation in the Comet assay is the percentage of DNA in tail (%DNA in tail) (**Note 20**).

3. For alkaline Comet assay, count the number of "droplet" (Fig. 2) per slide (**Note 21**). The percentage of "droplet" for each sample (i.e., per concentration of chemical) should be calculated (Table 5).

4. For the neutral Comet assay, count the number of "elongated comet" or "comets" per slide (Fig. 2). The percentage of "elongated comet" or "comets" for each sample should be calculated.

3.10 Statistical Analysis

1. DNA damage is assessed by the software system by measuring %DNA in tail, which is the percentage of DNA fragments present in the tail.

2. The mean and standard deviation (SD) can be calculated from three or more independent experiments.

Negative control | Droplet | Elongated comet

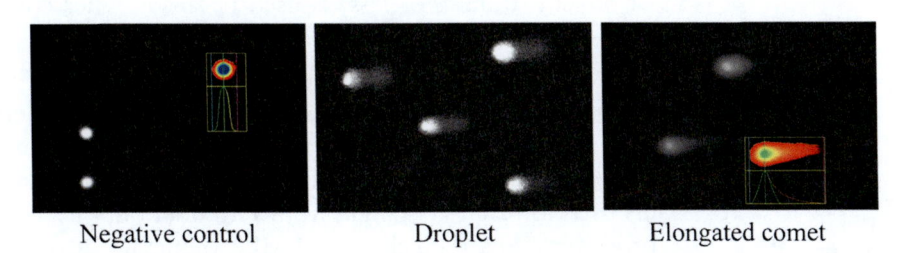

Fig. 2 Examples of comet images

Table 5
Slide analysis record

Study title or No. _____ Date: _____
Microscope: _____ Microscope magnification: _____
Slides prepared date: _____ Stain or dye: _____
Alkaline electrophoresis condition: _____
Neutral electrophoresis condition: _____

Slide ID[a]	CN	CD	SB	Number of "Droplets"/ "Comets" in total of 50 cells	%"Droplets"/ %"Comets"	Comments
A-1						
A-2						
B-1						
B-2						
C-1						
C-2						
…						
…						
X-1						
X-2						

CN cell number (0—good number to count; 1—too many; 2—not enough to count; 3—no cells), *CD* cell damage (0—no damage; 1—small damage, countable; 2—large damage, countable; 3—ghost; unable to count), *SB* slide background (0—no noise; 1—some noise; countable; 2—too dirty, unable to count)
[a]Coded number for negative control, positive control, and samples

3. ANOVA and paired *t*-tests (if just two doses) and their nonparametric equivalents can be used for data analysis (**Note 22**).

4 Notes

1. Trypsin is thawed immediately before use. Low concentration of trypsin (0.05 % or less) is used because higher concentrations may increase DNA damage.

2. The concentration of agarose in NMPA and LMPA may be different each time the Comet assay is conducted. Generally, the concentrations of NMPA and LMPA are varied from 0.5 to 1.0 % and 1.0–1.5 %, respectively. The number of agarose layers used for each gel ranges from 1 to 3. (1) For a single layer procedure, mix the cells with LMPA (generally at 37 °C) and place the mixture directly on a microscope slide. (2) For two-

layer method, pre-coat the microscope slides with a layer of NMPA (first layer), and then add the cells mixed with LMPA to directly spread onto the pre-coated slides (second layer). (3) In the three-layer procedure, the pre-coated microscope slides will have two layers as specified in (2) above and then another layer of LMPA is added to the second layer for increasing the distance between cells and the gel surface. A cover slip with an appropriate size is often used to flatten out molten agarose layer. The slides are often placed at 4 °C to enhance gelling of the agarose during the process. The pre-coated slides developed specifically for the Comet assay are commercially available (CometSlide, Trevigen).

3. The commercial lysis solution (Cat. #4259-050-01, Trevigen) can be used for the neutral and alkaline Comet assay.

4. The purpose of adding DMSO into the lysis solution is to remove radicals generated by the iron released from hemoglobin in blood cells or samples used. If slides are kept in the lysis solution for only a brief time, DMSO is not required in the lysis solution.

5. SYBR Gold or SYBR Green shows an excitation from 465 to 505 nm (blue light), with an emission from 515 to 565 nm (green light). Also, other DNA stains such as ethidium bromide, propidium iodide (PI), 4′6-diamidine-2-phenylindol dihydrochloride (DAPI), YOYO-1, TOTO, and silver stain (non-fluorescent staining) can be used in this assay.

6. For L5178Y cells, the growth medium is Fischer's medium or RPMI-1640 supplemented with 10 % heat-inactivated horse serum, sodium pyruvate (1 mM), penicillin (100 U/ml), and streptomycin (100 μg/ml). Generally, short-time treatment (3–6 h) with chemicals should be conducted in the absence and presence of S9. In some cases, multiple S9 formulae and/or concentrations may be used in order to adequately test the genotoxicity of a chemical.

7. The positive control used in the Comet assay should show a significant biological response such as DNA SSB or DSB (H_2O_2 and 4-NQO) compared to a negative control. To assess the extent of DNA damage in the Comet assay, different control cells commercially available from Trevigen can be used (Cat: #4257-010-NC0, -NC1, -NC2, -NC3 for neutral Comet assay; Cat: #4256-010-CC0, -CC1, -CC2, -CC3 for alkaline Comet assay). These suspension cells contain different levels of DNA damage and can be used under defined electrophoresis conditions as controls with Trevigen's Comet Assay kits.

8. For adherent cells, remove the media including the chemical and S9 mix, and then add 1–2 ml of 0.05 % trypsin to the cells. Keep the cells at 37 °C for 2–3 min to detach cells. Add an equal

amount of medium (with FBS) to quench trypsin, then remove the media including trypsin by centrifugation at 200 g for 10 min, wash the cells once with cold PBS, and resuspend the cells in cold PBS as single cell suspension.

9. One of the major problems during the preparation of Comet slides is detachment of agarose gel from the slides at some point. Pre-coating the slides with NMPA can prevent this problem. There should not be any bubbles and cracks in the agarose film on the slide. In addition, if it is humid inside the laboratory, the agarose gels may not stick to the slides as well as they stick under a dry condition. Generally, if the cell suspension contains a trace of culture medium with serum, the gel will fall off easily when the slides are in lysis solution. Using plastic films instead of glass slides can effectively prevent the gel detachment from slides but the film costs more than slides.

10. Three cytotoxicity assays are commonly used for the in vitro Comet assay: (1) trypan blue dye exclusion assay, (2) neutral diffusion assay, and (3) relative cell growth for 24 h after the treatment [30]. In addition, other tests, such as 3-(4,5-dimethylthiazol-2-yl)-2,5-diphenyltetrazolium bromide (MTT), or 2,3-bis-(2-methoxy-4-nitro-5-sulfophenyl)-2H-tetrazolium-5-carboxanilide (XTT), or 3-(4,5-dimethylthiazol-2-yl)-5-(3-carboxymethoxyphenyl)-2-(4-sulfophenyl)-2H-tetrazolium (MTS) assay, water soluble tetrazolium salt (WST) assay, adenosine triphosphate (ATP) luminescent assay, Alamar blue (redox dye resazurin) detection assay, and lactate dehydrogenase (LDH) assay, can also be used for detection of the cell viability. Before conducting the Comet assay, the cell viability for the treated cells should be tested and the cell viability should not show below 70 % compared to the concurrent control cells [31].

11. It is critical to ensure that the agarose is completely dissolved. To produce minimal bubbling and evaporation during dissolving the agarose, it is preferred to use a microwave oven at low power. The homogeneity of the agarose is also important factor for making the Comet slides, which depends on the temperature during gelling [32]. Therefore, this step should be performed in a way that can be reproduced easily, such as maintaining a constant time and temperature of gelling before immersing the slides into lysis solution.

12. It is one of the key steps to prepare the agarose with cell suspension. The cell number per slide can interfere with the image capture and analysis. To obtain optimal results, too many or too few cells per slide should be avoided so that adequate number of cells will be collected for the analysis. Therefore, appropriate cell number (cell density) or appropriate volume used to

spread onto the slide will help obtaining optimal results. For a 2-well CometSlide, 50 μl of a cell suspension (1×10^4 cells/ml) in agarose should be transferred onto the slide. For most microscope slides, 75–200 μl of a cell suspension ($1-5 \times 10^5$ cells/ml) in agarose should be transferred, depending on the area of the slide (the larger surface area of a slide allows more volume of cell suspension).

13. The optimal time for alkaline lysis is also an important factor to obtain valid and reproducible results for each cell line. The treatment conditions for both treated and control samples should be same. The temperature during alkaline lysis can significantly affect the amount of DNA migration. Therefore, it is important to establish stable and reproducible conditions for the laboratory.

14. The efficiency of enzymes is most important for this application. Therefore, it is required to use higher efficiency enzymes and sometimes enzyme purification may be required to improve efficiency. Endo III, FPG, and hOGG1 are commercially available from Trevigen. The concentration of enzymes is also important in order to obtain maximum numbers of breaks at sites of base damage. It is worth mentioning that using this method may decrease the sensitivity of base damage detection when a large number of DNA strand breaks occurs [21, 22].

15. In practice, any horizontal standard flat bed gel electrophoresis chamber can be used for this assay as long as the gel electrophoresis chamber with a power source is able to supply a constant voltage. The size of electrophoresis chamber is not critical. It is better to have a larger chamber that can run more slides at a time than a small chamber and running multiple times. The volume of electrophoresis buffer used for electrophoresis depends on the size of the chamber.

16. In the Comet assay, the voltage during electrophoresis determines migration of DNA fragment in the gel. The choices of voltage and duration of electrophoresis time are largely arbitrary if the comet size obtained is suitable to conduct image analysis. A typical voltage for electrophoresis is determined by measuring the perpendicular distance (cm) between the anode and cathode in the electrophoresis chamber and multiplying this distance by 0.6–1.0, i.e., V = perpendicular distance (cm) between the anode and cathode × (0.6–1.0). Generally, for the alkaline electrophoresis, the current should be adjusted to be less than 300 mA by raising or lowering the buffer level. Under this voltage, electrophoresis for 25–40 min can provide a useful image for comet analysis.

17. The optimal length of electrophoresis time is another key factor to obtain valid and reproducible results for each cell line. In

addition, the temperature during electrophoresis significantly influences the amount of DNA migration [33]. Therefore, it is required to establish stable and reproducible conditions, such as running electrophoresis in a cold room for a constant length of time.

18. It is critical to check the background for each slide, in order to estimate whether or not the slide is appropriate to score. For example, if the slide background is too dirty (high noise), the slide is not scorable. Generally, the cell number (CN) is used to determine if the cells on each slide are too close together, and cell damage (CD) based on the comet tail length is used to estimate the overall cell damage (this observation can be recorded on Table 5).

19. The light intensity under the microscope must be adjusted for even illumination across the field. The intensity and duration of illumination should not produce any noticeable bleaching of the image. During the image capture, analyses of the cell images at edges, and the images of two or more overlapping should be avoided. In addition, when (1) software can not differentiate between the head and the tail of a cell, (2) the staining of the nucleus of a cell is considered poor, and (3) a cell (heavily damaged) has 90 % or more DNA in the tail, these cells should not be scored during the image analysis.

20. These parameters can also be collected during image analysis, such as the total image intensity (DNA content), comet length or tail length, head diameter/tail length, percentage of DNA in tail (%DNA in tail), and tail moment.

21. In practice, the images like "clouds" or "hedgehogs" or "ghosts" are often observed; and these images (more than 90 % DNA in tail) are not scored during the image capture. The "clouds", "hedgehogs", or "ghosts", are morphologically indicative of highly damaged cells that are often associated with severe genotoxicity, necrosis, or apoptosis. The "cloud" is formed when almost the entire cell DNA is in the tail of the comet and the head is reduced in size or is almost nonexistent [12].

22. For the data analysis of the Comet assay, there is no optimal or standard statistical method [13]. Different statistical methods are used in the different laboratories and this may result in different interpretations of results, especially when a positive response is obtained using a lab-owned method. The biological significance or relevance of a positive response should also be considered [34]. In addition, historical control data (including negative and positive controls) may help interpreting these results.

Acknowledgements

The authors thank Dayton Petibone and Yan Li for the critical review of the manuscript. Use of trade names is for informational purpose only and in no way implies endorsement by the U.S. Food and Drug Administration (FDA). The views presented in this article are those of the authors and do not necessarily reflect those of the U.S. FDA.

References

1. Ostling O, Johanson KJ (1984) Microelectrophoretic study of radiation-induced DNA damages in individual mammalian cells. Biochem Biophys Res Commun 123:291–298
2. Singh NP, McCoy MT, Tice RR, Schneider EL (1988) A simple technique for quantitation of low levels of DNA damage in individual cells. Exp Cell Res 175:184–191
3. Olive PL, Banath JP, Durand RE (1990) Heterogeneity in radiation-induced DNA damage and repair in tumor and normal cells measured using the "comet" assay. Radiat Res 122:86–94
4. Olive PL, Banath JP (2006) The comet assay: a method to measure DNA damage in individual cells. Nat Protoc 1:23–29
5. Wood DK, Weingeist DM, Bhatia SN, Engelward BP (2010) Single cell trapping and DNA damage analysis using microwell arrays. Proc Natl Acad Sci USA 107:10008–10013
6. Witte I, Plappert U, de Wall H, Hartmann A (2007) Genetic toxicity assessment: employing the best science for human safety evaluation part III: the comet assay as an alternative to in vitro clastogenicity tests for early drug candidate selection. Toxicol Sci 97:21–26
7. Dusinska M, Collins AR (2008) The comet assay in human biomonitoring: gene-environment interactions. Mutagenesis 23:191–205
8. Valverde M, Rojas E (2009) Environmental and occupational biomonitoring using the Comet assay. Mutat Res 681:93–109
9. McKenna DJ, McKeown SR, McKelvey-Martin VJ (2008) Potential use of the comet assay in the clinical management of cancer. Mutagenesis 23:183–190
10. Dhawan A, Mathur N, Seth PK (2001) The effect of smoking and eating habits on DNA damage in Indian population as measured in the Comet assay. Mutat Res 474:121–128
11. Cotelle S, Ferard JF (1999) Comet assay in genetic ecotoxicology: a review. Environ Mol Mutagen 34:246–255
12. Collins AR (2004) The comet assay for DNA damage and repair: principles, applications, and limitations. Mol Biotechnol 26:249–261
13. Tice RR, Agurell E, Anderson D et al (2000) Single cell gel/comet assay: guidelines for in vitro and in vivo genetic toxicology testing. Environ Mol Mutagen 35:206–221
14. Burlinson B, Tice RR, Speit G et al (2007) Fourth International Workgroup on Genotoxicity testing: results of the in vivo Comet assay workgroup. Mutat Res 627:31–35
15. Hartmann A, Agurell E, Beevers C et al (2003) Recommendations for conducting the in vivo alkaline Comet assay. 4th International Comet Assay Workshop. Mutagenesis 18:45–51
16. Burlinson B (2012) The in vitro and in vivo comet assays. Methods Mol Biol 817:143–163
17. Liao W, McNutt MA, Zhu WG (2009) The comet assay: a sensitive method for detecting DNA damage in individual cells. Methods 48:46–53
18. Speit G, Hartmann A (2006) The comet assay: a sensitive genotoxicity test for the detection of DNA damage and repair. Methods Mol Biol 314:275–286
19. Rothfuss A, O'Donovan M, De Boeck M et al (2010) Collaborative study on fifteen compounds in the rat-liver Comet assay integrated into 2- and 4-week repeat-dose studies. Mutat Res 702:40–69
20. ICH (2011) Guidance on genotoxicity testing and data interpretation for pharmaceuticals intened for human use S2(R1). Accessed Nov 2011. http://www.ich.org/fileadmin/Public_Web_Site/ICH_Products/Guidelines/Safety/S2_R1/Step4/S2R1_Step4.pdf

21. Collins AR, Duthie SJ, Dobson VL (1993) Direct enzymic detection of endogenous oxidative base damage in human lymphocyte DNA. Carcinogenesis 14:1733–1735
22. Collins AR, Dusinska M, Gedik CM, Stetina R (1996) Oxidative damage to DNA: do we have a reliable biomarker? Environ Health Perspect 104(Suppl 3):465–469
23. Smith CC, O'Donovan MR, Martin EA (2006) hOGG1 recognizes oxidative damage using the comet assay with greater specificity than FPG or ENDOIII. Mutagenesis 21:185–190
24. Klaude M, Eriksson S, Nygren J, Ahnstrom G (1996) The comet assay: mechanisms and technical considerations. Mutat Res 363:89–96
25. Santos SJ, Singh NP, Natarajan AT (1997) Fluorescence in situ hybridization with comets. Exp Cell Res 232:407–411
26. Kiraly O, Wood D, Weingeist D et al (2012) Recombomice and CometChip technology shed light on gene-exposure interactions that impact genomic stability. Environ Mol Mutagen 53:S14
27. Spivak G (2010) The Comet-FISH assay for the analysis of DNA damage and repair. Methods Mol Biol 659:129–145
28. Mladinic M, Zeljezic D, Shaposhnikov SA, Collins AR (2012) The use of FISH-comet to detect c-Myc and TP 53 damage in extended-term lymphocyte cultures treated with terbuthylazine and carbofuran. Toxicol Lett 211:62–69
29. Woods JA, O'Leary KA, McCarthy RP, O'Brien NM (1999) Preservation of comet assay slides: comparison with fresh slides. Mutat Res 429:181–187
30. Olive PL, Banath JP (1992) Growth fraction measured using the comet assay. Cell Prolif 25:447–457
31. Anderson D, Yu TW, McGregor DB (1998) Comet assay responses as indicators of carcinogen exposure. Mutagenesis 13:539–555
32. Kusukawa N, Ostrovsky MV, Garner MM (1999) Effect of gelation conditions on the gel structure and resolving power of agarose-based DNA sequencing gels. Electrophoresis 20:1455–1461
33. Speit G, Trenz K, Schutz P, Rothfuss A, Merk O (1999) The influence of temperature during alkaline treatment and electrophoresis on results obtained with the comet assay. Toxicol Lett 110:73–78
34. Lovell DP, Omori T (2008) Statistical issues in the use of the comet assay. Mutagenesis 23:171–182

Chapter 32

Assessing DNA Damage Using a Reporter Gene System

Michael Biss and Wei Xiao

Abstract

It is of extreme importance to determine the genotoxicity of potential pharmaceutical products as it can drastically affect the potential use of those compounds. Described in this chapter is a system based in eukaryotic yeast cells that utilizes an endogenous DNA damage-responsive gene promoter and a reporter gene fusion to assess the ability of the test compounds to damage DNA. This system has been demonstrated to identify a broad range of DNA-damaging agents that show high correlation to rodent carcinogens. Furthermore, the system offers a cost effective, safe, reliable and rapid screen for genotoxic agents.

Key words Genotoxicity test, Gene expression, Budding yeast, *RNR3-LacZ*

1 Introduction

Determining the genotoxicity of a potential pharmaceutical compound is essential for determining the viability of that compound's practical use. Most pharmaceutical compounds are expected to have little or no genotoxicity if they are to be used commercially; however, there are circumstances in which genotoxicity may be desirable. It may be advantageous in the case of anticancer drugs to have genotoxic characteristics to selectively kill tumor cells that escape cell-cycle regulation and grow more quickly than normal cells, and thus are more susceptible to killing by such agents. In order to quantitate the genotoxicity of various compounds, bacteria-based test systems were developed to assess a compound's potential hazard [1–3]. In recent years, yeast-based genotoxicity screening systems have been developed and refined to surpass the risk assessment of the bacteria-based systems [4–8].

There are key benefits to utilizing a yeast-based system over that of a bacterial system. The budding yeast *S. cerevisiae* is a unicellular organism, which has the rapid growth advantage and ease of manipulation also available in bacterial systems. Furthermore, yeast are the simplest eukaryotic organism, sharing many basic metabolic processes, including similar responses to DNA-damaging and genotoxic agents to those of higher eukaryotes, including humans. This is especially important in the assessment of compounds as they may require a level of bioactivation to become hazardous to the cell. There is a wealth of knowledge available in regards to budding yeast. It was the first organism to have its genome sequenced and has been extensively studied. The *Saccharomyces Genome Database* (http://www.yeastgenome.org/) provides user-friendly access to this information. Finally, the use of budding yeast as a tool for genotoxic testing offers few potential hazards to the environment and public health, as it is already accepted for use in the food and beverage industries.

The reporter-based system described in this chapter utilizes the yeast gene *RNR3* as a sensor for DNA damage and thus genotoxicity. The yeast gene *RNR3* encodes a large subunit of ribonucleotide reductase, which is the most highly transcriptionally-upregulated damage-response gene upon treatment with DNA-damaging agents [5]. This gene was chosen not only for its sensitivity in detecting DNA damage but also for its well characterized regulatory mechanism [9–11]. In addition, *RNR3* has the ability to respond selectively to a variety of mutagenic and non-mutagenic genotoxic agents [5]. The second component of the system involves the bacterial *LacZ* gene, which acts as the reporter element of the system, under the control of the *RNR3* promoter. The *LacZ* gene encodes β-galactosidase whose enzymatic activity may be easily quantitated [12]. It has been determined that a centromeric-based, single-copy plasmid P_{RNR3}-*LacZ*, containing the promoter of *RNR3* fused to *LacZ* faithfully reflects the steady-state transcript level in both untreated and treated cells [13]. This permits use of a plasmid-based copy of the system to faithfully represent genomic *RNR3* expression. A stable genomic integration of the P_{RNR3}-*LacZ* fusion was also created to simplify the use of the reporter system [5].

To improve the yeast-based genotoxic testing system, genomic deletions of various genes involved were assessed for their ability to increase sensitivity of the system to a variety of agents. It was found that the deletion of the transcription factor *YAP1*, responsible for the activation of many genes involved in stress responses—most notably to oxidative damage, increased the detection sensitivity to many oxidative agents and other non-oxidative agents as well [7] (Fig. 1). A hyperpermeable yeast strain was also created by deleting the cell wall protein (*CWP*) genes as well as some pleotropic drug-resistant (*PDR*) genes involved in the activation of multidrug efflux pumps [6, 8]. These two deletion strain backgrounds increased the sensitivity of the system allowing for a wider range of detection [6, 7]. In addition they also were able to accurately identify compounds such as phleomycin and paraquat as genotoxic, whereas bacteria-based systems such as the Ames test have failed to do so [6, 7].

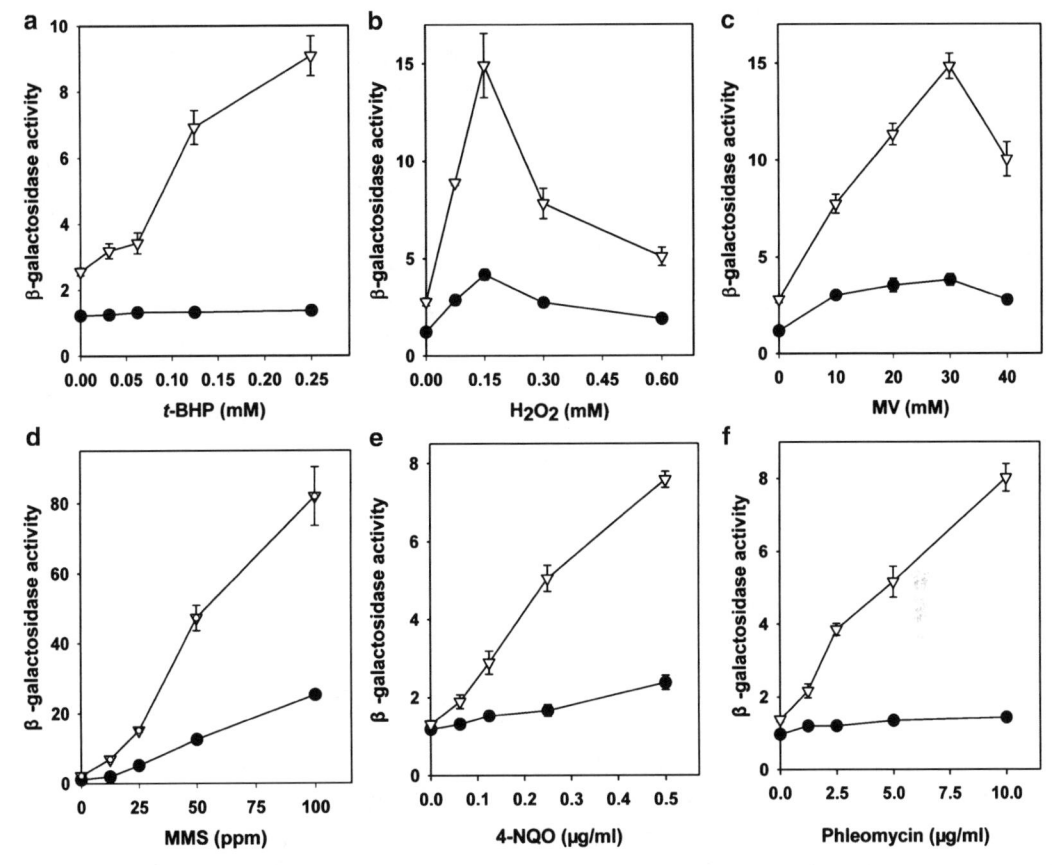

Fig. 1 Increased sensitivity of *yap1* mutation to representative genotoxic and oxidative agents. (**a**) tert-butyl hydroperoxide (*t*-BHP); (**b**) hydrogen peroxide (H$_2$O$_2$); (**c**) methyl viologen (MV); (**d**) MMS; (**e**) 4-nitroquinoline-N-oxide (4-NQO) and (**f**) phleomycin. Yeast strains used: (*filled circle*) BY4741 *trp1Δ* (wild type); (*open reverse triangle*) BY4741 *yap1Δ* mutant. Data adapted from [7] with permission

2 Materials

1. Buffer Z (60 mM Na$_2$HPO$_4$·7H$_2$O, 40 mM NaH$_2$PO$_4$·H$_2$O, 10 mM KCl, 1 mM MgSO$_4$·7H$_2$O, 40 mM β-mercaptoethanol, pH 7.0)
2. 0.1 % sodium dodecyl sulfate (SDS) solution
3. Chloroform
4. 4 mg/mL orthonitrophenyl-β-galactoside (ONPG), dissolved in sterile distilled water (note: light sensitive, dissolves at RT in a few hours)
5. 1 M Na$_2$CO$_3$
6. Yeast reporter strain: WXY1111, derivative of DBY747 carrying a stable integration of the P$_{RNR3}$-*LacZ* cassette [5]

7. Yeast extract-peptone-dextrose (YPD) liquid culture medium: 1 % (w/v) Bacto-yeast extract, 2 % (w/v) Bacto-peptone, and 2 % (w/v) glucose dissolved in distilled water, autoclave-sterilized, and stored at room temperature

8. YPD agar plates: YPD liquid medium plus 2 % (w/v) agar dissolved in water, autoclave-sterilized, poured in sterile Petri dishes and stored at 4 °C for up to 3 months

9. Synthetic dextrose (SD) liquid medium: 0.67 % (w/v) yeast nitrogen base, 2 % (w/v) glucose, add amino acids and nucleotides at working concentrations as recommended in Chap. 1

10. Synthetic dextrose agar plates: SD liquid medium plus 2 % (w/v) agar

11. Spectrophotometer capable of measuring optical density (OD) at 600 and 420 nm

12. 30 °C shaking incubator

3 Methods

3.1 Creation of Yeast Reporter Strain

The WXY1111 strain was created by integrating the P_{RNR3}-LacZ cassette into the yeast genome. The 0.88-kb promoter fragment of RNR3 fused to full length LacZ was extracted from the pZZ2 plasmid. This fragment was inserted into the genome at the HO locus with the aid of endogenous homologous recombination [14, 15].

For the use of the reporter gene system in the yap1Δ and the hyperpermeable yeast strain background the pZZ2 plasmid may be directly used [13]. The pZZ2 plasmid contains the URA3 selectable marker. To maintain the plasmid, cells are cultured in SD minimal media lacking uracil.

3.2 Yeast Cell Culture and Storage

Yeast cells (see **Note 1**) may be grown at 30 °C in YPD liquid media or on a YPD plate. It is recommended that cultures used for this study be subcultured from an overnight culture grown to saturation into fresh liquid YPD at a ratio of 1:30.

Yeast cells may be viably stored for many years. Cells are grown on YPD plates for 2 days and transferred into 1 mL of 15 % (v/v) glycerol. Cells may also be grown in liquid YPD. Spin 2 mL of liquid culture down and resuspend in 1 mL of 15 % (v/v) glycerol. Store the cells at −70 °C.

3.3 Optimization of Drug Dose

Optimization of drug dose is essential as most genotoxic agents display a dose-dependent increase in RNR3-LacZ expression, which peaks and is followed by a decrease in activity. This is most likely due to extensive killing of cells by the agent past a certain drug concentration.

1. Culture reporter strain overnight in YPD liquid media.
2. Subculture overnight culture into fresh YPD liquid media at a ratio of 1:30 and let incubate at 30 °C with shaking at 150–200 rpm for 2 h.
3. Separate the culture into 2-mL aliquots in test tubes.
4. Make serial dilutions of the drug to be tested and add the drug to the cultures to achieve the desired range of test concentrations.
5. Incubate at 30 °C with shaking for 4 h.
6. Transfer 1 mL of each culture to a separate microcentrifuge tube and spin at 15,000 rpm for 30 s. Dump the supernatant and wash once with sterile distilled water. Resuspend the cells in 1 mL sterile distilled water.
7. Make serial dilutions of each test and spread 100 µL onto YPD plates.
8. Incubate plates for 2 days at 30° and record colony growth.

The cell survival rate is determined by comparing the colony-forming units (cfu) on the drug-treated samples to the untreated control sample. It is recommended that a drug dose that yields between 10 and 90 % survival be used to achieve the optimal *RNR3-lacZ* genotoxicity test.

3.4 β-Galactosidase (β-Gal) Activity Assay

This method is adapted from the previously-described assay in reference 16. Each drug to be tested must be accompanied with an untreated control, which is comprised of the addition of only the solvent used to dissolve the drug. Each test should be performed in triplicate to ensure statistical relevance.

1. The reporter strain is cultured overnight in 2 mL of YPD liquid medium.
2. Subculture 1 mL of overnight culture into 30 mL of fresh YPD liquid media to a cell density of approximately $OD_{600nm} = 0.1$. The final volume may be altered to accommodate the number of assays to be performed. 3 mL is required per individual test.
3. Incubate cells at 30 °C with shaking for 2 h or until the cell density has reached approximately 1×10^7 cells/mL or an $OD_{600nm} = 0.2$–0.3.
4. Aliquot 3 mL of culture into one sterile culture tube per individual test.
5. Prepare drug to be tested by dissolving in appropriate solvent to the desired concentration and add to above culture tubes to achieve the desired final concentrations (*see* **Note 2**).
6. Incubate at 30 °C with shaking for 4 h (*see* **Note 3**).
7. Remove 1 mL per tube to determine the OD_{600nm} (*see* **Note 4**).

8. Centrifuge the remaining 2 mL of sample at 3,500 rpm for 4 min and remove the supernatant.
9. Resuspend the pellet in 1 mL of Buffer Z.
10. Add 50 μL of 0.1 % SDS and 50 μL of chloroform to the samples. Vortex each sample on top speed for 10 s to permeabilize cells.
11. Add 200 μL of ONPG to initiate the reaction and incubate at 30 °C with shaking for 20 min.
12. Add 500 μL of 1 M Na_2CO_3 to quench the reaction.
13. Centrifuge at 3,500 rpm for 4 min.
14. Remove the supernatant taking care not to disturb the pellet to measure the OD_{420nm} (*see* Note 5).
15. Determine the β-gal specific activity ($SA\beta_{-gal}$) using the following equation:

$$SA_{\beta\ gal} = \frac{1000 \cdot OD_{420nm}}{Reaction time (\min) \cdot Reaction Volume (mL) \cdot OD_{600nm}}$$

The recommended reaction time is 20 min. The recommended culture volume is 2 mL or less. β-gal activity is measured in Miller units [16].

3.5 Data Analysis

Expression of *RNR3-LacZ* is measured as a ratio of induction of β-gal activity for treated versus untreated cells assayed in the same experiment. It is suggested that a minimum of three independent experiments be used in determining the average activity for a specific treatment. Typical standard deviation between replicates of the same experiment is expected to be within 10 % as the assay is highly reproducible (Fig. 1). A compound, which produces a greater than twofold increase in β-gal activity, is considered to offer a positive result of genotoxicity.

4 Notes

1. If the WXY1111 strain is unsuitable for your experiment, a different strain background may be used with the plasmid pZZ2 transformed into the strain of interest [17]. The plasmid must be maintained by culturing cells in SD liquid medium lacking uracil.
2. The untreated control test cultures should be treated with the same volume of solvent used to dissolve the drug.
3. This drug dosage time is based on the determined optimal dosage time for methyl methanesulfonate. The dosage time may be changed to achieve optimal effect for different compounds.

4. It is important for the cells to be grown to an $OD_{600nm} = 0.4$–0.5, or cell density of 2×10^7 cells/mL representing active cell division in the culture. It has been shown that active cell division is required for the induction of *RNR3* in response to damaging agents.

5. To reduce background β-gal activity, a strain isogenic to your test strain but which does not contain the *RNR3-LacZ* cassette may be cultured under identical conditions to zero the spectrophotometer at OD_{420nm}.

References

1. Quillardet P, Huisman O, D'Ari R, Hofnung M (1982) The SOS chromotest: direct assay of the expression of gene sfiA as a measure of genotoxicity of chemicals. Biochimie 64:797–801
2. Oda Y, Nakamura S, Oki I, Kato T, Shinagawa H (1985) Evaluation of the new system (umu-test) for the detection of environmental mutagens and carcinogens. Mutat Res 147:219–229
3. el Mzibri M, De Meo MP, Laget M, Guiraud H, Seree E, Barra Y, Dumenil G (1996) The salmonella sulA-test: a new in vitro system to detect genotoxins. Mutat Res 369:195–208
4. Jia X, Xiao W (2003) Compromised DNA repair enhances sensitivity of the yeast RNR3-lacZ genotoxicity testing system. Toxicol Sci 75:82–88
5. Jia X, Zhu Y, Xiao W (2002) A stable and sensitive genotoxic testing system based on DNA damage induced gene expression in Saccharomyces cerevisiae. Mutat Res 519:83–92
6. Zhang M, Hanna M, Li J, Butcher S, Dai H, Xiao W (2010) Creation of a hyperpermeable yeast strain to genotoxic agents through combined inactivation of PDR and CWP genes. Toxicol Sci 113:401–411
7. Zhang M, Zhang C, Li J, Hanna M, Zhang X, Dai H, Xiao W (2011) Inactivation of YAP1 enhances sensitivity of the yeast RNR3-lacZ genotoxicity testing system to a broad range of DNA-damaging agents. Toxicol Sci 120:310–321
8. Zhang M, Liang Y, Zhang X, Xu Y, Dai H, Xiao W (2008) Deletion of yeast CWP genes enhances cell permeability to genotoxic agents. Toxicol Sci 103:68–76
9. Zhou Z, Elledge SJ (1993) DUN1 encodes a protein kinase that controls the DNA damage response in yeast. Cell 75:1119–1127
10. Huang M, Zhou Z, Elledge SJ (1998) The DNA replication and damage checkpoint pathways induce transcription by inhibition of the Crt1 repressor. Cell 94:595–605
11. Fu Y, Xiao W (2006) Identification and characterization of CRT10 as a novel regulator of Saccharomyces cerevisiae ribonucleotide reductase genes. Nucleic Acids Res 34:1876–1883
12. Beck CF (1979) A genetic approach to analysis of transposons. Proc Natl Acad Sci USA 76:2376–2380
13. Zhou Z, Elledge SJ (1992) Isolation of crt mutants constitutive for transcription of the DNA damage inducible gene RNR3 in Saccharomyces cerevisiae. Genetics 131:851–866
14. Voth WP, Richards JD, Shaw JM, Stillman DJ (2001) Yeast vectors for integration at the HO locus. Nucleic Acids Res 29:e59
15. Boeke JD, Trueheart J, Natsoulis G, Fink GR (1987) 5-Fluoroorotic acid as a selective agent in yeast molecular genetics. Methods Enzymol 154:164–175
16. Guarente L (1983) Yeast promoters and lacZ fusions designed to study expression of cloned genes in yeast. Methods Enzymol 101:181–191
17. Mount RC, Jordan BE, Hadfield C (1996) Transformation of lithium-treated yeast cells and the selection of auxotrophic and dominant markers. Methods Mol Biol 53:139–145

Chapter 33

Improved AMES Test for Genotoxicity Assessment of Drugs: Preincubation Assay Using a Low Concentration of Dimethyl Sulfoxide

Atsushi Hakura

Abstract

The *Salmonella*/microsome bacterial mutagenicity test (Ames test) is used worldwide as a simple and rapid mutagenicity testing system. Several modified versions of the Ames test have been developed, subsequent to the original "plate incorporation assay" using the *Salmonella* bacterial tester strains, and rat liver homogenate fraction (S9) for generating reactive metabolites of test compounds. Among the modifications, Ames test with a modified procedure of preincubation (called a preincubation assay) has been the most frequently used. In this assay, dimethyl sulfoxide (DMSO) has been often used at a concentration of 7 or 14 % (particularly in Japan) in the preincubation mixture (0.05 or 0.1 mL of DMSO in 0.7 mL of the mixture), as a vehicle to dissolve a wide range of chemicals. However, DMSO is known to inhibit several kinds of drug-metabolizing enzymes including CYPs involved in the metabolic activation or detoxification of chemical mutagens, even at low concentrations of 1 % or less. Therefore, an improved preincubation assay using a low concentration (*e.g.*, 1 %) of DMSO is recommended as one option for modification of the Ames test to improve its sensitivity towards the detection of promutagens that require metabolic activation by reducing the inhibiting effect of DMSO on drug-metabolizing enzymes. In genotoxicity tests using mammalian cells, a DMSO concentration of 1 % is often used. In this chapter, a detailed protocol for this slightly modified preincubation assay is presented, together with data useful for the performance of the Ames test and the interpretation of the results.

Key words Ames test, Preincubation assay, Revertant, Dimethyl sulfoxide, Drug metabolism, Cytotoxicity, Bacterial background lawn

1 Introduction

The *Salmonella*/microsome bacterial mutagenicity test (called Ames test) is used worldwide as a simple and rapid testing system for detecting mutagens and possible carcinogens [1, 2]. Several modified versions of the Ames test have been developed, subsequent to the original "plate incorporation assay" using the *Salmonella* bacterial test strains, and rat liver homogenate fraction (S9) so as to assess the mutagenicity of reactive metabolites generated from test compounds.

The main purpose of these modifications is to raise the sensitivity of the Ames test towards detecting mutagens and to understand the mechanisms of mutagenesis. The modifications include the use of genetically-engineered bacterial strains (*e.g.*, drug-metabolizing enzyme-overexpressing strains and DNA repair-deficient strains) [3–7], the use of liver S9 fraction prepared from other kinds of animals (*e.g.*, hamster) [8, 9] or the addition of supplements (*e.g.*, addition of reducing agents such as riboflavin to detect the mutagenicity of azo compounds) for more efficient, metabolic activation [10–12], and modified procedures of the original "plate incorporation assay" (*e.g.*, preincubation assay, and desiccator assay for volatile chemicals) [1, 2, 12–15]. This chapter deals with a preincubation assay using a low concentration of dimethyl sulfoxide (DMSO) [16].

The preincubation assay is generally considered to be equal to or more sensitive than the plate incorporation assay, with a few exceptions. This is probably because short-lived mutagenic metabolites may have a better chance reacting with tester strains at a smaller volume or at a higher concentration of the preincubation mixture (treatment mixture) in test tubes, and the higher concentration of S9 fraction is achievable in the preincubation mixture than on the plate [2, 14, 17].

DMSO has been often used as a good vehicle to dissolve a wide range of chemicals in the Ames test, because: (1) it is completely miscible with molten top agar, (2) it has nearly universal capability to dissolve organic chemicals, and (3) it is relatively nontoxic to bacteria [1, 2, 18]. In the preincubation assay, DMSO has been usually used at concentrations of 7 or 14 % (particularly in Japan) in the preincubation mixture (0.05 or 0.1 mL of DMSO in 0.7 mL of the mixture) [2, 19]. However, DMSO is known to inhibit several kinds of drug-metabolizing enzyme such as CYPs involved in the metabolic activation or detoxification of chemical mutagens [20, 21], even at low concentrations of 1 % or less [22–26]. Recently, we showed the inhibitory effect of DMSO on the mutagenicity of several kinds of promutagens that require metabolic activation; DMSO reduced the mutagenic activity of several promutagens at a concentration of 14 % (corresponding to the addition of 0.1 mL of DMSO), as compared to that of 1 % (corresponding to its addition of 0.07 mL) or 0 % (no addition of DMSO). Therefore, an improved preincubation assay using a low concentration (*e.g.*, 1 %) of DMSO is recommended as one option for modification to raise the sensitivity of the Ames test [16]. In genotoxicity tests using mammalian cells, a DMSO concentration of 1 % is often used. Thus, this simple modification may provide us better approximated-metabolic profiles to *in vivo* animal systems and allow us to directly compare the results from *in vitro* mammalian genotoxicity tests [27]. More importantly, reduction in the risk of misjudgement as false-negative results may be expected, particularly at a relatively low dose level. Butter yellow

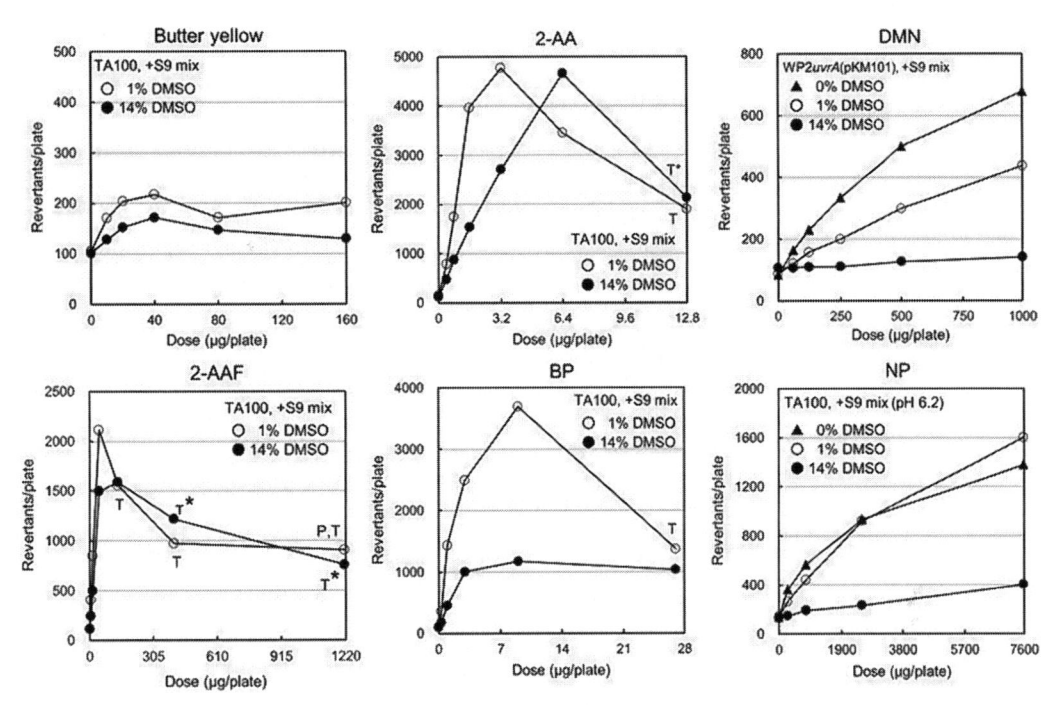

Fig. 1 Dose-response curves of the mutagenicities of chemicals in the presence of DMSO concentrations of 1 % (*open circle*) and 14 % (*filled circle*). Butter yellow and 2-AAF showed no significant differences in the mutagenicity between 1 and 14 %. In contrast, moderate (2-AA and BP) or marked (DMN and NP) differences were found. DMN and NP were also tested in the absence of DMSO (*filled triangle*). The symbols, "T", "T*", and "P" indicate 'toxic at 1 % DMSO', 'toxic at 14 % DMSO', and 'precipitation of test article', respectively. Mutagenicity was assayed by the preincubation assay (37 °C, 20 min) in the presence of PB/BF-induced rat liver S9 mix (10 % S9). Assays were conducted in duplicate, and mean values were plotted. Abbreviations of mutagens tested: butter yellow; *N,N*-dimethyl-4-aminoazobenzene, *2-AAF* 2-acetylaminofluorene, *2-AA* 2-aminoanthracene, *BP* benzo[*a*]pyrene, *DMN* dimethylnitrosamine, *NP* N-nitrosopyrrolidine. Modified from [16]

may be a good example; it shows a more significant dose-response relationship with a twofold increase in the number of revertants over the vehicle control at a DMSO concentration of 1 % as compared with 14 %, being judged as equivocal at 14 %, but as positive at 1 % DMSO [16] (*see* Fig. 1).

2 Materials

1. 0.5 mM Histidine/0.5 mM biotin solution: D-Biotin (122 mg) and L-histidine·HCl monohydrate (105 mg) are dissolved in purified water (1,000 mL), and the solution is filtered through a 0.45-μm Millipore filter for sterilization. Stock solution stored in a refrigerator (4 °C) can be used for at least 6 months.

2. 0.5 mM Tryptophan solution: L-Tryptophan (51 mg) is dissolved in purified water (500 mL), and the solution is filtered

through a 0.45-μm Millipore filter for sterilization. Stock solution stored in a refrigerator (4 °C) can be used for at least 6 months.

3. 100 mM Sodium phosphate buffer (pH 7.4): Solution II (3.6 g of NaH_2PO_4 in 300 mL of purified water) is gradually added to Solution I (14.2 g of Na_2HPO_4 in 1,000 mL of purified water) to adjust the pH to 7.4. The buffer is then distributed to bottles and autoclaved at 121 °C for 15 min. Stock solution of the phosphate buffer stored in a refrigerator can be used for at least 1 year.

4. 0.05 mM Histidine/0.05 mM biotin-top agar: 0.5 mM Histidine/0.5 mM biotin solution (20 mL), Bacto Agar (Difco, 1.2 g), and NaCl (1.0 g) are added to purified water (180 mL). The mixture (200 mL) is autoclaved at 121 °C for 15 min prior to use. Molten top agar is made homogeneous by rotary shaking, and then maintained at 45 °C during use.

5. 0.05 mM Tryptophan-top agar: 0.5 mM Tryptophan solution (20 mL), Bacto Agar (Difco, 1.2 g), and NaCl solution (1.0 g) are added to purified water (180 mL). The mixture (200 mL) is autoclaved at 121 °C for 15 min prior to use. Molten top agar is made homogeneous by rotary shaking, and then maintained at 45 °C during use.

6. Nutrient broth liquid medium (NB): Oxoid Nutrient No. 2 (Unipath, 2.5 g) is dissolved in purified water (100 mL) and autoclaved at 121 °C for 15 min. NB stored at room temperature in the dark can be used for at least 1 month (*see* **Note 1**).

7. Minimal-glucose agar medium (plate): The medium is commercially available (*e.g.*, CLIMEDIA AM-N, Oriental Yeast Co., Ltd., Tokyo). For preparation, Solution A (minimal medium: 0.2 g of $MgSO_4 \cdot 7H_2O$, 2 g of citrate·H_2O, 10 g of K_2HPO_4, 1.92 g of $(NH_4)_2HPO_4$, and 0.66 g of NaOH in 200 mL of purified water), Solution B (20 % glucose solution: 20 g of glucose in 100 mL of purified water), and Solution C (agar solution: 15 g of agar in 700 mL of purified water) are separately prepared, autoclaved at 121 °C for 15 min, and mixed. The prepared medium is distributed to plastic plates (diameter: 86 mm) at a volume of 30 mL. After the agar has hardened, the plates are stored in sealed plastic bags to prevent drying at room temperature. They can be used for at least 6 months.

8. Rat liver S9 fraction: Rat liver S9 fraction (supernatant from a 9,000g centrifugation of liver homogenate) including CYP enzymes is prepared from male Sprague-Dawley rats intraperitoneally treated with phenobarbital (PB) at doses of 30 mg/kg/day (96 h before sacrifice) and 60 mg/kg/day (24, 48, and 72 h before sacrifice) and with 5,6-benzoflavone (BF) at a

dose of 80 mg/kg/day (48 h before sacrifice). It is commercially available, for example, from Oriental Yeast Co., Ltd.

9. S9 mix: CYPs present in S9 fraction are largely involved in the metabolic activation or detoxification of chemical mutagens, and therefore, S9 mix usually consists of S9 fraction and cofactors for CYPs. The cofactors are commercially available (*e.g.*, Cofactor I, Oriental Yeast Co., Ltd.). One vial (for 10 mL of S9 mix) of Cofactor I contains 8 μmol of $MgCl_2 \cdot 6H_2O$, 33 μmol of KCl, 5 μmol of glucose-6-phosphate, 4 μmol of NADPH, 4 μmol of NADH, and 100 μmol of $Na_2HPO_4/NaH_2PO_4 \cdot 2H_2O$ per mL solution, and it can be used for at least 1 year when stored in a refrigerator. The cofactor solution is prepared by adding 9 mL of purified water into a vial containing cofactors, followed by filtration through a 0.45-μm Millipore filter. The solution is combined with thawed S9 fraction (often 1 mL of S9 fraction to prepare 10 % v/v S9 fraction in the S9 mix, corresponding to a content of *ca.* 1 mg S9 protein per plate), mixed, and immediately placed on ice to maintain S9 enzyme activity.

10. Bacterial tester strains: Bacterial tester strains (*Salmonella typhimurium* TA100, TA1535, TA98, and TA1537, and *Escherichia coli* WP2*uvrA*(pKM101)) should be confirmed for genetic characteristics (histidine or tryptophan dependence, *rfa* marker, *uvrA* or *uvrB* deletion, and pKM101 plasmid), and His⁺ or Trp⁺ negative and positive control values before use. The genetic characteristics of these strains are listed in Table 1, and Table 2 lists in-house background data for negative (spontaneous) and positive control values of revertants [1, 2, 28, 29].

Table 1
Genetic analysis of the bacterial tester strains used in the Ames test [1, 2, 28, 29]

Strain	Amino acid dependency for mutation	Crystal violet sensitivity for *rfa* marker	UV sensitivity for *uvr* deletion	Ampicillin sensitivity for pKM101
TA1535	His	S	S	S
TA100	His	S	S	R
TA1537	His	S	S	S
TA98	His	S	S	R
WP2*uvrA* (pKM101)	Trp	R	S	R

His histidine, *Trp* tryptophan, *S* sensitive, *R* resistance, *UV* ultraviolet

Table 2
Our background data for negative (spontaneous) and positive revertant control values[a]

	No. of revertants/plate			
	Negative control		Positive control	
Strain	−S9 mix	+S9 mix	−S9 mix	+S9 mix
TA1535	3–33 (Water)	3–25 (DMSO)	381–679 (SA, 0.5 μg/plate)	150–351 (2-AA, 2 μg/plate)
TA100	84–155 (DMSO)	87–182 (DMSO)	320–690 (AF-2, 0.01 μg/plate)	752–1,764 (2-AA, 1 μg/plate)
TA1537	2–37 (DMSO)	2–32 (DMSO)	195–1,152 (9-AA, 80 μg/plate)	90–286 (2-AA, 2 μg/plate)
TA98	13–54 (DMSO)	16–54 (DMSO)	497–916 (AF-2, 0.1 μg/plate)	362–925 (2-AA, 0.5 μg/plate)
WP2*uvrA* (pKM101)	45–173 (DMSO)	53–245 (DMSO)	569–1,610 (AF-2, 0.005 μg/plate)	721–2,150 (2-AA, 2 μg/plate)

DMSO dimethyl sulfoxide, *SA* sodium azide, *2-AA* 2-aminoanthracene, *AF-2* 2-(2-furyl)-3-(5-nitro-2-furyl)acrylamide, *9-AA* 9-aminoacridine

[a]Minimum-maximum revertants yielded in the preincubation assay at a DMSO concentration of 10 % in the preincubation mixture

3 Methods

3.1 Preparation of Frozen Permanent Cultures

Each frozen permanent culture should be prepared by cultivating a single colony of bacterial tester strain, and selected or confirmed for subsequent use after the completion of the strain check in terms of phenotypic characterization, spontaneous mutation induction, and sensitivity to positive control articles (*see* Subheading 2.10).

1. Each single colony obtained is inoculated to a glass conical flask (100 mL) containing 20 mL of NB, and the flask is incubated at 37 °C with shaking (140 rpm) for 10–12 h to obtain bacterial culture in the early stationary phase of the strain (over 1×10^9 cells/mL). The cell density may be monitored by measurement of a portion of cell culture at 660 nm using a spectrophotometer.

2. DMSO (0.7 mL) from freshly opened bottle (sterilization is not necessary) is placed in a test tube and combined with 8 mL of bacterial culture obtained. The rest is used for selection and confirmation of each bacterial culture of the strains to check phenotypic characterization, spontaneous mutation induction, and sensitivity to positive control articles [1, 2, 28, 29] (*see* Tables 1 and 2).

3. Bacterial culture to be used for frozen permanent stock is dispensed in 200-μL aliquots to γ-irradiated Assist tubes. After being tightly capped, the tubes are quickly frozen in liquid nitrogen and stored in a deep freezer (−80 °C). Freshly frozen permanent cultures can be used for preparation of frozen working cultures for at least 5 years.

3.2 Preparation of Frozen Working Cultures

Frozen working cultures should be prepared from selected and confirmed frozen permanent culture stocks, and likewise selected or confirmed for subsequent use after the completion of the strain check in terms of phenotypic characterization, spontaneous mutation induction, and sensitivity to positive control articles (*see* Subheading 2.10).

1. Frozen permanent culture is taken from the deep freezer and allowed to thaw at room temperature.

2. Thawed permanent culture stock (20 μL) is transferred to a glass conical flask (100 mL) containing 20 mL of NB, and the flask is incubated at 37 °C with shaking (140 rpm) for 10 h to obtain bacterial culture in the early stationary phase of the tester strains ($2–5 \times 10^9$ cells/mL). The cell density may be monitored by measurement of a portion of cell culture at 660 nm using a spectrophotometer.

3. DMSO (0.7 mL) from freshly opened bottle (sterilization is not necessary) is placed in a test tube and combined with 8 mL of bacterial culture obtained. The rest is used for selection and confirmation of each bacterial culture of the strains to check phenotypic characterization, spontaneous mutation induction, and sensitivity to positive control articles [1, 2, 28, 29] (*see* Tables 1 and 2).

4. Bacterial culture for stock is dispensed in 200-μL aliquots to γ-irradiated Assist tubes. After being tightly capped, the tubes are quickly frozen in liquid nitrogen and stored at a deep freezer (−80 °C) until use. Freshly frozen working cultures can be used for mutagenicity testing for at least 1 year.

3.3 Preparation of Bacterial Cultures

1. Frozen working culture is taken from the deep freezer and allowed to thaw at room temperature.

2. A portion of the thawed working culture is added to a glass conical flask (100 mL) containing 20 mL of NB (*see* **Note 2**). The number of bacterial cells transferred is 8×10^7 for the *S. typhimurium* strains and 2×10^7 for the *E. coli* strain. The volume of the working culture should be determined in advance, based on the cell number of the frozen working cultures and the growth curve of each bacterial tester strain in NB. The remainder of the thawed working culture should be discarded.

3. The flask is placed at 4 °C in a shaking incubator equipped with a timer (Bioshaker BR–40LF, Taitec). The incubator is set up for the first 6 h (depending on the time when bacterial culture is needed for treatment) at 4 °C, followed by shaking (140 rpm) of the flask for 10 h so as to obtain bacterial culture in the early stationary phase of the tester strains (*see* **Note 3**).

4. A portion of bacterial culture obtained is diluted tenfold with NB, and the optical density of the diluted cell suspension is measured at 660 nm by a spectrophotometer. The cell number should be confirmed to be $2–5 \times 10^9$ cells/mL (*see* **Note 4**), based on the pre-made working curve of optical density *vs* cell number (*see* **Note 5**). Unless the cell number satisfies the above criterion, the bacterial culture should not be used.

5. The bacterial culture is transferred to a 50-mL plastic tube, and placed at room temperature or on ice (*see* **Note 6**).

3.4 Mutagenicity Testing

1. S9 mix or sodium phosphate buffer (0.5 mL) is added to each test tube (105×16.5 mm) using an Eppendorf dispenser.

2. Bacterial culture (0.1 mL) is immediately added to each test tube and mixed by a touch mixer for 1 s.

3. Test article solution dissolved in DMSO (6 µL) is then immediately added to each test tube (*see* **Note 7**) and mixed by a touch mixer for 1 s.

4. After molten caps are put on the test tubes, the tubes are incubated for 20 min at 37 °C in a shaking (120 rpm) water bath (water bath shaker MM-10, Taitec) (*see* **Note 8**).

5. After shaking (*see* **Note 9**), 2 mL of molten 0.05 mM histidine/0.05 mM biotin-top agar (for the *Salmonella* strains) or 0.05 mM tryptophan-top agar (for the *E. coli* strain) (*see* **Note 10**), maintained at 45 °C (Dry Thermo Unit, DTU-1C, Taitec) is added to each test tube and mixed by a touch mixer for 1–2 s. The contents are immediately poured onto the surface of minimal-glucose agar plates (two plates per each dose).

6. Within about 5 min after the top agar has hardened (1–2 min), plates are inverted and placed in an incubator at 37 °C for about 48 h (*see* **Note 11**).

7. The number of revertants that has appeared on each plate is counted, together with observation of bacterial background lawn as an indicator for cytotoxicity of test chemical (*see* **Note 12**).

3.5 Assay Acceptance Criteria

This test is considered valid if the following criteria are satisfied:

1. No bacterial contamination is observed.

2. The numbers of revertants for the negative and positive controls are similar to the historical or reference data (*see* Table 2).

3.6 Assay Evaluation Criteria

Test article is considered to be mutagenic if the following criteria are satisfied:

1. There is a dose-related increase in the number of revertants.
2. The number of revertants at one or more treatment doses is twofold or more over the negative control (*see* **Note 13**).
3. Results are reproducible.

4 Notes

1. Source or lot of nutrient broth is important. They might greatly affect the growth rate of bacteria (we have had the rare experience of no growth of bacteria). Therefore, we should obtain data regarding the growth curve of each bacterial tester strain in each lot of nutrient broth.

2. For preparation of bacterial culture of TA100, TA98, and WP2*uvr*A(pKM101), which carry the pKM101 plasmid, ampicillin might be added to NB to prevent its possible loss. However, its addition is not necessary for preparation of bacterial cultures to be used in the mutagenicity testing. Because the Ames test results are not affected by its presence or absence.

3. Because the experimental conditions (*e.g.*, aeration and incubation period) for preparation of bacterial culture are known to affect the test results to a large extent, it is important to determine the best conditions matched to the equipment and supplies used in each laboratory in advance. Aerobic conditions are recommended. The use of bacterial culture in the late stationary phase, obtained by an excessive period of incubation should be avoided to prevent possible reduction of the test sensitivity to the mutagenicity of chemicals. AF-2 is exemplified in Fig. 2; its mutagenicity is largely affected by the culture conditions of aeration [30].

4. The cell number of bacterial culture is dependent on the bacterial tester strain and the source or lot of nutrient broth. The cell number of the *E. coli* strain is slightly larger than that of the *Salmonella* strains [31].

5. The working curve of optical density *vs.* cell number is recommended to be prepared with each tester strain, because their relationships are slightly different from each other.

6. Bacterial culture can be used within 3 h after the end of the cultivation.

7. If test articles are soluble in water, it is preferable for use as a vehicle to avoid the inhibitory effect of DMSO. In addition, our previous study showed that in the preincubation assay, the number of bacteria was reduced by treatment with DMSO at a

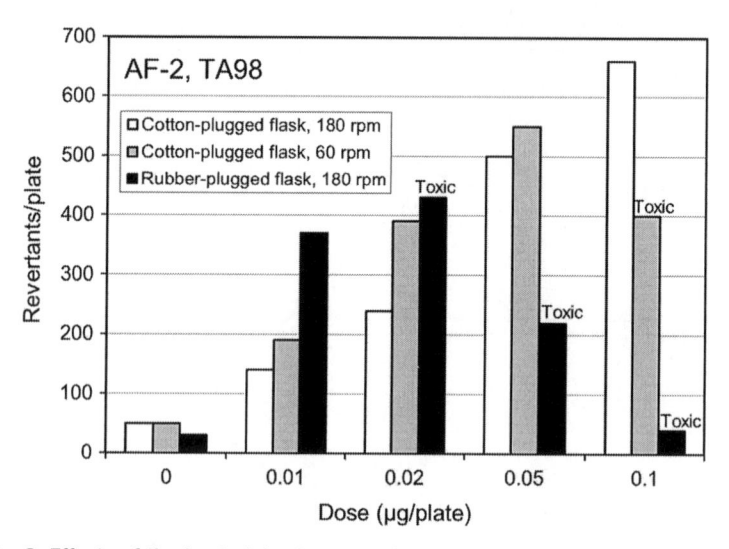

Fig. 2 Effects of the bacterial culture conditions employed in the preincubation assay on the mutagenicity of AF-2 towards the TA98 strain. Three different conditions to affect the oxygen concentration of culture medium were examined; (1) culture in a cotton-plugged flask by shaking at 180 rpm (the most aerobic condition), (2) culture in a cotton-plugged flask by shaking at 60 rpm, and (3) culture in a rubber-plugged flask shaking at 180 rpm (the most anaerobic condition). The TA98 strain shows different responses to the mutagenicity and cytotoxicity of AF-2 under the different culture conditions. These different responses are probably attributable to the different activities of nitroreductase induced by the different conditions, which is responsible for the metabolic activation and detoxification in its mutagenicity and cytotoxicity. Modified from [30]

concentration of 14 % in the preincubation mixture, with no notable diminution in bacterial background lawn [31]. This cytotoxic effect of DMSO is observed particularly with the TA100 and TA1535 strains among the recommended bacterial tester strains by guideline [17, 19]. However, importantly, the reduction in the number of surviving bacterial cells (reduction of 60–90 %, depending on the bacterial tester strains and the presence or absence of S9 mix) is likely not to largely affect the sensitivity of the Ames test towards detecting the mutagenicity of chemicals, as shown in Fig. 3 with a direct mutagen, methyl methanesulfonate (MMS) [31]. The reason why the number of revertants was not markedly reduced in spite of the low cell survival rate is not clear. But, one possible reason is an increase in the mutation frequency of His⁻ to His⁺, accompanied by an increase in cell divisions with continuous exposure to a relatively long-lived MMS [32–34], in proportion to the increasing amount of histidine per surviving cell; such a mechanism would be analogous to the finding that a proportional correlation does not exist between the number of His⁺ revertants produced and the number of cells plated (*see* Fig. 4a).

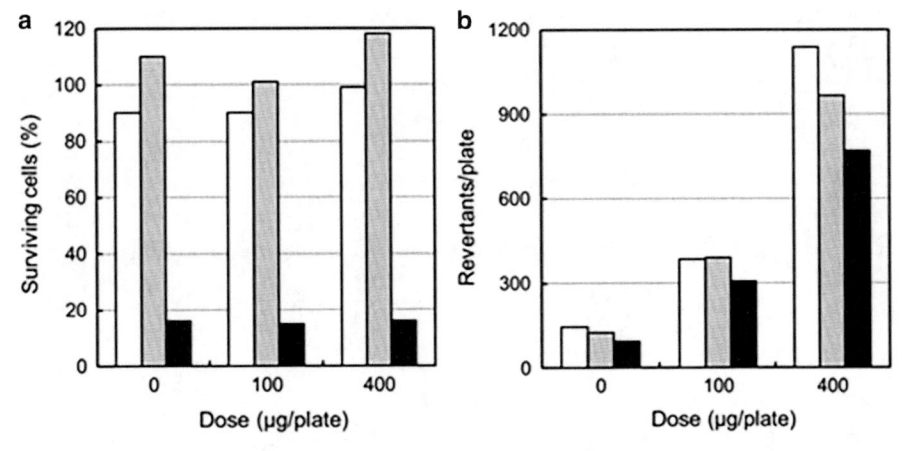

Fig. 3 Cell survivors (**a**) and revertant colonies (**b**) following treatment of TA100 with MMS in a preincubation mixture with a DMSO concentration of 0 % (*white bars*), 1 % (*gray bars*), or 14 % (*black bars*) in the preincubation assay. Surviving cells are indicated as the percent ratio compared with those before preincubation in the preincubation mixture. The numbers of surviving colonies and revertant colonies were determined using samples taken from the same treatment mixture. MMS is a direct mutagen, and was, hence, selected as a representative mutagen so as to remove the possible effects of DMSO on the metabolic activation or deactivation of mutagens. In spite of the marked reduction in cell survivors (a cell survival rate of about 20 %) at a DMSO concentration of 14 % (**a**), the number of revertant colonies was slightly lower than that produced with a negative control (water) with no notable diminution in bacterial background lawn (**b**). Modified from [31]

However, DMSO might still lower the yield of revertants or cause the preincubation assay to be less sensitive as a result of the decrease in cell survival arising from DMSO cytotoxicity.

8. The plate incorporation assay (original procedure for the Ames test) lacks this step. Bacterial tester strains can be additionally exposed to mutagens after being plated on plates, if mutagens are relatively long-lived.

9. It is important to prevent bacterial contamination by blotting water attached to outside the test tubes with absorbent paper or cotton sheets dipped in 70 % ethanol prior to the addition of top agar to test tubes.

10. The trace of the amino acid in the top agar allows all the bacteria on the plate to undergo several divisions for fixation of mutational lesions. Increase in the amount of the amino acid on the plate enhances mutagenesis, but also causes heavy growth of the background lawn that obscures the revertants. The deletion through the *uvrB* gene includes the biotin gene, and therefore biotin is necessary for growth of the *Salmonella* tester strains.

11. The same incubation period is recommended to be always used between 48 and 72 h for the formation of revertant colonies (we always incubate plates for approximately 48 h), because

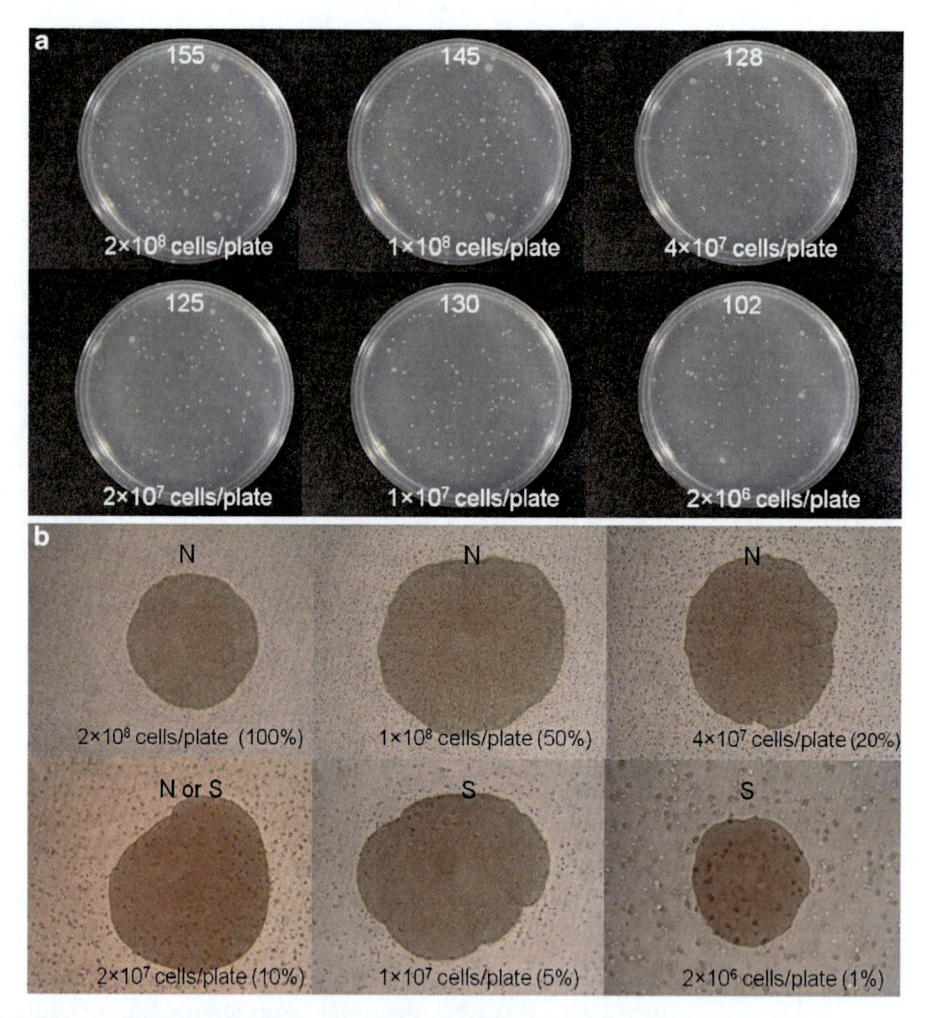

Fig. 4 Representative views of TA100 revertant colonies that spontaneously appeared after incubation of a content containing 2 mL of 0.05 mM histidine/0.05 mM biotin-top agar and 0.1 mL of a serially diluted bacterial culture. The number of His⁻ bacterial cells plated are 2×10^8, 1×10^8, 4×10^7, 2×10^7, 1×10^7, and 2×10^6 cells/plate. The figures in parentheses (100–1 %) are the ratios of the number of survivors expressed as a percentage of the number of those from non-diluted bacterial culture (2×10^8 cells/plate). The mean number of revertants/plate in duplicate is shown in the upper side (**a**). A proportional correlation does not exist between the number of His⁺ revertants produced and the number of cells plated. Their magnified views are shown in (**b**). A large colony located in the center of each view is a revertant one. Numerous microcolonies visible around the revertant colony are non-revertant ones. The judgment of bacterial background lawn by the naked eye is shown in the *upper* side; "N" (normal or no diminution) or "S" (slight but significant diminution). A slight but significant diminution of bacterial background lawn is noted at *ca.* 10 % or less of the survivors in the TA100 strain. Its significant diminution may be noted at *ca.* 20 % of the survivors by microscopic (40×) examination. Reduced survivors resulted in more histidine being available to the surviving His⁻ bacteria on a per cell basis. Therefore, these bacteria can undergo additional cell divisions until the depletion of the histidine, forming a larger colony

incubation of a long period of over 48 h may increase the occurrence of His⁺ or Trp⁺ small colonies induced by external suppressor mutations in addition to true reversions, depending on mutagens. Suppressor mutations are capable of overcoming at a single step auxotrophies at several loci; however, histidine- or tryptophan-synthesizing enzymes generated by suppressor mutations may have several kinds of amino acids different from those found in prototroph in the active center, resulting in reduced enzyme activity [35–37]. Thus, suppressor mutation-induced colonies may grow more slowly than true mutation-induced colonies. His⁺ or Trp⁺ small colonies derived from true or suppressor mutations may also occur due to their delayed or slow growth induced by test chemicals. The increased occurrence of such small-sized revertant colonies may lead to an increased sensitivity of the Ames testing system but also may result in less reproducibility between inter- or intra-laboratories.

12. Colonies, accompanied with His⁻ or Trp⁻ bacterial background lawn, should be counted as His⁺ or Trp⁺ revertants (*see* Fig. 4a). Small colonies (may be called pinpoint colonies), not accompanied with bacterial background lawn, are not revertants. The presence or absence of bacterial background lawn can be judged by the naked eye or microscopic (40×) examination. The judgment of bacterial background lawn on plates following treatment of bacteria with test chemicals is easily achievable by comparison with the contrasts of agar medium of fresh (without bacteria) and negative control (with bacteria exposed to vehicle) plates. A slight but significant diminution of the bacterial background lawn is noted at *ca.* 10 % or less of the survivors in the TA100 strain (*see* Fig. 4b). Its significant diminution may be noted at *ca.* 20 % of the survivors by microscopic (40×) examination. Thus, if significant diminution of the bacterial background lawn is observed following treatment of bacteria with chemicals at certain doses, then the doses will be noted as toxic ones. Majority of mutagens yield maximum revertant colonies around the least toxic dose or at slightly lower doses, dependent on chemicals. Alkyl methanesulfonates, which are potent mutagens, produces maximum revertant colonies around the values of LD_{50} (dose reducing survivors by 50 %) [32].

13. Twofold rule may be too insensitive for strains with relatively high reversion frequencies, such as TA100 and WP2*uvrA*(pkM101), and too sensitive for strains with low reversion frequencies, such as TA1535 and TA1537 (*see* Table 2). Hence, this evaluation method may be slightly modified, taking into consideration each laboratory's background data for negative control values; *i.e.*, setting a minimum fold increase from 2-fold to 1.5–2.0-fold for the former strains, and to 2.5–3.0-fold for the latter strains [1, 2, 38].

References

1. Maron DM, Ames BN (1983) Revised methods for the *Salmonella* mutagenicity test. Mutat Res 113(3–4):173–215
2. Mortelmans K, Zeiger E (2000) The Ames *Salmonella*/microsome mutagenicity test. Mutat Res 455(1–2):29–60
3. Watanabe M, Ishidate M Jr, Nohmi T (1990) Sensitive method for the detection of mutagenic nitroarenes and aromatic amines: new derivatives of *Salmonella typhimurium* tester strains possessing elevated O-acetyltransferase levels. Mutat Res 234(5):337–348
4. Kushida H, Fujita K, Suzuki A, Yamada M, Endo T, Nohmi T, Kamataki T (2000) Metabolic activation of N-alkylnitrosamines in genetically engineered *Salmonella typhimurium* expressing CYP2E1 or CYP2A6 together with human NADPH-cytochrome P450 reductase. Carcinogenesis 21(6):1227–1232
5. Cooper MT, Porter TD (2000) Mutagenicity of nitrosamines in methyltransferase-deficient strains of *Salmonella typhimurium* coexpressing human cytochrome P450 2E1 and reductase. Mutat Res 454(1–2):45–52
6. Yamada M, Matsui K, Sofuni T, Nohmi T (1997) New tester strains of *Salmonella typhimurium* lacking O^6-methylguanine DNA methyltransferases and highly sensitive to mutagenic alkylating agents. Mutat Res 381(1):15–24
7. Suzuki M, Matsui K, Yamada M, Kasai H, Sofuni T, Nohmi T (1997) Construction of mutants of *Salmonella typhimurium* deficient in 8-hydroxyguanine DNA glycosylase and their sensitivities to oxidative mutagens and nitro compounds. Mutat Res 393(3):233–246
8. Lijinsky W, Andrews AW (1983) The superiority of hamster liver microsomal fraction of activating nitrosamines to mutagens in *Salmonella typhimurium*. Mutat Res 111(2):135–144
9. Nohmi T, Mizokami K, Kawano S, Fukuhara M, Ishidate M Jr (1987) Metabolic activation of phenacetin and phenetidine by several forms of cytochrome P-450 purified from liver microsomes of rats and hamsters. Jpn J Cancer Res 78(2):153–161
10. Matsushima T, Sugimura T, Nagao M, Yahagi T, Shirai A, Sawamura M (1980) Factors modulating mutagenicity microbial tests. In: Norpoth KH, Garner RC (eds) Short-term test systems for detecting carcinogens. Springer, Berlin, pp 273–285
11. Prival MJ, Mitchell VD (1982) Analysis of a method for testing azo dyes for mutagenic activity in *Salmonella typhimurium* in the presence of flavin mononucleotide and hamster liver S9. Mutat Res 97(2):103–116
12. Yahagi T, Degawa M, Seino Y, Matsushima T, Nagao M, Sugimura T, Hashimoto Y (1975) Mutagenicity of carcinogenic azo dyes and their derivatives. Cancer Lett 1(2):91–96
13. Yahagi T, Nagao M, Seino Y, Matsushima T, Sugimura T, Okada M (1977) Mutagenicities of N-nitrosamines on *Salmonella*. Mutat Res 48(2):121–129
14. Araki A (1997) Mutagenicity and carcinogenicity of agricultural and medicinal chemicals. Environ Mutagen Res 19(1–2):55–61 (in Japanese)
15. Araki A, Noguchi T, Kato F, Matsushima T (1994) Improved method for mutagenicity testing of gaseous compounds by using a gas sampling bag. Mutat Res 307(1):335–344
16. Hakura A, Hori Y, Uchida K, Sawada S, Suganuma A, Aoki T, Tsukidate K (2010) Inhibitory effect of dimethyl sulfoxide on the mutagenicity of promutagens in the Ames test. Genes Environ 32(3):53–60
17. Gatehouse D, Haworth S, Cebula T, Goche E, Kier L, Matsushima T, Melcion C, Nohmi T, Ohta T, Venitt S, Zeiger E (1994) Recommendations for the performance of bacterial mutation assays. Mutat Res 312(3):217–233
18. Maron DM, Katzenellenbogen J, Ames BN (1981) Compatibility of organic solvents with the *Salmonella*/microsome test. Mutat Res 88(4):343–350
19. OECD (Organisation for Economic Co-operation and Development) (1997) Guideline for testing of chemicals, test guideline 471: bacterial reverse mutation test. OECD, Paris, France
20. Guengerich FP (2000) Metabolism of chemical carcinogens. Carcinogenesis 21(3):345–351
21. Shimada T (2006) Xenobiotic-metabolizing enzymes involved in activation and detoxification of carcinogenic polycyclic aromatic hydrocarbons. Drug Metab Pharmacokinet 21(4):257–276
22. Kawalek JC, Andrews AW (1980) The effect of solvents on drug metabolism *in vitro*. Drug Metab Dispos 8(6):380–384
23. Chauret N, Gauthier A, Nicoll-Griffith DA (1998) Effect of common organic solvents on *in vitro* cytochrome P450-mediated metabolic activities in human liver microsomes. Drug Metab Dispos 26(1):1–4
24. Hickman D, Wang J-P, Wang Y, Unadkat JD (1998) Evaluation of the selectivity of *in vitro* probes and suitability of organic solvents for the measurement of human cytochrome P450 monooxygenase activities. Drug Metab Dispos 26(3):207–215

25. Busby WF, Busby JR, Ackermann JM, Crespi CL (1999) Effect of methanol, ethanol, dimethyl sulfoxide, and acetonitrile on *in vitro* activities of cDNA-expressed human cytochromes P450. Drug Metab Dispos 27(2):246–249
26. Easterbrook J, Lu C, Sakai Y, Li AP (2001) Effects of organic solvents on the activities of cytochrome P450 isoforms, UDP-dependent glucuronyl transferase, and phenol sulfotransferase in human hepatocytes. Drug Metab Dispos 29(2):141–144
27. Ku WW, Bigger A, Brambilla G, Glatt H, Gocke E, Guzzie PJ, Hakura A, Honma M, Martus H-J, Obach RS, Roberts S (2007) Strategy for genotoxicity testing –metabolic considerations. Mutat Res 627(1):59–77
28. Green MHL, Muriel WJ (1976) Mutagen testing using Trp⁺ reversion in *Escherichia coli*. Mutat Res 38(1):3–32
29. Mortelmans K, Riccio ES (2000) The bacterial tryptophan reverse mutation assay with *Escherichia coli* WP2. Mutat Res 455(1–2):61–69
30. Matsushima T (1991) Utility and problems of short-term assay—microbial mutation assay. Environ Mutagen Res 13(3):279–283 (in Japanese)
31. Hakura A, Sugihara T, Uchida K, Sawada S, Suganuma A, Aoki T, Tsukidate K (2010) Cytotoxic effect of dimethyl sulfoxide on the Ames test. Genes Environ 32(1):1–6
32. Hakura A, Ninomiya S, Kohda K, Kawazoe Y (1984) Studies on chemical carcinogens and mutagens. XXVI. Chemical properties and mutagenicity of alkyl alkanesulfonates on *Salmonella typhimurium* TA100. Chem Pharm Bull 32(9):3626–3635
33. Hakura A, Mizuno Y, Goto M, Kawazoe Y (1986) Studies on chemical carcinogens and mutagens. XXXV. Standardization of mutagenic capacities of several common alkylating agents based on the concentration-time integrated dose. Chem Pharm Bull 34(2):775–780
34. Hakura A, Kawazoe Y (1986) Studies on chemical carcinogens and mutagens. XXXVI. Apparent activation energy for mutagenic modification induced in *E. coli* by alkylating agents. Estimation from mutation frequency. Chem Pharm Bull 34(4):1728–1734
35. Bridges BA, Dennis RE, Munson RJ (1967) Differential induction and repair of ultraviolet damage leading to true reversions and external suppressor mutations of an ochre codon in *Escherichia coli* B/r WP2. Genetics 57(4):897–908
36. Ohta T, Tokishita S, Tsunoi R, Ohmae S, Yamagata H (2002) Characterization of Trp⁺ reversions in *Escherichia coli* strain WP2*uvrA*. Mutagenesis 17(4):313–316
37. Qian Q, Li JN, Zhao H, Hagervall TG, Farabaugh PJ, Björk GR (1998) A new model for phenotypic suppression of frameshift mutations by mutant tRNAs. Mol Cell 1(4):471–482
38. Cariello NF, Piegorsch WW (1996) The Ames test: the two-fold rule revised. Mutat Res 369(1–2):23–31

Chapter 34

Methods for Using the Mouse Lymphoma Assay to Screen for Chemical Mutagenicity and Photo-Mutagenicity

Nan Mei, Xiaoqing Guo, and Martha M. Moore

Abstract

The mouse lymphoma assay (MLA) quantifies genetic alterations affecting expression of the thymidine kinase (*Tk*) gene which is located on chromosome 11. This assay is widely used for evaluating the genotoxic potential of various agents. It can also be used for photo-mutagenicity. The *Tk*-deficient mutants in the MLA result not only from point mutations but also from gross structural and numerical changes at the chromosomal level. The MLA has been recommended as one component of the genotoxicity test battery in many regulatory authorities, countries, and international organizations. The protocol for the MLA has been optimized and adopted by many laboratories around the world. In this chapter, the authors provide an example of the application of the MLA by detailing the procedures that are used for performing the assay in their laboratory and analyzing and interpreting the data generated from both the soft-agar and microwell versions of the assay. In addition, they provide information for using the assay for the analysis of potential photo-mutagenicity.

Key words Mouse lymphoma assay, Gene mutation, Chromosomal mutation, Mutagenicity

1 Introduction

The mouse lymphoma assay (MLA), which quantifies genetic alterations involving the thymidine kinase (*Tk*) gene, uses the $Tk^{+/-}$-3.7.2C clone of the L5178Y mouse lymphoma cell line. It was originally developed by Dr. Donald Clive and his colleagues [1], who used a soft-agar method to enumerate mutants. In 1983, an alternative method using liquid media and 96-well microtiter plates was developed [2]. Both procedures for conducting the MLA are equally acceptable methodologies, although there are some differences in the counting and sizing of mutant colonies and the calculation of mutant frequency (MF) (see Sect. 3.16). Along with other *in vitro* mammalian cell genotoxicity assays, the MLA has the advantage that it can be conducted relatively quickly and inexpensively. Because the MLA is capable of evaluating the ability of mutagens to induce a variety of mutational events, including point mutations, large scale chromosomal mutations, recombination, and mitotic nondisjunction [3–7], it is generally recommended as an *in vitro* mammalian gene mutation assay in regulatory test batteries including the U.S. Food and Drug Administration (FDA)/Center for Food Safety and Applied Nutrition (CFSAN) [8], the U.S. Environmental Protection Agency (EPA) [9], and the International Committee on Harmonization (ICH) [10–12]. Currently Test Guideline 476 of the Organization for Economic Co-operation and Development (OECD) [13] provides some guidance for conducting the assay. An OECD workgroup is currently writing a new test guideline for the MLA that provides specific guidance and it should be adopted in the future. This new test guideline incorporates recommendations from the MLA Expert Workgroup of the International Workshop for Genotoxicity Testing (IWGT). Beginning in 1999, this workgroup has held six meetings and reached a consensus on a number of important issues concerning the conduct of the assay, acceptance criteria for both the microwell and soft-agar versions of the MLA, and the interpretation of the data [14–19]. We incorporated these recommendations into this chapter.

The MLA uses a cell line that is heterozygous ($Tk^{+/-}$) and it detects forward mutation of the wild-type *Tk*1 allele located on chromosome 11. The two chromosomes 11 in this cell line can be distinguished by a difference in centromere size [20, 21]. The functional allele *Tk*1 is on chromosome 11b and the nonfunctional allele is on chromosome 11a. Because a functional *Tk* gene is not essential for cell survival, the *Tk*-deficient mutants ($Tk^{-/-}$ or $Tk^{0/-}$) are viable and can be selected by using the pyrimidine analogue trifluorothymindine (TFT). TFT inhibits the growth of normal cells (*Tk* proficient cells that are either $Tk^{+/-}$ or $Tk^{+/+}$) and because the *Tk* mutant cells do not have a functional pyrimidine salvage pathway they can survive in media containing TFT (Fig. 1). It should be noted that the assay was developed using bromodeoxyuridine as the selective agent [1] but it was subsequently shown that TFT is a superior selective agent [22] and bromodeoxyuridine

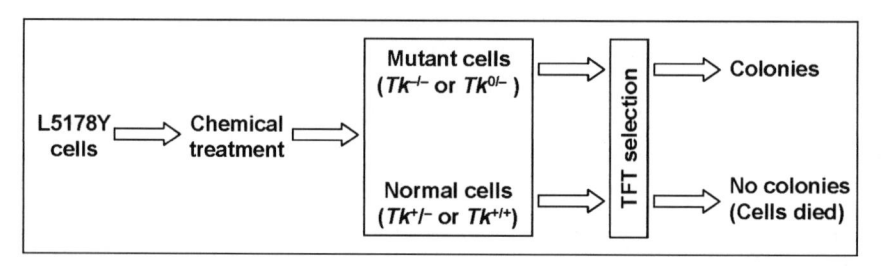

Fig. 1 Principle of the mouse lymphoma assay

is no longer acceptable. Earlier studies using bromodeoxyuridine as the selective agent should generally be considered unacceptable, particularly if the test chemical was found to be nonmutagenic.

The observation that TFT-resistant colonies grow at different rates was first described in the soft-agar version of the assay as a bimodal distribution [23]. Colonies can be classified as small (slow growing) or large (normal growing) [24]. Mutagenic chemicals induce different frequencies of small and large colonies [25]. Generally, chemicals acting primarily as clastogens induce a relatively higher proportion of small colony *Tk* mutants. Molecular analysis indicates that small colony mutations result predominantly from loss of heterozygosity (LOH). Chemical compounds that primarily induce point mutations result in a relatively higher proportion of large colony mutants which generally do not result from LOH [3]. *Tk* mutants can be characterized using molecular and cytogenetic techniques to determine the types of mutation induced by a specific mutagen and to provide some insight into the possible mode of action for the induction of mutations [7].

Because the L5178Y $Tk^{+/-}$-3.7.2C mouse lymphoma cells have a mutation in both of the *Trp53* alleles (located on chromosome 11) and no wild-type *Trp53* allele in either $Tk^{+/-}$ cells or $Tk^{-/-}$ mutants [26], mouse lymphoma cells do not have a large rate of apoptosis and therefore mutations that might be lethal in cell lines with wild-type p53 can be detected. This coupled with the wide spectrum of genetic events detected by the assay makes the MLA the most sensitive *in vitro* mammalian cell gene mutation assay and the most recommended gene mutation assay.

In addition to its use screening chemicals for their mutagenic potential, the MLA can be used for photo-mutagenicity evaluation. Because some compounds can become activated after the absorption of ultraviolet (UV) or visible light energy, there is a need to test pharmaceuticals and cosmetic products for photochemical genotoxicity. The existing methods for several genotoxicity tests have been adapted to use concurrent UV-visible light irradiation for the assessment of photomutagenicity [27]. We have evaluated a number of chemicals using the photo-MLA [28–30]. Considering the first law of photochemistry, photochemical reactions, and subsequent phototoxic actions are not possible without sufficient absorption of photons [31]. Therefore, the selection of the irradiation dose for combination chemical and light exposure is important and should be done specifically for the MLA [32].

In this chapter, we describe the procedures for both soft-agar and microwell versions of the MLA used in our lab (Fig. 2). In addition to standard chemical exposures, we provide information as to how we conduct photomutagenicity experiments. We include specific details as to how to set up the experiments, perform the assay, and analyze and interpret the data.

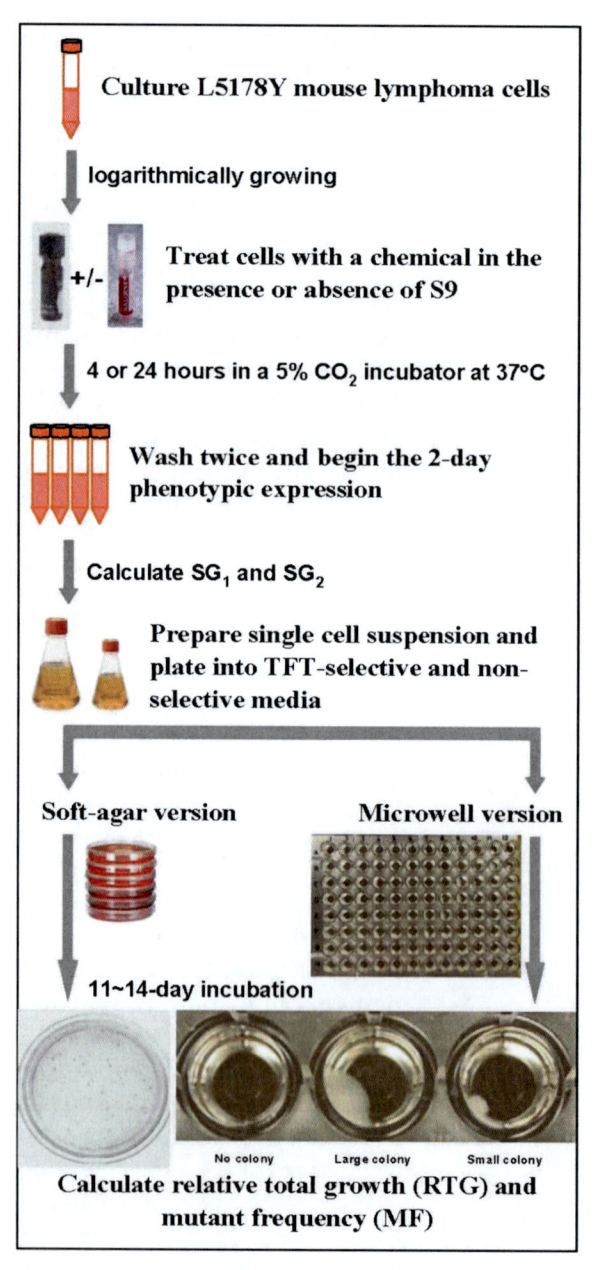

Fig. 2 Schematic for conducting the mouse lymphoma assay

2 Materials

2.1 Equipment

1. A laminar flow hood, 6 ft (any brand).
2. Humidified incubators at 37 °C in presence of 5 % CO_2 (any brand).
3. Multipurpose centrifuges, with tube adapters for 15 and 50 mL tubes (any brand).

4. Z1 Coulter counter, dual threshold analyzer, Cat. #66005699 (Beckman Coulter, Inc.; Bera, CA) and/or a hemocytometer.

5. Matrix electronic multichannel pipette, 1250 µL, #2004 (Thermo Scientific Inc.; Hudson, NH).

6. ProtoCOL automated colony counter (Microbiology International; Frederick, MD)

7. Incubator shaker, Innova 43 (New Brunswick Scientific; Enfield, CT).

8. Roller drum, digital display low profile base and 128P roller drum/30 mm tubes, #7736-11115 and #7736-20050 (Bellco Glass, Inc.; Vineland, NJ).

2.2 Cell Culture

1. Fischer's medium, 500 mL/bottle, #112-032-101 (Quality Biological Inc.; Gaithersburg, MD).

2. Horse serum, 500 mL/bottle, #16050-122 (Life Technologies Co.; Carlsbad, CA).

3. Penicillin-Streptomycin, 100×, 100 mL/bottle, #15140-122 (Life Technologies).

4. Pluronic F-68, 10 % (100×), 100 mL/bottle, #24040-032 (Life Technologies).

5. Sodium pyruvate, 100 mM (100×), 100 mL/bottle, #11360-070 (Life Technologies).

6. Difco Noble agar, 500 g, #214230 (BD; Franklin, NJ)

7. Aroclor 1254 induced male Sprague-Dawley rat liver post-mitochondrial fraction (S9), #11-101 (Molecular Toxicology Inc.; Boone, NC).

8. Trifluorothymidine (TFT), 100 mg, #T2255 (Sigma-Aldrich Co.; St. Louis, MO).

9. 4-Nitroquinoline N-oxide (4-NQO), 250 mg, #N8141 (Sigma-Aldrich).

10. Benzo[a]pyrene (BP), 100 mg, #B1760 (Sigma-Aldrich).

11. Dimethyl sulfoxide (DMSO), 50–500 mL/bottle, #D8418 (Sigma-Aldrich).

12. D-Glucose 6-phosphate sodium salt (G-6-P), #G7879 (Sigma-Aldrich).

13. β-Nicotinamide adenine dinucleotide phosphate sodium salt hydrate (NADP), #N0505 (Sigma-Aldrich).

14. Thymidine, 1 g, #T1895 (Sigma-Aldrich).

15. Hypoxanthine, 1 g, #H9636 (Sigma-Aldrich).

16. Glycine, 100 g, #G8790 (Sigma-Aldrich).

17. Methotrexate, 100 mg, #A6770 (Sigma-Aldrich).

18. ImmuMark™ MycoTest™ Kit, #30200 (MP Biomedicals; Solon, OH).

2.3 Supplies

1. Corning cell culture plates, 96-well, 50/case, #07-200-87 (Fisher Scientific Inc.; Waltham, MA).
2. Extra-deep petri dishes, 100×25 mm, 500/case, #08-757-11 (Fisher).
3. Cell culture flasks, 25 cm^2, vented cap, 200/case, #10-126-28 (Fisher).
4. Cell culture flasks, 75 cm^2, vented cap, 100/case, #10-126-31 (Fisher).
5. Erlenmeyer flasks, 125 mL, standard screw cap, 50/case, #10-041-8 (Fisher).
6. Polystyrene pipettes, 5 mL, individual, 200/case, #13-678-11D (Fisher).
7. Polystyrene pipettes, 10 mL, individual, 200/case, #13-678-11E (Fisher).
8. Polystyrene pipettes, 25 mL, individual, 200/case, #13-678-11 (Fisher).
9. Centrifuge tubes, 15 mL, 50/rack, 500/case, #05-538-51 (Fisher).
10. Centrifuge tubes, 50 mL, 25/rack, 500/case, #05-538-49 (Fisher).
11. Reagent reservoirs, 25 mL, individually wrapped, 50/case, #14-387-070 (Fisher).
12. Reagent reservoirs, 100 mL, individually wrapped, 50/case, #14-387-068 (Fisher).
13. Flat-cap PCR tubes, 0.2 mL, 1000/case, #14-230-225 (Fisher).
14. Matrix Pipette Tips, 1250 µL, 10 racks/case, #8041-11 (Thermo).

3 Methods

3.1 Medium and Culture Conditions

1. The basic medium is Fischer's medium for leukemic cells of mice (**Note 1**) with L-glutamine (illustrated in this chapter) supplemented with pluronic F-68 (0.1 %) (**Note 2**), sodium pyruvate (1 mM), penicillin (100 U/mL), and streptomycin (100 µg/mL).
2. The treatment medium (F_{5p}) (**Note 3**), growth medium (F_{10p}), and cloning medium (F_{20p}) are the basic medium supplemented with 5 %, 10 %, and 20 % heat-inactivated horse serum (**Note 4**), respectively.
3. The cells are (1) suspended in cell culture flasks and placed in a humidified incubator at 37 °C in the presence of 5 % CO_2

(stationary culture); or (2) suspended in Erlenmeyer flasks, gassed with 5 % (v/v) CO_2 in air, and maintained in an incubator shaker at 37 °C at 200 orbits/min (shaking culture). In our lab, we usually use shaking culture to grow cells.

3.2 Cells and Cell Growth

1. L5178Y $Tk^{+/-}$ clone 3.7.2C mouse lymphoma cell line was derived from a 3-methylcholanththrene-induced thymic lymphoma from a DBA-2 mouse. We use cells obtained from Dr. Donald Clive's laboratory in the 1980s and maintained first in the EPA laboratory of Dr. Martha Moore and subsequently moved to our laboratory at the FDA/National Center for Toxicological Research (Jefferson, AR). It should be noted that there is currently an ILSI/HESI effort to establish a repository of all the cell lines used in regulatory genetic toxicology laboratories and this should, in the future, be the preferred source for all laboratories starting to use the assay and established laboratories wishing to obtain a new cell stock [33]. If possible, karyotype analysis should be carried out when cryopreserving a master stock and chromosome painting is also useful to confirm normal chromosome 11.

2. Take a vial of frozen stocks of the mouse lymphoma cells from the liquid nitrogen freezer (**Note 5**).

3. Thaw the cells, by placing the vial in a warm-water bath.

4. Gently pipet the cells into a 15-mL centrifuge tube and wash the cells once using F_{10p} medium (**Note 6**).

5. Resuspend the cell pellet in an Erlenmeyer flask with 20 mL F_{10p} medium, loosen the flask cap, and place in a CO_2 incubator for 1 day.

6. After 1-day incubation, gas with 5 % (v/v) CO_2 in air, tighten the cap, and place in the incubator shaker at 37 °C.

7. Maintain cells in log-phase with cell densities between 2×10^5 cells/mL and 10×10^5 cells/mL and with a population doubling time of 8–10 h (**Note 7**). Determine cell densities using a Coulter counter or hemocytometer.

8. During the working week, determine the cell density and dilute cells each day to 2×10^5 cells/mL.

9. For the weekend, dilute cells to 0.05×10^5 cells/mL on Friday and culture until Monday morning (**Note 8**).

10. Historically we have often maintained a stock culture for up to 3 months when it was growing normally. More recently we have cleansed a large culture of preexising mutants (see Sect. 3.4), frozen a large number of vials, thawed a vial prior to each experiment, allowed the cells to acclimate and then used those cells for a single or small number of experiments.

3.3 Mycoplasma Detection in Cell Culture

It is important to routinely check cells for mycoplasma contamination. Only cells free of mycoplasma should be used because their presence can induce alteration in metabolism, cell growth rate, and morphology; disrupt DNA and RNA synthesis; cause chromosomal abnormalities; and consequently compromise the experimental results and conclusions. A variety of molecular methods is available for mycoplasma detection, including fluorescent staining of DNA, ELISA, autoradiography, immunostaining, and direct or nested PCR. A simple and sensitive method is the direct immunofluorescence test with the capability of detecting the broad range of mycoplasma species that account for more than 96 % of cell culture infections. In our lab, we use a commercially available kit, and the procedure is as follows.

1. Place 2×10^4 cells in a volume of 20–30 µL culture medium on a coated glass microscope slide.
2. Dry the sample at 50 °C for 45 min.
3. Fix in −20 °C cold 70 % ethanol for 60 s and allow the slide to air dry at room temperature.
4. Add one drop of the FLUOS-labelled monoclonal antibody to the fixed cell preparation.
5. Incubate for 20 min at room temperature.
6. Carefully rinse the slide twice for a total of 2 min in a bath of phosphate buffered saline (PBS) and dry at room temperature.
7. Add one drop of the Goat Anti-Mouse Fluorescein Conjugate to the fixed cell preparation.
8. Incubate for 20 min at room temperature.
9. Wash the slide twice in PBS for a total of 2 min and dry at room temperature.
10. Place one drop of mounting medium and a cover slip over the mounting fluid. Avoid trapping air bubbles.
11. Scan the cells on the slide using a fluorescence microscope (maximum excitation wavelength 490 nm, mean emission wavelength 520 nm). The presence of stained mycoplasma should be easily visible at ×400 to ×600 magnification (**Note 9**).

3.4 Cleansing Mouse Lymphoma Cells

Before freezing the working cell stocks or performing experiments, the mouse lymphoma cells are purged of preexisting $Tk^{-/-}$ mutants, in order to lower the background MF. The cells should be treated with THMG for 1 day and then THG for 2 days. Table 1 shows how to prepare THG and THMG solutions.

1. Count the cells using either a Coulter counter or hemocytometer and adjust cell density to 2×10^5 cells/mL (**Note 10**).
2. Add 0.5 mL of THMG stock (100×) to a 49.5 mL F_{10p} cell culture (about 100×10^5 of total cells).

Table 1
THMG and THG solutions

	100X stock Concentration (μg/mL)	THG[a] Weight (mg)	THG[b] 50 mL	THMG 50 mL of THG	THMG[b] 50 mL
Thymidine	300	30	Filtered, dispense in 0.6 mL		Filtered, dispense in 0.6 mL per tube
Hypoxanthine	500	50			
Glycine	750	75			
Methotrexate	10			0.5 mg[c]	
F_{0p} medium		100 mL			

[a]Warm F_{0p} medium (37 °C) and mix on magnetic stirrer
[b]Filter sterilize, dispense 50 mL into 0.6 mL portions in 1.5-mL sterile tubes, and store at −20 °C
[c]Mix or 0.5 mL of (10 mg Methotrexate + 10 mL PBS = 1 mg/mL)

3. Mix and gas the culture with 5 % CO_2-in-air and place in an incubator shaker for 24 h (**Note 11**).

4. After 24-h incubation, count and centrifuge 100×10^5 cells at $200 \times g$ for 10 min.

5. Resuspend the cell pellet in 20 mL F_{10p}, and add an additional 29.5 mL of F_{10p} (total volume is 49.5 mL).

6. Add 0.5 mL of THG stock (100×) to the 49.5 mL F_{10p} cell culture.

7. Mix and gas the culture with 5 % CO_2-in-air, and then place in an incubator shaker for 24 h.

8. After 24-h incubation, adjust the cell concentration to 2×10^5 cells/mL, using medium containing THG.

9. After additional 24-h incubation, count and centrifuge 100×10^5 cells at $200 \times g$ for 10 min.

10. Resuspend the cell pellet in 20 mL F_{10p}, then add an additional 30 mL F_{10p} (total volume is 50 mL).

11. Culture the cleansed cells as indicated in Sect. 3.2 (**steps 5–7**).

3.5 Preparation of S9-mix

For general screening, all chemicals should be tested both with and without metabolic activation (S9). The S9 fraction of livers from male Sprague-Dawley rats treated with Aroclor 1254 is mixed with a nicotinamide adenine dinucleotide phosphate (NADPH)-generating system. Each batch of S9 is checked by the manufacturer (Moltox) for sterility, protein content, some CYP-catalyzed enzyme activities, and promutagen activation; and reported on the quality control and production certificate. Table 2 shows different S9 cofactor formulae which have been used in the MLA.

Table 2
Preparation of S9 mix for each sample[a]

Chemical	Formula 1 [34] Stock conc. (mM)	Formula 1 Volume (mL)	Formula 2 [4] Stock conc. (mM)	Formula 2 Volume (mL)	Formula 3 [35] Stock conc. (mM)	Formula 3 Volume (mL)
S9		0.1–0.2		0.1–0.2		0.1–0.2
Sodium phosphate buffer, pH 8.0	125	0.8				
KCl	60	1	150	0.1		
G$_6$P	10		638	0.1		
NADP	8		33	0.1	75	0.4
CaCl$_2$	20					
MgCl$_2 \cdot$ 6H$_2$O	20					
C$_6$H$_5$O$_7$Na$_3$					375	0.4
Total[b]		2		0.5		1
Cell culture		8		9.5		9
Final S9%		1–2 %		1–2 %		1–2 %

[a]Multiply these numbers in the Table by the number of samples plus 1 (n + 1) for each experiment. The cofactors are prepared in dH$_2$O or serum-free medium, pH adjusted to ~7.0 with 1 N NaOH, and filter sterilized before use. Keep all ingredients chilled during preparation, and add the S9 fraction last

[b]When using less volume of S9, add dH$_2$O or medium to the final volume of each formula

The choice of S9-cofactor formula may depend on the property of the test chemical. In some cases, multiple formulae and/or concentrations may be used in order to adequately evaluate a test chemical. In all cases, including the vehicle control and positive control (see Sect. 3.7), the final S9 concentration in the treatment medium should be less than 2 % (**Note 12**). In our lab, we generally use formula 2 and the final S9 concentration usually is 1 %. The volume of S9 may be calculated based on "mg of protein per mL".

3.6 Preparation of TFT and Selection Medium

TFT is used to inhibit the growth of cells containing the TK enzyme, yet allowing the growth of cells that are *Tk*-deficient [22]. TFT is light sensitive and has a short half-life in medium. Therefore, appropriate precautions must be taken when handling TFT.

1. To make the stock TFT solution of 3 mg/mL, dissolve 100 mg of TFT in 33.3 mL of distilled water.

2. Filter-sterilize this stock solution through a 0.22 μm filter.

3. Aliquot 0.5, 1, and 1.5 mL of the filtered stock solution into 1.5-mL microcentrifuge tubes. This provides various volumes that can be used depending upon the number of cultures (**Note 13**). Keep all tubes in a storage box at −20 °C. Avoid multiple freeze-thaw cycles—thaw each tube only once and use immediately.

4. In the microwell version, for each sample, add 90 μL of the stock solution into 90 mL of the cloning medium to obtain a final concentration of 3 μg/mL TFT for mutant selection.

5. In the soft-agar version, for each sample, add 34 μL of the stock solution into 100 mL of the cloning medium to obtain a final concentration of 1 μg/mL TFT for mutant selection.

3.7 Positive Control

To demonstrate that the assay is working properly, each experiment must include a positive control. There are several chemicals that can be used as positive controls. The selection is based on whether the specific experiment is being performed in the presence or absence of exogenous metabolic activation (S9 mix). We usually use 0.1 μg/mL 4-nitroquinoline *N*-oxide (4-NQO) or 10 μg/mL methylmethansulfonate (MMS) in experiments without S9 mix, and 0.3 μg/mL benzo[a]pyrene (BP) or 3 μg/mL cyclophosphamide (CP) in experiments with S9 mix. MMS and CP stocks should be freshly prepared, while NQO and BP can be prepared in DMSO at a 100-fold higher concentration and stored as frozen aliquots at −20 °C. In this chapter, we will describe the preparation of 4-NQO solution as an example.

1. Dissolve 10 mg of 4-NQO in 1 mL of DMSO to make a stock solution containing 10 mg/mL.

2. Aliquot 20–50 μL of the stock solution into 0.2-mL microcentrifuge tubes; keep all tubes in a storage box and store the box at −20 °C. Avoid multiple freeze-thaw cycles.

3. Transfer 10 μL of 10 mg/mL stock solution into 10 mL of DMSO to make the 100× working solution of 10 μg/mL.

4. Aliquot 110 μL of the working solution into 0.2-mL microcentrifuge tubes; keep all tubes in a storage box, store at −20 °C.

5. When performing the experiments without S9 mix, add 100 μL of the working solution (from *step 4*) into 10 mL of the medium (1:100 dilutions) as the positive control. The final concentration of 4-NQO is 0.1 μg/mL (0.53 μM).

3.8 Test Chemicals

1. The working solution (100×) for each chemical is made by dissolving an appropriate amount of the chemical in medium, water, saline, or DMSO based on its solubility. Test stocks should be prepared just prior to use by a series of dilutions from the original stocks. If DMSO is used, the final concentration of DMSO in the medium is 1 % in all the test cultures and in the vehicle control (**Note 14**).

2. The cytotoxicity (see Sect. 3.16) and solubility of the test chemical, and any changes in pH, osmolality, and precipitate after adding the test chemical in the medium, are always considered for each culture. The highest concentration is based on one of these factors and is generally cytotoxicity.

3. For relatively non-cytotoxic compounds, the maximum concentration should be 2 mg/mL or 10 mM, whichever is lower for all chemicals except pharmaceuticals intended for human use. The last revision of the ICH which provides guidance for pharmaceuticals intended for human use recommends that the maximum top concentration be 0.5 mg/mL or 1 mM, whichever is lower [12]. However, for those pharmaceuticals with low molecular weight (e.g., less than 200), the ICH recommends that higher test concentrations should be considered.

4. For the majority of test chemicals there will be cytotoxicity, and the test concentrations should cover a range from 10 to 20 % relative total growth (RTG) to little or no toxicity (100 % RTG).

5. Because mouse lymphoma cells grow in suspension, it is possible that precipitates can interfere with the assay, i.e., it is difficult to separate the cells and the precipitates by centrifugation and wash after the treatments. Therefore, it is generally recommended that the lowest dose showing precipitate should be the highest dose tested. The solubility of a test chemical in the cultures is checked by naked eye, both at the beginning and end of treatment (**Note 15**).

6. Non-physiological pH may affect the results by mechanisms unrelated to the test chemicals. The pH of Fisher's medium is about 7.2 ± 0.2. If the pH of the medium shifts from neutrality after adding the test chemical, the treatment cultures should be pH adjusted in the medium to the range of neutrality.

7. The osmolality of Fisher's medium is about 300 ± 5 % mOsm/kg of water. The treatment medium (F_{5p}) with 1 % DMSO is around 425–450 mOsm/kg. Usually, the osmolality of treatment cultures should be less than 50 mOsm/kg over the osmolality of the vehicle control. Assessing osmolality is important when high amounts of test substance are used.

8. Before the main experiment, one or more dose range finding tests are performed with 6–10 concentrations of the test material. In the absence of any information, the dose-range experiment should start with 10 mM or 2 mg/mL (whichever is lower) and cover a range of concentrations to perhaps as low as 1 μg/mL.

9. Based on the dose range finding tests, 6–8 concentrations are selected for the main experiment which can be conducted using single, double or more cultures for each test concentration (**Note 16**).

3.9 Period of Treatment

1. The standard MLA usually employs a short treatment period, i.e., 3 or 4 h, in the presence or absence of metabolic activation (S9). We use a 4-h treatment and that is illustrated in this chapter.

For routine chemical screening it is necessary to conduct experiments both with and without S9 (**Note 17**).

2. When negative results are obtained for the short treatment with and without S9, a continuous treatment without metabolic activation for 24 h is recommended by the ICH for pharmaceuticals intended for human use. A 24-h treatment should also be considered for chemicals that have issues with solubility and that cannot be adequately tested for the 3 or 4-h treatments (**Note 18**).

3.10 Treatment

Mouse lymphoma cells are suspended in 50-mL centrifuge tubes containing 6×10^6 cells in 10 mL of F_{5p} treatment medium (**Note 3**). One hundred microliters of the working solutions (100×) of the test chemical and, if necessary, 250 μL of S9 mix (we generally use formula 2 in Table 2) are added to the cell cultures, which are then gassed with 5 % (v/v) CO_2-in-air and placed on a roller drum (15 rpm) in a 37 °C incubator for 4 h. Duplicate vehicle control cultures and one positive control culture with or without S9 should be used. After the 4-h treatment, the cells are centrifuged and washed twice with fresh medium, and then resuspended in 20 mL of growth medium. The culture tubes are placed on a roller drum in a 37 °C incubator to begin the 2-day phenotypic expression (**Note 19**).

1. One day before the treatment, subculture the cell culture at 2×10^5 cells/mL in Erlenmeyer flasks (usually 2 flasks, 45–50 mL in each flask and 90–100 mL in total).

2. On the day of the treatment, count the cells and determine if there are a sufficient number of cells for the treatment (**Note 20**).

3. Transfer the cells from the Erlenmeyer flasks into two 50-mL tubes, centrifuge at $200 \times g$ for 5–10 min, and resuspend the cells in 40 mL F_{5p} medium.

4. Re-count the cells and adjust to 2×10^6 cells/mL.

5. For each test culture, transfer 3 mL of the cell culture and 7 mL F_{5p} medium into a 50-mL centrifuge tube.

6. Using previously prepared chemical stock (see Sect. 3.8), add 100 μL chemical stock for each test concentration (if necessary, add S9 mix) into each culture tube, mix, gas with 5 % (v/v) CO_2-in-air, and place the sealed tube in a roller drum inside a 37 °C incubator for 4 h.

7. After the 4-h treatment, centrifuge the tubes containing the cells and test chemical at $200 \times g$ for 10 min.

8. Decant the supernatant and resuspend in 10 mL fresh F_{0p} medium (or F_{10p}). Repeat twice to wash the cells.

9. After the media washes, resuspend the cells in 20 mL of fresh F_{10p} medium, and incubate for approximately 24 h from the start of the treatment (**Note 19**).

10. After a 24 h incubation (day 1), count the cells in each culture, and dilute the cell density to 2×10^5 cells/mL with F_{10p}. Use the cell counts to calculate the day 1 suspension growth (SG1) (see Sect. 3.16).

11. After an additional 24 h (day 2), count the cells and adjust the cell density to 3×10^5 cells/mL with F_{10p}, and calculate day 2 the suspension growth (SG2) (**Note 21**).

12. Clone for plating efficiency and mutant enumeration (see Sects. 3.12 and 3.13).

3.11 Treatment for the Photo-MLA Assay

Some pharmaceuticals and cosmetics may induce genotoxic effects when they are irradiated with UV-visible light. The MLA and other existing methods have been adapted to assess the potential for photochemical genotoxicity [27]. Usually, photogenotoxicity testing for hazard identification focuses on visible (390–750 nm)-, UVA (315–400 nm)-, and UVB (280–315 nm)-irradiation. In the authors' lab, the photo-MLA has been performed with an adapted protocol using concurrent UVA irradiation during substance treatment [28–30]. Photochemical reactions and subsequent phototoxic actions are not possible without sufficient absorption of photons from the irradiation source [31]. There is, however, no specific irradiation dose that can be generally applied to different photogenotoxicity testing systems. The selection of the irradiation dose should be performed individually for each test system on the basis of the biological characteristics of the system and the specific question being addressed [27, 28, 31, 32]. Since the only difference between the MLA and photo-MLA is the treatment procedures, in this section, the authors describe the procedures for the cell treatment with retinyl palmitate and UVA light as an example [28].

1. The UVA light box is made using four UVA lamps (National Biologics; Twinsburg, OH). The irradiance of the light box is determined using an Optronics OL754 Spectroradiometer (Optronics Laboratories; Orlando, FL), and the light dose is routinely measured using a Solar Light PMA-2110 UVA detector (Solar Light Inc.; Philadelphia, PA). The maximum emission of the UVA light box is 350–352 nm with 98.9 % UVA.

2. The working solution (100×) for a test chemical (e.g., retinyl palmitate) is prepared just prior to use. A serial dilution from the working solution is prepared so that it can be added to the cell culture dishes (*step 3* below) to provide the desired number of cultures to be treated at the appropriate test chemical concentrations.

3. The cells are suspended in 100-mm diameter tissue culture dishes at a concentration of 6×10^6 cells in 10 mL of treatment medium. In all cases, the cells are treated with (1) test chemical alone (100 µL of the chemical working solutions is added),

(2) UVA irradiation alone (the dose selected from *step 4*), and (3) test chemical and UVA.

4. For the UVA irradiation alone, the cells in 100-mm diameter tissue culture dish are exposed to UVA light at a irradiation rate for various times (e.g., 82.8 mJ/cm^2/min from 5 to 45 min). One or more doses can be chosen for the treatment with a test chemical.

5. Cells are treated with different concentrations of a test chemical (e.g., retinyl palmitate) and exposed to UVA light (e.g., 2.48 J/cm^2 of UVA) during a period of 30 min.

6. The treated cultures are then incubated at 37 °C (without UVA light irradiation) for an additional 3.5 h.

7. After 4 h treatment, the cell suspension is removed from the tissue cultures dishes and centrifuged and washed twice with fresh medium. Cells are then resuspended in growth medium to begin the 2-day phenotypic expression (see Sect. 3.10, **steps 9–12**).

3.12 Mutant Enumeration Using the Soft-Agar Version of the MLA

Three million cells are suspended in 100 mL of 0.27 % soft agar cloning medium containing 1 μg/mL of TFT to enumerate the mutants, and six hundred cells are suspended in 100 mL of 0.27 % soft agar cloning medium for determining plating efficiency. After 11–14 days of incubation, colonies are counted using an automatic colony counter. Mutant colonies are categorized as small or large (see Sect. 3.14), and then MFs are calculated (see Sect. 3.16).

1. Prepare F_{10p} and F_{20p} media by warming in a 37 °C water bath.

2. Make a 3 % Difco Nobel agar solution in water (about 300 mL in a 500-mL bottle), autoclave the bottle for 30 min, then place it in a 56 °C water bath for cooling down. Each culture will need 18 mL of the agar solution for cloning.

3. Prepare two 125-mL flasks for each test sample with different labels (**Note 22**); one for non-selection medium (plating efficiency) and the other for selection medium with TFT (mutant determination) (see Table 3).

4. Gently pipette each test culture using 10 mL pipettes to form single-cell suspension.

5. (See Sect. 3.10, *step 11*) Determine the cell density and dilute cells to 3×10^5 cells/mL using F_{10p}. Allow cells to acclimate in the incubator while preparing the media for cloning.

6. Transfer 81 mL F_{20p} medium into each flask with selection label and 90 mL F_{20p} medium in each flask with non-selection label.

7. Transfer 9 mL 3 % Difco Nobel agar (*Step 2* of this section) to 125-mL flask containing either 81 or 90 mL of medium, mix well, and keep in the 37 °C incubator shaker.

Table 3
Preparation of cloning medium[a]

	Soft-agar version		Microwell version	
	Selection	Non-selection	Selection	Non-selection
	125-mL flask	125-mL flask	125-mL flask	50-mL tubes
F_{20p} medium (mL)	81	90	87	45
3 % Difco agar (mL)[c]	9	9	–	–
Cells to be added (mL)	10	1	3	36 µL
TFT (3 mg/mL)	34 µL[b]	–	90 µL[b]	–
Total volume (mL)	100	100	90	45
	Mix well			
Cell density (cells/mL)	3×10^4	6	1×10^4	8
	Pour into dishes		Dispense into plates	
100-mm dishes	3	3	–	–
96-well plates	–	–	4	2

[a]The cloning medium is maintained in the 37 °C shaker
[b]Final TFT concentration is 1 µg/mL in the soft-agar version and 3 µg/mL in the microwell version
[c]Final agar concentration is 0.27 %

8. Transfer 10 mL of cells in media (*Step 5* of this section) to selection flasks for each sample, mix well (3×10^4 cells/mL for 100 mL), and keep in the 37 °C incubator shaker.

9. Before adding TFT, transfer 1 mL cell culture (*Step 8 in this section*) to a 15-mL tube containing 9 mL F_{10p} (1:10 dilution, the first dilution) and mix well; then take 1 mL of this cell culture to a 15-mL tube containing 4 mL F_{10p} (1:5 dilution, the second dilution) and mix well; finally transfer 1 mL of this diluted cell culture to the flask that will be used (non-selection) for plating efficiency for each sample, and mix well (6 cells/mL for 100 mL).

10. Add 34 µL of the 3 mg/mL TFT stock solution into each 125-mL flask containing 3 million cells (selection solution, *step 8* of this section) in 100 mL F_{20p} (final TFT concentration is 1 µg/mL), mix well, and allow to incubate for 15 min in the 37 °C incubator shaker.

11. Pour the plating efficiency culture (non-selection solution, *step 9* of this section) into three 100-mm dishes (about 33 mL for each dish).

12. Pour the TFT selection culture (**step 10** of this section) into three 100-mm dishes (about 33 mL for each dish).
13. Chill all plates (**Steps 11** and **12** of this section) at 4 °C for 12–15 min, then move to a 5 % CO_2 incubator.
14. Incubate at 37 °C for 11–14 days.

3.13 Mutant Enumeration Using the Microwell Version of the MLA

The cultures to which 3 μg/mL of TFT will be added for mutant enumeration are adjusted to 1×10^4 cells/mL, and the cultures used for the determination of plating efficiency are adjusted to 8 cells/mL medium. The cells for plating efficiency are seeded into two 96-well flat-bottom microtiter plates and the cells for mutant selection are seeded into four 96-well plates (200 μL of cell suspension per well). After 11–14 days of incubation, colony counting is done visually (see Sect. 3.15). MFs are calculated using the Poisson distribution (see Sect. 3.16).

1. After 2-day expression period, on the day of plating, warm up F_{10p} and F_{20p} medium in a 37 °C water bath.
2. Prepare 125-mL flasks and 50-mL tubes, one for each sample (see Table 3).
3. Transfer 87 mL F_{20p} medium in each 125-mL flask and 45 mL F_{20p} medium in each 50-mL tube.
4. Pipette each sample using 10 mL pipettes to form single-cell suspension.
5. (See Sect. 3.10, *step 11*) Count the cells and dilute the cell density to 3×10^5 cells/mL using F_{10p}.
6. Transfer 3 mL of cells in media to 125-mL flask for each sample, and mix well (1×10^4 cells/mL for 90 mL).
7. Transfer 36 μL of 1×10^4 cells/mL from 125-mL flask (**step 6** of this section) into 50-mL tube containing 45 mL of F_{20p} medium (**Note 23**), mix well, and keep in the 37 °C incubator shaker.
8. Add 90 μL of TFT solution (3 mg/mL) into 125-mL flasks (**step 6** of this section) for all samples, mix well (final TFT concentration is 3 μg/mL), and keep in the 37 °C incubator shaker.
9. For the non-selection solution (**step 7** of this section), using an electronic 8-channel transfer pipette (1,250 μL), dispense 200 μL into each well of two 96-well flat-bottom plates.
10. For the selection solution with TFT (**step 8** of this section), using an electronic 8-channel transfer pipette, dispense 200 μL into each well of four 96-well plates.
11. Incubate at 37 °C in a 5 % CO_2 incubator for 11–14 days.

3.14 Colony Counting for the Soft-Agar Version of the MLA

After 11–14-day incubation, the colonies can be counted using an automatic colony counter that has adequate resolution to conduct colony sizing and provides general information on colony size. We use a ProtoCOL colony counter that has software developed specifically for the MLA.

1. (**Steps 1–5** for colony counting) Open the program ProtoCOL on the computer.
2. Create a new file with a name identifying the experiment and enter the parameters specific to the experiment including identifying information for each plate.
3. Read a plate, check sensitivity (keep low), and provide an identifier for the plate.
4. Read all the plates (remember each sample has three TFT-selection plates and three non-selection plates).
5. Save all data.
6. (**Steps 6–13** for colony size distribution) Select #1 TFT-plate of one sample, view size distribution.
7. Select #2 TFT-plate of same sample, add to size distribution.
8. Select #3 TFT-plate of same sample, add to size distribution.
9. View the size distribution of all three plates from the same sample.
10. Change the size limit selection (for example, 0–3 mm and 30 intervals).
11. Transfer the data to Excel and save it.
12. Click window and go to analysis view again.
13. Repeat **steps 6–12** for all samples.
14. (**Steps 14–20** for calculating small and large colonies) Go back to excel file for each sample.
15. Select a cutoff based on the dip in the middle of the bimodal size distribution. The left side of the size distribution is used to calculate the number of small colony mutants (**Note 24**). The right side of the size distribution is used to calculate the number of large colony mutants.
16. Sum the numbers of small colony mutants for each of the three TFT-selection plates.
17. Sum the total numbers of small colony mutants from **step 16** for all three plates.
18. Repeat for the large colony mutants.
19. Calculate the percentage of small colonies (see Sect. 3.16).
20. Repeat **steps 14–19** for all samples.

3.15 Colony Counting for the Microwell Version of the MLA

1. After the incubation of 11–14 days, the colonies on the plates are identified by naked eye using background illumination (**Note 25**).

2. For the non-selection plates (usually two plates), the empty wells will be counted.

3. For the TFT-selection plates (usually four plates), small and large colonies will be counted separately. Small colonies are defined as less than a quarter of the diameter of the well, while large colonies are more than a quarter of the diameter of the well. The number of empty wells should also be calculated as this is used in the calculation of the mutant frequency as well as the small and large colony mutant frequencies (see Sect. 3.16).

3.16 Data Calculations

Steps 1–4 in this section are performed during the suspension growth phase of the assay (before the cloning phase) and are the same for both the soft-agar version and microwell version of the MLA. **Steps 5–12** which deal with the calculation of mutant frequency are used for the soft-agar version, while **Steps 13–21** are for the microwell version.

1. Suspension growth (SG) is a measure of cell growth during treatment and the expression period. SG_1 is the suspension growth rate between day 0 (usually starting with 3×10^5 cells/mL) and day 1. It includes the 4 h treatment and about 20 h of incubation after the treatment, and is expressed as the fold increase in cell density. On the first day (day 1) following the treatments, the cell density for each culture is readjusted generally to 2×10^5 cells/mL. SG_2 is defined as the suspension growth rate between day 1 (after subculture) and day 2 (before the initiation of the cloning procedure).

2. Calculate $SG_{1(control)} \times SG_{2(control)}$ for the control sample. If the experiments have two vehicle controls, calculate the average of two $SG_{1(control)} \times SG_{2(control)}$.

3. Calculate $SG_{1(test)} \times SG_{2(test)}$ for the positive control and each concentration of test chemical (**Note 26**).

4. Relative suspension growth (RSG) is the relative total 2-day suspension growth of the test culture compared to the total 2-day suspension growth of the vehicle control, and is calculated for the positive control and each concentration of the test chemical. The RSG for the vehicle control will be defined as 100 %. The RSG for all other samples in this experiment can be calculated by [RSG (%) = $(SG_{1(test)} \times SG_{2(test)}) / (SG_{1(control)} \times SG_{2(control)}) \times 100$] (**Note 27**).

5. (**Steps 5–12** *for the soft-agar version*) The cloning efficiency (CEv) from the non-selection plates for all samples (including vehicle and positive controls, and test chemical) is calculated

by [CEv = (the total colony number of three non-selection plates) ÷ 600] (see Table 3, the cells in 100 mL of non-selection medium).

6. The cloning efficiency for mutant colonies (CEm) from the TFT-selection plates for all samples is calculated by [CEm = (the total colony number from three TFT-selection dishes) ÷ 3,000,000] (see Table 3, the cells in 100 mL TFT-selection medium).

7. Mutant frequency (MF) is calculated using the number of colonies on the TFT-selection plates divided by the cloning efficiency of the non-selection plates from the same sample, i.e., [MF = CEm ÷ CEv]. Usually, MF is expressed as mutants per million ($\times 10^{-6}$) viable cells.

8. From Sect. 3.14, the number of small colonies in TFT-selection medium for each sample is known. The percentage of small colony (SC) can be calculated by the number of small colony (#SC) divided by the total colony number (#TC) of the three TFT-selection plates, i.e., [SC% = #SC ÷ #TC × 100].

9. The small colony MF is calculated by multiplying the percentage of small colony (SC%) by the total MF (**Step 7**), i.e., [MF_{SC} = MF × SC%].

10. The large colony MF is calculated by subtracting the SC-induced MF (*Step 9*) from the total MF (**Step 7**), i.e., [MF_{LC} = MF - MF_{SC}].

11. Using the cloning efficiency of viable cells (CEv) in non-selection medium for each sample (*Step 5*), the relative cloning efficiency (RCE) can be calculated by [RCE (%) = $CEv_{(test)}$ ÷ $CEv_{(control)}$ × 100]. If the experiments have two vehicle controls, the average of two $CEv_{(control)}$ will be used in the calculation.

12. Relative total growth (RTG) is used as a parameter to define cytotoxicity and takes into account all cell growth and cell loss during the treatment period (4 or 24 h) and the 2-day expression period (RSG), as well as the cells' ability to clone 2 days after treatment (RCE, viability after 11–14-day incubation). RTG is calculated by multiplying RSG and RCE, i.e., [RTG% = RSG% × RCE% × 100].

13. (**Steps 13–21** *for the microwell version*) The microwell version of the assay uses 96-well cell culture plates for mutant selection. The plating efficiency (PE) in non-selection and TFT-selection plates is calculated using probability theory—from the zero term of the Poisson distribution (**Note 28**). After counting the number of empty wells in which a colony has not grown after 11–14 day incubation, the probable number of clones or wells (P_0, the mean of the distribution) on 96-well plates is calculated by [P_0 = −ln(empty wells ÷ total wells)].

Therefore, the plating efficiency (PE) is calculated by the probable number of wells divided by the average number of cells in each well (#cells), i.e., [$PEv = P_0 \div$ #cells].

14. Usually, two plates (for a total of 96 × 2 wells) are used for plating efficiency (non-selection medium) for all samples (including vehicle and positive controls, and test chemical) with the cell concentration of 8 cells/mL (i.e., an average of 1.6 cells per well; see Table 3). The PE is calculated by [$PEv = (-\ln(\text{empty wells} \div 192)) \div 1.6$].

15. Usually, four plates (a total of 96 × 4 wells) are used for TFT-mutant selection for all samples with a cell concentration of 1×10^4 cells/mL (i.e., an average of 2,000 cell per well; see Table 3). The numbers of small and large colonies are counted (Sect. 3.15). The number of empty wells, therefore, is 384 minus the number of wells containing either a small or a large colony. The PE of the TFT-selection plates is calculated by [$PEm = (-\ln(\text{empty wells} \div 384)) \div 2,000$].

16. MF is calculated using the plating efficiency of the TFT-selection plates divided by the plating efficiency of the non-selection plates for each sample, i.e., [$MF = PEm \div PEv$]. Usually, MF is expressed as mutants per million ($\times 10^{-6}$) cells.

17. Based on the colony counting in Sect. 3.15, the number of small colony (SC) and large colony (LC) TFT-resistant mutants is used to calculate the percentage of SC mutants (**Note 29**). The percentage of SC mutants in the four TFT-selection plates can be calculated by the number (#) of SC mutants divided by the total number of mutant colonies (#SC + #LC), i.e., [$SC\% = \#SC \div (\#SC + \#LC) \times 100$].

18. The small colony MF is calculated by multiplying the percentage of small colonies (SC%) by the total MF (*Step 16*), i.e., [$MF_{SC} = MF \times SC\%$].

19. The large colony MF is calculated by the total MF minus the SC-MF, i.e., [$MF_{LC} = MF - MF_{SC}$].

20. The relative plating efficiency (RPE) is calculated using the plating efficiency for the test culture relative to that of the vehicle control culture [$RPE(\%) = PEv_{(test)} \div PEv_{(control)} \times 100$]. If the experiment has two vehicle controls, the average of the two $PEv_{(control)}$ will be used in the calculation.

21. (*see* **Step 12**) RTG for each sample in the microwell version is calculated by multiplying RSG and RPE, i.e., [$RTG\% = RSG\% \times RPE\% \times 100$].

3.17 Data Evaluation

The acceptance criteria for individual experiments and the data evaluation criteria developed by the MLA Expert Workgroup of the IWGT are used to determine whether an assay is acceptable and whether a

Table 4
The acceptance criteria for the negative control of the MLA

Parameter	Soft-agar version	Microwell version
Mutant frequency (MF)[a]	$(35-140) \times 10^{-6}$	$(50-170) \times 10^{-6}$
Cloning or plating efficiency[b]	$(65-120)\%$	$(65-120)\%$
Suspension growth[c]	8–32	8–32

[a]See Sect. 3.16, Steps 7 and 16
[b]See Sect. 3.16, Steps 5 and 14
[c]See Sect. 3.16, Step 2. Here is for 4-h treatment. It will be 32–180 following 24-h treatment

specific treatment condition is positive [15, 17]. Appropriate cell growth for the vehicle/solvent controls is demonstrated by a cell suspension growth during the 2-day expression period that results in an 8–32 fold increase. The cloning efficiency or plating efficiency of the vehicle/solvent control should be 65–120 %; and the spontaneous MF should be within 50–170 mutants per 10^6 cells for the microwell version or 35–140 mutants per 10^6 cells for the soft-agar version (Table 4).

The positive controls must demonstrate adequate small colony mutant recovery. This can be done in one of two ways. (1) There is an induced small colony mutant frequency of at least 150 per 10^6 cells. That is, the small colony mutant frequency of the treated culture must be at least 150×10^{-6} greater than that of the concurrent vehicle/solvent control. (2) There is an induced total mutant frequency of at least 300×10^{-6} and at least 40 % of that total mutant frequency is from small colony mutants. That is, the total mutant frequency of the treated culture must be at least 300×10^{-6} greater than that of the concurrent vehicle/solvent control. So if a test culture shows an induced (mutant frequency above background) mutant frequency that is 500×10^{-6}, the small colony mutant frequency must be at least 200×10^{-6}.

Positive responses are defined as those where the induced MF in one or more treated cultures exceed the global evaluation factor (GEF) of 126 mutants per 10^6 cells for the microwell version or 90 mutants per 10^6 cells for the soft-agar version (**Note 30**) and where there is also a dose-related increase with MF [15, 17]. Negative responses are defined as those where the induced MF is less than the GEF and a sufficient level of cytotoxicity is attained.

Appropriate dose selection includes concentrations that cover the cytotoxicity range between 10 and 20 % RTG (highest dose) and 100 % RTG (lowest dose). If the test chemical is positive at cytotoxicity levels greater than 10–20 %, it is not necessary to have concentrations covering the entire range. It is necessary to have at least one culture yielding between 10 and 20 % RTG in order to call the response negative. Positive responses seen only in cultures with RTGs less than 10 % should not be considered biologically relevant.

It is worth noting that the concept of RTG tends to confuse people particularly when comparing the cytotoxicity seen in the MLA to the cytotoxicity seen in other *in vitro* mammalian assays. For example, it is agreed that the top cytotoxicity level for the *in vitro* cytogenetic assays (both for metaphase chromosome aberration analysis and for micronucleus analysis) should not exceed a reduction of about 50 % in cell growth [12]. However, for the MLA it is recommended that the top dose should have an RTG of 20–10 %, i.e., 80–90 % cytotoxicity. The difference between the assays is based on the fact that different parameters are used. The *in vitro* cytogenetic assays use relative increase in cell growth or relative population doubling, determined approximately 24 h after the initiation of treatment. On the other hand, the MLA uses RTG. This measure, first described by Donald Clive (the developer of the MLA), includes a measure of cell growth during the treatment (4 or 24 h), expression (2 days), and cloning (11–14 days). It is calculated by multiplying RSG (which covers treatment and expression) and RPE (which covers the 11–14-day colony formation) (see Sect. 3.16 in detail). The rationale for using RTG is based on the fact that some chemicals cause delayed toxicity following treatment and factoring in this cytotoxicity using the growth during the 2-day expression and the cloning phase of the assay captures this effect.

4 Data Sheets

Every laboratory should develop their own procedures for documenting each experiment. Here, we show a set of data sheets that we use for the MLA. For convenience, we convert these tables to Excel worksheets which include the formulae for the calculations. Tables 5, 6 (both A and B), and 8 can be used for the soft-agar version of the MLA. For the microwell version, we use Table 7 (both A and B) instead of Table 6. For the dose range finding tests, Tables 5 and 6A (or Table 7A) can be used to evaluate the cytotoxicity (RSG) after 2-day expression. This RSG information from the dose ranging experiment will be used to select doses for the mutation experiments.

Table 5
Cell counting

Day 1

No.	Sample	1:1,000 dilution Count #1	#2	#3	Average	Cell Concentration (×10⁵)	Suspension growth (SG1)	Subculture for day 2 (2×10^5 cells/mL, >10 mL) Cells (mL)	F10p (mL)
1	NC1	A	B	C	D=	$E = D \div 100$	$F = E \div 3$		
2	NC2				$(A+B+C) \div 3$				
3	PC								
4	Dose 1								
→	→								
13	Dose 10								

Day 2

No.	Sample	1:1,000 dilution Count #1	#2	#3	Average	Cell concentration (×10⁵)	Suspension growth (SG2)	Adjust concentration (3×10^5 cells/mL) Cells (mL)	F20p (mL)
1	NC1	G	H	I	J=	$K = J \div 100$	$L = K \div 2$		
2	NC2				$(G+H+I) \div 3$				
3	PC								
4	Dose 1								
→	→								
13	Dose 10								

Table 6
A, B: Mutant frequency (soft-agar version)

A

No.	Sample	SG=SG1×SG2			RSG[a] (%)	Cloning efficiency					CEv=T/600	RCE[b] (%)	RSG×RCE RTG (%)
		SG1	SG2	SG		Number of colonies			Total				
						Plate 1	Plate 2	Plate 3					
1	NC1	F	L	M=	N	Q	R	S	T=		U=	V	W=
2	NC2			F×L					Q+R+S		T÷600		N×V/100
3	PC												
4	Dose 1												
→	→												
13	Dose 10												

B

No.	Sample	CEv	Mutant selection				CEm		MF (×10⁻⁶)	Number of small colony[c]	SC (%) SC/TC	Small MF (×10⁻⁶)
			Number of colonies									
			Plate 1	Plate 2	Plate 3	Total	TC/3×10⁶	CEm/CEv				
1	NC1	U	X	Y	Z	a=	b=	c=		d	e=	f=e×c
2	NC2					X+Y+Z	a÷(3×10⁶)	b÷U×10⁶			d÷a×100	
3	PC											
4	Dose 1											
→	→											
13	Dose 10											

[a]RSG (%) = $(SG_{1(test)} \times SG_{2(test)}) / (SG_{1(control)} \times SG_{2(control)}) \times 100$ (see Sect. 3.16)
[b]RCE (%) = $CEv_{(test)} \div CEv_{(control)} \times 100$ (see Sect. 3.16)
[c]Data from Sect. 3.14

Table 7
A, B: Mutant frequency (microwell version)

A

No.	Sample	SG = SG1 × SG2			RSG[a] (%)	Plating efficiency					RSG × RPE[T]
						Number of wells		Colony	$P = -\ln(EW/TW)$	RPE[b] (%)	RTG (%)
		SG1	SG2	SG		Total	Empty		PEv = P/1.6		
1	NC1	F	L	M=	N	192	Q	R=	S=	T	U=
				F × L							
2	NC2					192		192-Q	$(-\ln(Q/192)) \div 1.6$		N × T/100
3	PC					192					
4	Dose 1					192					
→	→					↓					
13	Dose 10					192					

B

No.	Sample	PEv	Mutant selection					MF (×10⁻⁶)	SC (%)	(×10⁻⁶)	
			Number of wells			$P = -\ln(EW/TW)$				Small MF	Large MF
			Total	Large	Small	Empty	PEm = P/2000	PEm/PEv	SC/TC		
1	NC1	S	384	V	W	X=	Y=	Z=	a=	b=	c=
2	NC2		384			384-V-W	$(-\ln(X/384)) \div 2000$	Y ÷ S × 10⁶	W ÷ (V+W) × 100	a × Z	Z-b
3	PC		384								
4	Dose 1		384								
→	→		↓								
13	Dose 10		384								

[a] RSG (%) = (SG$_{1(test)}$ × SG$_{2(test)}$) / (SG$_{1(control)}$ × SG$_{2(control)}$) × 100 (see Sect. 3.16)
[b] RPE (%) = PEv$_{(test)}$ ÷ PEv$_{(control)}$ × 100 (see Sect. 3.16)

Table 8
Summary of experiment

No.	Sample	Soft-agar version[a]				Microwell version[b]			
		CEv (%)	RTG (%)	MF (×10⁻⁶)	SC (%)	PEv (%)	RTG (%)	MF (×10⁻⁶)	SC (%)
1	NC1	U	100	c	e	S	100	Z	a
2	NC2		100				100		
3	PC								
4	Dose 1								
↓	↓								
13	Dose 10								

[a]Data from Table 6 (A and B)
[b]Data from Table 7 (A and B)

5 Notes

1. RPMI 1640 has also been used successfully and in some laboratories it is the preferred medium for all or parts of the assay.

2. Pluronic F-68 is used to prevent mechanical disruption of the cells during the shaking. While it is not absolutely necessary (and some laboratories do not use it) we use it because the assay was originally developed using pluronic F-68.

3. A reduced serum level (5 %) is used for the treatment phase of the assay, because the presence of serum may interfere with the activity of some mutagens.

4. Horse serum is heat inactivated by warming at 56 °C for 30 min, in order to eliminate a factor which may degrade TFT. It is important that the horse serum should be at a temperature of 56 °C prior to starting the 30 min incubation.

5. Historically, in our lab, the mouse lymphoma cells from a vial of frozen stocks were used up to 3 months. During this period, we usually perform mycoplasma detection in the cell culture one or two times, and cleanse the culture of spontaneous *Tk* mutants at least three times.

6. Usually, freshly thawed cells need 3–7 days for recovery. Therefore, if one thaws a new culture on Friday, it is not necessary to check it during the weekend.

7. Avoid overgrowth of cells, i.e., more than 10^6 cells/mL.

8. In case of a holiday weekend, the cells can be diluted to 0.5×10^5 cells/mL and cultured for 2 days, and then be subcultured to 0.5×10^5 cells/mL again for another 2 days.

9. If the cultures are infected with mycoplasma, yellow-green fluorescence is visible on the cell membranes of the infected cells. In many cases, mycoplasmas are crowded on a spot on the cell's surface. Depending on the mycoplasma species present in the samples, the shape of the stained bacteria may vary from very small, coccid bodies with bright fluorescence to short, more diffusely filaments. It is critical that any infected cells be disposed of and measures taken to assure that the laboratory is properly disinfected so that the contamination does not spread to the stock cultures.

10. Usually, the cleansing is started on Monday. Total media volume is generally 50 or 100 mL (99 + 1 mL THMG), and starting cell concentration is generally 3×10^5 cells/mL (or 2×10^5 cells/mL).

11. The cells can be expected to grow at longer doubling times during the cleansing.

12. There are several different recipes for the S9 mix and for the final concentration of S9 in the test culture. It can be as high as 10 %, depending upon the batch of S9 and the class of test chemical.

13. For the soft-agar version, 0.5 mL of TFT stock solution will be sufficient for up to 14 samples. For the microwell version, 1 and 1.5 mL of TFT stock solution will be sufficient for up to 11 and 16 samples, respectively. If both selection methods are used in a single cloning, 1.5 mL of TFT stock solution will be sufficient for up to 12 samples.

14. Usually, each experiment has two vehicle (solvent) controls and one positive control.

15. It should be noted that nanomaterials tested in the MLA likely stay with the cell pellet after treatment and their presence cannot be readily identified by eye. The exposure aspects of nanomaterial evaluation require methodologies to characterize the nanomaterials both prior to and during cell treatment. The discussion of these methods is beyond the scope of this chapter.

16. Test cultures may be in duplicate and in this case, a minimum of four doses will be used.

17. The toxicity of a test chemical may well be different in the two metabolic activation conditions (with and without S9). The range finding tests for both conditions may be performed at the same time. It is also advisable to conduct the with and without S9 assays in the same experiment depending upon the number of cultures used and the capability of the staff to manage the number of cultures without negatively impacting assay performance.

18. It has been suggested that such short (3 or 4 h) treatments may be insufficient for detecting some clastogens and spindle poisons [36]. The ICH requires that a 24-h treatment be used if the short treatment is negative when evaluating pharmaceuticals intended for human use.

19. Phenotypic expression time is the time after treatment during which the genetic alteration is fixed within the genome and all residual thymidine kinase enzyme is removed from the newly induced $Tk^{-/-}$ cell, either by cell division or enzyme degradation. In the MLA, the period for expression of mutation at the Tk gene is 2 days.

20. Each treatment culture will require 6×10^6 cells. Therefore, a 100 mL cell culture should be adequate for up to 13 test cultures.

21. Using SG1 and SG2, calculate the relative suspension growth (RSG) for each sample. Any test culture with an RSG less than 20 % will not be cloned for mutant selection, because it is likely to have an RTG below 10 %.

22. It is also acceptable to use 125-mL glass bottles (Wheaton, Cat. #219815; Millville, NJ) instead of 125-mL Erlenmeyer flasks. These glass bottles can be washed, autoclaved, and reused. It is, however, critical that any glassware be carefully and thoroughly washed using tissue culture grade detergent. Any glassware that is not properly cleaned can be the source of poor cell growth.

23. Two-step dilutions can be used. Transfer 1 mL of cell culture (*Step 6*, before adding TFT) to a 15-mL tube containing 9 mL F_{10p} (1:10 dilution, the first dilution) and mix well; then transfer 6 mL of this solution to another 15-mL tube containing 4 mL F_{10p} (1:1.67 dilution, the second dilution) and mix well; finally transfer 0.6 mL of this diluted solution to non-selection flasks for each sample and mix well (8 cells/mL for 45 mL).

24. Based on the OECD recommendation, if the test chemical is positive in the MLA, colony sizing should be performed on at least one of the test cultures (the highest positive concentration) and on the vehicle and positive controls. In our lab, we generally conduct colony sizing for all cultures.

25. The colonies may also be identified by low-power microscope.

26. If a 24-h treatment is used, the 2-day expression period will begin following treatment. RSG should be calculated by $SG_T \times SG_1 \times SG_2$ (SG_T means suspension growth during 24-h treatment).

27. RSG should be calculated and used to select cultures for cloning. If the RSG for some of the higher doses of test chemical is less than 20 %, these cell cultures should not be cloned for mutant selection because the samples having RSG less than

20 % will most likely result in an RTG of less than 10 %. This will save supplies and labor.

28. A Poisson distribution is the probability distribution of a Poisson random variable. $P(x; \lambda) = (e^{\lambda})(\lambda^x)/x!$, where e is the base of the natural logarithm; x is the actual number of occurrences; λ is the mean number of occurrences. For the MLA, one should count the number of empty wells, that is, the wells which do not show any colony growth after 11–14 day incubation, i.e., x = 0. Therefore, $P(0; \lambda) = EW/TW = (e^{\lambda})(\lambda^0)/0! = (e^{\lambda})$, where EW and TW are the number of empty wells and total wells, respectively. The mean of the distribution $\lambda = P_0 = -\ln(EW/TW)$.

29. Alternatively, the percentage of small colony mutants (SC%) can be calculated by the small colony MF divided by the total MF. Small colony MF is calculated using the relevant number of empty wells for small colony, i.e., the zero term of the Poisson distribution.

30. The GEF was defined by the MLA Expert Workgroup of the IWGT [17]. It was based on the distribution of vehicle MFs from ten laboratories (4 using the agar version and 6 using the microwell version). It is the mean of that distribution plus one standard deviation. A separate value was calculated for the agar and microwell version because the distribution is slightly different between the two versions.

Acknowledgements

The authors thank Drs. Mugimane G. Manjanatha and Haixia Lin for the critical review of the manuscript. Use of trade names is for informational purposes only and in no way implies endorsement by the U.S. Food and Drug Administration (FDA). The views presented in this paper do not necessarily reflect those of the U.S. FDA.

References

1. Clive D, Flamm WG, Machesko MR, Bernheim NJ (1972) A mutational assay system using the thymidine kinase locus in mouse lymphoma cells. Mutat Res 16:77–87
2. Cole J, Arlett CF, Green MH, Lowe J, Muriel W (1983) A comparison of the agar cloning and microtitration techniques for assaying cell survival and mutation frequency in L5178Y mouse lymphoma cells. Mutat Res 111:371–386
3. Applegate ML, Moore MM, Broder CB et al (1990) Molecular dissection of mutations at the heterozygous thymidine kinase locus in mouse lymphoma cells. Proc Natl Acad Sci USA 87:51–55
4. Clements J (2000) The mouse lymphoma assay. Mutat Res 455:97–110
5. Liechty MC, Scalzi JM, Sims KR et al (1998) Analysis of large and small colony L5178Y tk-/- mouse lymphoma mutants by loss of heterozygosity (LOH) and by whole chromosome 11 painting: detection of recombination. Mutagenesis 13:461–474
6. Honma M, Momose M, Sakamoto H, Sofuni T, Hayashi M (2001) Spindle poisons induce allelic loss in mouse lymphoma cells through

mitotic non-disjunction. Mutat Res 493: 101–114

7. Wang J, Sawyer JR, Chen L et al (2009) The mouse lymphoma assay detects recombination, deletion, and aneuploidy. Toxicol Sci 109: 96–105

8. FDA (2000) Redbook 2000: IV.C.1.c mouse lymphoma thymidine kinase gene mutation assay. http://wwwfdagov/Food/Guidance ComplianceRegulatoryInformation/Guidance Documents/FoodIngredientsandPackaging/ Redbook/ucm078336htm

9. Dearfield KL, Auletta AE, Cimino MC, Moore MM (1991) Considerations in the U.S. Environmental Protection Agency's testing approach for mutagenicity. Mutat Res 258: 259–283

10. ICH (1995) Topic S2A genotoxicity: guidance on specific aspects of regulatory genotoxicity tests for pharmaceuticals. In: International conference on harmonisation of technical requirements for registration of pharmaceuticals for human use, harmonised tripartite guideline CPMP/ICH/141/95. Approved Sept 1995. http://www.ifpma.org/ich1.html

11. Muller L, Kikuchi Y, Probst G et al (1999) ICH-harmonised guidances on genotoxicity testing of pharmaceuticals: evolution, reasoning and impact. Mutat Res 436:195–225

12. ICH (2011) Guidance on genotoxicity testing and data interpretation for pharmaceuticals intended for human use S2(R1). Approved Nov 2011. http://www.ich.org/fileadmin/ Public_Web_Site/ICH_Products/Guidelines/ Safety/S2_R1/Step4/S2R1_Step4.pdf

13. OECD (1997) In vitro mammalian cell gene mutation test, OECD guideline for testing of chemicals, no. 476. http://wwwoecdorg/ chemicalsafety/assessmentofchemicals/19484 26pdf

14. Moore MM, Honma M, Clements J et al (2000) Mouse lymphoma thymidine kinase locus gene mutation assay: International Workshop on Genotoxicity Test Procedures Workgroup Report. Environ Mol Mutagen 35:185–190

15. Moore MM, Honma M, Clements J et al (2002) Mouse lymphoma thymidine kinase gene mutation assay: follow-up International Workshop on Genotoxicity Test Procedures, New Orleans, Louisiana, April 2000. Environ Mol Mutagen 40:292–299

16. Moore MM, Honma M, Clements J et al (2003) Mouse lymphoma thymidine kinase gene mutation assay: International Workshop on Genotoxicity Tests Workgroup report– Plymouth, UK 2002. Mutat Res 540:127–140

17. Moore MM, Honma M, Clements J et al (2006) Mouse lymphoma thymidine kinase gene mutation assay: follow-up meeting of the International Workshop on Genotoxicity Testing–Aberdeen, Scotland, 2003–assay acceptance criteria, positive controls, and data evaluation. Environ Mol Mutagen 47:1–5

18. Moore MM, Honma M, Clements J et al (2007) Mouse lymphoma thymidine kinase gene mutation assay: meeting of the International Workshop on Genotoxicity Testing, San Francisco, 2005, recommendations for 24-h treatment. Mutat Res 627:36–40

19. Moore MM, Honma M, Clements J et al (2011) Suitable top concentration for tests with mammalian cells: mouse lymphoma assay workgroup. Mutat Res 723:84–86

20. Hozier J, Sawyer J, Clive D, Moore M (1982) Cytogenetic distinction between the TK$^+$ and TK$^-$ chromosomes in the L5178Y TK$^{+/-}$ 3.7.2C mouse-lymphoma cell line. Mutat Res 105:451–456

21. Sawyer J, Moore MM, Clive D, Hozier J (1985) Cytogenetic characterization of the L5178Y TK$^{+/-}$-3.7.2C mouse lymphoma cell line. Mutat Res 147:243–253

22. Moore-Brown MM, Clive D, Howard BE, Batson AG, Johnson KO (1981) The utilization of trifluorothymidine (TFT) to select for thymidine kinase-deficient (TK$^{-/-}$) mutants from L5178Y/TK$^{+/-}$ mouse lymphoma cells. Mutat Res 85:363–378

23. Clive D, Johnson KO, Spector JF, Batson AG, Brown MM (1979) Validation and characterization of the L5178Y/TK$^{+/-}$ mouse lymphoma mutagen assay system. Mutat Res 59:61–108

24. Moore MM, Clive D, Hozier JC et al (1985) Analysis of trifluorothymidine-resistant (TFTr) mutants of L5178Y/TK$^{+/-}$ mouse lymphoma cells. Mutat Res 151:161–174

25. Moore MM, Clive D, Howard BE, Batson AG, Turner NT (1985) In situ analysis of trifluorothymidine-resistant (TFTr) mutants of L5178Y/TK$^{+/-}$ mouse lymphoma cells. Mutat Res 151:147–159

26. Clark LS, Harrington-Brock K, Wang J et al (2004) Loss of P53 heterozygosity is not responsible for the small colony thymidine kinase mutant phenotype in L5178Y mouse lymphoma cells. Mutagenesis 19:263–268

27. Brendler-Schwaab S, Czich A, Epe B et al (2004) Photochemical genotoxicity: principles and test methods. Report of a GUM task force. Mutat Res 566:65–91

28. Mei N, Xia Q, Chen L, Moore MM, Fu PP, Chen T (2005) Photomutagenicity of retinyl

palmitate by ultraviolet a irradiation in mouse lymphoma cells. Toxicol Sci 88:142–149

29. Mei N, Xia Q, Chen L, Moore MM, Chen T, Fu PP (2006) Photomutagenicity of anhydroretinol and 5,6-epoxyretinyl palmitate in mouse lymphoma cells. Chem Res Toxicol 19:1435–1440

30. Mei N, Hu J, Xia Q, Fu PP, Moore MM, Chen T (2010) Cytotoxicity and mutagenicity of retinol with ultraviolet A irradiation in mouse lymphoma cells. Toxicol In Vitro 24:439–444

31. Gocke E, Muller L, Guzzie PJ et al (2000) Considerations on photochemical genotoxicity: report of the International Workshop on Genotoxicity Test Procedures Working Group. Environ Mol Mutagen 35:173–184

32. Mei N, Chen T, Godar DE, Moore MM. UVA-induced photomutagenicity of retinyl palmitate. Mutat Res 2009;677:105–106; author reply 107–108

33. Lorge E (2013) Recommendations for good cell culture practices in genotoxicity testing. Genetic Toxicology Technical Committee, ILSI-HESI, Washington, DC (Personal Communication)

34. Machanoff R, O'Neill JP, Hsie AW (1981) Quantitative analysis of cytotoxicity and mutagenicity of benzo[a]pyrene in mammalian cells (CHO/HGPRT system). Chem Biol Interact 34:1–10

35. Majeska JB, Matheson DW (1990) Development of an optimal S9 activation mixture for the L5178Y TK$^{+/-}$ mouse lymphoma mutation assay. Environ Mol Mutagen 16:311–319

36. Honma M, Zhang LZ, Sakamoto H et al (1999) The need for long-term treatment in the mouse lymphoma assay. Mutagenesis 14:23–29

INDEX

A

ABC. *See* ATP-binding cassette (ABC)
Absorption, distribution, metabolism and excretion
    (ADME) 1, 6, 41, 101, 337, 441
Acyl
    glucuronides 118, 133,
        134, 505–515
    migration 509, 510, 514
Acylcarnitine profiling 461–474
ADME. *See* Absorption, distribution, metabolism and
    excretion (ADME)
Alamethicin 90, 92–94,
    101–114, 123, 125, 126, 133, 144, 165
Ames test 538, 545–557
Aryl hydrocarbon (Ah) receptor 207–217,
    221–233, 237
ATPASE assay 371–377, 398, 399
ATP-binding cassette (ABC) 337, 353, 369, 371
    ABCB1 337

B

Bacterial
    background lawn 552, 554–557
    mutagenicity test 518, 545
BCRP. *See* Breast cancer resistance protein (BCRP)
BCS. *See* Biopharmaceutical classification system (BCS)
β-Galactosidase (β-Gal) activity assay 541–542
Bioavailability and bioequivalence studies 3, 78
Biopharmaceutical classification system
    (BCS) 2, 3, 51, 78
Biopharmaceutics
    dissociation constants (Ksp) 2, 5
    dissolution rates 2, 5
    hydrophobicity 2, 5
    ionization potential (pKa) 2, 5
    lipophilic partition coefficients (Log P) 2, 5
    stability 2, 5, 445
Blood to plasma ratio 39–41, 43–46
Breast cancer resistance protein (BCRP) 51, 341,
    353–366, 388, 392, 400
Budding yeast 537, 538
Butyric acid 52, 53, 72
Bypassing the stomach 77

C

Caco-2 cells
    2/15 50
    AQ 50
    BBE 1&2 50
    5-day cultured, cell monolayers 49–74
    HTB-37 .... 50, 52–54, 56, 58
    TC7 .... 50, 53
    21-day cultured, cell monolayers 54
Calibration standards 34, 437–439,
    441, 451, 462, 466, 474, 513
CALUX. *See* Chemically activated luciferase expression
    (CALUX)
Carrier-mediated 50, 54
Cell
    culture 5, 50, 53, 55, 58–59,
        197, 208, 209, 211, 224, 225, 233, 340–342,
        355–357, 385, 388, 390, 406–408, 419, 420,
        429, 518, 521–522, 524, 526, 550, 551, 565,
        566, 568, 570, 573, 574, 576, 580, 587, 589
    viability 44, 95,
        98, 175, 178, 181–186, 189, 202, 266, 273, 523,
        526, 532
Chemical degradation half-life 509–512,
    514, 515
Chemically activated luciferase expression
    (CALUX) 222, 223
Chinese hamster ovary (CHO) 338, 418,
    419, 521, 522
Chromosomal mutation 562
Co-activating 101–114
Co-factor
    NADPH 89, 138, 151
    UDPGA 89, 456
Comet assay 517–534
Constitutive androstane receptor
    (CAR, NR1I3) 195–204, 237
Covalent binding 225, 318,
    506, 512–514
Cytochrome P450 (CYP)
    1A2 .... 104, 111,
        148, 150–152, 160, 161, 238, 242–244, 247,
        251, 283, 289, 306, 316, 324, 329–331, 333

Index

Cytochrome P450 (CYP) (cont.)
  3A4 .... 72, 104, 111,
    119, 148, 150–152, 155, 156, 158–161, 163,
    173–176, 181–183, 185, 186, 190, 191, 238,
    242–244, 247, 251, 255–267, 283, 297–301,
    311, 316, 324, 329–331, 333
  2C9 .... 104, 111, 139, 148,
    150, 152, 160, 161, 174, 242–244, 247, 252, 283, 299
  2C19 .... 94, 101, 114,
    148, 150–152, 160, 161, 174, 242–244, 247, 251,
    283, 291, 295, 316, 324, 329, 330, 333
  2D6 .... 104, 111, 118, 139,
    148, 150, 152, 158–161, 163, 283, 290, 291, 316, 324,
    329, 330, 333
Cytosol 103, 208, 210–213, 215, 417
Cytotoxicity 181, 183–186, 189, 200,
  242–243, 249, 251, 266, 523, 532, 552, 554, 555, 571,
  572, 580, 582, 583
  LDH 242, 251

D

Dansyl glutathione 493–494, 498–499, 502
Data mining 453, 457
DDI. *See* Drug-drug interactions (DDI)
Detection
  fluorescence 124
  HPLC 13
  LC/MS/MS 88
  radioactive 495
  UV 122, 124
Dimethyl sulfoxide (DMSO) 4, 8, 41,
  56–58, 62, 63, 72–74, 78, 80, 81, 84, 91, 92, 94, 96,
  104, 106, 107, 113, 133, 142, 143, 146, 151, 154,
  157, 162, 164, 175, 178–181, 183–186, 188, 189,
  200, 201, 203, 215, 216, 225, 226, 230, 239, 242,
  243, 251, 259, 262, 264, 267, 272, 273, 289–295,
  300, 308–311, 322, 325, 333, 372–374, 379, 398,
  429, 481, 506, 508, 521, 523, 524, 531, 545–557, 565
DMPK. *See* Drug metabolism/pharmacokinetic (DMPK)
DMSO. *See* Dimethyl sulfoxide (DMSO)
DNA
  binding analysis 208, 214–216
  damage 517, 518, 523,
    525, 527, 529–531, 537–543
Dried blood spots (DBS) 258, 262, 461–476
Drug discovery 1–18, 21–37,
  50, 51, 78, 97, 101, 102, 139–141, 148, 153, 159, 164,
  167, 173, 190, 191, 240, 255, 270, 272, 281, 283, 288,
  307, 308, 312, 316, 321, 445–457, 461–474, 478
Drug-drug interactions (DDI) 22, 78, 101,
  118, 137, 138, 141, 172, 188, 190, 191, 195–196, 237,
  249, 255, 256, 267, 269, 271, 272, 283, 284, 286–288,
  307, 309, 315, 319, 337, 353, 354, 370, 392, 405, 418

Drug metabolism/pharmacokinetic (DMPK) 50, 97
Dual-luciferase reporter assay 198–201

E

Efflux 50–52, 69–71,
  173, 195, 256, 337–339, 343, 344, 346–350, 354,
  357–363, 365, 370, 383, 387, 388, 391–397, 400,
  401, 417–419, 538
EMA. *See* European Medicines Agency (EMA)
Endoplasmic reticulum (ER) 88, 102, 103, 138
Enhanced green fluorescent protein (EGFP) 223, 225
Environmental protection agency (EPA) 562, 567
Enzymes
  aldehyde oxidase 88, 89
  glutathione transferases 88
  methyltransferases 88
  N-acetyl-transferases 88
  sulfotransferases 88
  xanthine oxidase 88
EPA. *See* Environmental protection agency (EPA)
Epithelium membrane 3, 49–51
Equilibrium dialysis
  classical equilibrium dialysis (CED) 22, 23,
    25–27, 29, 37
  rapid equilibrium dialysis (RED) 22, 23,
    28–32, 37
  ultrafiltration methods 22, 23, 33–34
European Medicines Agency (EMA) 138, 237,
  249–251, 338, 344, 346, 347, 354, 359, 363, 364,
  405, 418

F

FBS. *See* Fetal bovine serum (FBS)
FDA. *See* Food and Drug Administration (FDA)
Fetal bovine serum (FBS) 52–54, 57,
  175, 178, 198, 225, 230, 259, 261, 273, 340, 342, 355,
  357, 372, 384, 389, 392, 393, 407, 408, 419, 522
Flavin-containing monooxygenases
  (FMOs) 97, 138, 141, 146, 149–153
Fluorometric
  analyses 281–301
  detection 407
FMOs. *See* Flavin-containing monooxygenases (FMOs)
Food and Drug Administration (FDA) 3, 51,
  78, 84, 85, 138, 237, 249–251, 338, 343, 345,
  347, 350, 354, 358, 361, 362, 366, 405, 406, 418,
  441, 535, 562, 567
Formulation
  complexants 7, 9, 10
  cyclodextrins 6
  decision tree 2, 7, 9
  5 % dextrose (D5W) 7
  hydroxylethyl cellulose (HEC) 6

hydroxypropyl cellulose (HPC) 6
screening methyl cellulose (MC) 6
surfactants.. 6, 7, 9–11, 15
Fraction unbound (*f*u) 22, 26, 31, 37

G

Gel retardation assay (GRA) 207–217
Gene
    expression 173, 209, 222–225,
        229–231, 233, 247
    mutation 562, 563
Genotoxicity............................ 517–534, 537, 538,
    541, 542, 545–557, 561–563, 574
Glucuronidation 102, 103,
    109–113, 117–134, 141, 145, 165, 257, 432, 505
Glutathione 88, 286, 370,
    379, 446, 449, 452, 456, 478–481, 485, 486,
    490–495, 498–502
GRA. *See* Gel retardation assay (GRA)

H

Harvesting cells 199, 200
HEK. *See* Human embryonic kidney (HEK)
Hematocrit...............................43, 46, 462, 464, 474
Hepatic clearance 87, 88, 90, 97,
    102, 103, 112
Hepatocytes
    cryopreserved................................... 88, 92, 95,
        240–241, 250, 259–261, 271–274, 393, 394, 446
    plated 88, 140, 255–267, 271, 456
    sandwich-cultured 339, 391–397
    suspension................................87, 88, 92–93, 95–98, 140
HILIC. *See* Hydrophilic interaction liquid chromatography (HILIC)
Human colorectal carcinoma intestinal cell line (Caco-2)
    cells 5, 49–74, 80, 339,
        341, 343, 349, 350, 354, 356–361, 388–391
Human embryonic kidney (HEK)..................... 339, 418, 419
Hydrophilic interaction liquid chromatography
    (HILIC) 435, 463, 465, 472

I

IC$_{50}$
    shift... 256, 270, 287,
        307–313, 315–334
    values 262–264, 266, 271, 286, 288,
        289, 293, 295–299, 307, 308, 312, 313, 318, 320, 321,
        323, 327, 329, 330, 333, 371, 376, 378, 398, 400, 427
Induction
    CYP3A4............................ 185, 186, 190, 191
    hepatic 172, 174
    Pgp 173
    PXR.. 185, 187, 191

In situ single rat perfusion 77–85
International Committee on Harmonization
    (ICH) 518, 562, 572, 573, 589
Intestinal
    apical membrane..353
    basolateral membrane 3, 49
Irreversible inhibitors.................................283, 288,
    296–298, 316

J

Jejunum 78, 81, 82, 85, 370, 388

K

Kinetics
    comparative analysis 98, 129
    CYP......................139, 148, 153–155, 161,
        165, 284, 285, 316–319
    enzyme 129–130
    FMO 153
    irreversible 285
    parameters 129, 145, 148,
        153, 183, 186, 256, 257, 272, 276–278, 286, 299, 312,
        317, 318, 410, 422
    reversible...................................... 284, 293, 316

L

Lactate dehydrogenase (LDH)................. 242, 243, 251, 532
LC/MS/MS analyses
    calibration standards 437–439,
        441, 451, 462, 466, 513
    qualitative 139–140, 450–452, 457
    quantitative..................................... 67, 68, 258,
        328, 450–451
LDH. *See* Lactate dehydrogenase (LDH)
LDH assay...................................... 243, 251, 532
Lipid solutions
    lipid dispersions 6
    lipid emulsions 6
    self-emulsifying drug delivery systems
        (SEDDS)...6
    self-microemulsifying drug delivery systems
        (SMEDDS)..6
Liquid chromatography (LC)............................... 10, 24, 40,
    55, 80, 88, 104, 139, 243, 258, 273, 285, 309,
    318, 345, 359, 381, 408, 422, 431, 446, 462, 478,
    491, 506
Liver microsomes 87–99, 101–114,
    121, 129, 140, 144, 146, 148, 149, 154, 155,
    163, 256, 271, 281–301, 305–313, 318, 319,
    323, 456, 481, 482, 490, 492–494, 500,
    506, 508
Luciferase 176, 181, 183–186, 188,
    189, 196–201, 223, 225–227, 230–233
Luminal fluid...3, 50

Index

M

Madin-Darby canine kidney (MDCK) 339, 341–343, 348, 349, 354–361, 365, 418
    II cells overexpressing MRP2 (MDCKII-MRP2) 383–388
Mass spectrometry (MS) 52, 64, 65, 67, 127, 130, 133, 285, 315–334, 431, 432, 436–437, 448, 453, 454, 457, 462, 477–487, 490, 491, 499
MBI. *See* Mechanism based inhibition (MBI)
MDCK. *See* Madin-Darby canine kidney (MDCK)
Mechanism based inhibition (MBI) 283–285, 288, 296, 298, 300, 301, 306–308, 312, 316, 318, 331
Metabolic stability 87–99, 106, 108, 110, 112, 140, 148, 209, 221, 222, 445, 463
Metabolism
    Phase I 88, 96, 102, 103, 111, 138, 139, 141, 451, 456, 457, 489
    Phase II 102, 111, 141, 142, 144, 223, 457
    reactivity 478, 505
Metabolite intermediate complex (MIC) 283, 306, 318
Microsomes 88–90, 94, 97, 98, 102–107, 111, 113, 124, 126, 129, 133, 140, 144, 151, 154–156, 159–163, 295, 323, 324, 327, 332, 446, 449, 450, 456, 545
Mouse lymphoma assay 519, 561–590
mRNA levels 52, 224, 233, 288
Multidrug resistance-associated protein 2 (MRP2; ABCC2)
    hepatic 369–401
    intestinal 369–401
Mutagenicity 518, 545–547, 551–554, 561–590

N

Nicotinamide adenine dinucleotide phosphate (NADPH) 88, 89, 92–95, 97, 98, 102–109, 111, 113, 138, 139, 141–143, 145–147, 149, 151–153, 161, 165, 166, 172, 288, 290–291, 306–311, 319–323, 325–330, 332–334, 446, 449, 450, 482, 486, 490, 492, 493, 495, 498, 502, 549, 569
Non-specific binding 23, 25, 28, 31, 33, 34, 37, 51, 88, 90, 97, 165, 166, 249–250, 286, 288, 296, 319, 350, 366, 396, 400, 464
Nuclear receptor
    AhR 173, 195, 207–217, 221–234, 237
    CAR 173, 195–204, 237
    PXR 173, 174, 176, 178–191, 195, 196, 237
Nuclear translocation 197, 201–204, 215, 222, 223

O

OECD. *See* Organization for Economic Co-operation and Development (OECD)
Oligonucleotide annealing 212–213
Oral absorption 1, 3, 6, 49, 50, 388–391
Organic anion transporting polypeptides (OATPs)
    OATP1B1 405–416
    OATP1B3 405–416
Organization for Economic Co-operation and Development (OECD) 562, 589

P

PAGE. *See* Polyacrylamide gel electrophoresis (PAGE)
Paracellular 50, 51, 54, 62, 69, 70
Parallel artificial membrane assay (PAMPA) 5
Passive transcellular drug flux 50
Permeability 2–6, 49–74, 80–84, 270, 341–343, 345, 349, 356, 357, 360, 361, 365, 370, 378, 383, 385, 387, 389–391, 400, 432
    intestinal 50, 51, 77, 78, 83
P-glycoprotein (P-gp/ABCB1) 173, 337–350
Pharmacokinetics (PK) 1, 4, 6, 10, 21, 25, 29, 39, 41, 88, 90, 97, 102, 112, 164, 172, 195, 269, 271, 286, 319, 334, 337, 340, 354, 370, 406, 416, 429, 446, 451, 462, 463
Pharmacokinetics-pharmacodynamics (PK-PD) 22
Phenotyping 117–134, 137–167
Plasma protein binding 21–37, 39
Polyacrylamide gel electrophoresis (PAGE) 207, 208, 210, 213–214, 217, 434
Potential drug interactions 22, 78, 118, 137, 190, 255
Pregnane X receptor (PXR) 171–191, 195, 196, 237
Pre-incubation 43, 110, 256, 257, 260–265, 267, 274–277, 286, 287, 297, 298, 307–313, 319, 321, 323, 326, 327, 330, 344–346, 350, 359–361, 363, 366, 385, 426
Protein-DNA interactions 207, 208, 215, 217
PXR. *See* Pregnane X receptor (PXR)

Q

Quantitative
    LC/MS/MS 67, 127, 130, 258, 436
    real time PCR 127, 224, 225, 228–229, 233, 247
Quasi-irreversible 255, 283, 286, 305, 306, 318
Quaternary ammonium glutathione 494–495, 499–591

R

Reactive metabolite 172, 255, 288, 297, 306, 312, 432, 446, 457, 477–487, 489–502, 545
Recombinant expressed
    CYPs 126
    UGTs 126
Red cell partitioning 41, 46
Reporter gene bioassays 225–228, 230–233
Reversible inhibition 256, 283, 284, 286, 297, 314–319, 321
RNR3-LacZ 540–543

S

Salmonella 545, 549, 552, 553, 555
Screening of reactive metabolites 477–487
S9 fractions 522, 546, 548, 549, 569, 570
Simcyp 159, 271, 272
Simulating intestinal fluid
    fasted-state simulating intestinal fluid
        (FaSSIF) 5
    fed-state simulating intestinal fluid
        (FeSSIF) 5
Single cell gel electrophoresis (SCGE) 517, 518
Solubility 2–8, 10–12, 14, 15, 17,
    50, 51, 58, 77, 80, 83–85, 106, 133, 162, 164, 166,
    180, 189, 242, 249, 273, 308, 334, 350, 366, 387,
    391, 398, 416, 429, 448, 466, 571–573
Solubilization
    buffers 6, 7
    complexants 6, 7
    cosolvents 6, 7
    lipids 6, 7
    surfactants 6, 7
Solute carrier (SLC) superfamily of transporters
    OAT1 (SLC22A6) 417
    OAT3 (SLC22A8) 417
    OCT2 (SLC22A2) 417
Stable isotopic labeled 315–334, 452
Structure elucidation of metabolites 453–456
Substrates 69, 102, 117,
    137, 182, 227, 243, 256, 276, 283, 308, 315, 337, 354,
    369, 405, 417, 434, 454

T

Tandem mass spectrometry 285, 318,
    432, 448, 451, 457, 462, 478, 499
TCI. *See* Time-dependent inhibition/inactivation (TDI)
Thymidine kinase (*Tk*) gene 561, 562

Tiered approach 4, 281–288, 298, 370
Time-dependent inhibition/inactivation
    (TDI) 255–267, 269–278,
    281–301, 305–313, 315–334
Toxicity 2, 22, 101, 102,
    184, 221, 222, 227, 231, 250, 269, 270, 288, 418, 446,
    477, 489, 490, 500, 501, 572, 583, 588
    surfactants 6
Toxicokinetics (TK) 1, 10, 462, 463, 570
Transcellular drug flux 3, 5, 50
Transcytotic pathways 50
Transporter
    OAT1 417–429
    OAT3 417–429
    OATP 405, 409, 410
    OATP1B1 405–416
    OATP1B3 405–416
    OCT2 417–429
    solute carrier (SLC) family 405, 417
Trapping agents 478, 480–481, 485,
    490, 492–498, 501, 502

U

UDP-glucuronsyltransferase 88, 102, 103,
    117–134, 195
Uridine-diphosphate-glucuronic acid
    (UDPGA) 88, 89,
    92–95, 102, 104–107, 109–111, 113, 117,
    125, 126, 129, 141, 142, 144, 145, 165, 446,
    449, 456, 506, 508

V

Vesicular transport assays 378–382

Y

Yeast-based genotoxicity 537, 538

Printed by Publishers' Graphics LLC